Daniel J. Keyser, Ph.D.
Richard C. Sweetland, Ph.D.

General Editors

TEST CRITIQUES
Volume VI

TEST CORPORATION OF AMERICA

LC 84-26895

ISBN 0-9611286-6-6 (v. 1)   ISBN 0-933701-14-4 (v. 1 softcover)
ISBN 0-9611286-7-4 (v. 2)   ISBN 0-933701-15-2 (v. 2 softcover)
ISBN 0-9611286-8-2 (v. 3)   ISBN 0-933701-16-0 (v. 3 softcover)
ISBN 0-933701-02-0 (v. 4)   ISBN 0-933701-17-9 (v. 4 softcover)
ISBN 0-933701-04-7 (v. 5)   ISBN 0-933701-18-7 (v. 5 softcover)
ISBN 0-933701-10-1 (v. 6)   ISBN 0-933701-19-5 (v. 6 softcover)
                            ISBN 0-933701-13-6 (softcover set)

Printed in the United States of America

CONTENTS

ACKNOWLEDGEMENTS

The editors wish to acknowledge the special contributions of our test reviewers. They have done an outstanding job. Our thanks extend from our deep pleasure and gratitude over their participation and the quality of their work. We know many of the contributing reviewers were as "caught up" in this project as we, and are now writing additional reviews for subsequent volumes. And, thanks also go to the test publishers themselves who released information to the reviewers in an expeditious manner.

We also wish to express thanks to the staff members at Test Corporation of America who were involved in this project: Jane Doyle Guthrie, Steve Poole, Kelly Scanlon, Marletta McCarty, and Shannon Moening. Eugene Strauss and Leonard Strauss, directors of Westport Publishers, Inc., have given freely and generously their support, encouragement, and business advice. Our indebtedness to both gentlemen is legion.

Finally, we want to express our warmest thanks to our readers. It is their use of *Test Critiques* that gives a final validity to this project. It is our sincerest desire that *Test Critiques* will have a true application for them.

INTRODUCTION

Test Critiques is a fulfillment of a goal of the editors and a continuation of a task begun with the publication of *Tests: A Comprehensive Reference for Assessments in Psychology, Education and Business* (1983), its *Supplement* (1984), and *Tests: Second Edition* (1986). With the *Test Critiques* series, we believe that we have moved into the final phase of this project—to include those vital parts that were not appropriate for our directory. With *Tests: Second Edition* and the *Test Critiques* series, the reader will have a full spectrum of current test information.

When *Tests* was published, a decision was made to leave out important psychometric information relating to reliability, validity, and normative development. Normative data and questions of reliability and validity were considered simply too complex to be reduced to the "quick-scanning" desk reference format desired. It was also apparent to the editors that a fair treatment of these topics would unnecessarily burden less sophisticated readers. More learned readers were familiar with other source books where such information could be obtained. The editors were aware, however, that a fuller treatment of each test was needed. These complex issues, along with other equally important aspects of tests, deserved scholarly treatment compatible with our full range of readers.

The selections for each volume were in no way arbitrarily made by the editors. The editorial staff researched what were considered to be the most frequently used psychological, educational, and business tests. In addition, questionnaires were sent to members of various professional organizations and their views were solicited as to which tests should be critiqued. After careful study of the survey results, the staff selected what was felt to be a good balance for each of the several volumes of critiques and selection lists were prepared for invited reviewers. Each reviewer chose the area and test to be critiqued and as can be noted in each volume's table of contents, some reviewers suggested new tests that had not been treated to extensive reviews. As test specialists, some reviewers chose to review tests that they had extensively researched or were familiar with as users; some chose to review instruments that they were interested in but had never had the opportunity to explore. Needless to say, the availability of writers, their timetables, and the matching of tests and writers were significant variables.

Though the reviewers were on their own in making their judgments, we felt that their work should be straightforward and readable as well as comprehensive. Each test critique would follow a simple plan or outline. Technical terms when used would be explained, so that each critique would be meaningful to all readers—professors, clinicians, and students alike. Furthermore, not only would the questions of reliability and validity along with other aspects of test construction be handled in depth, but each critique would be written to provide practical, helpful information not contained in other reference works. *Test Critiques* would be useful both as a library reference tool containing the best of scholarship but also useful as a practical, field-oriented book, valued as a reference for the desks of all professionals involved in human assessments.

It might be helpful to review for the reader the outline design for each critique contained in this series. However, it must be stressed that we communicated with each critique writer and urged that scholarship and professional creativity not be sacrificed through total compliance to the proposed structure. To each reviewer we wrote, ". . . the test(s) which you are reviewing may in fact require small to major modifications of the outline. The important point for you to bear in mind is that your critique will appear in what may well become a standard reference book on human assessment; therefore, your judgment regarding the quality of your critique always supercedes the outline. Be mindful of the spirit of the project, which is to make the critique practical, straightforward, and of value to all users—graduate students, undergraduates, teachers, attorneys, professional psychologists, educators, and others."

The editors' outline for the critiques consisted of three major divisions and numerous subdivisions. The major divisions were Introduction, Practical Applications/Uses, and Technical Aspects, followed by the Critique section. In the Introduction the test is described in detail with relevant developmental background, to place the instrument in an historical context as well as to provide student users the opportunity to absorb the patterns and standards of test development. Practical Applications/Uses gives the reader information from a "user" standpoint—setting(s) in which the test is used, appropriate as well as inappropriate subjects, and administration, scoring, and interpretation guidelines. The section on Technical Aspects cites validity and reliability studies, test and retest situations, as well as what other experts have said about the test. Each review closes with an overall critique.

The reader may note in studying the various critiques in each volume that some authors departed from the suggested outline rather freely. In so doing they complied with their need for congruence and creativity—as was the editors' desire. Some tests, particularly brief and/or highly specialized instruments, simply did not lend themselves easily to our outline.

Instituted in Volume III, an updated cumulative subject index has been included in this volume. Each test has been given a primary classification within the focused assessment area under the main sections of psychology, education, and business. The subject index has been keyed to correspond with *Tests: Second Edition*.

It is the editors' hope that this series will prove to be a vital component within the available array of test review resources—*The Mental Measurements Yearbooks*, the online computer services for the Buros Institute database, *Psychological Abstracts*, professional measurement journals, the forthcoming *A Consumer's Guide to Tests in Print* by Hammill, Brown, and Bryant, etc. To summarize the goals of the current volume, the editors had in mind the production of a comprehensive, scholarly reference volume that would have varied but practical uses. *Test Critiques* in content and scholarship represents the best of efforts of the reviewers, the editors, and the Test Corporation of America staff.

TEST CRITIQUES

Ellis D. Evans, Ed.D.

Professor of Educational Psychology, College of Education, University of Washington, Seattle, Washington.

ADOLESCENT ALIENATION INDEX

F. K. Heussenstaumm. Hollywood, California: Monitor.

Introduction

The Adolescent Alienation Index (AAI; Heussenstaumm, 1971) is a brief, paper-and-pencil, self-report measure of alienation for use with adolescents aged 12-19. Keniston's (1965) definition of alienation as an individual's conscious rejection or repudiation of expected roles represented by societal values frames this measure. However, specific item content is derived from Seeman's (1959) analysis of alienation in terms of five major dimensions of subjective experience: normlessness, meaninglessness, powerlessness, self-estrangement, and social isolation. The result is a 41-item scale crafted according to the forced-choice technique.

The AAI was developed by Frances K. Heussenstaumm, Ph.D., apparently in connection with a doctoral dissertation about alienation and creativity. He earned his doctorate from the University of Southern California in 1968. Subsequently, Heussenstaumm taught for several years at Columbia University and more recently has been a practicing psychologist in Santa Monica, California. A preliminary version of the AAI, consisting of 47 forced-choice items, was piloted with a sample of 75 lower-division college students. This pilot work included an attempt to establish concurrent validity for the test by correlating its scores with scores on the original Rotter I-E Scale (Rotter, 1966) from this sample. Nonfunctional items were culled from the original measure, resulting in a revision to 41 items. The final version then was normed on 1,228 adolescents of both sexes differentiated by demographics such as race, geography, and school status. No further revision of the AAI has occurred since publication of this final version. The measure is available only in standard English for use with literate subjects.

The AAI items are 41 pairs of self-descriptive statements presented in two booklets, Form A and Form C. No rationale is given for the existence of these two different forms. They are not parallel forms (i.e., forms consisting of two sets of similar items drawn from the same item population). They contain identical items, differing only in terms of the material upon which respondents record their answers. Form A, which is reusable, is accompanied by a separate one-page answer sheet printed in the standard "mark-sense" tradition. For ease of scoring, an administrator's scoring mask is provided. Form C is a consumable booklet that requires respondents to circle their choices in the booklet itself.

All of the item statements are written in one-sentence declarative form. Respondents using either test form must select one statement from each item pair, A or B, that best applies to their current feelings or perceptions. Most of the item content deals with sentiment and beliefs about school experience (e.g., academic studies,

3

school regulations, and teachers), with a sprinkling of items about people and life in general. Basically, the two statements in each item contrast a positive, self-confident, socially responsible attitude with a negative, self-doubting, passive, and irresponsible attitude (Friedes, 1978).

Examiner participation in the measurement process is relatively simple but straightforward. Instructions for responding are stated clearly on the booklets and include a sample item to illustrate the format. Items are written suitably for easy comprehension by adolescents in the intended age range of 12-19 years. Administration normally requires only about 10-15 minutes. The AAI yields one score, which is intended to summarize an individual's general sense of alienation or estrangement. The higher one's score, the greater one's expression of alienation.

Practical Applications/Uses

The zeitgeist in which the AAI was developed reflected much concern and apprehension about the psychological and social effects of widespread alienation among youth (Wynne, 1976). Indeed, such concern has continued in various quarters of the scholarly community committed to values of self-actualization and a positive sense of community (Bronfenbrenner, 1986). Contemporary scholars (i.e., Newmann, 1981) periodically echo cries issued during the turbulent period of social unrest (late 1960s and early 1970s), namely, that student alienation can adversely affect the qualities of school life and feed problems such as violence, vandalism, and poor achievement. Having identified with these persistent and long-standing concerns, Heussenstaumm (1971) claims three points of significance for his exemplar of alienation measurement. First, the AAI is said to have theoretical significance; that is, it provides a means for operationalizing the complex alienation construct for use in tests of theory. Second, because of its efficient and economical use in identifying emergent or developing alienation in adolescents and in studying the degree of its seriousness, the AAI technique is said to have methodological significance. Third, Heussenstaumm (1971) values the AAI for its practical significance in the schools; that is, he believes that the measure is useful to teachers, counselors, and others upon whom the "responsibility for devising and implementing counter measures to alienation must ultimately rest" (p. 4). More specifically, timely use of the AAI might help educators probe more easily for "incipient estrangement" in order to anticipate and ward off more fully developed behavioral expressions of alienation.

In short, Heussenstaumm envisions a variety of uses for the AAI. Its function as a tool for basic empirical research is clear. More speculatively, the AAI can be seen as a means of 1) identifying individuals or groups of adolescents needing special educational or counseling services, 2) generating clues for the development of therapeutic curriculum experiences in the secondary school setting, and 3) providing data for the evaluation of such interventions.

In light of these preferred functions and the item content itself, the AAI is best suited for use with its intended subjects: "normal" secondary school and early college-level students representing the 12-19-year-old age range. No accommodations to other types of subjects are reported in the manual. Considering the norm sample for this measure, score interpretations would be problematical for exceptional students.

No special training is required of the administrator of the AAI. Heussenstaumm suggests that the test can be administered by teachers, psychologists, sociologists, or even students themselves, in either small or large group settings. Directions printed on the front page of the instrumental protocol sheet are sufficient for self-administration. It seems preferable, however, for an administrator to introduce the AAI in general terms, without mentioning the word "alienation." Heussenstaumm fears that to do so may arouse an undesirable response set in examinees. For Form A, the AAI booklet and one answer sheet are distributed to each respondent. Verbal instructions are given to ensure that respondents supply vital information (full name, gender, age in years) on the answer sheet. For Form C, the AAI test booklets are distributed, and examinees provide appropriate background information on the instruction booklet itself. After distributing the AAI materials to examinees, an administrator can elect either to read the instructions aloud, if appropriate, or to allow examinees to read them silently. Questions from examinees should be answered only by paraphrasing the stated instructions as necessary.

Under normal conditions, the AAI takes about 15-20 minutes to complete. The measure is untimed, however, and every opportunity should be given to allow each examinee to complete the instrument. To maintain a reasonable pace, the administrator is advised to monitor the assessment process. After about 10 minutes, the examiner should remind examinees that they should have completed the first page. Later, when several respondents appear to have completed their work, another prompt, such as a recommendation to recheck one's answers, is suggested. No specifications for modifying or adjusting AAI administration are included in the manual.

All scoring is simple, objective, and performed manually. No computer scoring provisions accompany this measure; however, the scoring procedure could easily be converted to machine scoring using standard mark-sense forms. For manual scoring, layover stencils are provided. The Form A stencil fits the AAI answer sheet; Form C requires scoring directly on the completed test booklet. Using the scoring stencil for Form A, a scorer must mark and sum the keyed answers to arrive at a total raw score. This answer sheet serves as the score record. For Form C, the score is recorded directly on the test booklet.

Raw scores provide the basis for score interpretation. Scores can range from 0 to 41, with higher scores inferring greater alienation in a relative sense. A table containing six sets of percentile norms that correspond to this range of possible scores is presented in the manual. No other standard score conversions are included. The principal percentile set includes norm conversions for the total sample of 1,262 individuals upon which the AAI was standardized. In addition, percentiles are reported for five subdivisions of this norm sample: 1) Group 1, Caucasian suburban high school students of both sexes (n = 134); 2) Group 2, black male Job Corps enrollees (n = 65); 3) Group 3, black male and female urban high school students (n = 681); 4) Group 4, a combined group of male and female students from junior high and high school settings (n = 221); and 5) Group 5, Mexican-American rural high school students of both sexes (n = 161). A breakdown of ages, gender distribution, and socioeconomic status is lacking for these subgroups as well as for the total norm group. The various subgroups were selected on the basis of hypotheses generated from the pre-1971 literature about alienation and its correlates. Heussenstaumm anticipated that mean score differences would function as validity data, as discussed later in this review.

To assist interpretation, ranges of percentile norms for the total sample and all five subgroups are organized into categories that express degree of alienation from low to high: "minimal," "little," "typical," "moderate," and "extreme" alienation. Adolescents whose percentile scores place them in the latter two categories are presumably prime candidates for some kind of attention by helping adults. It is important to note, however, that this scaling is relative: no independent rationale, theoretical or otherwise, is offered to explain these score category designations.

The manual also presents a limited array of conventional descriptive statistics that may aid score interpretation further. These statistics include means and standard deviations for each of the five aforementioned subsamples, correlations between AAI scores and various demographic characteristics obtained for the norm group (e.g., gender, age, part-time work hours, estimated grade point average, paternal educational level, marital status of parents), and z-test results to compare AAI scores for four of the subgroups.

Technical Aspects

Validity questions about the AAI concern, first, the extent to which item content reflects the core meaning of alienation as a theoretical construct and, second, empirical relationship of AAI scores to adolescent status and behavior as projected from theory. As for the first concern, content, items were written by Heussenstaumm (1971) with reference to Seeman's (1959) five dimensions of alienation, presented in the Introduction to this review. At issue is accuracy in the translation of these dimensions into terms that adolescents comprehend. On the face of it, their translation appears sensible, although no indications are given about which and how many items are generated from each of Seeman's five dimensions.

As for the second concern, the original norming study represents the basis upon which to judge the empirical validity of the AAI. For this study, the adolescents comprising the five subgroups described previously seemingly were chosen for convenience and from expectations about how selected demographics would differentiate subgroups of subjects in terms of alienation theory. Eleven specific hypotheses, largely correlational in nature, were tested and produced mixed results. The demographic characteristics of adolescents showing the most consistent and statistically significant correlations with AAI scores are 1) self-reported grade point average, 2) self-estimate of cohesiveness or solidarity with peers, and 3) self-estimates of parental approval of friends. These three sets of correlations vary in their magnitude across subsamples. None exceeds $r = -.33$, and the patterns of correlations are not entirely consistent with theoretical inference. For example, father's educational level, the primary socioeconomic indicator in the validation study, related significantly with AAI scores only for suburban Caucasian and black male Job Corps students. Moreover, neither parental marital status nor family transiency (as measured by number of moves in the past 5 years) correlated with AAI scores in any meaningful way.

Perhaps more telling was the failure of test data to support hypotheses about gender and age differences in alienation. Contrary to expectations, older students showed no progressive trend for alienation, and males indicated no greater vulnerability to alienation experience in the school setting than their female counterparts. A predicted tendency for ethnic, minority-group adolescents to express somewhat higher AAI scores was confirmed. Finally, subsample group means reveal an AAI score hierarchy of uncertain relevance for measurement validity. The rank order of means, from high to low, is as follows: urban junior and senior high school students (14.17), black Job Corps enrollees (13.42), rural Mexican-American high school students (12.45); urban high school students (11.89), and suburban Caucasian high school students (10.83) (Heussenstaumm, 1971, p. 5). The validity puzzle concerns the top scoring group, which is identified only as subjects from target schools selected for a Coro Foundation teacher education project. No explanation of this project appears in the manual. AAI score variability was remarkably similar across four subgroups (standard deviations from 5.24 to 5.55), the lone outlier being the black male Job Corps groups ($sd = 7.91$). These data add little to any validity interpretation. Regardless, the percentile norm score categorization indicates that mean alienation scores for all subsamples reflect "typical alienation" (Heussenstaumm, 1971, p. 5).

The aforementioned subsample score dispersion pattern is paralleled by Kuder-Richardson Formula 20 reliability estimates for the five groups. AAI scores for the black Job Corps enrollees show strong reliability ($r = .89$) with lower, but still strong, estimates for the remaining groups ($r = .75-.79$). These estimates address internal consistency reliability only. No stability coefficient is reported. Because the AAI is a single scale with no parallel form, an equivalence coefficient cannot be stated.

Critique

The AAI would seem to appeal most strongly to professionals involved with adolescents who are deviant or otherwise experiencing difficulty with their transition to adulthood, especially as schooling may be implicated. Two features of the AAI are attractive. First is the intuitive appeal of the alienation construct upon which the AAI is based. This construct has a respectable niche in the formal scholarly literature, beginning with Durkheim's 19th-century writings and extending through and beyond Merton's (1957) work on social theory and structure. Second, the AAI provides a practical operational device with which to fan away some of the existential haze that has engulfed the alienation literature since its inception.

Beyond first impressions, however, the AAI reveals a number of shortcomings that themselves create a mist of uncertainty. Because it originally was published in 1971, has not undergone revision, and has a virtually unknown research history, it is not surprising that previous reviewers (Friedes, 1978; Hogan, 1978) have probed this instrument with a critical lance. Friedes, for example, writes unequivocally that the AAI should not be used for counseling or clinical work because of its weak psychometric properties, especially validity. He correctly observes that the AAI was published commercially and offered for school use in the absence of any case study material or test of peer-reviewed journal publication. Friedes questions the AAI item integrity as well. Just what does this scale measure? Scale items seem generally to reflect a continuum of positive-to-negative social attitudes that could show considerable overlap with personality variables or affect states such as self-esteem, anxiety, depression, and generalized school sentiment (Tolor & LeBlanc, 1971).

As Friedes (1978) suggests, a more definitive critical and factorial analysis of the original data, together with correlational data of the AAI with other established measures to address both discriminant and convergent validity, is needed. In the manual, the sole indication of such correlation with other measures is a .61 correlation between AAI scores and the original Rotter I-E Scale (Rotter, 1966), obtained from the pilot version of the AAI used with 75 lower-division college subjects. Apparently, Heussenstaumm's only additional contribution to the published alienation literature is a paper that restates findings from the original norming study (Heussenstaumm & Hoepfner, 1972). In view of these facts, a yellow flag of caution can be raised even for those considering research use of the AAI.

Hogan's (1978) concurrent criticism is mellowed somewhat by an affinity for the alienation concept and theoretical rationale for AAI construction, especially Heussenstaumm's (1971) justification of scale variables according to Seeman's (1959) analysis. Nonetheless, Hogan is critical about inadequate validation data, ambiguity in norm group specifications, failure to provide separate norms by gender in light of sex differences in alienation, occasionally stilted item wording that may "put off" working-class students, and some careless errors in the test manual, such as misspellings and missing bibliographic citations. For such weaknesses, Monitor, the publisher of the AAI, also must take responsibility.

These collective shortcomings underscore the need for a revision of the material in addition to the factorial analysis and correlational studies suggested by Friedes. Fundamental to a revision should be a renorming that includes a broad sampling of contemporary adolescents, a study that seems particularly crucial in view of the societal differences between the late 1960s and the late 1980s. In addition, there is a critical need for evidence about the behavioral correlates of AAI score levels.

Any revision, however, may presuppose, to some degree, the use of the AAI by researchers and practitioners. It is difficult to project any future use of the AAI in the absence of past sales figures for the original measure. Sales figures aside, past use indications are not strong. An extensive review of alienation literature by this reviewer has not revealed evidence of published research using the AAI. Rather, variations on the Dean (1961) Alienation Scale have appeared with some frequency (e.g., Baker & Siryk, 1980; Calabrese, Miller, & Dooley, 1987). Still other more visible measures of alienation are notable for their research use as well, including an adolescent alienation and involvement battery created by Kulka, Kahle, and Klingel (1982) and, especially, the alienation scales developed by Mackey and Ahlgren (1977). Taken together, these efforts signal continued interest in alienation theory and research, as well as concern for reducing levels of alienation among deviant adolescents (Calabrese & Schumer, 1986). The AAI does not seem to have figured strongly in this mainstream of alienation research activity. Personal correspondence between this reviewer and the test author has revealed Heussenstaumm's continued interest in the alienation topic, although his focus is now the clinical treatment of alienated individuals and issues of therapy efficacy (F. K. Heussenstaumm, personal communication, June, 1987). Unfortunately, this correspondence also revealed that the AAI publisher's records, which would reveal past usage, are irretrievable. Thus, in the absence of additional validation data and accessible clinical-use reports about the AAI, the data base for informed use of this scale seems no greater than when the AAI first was published in 1971.

References

Baker, R. W., & Siryk, B. (1980, September). Alienation and freshman transition into college. *Journal of College Student Personnel*, 437-442.

Bronfenbrenner, U. (1986). Alienation and four worlds of childhood. *Phi Delta Kappan*, *67*(6), 430-436.

Calabrese, R. L., Miller, J. W., & Dooley, B. (1987). The identification of alienated parents and children: Implications for school psychologists. *Psychology in the Schools, 24*, 145-150.

Calabrese, R. L., & Schumer, H. (1986). The effects of service activities on adolescent alienation. *Adolescence, 21*(83), 675-687.

Dean, D. (1961). Alienation: Its meaning and measurement. *American Sociological Review, 26*, 753-758.

Friedes, D. (1978). Review of Adolescent Alienation Index. In O. K. Buros (Ed.), *The eighth mental measurements yearbook* (pp. 691-692). Highland Park, NJ: The Gryphon Press.

Heussenstaumm, F. K. (1971). *Adolescent Alienation Index*. Hollywood, CA: Monitor.

Heussenstaumm, F. K., & Hoepfner, R. (1972). Black, white, and brown adolescent alienation. *Educational Leadership, 30*(3), 241-244.

Hogan, R. (1978). Review of Adolescent Alienation Index. In O. K. Buros (Ed.), *The eighth mental measurements yearbook* (pp. 692-693). Highland Park, NJ: The Gryphon Press.

Keniston, K. (1965). *The uncommitted: Alienated youth in American society*. New York: Harcourt, Brace and World.

Kulka, R. A., Kahle, L. R., & Klingel, D. M. (1982). Aggression, deviance, and personality adaptation in antecedents and consequences of alienation and involvement in high school. *Journal of Youth and Adolescence, 11*(3), 261-279.

Mackey, J. A., & Ahlgren, A. (1977). Dimensions of adolescent alienation. *Applied Psychological Measurement, 1*(2), 219-232.

Merton, R. (1957). *Social theory and social structure* (rev. ed.). New York: The Free Press of Glencoe.

Newman, F. M. (1981). Reducing student alienation in high schools: Implications of theory. *Harvard Educational Review, 51*(4), 546-564.

Rotter, J. R. (1966). Generalized expectations for control of reinforcement. *Psychological Monographs, 80*, 7(609), 28.

Seeman, M. (1959). On the meaning of alienation. *American Sociological Review, 24*, 783-791.

Tolor, A., & LeBlanc, R. F. (1971). Personality correlates of alienation. *Journal of Consulting and Clinical Psychology, 37*(3), 444.

Wynne, E. (1976). Adolescent alienation and youth policy. *Teachers College Record, 78*(1), 23-40.

Andrew F. Newcomb, Ph.D.
Associate Professor of Psychology, University of Richmond, Richmond, Virginia.

ADOLESCENT-COPING ORIENTATION FOR PROBLEM EXPERIENCES

Joan M. Patterson and Hamilton I. McCubbin. Madison, Wisconsin: Family Stress, Coping and Health Project.

Introduction

The Adolescent-Coping Orientation for Problem Experiences (A-COPE) is a self-report inventory designed to assess adolescent coping style and behavior. The 54 A-COPE items have been collapsed to provide the user with 12 scales that assess different coping patterns among adolescents between 13 and 18 years of age. The objective of this questionnaire is to provide an assessment of how the adolescent manages the developmental tasks confronted during the transition from childhood to young adulthood. The test authors suggest that the coping style that develops during this time has significant implications for future adult adjustment.

The A-COPE was developed by Joan Patterson and Hamilton McCubbin as part of the Family Stress, Coping and Health Project directed by Hamilton McCubbin (McCubbin & Thompson, 1987). The Family Stress, Coping and Health Project had its origins at the University of Minnesota and is currently located at the University of Wisconsin-Madison. In the past 6 years, McCubbin and his colleagues have utilized the Double ABCX-Model of Family Adjustment and Adaptation (McCubbin & Patterson, 1983a, 1983b) to examine how families make the transition and adjust to major life changes and illness and as the conceptual basis for developing a battery of instruments to assess how the family system responds to this life stress and transition. Most recently, this conceptual framework has been expanded to include a consideration of the role of family types (i.e., Balanced, Regenerative, Resilient, Rhythmic, and Traditionalistic) in the coping and adaptation process (McCubbin & McCubbin, 1987).

A unique characteristic of the A-COPE inventory is that the theoretical conceptualization underlying the instrument is derived from an integration of individual coping theory and family stress theory (Moos & Billings, 1982). As a consequence, the adolescent is conceived of as needing to manage both individual demands and those related to the family and the community. Successful coping is achieved when the adolescent is simultaneously able to fit into the family and the community, which consists of peers, school, and so on. Patterson and McCubbin (in press) propose that a "fit" occurs when there is a reciprocal balance between each system's demands and capabilities. Development is considered to take place when demands exceed the individual's existing capacities and new coping responses are acquired to reobtain balance.

The development of the A-COPE began by having a group of thirty 10th-, 11th-, and 12th-graders complete the Adolescent-Family Inventory of Life Events and Changes (A-FILE; McCubbin, Patterson, Bauman, & Harris, 1981). The answers to this inventory were used as a stimulus for interviewing each respondent. In the interviews, the adolescents were asked how they managed their most difficult personal life stress, the most difficult life stress faced by a family member, and difficult life events in general. These responses were used to generate 95 items for the initial revision of the A-COPE inventory. The test authors describe these items as reflecting both desirable and undesirable behavior and as representing the three primary coping functions: 1) direct action (problem-focused coping), 2) altering meaning (appraisal-focused coping), and 3) managing tension (emotion-focused coping).

As the next step in instrument development, these 95 items were administered to a group of 467 junior and senior high school students. These students were asked to complete a questionnaire in which they responded, on a 5-point scale, to how often they used each of the 95 behaviors when they felt tense or were confronted with difficulties. On the basis of this sample's responses, 27 items were removed from the scale either because of infrequent use or minimal variance. The remaining 68 items were factor analyzed, resulting in 54 items with factor loadings of .40 on 12 scales; each scale had an eigenvalue of 1.0 or greater. A description of the 12 scales follows.

Ventilating Feelings: a 6-item scale focused on expression of feelings through means such as yelling at people, complaining to family or friends, or swearing.

Seeking Diversions: an 8-item scale that includes primarily recreational activities (e.g., going to the movies, going shopping, working on a hobby) or sedentary activities (e.g., sleeping, watching T.V., reading, using prescription drugs).

Developing Self-Reliance: a 6-item scale that is focused on utilizing personal resources to handle life's circumstances (e.g., organizing your life and what you have to do).

Developing Social-Support: a 6-item scale that emphasizes means of maintaining social support networks; one item (i.e., cry) appears less directly related to the other items in the scale.

Solving Family Problems: a 6-item scale that measures the degree to which problem solving and discussion is undertaken with parents and siblings.

Avoiding Problems: a 5-item scale that includes items on drug, alcohol, and cigarette use as well as on avoidance of problems.

Seeking Spiritual Support: a 3-item scale in which religious activities (e.g., going to church or talking with clergy) are emphasized.

Investing in Close Friends: a 2-item scale focused on being with "someone you care about" or being with "a boyfriend or girlfriend."

Seeking Professional Support: a 2-item scale in which the assistance of professional and school counselors is sought.

Engaging in Demanding Activity: a 4-item scale that assesses involvement in activities (e.g., physical activity and school work) as a means of coping.

Being Humorous: a 2-item scale that measures the degree to which humor (e.g., joking or making light of a situation) is used as a coping behavior.

Relaxing: a 4-item scale focused on activities that may be viewed as being relaxing (i.e., daydreaming, listening to music, riding in the car, and eating).

Practical Applications/Uses

The purpose of the Adolescent-Coping Orientation for Problem Experiences is to assess how often adolescents use different behaviors in situations where they "face difficulties or feel tense" (Patterson & McCubbin, 1987). The test authors suggest two likely applications of the A-COPE inventory. First, the instrument has the potential to help adolescents better identify the types of coping behaviors and patterns they currently utilize. This self-education function may be quite useful to guidance counselors and other mental health professionals as a means to stimulate discussion.

Second, the A-COPE has been suggested as a possible pre/post-assessment instrument for intervention programs designed to facilitate adolescent coping with life stress and/or developmental tasks. The value of the A-COPE as a program evaluation tool must be carefully balanced against the current evidence for the reliability and validity of the instrument. The potential utility of the A-COPE as a research instrument would appear to be very good. Although the instrument clearly needs further development, the conceptual rationale underlying the A-COPE is well articulated and the need to examine adolescent coping style and behavior is a high priority.

As a guidance and research tool, the A-COPE is designed to be easily self-administered by adolescents between the ages of 13 and 18. Instructions for the administration of the A-COPE are provided directly on the answer sheet. The instructions are sufficiently clear that eighth-graders with average reading ability would have little difficulty completing this self-report instrument.

The authors begin the assessment by providing the respondent with a description of the purpose of the questionnaire and a definition of coping. In particular, coping is operationalized for the adolescent as "individual or group behavior used to manage the hardships and relieve the discomfort associated with life changes or difficult events" (Patterson & McCubbin, 1987). The directions that follow are simple and straightforward; the respondents are to decide how often they use the specified behaviors when they are confronted with life stresses. Each of the items on the A-COPE has five alternative responses: 1) never, 2) hardly ever, 3) sometimes, 4) often, and 5) most of the time. The respondent is simply asked to circle a response for each statement and to make sure all items are answered. A note is provided to assure the respondent that terms such as parent and stepparent are interchangeable.

Although the test authors do not discuss group and individual administration, the A-COPE appears appropriate for either use. The instrument could be read aloud to an individual or group if the reading level of the respondent(s) was in question. Overall, the questionnaire should take little more than 15 minutes to complete.

A separate test manual for the A-COPE has not been developed. Instead, the user should consult both a research article in the *Journal of Adolescence* (Patterson & McCubbin, in press) and McCubbin's compilation of family assessment inventories

(Patterson & McCubbin, 1987) for further information on the development and psychometric properties of the instrument.

Scoring procedures are described by Patterson and McCubbin (1987). The user is advised that the respondent's score on each of the 12 scales is obtained by simply summing responses to each item in the scale. In completing this process, nine items (7, 8, 19, 24, 26, 28, 42, 46, and 49) must be reverse scored (i.e., 5=1, 4=2, 3=3, 2=4, 1=5). Inasmuch as no templates are provided for these purposes, the scoring process can be somewhat tedious. Frequent users of this instrument would be advised to develop their own templates. In contrast, researchers using the A-COPE should be minimally inconvenienced except for the time required to complete data entry for later analysis.

Interpretation of A-COPE results will vary depending on the intended use of the instrument. If the A-COPE is used for guidance purposes, the user will find that normative comparison data are very limited. Means and standard deviations are reported separately for the 185 boys and 241 girls included in the sample that was used to determine the factor structure of the A-COPE. The mean age of this sample was 15.9 years, and the children were recruited from a midwestern, suburban school district in which families were predominantly of middle to upper-middle socioeconomic status. In a second comparison sample described by Patterson and McCubbin (1987), the A-COPE means and standard deviations for black and white adolescents ($N = 203$) in residential treatment for deviancy and social adjustment problems are reported. The scores are divided on the basis of whether the adolescents are from single parent or nuclear families. No further information is provided.

The guidance counselor or mental health professional should use extreme caution in employing these data for comparison purposes. First, limited description is provided for the samples, and the authors make no claim that these data are representative of any particular population or generalizable to the general population. Second, cutoffs for extreme scores and descriptions of the meanings of extreme scores are not provided. And last, inasmuch as the scales have different numbers of items and an individual's score on each scale is the sum of responses, no relative comparison can be made between scale scores. Instead, the clinical use of this instrument should be limited to situations in which responses could be utilized to promote further discussion with the adolescent and in which clinical judgment is an acceptable evaluation criterion.

The clinical use of the A-COPE is further compromised by the limited reliability and validity data currently available for the instrument. The empirical investigator may also be hindered by this small reliability and validity data base; however, the A-COPE may prove to be a valuable research tool in some instances. In particular, if the investigator has a large sample size that will allow for the verification of the proposed factor structure and a design that will permit the establishment of test-retest reliability, then the A-COPE has the potential to provide useful information on adolescent coping style and behavior.

Technical Aspects

The only reliability data available for the Adolescent-Coping Orientation for Problem Experiences are estimates of internal consistency for each of the 12 factor scales. These alpha estimates range from .50 to .76, with a mean of .70. Generally, a minimum alpha is considered to be .60. Using this criterion, the internal consistency of the Seeking Professional Support factor (alpha = .50) is unacceptable, and the internal consistency of the other scales is adequate but not particularly impressive. No additional reliability information is currently available.

Inasmuch as an instrument can be no more valid than it is reliable, the validity data on the A-COPE must be seriously questioned. The report of concurrent validity includes an examination of the relationship between 8 of the 12 coping scales and the examinee's reported use of cigarettes, beer, wine, liquor, and marijuana. A significant pattern of correlations supporting concurrent validity were obtained; however, the magnitude of these relationships was relatively small (range of $r = -.21$ to .25). The data from a second validity study with adolescents having cystic fibrosis ($N = 17$) are reported by Patterson and McCubbin (1987), yet no statistical comparison is made with a normal sample.

Critique

In light of the fact that the Adolescent-Coping Orientation for Problem Experiences has a number of limitations at this time, the authors of the instrument make only qualified suggestions for its use. If the constraints of the inventory are understood, the A-COPE can be a helpful guidance tool when combined with clinical judgment or can be a useful research tool given particular experimental designs. Beyond these uses, the A-COPE is an instrument that needs further development.

At the core of any effort to improve the instrument is the need to continue to build on the impressive theoretical rationale that underlies the measure. First, the relationship among scales should be examined and scales should be collapsed as appropriate, the number of items in several scales needs to be increased, and the internal consistency of each scale should be improved. Second, the test-retest reliability of the measure needs to be established. Third, a more powerful demonstration of validity is warranted. And finally, if this instrument is to be used for clinical comparison purposes, a careful normative comparison study will be necessary.

In the process of reducing and strengthening the A-COPE scales, greater care should be taken in describing the meaning of the resulting scales. Currently, the description of the items in the scales suggests connections between items that are based on supposition and not actual data. For example, test items such as "use drugs (not prescribed by doctor)" and "drink beer, wine, liquor" have been interpreted as "the use of subtances as a way to escape" in the Avoiding Problems scale. Similarly, in the Investing in Close Friends scale, the item "be close with someone you care about" is interpreted as including only peers; this interpretation is presumably made to be consistent with the other item in the scale, "be with a boyfriend or girlfriend." Instead, the respondent might also be considering other social support persons when answering this question. The problem of overinterpretation is also evident in the Developing Social Support scale, where the rationale of the respondent is interpreted from the scale items without justification.

As interest in adolescent coping and stress continues to grow and investigators continue to address the more complex intersystem relations (e.g., individual coping and family functioning), the need for instruments like the A-COPE will increase. Currently, the A-COPE provides an example of a good self-report inventory at an early stage of development. As further refinements of the instrument are completed, the A-COPE has the potential to be a valuable research and clinical tool.

References

McCubbin, H. I., & Patterson, J. M. (1983a). Family stress and adaptation to crisis: A double ABCX model of family behavior. In D. Olson & B. Miller (Eds.), *Family studies review yearbook* (pp. 125-135). Beverly Hills, CA: Sage.

McCubbin, H. I., & Patterson, J. M. (1983b). The family stress process: The double ABCX model of adjustment and adaptation. In M. Sussman, H. McCubbin, and J. Patterson (Eds.), *Social stress and the family: Advances and developments in family stress theory and research* (pp. 7-37). New York: Haworth.

McCubbin, H. I., Patterson, J. M., Bauman, E., & Harris, L. (1981). *Adolescent-Family Inventory of Life Events and Changes (A-FILE).* Madison, WI: University of Wisconsin-Madison.

McCubbin, H. I., & Thompson, A. I. (1987). *Family assessment inventories for research and practice.* Madison, WI: The University of Wisconsin.

McCubbin, M. A., & McCubbin, H. I. (1987). Family stress theory and assessment: The T-double ABCX model of family adjustment and adaptation. In H. I. McCubbin & A. I. Thompson (Eds.), *Family assessment inventories for research and practice* (pp. 3-32). Madison, WI: The University of Wisconsin.

Moos, R. H., & Billings, A. G. (1982). Conceptualizing and measuring coping resources and processes. In L. Goldberger & S. Breznitz (Eds.), *Handbook of stress* (pp. 212-230). New York: Free Press.

Patterson, J. M., & McCubbin, H. I. (1987). A-COPE: Adolescent Coping Orientation for Problem Experiences. In H. I. McCubbin & A.I. Thompson (Eds.), *Family assessment inventories for research and practice* (pp. 225-243). Madison, WI: The University of Wisconsin.

Patterson, J. M., & McCubbin, H. I. (in press). Adolescent coping style and behaviors: Conceptualization and measurement. *Journal of Adolescence.*

Nancy A. Busch-Rossnagel, Ph.D.

Associate Professor of Psychology, Fordham University, Bronx, New York.

ADOLESCENT-FAMILY INVENTORY OF LIFE EVENTS AND CHANGES

Hamilton I. McCubbin, Joan M. Patterson, Edward Bauman, and Linda Hall Harris. Madison, Wisconsin: Family Stress, Coping and Health Project.

Introduction

The Adolescent-Family Inventory of Life Events and Changes (A-FILE) is a self-report instrument assessing adolescents perceptions of life events and changes. Developed from a family systems perspective, the A-FILE assesses the adolescent's vulnerability as a result of the "pile-up" of events affecting any family member. The instrument contains 50 items tapping both normative and nonnormative life events and changes experienced by the family over the past 12 months. A 27-item subset of the 50 items is rated for experiences previous to the past year because those 27 events require longer periods for adaptation or have chronic effects. Designed for use with adolescents from 12 to 18 years of age, the A-FILE represents an extension of the assessment available from the Family Inventory of Life Events and Changes (FILE; McCubbin, Patterson, & Wilson, 1983).

The A-FILE is one of the family assessment inventories developed by Hamilton McCubbin and his colleagues. McCubbin began his work with family assessment at the University of Minnesota and has continued it at the University of Wisconsin-Madison, where he currently is the dean of the School of Family Resources and Consumer Sciences and the director of the Family Stress, Coping and Health Project.

As part of his focus on family strengths and well-being, McCubbin and his colleagues developed the T-Double ABCX model of Family Adjustment and Adaptation. McCubbin's model elaborates on Hill's (1949) ABCX family crisis model in two ways. First, it notes that there are two distinct phases of response to family crisis, namely adjustment and adaptation. Second, it includes family types and levels of vulnerability as influences on the family's coping behavior. The FILE and the A-FILE were developed as measures of the vulnerability of the family.

Items in the initial version of A-FILE included events from the FILE that particularly would impact the adolescent as well as items from a previous scale of adolescent life changes (Coddington, 1972). The first pretest sample included 30 suburban 11th graders who provided feedback about additional stressful life events. A second pretest sample of 50 students in the seventh, tenth, and eleventh grades was used to examine wording, administration, and variability. These two pretests resulted in the 73-item Form A of the A-FILE. Form B (50 items) resulted from the analysis of

16

the responses of 500 junior and senior high school students. Three criteria guided the selection of items to be deleted or retained: 1) construct validity, internal consistency, and test-retest reliability; 2) theory and research on family life changes, and 3) frequency of occurrence. The test authors note, however, that they retained some items with low frequencies because the events described by the items were identified as major stressors (e.g., death of a parent).

A-FILE consists of a single 8½" ×11" sheet with directions and background information on the front and the 50 items on the back. Examinees respond to each item by checking yes or no. There are two columns for responses, one indicating the last 12 months and the second indicating time before the last 12 months. The items are grouped into six conceptual dimensions: Transitions (14 items), Sexuality (4 items), Losses (7 items), Responsibilities and Strains (19 items), Substance Use (4 items), and Legal Conflict (2 items).

Practical Applications/Uses

The Adolescent-Family Inventory of Life Events and Changes measures the "pile-up" of family events that might affect an adolescent. "Pile-up" is defined as "the sum of normative and non-normative stressors and intra-family strains" (McCubbin & Patterson, 1987) and is considered an index of the Family Vulnerability, or V, factor of the T-Double ABCX model. In this model, McCubbin and his colleagues suggest that family vulnerability, or pile-up, may be one reason why some families are unable to cope with a single stressor: If the family is already overburdened from other life changes, the members may have difficulty dealing with additional life events. Thus, this model views family changes as additive.

Consistent with a family systems perspective that an event that happens to one family member affects every member of the family, the A-FILE indicates the adolescent's perception of family stress, rather than just stress on the individual. From a research perspective, both the total stress score and the weighted stress score can be used as one predictor of a variety of dependent variables, such as adolescent adjustment or family adaptation. Clinical uses of the A-FILE range from assessing the impact of a single life event on adolescents' perception of stress to evaluation of intervention programs designed to alleviate stress. The clinician also can gain a fuller understanding of the family system by comparing the adolescent's responses on A-FILE to the parents' responses on FILE. Given the link between stress and physical health, health care practitioners can use the A-FILE to identify adolescents vulnerable to health problems that result from a high level of stress.

This test is appropriate for adolescents ages 12 to 18 who live in a family setting. The wording of the items and format are fairly straightforward, although the size of the type is fairly small. Adolescents attending school will find the format similar to standardized tests.

The manual for the A-FILE is a chapter from Family Assessment Inventories for Research and Practice (McCubbin & Thompson, 1987). The manual gives no information about administration. However, it appears that the test is designed as a self-report measure that does not require the presence of the examiner; therefore, it could be mailed to respondents for administration in their homes. When an examiner is present, the A-FILE can be administered individually or to groups. The test requires less than 10 minutes to complete.

There are two methods of scoring the A-FILE. To derive the Total score, a "yes" response (indicating that the event or change has occurred) is coded as "1" and a "no" is coded as "0". The Total score is the sum of the "yes" responses. Total scores may be computed for recent life events, past life changes, and the six dimensions. The Total Recent Life Changes score measures the current pile-up, and the Total Past Life Changes score indicates the degree of persistent stress in the family system, with high scores indicating high stress.

The second method of scoring the A-FILE is to obtain weighted scores for Total Recent Life Changes and for Total Past Life Changes. Using an approach similar to Holmes and Rahe (1967), the weights assigned to each life event reflect the degree of readjustment the event requires. The weights for the A-FILE were obtained from 88 students in the tenth and eleventh grades who rated each event in terms of the change required for an average family. The means of the ratings are presented as standardized weights on the A-FILE instrument. To obtain this weighted score, the weights for each event checked by the adolescent are summed. Most users will find that although this procedure requires a calculator, it takes only a few minutes.

The clinical interpretation of these scores is not straightforward because the meaning of high versus low stress is relative. For the first-time user, scores could be compared to the norms provided, but, as will be discussed later, these norms have serious limitations. Appropriate interpretations probably come only with repeated experience with the A-FILE for specific samples.

Technical Aspects

Although the authors present both internal consistency and test-retest reliability estimates for some of the Adolescent-Family Inventory of Life Events and Changes scores, they give no information about the sample used to determine these measures of reliability. Cronbach's alpha for the Total Recent Life Changes score is .69. Because of the heterogeneity in the frequency of occurrence for some of the items, the authors present alphas only for the scales of Responsibilities and Strains (.67) and Legal Conflict (.89).

Using a sample of 74 junior and senior high school students, 2-week test-retest reliabilities were calculated for the six scales, the Total Recent Life Changes score, and the Total Past Life Changes score. With the exception of Responsibilities and Strains [$r(73) = .69$], the reliabilities of the scales are high, above $r(73) = .80$. The reliability of the Total Recent Life Changes score is .82, and for Total Past Life Changes, $r(73) = .84$.

By providing the loadings for the hypothesized factor, the authors partially present the results of a factor analysis to support the construct validity of the A-FILE. Although the scales do seem to be well defined by several items with loadings above .30, the simple structure of the factors cannot be evaluated from the data presented. It is not clear why the items with very low loadings were retained because most of these items are not the infrequent events considered to be major stressors.

Criterion-related validity information also is presented. The authors had hypothesized that adolescent family life changes would be related significantly to substance use. No information is provided about the sample used to test this hypothesis, but the authors report significant correlations between alcohol use and marijuana use for the scales of Sexuality, Responsibilities and Strains, and Substance Abuse and for the Total Recent Life Changes score. Sexuality and Substance Use scale scores were related significantly to cigarette use during the past month and the past year; the Total Recent Life Changes score was related to cigarette use during the past year only.

A second predictive validity analysis examined the relationship between the A-FILE scores and health locus of control. The Total Recent Life Changes score was related negatively to an internal locus of control, and high scores on the Responsibilities and Strains scale predicted a belief in the control of powerful others. Once again, no information is given about the size of the sample used to test these relationships.

The test authors present normative data for two samples; however, except for the size of sample, no characteristics about the subjects in the sample are provided. The means and standard deviations for each scale and for the two total scores from a sample of 500 junior and senior high school students are presented with notes that there were significant differences between junior and senior high school students in frequency of occurrence. However, the data are not broken down by age of subjects. For Total Recent Life Changes, the mean score was 7, with a standard deviation of 4; the mean for Total Past Life Changes was 8, with a standard deviation of 6.

Separate norms for males (n = 197) and females (n = 206) are presented, but these data are confusing. No indication about which raw scores are being used is given, leaving the user to assume that the norms are for the Total Recent Life Changes scores. The raw scores listed range from 1 (low stress) to 15 (high stress), but the range for these scores for these samples is 45 for the males and 25 for the females. In addition, no indication is given as to what type of score the norms are, although the range from 99 (low stress) to 10 (high stress) suggests percentiles.

Critique

McCubbin et al. appear to have taken Anastasi's suggestion that "empiricism need not be blind" (1986, p. 6) to heart. The T-Double ABCX model, which is the theoretical rationale behind the creation of the series of family assessment measures of which the A-FILE is a part, is well developed and currently generating much research. Thus, the authors have the potential for developing an important tool for assessing adolescents' perceptions of their families.

However, the A-FILE as it currently is presented needs more attention to make it "user friendly." Its major problems are a lack of adequate information in the manual and inadequate attention to the details of the standardization sample for the norms. The lack of information in the manual has been noted throughout this review. No information is given about administration, and not enough data from the factor analysis is presented to evaluate construct validity. All of the psychometric information suffers from a lack of precise information about the samples on which it was collected and often extends to specifying sample size.

Without the necessary information about the sample populations, the norms are almost meaningless, even if the user assumes that they are percentiles for the Total Recent Life Changes scores. More information about the norms will not overcome all of the problems, however. The authors themselves present information indicating that there are significant differences between two age groups, but fail to break the norms down accordingly, which makes it very difficult for a practitioner to use the A-FILE in the manner suggested by the authors (e.g., to identify adolescents who may be at risk for problems due to high levels of stress).

The A-FILE is more appropriate for the researcher. The measures of internal consistency and test-retest reliabilities for the Total Recent Life Changes score document that this score may be considered reliable. Likewise, the test-retest reliabilities for the scale scores indicate high reliability; however, the factor analysis, as presented, does not confirm the construct validity of the scales. Using Anastasi's (1986) suggestion of a multistage process of validation, the A-FILE does have a well-developed theoretical base. However, more attention needs to be given to construct validity. Assuming that the T-Double ABCX model continues to generate research, more criterion-related validity information should be available in the future.

In summary, the A-FILE is currently a tool best used by the researcher rather than the practitioner. Although the potential clinical use of the A-FILE has been indicated by the test authors, practitioners cannot rely on the norms to aid in their interpretation of scores. Only by second guessing the authors about the psychometric information missing from the manual (e.g., assuming that the norms are percentiles) can practitioners even use the norms. Many research questions will not require referral to norms, and the reliability and theoretical validity of the A-FILE appear adequate, justifying its use in research endeavors.

References

Anastasi, A. (1986). Evolving concepts of test validation. *Annual Review of Psychology, 37,* 1-15.

Coddington, R. D. (1972). *Life Events Scales for Children and Adolescents.* New Orleans: Stress Research Co.

Hill, R. (1949). *Families under stress.* New York: Harper & Row.

Holmes, T. H., & Rahe, R. H. (1967). The Social Readjustment Rating Scale. *Journal of Psychomatic Research, 11,* 213-218.

McCubbin, H. I., & Patterson, J. M. (1987). Adolescent-Family Inventory of Life Events and Changes. In H. I. McCubbin & A. I. Thompson (Eds.), *Family assessment inventories for research and practice.* Madison: University of Wisconsin-Madison.

McCubbin, H. I., Patterson, J. M., & Wilson, L. R. (1983). *Family Inventory of Life Events and Changes (FILE).* Madison: University of Wisconsin, Family Stress, Coping and Health Project.

McCubbin, H. I., & Thompson, A. I. (Eds.). (1987). *Family assessment inventories for research and practice.* Madison: University of Wisconsin-Madison.

Robert J. Drummond, Ed.D.
Program Leader, Counselor Education, and Interim Chairperson,
Division of Educational Services and Research, University of North
Florida, Jacksonville, Florida.

ADULT PERSONALITY INVENTORY

Samuel E. Krug. Champaign, Illinois: Institute for Personality
and Ability Testing.

Introduction

The Adult Personality Inventory (API; Krug, 1984) is a 324-item paper-and-pencil instrument designed to analyze the individual differences in personality, interpersonal style, and career and life-style preferences of adults ages 16 and older.

The test author, Samuel E. Krug, Ph.D., has published widely in the field of personality assessment and computer-based interpretation. Some of his major publications are *Interpreting 16PF Profile Patterns* (1981), the *Manual for the IPAT Depression Scale* (Krug & Laughlin, 1976) and the *Handbook for the IPAT Anxiety Scale* (Krug, Scheier, & Cattell, 1976). He recently has compiled and edited the 1987-1988 *Psychware Sourcebook*.

The API is rooted in the work on personality assessment conducted by Raymond B. Cattell and his associates. Their studies led to the development of the 16 Personality Factor Questionnaire and other related instruments. The API was developed from the 16PF's pool of 564 items, which underwent factor analytic and correlational analysis in order to:

1. identify items within the pool that most validly and reliably measured the primary personality dimensions underlying each trait scale;
2. recast the item content in shorter, more easily readable form, and
3. present the final item selection in a standard format that would simplify and shorten test administration. (Krug, 1984, pp. 45-46)

The items that resulted from these analyses were used to develop the pool of primary source traits. The new revised item format also was correlated with Form A of the 16PF.

The API test booklet contains five parts. In the first part (159 items), the examinee uses a 3-point scale (generally true, uncertain, and generally false) to respond to briefly and simply stated items focusing on the examinee's personality characteristics, interpersonal style, and career and lifestyle (e.g., I like to watch football games). Part II consists of 10 verbal items, and Part III consists of 10 numerical items involving numbers and number operations. In both Part II and Part III, the examinee chooses the correct answer from among three options for each item. Part IV consists of 10 reasoning items. Three options are presented, and the examinee must identify the one option that is different from the other two. In Part V (135 items), the examinee uses the 3-point rating scale employed in Part I to indicate his or her responses to items focusing on personality dimensions.

The API contains 21 trait scales. The scales were derived using multivariate analysis techniques, such as multiple discriminant analysis and multiple regression. The procedures for developing the scales are not discussed fully in the manual. Seven scales deal with personal characteristics: 1) Extraverted (outgoing, prefers to be with other people), 2) Adjusted (stable, calm, unfrustrated, functions well under stress, and coldly objective), 3) Tough minded (approaches problems rationally), 4) Independent (self-directed, stubborn, and willful), 5) Disciplined (controlled, careful, organized), 6) Creative (imaginative, unconventional, sensitive), and 7) Enterprising (adventurous, ambitious, motivated to succeed).

Eight scales describe the interpersonal style of the examinee: 1) Caring (warm, trusting, accepts others openly), 2) Adapting (dependent, self-effacing, meek), 3) Withdrawn (quiet, bashful, shy, unhappy), 4) Submissive (insecure, tense, needs support and approval from others), 5) Hostile (angry, jealous, exploitative), 6) Rebellious (critical, unconforming, radical), 7) Sociable (open, cheerful, jovial), and 8) Assertive (forceful, dominant, take-charge people).

The examinee's career and life-style preferences are measured by six scales that are similar to John Holland's six occupational themes: 1) Practical (cf. Holland's Realistic), 2) Scientific (cf. Holland's Investigative), 3) Aesthetic (cf. Holland's Artistic), 4) Social (cf. Holland's Social), 5) Competitive (cf. Holland's Enterprising), and 6) Structured (cf. Holland's Conventional).

There are also four validity scales. The Good Impression scale indicates the extent to which faking good tendencies may have influenced test results. The Bad Impression scale measures the extent to which faking bad tendencies may have influenced test results. The Infrequency scale consists of items that had low endorsement frequencies (< 10%) in the derivation samples. The Uncertainty scale measures the number of "middle" or "uncertain" responses chosen.

The test can be self- or group-administered. According to the test author, a fourth-grade reading level is required. The directions are stated clearly on the front of the test booklet, and examples are provided on the separate answer sheet, which is nine columns wide and allows space for the examinee to "bubble" in responses to 187 questions per side. Although the test is not timed, it usually requires approximately 1 hour to complete. Computerized scoring is available from the publisher. In addition, the Decision-Making Worksheet, which requires the examinee to consider only two scales at a time and to determine which of the two is more important, is available.

Practical Applications/Uses

According to Krug (1984, p. 1), the API is an inventory "intended to facilitate selection and placement decisions in industry." The API also is suggested for use in individual counseling because the scales reflect major personality dimensions and help clients and counselors to an increased awareness of underlying personality dynamics. The Interpersonal scales can be used as a "simple, understandable model for explaining how the individual relates to others and how relationships between people are affected by differences in primary styles" (Krug, 1984, p. 1). The API also is suggested for use in marriage and family counseling, in relationship training, and in personal growth programs. Krug feels that the test results can be of value to decision makers in industry, public service, health care, and education. In addition, the test may be valuable as a research tool to study dimensions of adult personality.

The API can be administered on an individual or group basis to individuals ages 16 and older. Because it requires only a fourth-grade reading level, the test can be given to a wide range of individuals. However, physically or visually handicapped adults may have problems reading the items due to the small type size used in the test booklet and "bubbling" in answers on the separate answer sheet.

No special training is necessary for administering the test. Administration procedures are discussed in the manual, and the test booklet and answer sheet contain directions as well. The examiner is directed to read the instructions on the front of the test booklet to the examinees. (The instructions are typical of most personality inventories.) The examinee is told to answer each question as quickly as possible. Because some of the items measure Factor B (Intelligence), the examinees are instructed that there is only one correct answer for each item. Although the test is written at a fourth-grade level, some words may be difficult for examinees; therefore, an examiner's presence would be helpful during testing.

The API must be sent to the Institute for Personality and Ability Testing (IPAT), the test publisher, for computer scoring. The manual does describe the true/false-keyed items, but because the scoring requires sophisticated statistical procedures, the test would be difficult to hand score. Krug explains in the manual that the test is computer scored to insure maximum accuracy, as well as speed, convenience, and economy (p. 2).

The Individual Assessment Report is a computer-generated report that organizes the information contained in the trait scores into a format that is attractive and easy for the test user to read and understand. The test presents the individual's results on the 21 report scales both in words and graphically. Profiles also are presented on each of the three major dimensions (Personality, Interpersonal, and Career/Life Style). The norms are based upon a sample of over 1,000 adults ranging in age from 16 to 70 and from all 50 states. Although in the manual Krug describes the data base as extensive and representative (p. 62), he does not provide any breakdown by age or geographic region. The user needs to be familiar with the conceptual model of Cattell and his associates as well as factor analytic techniques to have the background to use the API properly. The manual presents a sample case and a section on sharing the report with the client.

Technical Aspects

As evidence for the construct validity of the Adult Personality Inventory, the test manual presents the primary factor pattern matrix for the trait scores as well as primary factor correlations. In the description of the scales, Krug refers to the positive and negative test correlations of each of the API scales with other established personality instruments such as the Eysenck Personality Inventory, the 16PF, Thorndike's Dimensions of Temperament, and others. Intercorrelations of the scales within areas (e.g., Career/Life Style) are presented. The Career/Life Style scales tend to fit Holland's theoretical framework. Although Krug suggests that the test can be used for placement and selection decisions, little, if any, criterion-referenced validity evidence is presented. Intercorrelations of the 21 scales with each other are not presented in the manual.

Three studies concerning the reliability of the scales are presented in the manual (Krug, 1984, p. 61). The first and third study provide internal consistency coefficients, the second test-retest coefficients. The first study is based upon a sample of 281 adults; the second study, 31; and the third study, 612. The lowest reliability coefficients are found on three of the validity scales: Good Impression, Bad Impression, and Infrequency. The second lowest set are the Career/Life Style scales, which range in Study 3 from a low of .44 on Practical to a high of .84 on Scientific. In the same study, the Interpersonal scales range from .77 on Adapting to .90 on Assertive. The highest Personality scale was Adjusted with coefficients of .91, .74, and .90 for the first, second, and third studies, respectively. The lowest was Enterprising with coefficients of .51, .66, and .72, respectively. There is a need for additional reliability information about the API. For example, the test-retest study is based on a sample of only 31 adults. The standard error of measurement is not reported for any of the scales.

Critique

The manual, test booklet and answer sheet, and test report for the Adult Personality Inventory are user friendly. They are easy to read and present clear and understandable information about the scale. However, the examples provided in the manual to demonstrate how to bubble in responses on the answer sheet do not match the options given on the test.

In addition, the intercorrelations of all the API scales and the demographic description of the norming group are not reported. In his review of the API in the *Ninth Mental Measurements Yearbook,* Brian Bolton (1985) states that although the test has a good conceptual base and might be considered a modern version of the 16PF designed to assess 21 normal personality traits of adults, the derivation and construction of the Interpersonal scales are not explained adequately and that, overall, a technical supplement describing test construction procedures and presenting the psychometric characteristics of the instrument would be useful. The test does have a Decision Making Worksheet designed to help facilitate profile analysis for personnel classification and decision making. Because more evidence about the criterion-referenced validity of the test is needed, this form should be used cautiously.

This reviewer feels that, overall, the API has excellent potential for use in counseling situations and in research studies on the dynamics and structure of the personality of adults. As users become familiar with the API, more evidence of the validity and reliability of the scales will accumulate. This reviewer feels that the test covers important personality constructs and has an excellent, user-friendly computerized interpretative report. As such, the API is one of the best new personality tests for normal adults available.

References

Bolton, B. (1985). The Adult Personality Inventory. In J. V. Mitchell, Jr. (Ed.). *The ninth mental measurements yearbook* (pp. 55-56). Lincoln, NE: The Buros Institute of Mental Measurements.

Krug, S. E. (1981). *Interpreting 16PF profile patterns.* Champaign, IL: Institute for Personality and Ability Testing.

Krug, S. E. (1984). *The Adult Personality Inventory.* Champaign, IL: Institute for Personality and Ability Testing.

Krug, S. E. (Ed.). (1987). *Psychware Sourcebook.* Kansas City, MO: Test Corporation.

Krug, S. E., & Laughlin, J. E. (1976). *Manual for the IPAT Depression Scale.* Champaign, IL: Institute for Personality and Ability Testing.

Krug, S. E., Scheier, I. H., & Cattell, R. B. (1976). *Handbook for the IPAT Anxiety Scale.* Champaign, IL: Institute for Personality and Ability Testing.

Zoli Zlotogorski, Ph.D.
Professor of Psychology, The Hebrew University of Jerusalem, Mount Scopus, and Chief Neuropsychologist, Department of Psychiatry, Shaare Zedek Medical Center, Jerusalem, Israel.

Judith Guedalia, Ph.D.
Director, Pediatric Neuropsychology Clinic, Shaare Zedek Medical Center, Jerusalem, Israel.

ANN ARBOR LEARNING INVENTORY AND REMEDIATION PROGRAM
Waneta B. Bullock and Barbara Meister Vitale. Naples, Florida: Ann Arbor Publishers, Inc.

Introduction

The Ann Arbor Learning Inventory (AALI) and Remediation Program was designed to screen for central processing difficulties. It is a group test administered by teachers to form a "comprehensive picture of a student's specific disabilities" (Bullock & Vitale, 1982, p. i). The test's authors, Waneta B. Bullock and Barbara Meister Vitale, claim that it is useful in screening for learning disabilities on a large scale and in selecting proper learning programs relative to performance. According to the test authors, the AALI is "not for labeling but rather an instrument that gives the in-depth information on a group of students so that the teacher can teach to modality strengths and remediate specific central processing deficits" (Bullock & Vitale, 1982, p. i).

The test consists of two skill levels, A and B. At each level, the authors provide a student assessment booklet and its corresponding manual. Teachers are provided with fairly clear directions and instructions. Students are instructed to mark their answers in their booklets after receiving oral directions from their teachers. Time constraints are not listed, but the authors suggest a total test time of 60-90 minutes. After completing the test, a child's performance is evaluated in terms of level of task competency. An itemized list of scoring information is provided for each subtest. Finally, descriptive definitions of areas of weakness and lists of appropriate remediation techniques are provided. In this manner, the test authors claim that the AALI can identify the student's modality strength and adapt the student's curriculum to the modality.

Skill Level A (Grades K-1) is designed to measure seven elements: 1) body image, 2) visual discrimination skills, 3) visual motor coordination skills, 4) visual sequential memory skills, 5) aural discrimination skills, 6) aural sequential memory skills, and 7) aural conceptual skills.

Each of these elements is divided into subtests designed to assess processing skills. In the body image section, children are required to draw themselves. The

26

visual discrimination skills section is made up of six subtests, including those requiring the child to find differences and likenesses among shapes, letters, numbers, and objects. Items in the six subtests are presented both horizontally and vertically. The visual motor coordination skills section consists of five subtests that test near-point and far-point copying skills. The items in these five subtests also are presented both horizontally and vertically. The visual sequential memory skills section contains three subtests requiring the child to match numbers and letters and words and complex numbers. Target items are presented orally by the teacher. In the third subtest in the section, the child is requested to write the number or letter sequence after the teacher says it. The aural discrimination section also has three subtests. In this section, the child is asked to discriminate likenesses and differences in words by circling either a happy face or a sad face. In the Rhyming subtest, the teacher orally presents the child with a pair of words that may or may not rhyme. The child responds by circling either a happy or a sad face. The aural sequential memory skills section requires the child to recall a string of numbers, letters, and words. Finally, the aural conceptual skills section is designed to assess the child's ability to understand and follow directions.

Skill Level B (Grades 2-4) is composed of five sections: 1) visual discrimination skills, 2) visual motor coordination skills, 3) sequential memory skills, 4) auditory discrimination skills, and 5) comprehension skills. Essentially, the Skill Level B subtests are directly comparable to the Skill Level A subtests described previously. Score sheets are provided to summarize ideographic data as well as performance scores.

Practical Applications/Uses

The AALI is designed by its authors to help the teacher determine "each student's task competencies and deficiencies and select the proper text materials or series of texts to ease their difficulties" (Bullock & Vitale, 1982, p. i). The test is group-administered to kindergarten through elementary level students or "older students whose proficiency levels have not met basic requirement for their grade levels" (Bullock & Vitale, 1982, p. i). However, it can be administered individually. The teacher scores the test by hand directly from the pupil booklets. The manual provides nominal standards for judging correct and incorrect responses. There are scant criterion provided for evaluating the quality of the response. The evaluation form contains an "errors allowed" section, and, based on the child's performance (a few or a lot of errors), remedial suggestions are made. No normative data are provided, nor do the authors make any serious effort to provide developmental norms for the various skills tested.

Technical Aspects

Despite the test author's rather ambitious claims for the utility of the AALI, no reliability or validity data are provided. In addition, no references are provided in support of the claim that the AALI is a criterion-referenced test. In the present exhaustive review of the literature, no empirical evidence for either the psychometric properties or the utility of the instrument were found. In addition, the manuals are written in vernacular that seems to patronize the test administrator.

Critique

The Ann Arbor Learning Inventory and Remediation Program was designed as a screening and remediation device for central processing disorders. Many of the test items seem to be strikingly similar to those of the Detroit Tests of Learning Aptitude, the Auditory Discrimination Test, the Wide Range Achievement Test, the Illinois Test of Psycholinguistic Abilities, the Bender Visual Motor Gestalt Test, and others. No mention of these sources are found anywhere in the test material.

It seems that a fully trained learning disabilities specialist might find the AALI a useful device for screening learning disabilities in a large group, based on a qualitative analysis of the results. However, suffice it to say that the use of standardized, reliable, and valid screening tests in this important assessment area is highly recommended and tests lacking those psychometric properties should be discouraged strongly.

References

This list includes text citations and suggested additional reading.

Bullock, W. B., and Vitale, B. M. (1977). *Ann Arbor Learning Inventory, manual skill level A.* Naples, FL: Ann Arbor.

Bullock, W. B., & Vitale, B. M. (1977). *Ann Arbor Learning Inventory, manual skill level B.* Naples, FL: Ann Arbor.

Bullock, W. B., & Vitale, B. M. (1982). *Ann Arbor Learning Inventory, manual skill level A* (rev. ed.). Naples, FL: Ann Arbor.

Bullock, W. B., & Vitale, B. M. (1982). *Ann Arbor Learning Inventory, manual skill level B* (rev. ed.). Naples, FL: Ann Arbor.

Louis M. Hsu, Ph.D.
Professor of Psychology, Fairleigh Dickinson University, Teaneck, New Jersey.

ASSESSMENT OF SUICIDE POTENTIAL (RESEARCH EDITION)

Robert I. Yufit and Bonnie Benzies. Palo Alto, California: Consulting Psychologists Press, Inc.

Introduction

The Assessment of Suicide Potential (Research Edition) (Time Questionnaire; TQ) is a 39-item, semiprojective (and semi-objective) instrument for the measurement of "suicide potential." Yufit and Benzies (1979) describe "suicide potential" as the "probability that a person will consciously commit an act destined to destroy his life" (p. 9). The major rationale of the TQ is that persons who have pervasive conscious thoughts about suicide will a) not have a developed future perspective (p. 2); b) have a negative present self-appraisal; and c) have conflicted, guilt-ridden recall of the past (p. 3). Reflecting this rationale, the TQ is divided into three sections: Future (17 items), Present (15 items), and Past (7 items). Five principal variables, each of which is expected to be negatively related to suicide potential, are defined in terms of the 39 item responses: a Future score, a Present score, a Past score, a Total score, and a "Number of Years Projected" score. The last of these scores is determined from the first item in the Future section, which asks the respondent to select some year in the future and answer a set of questions as if he or she were living in that future year. The Number of Years Projected score is simply the difference between the year selected by the respondent and the current year. Four additional scores are determined: an F (Falsification) score, which is considered to measure truthfulness or denial; an O (Omissions) score, which is considered to reflect evasive, apathetic, and uncooperative attitudes, but which is also viewed as an indicator of suicide risk; a B (Bizarre) score, which is considered to reflect psychotic processes, sarcasm, and hostility; and a U (Unscorable) score, which is considered to indicate lack of adherence to the directions.

Yufit and Benzies's motivation for developing the TQ included a) dissatisfaction with what they viewed as the "generally unproductive . . . traditional psychological techniques" (Yufit & Benzies, 1979, p. 1) for the estimation of suicide intent; b) the senior author's determination, from analysis of psychiatric patients' autobiographies, of a relation between future time-perspective and suicidality; and c) the lack of attention that time-perspective had received in the psychological literature—the authors note that time-perspective has only been studied systematically in philosophical essays.

For many years, the senior author included an autobiographical technique as part of a diagnostic test battery in his work with psychiatric patients. This technique was formalized by asking patients to project themselves into the future and write a

"Future Autobiography" (Yufit & Benzies, 1979, p. 1). The technique was then modified by providing patients with an "Ideal Future" form, in which they were asked to select a future year and then to answer questions relating to their status and activities in that year. The 1979 revision of the questionnaire contains 25 multiple-choice and 14 open-ended questions dealing with the present and past as well as with the future. According to the authors, the current (i.e., 1979) version of the TQ "represents a decade of evolution from a purely clinical technique to a semi-objective, semi-projective instrument for suicide assessment" (Yufit & Benzies, 1979, p. 2).

Yufit's earlier published works in the area of suicide risk measurement include a chapter in *Suicide and Bereavement* (cited in Yufit & Benzies, 1979) and two papers, one in the *Archives of General Psychology* (1970) and the other in the *Journal of Life Threatening Behavior* (1973). Two other relevant papers are listed in the reference section of the manual as "in preparation"; however, a computer search (BRS, Information Technologies) of *Psychological Abstracts* and *Index Medicus* did not list any post-1979 publications by the senior or junior authors.

The TQ is printed on four 8½" × 11" sheets. The first sheet lists the 15 items of the Present section; the second sheet lists the 17 items of the Future section; and the third sheet lists the 7 scorable items of the Past section, together with an additional three items that the authors indicate should be evaluated "qualitatively." These last three items are attributed to Erikson and Winnicott (Yufit & Benzies, 1979, p. 22). The last sheet is a Personal Information Sheet, which the authors indicate provides "a baseline for scoring the Future section" (Yufit & Benzies, 1979, p. 22). The authors indicate that the TQ has been administered to persons from the ages of 16 to 81, but they note that it is not considered useful with grossly psychotic, organically brain damaged, or mentally retarded persons. Although there is no time limit, the instructions on the first sheet urge the examinee to answer the questions quickly. The authors note that most people need 10 to 15 minutes, but that very depressed persons may take longer, although usually not more than 25 minutes. They also note that few people have difficulty following the directions.

Practical Applications/Uses

Yufit and Benzies (1973) state that "the TQ consistently and significantly differentiates suicidal high risk patients from both nonclinical controls and from such clinical comparison groups as nonsuicidal psychiatric patients and suicidal low risk patients" (p. 270). This would seem to suggest that the TQ may be used in clinical settings for the diagnosis of suicidal intent. However, the publisher of the TQ includes on the title page of the manual the following cautionary note: "The [TQ] is an experimental assessment technique for use by the clinically sophisticated psychologist in research investigations for evaluating suicide potential. At present it is not recommended for clinical use" (Yufit & Benzies, 1979). This reviewer's evaluation of the TQ suggests that, at its present stage of development, it likely is to be primarily of use to researchers who a) are interested in the TQ authors' theory concerning the relation of time-perspective to suicidality or b) wish to supplement their batteries of suicide intent predictors, with the seldom used time-perspective variables measured by the TQ. For reasons to be presented later, this reviewer completely agrees that the TQ is not ready for clinical use.

The TQ may be administered orally or in written form by "properly trained" examiners (Yufit & Benzies, 1979, p. 4). Oral presentation is recommended for very depressed patients, but in general, the written format is preferred, both because the concurrent, overt interaction with the examiner is minimized and because the current norms are based on the self-administration of the TQ. The authors consider that the proper training of examiners consists of at least one graduate course in testing and practical experience under the supervision of a registered or licensed psychologist. The different sections of the TQ should be presented in the order in which they appear on the printed form.

Approximately 14 of the 23 pages of the manual are devoted to descriptions of various aspects of scoring. The description of scoring multiple-choice items is clear, albeit very lengthy when compared to other instruments of this type. There should be no interexaminer disagreement in scoring of these items. The description of scoring the open-ended items is also very lengthy but need not result in interexaminer agreement because subjective examiner judgments are involved. Fourteen of the 39 scorable TQ items involve subjective judgments. Although measurement errors associated with open-ended questions may be unavoidable, it should be noted that some of the scoring instructions introduce the possibility of measurement errors that could have been avoided. For example, the manual calls for scoring omitted items first, using the table of weights for omitted items. Then, the examiner is instructed to score the multiple-choice items and the open-ended items, using the lengthy instructions provided in the manual. A reasonable strategy then would be to score items the responder omitted and then to score the remaining items. However, this could lead to scoring errors. For example, Item 4 in the Future section should be scored "-2" if omitted, according to Table 6 on page 10, which provides information about scoring of omitted items. However, five pages later, the manual instructs the examiner to score the same item "0" if the omission comes from an unmarried responder.

Another possible source of measurement error comes from the authors' description of how the total scores should be determined. Yufit and Benzies (1979) instruct the examiner to subtract the negative weights from the positive in order to obtain the totals. Most anyone who has had a high school algebra course would interpret this to mean that the signs of the negative weights are to be changed to +'s and that all weights should then be added. But this is clearly not what the authors had in mind! Another unnecessary source of difficulty in the instructions for scoring comes from the authors' referral of the examiner to nonexisting pages for determination of scoring weights. On page 9, the authors direct the examiner to "pages 00 and 00" for multiple-choice item scoring weights, and on page 10, they direct the examiner to the same nonexistent pages for open-ended scoring weights.

The authors warn that "the interpretation of scores is often subjective and inferential" (p. 3). Certainly, the absence of any clearly identified norms in the manual would be consistent with that view. On the title page of the manual, the publisher states that ". . . while strongly negative scores suggest the likelihood of high suicide potential, the absence of such high negative scores is no assurance that high suicide potential is not present." Some of the evidence related to this belief will be critically evaluated in the discussion of validity in the next section of this critique.

Technical Aspects

Four types of reliability coefficients are listed in the manual: 1) test-retest correlations, 2) pairwise correlations of the five principal scores, 3) (so-called) interrater reliability by groups, and 4) interrater reliability by items. The largest reliability coefficients reported (with a couple of exceptions) are of the third type. However, this reviewer will argue that the sizes of these coefficients probably primarily reflect scaling characteristics of the items and not interrater reliability. Reliability information provided in the manual will be described and evaluated in the order listed above. It should be noted that no information is provided in the reliability section of the manual about the source of the statistics reported therein. In fact, all of these statistics come from Yufit and Benzies (1973).

Five 1-week test-retest reliability coefficients are reported in the manual, one for each of the five principal scores. These coefficients range from .54 (for the Past scores) to .78 (for the Future scores) and were computed on scores obtained for "25 hospitalized patients, most of whom were suicidal" (Yufit & Benzies, 1979, p. 6). However, the reliability section of the manual provides no information about a) how or why these patients were selected, or b) these patients' demographic characteristics. From the information provided in the manual and in Yufit and Benzies (1973), it would appear likely that these patients constituted a homogeneous group, as compared to the combined patient and normal groups tested by the author. It is therefore reasonable to expect that, due to range restriction effects on the reported reliability coefficients, the 1-week test-retest reliability coefficients for the combined groups would be higher than those reported in the manual. In addition, the authors provide virtually no information about how they established that "most" of the 25 patients were suicidal. And the difference between the largest and smallest reported reliability coefficients may, in part, reflect the effects of the differences in the lengths of the corresponding scales (the Future section, for example, has 2.43 times as many items as the Past section).

Intercorrelation of the five principal scores of the TQ are reported in the manual for the 25 patients used in the test-retest study. One set of correlation coefficients was determined from the admission scores, and the other from the retest scores. The authors report that

> for the admission TQ's the [Present, Future, Past, and Total] scores intercorrelated at the .05 level or better, but did not correlate with the number of years projected. For the retest Questionnaires, however, all five variables intercorrelated at the .01 level of significance with the exception of the Number of Years Projected versus the Present Section ($p < .05$) and the Number of years Projected versus the Past Section [N.S.]. (Yufit & Benzies, 1979, p. 6)

However, it is incorrect to apply ordinary tests of significance of Pearson rs (which were apparently used to obtain the p-values reported in Table 2 of the manual) to rs of pairs of variables that share items because of the spurious effects of the overlapping items on these correlations. It should be noted that, out of the 10 pairwise correlations in each set, 3 involve overlapping items keyed in the same way (viz., correlations of Present, Future, and Past scores with Total scores) and that with one exception it is these 6 correlations that are the largest in their respective sets. In fact, the two highest correlations (.93 and .96) were obtained for the pair of scales that share the most items, the Future and Total scales, which share 17.

This reviewer subjected the intercorrelation of the admissions scores, as well as of the retest scores (leaving out Total scores) to principal-components analyses (PCAs), to determine if there was evidence of multidimensionality of the construct measured by the four scales of the TQ. Neither the PCA for the admissions data nor that for the retest data yielded more than one component with an eigenvalue greater than 1.00. Therefore, the interscore correlations reported in the manual are consistent with the view that a single construct (which could perhaps be called *time-perspective*) is measured by all scales of the TQ.

The manual reports what are described as interrater reliability coefficients by groups. Specifically, eight completed TQs from each of four groups (a nonclinical control group and three psychiatric groups reported to differ in degree of suicidal tendencies) were selected at random for scoring. Only the 14 items requiring some degree of subjective judgment were analyzed. The raters consisted of the two principal investigators, two experienced raters, and three "lesser experienced" raters. The data were analyzed by group, "between each pair of raters for the eight TQ's in each group, giving a Pearson *r* based on 112 items" (p. 7). The means of the resulting correlations are reported in Table 3 of the manual. Two things should be noted about these correlations. First, the authors apparently calculated, for each pair of raters and each group, the correlation of two columns of item scores. In column 1 were the 112 scores assigned to the subjects on the subjective items by the first rater; in column 2 were the scores assigned to these subjects by the second rater. This reviewer would argue that such correlations are quite meaningless because of the scoring system used for the 14 subjective items in the TQ. More specifically, the range of scores that can be obtained for any subjective item differ from item to item. For example, the maximum score permitted by the scoring instructions for item 1 of the Future section is +4, but the maximum score for item 2 in that section is +1. Therefore, the size of the correlation reflects, to an unspecified degree, differences in allowable scores of different items. A more extreme example may further clarify this point. If one item had a permissible scoring range extending from +5 to +10 and another had a permissible range of -10 to -5, then the interrater correlation of the item scores of a group of *N* subjects would have to be large and positive, even if both raters assigned their scores completely at random. Second, the manual provides no information about how interrater reliability coefficients differ as a function of level of training or experience of the examiners. However, some information on this subject is provided in Yufit and Benzies (1973). The authors note that "most of the low coefficients involved the principal investigator who has been least involved in the routine scoring of TQ's. When these results were discussed by the team of raters, it emerged that the principal investigator was using a more intuitive and subjective approach when scoring, and not adhering to the scoring manual as were the other raters" (pp. 280-281).

Mean interrater reliability coefficients are reported by item (for the (14)x(8) subjects described in the preceding paragraph) in Table 4 of the manual. These means are described as ranging from .43 to 1.00, with an average value of .86. The authors' description of this portion of the reliability study appears, to this reviewer, to involve several factual inconsistencies. However, the relative lack of importance of the interrater item reliabilities does not warrant allowing space to any further discussion of these inconsistencies.

The information that the manual provides concerning the validity of the TQ can be categorized as follows: a) TQ scores of persons who actually committed suicide, b) information available in studies listed in the manual's reference section, c) criterion-group validity data, and d) what the authors describe as construct validation data.

The information provided by the manual on how well the TQ predicts actual successful suicides is extremely limited. The manual notes that "of 1190 patients and controls tested by the authors (to 1976) five patients are known to have committed suicide. All five had very high negative total scores (greater than -30)" (p. 8).

The manual of any technique developed to predict suicide should report information about the selectivity, sensitivity, and hit rates of the technique as precisely and fully as possible. If we interpret the preceding quotation to mean that up to 1976 the authors knew of only five suicide cases for whom TQ scores were available, then we could infer that the sensitivity of the TQ is 100%, given that a total score of -30 is adopted as the critical cutoff score; that is, adopting a cutoff of -30 for the diagnosis of suicide would have resulted in 100% correct diagnoses for those who actually committed suicide. However, the information provided in the quotation does not permit calculation of the selectivity (i.e., the reader has absolutely no data about the percentage of nonsuicide cases who would have been diagnosed as nonsuicidal by the same diagnostic rule). There is a more serious problem, however. Upon reading one of the articles listed in the manual's references (Yufit, Benzies, Fonte, & Fawcett, 1970), this reviewer found a table summarizing information of seven case studies, which the authors describe as "a subsample of patients manifesting serious suicidal behavior" (p. 161). Of these seven cases, two are listed as having died. One died 24 hours after ingesting a bottle of formaldehyde, and the second died after ingesting strychnine. Neither of these individuals had negative TQ total scores! One had a total score of 0, and the other a total score of +19. This reviewer cannot reconcile this information with the authors' statements in the manual that "of 1190 patients and controls tested by the authors (to 1976) five patients are known to have committed suicide" and that "all five had very high negative total scores (greater than -30)." It might also be noted that none of the other five "patients manifesting serious suicidal behavior" (in the article by Yufit et al., 1970) had negative scores. Three of these patients were listed as comatose following suicide attempts. Of these three, one had a total TQ score of +15, another a score of +13, and the third had a score of +6 (as determined from the table) and +4 (as determined from the text).

The description of the findings of two validity studies listed in the manual's (Yufit & Benzies, 1979) reference section is very brief:

> A . . . study [i.e., an unpublished (?) 1974 dissertation by Flynn] has been completed comparing the TQ with another instrument relating to time orientation. The study reveals that time perspective does shift when the suicidal crisis is judged to be past. However, when the suicidal wish is present . . . the time perspective appears to remain relatively constant with low future perspective and high orientation to the past. TQ scores are also correlated with some measures of depression (Yufit et al., 1970). (p. 8)

The manual provides no statistics to support the statements in this quotation. Without access to Flynn's dissertation, this reviewer was unable to evaluate the authors' statements in the first part of the previous quotation. Concerning the statement about the correlation of TQ scores with "measures" of depression, this reviewer found in the Yufit et al. (1970) paper one correlation ($r = .50$) of the TQ total score with the Zung Self-Rating Depression Scale (Zung, 1965).

The principal validity statistics presented in the manual consist of means and standard deviations of Total scores, Future scores, and Years Projected for four criterion groups (inpatients, outpatients, staff, and college groups). These statistics, together with group sizes (44, 26, 35, & 57), are reported in Table 5 of the manual. The authors note that

> mean scores of "normal" and patient groups show large differences in the expected direction, and inpatient groups obtain more suicide-prone scores than outpatients. This progression of scores applies to Total score, Future score, and Number of Years Projected. (Yufit & Benzies, 1979, p. 9)

No information is provided in the manual concerning the source of the statistics in this table. And no information (other than group labels) is provided concerning the characteristics of subjects in the criterion groups. Upon reading the Yufit et al. (1970) article, this reviewer found that Table 5 in the manual is identical to Table 2 in the article, even though this fact is not reported in the manual. Another table that appears in Yufit et al. (1970) shows that subjects in the inpatient group were markedly older (mean age = 43) than subjects in the outpatient group (mean age = 29) and than subjects in the "normal" groups (mean ages = 28 and 33). This fact is not reported in the manual. An attempt is made by Yufit et al. (1970) to control for possible age effects; groups were dichotomized at age 35, and clinical groups were then compared to control groups on the TQ variables within each of the two age categories. Yufit et al. (1970) report generally significant differences between these contrasted means. This procedure is incorrect. There is no assurance that when two groups differ in locations of their age distributions, dichotomizing these groups at some age value will result in the elimination of age differences between subgroups belonging to the same category of the age dichotomy. In fact, one could be virtually certain that it will not. A much better procedure would have been to test for group differences in TQ means that have been adjusted for age effects by means of analysis of covariance. Perhaps the greatest limitation of the information in Table 5 is that it is not at all clear what differences in group means (adjusted or not for age effect) on these variables indicate about validity of the TQ variables as predictors of *suicide*. The information provided by Yufit et al. (1970) concerning the groups is that

> the clinical group ($N = 70$) consisted of two subgroups of persons with serious psychological problems, one group being hospitalized on a Depression and Suicide Prevention Research Unit . . . ($N = 44$) the other group consisting of persons seeking, or already in individual psychotherapy as clinic or private outpatients ($N = 26$). Both clinical groups consisted of diagnostically heterogeneous patients but with an emphasis on classification of either primary depressive syndrome (with or without suicidal potential) or schizophrenic with depressive features. (p. 159)

The clinical groups are later referred to as "suicide prone" and the comparison groups are later referred to as "not suicide prone" in the article. However, no information is provided in Yufit et al. (1970) or in the manual concerning what fraction of each group are actually suicide prone as opposed to being suspected of being suicide prone by virtue of their group membership. In other words, it is not clear that group TQ variable mean differences, if such exist when effects of age and possibly other nuisance variables are *correctly* partialled out, reflect the ability of these variables to discriminate between suicidal and nonsuicidal persons, as opposed to between disturbed and nondisturbed.

The last type of validity information provided in the manual consists of a brief statement by the authors: "Efforts to establish construct validity are being made by comparing the degree of agreement among psychotherapists' ratings of suicide potential and the TQ scores of the same patients. There is 72% agreement between such clinical judgments and TQ scores" (Yufit & Benzies, 1979, p. 9). Three things should be noted here. First, psychotherapists' judgments of suicide potential are, at best, of questionable validity (see, for example, Arbeit & Blatt, 1973). Second, the authors provide no information about sources in which further information about this work may be obtained, and no information is provided about numbers or characteristics of the psychotherapists, of the patients, or of the TQ scores in this study. Last, in light of this, it is totally unclear what the "72% agreement between . . . clinical judgments and TQ scores" means.

Critique

Yufit and Benzies developed the Time Questionnaire because of their dissatisfaction with what they described as the "unproductive . . . traditional psychological techniques" (1979, p. 1) for the estimation of suicide intent. They focused on a single construct (time-perspective), which they must have felt, in the context of their theory and clinical observations, would be more productive than or would contribute to the traditional methods. The objectives of the TQ are therefore clear: 1) the five principal scales of the TQ attempt to tap different facets of the time-perspective construct, and 2) the TQ scale scores attempt to predict suicide. Given these objectives, the criteria for evaluation of the TQ appear to be equally clear. First, to what extent are the two goals attained in an absolute sense (i.e., without comparison to other techniques)? Second, in view of the fact that the TQ was developed because of dissatisfaction with traditional techniques for predicting suicide, a) how well does the TQ achieve its second goal, in comparison to more traditional techniques, and b) how much of an incremental increase in predictive validity would result from use of the TQ in conjunction with the traditional methods? The soundness of the single time-perspective construct theory, relative to, say, multiconstruct models of suicidal tendencies (e.g., Beck, Weisman, Lester, & Trexler, 1974; Blatt & Ritzler, 1974; Maris, 1981) or purely empirical approaches to suicide prediction (e.g., Pierce, 1981) could be viewed as a third criterion for evaluation of the TQ, but this criterion would not appear to have the importance of the first two, especially for practicing clinicians.

Evaluation of the TQ in terms of the second criterion would require comparison of the predictive and/or incremental predictive validity statistics of the TQ with those of more traditional methods when the competing procedures have been applied to the same groups of subjects. Virtually no statistics of this type are reported in the TQ manual or appear to be available anywhere else (to the best of this reviewer's knowledge).

Evaluation of the TQ in terms of the first criterion can be based on the current manual's information concerning the TQ's reliability and validity. Some of the reliability coefficients reported in the manual are large. Unfortunately, as was noted previously, most of the large reliability coefficients could be large for the wrong reasons. More specifically, large interscale coefficients could primarily reflect item overlap, and large interrater reliability coefficients could be primarily artifacts of item scaling. Concerning the validity information, it was noted that the group TQ variable mean differences reported in the two principal criterion-group validation studies could reflect differences in such factors as level of psychopathology (for the 1970 study) or psychotherapists' judgments of suicide risk (for the 1973 study) rather than differences in proneness to suicide. Improper controls of effects of nuisance variables were also noted. Thus, these studies only provide questionable indirect evidence of validity of the TQ. The absence of information about the selectivity of the TQ scales and the apparent inconsistency (between the manual and the 1973 study) of information concerning TQ characteristics of the few individuals who are reported to have committed suicide would further suggest that the TQ is not ready for clinical applications. Nevertheless, the plausibility of the authors' theory concerning the relation of time-perspective to suicide potential, the care manifested in selection of items consistent with this theory, and some existing evidence (albeit mostly indirect and somewhat ambiguous) of the test's validity suggest that the TQ should be studied further as a possible indicator of suicide proneness. Because of the limited and different focus (time-perspective) of this instrument, studies of its incremental predictive validity, when combined with other indicators of suicide such as those described by Beck et al. (1974), Maris (1981), and Pierce (1981), would seem particularly desirable.

References

Arbeit, S. A., & Blatt, S. J. (1973). The differentiation of simulated and genuine suicide notes. *Psychological Reports, 33*, 283-297.

Beck, A. T., Weissman, A., Lester, D. & Trexler, L. (1974). The measurement of pessimism: The hopelessness scale. *Journal of Consulting and Clinical Psychology, 42*, 861-865.

Blatt, S. J., & Ritzler, B. A. (1974). Suicide and the representation of transparency and cross-sections on the Rorschach. *Journal of Consulting and Clinical Psychology, 42*, 280-287.

Maris, R. W. (1981). *Pathways to suicide: A survey of self-destructive behaviors*. Baltimore: The Johns Hopkins Press.

Pierce, D. (1981). The predictive validation of a suicide intent scale: A five year follow-up. *British Journal of Psychiatry, 139*, 391-396.

Yufit, R. I., & Benzies, B. (1973). Assessing suicide potential by time perspective. *Journal of Life Threatening Behavior, 3*, 270-283.

Yufit, R. I., & Benzies, B. (1979). *Preliminary manual: Time Questionnaire: Assessing suicide potential*. Palo Alto, CA: Consulting Psychologists Press.

Yufit, R. I., Benzies, B., Fonte, M. E., & Fawcett, J.A. (1970). Time perspective and suicide potential. *Archives of General Psychology, 23*, 158-163.

Zung, W. W. (1965). A self-rating depression scale. *Archives of General Psychiatry, 12*, 63-70.

Judith L. Whatley, Ph.D.
Assistant Professor of Clinical Pediatrics and Psychiatry, University of Texas Health Science Center, San Antonio, Texas.

BAYLEY SCALES OF INFANT DEVELOPMENT

Nancy Bayley. San Antonio, Texas: The Psychological Corporation.

Introduction

The Bayley Scales of Infant Development (BSID) provide a three-component approach to the assessment of children's developmental status from 2 months to 2½ years of life. Two scales, the Mental Scale and the Motor Scale, result in quantitative standard scores. The Mental Scale is intended to assess sensory-perceptual abilities, object constancy, memory, learning and problem-solving ability, communication and verbal skills, and early abstracting ability. The Motor Scale is intended to measure gross and fine motor skills and control of the body. The third component, the Infant Behavior Record, provides an assessment of social and objective orientation toward the environment. Designed for use in both research and clinical practice, the scales provide an assessment of a child's current developmental status in comparison with normatively based expectations. In practice, the Mental and Motor Scales are more often employed, both by researchers and clinicians, than is the Infant Behavior Record.

The BSID culminate more than 40 years of research on the part of Nancy Bayley and her colleagues. Bayley was motivated to develop an assessment tool for infant development for which the standardization sample would be representative of normal children. Hers was a reaction against the relatively small samples of institutionalized children upon which many tests were based at the time her work began.

Bayley (1969) acknowledges the contribution of the California First-Year Mental Scale (Bayley, 1933), the California Preschool Mental Scale (Jaffa, 1934), and the California Infant Scale of Motor Development (Bayley, 1936) in the development of the BSID. An unpublished 1958 version of the BSID covered the first 15 months of life and was employed in a research program sponsored by the National Institute of Neurological Diseases and Blindness. These scales were then expanded to include the second year of life by taking items from the California Preschool Mental Scale and developing new items. This 1958-60 version was subsequently used in research sponsored by the National Institutes of Health. Using data from approximately 1,400 children aged 1 to 15 months and a sample of 160 children aged 18 to 30 months, items were selected that became the Mental and Motor Scales of the current edition.

The scales were standardized on a stratified sample of non-institutionalized children selected on the basis of the 1960 United States Census. The norms for the BSID are based upon this normal, English-speaking sample. Norms and standardized procedures have not been published for non-English-speaking populations, specific handicapped groups, or premature infants, although the scales have been used with such groups.

The Infant Behavior Record was developed from rating scales originally used with the sample from which the earlier California Mental and Motor Scales (Bayley, 1933, 1936) were developed. A 1958 unpublished version of the rating scales was administered to approximately 1,350 children; this research resulted in the present edition.

The BSID consist of 163 items on the Mental Scale and 81 items on the Motor Scale. Only those items within the range of performance, basal to ceiling, of a given child are administered. Items in the 2-to-5-month range are normed at half-month intervals; those in the 6-to-30-month range are normed at 1-month intervals. In general, test materials are presented to the child in a specified way, and the child's reactions to or behavior with the stimulus is observed and scored by the examiner. Other spontaneous behaviors that may be expected of a child in a given age range are also scored for their occurrence or nonoccurrence.

Materials for administration of the examination include a set of standardized, manipulable objects, as well as objects for visual and auditory presentation: a set of 1" cubes, a red ring with attached string, cups, a saucer, red crayons, a rattle, a red ball, sugar pellets, a bell, a whistle doll, a picture book, puzzles, a jointed doll, a small broken doll, picture cards, a pull toy, spoons, and a non-breakable mirror. The manual also includes building specifications for a set of stairs and a walking board for assessment of motor skills for children in the older age ranges. (These standardized materials are available through The Psychological Corporation.) Other materials that must be supplied by the examiner include a table, a crib for the testing of young infants, white paper, and tissue.

The Infant Behavior Record consists of a set of 30 descriptive rating scales for behaviors characteristic of children in the first 2½ years of life. Twenty of these employ a 9-point scale, six use a 5-point scale, three use a yes/no format, and one asks for a Normal/Exceptional general evaluation. These rating scales focus on the child's social orientation, emotional tone, object orientation, attention span, goal directedness, interest focus, energy, overall evaluation of the child's performance, and representativeness of test performance. The Infant Behavior Record is based upon observations during the assessment and is completed after the exam.

Practical Applications/Uses

The Bayley Scales of Infant Development are used primarily in research settings and in clinical practice when there is a question about developmental status or when a high-risk/special population is being followed. Because of the skill level and time involved in administration, it is less likely to be used as a routine screening instrument.

The BSID were standardized on a full-term, nonhandicapped population. Since its development, it has frequently been used with children whose development is in question due to premature birth or other birth-associated risk factors. Most commonly, when BSID scores are computed for premature infants, the age of the infant is corrected for the prematurity. Rhodes, Bayley, and Yow (1984) cite Hunt and Rhodes (1977) for further information about age correction.

Use of the BSID with handicapped children is difficult and often impossible because items on the Mental Scale often require visual and auditory abilities. At all but the earliest age ranges, required responses involve reaching, gesturing, or otherwise manipulating test materials. No standardized version of the BSID has been developed for use with handicapped, blind, or deaf children, although some researchers and clinicians have employed modifications of the standardized procedure (DuBose, 1977; Kierman & DuBose, 1974). Experienced clinicians may wish to use items from the BSID as part of an assessment with these children, but no valid score is obtainable.

General instructions as well as the administration procedure for each item on the Mental and Motor Scale are detailed in the manual (Bayley, 1969). Additional clarifications are provided in the recent manual supplement (Rhodes, Bayley, & Yow, 1984). In general, the instructions are sufficiently detailed to inform the examiner of the proper procedure for administering each item. (Videotapes for use in training examiners are available through The Psychological Corporation, and training films may be rented from the Extension Media Center, University of California, Berkeley.)

The BSID are best administered in a room large enough to contain an examining table, chairs for adults, a crib or youth chair (depending on the age of the child), and free floor space for testing children who locomote. The room should be pleasant but not distracting. Optimally, other children and unnecessary adults should not be present to distract the child being examined. Rhodes, Bayley, and Yow (1984) remind examiners that the test materials should be managed in such a way that only materials being employed at the moment be visible to the child and that the child not have access to the equipment kit and its materials.

The role of the examiner is a very active and integral part of the assessment. The examiner administers test items, motivates the child to perform, observes and scores administered items and scorable spontaneous behavior, and is a stimulus toward whom the child responds. The attendant parent, at the examiner's request and direction, may administer items as instructed.

BSID norms are based on testing in a laboratory setting; however, children may be tested in a clinic or at home provided that care to limit distractions is taken. Durham and Black (1978) have reported data on infants 16 to 21 months old that showed that infants assessed at home scored higher on the Mental Scale than when they were assessed in laboratory settings. Verbal performance appeared to be especially affected.

Administration of the BSID requires training and practice. Examiners need to be able to relate well and easily to infants and their parents and should also have some background in child development and psychometrics. Most often the BSID are administered by psychologists, child development specialists, and occupational and physical therapists. The BSID are not difficult to administer, but neither can they be considered easy. This is not a test that can be administered as one reads along in the manual.

The scales are to be administered with the parent or appropriate substitute present, generally someone with whom the child is at ease. Bayley recommends that the examiner sit across from the infant rather than to the side in order to facilitate observation. Items are administered by the examiner but may be administered by the parent at the examiner's request and instruction.

The Mental Scale is to be administered before the Motor Scale due to the change of pace involved. However, many of the early items of the Motor Scale may be observed incidentally, or interspersed among items of the Mental Scale. In both the Mental and Motor Scales there is no set order for administration of items, and administration should be guided by the child's interest, energy, and attention. Basal and ceiling level criteria consist of 10 successive items passed or failed on the Mental Scale and 6 on the Motor Scale.

The Infant Behavior Record is completed following administration of the Mental and Motor Scales. The examiner selects the one most descriptive statement for each of the rating scales and may also note additional observations. Except for three items, ratings are characterizations of the child's behavior without reference to the normal distribution of the characteristic. For the three specific items, ratings are to be estimated with respect to other children of the child's age.

The time for testing will vary depending on the age and characteristics of the child. Bayley (1969) has estimated 45 minutes as the average amount of time to administer both the Mental and Motor Scales, with only 10% of cases requiring as much as 75 minutes. Completion of the Infant Behavior Record and computation of scores requires additional time.

Scoring of the BSID uses three forms, one for each of the two scales and one for the Infant Behavior Record. The Mental and Motor Scale forms provide descriptive phrases for each item and a space for noting "Pass" or "Fail" status or other notation. In addition, the record forms for these scales indicate the age at which 50% of children demonstrate an item and the range of ages that 5-95% of children pass an item. There are also situation codes noted on the form that facilitate the examiner's orderly administration of the examination. These codes draw the examiner's attention to the use of the same stimulus material or procedure for multiple items at different age levels that can be scored in one or a few presentations. The Infant Behavior Record form contains the rating scales and space for additional observations the examiner might wish to add. All three forms provide space for background information, computation of age at testing, and conversion of raw scores to standard scores.

Items on the Mental and Motor Scales are scored on a Pass/Fail basis, and only responses observed by the examiner may be scored as a "Pass." Even though there is a space provided to note responses reported by the parent, credit is not given for these responses. Scoring is intimately related to the administration of the test. Each item is scored as it is administered or observed; additionally, any scorable response is scored at any time that it is observed. Raw scores are obtained by summing items passed on the Mental and Motor Scales and adding this number to the basal level score for each scale. Using tables in the manual, raw scores are converted into standard scores known as the Mental Index and the Psychomotor Index. The child's date of birth is used as the reference from which conversions are made. Age tables making the conversion are broken down by 2-week intervals for the age range 2 to 6 months and by monthly intervals for the 6-to-30 month range. The Mental Index and the Psychomotor Index have a mean of 100 and a standard deviation of 16. Bayley (1969) cautions that these are developmental quotients and not IQs and offers instructions for individuals wishing to convert the raw score into an "age equivalent" rather than a standard score. An age equivalent score, however, does not offer the psychometric advantages of the standard score. No subtest scores are derivable from the two scales. Ratings from the Infant Behavior Record and the comparison percentages for these ratings add to the clinical evaluation of the child's social, behavioral, and emotional functioning during the test situation. These ratings may be examined using norm-referenced tables provided in the manual. The sample size for the Infant Behavior Record was relatively small for children 15 months and younger, and Bayley urges caution when interpreting these ratings (Bayley, 1969).

The BSID offer the advantages of a representative, normative sample and standardized scores with statistical properties useful for interpreting scores. Interpretation is based on reference to the norms established for the test. Clinical judgment plays a role in decision-making concerning the extent to which the child's performance is a fair representation of his or her ability. Such judgment is also called into play in developing hypotheses concerning the pattern of results that make up a given score, differences encountered between Mental and Psychomotor Indices, and the meaning of a score for a child from certain population groups. Clinical experience with children and an understanding of child development are also necessary to understand the meaning of a given score. Unfortunately, the manual offers little to educate or facilitate interpretation of scoring beyond statistical and standardization data.

Interpretation of the Infant Behavior Record makes use of percentage tables for individual rating scales. However, use of the rating scales and their interpretation calls upon the clinical skill and judgment of the examiner even more than do the Mental and Motor Scales.

Technical Aspects

The advantage of the Bayley Scales of Infant Development in comparison to other tests of infant development has been and continues to be its standardization and statistical properties. Norms of the BSID are derived from a stratified sample based upon the 1960 U.S. Census, controlling for sex, color within age group, urban/rural residence, and education of head of household. Bayley acknowledges that rural children are somewhat underrepresented but considers the effect on the norms to be negligible. The children making up the sample were tested at 14 ages: 2, 3, 4, 5, 6, 8, 10, 12, 15, 18, 21, 24, 27, and 30 months. These children were located through hospitals, well-baby clinics, municipal birth records, and social agencies, but included only "normal" children living at home. Excluded were institutional children with severe behavioral or emotional problems, those born more than 1 month prematurely, and children over 12 months of age from bilingual homes who showed significant difficulty using English. In all, 1,262 children were tested.

Bayley (1965) examined the effects of socioeconomic variables on scores using the 1958-60 precursor of the current scales. Although the difference was small, she noted significantly higher scores on the Motor Scales by black infants at all ages from 3 through 14 months. No other differences due to sex, birth order, geographic location, or parental education were found on either the Mental or Motor Scales.

Statistical properties are presented in the manual for the current scales and for research on earlier versions. Split-half reliabilities were computed for both the Mental and Motor Scales. Mutually dependent items were grouped on the same half of each scale, and reliabilities were computed separately for each age group in the standardization sample. Coefficients corrected by the Spearman-Brown formula estimate the reliability of the full-length scales. Resulting reliability coefficients for the Mental Scale ranged from .81 to .93, with a median value of .88. For the Motor Scale, values ranged from .68 to .92, with a median value of .84. The lower reliabilities were obtained on the Motor Scale when testing infants under 6 months of age. Overall, the reliability values compare quite favorably with other infant tests.

Bayley also reports standard error of measurement values for each scale by age group tested. For the Mental Scale, these range from 4.2 to 6.9. On the Motor Scale, the values range from 4.6 to 9.0.

Werner and Bayley (1966) report tester-observer and test-retest reliabilities (for a 1-week time difference) for a sample of 8-month-old infants tested with the 1958-1960 precursor of the BSID. Mean percentage of agreement between observers on the Mental Scale was 89.4 with a standard deviation of 7.1; on the Motor Scale, mean percentage agreement was 93.4 with a standard deviation of 3.2. These figures document a high level of agreement but are unfortunately limited to a restricted age range and are not based on the current version of the BSID. Test-retest reliabilities for a subsample of these babies showed mean percentage agreement on the Mental Scale of 76.4; however, the same limitations pertain. Reliability estimates are not provided in the manual for Infant Behavior Record ratings.

Reliability across test setting has been examined in a study by Durham and Black (1978) with a sample of 16- to 21-month-olds. These researchers found that children were likely to score higher when tested in the home compared to testing in a laboratory setting. Verbal performance appeared to be especially affected.

Another issue addressed in the manual is the relationship between the Mental and Motor Scales. Bayley (1969) reports that correlation coefficients by age between raw scores on these scales range from .24 to .78; the range is from .18 to .75 between standard scores. The median coefficient is .46 for both raw and standard scores. Overall, the relationship between mental and motor scores decreases with age. Bayley interprets this as being due to the increasing differentiation between mental and motor skills with development.

Researchers generally have found that infant intelligence tests in the first year of life do not accurately predict IQ performance later in childhood (Stott & Ball, 1965; McCall, Hogarty, & Hurlburt, 1972). These results are not inconsistent with Bayley's view of mental development—that it is emergent and may take different forms at different ages. Attempts to relate BSID scores to measures of intelligence in childhood have generally followed this pattern, especially for infants scoring within the normal range of functioning.

Bayley (1969) presents data correlating scores on the Mental Development Index and the IQ obtained from the Stanford-Binet Intelligence Scale for 120 children who were 24, 27, and 30 months of age. The sample was limited to children who earned basal scores on the Stanford-Binet. The coefficient of correlation obtained was .57.

Although Bayley argues that this is substantial given the limited range of scores on the Stanford-Binet, the correlation still leaves much unexplained.

Cohen and Parmalee (1983) found only a moderate relation between BSID scores for preterm infants at 25 months of age and their Stanford-Binet scores at age 5 years ($r = .65$, $p < .05$). Their analysis supported social factors as being more important than any other set of variables in predicting outcome at age 5. McCall, Hogarty, and Hurlburt (1972), looking at Bayley's 1949 data, concluded that prediction of intellectual functioning in childhood was poor, although the later in infancy the developmental assessment was made and the closer in time the two assessments were made, the better the predictive ability.

The picture is somewhat different when attempting to predict the future functioning of children who receive BSID scores that are significantly lower than average. Vander Veer and Schweid (1974) related BSID scores at 18 to 30 months to tests 1 to 3 years later and found that 75% of the infants judged to be moderately to profoundly retarded remained so classified at the later testing. Additionally, none of the original group that were classified as retarded were scoring in the normal range at the subsequent testing. Ireton, Thwing, and Gravem (1970) have reported similar findings of greater predictability for low scores on the BSID. Thus, while the BSID may offer little predictability for children in the normal range or better, it does appear to be useful in predicting the course of development for those children scoring very poorly early in life.

Fewer efforts have attempted to look at the predictive value of the Infant Behavior Record. McGowan, Johnson, and Maxwell (1981) collected BSID Mental Scale scores and infant behavior ratings at 12 and 24 months and Stanford-Binet scores at 36 months on a sample of Mexican-American babies. They concluded that the behavior ratings were not helpful in adding to the predictive power of BSID Mental Scores at 12 months.

Other researchers have looked at infant behavior ratings over time, looking at the distributions of ratings and at the underlying factor structure of these ratings. Several sources (Dolan, Matheny, & Wilson, 1974; Matheny, 1983; Bayley, 1969) have noted developmental as well as individual difference changes in the distribution of ratings over the age span covered by the record. Research with twins has related these changes to genetic influences (Matheny, 1980; Matheny, Dolan, & Wilson, 1976). Results concerning factors underlying the Infant Behavior Record are mixed. Matheny (1980) has reported three factors to be recurrent across the first two years: task orientation, test affect-extraversion, and activity. McGowan, Johnson, and Maxwell (1981) failed to find the same factor structure in a study of Mexican-American 12-month-olds. It should be noted that attempts to identify factors underlying the abilities measured by the BSID Mental and Motor Scales, such as that by Hofstaetter (1954), have not been clearly successful (see Cronbach, 1967, for a critique).

Critique

Almost 20 years after their publication, the Bayley Scales of Infant Development remain one of the most used and useful assessment instruments for infant development. Bayley's original aim, to offer a well-standardized assessment making use of a representative sample of non-institutionalized children for the determination of current developmental status, has been achieved. This, indeed, is one of the greatest strengths that the BSID have to offer in contrast to many other infant assessment techniques. Additionally, the BSID offer a set of standardized scores and norms for both mental and motor development based on the same group of children. Directions for administration are clearly presented in the manual, and the more recent 1984 supplement offers additional clarification to facilitate reliable and standardized administration.

The BSID are useful in identifying infants whose development is significantly below average, and for these children it offers good predictive value for the future. The specific scale scores and the observations the test facilitates help identify neuromuscular deficits.

The BSID are also very useful in documenting, especially for research purposes, that a sample of children is developmentally within the range expected for their age. Such information provides an important base when attempting to interpret other observations made on those children. An additional advantage of the BSID is that they are one of the most frequently used infant tests and have consequently accumulated a background of data. This includes research by Bayley as well as many other researchers.

What the BSID do not do is predict future intellectual development, especially 1) within the normal range and above and 2) from the first year of life to later childhood and beyond. This shortcoming is not just a characteristic of this particular test but reflects the state of our understanding of development. Bayley and others discuss the emergent nature of intelligence and see it taking different forms at different times. However, we continue to ask, "What will this baby be like when he or she is older?" and we are not able to answer with confidence.

A second shortcoming of the BSID is the lack of reliable and valid subscores. Children appear to have strengths and deficits in particular areas, but these areas of ability are not reliably defined or measured. Individual clinicians make observations about verbal skills, problem-solving skills, fine motor skills, and so on, but no reliable or standardized subscales are available to facilitate such observations.

A related difficulty with the BSID is the lack of independence between the Mental and Motor Scales. Clearly, children with significant motor deficits will score less well on the Mental Scale as we measure "mental abilities" as manifested in motor performance.

Another problem concerns the Motor Scale at the older age range where it focuses on gross motor skills—climbing, walking, jumping, and so on. Children with fine motor deficits will fail to be identified formally by the Motor Scale as having motor problems. Clinical observations from performance on the Mental Scale can help identify such cases, but the test, and especially the Psychomotor Index, will fail to reflect this observation directly.

Finally, the manual offers little information to facilitate clinical interpretation of BSID results beyond the statistical properties of the test. Neither does it offer therapeutic or remedial interventions related to test performance or particular difficulties that a child might display. Those using the BSID in clinical settings would benefit from information concerning both these areas.

The fact that the BSID offers so much leads us to want it to do more. It does not claim to be standardized on handicapped or preterm infants, but we want to use it for these very populations and do so at the risk of misinterpreting our findings. These cannot be said to be shortcomings of the test but rather a need for better such instruments in the field of assessment, instruments for which the BSID is a model.

References

Bayley, N. (1933). *The California First-Year Mental Scale.* Berkeley: University of California Press.

Bayley, N. (1936). *The California Infant Scale of Motor Development.* Berkeley: University of California Press.

Bayley, N. (1965). Comparison of mental and motor test scores for ages 1–15 months by sex, birth, order, race, geographical location, and education of parents. *Child Development, 36,* 379–411.

Bayley, N. (1969). *Bayley scales of infant development.* San Antonio, TX: The Psychological Corporation.

Cohen, S. E., & Parmalee, A. H. (1983). Prediction of five-year Stanford-Binet scores in preterm infants. *Child Development, 54,* 1242–1253.

Cronbach, L. J. (1967). Year-to-year correlations of mental tests: A review of the Hofstaetter analysis. *Child Development, 38,* 283–290.

Dolan, A. B., Matheny, A. P., Jr., & Wilson, R. S. (1974). Bayley's Infant Behavior Record: Age trends, sex differences, and behavioral correlations. *JSAS Catalog of Selected Documents in Psychology, 4,* 9.

DuBose, R. F. (1977). Predictive value of infant intelligence scales with multiply handicapped children. *American Journal of Mental Deficiency, 81,* 388–390.

Durham, M., & Black, K. (1978). The test performance of 16 to 21 month-olds in home and laboratory settings. *Infant Behavior and Development, 1,* 216–223.

Hofstaetter, P. R. (1954). The changing composition of "intelligence": A study in t-technique. *The Journal of Genetic Psychology, 85,* 159–164.

Hunt, J. V., & Rhodes, L. (1977). Mental development of preterm infants during the first year. *Child Development, 48,* 204–210.

Ireton, H., Thwing, E., & Gravem, H. (1970). Infant mental health development and neurological status, family socioeconomic status, and intelligence at age four. *Child Development, 41,* 937–946.

Jaffa, A. S. (1934). *The California Preschool Mental Scale.* Berkeley: University of California Press.

Kierman, D. W., & DuBose, R. F. (1974). Assessing the cognitive development of preschool deaf-blind children. *Education of the Visually Handicapped, 6,* 103–105.

Matheny, A. P., Jr. (1980). Bayley's Infant Behavior Record: Behavioral components and twin analyses. *Child Development, 51,* 1157–1167.

Matheny, A. P., Jr. (1983). A longitudinal study of stability of components from Bayley's Infant Behavior Record. *Child Development, 54,* 356–360.

Matheny, A. P., Jr., Dolan, A. B., & Wilson, R. S. (1976). Twins: Within-pair similarity on Bayley's Infant Behavior Record. *Journal of Genetic Psychology, 128,* 263–270.

McCall, R. B., Hogarty, P. S., & Hurlburt, N. (1972). Transitions in infant sensorimotor development and the prediction of childhood IQ. *American Psychologist, 27,* 728–748.

McGowan, R. J., Johnson, D. L., & Maxwell, S. E. (1981). Relations between infant behavior ratings and concurrent and subsequent mental test scores. *Developmental Psychology, 17,* 542–553.

Rhodes, L., Bayley, N., & Yow, B. C. (1984). *Supplement to the manual for the Bayley Scales of Infant Development.* San Antonio, TX: The Psychological Corporation.

Stott, L. H., & Ball, R. S. (1965). Infant and preschool mental tests: Review and evaluation. *Monographs of the Society for Research in Child Development, 30* (3, Serial No. 101).

Vander Veer, B., & Schweid, E. (1974). Infant assessment: Stability of mental functioning in young retarded children. *American Journal of Mental Deficiency, 79,* 1–4.

Werner, E. E., & Bayley, N. (1966). The reliability of Bayley's revised scale of mental and motor development during the first year of life. *Child Development, 37,* 39–50.

Michael K. Gardner, Ph.D.

Assistant Professor of Educational Psychology, University of Utah, Salt Lake City, Utah.

BLOOM ANALOGIES TEST

Philip Bloom. Brooklyn, New York: Philip Bloom.

Introduction

The Bloom Analogies Test (BAT; Bloom, 1981) is a multiple-choice analogy completion test designed to measure the general mental aptitude of adults in the extremely high ability range (98th percentile and above). The test was developed by Philip Bloom, who holds a bachelor's degree in psychology and a master's degree in physiology. He has spent over 15 years working with and testing for high-level IQ groups such as MEGA, the Prometheus Society, the Triple Nine Society, and the International Society for Philosophical Inquiry. The Bloom Analogies Test was developed as a screening tool for these and other similar groups.

Approximately 4,000 individuals have taken the BAT during its 15-year history, including MENSA members, applicants to the societies previously mentioned, and college students tested as part of a class exercise. In addition, some people of high ability were referred for testing by teachers and fellow students. No formal norming studies have been performed for this test, presumably because of the difficulty in finding a sufficient number of individuals at such a high ability level to form a representative sample. To some extent, the pool of previous examinees serves as a norming sample, although obviously an imperfect one. The test has not been revised formally since its inception, but Bloom continues to evaluate the items to determine which ones contribute significantly to the prediction of high-level mental ability.

The test exists in both a timed form and an untimed form. The timed form has a 15-minute time limit. The untimed form has no time limit; however, the examinee is required to complete the test in a single session. In order to equate means across the two forms of the test, the raw score mean for the untimed form is reduced by two.

The test consists of a single 8½" ×14" sheet of paper folded in half, which yields four 8½" ×7" pages. The first page contains instructions and four sample analogies. Pages 2 and 3 (the interior of the folded sheet) contain 50 multiple-choice analogies that comprise the body of the test. The last page contains 1) a statement, which the examinee must sign, concerning testing conditions and 2) space for the examinee's name and address.

Because the test is self-administered, there is no direct examiner involvement in the testing procedure. Testing is conducted on the "honor system." Answer forms are not available for this instrument: all scoring is done by Mr. Bloom. As a result, examinees usually must pay a fee to cover scoring and handling.

BAT subjects are high-ability adults who usually are being tested voluntarily, as

48

when applying for membership in a high-IQ society. As one would expect, the test is quite difficult. The content is entirely verbal analogies, and successful performance requires a strong vocabulary and general knowledge background. The BAT is similar to the Miller Analogies Test in this way.

Practical Applications/Uses

As previously mentioned, the major users of the Bloom Analogies Test are high-IQ societies in need of a screening device to administer to applicants as part of the entrance requirements. The test measures general ability related to IQ, but is slanted toward verbal ability to the neglect of performance ability. The test really is not oriented toward professionals, but it might be of some interest to educators of the gifted. For example, the test might be useful as a screening device for identifying highly gifted older adolescents. Because the test items are very difficult, the BAT would not be appropriate for children or young adolescents (i.e., 11th grade and below); likewise, the test is not useful for assessing adults within the normal intelligence range. Other excellent measures (e.g., the WISC-R and the WAIS-R) exist for these groups.

The BAT is somewhat different from other IQ tests in that it is self-administered. No examiner is necessary. All that is needed is a quiet, undisturbed setting in which the examinee can complete the test. Comments on the test indicate that the normal range of completion times is 15 minutes to 2 hours. Instructions are provided on the face of the test. No manual is provided, nor is one available upon request.

The 50 analogies are presented in a multiple-choice format. The examinee is asked to select the best of four answer options to complete one of the analogy's terms and is instructed to guess when he or she is uncertain (the raw score is corrected for guessing). It is possible to return to difficult items later in the testing period. The final page contains some additional instructions (examinees are referred to this page from page 1) and a statement indicating the amount of time the examinee spent completing the test (the examinee fills this information in) and asserting that no references or memory aids were used to complete the test. The examinee signs the statement, fills in his or her name and address, and returns the test to the sponsoring organization for scoring.

Scoring procedures are unusual. The answer key is not released, even to professionals. Instead, the exams are scored by Mr. Bloom for a fee. Although this procedure insures the security of the test, it makes any type of an error analysis by the sponsoring agency or professional impossible. All aspects of the BAT, from test acquisition to scoring, are handled directly by Mr. Bloom.

The BAT yields a single raw score that has been corrected for guessing and time limits. This omnibus measure is intended to reflect general intellectual capacity or functioning. On the face of it, interpretation is simple enough. Higher raw scores reflect greater mental power. But beyond that, interpretation is problematic. In an unpublished manuscript, Bloom (1985) attempts to convert BAT raw scores into percentile ranks using both a regression procedure and an empirical procedure. The extrapolation into the extremely high percentile ranks (i.e., from the 99.9th percentile to the 99.9999th percentile) seems uncertain at best. Insufficient norming samples exist for using a straightforward empirical (i.e., sampling) approach, and Mr. Bloom's regression approach makes assumptions that seem questionable. The difficulty in interpreting a BAT score, therefore, depends upon the degree of confidence one places in Mr. Bloom's percentile conversions. If one believes them, a master's student should be able to interpret the test; if not (and this reviewer doesn't), anything beyond a ranking of candidates based on raw scores is likely to be deceiving.

Technical Aspects

The Bloom Analogies Test has not been the subject of serious reliability and validity studies. The only data on the test come from Mr. Bloom's collection of information on the 4,000 previous testees, most of whom are members of one or another high-IQ societies. Although individuals who perform well on the BAT also perform well on other standardized ability instruments, no systematic validity data exist. Likewise, perhaps because successful high-IQ society applicants may be unwilling to take the test a second time and risk obtaining a lower score, no test-retest reliability data exist. In addition, this reviewer knows of no internal consistency reliability data (e.g., split-halves or coefficient alpha) on the test. This disappointing state of affairs leaves one mostly to take the author's word that the test is a good one, which seems unacceptable for a test that is to be used for any sort of decision-making that might have important consequences. Systematic studies of the BAT clearly are needed.

Critique

The Bloom Analogies Test is essentially a homemade instrument. Although there is nothing wrong with homemade instruments, without validity, reliability, and norming studies it is difficult to recommend them to others. Some of the difficulties faced by the BAT are understandable. If a test is aimed toward a population whose occurrence is only one per thousand to one per million, a very large sample of the general population would be needed to find an adequate number of acceptable cases. However, other types of studies could have been performed. Concurrent validity can be established by administering measures such as the Ravens Advanced Matrices to BAT examinees and correlating their scores. In this reviewer's opinion, the lack of internal consistency reliability is unforgivable.

Other points need mentioning to potential BAT users. First, there is no test manual, presumably because of the test's limited audience and use within private societies. Although this may suit some, this reviewer strongly advises against using a test for which there is no manual.

Second, the answer key is restricted severely. Only Mr. Bloom has access to it, and he will score tests for a fee. Supposedly, this guarantees test security, but it is a condition potential test users should be aware of. Scoring costs can become quite high if the user tests a large number of individuals and pays the fee rather than the examinees. Further, no diagnostic information can by gleaned by examining the items testees miss.

Third, the test copy this reviewer received was of poor publication quality.

Although all of the items were readable, the test looked like a second- or third-generation photocopy. Although that particular copy may not be representative, its appearance seems unacceptable for a nationally used instrument.

Finally, the test consists entirely of verbal analogies. Test performance seems to be determined by 1) analogical reasoning ability, 2) strength of vocabulary, and 3) general knowledge base (some questions require a good deal of general knowledge to solve the analogy). There is no performance component, as one would find on most general intelligence tests. To the extent that performance ability is something in which a user is interested, the BAT would be a poor choice.

The mixture of items on the Bloom Analogies Test is strongly reminiscent of the Miller Analogies Test, which consists of 100 verbal analogy items and often is used for graduate school admission. As a final suggestion, those who must assess high-ability individuals should use the Miller Analogies Test. It is a well-established test with known psychometric properties. If the examinees correctly answer all the analogies on the Miller Analogies Test, the BAT may be in order.

References

Bloom, P. (1981). *Bloom Analogies Test*. Brooklyn, NY: Author.

Bloom, P. (1985). *Does the Bloom Analogies Test reach the top/million level?* Unpublished manuscript.

The Psychological Corporation. (1987). *Miller Analogies Test*. San Antonio, TX: Author.

Arthur MacNeill Horton, Jr., Ed.D., ABPP(CL), ABPN
*Chief, Neuropsychology Section, Psychology Service, Veterans
Administration Medical Center, and Department of Psychiatry,
University of Maryland Medical School, Baltimore, Maryland.*

BODER TEST OF READING-SPELLING PATTERNS

*Elena Boder and Sylvia Jarrico. Orlando, Florida: Grune &
Stratton, Inc.*

Introduction

The Boder Test of Reading-Spelling Patterns is intended as a diagnostic screening
test for subtypes of reading disability. Essentially, from an assessment of reading
and spelling skills, children with specific reading disability (diagnostic dyslexia)
can be distinguished from children with nonspecific reading disability and placed
accurately within a typology of reading problems, each with its own prognostic
and remedial implications.

The test authors are Elena Boder, M.D., and Sylvia Jarrico, M.A. Dr. Boder, a
pediatric neurologist, currently is a clinical professor of pediatrics in the Division of
Pediatric Neurology at the School of Medicine, University of California at Los
Angeles. Boder grew up in Mexico City, where her father was the United States
Ambassador to Mexico, a cross-cultural experience that, perhaps, made her particularly
sensitive to the peculiarities of language patterns. During the mid-1960s,
Boder worked clinically with reading-disabled children in neurology clinics in the
Los Angeles area. From the wealth of clinical experience she aquired there, she
began to conceptualize the Boder Test of Reading and Spelling Pattern (BTRSP). In
1973, she published her often-cited paper *Developmental Medicine and Child Neurology,*
which describes her diagnostic approach based on three atypical reading-
spelling patterns. Boder's work, which melded clinical sensitivity with impressive
scholarly knowledge to produce an exciting new conceptualization of reading
problems, received a great deal of professional attention. While other professionals
in the field still perceived reading problems as a unity disorder and studied hetero-
genous groups, Boder saw that a typology of homogenous grouping of children
with reading problems would be a major advance. Despite the clinical acceptance
of her insights, the development of the BTRSP as a standardized test stemmed from
Boder's meeting with Jarrico, a research psychologist from Los Angeles, California,
at a conference on neuropsychology and cognition sponsored by the North Atlan-
tic Treaty Organization (NATO) in Augusta, Georgia, in 1980. Both were invited
participants and decided to develop a standardized test based on Boder's insights.

The primary assumption of the BTRSP is that poor readers differ along two distinct dimensions, the visual gestalt function and the auditory analytic function.
The developers of the BTRSP saw developmental dyslexia as a reading disability in

which an individual's reading and spelling performance suggest difficulties with either the auditory analytic function, the visual gestalt function, or both. On the other hand, if an individual does not display evidence of problems in reading or spelling patterns then the reading problem is not neuropsychologically based. It is due, instead, to poor instruction, low motivation, lack of academic opportunity, or other noncognitive factors. Based on this theory, Boder and Jarrico identified the four groups of individuals with reading problems: 1) dysphometric dyslexics (poor phonic analysis), 2) dyseidetic dyslexics (poor visual gestalt analysis), 3) mixed dysphometric-dyseiditics (poor phonic analysis and visual gestalt analysis), and 4) nonspecific reading disability (poor reading ability but possessing adequate phonic analysis and visual gestalt analysis abilities).

The BTRSP is composed of two subtests, a word-recognition oral-reading test and a spelling-to-dictation test. Interestingly, the spelling words are selected based on the reading level achieved on the reading test. The reading test is made up of 13 separate, 20-word lists presented in sequential order of difficulty. Of the words on each list, half are phonetically regular and the other half are not. The word lists first are presented in a 1-second flash presentation and again in a more lengthy 10-second presentation. The reading level is calculated by taking the grade level at which half or more of the words can be read without error in the first short presentation with the addition of 2 months for each word correctly read in the short presentation of words above the grade level. A reading quotient (RQ) can be calculated by dividing the reading age (RA) , which is derived by adding the absolute number 5 to the RL, by the mental age and multiplying the result by $100(\frac{RA}{CA} \times 100 = RQ)$.

Practical Applications/Uses

The test is most appropriate for use with children, adolescents, and adults suspected of having reading problems. Perhaps the professionals most likely to use the test would be reading teachers and school psychologists who are referred school-aged children with reading difficulties. The clinically relevant typology of reading subtypes, which have straightforward implications for remediation strategies, are of great potential value to special educators and reading teachers. However, the BTRSP may be useful to professionals in a variety of other disciplines, including physicians, school nurses, reading specialists, and speech therapists. The settings in which the test could be used are primarily education related but may, in special situations, include pediatric neurology clinics and child-related private practices. Potential new uses of the test might be to investigate the influence of dementing and degenerative diseases upon reading ability. In recent years, there has been considerable interest in reading comprehension skills as a measure of premorbid mental ability. One wonders if the differential pattern of reading and spelling patterns might be instructive in gerontological settings.

Conversely, the BTRSP is not appropriate for general neuropsychological screening of patients suspected of having brain damage. Also, it is inappropriate for use as a general intellectual measure, for attempting to draw inferences regarding personality, or for administration to individuals not suspected of having reading problems. The BTRSP is to be administered on an individual basis. It requires about 30 minutes for administration. The manual indicates that examiners qualified to administer the test include teachers, reading specialists, psychologists, physicians, and speech therapists. The administration instructions appear somewhat complicated, and the manual could profit from a revision. The test seems much too difficult for someone with little training in educational and psychological measurement to adequately administer. To administer the test correctly, one would expect the examiner to at least hold a master's degree or possess a year or two of experience. To expect physicians or classroom teachers to use the test successfully based on the manual appears unwise and unrealistic. Also, the spelling lists are not uniform for similar-aged subjects, increasing the possibility of measurement errors as the actual words in each spelling list are left to the individual judgment of administrators.

Instructions for scoring are presented clearly. Depending upon the test users prior training and experience in educational and psychological measurement, scoring could take a few hours or a number of days. Once mastered, however, test scoring should not require much more time than that necessary to administer the test (i.e., about 30 minutes). According to the manual, machine scoring is not available.

One of the problems in interpreting the BTRSP is that little normative data are provided. For example, no journal item analyses were conducted on the reading lists, precluding understanding of whether the items are truly ordered in terms of reading difficulty. The BTRSP would be much enhanced by assembling a nationally stratified randomized sample subdivided on relevant demographic variables such as race, education, age, and socioeconomic status.

There are also problems with the reading level (RL) and reading age (RA) senses on the BTRSP. Essentially, these are Grade Equivalent (GE) scores and, thus, are prone to the many problems of GEs. Simply put, GEs presuppose equal intervals between academic skill acquisition between grades and equal rates of learning. Both assumptions are false, and the GE is more properly considered an ordinal, rather than an interval, scale. The Reading Quotient (RQ) on the BTRSP not only shares the GE problem of the RL and RA measurement but also has some unique problems of its own. Chief among these is the provision of alternate formulas to calculate the RQ, allowing examiners to produce different RQs from the exact same data. It would be better to provide standard scores for the BTRSP so that meaningful comparisons and analyses can be made.

Technical Aspects

There is reasonably good evidence for the validity of the BTRSP. However, there are age confounds in the bulk of the criterion-related validity studies that limit the degree of confidence one can place in the study. Evidence for construct validity appears quite good and studies have related the BTRSP to WISC-R Verbal Performance Splits (Smith, 1970; Ginn, 1979) and EEG patterns (Sklar, Hanley, & Simmons, 1972).

Test-retest reliability was assessed at 2-month and 1-year intervals. For the 2-month condition, an r of .98 for ages 6-9 and .96 for ages 10-15 were obtained for the RA score. These scores would have to be corrected for spurious correlation with age, in effect, lowering the test-retest reliability. For the 1-year condition, the RA was .81. Again, there would need to be an age correction that could significantly reduce the test-retest to a level below that necessary for individual cases. Moreover, the 1-year test-retest reliability is based on only 14 subjects. Split-half reliability for the RL was reported as .97. Once more, however, a significant age confound is present, and the estimate of internal consistency is clearly inflated.

Reynolds (1986) essentially lauded the clinical value of the BTRSP but severely criticized its psychometirc aspects, including its normative data and evidence for validity and reliability. In addition, all of the validity and reliability studies have been performed with school-aged children and adolescents. Any generalization of the results to adults or the elderly is questionable.

Critique

In terms of the bottom line, the test appears to have intuitive appeal and impressive conceptual insight. It may become an important instrument in future years. At present, due to psychometric flaws, it is not possible to be certain of the contribution the test will make. Still, absence of evidence is *not* evidence for absence!

References

Boder, E. (1973). Developmental dyslexia: A diagnostic approach based on three atypical reading-spelling patterns. *Developmental Medicine and Child Neurology, 15*, 663-687.

Ginn, R. (1979). *An analysis of various psychometric typologies of primary reading disability.* Unpublished doctoral dissertation, University of Southern California.

Reynolds, C.R. (1986). Clinical acumen but psychometric naivete in neuropsychological assessment of educational disorder. *Archives of Clinical Neuropsychology, 1*, 121-137.

Sklar, Hanley, J., & Simmons, W. (1972). An EEG experiment aimed toward identifying dyslexic children. *Nature, 240,* 414.

Smith, M. (1970). *Patterns of intellectual abilities in educationally handicapped children.* Unpublished doctoral dissertation, Claremont College, Claremont, CA.

Gene Schwarting, Ph.D.
Project Director, Preschool Handicapped Program, Omaha Public Schools, Omaha, Nebraska.

BOEHM TEST OF BASIC CONCEPTS-REVISED

Ann E. Boehm. San Antonio, Texas: The Psychological Corporation.

Introduction

The Boehm Test of Basic Concepts-Revised (Boehm-R; Boehm, 1986) was developed to measure the understanding of basic positional concepts of young children, that is, whether individuals in kindergarten, Grade 1, and Grade 2 can correctly identify a picture from among a choice of three when presented with verbal cues incorporating such terms as *over, least, left,* and so on. Based on these results, children can be identified as deficient in conceptual development, with resultant curricular implications.

The author of the instrument, Ann E. Boehm, has been involved in the assessment of young children for some time and is well known in the field of measurement. The 1986 Boehm-R is a revision of the Boehm Test of Basic Concepts (BTBC; Boehm, 1971) and was designed to re-evaluate the individual items for that instrument, incorporating antonyms and synonyms of the original items and introducing a new section that involves combinations of two concepts. The items from Forms A and B of the original instrument were evaluated as to their continued appropriateness, whether they could be depicted accurately through pictures, and on additional criteria. These included the frequency of the terms in printed materials as measured by their apearance in "most used" word lists generated by Thorndike and Lorge (1944) and Taylor, Frackenpohl, and White (1979), as well as in five major reading and five major math series for Grades K-2, and through an analysis of oral directions presented by 18 teachers to their classes in Grades K-5. Once selected, the concepts were used to develop pictorial, multiple-choice items that were field-tested. Then, three "experts" evaluated the items for possible bias, with the items not discriminating being eliminated. Of the items utilized on the Boehm-R, the majority were also on the BTBC, with some modification of the artwork.

A Spanish translation of this instrument is available. Although there are no versions of the test developed by the author for handicapped populations, she reports independent studies with the original BTBC on children diagnosed as blind, acoustically handicapped, and learning disabled.

The Boehm-R instrument has two alternate forms (C and D), both of which consist of 50 pictorial items arranged by increasing difficulty in two booklets, preceded in each case by three sample questions. The author estimates that each booklet will require 15 to 20 minutes to administer to kindergarten students, with older children probably requiring less time. She also notes that administration may need to be divided into two separate testing periods, and into smaller groups for younger children. Materials needed include the test booklets and a pencil and crayon for each child, who should be separated as far as possible to avoid copying. Directions are simply and clearly stated for the examiner, who reads a simple sentence concerning the picture stimulus for that item, emphasizing and repeating the positional concept involved. The child then marks with an X the correct one of three alternatives for each item.

The entire instrument may be administered to children in kindergarten, first grade, or second grade (a separate version has been developed for preschool children, 3 to 5 years of age). Both forms of the instrument consist of two 8-page consumable test booklets and a simple 8-page administration manual. A 65-page manual includes technical information, administration and scoring procedures, interpretation and instructional planning suggestions, and tables of scores. Also available are class record forms, a parent-teacher conference report form, and both test booklets and administration for the Applications Test. The Applications section of the instrument is similar to forms C and D in format, presentation, and length, but differs in that the items involve combinations of two concepts.

The tables presented include the percent of the norm group passing each item for C, D, and Applications by grade and three broad socioeconomic levels, at the beginning as well as at the end of the school year; percentile equivalents of raw scores by grade and socioeconomic levels at the beginning and end of the school year; and normal curve equivalent conversions.

Practical Applications/Uses

The major use of the Boehm-R would appear to take place within elementary schools, preschools, and Head Start centers, to provide information on young children's knowledge of the concepts of space, quantity, and time. The results could be utilized to make both group and individual decisions for instructional purposes. The instrument also would be of use to speech therapists or resource teachers working individually with young children. Psychologists might well use the Boehm-R to determine the child's understanding of these terms prior to administering other instruments that include such concepts in their directions.

Specific training in test administration is *not* required for this instrument, so the administration might be done by a paraprofessional or volunteer following minimal in-service training. The Boehm-R can be presented individually or in groups, the size of which is restricted only by the examiner's ability to monitor progress. Both sections of a test usually can be administered in 45 minutes or less, depending on the test-taking skills of the children involved.

Scoring must be done by hand, one page at a time, with scoring time estimated at 5 to 10 minutes per booklet. Once scored, the results can be translated into percentiles, which may be done by socioeconomic level (low or medium/high) or by the total group. As noted previously, norms are available for Kindergarten, Grade 1, and Grade 2 for both forms C and D, both at the beginning and end of each school year. In addition, means and standard deviations are provided for each group. Due to the simplicity of both scoring and translation of the scores, only minimal training should be required.

Technical Aspects

The norming of the Boehm-R occurred in 1983, involving approximately 10,000 children in public school districts selected as nationally representative by size of the district and geographic location. Socioeconomic status was estimated based on each building's percentage of students who qualified for subsidized lunches, *not* on the characteristics of the individual children. A number of the children tested at the end of the year for grade level were also tested at the beginning of the following school year, so some familiarity with the instrument was possible.

A number of measures of the reliability of the instrument are presented. Correlations reported between Forms C and D (N=625) are .82 for Kindergarten, .77 for Grade 1, and .65 for Grade 2. Correlations of Forms C and D with Form A of the BTBC (N=173) are .65 and .62, respectively. Split-half reliability coefficients and standard errors of measurement are presented by form, grade, and socioeconomic level. These relationships vary between .55 and .87, with the co-efficients being lower at Grade 2 than at Kindergarten and Grade 1. Stability was measured through test-retest with a 1-week interval (N=548) and varies between .55 and .88.

Validity is measured through the relationship of scores on the Boehm-R and various measures of academic achievement (the total batteries of the Comprehensive Test of Basic Skills, the California Achievement Test, and the Iowa Test of Basic Skills), as well as through reading level as measured on the Harcourt Brace Jovanovich Bookmark Reading Program. The coefficients of correlation varied from .24 (Grade 2 applications with ITBS) through .64 (Kindergarten Form D with CTBS), with a median of .44.

Critique

The Boehm-R is, for the most part, a continuation of the Boehm Test of Basic Concepts with updated normative data. Thus, those who utilize that instrument will feel comfortable with the new one. The test is easy to administer and score, has alternate forms, and has a large normative population. The class record sheet and parent-teacher forms are useful, and the new Applications section extends the complexity of traits measured. Overall design is good, as is most of the artwork.

The reader is reminded that the Boehm-R measures only receptive conceptual development and so provides but one piece of information, which would appear to be most useful for instructional planning in kindergarten and first grade (second grade reliability and validity are both lower than for the other grade levels). The norm group, while large, may be restricted; no information is presented regarding racial or cultural representation and the SES data are very general in nature. Indeed, the norms table emphasis on socioeconomic grouping is confusing, and its purpose in score interpretation rather questionable.

References

This list includes text citations and suggested additional reading.

Boehm, A. E. (1971). *Boehm Test of Basic Concepts manual*. New York: The Psychological Corporation.

Boehm, A. E. (1986). *Boehm Test of Basic Concepts-Revised manual.* New York: The Psychological Corporation.

Estes, G., Harris, J., Moers, F., & Wodrich, D. (1976). Predictive validity of the Boehm Test of Basic Concepts for achievement in first grade. *Educational and Psychological Measurement, 36,* 1031-1035.

Manning, B. (1984). Problem-solving instruction as an oral comprehension aid for reading disabled third graders. *Journal of Learning Disabilites, 17,* 457-461.

Piersal, W., Sterrett-Blake, B., Reynolds, C., & Harding, R. (1982). Bias in content validity on the Boehm Test of Basic Concepts for white and Mexican-American children. *Contemporary Educational Psychology, 7,* 181-189.

Powers, S., Rossman, M., & Douglas, P. (1986). Reliability of the Boehm Test of Basic Concepts for Hispanic and Nonhispanic kindergarten pupils. *Psychology in the Schools, 23,* 34-36.

Preddy, D., Boehm, A., & Shepherd, M. (1984). PBCB: A norming of the Spanish translation of the Boehm Test of Basic Concepts. *Journal of School Psychology, 22,* 407-413.

Sarachan-Deily, A., Hopkins, C., & De Vino, S. (1983). Correlating the DIAL and the BTBC. *Language, Speech, and Hearing Services in the Schools, 14,* 54-59.

Silverstein, A., Belger, K., & Morita, D. (1982). Social class differences on the Boehm Test of Basic Concepts: Are they due to bias? *Psychology in the Schools, 19,* 431-432.

Silverstein, A., Morita, D., & Belger, K. (1983). Sex differences and sex bias on the Boehm Test of Basic Concepts: Do they exist? *Psychology in the Schools, 20,* 269-270.

Smith, E. (1982). The Boehm Test of Basic Concepts: An English standardization. *British Journal of Educational Psychology, 56,* 197-200.

Smith, E. (1982). The validity of the Boehm Test of Basic Concepts. *British Journal of Educational Psychology, 56,* 332-345.

Steinbauer, E., & Heller, M. (1978). The Boehm Test of Basic Concepts as a predictor of academic achievement in grades 2 and 3. *Psychology in the Schools, 15,* 357-360.

Taylor, S., Franckenpohl, H., & White, C. (1979). A revised core vocabulary. In S. Taylor, H. Franckenpohl, C. White, B. Nicrorodo, C. Browning, & E. Birsher (Eds.), *EDL core vocabularies in reading, mathematics, sciences, and social studies.* New York: EDL/McGraw-Hill.

Thorndike, E., & Lorge, I. (1944). *The teachers workbook of 30,000 words.* New York: Columbia University, Teachers College, Bureau of Publications.

Larry R. Cochran, Ph.D.

Associate Professor of Counseling Psychology, The University of British Columbia, Vancouver, British Columbia.

THE CALIFORNIA LIFE GOALS EVALUATION SCHEDULES

Milton E. Hahn. Los Angeles: Western Psychological Services.

Introduction

Over 6 years in development and intended for people who are 15 years of age and older, the California Life Goals Evaluation Schedules (CLGES) comprise a 150-item test of future direction designed to measure the strength of 10 life goals: Esteem, Profit, Fame, Power, Leadership, Security, Social Service, Interesting Experiences, Self-Expression, and Independence. A life goal is conceived as a motivating attitude toward a desired future state. Unlike an interest, which is regarded as a general and enduring motivation that might exist throughout the life span, a life goal is more specific and changeable. For example, a person might be strongly directed toward the pursuit of profit during a particular period of life, but not necessarily over the entire course of life. Given this distinction, the CLGES can be characterized as a test of short-term future direction that is of potential value in helping individuals to plan future activities.

Milton E. Hahn, the author of the test, was associated early with the trait-factor approach in counseling psychology, an approach that stresses the use of tests in helping people to solve problems and make plans and decisions (Williamson & Hahn, 1940). Among his contributions to the field are books on clinical counseling (Hahn & McLean, 1950) and psychological assessment (Hahn, 1963). The development of the CLGES grew directly out of Hahn's long-standing involvement in helping people to solve problems of living and to make viable plans for the future.

Behind the development of the CLGES is the problem of change over the life span. People change as they age and acquire different statuses. In addition, the rate of social change seems to be accelerating. Old goals that organized meaningful living can lose vitality or viability. Increasingly, living a satisfying and productive life seemed to Hahn to depend upon one's ability to set new and stimulating goals with which to refresh one's pursuits. Although interest tests are valuable for career and life planning, what they measure is rather general and constant, and the problems that Hahn had in mind were more specific and fluctuating. The same life structure might not be satisfying for a lifetime, and, consequently, there was a need for a test to help people make these more short-term adjustments. The CLGES was designed largely for this purpose.

The 10 life goals were adapted primarily from Centers's (1949) studies of social class and mobility. Two sets of debatable statements were constructed that concern economic, social, and political attitudes. These statements were used as test items, partially to avoid overlap with the categories of items found on interest inventories and partially to reflect directly the kinds of attitudes that Centers and others had found to be important. For example, a hypothetical item could have been: "What one earns indicates one's worth and usefulness to society." Each debatable statement is rated on a 5-point scale ranging from "strongly agree" (4) to "strongly disagree" (0). Initially, these two sets of statements (Forms A and B) were administered to 348 University of California at Los Angeles freshmen in a beginning psychology class who, in turn, were able to obtain completed tests from 670 of their parents and 140 of their grandparents, a tri-generational sampling technique that allows for age comparisons. From this data pool, items were selected that were significantly 1) associated with the total score for their respective scales and 2) discriminating in the ratings obtained. These items comprised Form C, which was then administered to a validation sample of 100 UCLA freshmen (50 males and 50 females), 100 of their parents, and 50 of their grandparents. Form C was informally field tested in a variety of practical settings (e.g., a community counseling center) to obtain reports of its usefulness and general comments. Lastly, a Form D was constructed by rewriting Form C items in simpler language, aiming for a comprehension level of 10th-grade students. There are 15 items for each of the 10 scales on Forms C and D.

Norms for Form D were developed in two ways. First, the scores for the original sample of 1,158 people were converted in some unspecified way to Form D equivalents. Reasonable equivalence was checked by comparison with the Form C validation sample, although the details of this comparison were not reported. Second, Form D was administered to 618 students from five secondary schools in Santa Clara County, California. Percentile norms, which indicate the percentage of students scoring at or below different levels, were constructed for 352 12th-grade students and 146 10th-grade students, with males and females combined.

Testing materials include a four-page test booklet, an answer sheet, a personal profile form, and the CLGES manual. Individual scores are placed on the personal profile form. For each life goal, there are separate rows of raw scores for males and females. One places an "X" in the appropriate row to indicate about where one's raw score belongs and then connects the Xs by straight lines. Toward the top of the page are percentile norms that allow for a rough interpretation. These percentile norms are general and tentative, formed from the original and validation samples who took Forms A, B, and C. On the back of the profile form are brief definitions of the 10 life goals along with suggested occupations for those scoring high or low in each. For example, those who score low on the profit motive might consider nursing, teaching, or being a housewife. Those scoring high might consider banking, law, dentistry, architecture, or being an accountant.

Practical Applications/Uses

A variety of practical uses were envisioned for the California Life Goals Evaluation Schedules. For example, this measure might facilitate self-examination or stimulate group discussions about differences in attitudes. Potentially, it might be useful in employee selection. However, its primary use is to help normal individuals plan future directions and evaluate current pursuits. In this regard, the scope for application is quite broad, restricted mainly by age (from about Grade 10 to retirement). For example, people approaching retirement might use it to plan activities. Secondary and college students might use it to make career plans. Women returning to the work force, persons wishing to change careers, and terminated workers might all use it to help in planning. In addition, married couples might use the CLGES to better understand their similarities and differences.

At this time, however, probably the most important use for the CLGES is research. In the manual, Hahn poses a number of significant questions concerning the way life goals vary according to socioeconomic level, ethnic group, and levels of performance (e.g., academic achievement). Of perhaps even more significance, he was concerned with the way life goals vary across the life span and vary in relation to major events, such as getting married. At least since the work of Levinson (1978), the topic of adult change and development has become prominent in research. Do mid-life career changes or Levinson's life transitions involve shifts in life goals? What initiates shifts in life goals? How are life goals affected by mental health? In the instrument he developed and the questions he posed, Hahn was ahead of his time. The research context of the 1960s was not congruent with his work, which is much more at home in the present.

In practical use, the CLGES is easy to administer. No special qualifications are required by the administrator. Aside from normal requirements for testing, no special procedures are necessary. Scoring is a straightforward summing of raw scores for scale items.

Form D-S is self-scored, and Form D-M is scored manually through 10 scoring templates. Although Hahn stressed the importance of providing clear statements of what the test measures and the immediate purpose of testing (e.g., career planning), the test is self-administered. Few people have difficulty in following the directions, and if there is doubt, the author advocates reading the instructions aloud. The test is untimed; generally, it can be completed and scored within 45 minutes.

Although administration seems relatively unproblematic, interpretation is difficult due largely to the general and tentative nature of the norms. As the author frequently notes, the CLGES is still experimental rather than established. At least a master's level knowledge of psychological measurement and a thorough reading of the manual are required to interpret the test with proper reservations. Problems of interpretation will be discussed later.

Technical Aspects

Evidence for scale reliabilities is incomplete. Split-half and internal consistency reliabilities for the original sample were computed using the Spearman-Brown and Kuder-Richardson formulas, respectively. The coefficients obtained ranged from .87 to .98. For the Form C validation group, Kuder-Richardson reliabilities ranged from .82 to .89. Although these reliabilities are strong, Forms A, B, and C can merely encourage the expectation that Form D has strong reliability as well. In this regard, a test-retest study of 41 upper-division and graduate students over a 90-day interval yielded scale reliabilities ranging between .71 and .86 for Form D-M. At present, there are encouraging indications that the CLGES produces reliable scores, but reliability must be demonstrated using Form D with samples from groups who are likely to use the test (e.g., secondary students).

Evidence for validity is suggestive but weak. Content validity is perhaps the strongest. Hahn did not claim that the 10 life goals he drew from Centers are an adequate representation of the range of life goals, nor that the 15 items for each scale adequately sample the construct being measured. However, he did have 10 experts in measurement examine the test, and they apparently expressed a consensus of opinion that scale items reflected the scale. In addition, items were examined by students in several graduate seminars, who also concluded that items reflected constructs. No details of the nature or degree of consensus were reported.

Evidence for concurrent validity is presented in tables that show the average profiles of students in different majors and of parents in different occupations, but this evidence is inconclusive. Generally, the profiles are what one might expect and, in this sense, are encouraging—but not convincing. It would be desirable to find out if those who score high, for instance, on Power actually do seek power over others. No such direct evidence is available. Rather, there is only suggestive, indirect evidence.

There is no evidence for predictive validity. That is, there are no studies that use the CLGES scales to predict college major or future occupation. More importantly, there is no evidence that if one scored high on a scale and then pursued that life goal one would be happy or in some way benefit from having done so.

Neither is there evidence for construct validity. For instance, no comparisons of the CLGES with other tests have been conducted. There is no evidence that the esteem scale correlates well with another test of esteem, or the profit scale with another such test. Construct validity rests, in this case, upon content validity, which is inadequate.

Using scale scores, the internal structure of the CLGES was examined by factor analysis with varimax rotation. The aim of this procedure is to identify clusters of scales that go together. For parent samples, Esteem, Fame, Power, and Profit consistently clustered together, but in results for students they broke apart. However, Leadership and Interesting Experiences rather consistently went together. Using different groups, factor analyses did not result in consistent clusters but only in tendencies. Hahn divided the 10 scales into three groups for interpretation: Protestant ethic group (Esteem, Profit, Fame, Power); social group (Leadership, Security, Social Service); and creative group (Interesting Experiences, Self-Expression, and Independence). However, it is difficult to understand what justifies these groups. It would be more desirable to examine the internal structure through factor analysis of items rather than scales, particularly because the scales were, to a large extent, rationally developed. Also, it is essential that analyses be conducted on Form D rather than inferred from previous forms.

Norm groups need to be expanded and updated. For any test, it is essential that inferences about an individual be based upon a reference group to whom that individual belongs or can be reasonably compared. The reported norms are too few and too narrow, and all were developed close to or over 20 years ago.

Critique

Without resources or funding for extensive sampling and large-scale research projects, Hahn had to develop the California Life Goals Evaluation Schedules largely through his own resources. With considerable industry, he launched numerous studies over the years of test development, particularly small-scale practical and informal studies to judge whether the test information would be of value to clients. However, the CLGES is still not suitable for practical use of this nature, even with the precaution to interpret it within the context of other test information and personal history. It is not just that norms, reliabilities, and validities are tentative, but that interpretation can be too misleading.

The percentile norms were derived from the original sample of university students, parents, and grandparents. However, this norm group yields percentile norms that are quite different than secondary students. Older people are apparently more decisive in their judgments of debatable statements than youth, leading to a greater range of scores. For example, the raw scores that mark the 10th and 90th percentiles on Esteem are 15 and 52 for males in the general norm group, and 18 and 37 for males in Grade 12. The range of scores for secondary students is consistently attenuated in comparison to the general norm. Thus, if a young male, for instance, marked his score of 37 on the personal profile, it would indicate that he was a little above average (the 60th percentile) whereas he would be at the 90th percentile for his own age group! The difference in interpretation is considerable. Not only would the youth be led to believe his desire for high regard from significant others was only average rather than strong, but the interpretations on the back of the profile sheet would potentially mislead him further. The profile sheet indicates that high scores are indicative of bankers, journalists, lawyers, and army officers and that low scores are indicative of nonprofessional groups. If this youth had scored 28 on Esteem (the 50th percentile for his group), he would have scored very close to the low scores of nonprofessional groups.

There is no adequate evidence to support the interpretive claims on the back of the profile forms. That is, there is not enough presented evidence to justify directing a person with a low esteem score to the nonprofessional ranks. The danger of faulty interpretation seems prominent, particularly in view of the sizeable differences in norms. Further, there is some indication that high scores might mean different things at different ages. For example, 11th-grade males who were not planning to go on to college scored higher on the profit motive than any other secondary school group, yet the profile interpretation for high Profit scores points to highly educated professional groups such as architects, dentists, and lawyers. The interpretive link between scale scores and occupations seems inappropriate, partially because the number of occupations are so few and partially because a high score on a scale can indicate such diverse occupations. Both dentists and economists in the sample of parents scored high on Profit, but perhaps for different reasons. It would be better to link scale scores to characteristic values, actions, or attitudes.

The use of the CLGES as a practical instrument for helping people to make plans is premature. Hahn advocates caution in interpretation and indicates on the profile form that interpretations are suggestive and tentative, but it does not seem likely that clients would reasonably observe these qualifications. The visual impact of seeing a score on a scientific-looking profile seems more apt to outweigh cautionary notes. At present, there are other tests that measure many of the same constructs found on the CLGES, such as the Work Values Inventory (Super, 1970) and the Hall Occupational Orientation Inventory (Hall & Tarrier, 1976). If the CLGES is to be a serious rival to these tests, an effort to update and provide more solid grounding for the test must be made.

In general, the manual does not meet many current standards of the American Psychological Association for educational and psychological tests. Procedures are too briefly reported. For instance, the conversion of scores from Forms A and B into norms for Form D is not described well enough. Standard errors of estimate are not reported. The term "life goals" is not defined fully enough to make important distinctions or to identify more specific criteria for what a person with a given life goal would do or be like. The role of the validation sample was not specified clearly, nor was evidence systematically presented. There are too many intrusions of hypothetical interpretations and educational asides about testing in general that weigh against the coherency of the presentation. As Lundin (1972) concluded in a much earlier review, the CLGES "is still in its experimental phase" (p. 86), and so it has remained for 15 years.

Although one can place only limited confidence in the CLGES for practical application, it does seem to offer promise for research. The idea of measuring the strength of life goals is sound and valuable. Oddly enough, some of the reasons that weigh against the CLGES for practical application favor it for further research. For example, it may be that professionals in dentistry, medicine, and the like hold quite different life goals in secondary school than they manifest later. Perhaps there is a normative pattern of change. The possibility that non-college-bound students might have life goals that resemble those held by professionals is fascinating. Might not school drop-outs also resemble professionals? Perhaps the holding of certain life goals might be adaptive and beneficial at one age or in one setting but not so at another age or in another setting. After all, a strong profit motive would not have much of an outlet for students and could be experienced as a continual frustration. Hahn indicates that those males who have the least esteem, money, fame, or power (those at the lowest socioeconomic level) are those who have the highest desire for them. Could the maturational difference in life goals auger well for the success of career pursuits for one group and auger ill for another? These kinds of highly significant questions are raised either implicitly or explicitly in the manual, suggesting the promise of the CLGES for research investigations, and eventually, perhaps, for practice.

References

Centers, R. C. (1949). *The psychology of social classes*. Princeton, NJ: Princeton University Press.

Hahn, M. E. (1963). *Psychoevaluation*. New York: McGraw-Hill.

Hahn, M. E. (1969). *The California Life Goals Evaluation Schedules manual*. Los Angeles: Western Psychological Services.

Hahn, M. E., & McLean, M. S. (1950). *General clinical counseling in educational institutions*. New York: McGraw-Hill.

Hall, L. G., & Tarrier, R. B. (1976). *Hall Occupational Orientation Inventory*. Bensenville, IL: Scholastic Testing Service.

Levinson, D. J. (1978). *The seasons of a man's life*. New York: Alfred A. Knopf.

Lundin, R. W. (1972). The California Life Goals Evaluation Schedules. In O. K. Buros (Ed), *The seventh mental measurements yearbook* (pp. 86-87). Highland Park, NJ: The Gryphon Press.

Super, D. E. (1970). *Work Values Inventory*. Boston: Houghton Mifflin.

Williamson, E. G., & Hahn, M. E. (1940). *Introduction to high school counseling*. New York: McGraw-Hill.

Joanne Gallivan, Ph.D.

Associate Professor of Psychology, University College of Cape Breton, Sydney, Nova Scotia.

CANADIAN COMPREHENSIVE ASSESSMENT PROGRAM: ACHIEVEMENT SERIES

R. R. Danley, J. W. Wick, and J. K. Smith. Toronto, Ontario: Learning House—Distributed exclusively by Guidance Centre.

Introduction

The Canadian Comprehensive Assessment Program (CCAP) is a coordinated set of measures with Canadian content and norms covering ability (the Developing Cognitive Abilities Test), attitude (the School Attitude Measure), and achievement (the Achievement Series). The Achievement Series tests are designed to assess development of basic skills in the areas of reading, language, mathematics, and reference and study skills.

The CCAP Achievement Series was developed by an 18-member team of authors coordinated by Dr. Raymond R. Danley, who was at the time Measurement Coordinator at Gage Publishing, the original publisher of the battery. The authors were comprised of educational measurement and curriculum specialists and elementary and high school teachers. In addition, other teachers and curriculum specialists, as well as testing personnel, were consulted during item tryout. Standardization of the Achievement Series took place during October, 1980, and April, 1981, and the tests have been available since that time.

The Achievement Series tests are divided into Levels 4-12, roughly corresponding to the chronological age of their intended population. Levels 4-11 are intended for the assessment of children in prekindergarten to Grade 6, with Level 12 appropriate for the assessment of students in Grades 7 to 9. For all levels, a single test booklet contains all subtests, and all questions use a multiple-choice format. For Levels 4-8, students record their responses directly in the booklets, which are available in hand- and machine-scorable versions. Levels 9-12 have reusable booklets, and students mark their responses on hand- or machine-scorable answer sheets. All levels of the series contain subtests assessing three major content areas, Reading, Language, and Mathematics. In addition, Levels 9-12 contain subtests in the area of Reference and Study Skills. The tasks within each content area change across levels to reflect changing curriculum across the grades.

The Level 4 Reading area contains four subtests requiring the child to mark pictures matching: 1) visual stimuli (Visual Matching), 2) orally presented words (Auditory Attention), 3) words formed by blending sequences of orally presented sounds (Auditory Picture Closure), and 4) words rhyming with orally presented words (Auditory Picture Rhymes).

The Language section of Level 4 contains five subtests. Combining Information

66

requires the child to select the picture matching the combination of characteristics described in a spoken sentence. In Auditory Comprehension, the child must select the picture representing information that could logically follow a set of presented statements. In Silly Pictures, the task is to select which of a group of pictures is foolish (e.g., in a series of scenes of fishing, a picture of a fish holding a rod would be the inappropriate one). In Visual Oddities, the dissimilar item within a set must be identified. And finally, Generalization requires the student to choose, the item that is most similar to a pair of related targets.

The Developmental Mathematics subtest consists of items requiring the child to choose, from several pairs, the pair of symbols or pictures that are most alike, based on similarities in number or geometry.

The Level 5 Reading area contains a Letter Recognition subtest, a Picture-to-Picture Rhymes subtest, which requires the choice of a picture whose name rhymes with the name of a target picture, and a Sound-Picture Relations subtest, which requires the child to select a picture that has an orally presented sound in its name.

The Level 5 Language area contains two subtests (Visual Oddities and Generalization) similar to those at Level 4 and four others: Auditory Classification, which requires the child to choose the group of pictured objects that is related to orally presented word(s); Visual Synthesis, which requires the child to select the correct set of pieces that would form a letter, number, or picture; Sequencing, in which the child must choose the picture matching a spoken spatial or temporal ordering statement; and Negation, in which the child must choose the picture that matches a description containing negative information. The Level 5 Developmental Mathematics subtest has the same format as that of Level 4.

In Level 6, there are eight Reading subtests, only one of which is retained from Level 5, Letter Recognition. Auditory Word Rhymes requires the choice of a printed word that rhymes with an orally presented one. Sound-Letter Relations requires the student to select the letter corresponding to a sound. Adding Sounds requires the student to blend sounds and choose the picture representing the resulting word. In Taking Sounds Away, the student must remove a given sound from a stated word and pick the picture representing the new word that results. Blending requires the selection of a printed word that is formed by blending presented sounds. In Words in Sentences, which requires the completion of a sentence, the child must choose a nonsense word that is keyed to a pictorial code. In Reading Comprehension, students must answer questions measuring literal comprehension of and simple inference from two short passages.

The Level 6 Language subtests include Picture Opposites, in which the child must choose the picture that is opposite in meaning to the target; Visual Closure, which requires the student to identify which pieces would complete a figure; and Finding Causes, in which the child must select the picture that illustrates the cause of a pictured event.

Level 6 includes a Developmental Mathematics subtest similar to that used in Levels 4 and 5, as well as a Mathematics Achievement subtest that measures knowledge of basic mathematical computations and concepts.

In Levels 7-12, the Reading area includes Vocabulary and Reading Comprehension subtests; in addition, Levels 7 and 8 also contain a Word Attack subtest, which measures knowledge of consonants, vowels, word formation, and syllabication.

The three Language subtests for Levels 7-12 measure competence in Spelling, Capitalization and Punctuation, and Grammar and Usage. The mathematics tests in these levels are Computation, Concepts, and, with the exception of Level 7, Problem Solving. A fourth area of assessment, Reference and Study Skills, is included in Levels 9-12. This subtest assesses knowledge of library reference skills, such as alphabetization and dictionary use, and social science skills, such as the use of graphs, charts, and maps.

Practical Applications/Uses

The Achievement Series of the Canadian Comprehensive Assessment Program is intended to measure educational achievement in the basic areas of language, reading, mathematics, and reference and study skills. The tests may be used as either group indicators or individual assessment devices. The authors of the teacher's manual suggest that information from the subtests can be used to evaluate program effectiveness and plan for specific instructional improvements (Danley, Wick, & Smith, 1983). As individual assessment tools, the tests allow for the identification of pupil capabilities, which provides a basis for individual adaptations in methods and programs. The authors also suggest that the tests may be used to 1) monitor student progress and communicate that progress to parents and school personnel and 2) identify students with special needs, such as the gifted or those with learning problems.

The Achievement Series tests are specifically designed for classroom administration by teachers. Directions for administration are clear and thorough. The tasks in Levels 4-6 are examiner-paced without specific time limits; average administration time is about 90 minutes. Levels 7-12 each require 4 to 4½ hours administration time, which should be spread across 3 or 4 days. Machine scoring of tests is available, or the tests may be hand-scored using the scoring key and norms booklet.

Interpretation of scores is based on norms derived from a national sample of over 27,000 children from over 250 schools in every Canadian province except Prince Edward Island. Guidance in score interpretation is given in the teacher's manual. For the more technically minded, additional information on the scores and their derivations is available in the technical manual (Learning House, 1985).

Technical Aspects

Reliability coefficients for every subtest at every level are reported in the technical manual. These measures reflect internal consistency, that is, the consistency with which students in the norm group tended to respond on the items within each subtest. The coefficients ranged from .49 to .98, with the great majority above .60. These figures represent moderate to high levels of internal reliability for the subtests; certainly, these are acceptable levels for this type of test. The alternate forms for Levels 7-12 are essentially equivalent (i.e., will produce similar scores for the same individual in most cases). This equivalence was accomplished through a special equating study involving a sample of over 11,000 students. The information obtained was used to adjust the raw score conversion tables for Form B so that scores derived from the two forms are similar.

The validity of these tests must, in large part, be determined by informed judgment. Administrators must decide on the basis of curriculum goals and other considerations whether the tests provide valid and useful information for their particular purposes and students. However, the test developers did make several efforts to ensure the content validity of the test for Canadian schools. First, the authors reviewed textbooks, supplementary instructional materials, and curriculum guides to create a set of instructional objectives upon which the content of the test was based. These were reviewed and refined by teachers, testing personnel, and curriculum specialists. Test items were developed based on these objectives and subjected to three phases of tryouts involving large numbers of students. Only those items that satisfied criteria for difficulty, discriminability, and absence of ethnic or gender bias and that fulfilled readability requirements were retained.

Critique

The Achievement Series of the Canadian Comprehensive Assessment Program was developed to meet an admirable goal—the provision of a general achievement battery with Canadian content and norms. It is unique in that it provides for comparison of achievement with measures of aptitude and school attitude, the norms for which were established with the same group of students who served as the standardization sample for the Achievement Series. Thus, it is particularly appropriate in situations where direct comparison of achievement with aptitude and/or school attitude is desired. For use alone, its appropriateness as an achievement measure can probably best be judged by examining the list of instructional objectives upon which the content of each subtest is based; these are listed in detail in the teacher's manual. The time required for administration, particularly at higher levels, may present practical problems, especially in those cases in which testing materials must be shared. Despite this limitation, the CCAP Achievement Series is a valuable addition to Canadian educational assessment.

References

Danley, R. R., Wick, J. W., & Smith, J. K. (1983). *Canadian Comprehensive Assessment Program: Achievement Series teacher's manual.* Toronto: Learning House.

Learning House. (1985). *Canadian Comprehensive Assessment Program: Achievement Series technical manual.* Toronto: Author.

Joanne Gallivan, Ph.D.
Associate Professor of Psychology, University College of Cape Breton, Sydney, Nova Scotia.

CANADIAN COMPREHENSIVE ASSESSMENT PROGRAM: DEVELOPING COGNITIVE ABILITIES TEST

R. R. Danley, J. W. Wick, and J. K. Smith. Toronto, Ontario: Learning House—Distributed exclusively by Guidance Centre.

Introduction

The Developing Cognitive Abilities Test (DCAT) is the ability measure component of the Canadian Comprehensive Assessment Program (CCAP), which also includes an achievement series and a school attitude measure; both also are reviewed in this volume. The DCAT is intended to assess learning characteristics and abilities presumed to contribute to academic performance. The test is based on Bloom's (1956) taxonomy of objectives in the cognitive domain. Unlike many ability tests, the DCAT is founded on the assumption that learning characteristics are not static, but are affected by learning and instruction.

The DCAT was developed by a team of authors under the supervision of Dr. Raymond R. Danley, then Measurement Coordinator at Gage Publishing. The standardization of the complete CCAP was conducted during the 1980-81 school year, and all components of the program, including the DCAT, have been available since that time.

The DCAT consists of six levels suitable for use in Grades 2 to 9. Levels 2, 3, and 4, used in Grades 2, 3, and 4, respectively, are available in one form only. Level 5/6, for Grades 5 and 6, Level 7/8, for Grades 7 and 8, and Level 9/12, for Grade 9, have two alternate forms. For all levels, a single test booklet contains all test items. Each question uses a multiple-choice format. For Levels 2 and 3, answers are entered directly into the test booklets, which are machine or hand scorable. For the remaining levels, answers are marked on machine- or hand-scorable response sheets. All levels of the DCAT contain items assessing three content areas: Verbal Ability, Quantitative Ability, and Spatial Ability. In addition, except for Level 2, information is provided on performance on items categorized according to five of the six levels in Bloom's taxonomy. Those included are knowledge, comprehension, application, analysis, and synthesis. Evaluation was not included in this assessment because of the difficulty of measuring it using objective-type questions.

Each of the upper five levels of the DCAT contains 80 items arranged in a single test with one set of instructions. Level 2 contains 80 items grouped into nine subtests. At this level, all materials are visual rather than verbal, and instructions are given orally.

Four subtests comprise the measure of Verbal Abilities: in Picture Opposites, the child must select a picture representing the opposite of a given target; Excluding Information tests ability to use the word *or*; Finding Causes assesses the ability to recognize possible causes for pictured events; and Drawing Conclusions measures the ability to consider two pieces of information and use them to form valid conclusions.

There are three subtests assessing Spatial Abilities: Visual Analogies contains basic visual analogy items; Visual Reasoning is a general measure of nonverbal reasoning; and Pattern Extension measures basic abilities in understanding patterns by requiring students to select the correct continuation of a given sequence of stimuli.

There are two tests of Quantitative Abilities: Mathematics is a measure of quantitative reasoning assessing knowledge of quantitative concepts, relations, and problem solving; and Developmental Mathematics requires the child to select pairs of items that have something in common, based on similarities in size, shape, or number.

Practical Applications/Uses

The Developing Cognitive Abilities Test was developed to assess cognitive abilities contributing to academic performance, which themselves can be altered through instruction. Appropriate as either a group or an individual assessment device, the authors (Danley, Wick, & Smith, 1983) suggest that the DCAT has utility for program development or improvement and for individual curriculum planning. The test can be used in conjunction with the CCAP Achievement Series, which was normed on the same standardization group, to identify students whose actual achievement levels are discrepant with their abilities. This makes it particularly useful, the authors (Danley, Wick, & Smith, 1983) suggest, for identifying students with learning problems or those who might be considered gifted. In addition, it is suggested that the results of group testing can be used to identify areas of strength and weakness in the curriculum.

The DCAT is designed for classroom administration by teachers. Directions for administration are clear and thorough. The subtests in Level 2 are examiner-paced without specific time limits; it is recommended that the testing be spread over 3 or 4 days. For each of the remaining levels, total administration time is approximately 65 minutes.

Interpretation of the scores is based on norms derived from a national sample of over 13,000 students from 55 school districts in every Canadian province except Prince Edward Island. Norms are available for fall and spring administrations of the test. Guidance in score interpretation is given in the teacher's manual (Danley, Wick, & Smith, 1983). Direct comparison of students' scores with their CCAP Achievement Series performance can be obtained if machine scoring is used. In that case, the grade equivalent scores for the Achievement Series are marked to indicate whether they are significantly different from the predicted levels for individual students or groups of students.

Technical Aspects

Several steps were taken by the developers of the CCAP Developing Cognitive Abilities Test to ensure the content validity of the test for the Canadian school population. The set of objectives on which content was based and the actual test items were evaluated by several groups of educational specialists, including a panel of minority-group representatives. Additional comments were solicited from teachers during field testing of the items. Only those items that satisfied criteria for difficulty, discriminability, and absence of ethnic or gender bias were retained in the final version of the test.

Reliability coefficients for every content area at every level are reported in the technical manual (Gage, 1983). These measures reflect internal consistency, or, the consistency with which students in the norm group tended to respond on the items within each content area. Most of the correlations were moderate to high (.56 to .90) and represent acceptable values for this type of test. However, there were a few areas for which the reliability coefficients were rather low. These included the Spatial Abilities measure at Level 2 (.32 for fall; .40 for spring) and Level 3 (.47 for spring) and the Quantitative Abilities measure at Level 5/6 for the Grade 6 spring norm group (.46 for Form A; .47 for Form B). For those particular measures, some caution is warranted in their use.

Critique

The CCAP Developing Cognitive Abilities Test was developed to meet an admirable goal—the provision of a general abilities battery with Canadian content and norms. It is unique in that it is part of a comprehensive testing package that allows for direct comparison of aptitude with measures of achievement and school attitude normed on the same sample of school children. Another unique feature of the DCAT is the provision of scores keyed to Bloom taxonomy objectives, in addition to scores for content areas. This makes the DCAT particularly useful for educators who have used the Bloom taxonomy as a guide to curriculum development. All in all, the DCAT is a valuable addition to Canadian educational assessment.

References

Bloom, B. S. (Ed.). (1956). *Taxonomy of educational objectives: Handbook 1. Cognitive domain.* New York: David McKay.

Danley, R. R., Wick, J. W., & Smith, J. K. (1983). Developing Cognitive Abilities Test: Teacher's manual. Toronto: Gage.

Gage Publishing Co. (1983). Developing Cognitive Abilities Test: Technical manual and norms. Toronto: Author.

Joanne Gallivan, Ph.D.
Associate Professor of Psychology, University College of Cape Breton, Sydney, Nova Scotia.

CANADIAN COMPREHENSIVE ASSESSMENT PROGRAM: SCHOOL ATTITUDE MEASURE

R. R. Danley, J. W. Wick, and J. K. Smith. Toronto, Ontario: Learning House—Distributed exclusively by Guidance Centre.

Introduction

The School Attitude Measure (SAM) is a self-report inventory designed to assess five dimensions of students' attitudes and opinions concerning their school environment and their academic skills and achievement. It is the attitude component of the Canadian Comprehensive Assessment Program (CCAP), which also includes measures of achievement (Achievement Series) and aptitude (Developing Cognitive Abilities Test).

The SAM was developed by a team of authors under the supervision of Dr. Raymond Danley, the Measurement Coordinator at Gage Publishing in Toronto at the time of the test's development. Standardization of the complete CCAP was conducted during the 1980-81 school year, and all three components of the program, including the SAM, have been available since that time.

The SAM is available on three levels composed of five scales each. Level 4/6 is suitable for use with students in Grades 4-6; Level 7/8 for Grades 7-8; and Level 9/12 for Grade 9. A reusable test booklet available for each level contains all the items for that level, and responses are recorded on a machine-scorable answer sheet. The items are simple statements worded both positively and negatively, to which students respond on a 4-point scale (never agree, sometimes agree, usually agree, always agree).

The five scales provide information on students' attitude dimensions. Responses to the Motivation for Schooling scale reflect students' general attitudes about the importance, value, and meaningfulness of their school experience. Academic Self-Concept—Performance Based provides information on students' confidence in and feelings about their academic abilities and school performance. Academic Self-Concept—Reference Based assesses students' perceptions of others' opinions of their performance and ability. The Sense of Control over Performance Scale measures students' feelings of control over school situations and their willingness to take responsibility for personal performance. The final scale, Student's Instructional Mastery, involves self-assessment of students' mastery of school skills, such as task persistence and concentration.

Practical Applications/Uses

As described previously, the School Attitude Measure is intended to assess five dimensions of students' school-related attitudes. It may be used as either a group or an individual assessment device. The test authors suggest that the test be used for program development or improvement, individual educational planning, and student guidance (Danley, Wick, & Smith, 1982). The authors further suggest that the test is suitable for developing and evaluating effective objectives for education and for assessing the impact of innovative programs whose stated goals may be affective, as well as cognitive, in nature. Information from the test also may aid in the selection of students for particular education programs. Finally, because it provides comparable measures of attitude across a 6-year grade span (Grades 4-9), the instrument can be used in research concerning the effects of schooling on affective development.

The main value of the test seems to lie in its effectiveness as a tool for student counseling. The authors emphasize that test results may be helpful in the early identification of students with negative attitudes. Consequently, intervention methods can be implemented before serious problems arise. In addition, because scores are obtained for five separate attitude scales, test results may help teachers select the most appropriate intervention for individuals or groups of students. When used in conjunction with the results from aptitude and achievement measures such as the companion tests in the CCAP, the SAM may allow identification of students who show discrepancies in aptitude and achievement due to attitudinal problems.

The SAM is designed specifically for classroom administration by teachers. Directions for administration are clear and thorough. The test is examiner-paced, and although it has no specific time limits, administration time is approximately 35 minutes. Hand scoring is possible, but because test items are arranged inconveniently for that method, machine scoring is advised.

Standardization of the SAM was conducted in conjunction with the norming of the CCAP Achievement Series and the DCAT. Norms are based on a national sample of approximately 8,700 children from school districts across Canada. Guidance in score interpretation is provided in the teacher's manual (Danley et al., 1982).

Technical Aspects

Reliability coefficients for each attitude scale at each grade level are reported in the teacher's manual. The values, which range from .69 to .88, reflect the test's internal consistency; that is, the consistency with which students in the norm group tended to respond on the set of items comprising each scale. Reliability coefficients for the total test at each grade level ranged from .91 to .95. Finally, test-retest reliability measures for each level were obtained for administrations spaced 4 weeks apart. These values ranged from .80 to .89. Certainly, then, the evidence indicates that the SAM is a highly reliable measure.

To ensure the validity of the SAM, its developers conducted discussions with school personnel and extensively analyzed research literature on affective correlates of schooling. From these studies, they derived a potential set of school attitude dimensions, which formed the basis for sample items. The number of sample items was reduced on the basis of factor analyses, leading to the development of the current five basic scales. Next, as a guide to item development, content specifications were devised for each dimension. The test was subjected to two trial phases, each involving several hundred students at each test level. Only the items that satisfied criteria for readability, reliability, discriminability, and lack of sex, race, or social desirability bias were retained in the final version of the test. Several findings supporting the claim of validity for the test are described in the teacher's manual, although the actual data on which the statements are based are not reported. The manual reports that "correlations with measure [*sic*] of social desirability with the final version of the School Attitude Measure are low and insignificant" (Danley et al., 1982, p. 24) and that "validity studies suggest strong convergent validity of specific subscales with other instruments that test only one aspect of affective development" (Danley et al., 1982, p. 24). The authors also report that significant relationships have been found between SAM scale scores and parent and teacher ratings of students' attitudes.

Critique

Although additional information on the details of the validity studies referred to in the teacher's manual would be welcome, there is no question that the School Attitude Measure is psychometrically sound. Perhaps its only flaw is that all scores are grade-referenced; given the considerable variation in grade assignment across the country, this could limit the test's usefulness.

Probably the greatest barrier to classroom application of the test, though, is the likely perception of school personnel that students' school attitudes can be assessed by observing behavior and performance and that the time and expense involved in using such a test is, therefore, unneccessary. The School Attitude Measure is a suitable instrument for educators who do want to use standardized attitude assessments, but its primary application may be in research, for which it is also appropriate.

References

Danley, R. R., Wick, J. W., & Smith, J. K. (1982). *School Attitude Measure teacher's manual*. Toronto: Gage Publishing.

Terry Cicchelli, Ph.D.
Associate Professor and Coordinator of Elementary Preservice Program,
Fordham University, School of Education at Lincoln Center, New York,
New York.

CANFIELD INSTRUCTIONAL STYLES INVENTORY

Albert A. Canfield and Judith S. Canfield. La Crescenta,
California: Humanics Media.

Introduction

The Canfield Instructional Styles Inventory (CIS; Canfield & Canfield, 1975) was developed to measure an instructor's values and preferences relative to instructional style in four areas that include 17 dimensions:

Condition—the relative emphasis an instructor prefers in the teaching/learning relationship with regard to 1) peer affiliation, 2) organization, 3) goal setting, 4) competition, 5) instructor affiliation, 6) detail, 7) independence, and 8) authority.

Content—an instructor's comparative level of interest in four dimensions: 1) numeric, 2) qualitative, 3) inanimate, and 4) people.

Mode—one's preference for particular instructional procedures in 1) listening, 2) reading, 3) iconic experience, and 4) direct experience.

Influence (referred to as Influence in the manual [Canfield & Canfield, 1975, p. 6] and Responsibility and Expectancy on the answer sheet)—the comparative amount of responsibility that an instructor feels he or she has or will accept for the learning process while 1) taking primary responsibility, 2) taking most responsibility, 3) taking minor responsibility, and 4) taking primary to total responsibility.

The manual offers no theoretical grounding for the selection and use of the concepts, areas, or dimensions included in the instrument. Specifically, although the CIS includes three concepts—values, preference, and instructional style—no literature is provided to show relationships, psychological or educational, between or among these concepts. Further, instructional style is operationalized in four areas and 17 dimensions, with an absence of theoretical or practical support for their selection and utility to teaching and learning.

However, on a more positive note, there is a report of some testing of the CIS along with the Learning Styles Inventory (LSI; Canfield & Canfield, 1974) an instrument similar in design, whereby the dimensions of the LSI correspond with the CIS. In one study conducted with nursing students and faculty at the College of San Mateo, California, findings showed a significant difference in 11 of the 17 scales measured. Students had higher preferences than instructors in the areas of peer affiliation, goal setting, and direct experiences, whereas instructors had higher preferences than students in the areas of competition, independence, authority, and reading (Henderson, n.d.) Two other studies were conducted at Nova University by Davis (1979) and Raines (1976). Davis, in a regression analysis of instructional style on training style for developmental students at six institutions in the university system of Georgia, accounted for 36% of the variation in the model. Raines, in research with mathematics classes at a junior college, found that students with higher grade levels of achievement had learning styles more closely related to instructor's teaching styles than students receiving lower grades. Further, Llorens and Adams (1978), in their work at the University of Florida, found significant differences in instructor and student preferences, suggesting the need for altering the teaching/learning environment at the university.

Although the CIS has been tested, the studies reported in the manual (pp. 17-18) do not provide sufficient information regarding sample sizes, data collection procedures, and statistical approaches used. Seemingly, findings should be viewed with moderate caution at best.

Practical Applications/Uses

The Canfield Instructional Styles Inventory was designed to be used with postsecondary students, either in conjunction with or independently of the LSI. The CIS uses a 4-point Likert-type response format, ranging from most to least preferred. Each of its 25 items contains four response options ranked in order of their "fit" with the values and preferences of the examinee. Scores are obtained on 17 dimensions of instructional style. The CIS is a paper-and-pencil test and may be completed in 30 minutes. The test may be hand scored, and a clear description of scoring procedures relative to the scales is provided in the manual. Scores obtained from the summary of item values are directly transferable to profile sheets. The scores on the profile sheets can be interpreted to gain an appreciation of the kinds of conditions, content, modalities, and influence that characterize the examinee in relation to others in the norm group. Although a matrix relating techniques to the modality and conditions combination is provided, its utility in the absence of reported theory or practice is questionable.

In short, the examinee should not experience problems in reading difficulty of items or instructions. Similarly, the administrator of the inventory would find ease in administering the test and interpreting the results. However, using the results raises at least moderate concern for caution.

Technical Aspects

One reliability study is reported in the manual. Test-retest reliabilities for the 17 scales of the CIS for a sample of 62 college-level students with a 7-day interval between administrations ranged from .38 to .96, with a median of .89. Internal consistency coefficients for the scales ranged from .59 to .85, and, using a random sample of 200 subjects (100 males and 100 females), interscale correlations ranged from .63 to -.68. The authors note that correlations of total scores within ranking groups are forced toward negative values and are not appropriate indicators of scale relationships found using noninterdependent ratings.

A series of validity studies could be conducted to ensure that which the instrument purports to measure. In conducting these studies, the work of Cronbach (1980) and Messick (1975) may be reviewed, particularly as they provide a broad definition of construct validity—seemingly an appropriate focus for this instrument. Specifically, the authors may apply the work of Anastasi (1976) and her interpretation of construct validity. She states that a construct measured by a particular test can be adequately defined only in the light of data gathered in the process of validating that test. Further, she emphasizes that the definition should take into account the variables with which the test correlates significantly, including the conditions that affect its scores and the groups that differ significantly in such scores.

Critique

Although the basic intent of the Canfield Instructional Styles Inventory is noteworthy, that is, the identification of an instructor's values and preferences relative to instructional style, its limitations and deficiencies cannot be overlooked with regard to theoretical grounding for the use of the key variables in the four areas.

An exploration of the work related to the area of motivation and instruction found in the investigations of Weisz and Stipek (1982), Weiner (1979, 1983), Nord, Connelly, and Diagnault (1974), Wang and Stiles (1976), Wang (1983), and McCombs (1983) may provide a fruitful orientation to issues related to locus of control, perception, attribution, and learning. Additionally, the teacher-effects literature focusing on instructional patterns of behavior provided by Rosenshine (1976), Rosenshine and Berliner (1978), Good and Grouws (1979), Soar, (1973, 1975), and Cicchelli (1983) offers rich descriptions and operations of effective teacher behaviors in varying instructional styles. The reader also is encouraged to review the work emphasizing the relationship of teacher behavior to student learning conducted by Anderson, Evertson, and Brophy (1979), Crawford et al. (1976), Emmer, Sanford, Clements, and Martin (1982), Coladarci and Gage (1984), and Berliner (1980, 1982). Although all of these references are not equally important relative to the intent of the CIS, each may contribute in some way to any revisions in the instrument and manual undertaken by the authors or publisher.

In short, upon refinement of this inventory, the use of the CIS may be explored in contexts including secondary and postsecondary students with the intent of gathering data on instuctors' values and preferences relative to instructional style.

References

Anastasi, A. (1976). *Psychological testing* (4th ed.). New York: Macmillan Publishing Co.

Anderson, L., Evertson, C., & Brophy, J. (1979). An experimental study of effective teaching in first grade reading groups. *Elementary School Journal, 79,* 193-223.

Berliner, D. C. (1980). Using research on teaching for the improvement of classroom practice. *Theory into Practice, 29*(4), 1-4.

Berliner, D. C. (1982, April). *Executive functions of teaching.* Paper presented to the annual meeting of the American Educational Research Association, Toronto.

Canfield, A., & Canfield, J. (1974). *Canfield Learning Styles Inventory.* La Crescenta, CA: Humanics Media.

Canfield, A., & Canfield, J. (1975). *Canfield Instructional Styles Inventory (manual).* La Crescenta, CA: Humanics Media.

Cicchelli, T. (1983). Forms and functions of instruction patterns: Direct and nondirect. *Instructional Science, 12,* 343-353.

Coladarci, T., & Gage, H. (1984). Effects of a minimal intervention on teacher behavior and student achievement. *American Educational Research Journal, 21*(3), 539-555.

Crawford, J., Gage, N. L., Corno, L., Stayrook, N., Mitman, A., Schunk, D., Stallings, J., Baskin, E., Harvey, P., Austin, D., Cronin, D., & Newman, R. (1979). *An experiment on teaching effectiveness and parent-assisted instruction in the third grade.* Stanford, CA: Center for Educational Research.

Cronbach, L. J. (1980). Validity on parole: How can we go straight? *New Directions for Testing & Measurements, 5,* 99-108.

Davis, C. (1979). *A study of instructor styles and training.* Unpublished manuscript, Nova University, Fort Lauderdale, FL.

Emmer, E. T., Sanford, J. P., Clements, B. S., & Martin, J. (1982). *Improving classroom management and organization in junior high schools: An experimental investigation* (Research & Development Rep. No. 6153). The University of Texas at Austin, Austin Research and Development Center for Teacher Education.

Good, T. L., & Grouws, D. A. (1979). *Process product relationships in 4th grade mathematics classes.* Columbia, MO: University of Missouri, College of Education.

Henderson, F. (n.d.) *A study of nursing students and faculty regarding learning and instructional styles.* Unpublished manuscript, College of San Mateo, California.

Llorens, L., & Adams, S. (1978, March). A study of instructor and student preferences. *American Journal of Occupational Therapy.*

McCombs, B. L. (1983, April). *Motivational skills training: Helping students to adapt by taking personal responsibility and positive self-control.* Paper presented at the annual meeting of the American Educational Research Association, Montreal.

Messick, S. (1975). The standard problem: Meaning and values in measurement and evaluation. *American Psychologist, 30,* 955-966.

Nord, W. R., Connelly, F., & Diagnault, G. (1974). Laws of control and aptitude test scores as predictors of academic achievement. *Journal of Educational Psychology, 66,* 956-961.

Raines, R. (1976). *A study of mathematics classes at Manalet Junior College.* Unpublished manuscript, Nova University, Fort Lauderdale, FL.

Rosenshine, B. (1976). Recent research on teaching behaviors and student achievement. *Journal of Teacher Education, 27*(1), 61-64.

Rosenshine, B., & Berliner, D. C. (1978). Academic engaged time. *British Journal of Teacher Education, 4,* 3-16.

Soar, R. S. (1973). *Follow-through classroom process measurement and pupil growth (1970-1971)* (Final Report). Gainesville: University of Florida, College of Education.

Soar, R. S. (1975). *An integrative approach to classroom learning.* Bethesda, MD: National Institute of Mental Health.

Wang, M. C. (1983). Development and consequences of students' sense of personal control. In J. Levine & M. C. Wang (Eds.), *Teacher and student perceptions: Implications for learning.* Hillsdale, NJ: Lawrence Erlbaum.

Wang, M. C., & Stiles, B. (1976). An investigation of children's concept of self-responsibility for their school learning. *American Educational Research Journal, 13,* 159-179.

Weiner, B. (1979). A theory of motivation for some classroom experiences. *Journal of Educational Psychology, 71,* 3-25.

Weiner, B. (1983). Speculations regarding the role of affect in achievement-change programs guided by attributional principles. In J. Levine & M. C. Wang (Eds.), *Teacher and student perceptions: Implications for learning.* Hillsdale, NJ: Lawrence Erlbaum.

Weisz, J. R., & Stipek, D. (1982). Competence, contingency, and the development of perceived control. *Human Development, 28,* 250-281.

Thomas L. Layton, Ph.D.
Associate Professor of Speech and Hearing Sciences, University of North Carolina, Chapel Hill, North Carolina.

CARROW AUDITORY-VISUAL ABILITIES TEST

Elizabeth Carrow-Woolfolk. Allen, Texas: DLM Teaching Resources.

Introduction

The Carrow Auditory-Visual Abilities Test (CAVAT) is an individually administered, norm-referenced test that measures auditory and visual perceptual, motor, and memory abilities in children aged 4-10. The 14 subtests comprising the instrument are divided into two sections: the Visual Abilities battery (5 subtests) and the Auditory Abilities battery (9 subtests). The CAVAT also includes the Entry Test, which is used to determine which subtests should be administered. The entire test takes about 1 hour, 30 minutes, or about 5 minutes per subtest, to administer.

The CAVAT was based on a four-dimensional theoretical model developed by Elizabeth Carrow-Woolfolk and Joan I. Lynch (1981). Dr. Carrow-Woolfolk received her Ph.D. in speech-language pathology from Northwestern University with a minor in reading and remedial reading. Dr. Lynch received her Ed.D. in speech-language pathology from Teachers College of Columbia University. Their model describes how cognitive behavior, linguistic knowledge, language performance, and the communicative environment affects the language/learning abilities of an individual. According to the theory, cognitive behavior translates external information into internal representation, concepts, and symbols by means of perception and memory. Linguistic knowledge describes the pragmatic, semantic, morphosyntactic, phonological, and graphing systems of language. Language performance, on the other hand, is the transmission of information by linguistic codes: listening, speaking, reading, and writing. The communicative environment details the purposes, needs, and desires for relating to others. The CAVAT reportedly measures these behaviors in both normal and abnormal language/learning children.

There were 170 subjects used in the original pilot study on the CAVAT, whereas 1,032 children representing four regions or 17 states were included in the norming population. The final population was selected according to age, sex, and race. Ages ranged from 4-10 years, with at least 125 children representing each age level. Communities and schools that represented a range of socioeconomic levels were selected so that the norming sample would represent a national population. Approximately 85% of the children were tested by trained psychologists, and the remaining examiners were college graduates who had received special training on the CAVAT.

Latent-trait test theory was used to shorten the original CAVAT to its present form. Of the 14 subtests on the CAVAT, 7 were shortened because they fit the requirements necessary for latent-trait test theory (Rasch, 1960). The results of this reduction made the current form more parsimonious.

The Entry Test contains 40 items from the Visual Abilities battery and the Auditory Abilities battery. The items were selected based on their 1) high item discrimination value, 2) representativeness of the subtests, and 3) representativeness of the various categories in the first four visual subtests. Items from three subtests—Motor Speed, Auditory Discrimination in Quiet, and Auditory Discrimination in Noise)—were not included in the Entry Test because they required extensive training and their total score could be used only for diagnostic purposes. The Entry Test includes sample items from all the other subtests. The items are clustered in order to sample each of the main subtests. If a child fails any of the sampled clusters, he or she must be adminisered the complete CAVAT subtest from which the failed item clusters were drawn. The sample subtests also are grouped into an Auditory Abilities Cluster and a Visual Abilities cluster. If a child should fail either of these two clusters, he or she is required to take all of the CAVAT subtests in that area. Entry Test items are presented with no ceiling rule applied.

The Visual Discrimination Matching subtest (50 items) assesses the child's ability to recognize and discriminate attributes of a visual-graphic stimulus, such as its shape, size, or orientation. The test items are categorized by the type of perceptual skill involved: 1) form perception (nonmeaningful line drawings of figures or shapes), 2) number perception (nonmeaningful figures in groups of two, three, four, five and six), 3) picture perception (sets of drawings of objects in horizontal order), 4) size perception (drawings of forms and objects differing only in size), 5) directionality (forms differing only in spatial orientation), 6) closure (incomplete forms and complete forms), 7) figure-ground (line drawings of forms presented on a background of other figures and lines), 8) form-space relations (two or more figures in spatial relation to each other), 9) form-order pattern relations (sets of nonmeaningful series of forms in horizontal order), 10) letter-order pattern relations (sets of series of letters in horizontal order). There are five items per category. For each item, five figures are printed on a page: one figure at the top and four at the bottom. The child is required to point to the bottom item that is identical to the one at the top. The child receives 1 point for each correct response. Testing continues until the child scores a zero on any two items within each of the 10 categories. At least two items within each category are to be administered.

The Visual Discrimination Memory subtest (50 items) assesses the child's ability to remember visual-graphic stimulus similar to those tested in the Visual Discrimination Matching subtest. The same 10 categories are measured, five items are presented per category, and the scoring procedure is identical. The only difference between the two subtests is that the first page shown to the child contains only one figure rather than five. After the child views the page for 10 seconds, a second page, containing four figures, is shown. The child is required to remember the figure from the first page and, without a direct model, find the matching figure in the second page.

The Visual-Motor Copying subtest (27 items) evaluates the child's accuracy in copying a visual-graphic stimulus while the stimulus remains in the child's view. Six of the same categories (form perception, number perception, directionality, form-space relations, form-order pattern relations, and letter-order pattern relations) used in the first two subtests are used again here. There are five stimuli for every category except letter-order pattern relations, which has two stimuli. The child is given one figure containing one or more forms or letters and is asked to copy the figure in a Motor Response Booklet. Scoring is similar to that in the first two subtests, and complete instructions and examples are provided.

The Visual-Motor Memory subtest (27 items) evaluates the child's ability to remember and reproduce a stimulus after it has been removed from view. The stimuli for this subtest are similar to those for the Visual-Motor Copying subtest except that one figure is presented to the child for 10 seconds, and then it is covered. The child is required to draw the stimulus figure from memory. Scoring is identical to that in the Visual-Motor Copying Subtest.

The Motor Speed subtest (24 items) evaluates the child's ability to reproduce simple visual-graphic stimuli with accuracy and speed. The task consists of four rows of circles within circles. The child is asked to draw another circle between the inner and outer circles. The first two rows require only accuracy, but the last two stress speed as well as accuracy. A circle is scored correct if the child's line does not cross either the inner or the outer circle. Both time and accuracy of performance are scored on this subtest. There is no ceiling because all of the items are presented.

The Picture Memory subtest (14 items) assesses the child's ability to recall a series of unrelated words spoken by the examiner. The child is required to point to a set of pictures that corresponds to the sequence of spoken words. The subtest consists of seven pairs of items containing three to nine words per series. The examiner says a series of words at the rate of one word per second and discontinues when the child scores zero on both items in one pair.

The Picture Sequence Selection subtest (10 items) evaluates a child's ability to recall a series of unrelated words spoken by the examiner in a specific sequence. The pictures used in this subtest are the same as those used in the Picture Memory subtest except the stimuli appear on four plates containing several pictures per plate. The examiner says the stimuli at a rate of one word per second, and the child is required to point to the correct pictures in sequence. The procedure is continued until no improvement in memory is demonstrated. Both the pictures selected and the sequence of items are recorded. The total score includes a memory score, as well as a sequence score and a bonus score, if one is available.

The Digits Forward and the Digits Backward subtests (7 items each) assess a child's ability to recall and reproduce orally a series of numbers. Digits 1-9 are used as items, with a range of three to nine digits per item. In the Digits Forward subtest, the child repeats the digits in the order they are presented, and in the Digits Backward subtest, the child is required to recall them in reverse order. Testing is terminated when the child shows no improvement in the number of digits recalled. One point is awarded for each sequence of two digits recalled if there are no digits omitted between them and no digits inserted between them. A bonus is provided when a sequence of four digits (i.e., 1 bonus point), five digits (i.e., 2 bonus points), or six digits (i.e., 3 bonus points) is recalled. There is a possible score of 50 points per subtest.

The Sentence Repetition subtest (14 items) evaluates the child's ability to recall and reproduce complete sentences. The grammatic structure of the sentences increase in difficulty throughout the subtest; however, performance is based solely on recall of repetition and not on grammatical complexity. Scoring is based on a 3-point weight system whereby the child receives 3 points for no errors, 2 points for one error, 1 point for two errors, and 0 points for three or more errors.

The Word Repetition subtest (18 items) evaluates the effect that a sequence of words has on auditory memory, especially when the words are organized increasingly to approximate true English sentences. That is, as the sequences of words approximate English sentences, the word series become easier to recall. The word sequences, which were developed specifically for the CAVAT, consist of three sets of words reflecting three orders of approximation of English. Each of the three orders contain word sequences that range from three to eight items. Scoring consists of both the words recalled and the exact sequence of recall. The child also can receive a bonus score for sequences over three items long. There is a possible score of 216 points for this subtest.

The Auditory Blending subtest (29 items) assesses the child's ability to blend phonemes that make meaningful words. The subtest is divided into three sections. The first section requires the child to point to a corresponding picture after the examiner presents the stimuli. The last two sections require the child to blend words after the examiner presents the stimuli. In one of these sections, items are presented as phonemes (e.g., b-l-a-k), and in the other section, they are presented as syllables (e.g., sp-la-sh). One point is awarded for each correct response for a total of 29 points.

The Auditory Discrimination in Quiet subtest (30 items) and Auditory Discrimination in Noise subtest (20 items) assess the child's ability to discriminate between similar phonemes in contrasting words within sentences. The CAVAT uses minimal pairs of words for contrast, with the words embedded in sentences rather than in isolation as traditionally presented (Wepman, 1958; Goldman, Fristoe, & Woodcock, 1970). Before either subtest is administered, a training activity is presented to familiarize the child with the sentences, pictures, and words. The child is required to select the correct response from a picture plate that contains four pictures: one picture contains the stimulus sentence, one contains the contrasting minimal word, and two are foils.

Practical Applications/Uses

Carrow-Woolfolk (1981) states that essentially anyone trained in test administration is qualified to give the CAVAT; however, she cautions that psychologists, language/learning specialists, and educational diagnosticians may find the auditory sections difficult because knowledge of phonetics is necessary. Speech/language pathologists, on the other hand, should have no difficulty giving the CAVAT. The author further suggests that the CAVAT be administered five times in practice before administering it under true testing conditions.

The CAVAT raw scores can be converted to percentile ranks or T-scores. The raw scores are grouped into 15 different clusters or battery scores: 1) General Visual Memory, 2) General Auditory Memory, 3) Auditory Memory for Sequence, 4) Auditory Memory for Unrelated Stimuli, 5) Short Term Auditory Memory Span, 6) Visual Discrimination, 7) Auditory Discrimination, 8) General Reproduction, 9) Graphic Reproduction of Visual Stimuli, 10) Verbal Reproduction of

Auditory Stimuli, 11) Response by Indication-Auditory Stimuli, 12) Grammatical Organization, 13) General Visual Processing, 14) Auditory Synthesis, and 15) Perceptual-Cognitive Integration.

Technical Aspects

The reliability data are presented clearly in the manual and suggest that the instrument is basically stable except in a couple of areas. Stability measures were completed on 39 clinical children rather than on normals because Carrow-Woolfolk (1981) believed that clinical children more likely would be retested in real-life situations. The test-retest conditions were conducted at 6-8-week intervals. The majority of the test-retest scores for each subtest were either moderate to high, except for Digits Forward ($r = .48$), Digits Backward ($r = .35$), Auditory Discrimination in Quiet ($r = .59$), and Auditory Discrimination in Noise ($r = .34$). The Digits Backward subtest, according to Carrow-Woolfolk, reported low scores because of a floor effect; that is, subjects performed quite poor on that subtest, reducing the variability of the scores. The Auditory Discrimination in Noise subtest score was low due to a leveling off effect at age 5, which caused a ceiling effect that depressed the correlations. Overall, however, the CAVAT does appear to be a stable instrument with a correlation of .94.

Internal consistency was reported by using the Spearman-Brown split-half formula for all but the Motor Speed subtest. It was not used in that subtest because speed devalues the use of split-half correlations. The results on the split-half reliability indicated that all of the correlations were above .72, with most above .90. Thus, the internal consistency of the instrument is quite good.

According to Carrow-Woolfolk (1981), the CAVAT meets four measures of validity: content, construct, criterion-related, and sex and race-related validity. Content validity was met based on the general description, theory, and item-selection procedures outlined in the manual. The additional pilot work and the use of the latent-trait test theory helps to support the test's content validity. Construct validity was met in three ways: changes, groups, and correlations. By changes, Carrow-Woolfolk refers to the increase in scores across ages; that is, the CAVAT is developmental in nature because the older subjects performed better than the younger ones. In other words, there was a statistically significant difference in scores between each age level for each subtest. However, one exception occurred between the ages of 9 and 10 years. At these two age levels, 8 of the 14 subtests were not significantly different. These results suggest that the CAVAT is valid for use with children between the ages of 4 to 9 but is not valid for use with children between the ages of 9 and 10 or older.

Content validity for groups was completed on a sample of clinical children ($n = 141$), some of whom experienced auditory-perceptual difficulties, some visual-perceptual problems, and some both. The clinical children were administered the CAVAT, and the results as compared to those from the norming sample indicated clear differences between the clinical and norming population. Stated differently, the CAVAT clearly discriminated the clinical children from the nonclinical children in all areas.

Content validity consisted of intercorrelating the 14 subtests and the overall clusters on the CAVAT. The data from 133 subjects from the norming sample were used. The results indicated that the visual subtests correlated higher with the visual battery than with the auditory battery, and the auditory subtests correlated higher with the auditory battery, as was expected. Overall, there were only low to moderate correlations between the subtests, indicating that each of the CAVAT subtests measures a separate theoretical construct.

Criterion-related validity pertained to the CAVAT's performance with other similar tests. The problem with this, however, is that there are no other tests that are both valid and reliable for making comparisons with the CAVAT. Carrow-Woolfolk did select six subtests from the Detroit Tests of Learning Aptitude (DTLA; Baker & Leland, 1967) that closely paralled CAVAT constructs and correlated the results between the two. The findings indicated strong concurrent validity between the selected DTLA subtests and those of the CAVAT. Nevertheless, additional testing still is needed before concurrent validity can be conclusively supported for the CAVAT.

Carrow-Woolfolk (1981) also reported appropriateness of the CAVAT by comparing performances by sex and by race. The results indicated that both differences between sexes and various races were nonsignificant. That is, of the 432 different comparisons made, only 16 were found to be statistically significant. Carrow-Woolfolk interprets these results to mean that the CAVAT is valid for use with blacks and Hispanics, as well as for children of either sex.

No predictive validity measures were reported in the manual.

Critique

The CAVAT is a technically sound instrument for measuring auditory-visual abilities. However, because there is a basic lack of strong, related measures in the professional field, the CAVAT should be considered, at this point, only an experimental instrument requiring additional validity studies before it can be considered a fully developed instrument. Furthermore, the significant discriminatory results between the clinical group and matched normals make the CAVAT less valid because the ages for the clinical group were not reported and assumably were different. In addition, because the CAVAT's raw scores correlated with age, a study in which age and disorder are controlled is needed in order to determine whether the test truly does parcel out clinical populations. Currently, the CAVAT manual contains several recommended clusters for clinical subgrouping and for additional diagnosis. However, because the manual does not contain any support for these clusters, additional work is needed with clustering analysis to determine whether they are true clusters. If they are, the results, along with standardized measures, need to be reported. This added information would certainly strengthen the value of the CAVAT and make it a more powerful diagnostic instrument.

References

Baker, H., & Leland, B. (1967). *Detroit Tests of Learning Aptitude.* Indianapolis, IN: Bobbs-Merrill.

Carrow-Woolfolk, E. (1981). *Carrow Auditory-Visual Abilities Test.* Hingham, MA: DLM Teaching Resources.

Carrow-Woolfolk, E., & Lynch, J. (1981). *An integrative approach to language disorders in children.* New York: Grune & Stratton.

Goldman, R., Fristoe, M., & Woodcock, R. (1970). *Goldman-Fristoe-Woodcock Test of Auditory Discrimination.* Circle Pines, MN: American Guidance Service.

Rasch, G. (1960). *Probablistic models for some intelligence and attainment tests.* Copenhagen: Danish Institute for Educational Research.

Wepman, J. (1958). *Auditory Discrimination Test.* Los Angeles: Western Psychological Services.

Michele A. Paludi, Ph.D.
*Visiting Associate Professor of Psychology, Women's Studies
Department, Hunter College, New York, New York, and Associate
Professor of Psychology, Kent State University, Kent, Ohio.*

CHILD AND ADOLESCENT ADJUSTMENT PROFILE

*Robert E. Ellsworth. Palo Alto, California: Consulting
Psychologists Press, Inc.*

Introduction

The Child & Adolescent Adjustment Profile (CAAP) measures a child's or adolescent's adjustment in the areas of peer relations, dependency, hostility, productivity, and withdrawal. The CAAP is designed to be rated by teachers, parents, counselors, probation officers, and treatment staff who work with children and adolescents.

The CAAP was developed in 1981 by Robert B. Ellsworth. The following individuals submitted data for the scale's development: Frances Ricks, Ph.D., George Doherty, Joan Derenge, and Marjorie Pett. It underwent several developmental phases before the present form was obtained. In the initial phase of scale development, 147 children referred for mental health services and 115 nonreferred children were rated by a parent, using a true/false answer format, on 292 items selected from a review of published studies and clinicians' judgments. The 292 items were examined with respect to the following criteria: 1) the extent to which each item differentiated between the rated behavior of both groups of children, 2) the magnitude of the items' factor loading, and 3) the judgments of outpatient and residential treatment staff as to each behavior's cinical importance for evaluating children and adolescents. Of the initial pool of 292 items, 115 met the first two criteria and were evaluated by clinicians. Of these 115, 69 were judged by staff as highly clinically relevant in evaluating the adjustment of children and adolescents. Using a Varimax factor solution, the 69 items were analyzed statistically and 55 items that best measured six factors—Peer Relations, Dependency, Hostility, Productivity, Anxiety-Depression, and Withdrawal—were selected for the next stage of the scale's development.

In the second stage of development, the 55 items selected in the first phase were administered to children and adolescents by raters who had observed the behavior of the examinees during the previous month. The raters, who included parents of children referred to mental health services, parents of probationers, parents of "normal" children, probation officers supervising delinquents, and teachers, used a four-choice answer format (i.e., either never, rarely, sometimes, often, or rarely, sometimes, often, almost always) to record the observed behaviors. In this phase of the scale's development, the 55 items that successfully met two criteria—sensitivity to pre- and posttreatment differences on children and adolescents referred to mental health centers and sensitivity to group differences—were submitted to a series of factor analyses. Four factor analyses were performed in order to obtain factors that were consistent across four age and sex groups: ages 6-11, ages 12-18, girls, and boys. Factor consistency across these groups was considered essential inasmuch as it insured that the adjustment factors selected would be relevant for individuals in these categories. The Anxiety-Depression factor was dropped from further analyses because this factor did not appear to be easy for raters to evaluate in reliable ways. The final factor analysis yielded a sample of 20 items that were measured reliably by internal consistency (coefficient alpha) and test-retest ratings, as well as by the magnitude of factor loading on one of the five remaining factors: Peer Relations, Dependency, Hostility, Productivity, Withdrawal.

The Peer Relations factor is concerned with children and adolescents getting along with others, joining other children and adolescents freely, inviting peers to play, and laughing and smiling with ease. The Dependency factor is composed of items dealing with wanting help for things that could be done easily on one's own, discouragement over attempts to do things on one's own, and asking unnessesary questions. The Hostility factor is comprised of items that deal with whether adolescents or children flare up if they cannot have their own way, become upset if other children and adolescents do not agree with them, and are responsive to discipline. The Productivity factor is comprised of items that deal with children and adolescents working hard at tasks and working carefully. The items that comprise the Withdrawal factor deal with children and adolescents sitting and staring, doing nothing, and appearing indifferent.

Using four answer choices (rarely, sometimes, often, almost always), individuals rate children or adolescents on 20 items (4 in each factor) expressing behaviors they have observed in the examinees during the past month. An item relating to the Peer Realtions factor might question an individual's ability to get along with others. To assess dependency, an item might question whether an individual asks unnecessary questions or works well on his or her own. To test the Hostility factor, an item might question whether an individual responds to discipline. Productivity may be measured with a question regarding an individual's ability to stay with an assignment until it is completed. Finally, an item relating to the Withdrawal factor might question an individual's indifference towards or interest in things. The CAAP can be used to measure change in adjustment over time. Change scores provide a model for gauging children and adolescents' responses to a standard form of treatment.

Practical Applications/Uses

The CAAP can be used as a pre- and posttreatment measure with children and adolescents. In addition, behavior in the classroom, as rated by the teacher, can be compared with parent ratings of behavior in the home. Ellsworth has reported that in comparing teacher and parent CAAP ratings of normal children, the greatest differences, in descending order, were in the areas of hostility, withdrawal, productivity, and dependency (Ellsworth, 1981). In the classroom, children were clearly different in that they scored lower on peer relations, hostility, productivity, and dependency and higher on withdrawal than they did at home. These findings are consistent with expected student behaviors. However, the finding that productivity is lower in the classroom is unexpected, reflecting, perhaps, that classroom expectations of productivity are much higher.

The scale may be used by parents, teachers, and various other individuals who work with children and adolescents. Examiners are asked to read 20 items and rate whether the behavior expressed in each item applies rarely, sometimes, often, or almost always to the behavior displayed by the examinee in the last month. Raters also are encouraged to offer comments on the back of the rating sheet. In addition, the sheet allows room for the rater to provide demographic information (e.g., name, date, relationship to child or adolescent, sex of youngster, age of youngster, grade in school) about the examinee. There is no time limit for administration of the CAAP, although most individuals can rate the items on the scale in 20-30 minutes. Administrators of the CAAP do not have to be psychologists trained in clinical psychology. They must, however, be knowledgeable about the purpose and application of the CAAP's factors.

The CAAP is scored according to a 4-point scale. A rating of "often" receives a score of 3, a rating of "sometimes" a score of 2, and a rating of "never" a score of 1. The total scores for each of the factors is obtained by adding the scores assigned to each of the items comprising that factor. If an item is left blank, the rater must insert a missing value prior to obtaining a total score. If two or more questions are left unanswered, that area of adjustment should be left unscored.

After the CAAP is scored, the sums for each of the five factors of adjustment can be transferred to the profile sheet. Comparing the scores of the children used to establish the profile norms with the child or adolescent who was rated can help to determine whether the individual has serious problems in one or more areas of adjustment. When using the CAAP in follow-up ratings, the same rater should be used. A change in raters often will result in score changes despite no observable behavior changes in the child or adolescent. Follow-up ratings are made in the same way as the initial ratings. The totals for each area of adjustment can be entered on the same profile sheet used for pretreatment scores, and then a comparison between pre- and posttreatment adjustment can be made.

Technical Aspects

The CAAP factors are independent. The largest correlations between factors in parent ratings was .42 between Dependency and Hostility, suggesting that children and adolescents who do not do things on their own tend also to express anger and be unresponsive to discipline. However, it is important to bear in mind that the correlation of .42 accounts for 18% of the variance and, thus, the CAAP measures different aspects of adjustment as rated by parents. Teacher ratings were found to be more interrelated than the ratings obtained from parents. This finding has been explained as indicating that teachers do not know the child or adolescent as well as the parents and, therefore, a more global impression is reflected in the parents' ratings.

Coefficient alphas and test-retest reliability coefficients were high, ranging from

.78 to .90. Test-retest reliability involved ratings made 1 week apart by parents, probation officers, and teachers.

With respect to validity, parent ratings were obtained on various groups of children and adolescents: mental health center clients, probationers, and normals. The results indicated that the pretreatment mental health group was the least well adjusted of any of the groups in the areas of hostility and withdrawal. The normals were the most well adjusted of any group in regard to peer relations, productivity, and withdrawal. The group of probationers fell between the other two groups, but was rated by parents as being less dependent than the other groups. The posttreatment clinical group showed statistically significant improvement with regard to dependency, hostility, and withdrawal.

Further analyses revealed a significant agreement in the ratings of youthful offenders by parents and probation officers. In addition, there were no differences among boys and girls on the CAAP scale scores as rated by parents. In the teacher ratings, children in the lower grades had lower scores on peer relations than children in upper grades. Girls also were rated by teachers as more productive than boys.

Critique

The CAAP is a valuable psychological tool that, when used in conjunction with clinical diagnoses, helps to delineate problems in adjustment. This reviewer is concerned with the wording of many items in the scale as many adolescents exhibit the behaviors as a normative part of their identity process (e.g., sat and stared without doing anything, daydreamed, appeared indifferent and uninterested in things, picked quarrels with others). These are developmentally appropriate behaviors for adolescents.

In addition, this reviewer is concerned about the frequency of stereotypic responses given by parents and teachers because of their knowledge of the sex of the child or adolescent. In the scale booklet, Ellsworth interprets the finding that girls were perceived by teachers as more productive than boys as being "consistent with what one would expect" (p. 7). However, these findings do not indicate a problem in the adjustment of boys necessarily. The sex of teachers and parents and other raters would be important information to have given the research on the sex of the observer and the rater effects on the interpretations of the behavior of children, adolescents, and adults.

References

Ellsworth, R.B. (1981). *CAAP Scale: The measurement of child and adolescent adjustments.* Palo Alto, CA: Consulting Psychologists Press.

Elaine Clark, Ph.D.
Assistant Professor of Educational Psychology, University of Utah, Salt Lake City, Utah.

CHILD ASSESSMENT SCHEDULE

Kay Hodges. Durham, North Carolina: Kay Hodges, Ph.D.

Introduction

The Child Assessment Schedule (CAS) is a semistructured interview designed to assess a broad spectrum of psychological disorders in children aged 7-12. The CAS was developed by Kay Hodges, Ph.D., Associate Professor, Department of Psychiatry, Duke University Medical Center, in collaboration with Don McKnew, M.D., and Leon Cytryn, M.D., both with the Unit on Childhood Mental Illness, Biological Psychiatry Branch of the National Institute of Mental Health and Department of Psychiatry, George Washington University. The interview originally was developed in 1978 in response to a need for an instrument to study children with psychosomatic illnesses. The original version most closely resembles a traditional, open-ended clinical interview. The CAS has since undergone three revisions. The first, in 1983, consisted mainly of modifying existing items and adding questions regarding the onset and duration of dysfunctional behavior. In a more extensive revision in 1985, several items were added to supplement various diagnostic categories (e.g., Overanxious Disorder, Attention Deficit Disorder, and Conduct Disorder) and to generate DSM-III diagnoses (American Psychiatric Association, 1980). The result was a lengthier and more comprehensive structured interview. Other than the modification of certain items and a reduction of the response categories, few changes were made in the 1986 revision.

The CAS consists of three sections. The first section is a series of open-ended questions pertaining to school, friends, activities, family, fears, worries, self-image, mood, somatic concerns, anger, and symptoms of thought disturbance. The questions are ordered from least to most threatening so that rapport can be established more easily. The child's answers on the 1986 CAS are scored "yes," "no," "ambiguous," or "not scored"; in the 1985 version, the "not applicable" and "no response" categories are used instead of the "not scored" response category. The purpose of the second section is to evaluate the onset and duration of symptoms identified in the first section; in other words, problem behaviors that received a "yes" score. The use of specific probes allows the interviewer to gather information that is necessary to generate DSM-III diagnoses. The third section, used to record behavioral observations of the child, is completed following the interview and consists of items pertaining to the child's insight, grooming, motor coordination, activity level, estimated cognitive ability, and spontaneous physical movements, and to the quality of the child's verbal communication, emotional expression, and interpersonal interactions. Sample items for all three sections are provided in a chapter on assessment with the CAS by Dr. Hodges (Hodges, in press).

A parallel parent form of the CAS (P-CAS) was developed in 1983. Information that might otherwise be unobtainable from the child may be elicited using the P-CAS (e.g., the parent's perception of the child and information regarding onset and duration of problem behaviors). Verhulst, Berden, and Sanders-Woudstra (1985) found that information obtained on both the CAS and P-CAS contributed significantly to the final diagnosis, although the CAS contributed most. When used in conjunction with the CAS, the P-CAS may serve to confirm or disconfirm a child's report and provide a basis of comparison between items, scores, and diagnoses. This comparison can best be made when the 1985 version of the CAS is used. Changes in the 1986 version (e.g., items modified and numbered differently) make it more difficult to make item-to-item comparisons between the child and parent forms. The P-CAS was revised in 1984; however, the change was primarily in the wording of items (e.g., "do you think your child . . . ," rather than "has your child . . ."). The item content and item order have remained the same.

Practical Applications/Uses

The CAS was designed as a clinical tool for obtaining information that would be relevant for assessing children's coping skills, as well as for diagnosing and treating psychopathology. In addition, the instrument provides a method for generating quantifiable data for research purposes and a systematic means for training novice clinicians. The CAS was intended to be used with children aged 7-12; however, the author states that the instrument can be used with children as young as 5 years of age who have above-average cognitive abilities and communication skills. The questions in the interview are organized in such a way that children are more likely to perceive it as a conversation rather than a test. For example, items are grouped according to topics like school, friends, and hobbies. Nonetheless, regardless of his or her age, the child must have sufficient intellectual skill to comprehend and respond to the questions. The CAS in its present form would not be appropriate for mentally retarded children or children with significant brain damage or psychotic disorders. On the other hand, the CAS can be administered to children with physical handicaps, as long as the questions and answers can be communicated (e.g, signing with a hearing-impaired child or using a translator for non-English-speaking children). Special considerations also may be needed when assessing very young or very old children. It is recommended that children aged 5-7 be given the less lengthy and less complex 1978 version, which is still available from the test author. The 1986 version, containing 320 items of which 261 are administered directly to the child, generally takes 45-90 minutes to administer. Depending on the child's age, disturbance, and behaviors during testing this estimated time will vary; however, for very young children, this length of time would be too fatiguing. A number of professionals working with inpatient adolescent populations have complained that the questions are too "kid-like." When interviewing adolescents, examiners may wish to add more age-appropriate items (e.g., questions regarding sexual experiences and heterosexual relationships, and drug and alcohol use): However, Dr. Hodges feels that the majority of questions assess behaviors and diagnoses that are relevant to adolescents and are not infantilizing.

The CAS is to be administered by clinicians who are familiar with the instrument and have experience testing and interviewing children. The setting in which the interview is conducted is not specified; however, the test author recommends that the interviewer hold at least a master's degree in a mental health field and have expertise in childhood psychopathology. Professionals such as school and clinical psychologists, child psychiatrists, and social workers would likely have the educational and experiential background to conduct the interview.

The testing manual, *Manual for the Child Assessment Schedule* (Hodges, 1985), provides clear and detailed instructions for administering and scoring the CAS. The author not only provides the statistical formula to calculate the interrater reliability, but also has written a manual supplement, *Guidelines to Aid in Establishing Interrater Reliability with the Children's Assessment Schedule* (Hodges, 1983), that provides suggested ways to improve scorer reliability. Although it is important that the interview be conducted in such a way that the child perceives it as a discussion and not an inquisition, the interview must be conducted in a standardized fashion in order to yield reliable scores. To improve reliability further, criteria for scoring each response as "yes," "no," "ambiguous response," or "not scored" are specified. The CAS can be hand-scored or machine-scored. Hand-scoring is quite simple; however, individuals using the CAS for research purposes may prefer machine-scoring.

The CAS, as a clinical tool, is analyzed qualitatively. The presence of dysfunctional behaviors provide the information necessary to make differential diagnoses according to the DSM-III using either the CAS or P-CAS or both. The DSM-III diagnostic categories that can be generated using the CAS include attention deficit disorder, conduct disorder, anxiety disorder, schizoid disorder, oppositional disorder, functional enuresis and encopresis, major depressive episode, dysthymic disorder, phobic disorder, and obsessive compulsive disorder. Items referring to less frequent childhood disorders also are contained in the interview for preliminary screening, including eating disorders, stereotyped movement disorders, sleepwalking and sleep terror disorders, substance abuse disorders, panic disorders, bipolar affective disorders, and psychosis.

For research purposes, the instrument can be quantitatively scored for content areas (e.g., school, family, and activities), symptom complexes (e.g., attention deficit and anxiety disorders), and total psychopathology. The three scores can be used to make group comparisons. There are no normative data for individual comparisons; however, the scores can provide general information about overall dysfunction, areas of strength and weakness (e.g., school, home, activities), and the ways in which the disturbance seems to have manifested itself (e.g., overanxious, oppositional). This reviewer is still unclear as to how to interpret the "total pathology score."

Technical Aspects

A series of four reliability studies have been conducted to examine the interrater reliability of the CAS. Unfortunately, none of the studies examined test-retest stability of the CAS scores. The studies that were conducted, however, examined the CAS data independently from the P-CAS data. The studies were conducted at independent sites, and, with few exceptions, used experienced practitioners as interviewers and raters. The children who were interviewed for the studies ranged in age from 5 to 18 and were drawn from both normal and clinical populations (inpatient and outpatient facilities). Different versions of the CAS were studied. It is not always clear which revision is being evaluated; however, other than the 1978 CAS, there is considerable similarity among the forms.

In one study using a child psychiatric population, the mean kappa coefficients ranged from .47 to .61. Although this range is rather low, for three of the four pairs of raters, the level of agreement was more acceptable, near or exceeding .60. When percent of agreement was examined using interviews with children of normal and affectively disturbed mothers, the mean agreement for item-to-item comparisons across all response categories was .91 (the agreement ranged from .87 to .96 for all subjects). The mean interrater agreement for the content areas was .93 (a range from .86 to .98), and, for the symptom complexes, the mean was .93 (a range from .89 to .97). Other studies conducted with psychiatric populations have yielded similarly high reliability coefficients (e.g., intraclass coefficients ranging between .66 to .98 and a mean kappa coefficient of .71).

Nine validity studies have been conducted. These studies have primarily provided support for the concurrent and discriminant validity of the CAS. Studies that examined the concurrent validity of the instrument often did so by comparing contrast groups (e.g., normals, inpatients, and outpatients) and correspondence between the CAS and other measures (e.g., child report and parent report). Basically, the studies have demonstrated that the CAS is capable of discriminating among groups of children with differing degrees of psychopathology. In one study using discriminant function analyses, 67% of the groups were classified correctly according to CAS scores. Only one stymptom complex, Attention Deficit Without Hyperactivity, and two content areas, Fears and Worries, did not discriminate well among the groups. When combined with another measure, the Achenbach and Edelbrock's Child Behavior Checklist (CBC), the overall number of correctly classified cases increased; however, the CAS contributed most to the discrimination of inpatient populations. Clearly, there was a high correspondence between the child's perception of his or her own behavior via the CAS, and the mother's view of the child's behavior using the CBC. Other studies of diagnostic concordance have been quite favorable. When the child's report on the CAS was compared to the mother's report on the Schedule for Affective Disorders and Schizophrenia for School-Age Children (Puig-Antich & Chambers, 1983), there was a high level of agreement for conduct disorders, attention deficit disorders, and affective disorders when the diagnoses were based on parent interviews alone or a combination of the P-CAS and CAS. When diagnoses were made solely on the basis of child interviews, there was only a moderate to poor concordance. Furthermore, studies examining the correspondence between the CAS and other child self-report measures, the Child Depression Inventory (Kovacs, 1978) and Spielberger's (1973) State-Trait Anxiety Inventory for Children have demonstrated a high level of concurrence. A comprehensive review of these and other validity studies can be found in Dr. Hodge's chapter on assessing children with the CAS (Hodges, in press). Also included in the chapter is a description of the previously mentioned reliability studies.

Critique

The CAS provides a unique opportunity to interview children in a semistruc-tured fashion that allows one to collect relevant data for clinical diagnosis, treat-ment planning, and research. With the parallel parent form, the interview is comprehensive and provides a much needed comparison between parent and child perspectives. The fact that the CAS generates diagnoses according to the DSM-III, the classification system used most often in clinical settings, makes it a particularly relevant instrument for educating the novice clinician in assessment and childhood psychopathology. The CAS may be particularly valuable to professionals who rely on other classification systems in their particular settings. School psychologists and school social workers who more often use behavioral and educational classifica-tions but need to be conversant with traditional psychiatric diagnoses might find the CAS a practical means of acquiring this knowledge. Actually, those who work in mental health settings where the DSM-III is used routinely for diagnoses may be less interested in using the CAS. The more seasoned practitioner may be more likely to use a less structured, and perhaps less time-consuming method of inter-viewing. This is not to suggest that these individuals, and those with whom they work, would not benefit from the use of the CAS. The CAS has at least some empirical support for its reliability and validity.

Although the majority of reliability and validity studies have been conducted by the test author and those with whom she has collaborated, this is certainly not unusual for a relatively new test, especially one that is unpublished. Unfortu-nately, the fact that the CAS is not published is likely to reduce the amount of press it receives. Nonetheless, the studies performed to date have provided impressive data; hopefully, more researchers outside the "CAS community" will elect to use and evaluate the test's performance. Dr. Hodges undoubtedly has good reasons for not publishing the CAS. Perhaps, publication is not feasible because the instru-ment itself is in a constant state of flux, as is the American Psychiatric Association's (APA) diagnostic criteria. Since the 1986 CAS revision, the DSM-III has been revised and is now the DSM III-R (American Psychological Association, 1987). Changes in the classification code, in particular those pertaining to childhood diag-noses like attention deficit disorder and conduct disorder, will undoubtedly neces-sitate another CAS revision which, of course, also means revising the P-CAS, a valuable counterpart to the CAS. However, before the next revision is complete, it may be time to revise again; the expected publication date for the DSM-IV is 1988 or 1989.

References

This list includes text citations and suggested additional reading.

Achenbach, T.M., & Edelbrock, C. (1983). *Manual for the Child Behavior Checklist and Revised Child Behavior Profile.* Burlington, VT: Queen City Printers.
American Psychiatric Association. (1980). *Diagnostic and statistical manual of mental disorders* (3rd ed.). Washington, DC: Author.
American Psychiatric Association. (1987). *Diagnostic and statistical manual of mental disorders* (rev. 3rd ed.). Washington, DC: Author.

Hodges, K. (in press). Assessing children with a clinical research interview: The Child Assessment Schedule. In R. J. Prinz (Ed.), *Advances in behavioral assessment of children and families.* Greenwich, CT: JAI Press.

Hodges, K. (1985). *Manual for the Child Assessment Schedule (CAS).* (Available from Dr. Kay Hodges, Department of Psychiatry, Duke University Medical Center, Durham, North Carolina, 27710).

Hodges, K. (1983). *Guidelines to aid in establishing interrater reliability with the Child Assessment Schedule.* (Available from Dr. Kay Hodges, Department of Psychiatry, Duke University Medical Center, Durham, North Carolina, 27710).

Hodges, K., Kline, J., Stern, L., Cytryn, L., & McKnew, D. (1982). The development of a child assessment schedule for research and clinical use. *Journal of Abnormal Child Psychology, 10,* 173-189.

Hodges, K., McKnew, D., Burbach, D. J., & Roebuck, L. (in press). Diagnostic concordance between two structured child interviews, using lay examiners: The Child Assessment Schedule and the Kiddie-SADS. *Journal of American Academy of Child Psychiatry.*

Hodges, K., McKnew, D., Cytryn, L., Stern, L., & Kline, J. (1982). The Child Assessment Schedule (CAS) diagnostic interview: A report on reliability and validity. *Journal of American Academy of Child Psychiatry, 21,* 468-473.

Kovacs, M. (1978). *Children's Depression Inventory (CDI).* Unpublished manuscript, University of Pittsburgh.

Puig-Antich, J., & Chambers, W. (1978). *The Schedule for Affective Disorders and Schizophrenia for School-Age Children (Kiddie-SADS).* New York: New York State Psychiatric Institute.

Spielberger, C.D. (1973). *Preliminary manual for the State-Trait Anxiety Inventory for Children.* Palo Alto, CA: Consulting Psychologists Press.

Verhulst, F.C., Berden, G.F., & Sanders-Woudstra, J.A.R. (1985). Mental health in Dutch children: II. The prevalence of psychiatric disorder and relationship between measures. *Acta Psychiatrica Scandinavica, 72*(Supp. No. 324), 1-45.

Geoffrey F. Schultz, Ed.D.

Assistant Professor of Educational Psychology and Special Education,
Illinois Benedictine College, Lisle, Illinois.

Harvey N. Switzky, Ph.D.

Professor of Educational Psychology and Special Education,
Northern Illinois University, DeKalb, Illinois.

CLASSROOM ENVIRONMENT SCALE

Rudolf H. Moos and Edison J. Trickett. Palo Alto,
California: Consulting Psychologists Press, Inc.

Introduction

As with the other instruments within the Social Climate Scales, the Classroom Environment Scale (CES) purports to measure the "personalities" of environments, to evaluate the impact of different interventions, and to compare the judgments of staff and group members. The scale consists of 90 true/false items and yields nine easily computed scores that cover three major dimensions of one's social environment: 1) Relationship—the nature and intensity of personal relationships; 2) Personal Growth—personal growth and self-enhancement influences; and 3) System Maintenance and Change—aspects of organization of classrooms, teacher innovation, control, and role clarity influences. Each subscale consists of 10 items selected by internal consistency procedures.

The Real Form (Form R) is the standard form of the CES. Form R measures students' and teachers' perceptions of their current classrooms. All the other CES forms are adaptations of Form R. The scale can be adapted to measure the ideal classroom social environment (Form I) or the expectations one has of this environment (Form E). Also, the Short Form (Form S) of the CES was developed to obtain a rapid assessment of a classroom's social climate. The Short Form comprises the first 36 items of Form R, including four items from each of the nine subscales.

In addition, the CES has been adapted to 1) tap teachers' perceptions of their classes on 5-point rating scales (Weisz & Cowen, 1976), 2) focus on the overall school rather than the classroom (Felner, Ginter, & Primavera, 1962; Felner, Aber, Primavera, & Cauce, 1965; Marjoribanks, 1979), and 3) tap primary grade children's views of their classes (Parker, 1982; Toro, Cowen, Gesten, Weissberg, Rapkin, & Davidson, 1985).

Practical Applications/Uses

The Classroom Environment Scale was designed to assess the "atmosphere" in junior and senior high school classrooms. The CES offers a method of evaluating the effects of course content, teaching methods, teacher personality, and class com-

97

position. The CES is based upon the asssumption that nine basic characteristics describe a classroom environment:

Involvement: the extent to which students are attentive and interested in class activities, participate in discussions, and do additional work on their own.

Affiliation: the level of friendship students feel for each other, as expressed by getting to know each other, helping each other with homework, and enjoying working together.

Teacher Support: the amount of help and friendship the teacher manifests toward students; how much the teacher talks openly with students, trusts them, and is interested in their ideas.

Task Orientation: the amount of emphasis on completing planned activities and staying on the subject matter.

Competition: how much students compete with each other for grades and recognition and how hard it is to achieve good grades.

Order and Organization: the emphasis placed on students behaving in an orderly and polite manner and on the overall organization of assignments and classroom activities.

Rule Clarity: the emphasis on establishing and following a clear set of rules and on students knowing what the consequences will be if they do not follow them; the extent to which the teacher is consistent in dealing with students who break rules.

Teacher Control: how strict the teacher is in enforcing the rules, the severity of punishment for rule infractions, and how much students get into trouble in the class.

Innovation: How much students contribute to planning classroom activities, and the extent to which the teacher uses new techniques and encourages creative thinking.

The Relationship dimension is measured by the Involvement, Affiliation, and Teacher Support subscales; the Personal Growth dimension is measured by the Task Orientation and Competition subscales, and the System Maintenance and Change dimension is measured by the Order and Organization, Rule Clarity, Teacher Control, and Innovation subscales.

The CES has practical applications in program evaluation and clinical contexts, including 1) describing and contrasting educational programs, 2) promoting improvement in classrooms, 3) conducting formative evaluations, 4) monitoring the impact of intervention programs, and 5) formulating ecologically relevant case descriptions (Moos, 1979). The CES has been used to help parents select a school for their child (Weiler, 1976), to help teachers enhance the learning environment (DeYoung 1977a, 1977b), and to monitor program-related modifications in classroom environments (Felner, Ginter, & Primavera 1982; Wright & Cowens, 1985; Olton, 1985; Evans & Lovell, 1979; Sutter, 1984; Zempel, 1983).

The CES also has research and practical applications. One of its important investigative uses has been to describe and contrast classroom learning environments in different types of schools. Students' and teachers' views of classes have been compared, as have students' reports of their actual and preferred classroom settings (Moos, 1979; Adams, 1983). In addition, an empirical typology of classroom environments has been developed that allows investigators to select classes with some assurance that they are distinctive and represent important and reasonable representative types. Six classroom types have been identified that provided information on the range of variation within which replications of specific findings can be expected (Moos, 1980).

The CES may also be used to investigate how classroom climates develop in such disparate ways; that is, what variants lead to an emphasis on teacher support, or task orientation, or order and organization? The CES lends itself to identifying specific dynamic interplays of variables that are associated with classroom learning environment, and it may be useful in helping identify particular environmental qualities that contribute to positive or negative learning environments.

The scale can also be used to measure how the learning environment is influenced by other contextual factors (e.g., physical layouts, administrative policies, teacher characteristics, and aggregate student characteristics). It is assumed that contextual factors cultivate specific social climates; in turn, if contextual factors are manipulated, then what measurable effect on classroom social milieu is registered?

Most importantly, the CES may lend itself to addressing the controversial issue of underachievement in secondary school students. Lower achievement test scores and an increase in the number of secondary school students who lack reading, writing, and arithmetic skills have become growing concerns. Advocates of more innovative educational programming argue that overly controlled, very structured classes stifle critical thinking, lower self-concept, and sacrifice students' intrinsic interest and motivation to learn beyond the short-term gains of grades and high test scores. The CES can identify critical classroom variables that may contribute to a clearer understanding of underachievement and its correlates.

Technical Aspects

Norms for the Classroom Environment Scale are based on students in about 400 classrooms and 300 teachers from a wide range of schools representing a broad spectrum of socioeconomic backgrounds and geographic locations. The newly revised manual includes comparable data on public versus private schools and alternative versus traditional high schools. In addition to these overall norms, separate norms have been obtained for classrooms focusing on specific subject areas. These data include scale and subscale means for students and teachers in English, social studies, math, and business and technical classes. Scale and subscale means on the Short Form are also provided.

As reported in the manual, the subscale internal consistency coefficients are moderately high, ranging from .67 for Competition to .86 for Teacher Control. The average item-to-subscale correlations are quite high for all nine subscales. An intercorrelation of the nine subscale scores yields an average subscale intercorrelation of about .27, suggesting that the subscales measure distinct though somewhat related aspects of classroom environment. The test-retest reliability coefficients for the nine subscales are moderately high, ranging from .72 for Rule Clarity to .90 for Innovation.

At least four investigators have identified three factorial dimensions for the CES (Schultz, 1979; Hughes, 1984; Keyser & Barling, 1981; Moyano-Diaz, 1983). Their principal component analyses resulted in three factors: 1) a relationship factor, which covered involvement, affiliation, teacher support, and innovation; 2) an orderliness/achievement factor, composed primarily of task orientation and organization; and 3) a control factor, which included competition, clarity, and teacher control.

The validity of the CES has been assessed in a variety of ways. The manual indicates that content and face validity have been established through procedures used in scale construction. Items were chosen on the basis of empirical criteria, such as item intercorrelations, item-subscale correlations, and internal consistency analyses. Further support for content validity is based on a formulated rationale agreed upon by independent consultants.

Criterion-related validity information is adequate, with the manual reporting that classroom involvement, affiliation, and teacher support were positively associated with students' satisfaction with teachers and school as measured by the Quality of School Life Scale (Chiou, 1985). In conjunction with the research cited earlier indicating the relationship of the subscale measures to other related contextual variables of classroom environments, these findings provide some evidence for criterion-related validity of the CES scales.

Construct validity is established by the authors (Moos & Trickett, 1986) through monitoring of teachers' behavior in 38 classes representatively sampled from two suburban schools. Relationships were examined between the perceived learning environment as measured by the CES and objectively identifiable methods of teaching (e.g., availability of special projects, small group instructional time, free time allocation, use of reward-punishment contingencies, etc.). A strong association between the CES Teacher Control subscale and the frequency of use of specific rewards and punishments is reported. These two indices also were higher in classes seen as competitive and lower in classes where the teacher was seen as supportive. In addition, students saw their classes high in Affiliation when allowed to work in small groups. Likewise, students in classes measured as high in Innovation and low in Task Orientation and Order and Organization had more free time and were more likely to self-initiate special projects.

Critique

The greatest concern an investigator has when using a self-report assessment measure such as the Classroom Environment Scale is the ability of the instrument to accurately reflect a subject's point of view. Inherent with any self-report instrument of this type is the problem that the subjects will report inaccurately; that is, a teacher may communicate their perceptions in terms of what they believe "should be" or, in some instances, what they feel the investigator believes "should be." In an attempt to control for this problem, Moos and Trickett developed a modification of the CES, Form I, which provides an assessment of what teachers and/or students prefer in an ideal classroom setting. Nevertheless, even with the control form response bias Form I offers, the CES really has no definite way of identifying the teacher or student who contrives or distorts their supposedly "real" perception of social climate in a particular classroom environment.

Also, it is important to be aware of the nature of the relationship between the classroom climate factors assessed by the CES and actual classroom learning. By using terminology that implies a cause-and-effect relationship (e.g., "impact," "determinants," etc.), the manual implies that the CES and its various subscales are capable of identifying the significant characteristics that create academically successful classroom environments. Actually, the scale appears adequate in differentiating between classroom environments in terms of student background differences, subject matter variations, teacher style differences, and type of school (public vs. private). The "impact" of these variable differences, however, does not go beyond establishing their correlations to other hypothetical constructs believed to be contextually related to successful classroom learning experiences (e.g., teacher work climates, student satisfaction and morale, student adjustment and self-concept, and student intrinsic motivation to learning).

For the CES to be more than just a correlational measure related to other contextual variables associated with classroom environments, causal connections among the classroom climate constructs assessed and academic achievement still need to be established. When reviewing the research that has utilized the CES, one must note that the scale is primarily used to identify other related contextual classroom variables that correlate with the CES subscales. For example, the manual reports a number of studies that link CES factors (e.g., Teacher Support and Involvement) to other assessments of student motivation for academic success, but as yet, no significant cause-and-effect relationship has been documented by the authors between any of the subscales and actual classroom academic success. For this reason, it is just as reasonable to conclude that students who are both inherently better learners and more highly motivated toward academic success create more positive classroom climates (i.e., high teacher support and strong task orientation). At this point in time, the CES and other instruments like it must be viewed as measuring correlational variants rather than causes of positive classroom learning environments. Users of the CES need to be aware that they can with assurance interpret the information the scale provides as a function of a teacher-student interaction but that caution should be used in interpreting the CES subscale information as identifying what "determinants" cause and affect specific classroom environments.

References

Adams, J. (1983). Preferences for the ideal classroom environment: A comparison between gifted and nongifted students (Doctoral dissertation, University of Tennessee, Knoxville, 1982). *Dissertation Abstracts International, 43*, 2869A.

Chiou, G.F. (1985). Students' perceptions of classroom environment and quality of school life (Doctoral dissertation, University of Minnesota). *Dissertation Abstracts International, 46*, 1558A.

DeYoung, A. (1977a). Classroom climate and class success: A case study at the university level. *Journal of Educational Research, 70*, 252-257.

DeYoung, A. (1977b). Measuring and changing classroom climate. *New Directions in Teaching, 6*, 40-47.

Evans, G. & Lovell, B. (1979). Design modification in an open-plan school. *Journal of Educational Psychology, 71*, 41-48.

Felner, R., Aber, M., Primavera, J., & Cause, A. (1985). Adaptation and vulnerability in high risk adolescents: An examination of environmental mediators. *American Journal of Community Psychology, 13*, 365-379.

Felner, R., Ginter, M., & Primavera, J. (1982). Primary prevention during school transitions: Social support and environmental structure. *American Journal of Community Psychology, 10,* 277-290.

Hughes, G. (1984). The relationship of learning to congruence between students' ideal and perceived learning environments (Doctoral dissertation, University of Minnesota, St. Paul, 1983). *Dissertation Abstracts International, 44,* 3266A.

Keyser, V., & Barling, J. (1981). Determinants of children's self-efficacy beliefs in an academic environment. *Cognitive Therapy and Research, 5,* 29-40.

Marjoribanks, K. (1979). Family and school environmental correlates of intelligence, personality and school-related affective characteristics. *Genetic Psychology Monographs, 99,* 165-183.

Moos, R. (1979). *Evaluating educational environments: Procedures, methods, findings and policy implications.* San Francisco, CA: Jossey-Bass.

Moos, R. (1980). Evaluating classroom learning environments. *Studies in Educational Evaluation, 6,* 239-252.

Moos, R., & Trickett, E. (1986). *Classroom Environment Scale manual* (2nd ed.). Palo Alto, CA: Consulting Psychologists Press.

Moyano-Diaz, E. (1983). *Le climat social en education: Sa mesure, ses determinants, les strategies d'optimisation.* Unpublished doctoral dissertation, Catholic University of Louvain, Belgium.

Olton, A.L. (1985). The effect of locus of control and perceptions of school environment on outcome in three drug abuse prevention programs. *Journal of Drug Education, 15,* 157-169.

Parker, K. (1982). The relationship of person-environment fit and social climate in home and classroom to individual behavioral adjustment in first grade (Doctoral dissertation, Southern Illinois University, Carbondale, 1982). *Dissertation Abstracts International, 42,* 4204B.

Schultz, R. (1979). Student importance ratings as an indicator of the structure of actual and ideal sociopsychological climates. *Journal of Educational Psychology, 71,* 827-839.

Sutter, E.C. (1984). Exploring the effect on school climate of the "Let's Rap" Program at Oaklands's McChesney Junior High School (Doctoral dissertation, Wright Institute, Berkeley, CA, 1983). *Dissertation Abstracts International, 44,* 2600B.

Toro, P., Cowen, E., Gesten, E., Weissberg, R., Rapkin, B., & Davidson, E. (1985). Social environmental predictors of children's adjustment in elementary school classrooms. *American Journal of Community Psychology, 13,* 353-364.

Weiler, D. (1976). A public school voucher demonstration: The first year of Alum Rock, summary and conclusions. In G. Glass (Ed.), *Evaluation studies review annual* (Vol. 1). Beverly Hills, CA: Sage.

Weisz, P.V., & Cowen, E. (1976). Relationships between teachers' perceptions of classroom environments and school adjustment problems. *American Journal of Community Psychology, 4,* 181-187.

Wright, S., & Cowen, E. (1985). The effects of peer teaching on student perceptions of class environment, adjustment and academic performance. *American Journal of Community Psychology, 13,* 417-431.

Zempel, C. (1983). The effects of affective education on classroom environment, student self-esteem, and grades (Doctoral dissertation, University of Tennessee, Knoxville, 1982). *Dissertation Abstracts International, 43,* 2322B.

Mark Stone, Ed.D.

Professor of Psychology, Forest Institute of Professional Psychology, Des Plaines, Illinois.

COGNITIVE DIAGNOSTIC BATTERY

Stanley R. Kay. Odessa, Florida: Psychological Assessment Resources, Inc.

Introduction

The Cognitive Diagnostic Battery (CDB) was developed to assess the nature and degree of intellectual disorders in order to aid in differential diagnosis. The purposes of the CDB are to differentiate between mental subnormality and abnormality, to assess cognitive deficits due to impaired development as opposed to deficits resulting from regression due to a psychiatric condition, and to differentially diagnose mental retardation from psychosis.

The CDB is comprised of five tests that assess: 1) early conceptual maturation, 2) higher order concept utilization, 3) egocentric versus socialized thinking, 4) perceptual motor development, and 5) attention span. The CDB Kit includes materials for administering the five tests, a manual with directions, scoring procedures and norms, and individual scoring forms.

The CDB was designed to assist the practicing psychologist with a method for assessing and differentiating among aspects of cognitive dysfunction. Its purpose is to provide quantitative and qualitative information regarding cognitive functioning or dysfunctioning that prevail over many clinical conditions. The principal objective of the CDB is to distinguish whether such cognitive impairments can be attributed to a developmental delay or a psychiatric disturbance.

The rationale of the test assumes that with developmental delay cognitive behavior is immature but relatively stable over time, whereas a psychiatric disturbance presents a cognitive picture consisting of a network of related symptoms. The primary purpose of the CDB is to distinguish between intellectual subnormalities (i.e., between deficient intelligence and subnormal intelligence, which reflects idiopathic cognitive behavior). The CDB was conceived as a diagnostic instrument to help differentiate among the subcategories of mental retardation and psychiatric disorders and to specify the character and degree of cognitive dysfunction.

The five tests that comprise the CDB are purported to examine concept formation, symbolic thinking, socialization of thought, perceptual development, and temporal attention. The CDB is organized as follows:

Color Form Preference Test. This subtest is a 20-item similarity of judgment test that requires the subject to match a standard card with one of three comparison cards according to color, form, or neither cue. An analysis of the response pattern can be translated into one of four hierarchical stages of early cognitive development: 1) purposelessness, 2) random response, 3) classification by salient attributes of color, or 4) dominance by form. This scale examines the basis for perceiving relationships and preconceptual modes of thinking and facilitates the evaluation of arousal related to cognitive disturbance.

Color Form Representation Test (CFR). This subtest is an extension of the CFP and is a more advanced similarity of judgment test consisting of 20 items. The CFR introduces the further option of matching cards by figural representation. This test evaluates conceptualization by more complex and symbolic cues. The responses are evaluated as to whether they are formed more by perceptual variance or by semantic relevance.

Egocentricity of Thought Test (EOT). The EOT is a hierarchical series of three sets of questions following the four major phases of cognitive development postulated by Piaget (1952): preconceptual, egocentric, socialized, and objective. Performance on this subtest is evaluated by whether the responses are self-centered and subjective, reliant on social orientation, or relatively independent of subjective and interpersonal cues.

Progressive Figure Drawing Test (PFDT). This subtest evaluates the examinee's copying of seven simple designs that children can be expected to master at successive periods between 2 and 6 years of age. This subtest utilizes the test methodology and stimuli of Gesell (1940) and Terman and Merrill (1960).

Span of Attention (SOA). This subtest is a measure of concentration and distractibility based upon the evaluation of concentration given to a single rote-motor task similar to the coding or digit symbol subtests of the Wechsler scales.

Practical Applications/Uses

The design of the Cognitive Diagnostic Battery facilitates the psychological evaluation of a patient, particularly where the use of a standardized individual intelligence test, such as one of the Wechsler scales, is deemed inappropriate. This situation can occur whenever tests that have been normed on samples of largely average subjects are used with psychiatric groups or retarded persons. The skewed distribution of such clinical samples often exceeds the practical range of tests lacking sufficient items to sample lower-level functioning. As users of the Wechsler scales should know, many low-functioning patients are overestimated by a total score of 0 for each of the subtests because the subtests have an arbitrary lower limit or basal level. Furthermore, the content of a subtest may be poorly suited to the extreme levels of dysfunction encountered in a clinical population. The CDB specifically fills this void in psychometric measurement by providing scales that evaluate subaverage functioning and that determine cognitive dysfunction occurring at extremely low levels.

The CDB was normed on 383 subjects, of whom 198 were nonpsychotic individuals, 97 were undefined schizophrenics, and 88 were mentally retarded psychotics. The clinical adult inpatient samples were drawn from two general locales. The psychiatric patients ranged from ages 18 to 62, and the nonpsychotic sample ranged from 2 to 65. There was a fairly even distribution between sexes, and race was represented by black and white subgroups.

The clinical diagnoses for sample selection were determined by the consensus of two psychiatrists and/or a unit psychiatrist and psychologist. The nonpsychotic sample consisted of staff, paraprofessionals, and their acquaintances. The test author has provided a separate series of research reports on these samples (see Kay, 1977, 1979b, 1979c, 1980; Kay & Singh, 1974, 1975, 1979; Kay, Singh, & Smith, 1975). These reports should be consulted for more details.

Appropriate use of the CDB requires comprehensive training in clinical psychology and experience with individually administered psychological tests, psychopathology, and differential diagnosis. The CDB is appropriate for use solely by clinical psychologists, primarily because it utilizes qualitative aspects of assessment, which require a high degree of training and experience. Furthermore, the sample to which this test is appropriate is clearly a clinical one, typically reflecting severe psychopathology with unclear diagnoses. Therefore, the use of this test should be restricted to only those individuals with sufficient training and experience. The CDB can be used in any clinical situation requiring the assessment of cognitive behavior in the context of emotional disturbance. The procedures for administration, scoring, and interpretation are given clearly in the manual with appropriate clinical examples. After carefully studying the manual, a clinical psychologist should have no difficulty in learning how to appropriately administer and score the test.

Administration time for each of the five subtests is about 3 to 5 minutes, making a total testing time of approximately ½ hour. This reviewer used the test with several inpatients and found that a longer test administration time (approximately an hour) was required, due principally to the need of these patients for more support and encouragement.

The CDB answer form presents a very helpful arrangement of the items with ample opportunity to make clinical notes on the answer form. The manual provides clear directions for scoring and norm tables for select psychiatric samples. The manual also gives several examples to aid the user in understanding the scoring and interpretation process. It must be emphasized, however, that the manual cannot be relied upon as the sole basis for diagnosis or test interpretation—a trained clinical examiner is required to properly utilize these test materials.

The CDB is appropriate for the evaluation of the cognitive processing of psychiatric patients and the mentally retarded, as specified in the manual. The reviewer also used the CDB with hearing-impaired persons hospitalized for severe psychiatric conditions and found the instrument very useful in differential diagnosis.

Technical Aspects

The reliability and validity data for the CDB must be evaluated separately for each of the five subtests.

Color Form Preference Test. Split-half reliability coefficients for 52 schizophrenics, 72 retarded psychiatrics, and 101 nonpsychiatrics aged 18 to 65, corrected by the Spearman-Brown proficiency formula were .96, .87, and .86, respectively (Kay, 1982). A test-retest reliablity coefficient on the CFP was .67 for 25 hospitalized schizophrenics (Kay, 1982).

Evidence for validity of the CFP is based on significant chi-square values among psychiatric samples, such as aged schizophrenics, and mentally retarded psychiatric patients. Additional data are reported in Kay and Singh (1974, 1975, 1979).

Color Form Representation Test. Kay (1982) reports that split-half coefficients adjusted by the Spearman-Brown proficiency formula for 66 schizophrenics, 71 mentally retarded psychiatrics, and 42 nonpsychiatrics were .94, .67, and .81, respectively. A test-retest study on 38 hospitalized schizophrenics with two administrations 1-week apart was .87 (Kay, 1982). The validity of the CFR was evaluated by significant mean differences between scores of nonpsychiatric adults, schizophrenic children, and mentally retarded psychiatrics (Kay, 1982).

Egocentricity of Thought Test. A test-retest study of 50 untreated schizophrenics using test administrations 1 week apart yielded a corrected test reliability coefficient of .82 (Kay, 1982). Validity studies of the EOT yielded inconsistent results, but significant differences were found between the scores of schizophrenics and retarded persons.

Progressive Figure Drawing Test. The manual gives a corrected reliability coefficient of .93 for 79 mentally retarded psychiatrics evaluated by use of an odd/even split of items. A corrected test-retest reliablity coefficient of .96 was computed for 34 patients retested 9 to 24 months later. Validity of the drawing test was evaluated by comparing PFDT scores to the Draw-A-Person mental age. This coefficient was .83 (Kay, 1982).

Span of Attention. The reliability of this test was evaluated by a test-retest design following a 1-week interval yielding a corrected reliability coefficient of .82 (Kay, 1982). Validity was evaluated by comparing the means of five normative groups described in the manual for which a significant value was computed (Kay, 1982). Several evaluation studies on the distribution of attention spans of schizophrenics before and after neuroleptic treatment are described in the manual.

Inasmuch as the scales are generally made up of items of increasing difficulty, the corrected split-half reliability coefficient appears to be an appropriate statistic for several of these subtests. The test-retest index appears useful for the latter two subtests. The results reported in the manual and cited in additional references demonstrate the basic reliability of the subtests. Further work is necessary, however, to provide more complete data on the reliability of the CDB with more substantive samples. The demographics of each sample should be carefully delineated and presented with appropriate statistical information to enable the test user to clearly understand the study.

The validity data are much weaker, although it is clear from the studies that significant differences can be determined among clinical groups. A great deal more evidence is required to substantiate the appropriateness of these subtests with clinical samples in order to justify interpretation hypotheses. This reviewer is sympathetic to the problems of producing such studies, yet they are necessary.

Critique

This review of the Cognitive Diagnostic Battery and its use in several clinical settings suggests that it is a useful instrument and that there is great merit in conducting further studies using it with specific clinical samples. One would be naive, however, to expect that significant differences of scores can be established among all clinical subgroups. Nevertheless, this reviewer would recommend that continued work with clinical subgroups is useful in order to increase the value of the

CDB to clinicians. This work is particularly important as the instrument greatly relies on examiner expertise and background, a fact that needs to be reiterated to avoid the use of this instrument by untrained persons. Although test administration is not particularly difficult, it is recommended that even more samples for scoring and interpretation using additional test norms be given in the manual in order to provide examiners with thorough documentation. More scoring examples would assure conformity to standardized procedures and less subjective interpretation.

Overall, this reviewer would recommend that clinical psychologists involved in the diagnosis of psychiatric conditions and mental retardation give consideration to the CDB as a supplemental instrument to aid in the evaluation of cognitive functioning and mental disturbance.

References

Bearison, D.J., & Siegel, I.E. (1968). Hierarchical attributes for categorization. *Perceptual and Motor Skills, 27,* 147-153.

Brian, C.R., & Goodenough, F.L. (1929). The relative potency of color and form perception at various ages. *Journal of Experimental Psychology, 12,* 197-213.

Gesell, A. (1940). *The first five years of life.* New York: Harper & Row.

Kagan, J., & Lemkin, J. (1961). Form, color, and size in children's conceptual behavior. *Child Development, 32,* 25-28.

Kay, S.R. (1979a). Maturity of schizophrenic conceptual style in relation to intelligence. *Perceptual and Motor Skills, 48,* 1286.

Kay, S.R. (1979b). Schizophrenic WAIS pattern by diagnostic subtypes. *Perceptual and Motor Skills, 48,* 1241-1242.

Kay, S.R. (1979c). Significance of torque in retarded mental development and psychosis. *American Psychologist, 34,* 357-362.

Kay, S.R. (1980). Progressive figure drawings in the developmental assessment of mentally retarded psychotics. *Perceptual and Motor Skills, 50,* 583-590.

Kay, S.R. (1982). *Cognitive Diagnostic Battery.* Odessa, FL: Psychological Assessment Resources.

Kay, S.R., & Gang, R.G. (1973). Critical study of the Organic Integrity Test as a diagnostic and therapeutic index in schizophrenia. *Perceptual and Motor Skills, 37,* 827-833.

Kay, S.R., & Singh, M.M. (1974). A temporal measure of attention in schizophrenia and its clinical significance. *British Journal of Psychiatry, 125,* 146-151.

Kay, S.R., & Singh, M.M. (1975). A developmental approach to delineate components of cognitive dysfunction in schizophrenia. *British Journal of Social and Clinical Psychology, 14,* 387-399.

Kay, S.R., & Singh, M.M. (1979). Cognitive abnormality in schizophrenia: A dual process model. *Biological Psychiatry, 14,* 155-176.

Kay, S.R., Singh, M.M., & Smith, J.M. (1975). Color Form Representation Test: A developmental method for the study of cognition in schizophrenia. *British Journal of Social and Clinical Psychology, 14,* 401-411.

Payne, R.W. (1973). Cognitive abnormalities. In H.J. Eysenck (Ed.), *Handbook of abnormal psychology* (2nd ed., pp. 420-483). London: Pitman Medical Press.

Piaget, J. (1952). *The origins of intelligence in children.* New York: International Universities Press.

Schaie, K.W. (1968). Developmental changes in response differentiation on a color-arrangement task. *Journal of Consulting and Clinical Psychology, 32,* 233-235.

Sigel, I.E. (1953). Developmental trends in the abstraction ability of children. *Child Development, 24*, 131-144.

Smiley, S.S. (1973). Optional shift behavior as a function of age and dimensional dominance. *Journal of Experimental Child Psychology, 16*, 451-458.

Suchman, R.G., & Trabasso, T. (1966). Color and form preference in young children. *Journal of Experimental Child Psychology, 3*, 177-187.

Terman, L., & Merrill, M. (1960). *Stanford-Binet Intelligence Scale: Manual for the third revision Form L-M*. Boston: Houghton Mifflin.

Venables, P.H., & Wing, J.K. (1962). Level of arousal and the subclassification of schizophrenia. *Archives of General Psychiatry, 7*, 114-119.

Wahl, O., & Wishner, J. (1972). Schizophrenic thinking as measured by developmental tests. *Journal of Nervous and Mental Disease, 155*, 232-244.

Stephen L. Franzoi, Ph.D.
Assistant Professor of Psychology, Marquette University, Milwaukee, Wisconsin.

COMMUNICATING EMPATHY

John Milnes and Harvey Bertcher. San Diego, California: University Associates, Inc.

Introduction

Communicating Empathy is an audiotape-assisted assessment and training program that has three stated purposes: 1) to teach participants how to make appropriate verbal empathic responses when talking to others, 2) to motivate participants to develop empathic communication skills, and 3) to provide participants with a structured way to assess the appropriateness of their own or another person's verbal empathic responses. This learning program is to be run by a trained facilitator and has been developed for a wide range of individuals, including those who have had little or no empathic training as well as people with considerable training in human relations skills. It is suggested that training groups be set up so that they are homogeneous, with no mixing of novices and future facilitators.

Communicating Empathy was initially developed by John Milnes and Harvey Bertcher while Milnes was taking a special studies course from Bertcher at the School of Social Work, University of Michigan. John Milnes received his master's degree in social work from the University of Michigan and has worked extensively with the chronically emotionally disturbed and those with alcohol and drug problems. In addition, he maintains a private practice in marriage and family therapy. Harvey Bertcher is Professor of Social Work at University of Michigan's School of Social Work. He has considerable experience in social group work, including work in residential and day-treatment centers, with street gangs, in settlement houses and community centers, and with former psychiatric-hospital patients and handicapped children. He has served as a consultant to human-service agencies, including 3 years as a national consultant in social work to the Office of the Surgeon General, United States Air Force.

Milnes and Bertcher's main purpose in developing the Communicating Empathy program was to have on hand a clear model for empathy training, one that would provide consistency in definitions, modeling, and feedback. Following the first field test of the program on 12 graduate students, a second version was developed and tested on 18 graduate students (Kozma & Bertcher, 1974). Based on this field test, Milnes and Bertcher added one final segment to the program, which teaches trainees how to give accurate feedback to each other. The current revision is designed for a wider audience than the earlier version and has moved from a self-instructional format to one that requires a group facilitator.

Milnes and Bertcher believe that the Communicating Empathy program has the following advantages as a tool in empathy training: 1) it is easy to use, 2) it proceeds at the pace of the learner, 3) it provides models of good performance, 4) it creates a structure for assessment of the participant's performance, and 5) it can be adapted easily to a wide range of learning situations.

In its present form, Communicating Empathy consists of two audiotapes, a 30-page facilitator's guidebook, and a 6-page participant's response form. The guidebook states that the facilitator will need a cassette tape player and either a chalkboard and chalk or large newsprint pads with thick dark-colored, felt-tipped pens and masking tape. Each participant will need a pencil or pen to complete the response form.

The Communicating Empathy program is intended to be a 1-day session but was designed for flexibility and can be adapted to various needs. The program itself utilizes progressive learning steps to maximize successful learning experiences, beginning with a simple exercise and progressing to more complex ones as the program continues. Almost all instructions are given on the tape; however, the first two exercises and the last exercise are not tape-assisted and are conducted by the facilitator with the facilitator's guidebook and the participants' response forms.

There are nine different exercises in the program. The purpose of Exercise One, which takes approximately 30 minutes, is to help participants select "feeling words" and recognize and describe many subtle shades of feelings. The facilitator introduces the exercise and asks each of the participants to write privately 10 words that connote feeling on the response form. After everyone has completed a list, small groups are formed and members read their lists to their own groups. The groups must agree that at least 8 of the 10 words on each list connote feeling. Following this, each group brainstorms as many new words as possible within a 5-minute span. The facilitator concludes the exercise with a full-group discussion of feeling words.

The purpose of Exercise Two, which also takes about 30 minutes, is to help participants select synonyms that connote approximately the same meaning as the given feeling words. The facilitator reads a list of 10 feeling words and pauses long enough after each word for the participants to write on their response form three synonyms for each key word. Following this, small groups are formed in which members discuss the subtle differences between words that connote the same general feeling. Exercise Two concludes with a full-group discussion.

Exercise Three takes approximately 45 minutes and is the first exercise to utilize the audiotape. Its purpose is to help participants 1) identify the predominant feeling(s) expressed by a person who is describing a problem situation and 2) provide synonyms for these feeling(s). Feeling statements are not presented more than twice apiece in this exercise in order to teach participants the importance of responding both accurately and rapidly.

Exercise Four takes 30 minutes and is designed to help participants identify different levels of empathic understanding expressed by a listener to a speaker. The tape briefly discusses various levels of empathy and instructs participants to read the scale descriptions provided. Three examples illustrate varying degrees of empathy in response to the same spoken comment. Participants then hear a series of five statement/response pairs and evaluate the empathy level of each response.

The purpose of Exercise Five, which takes 15 minutes, is to help participants identify and describe the difference between statements referring to *thinking* and those referring to *feeling*. The tape briefly explains the difference between a thinking statement and a feeling statement, and then participants identify statements read to them as being either *thinking* or *feeling* in nature.

The purpose of Exercise Six is to help participants compose statements that are interchangeable with a speaker's feeling statements. The tape briefly reviews the elements that are crucial for good empathic responses; this is followed by the administration of five practice statements, which the participants must respond to in appropriate fashion. This exercise takes approximately 45 minutes.

Exercise Seven takes 30 minutes and is designed to help participants recognize and describe situations in which it is or is not appropriate to respond empathically. The tape provides a brief introduction describing five situations appropriate for empathic responses. The facilitator then stops the tape and leads a group discussion. Following this discussion, the facilitator starts the tape for a lecture on when not to use empathic responses. Finally, participants listen to five separate statements and indicate where in each statement they should make an empathic response. The correct answers are given after each statement.

The purpose of Exercise Eight is to help participants accurately rate, in writing, several dimensions of a listener's verbal empathic responses. The 45-minute tape contains three segments, each a series of interchanges between a speaker who has a problem and a listener who is attempting to respond empathically. Participants rate the listener in each segment, using the dimensions of empathic communication presented in previous exercises.

Finally, Exercise Nine is approximately 1 hour in length and is designed to help participants make verbal responses that are interchangeable with the feeling components of the speaker's statements. The facilitator instructs participants to form themselves into groups of three, with each member in turn being the "speaker," "listener," and "observer." Each role play lasts 5 minutes. After each role play each participant completes a rating form on which he or she judges his or her own performance. Once everyone has had one or two turns as speaker, listener, and observer, the trio discusses what they have learned. At the conclusion of Exercise Nine, the facilitator is instructed to review the entire program with the assistance of the participants.

Practical Applications/Uses

In response to the belief that accurate empathic understanding is one of the central interpersonal skills necessary for effective client-centered counseling and psychotherapy (e.g., Bergin & Garfield, 1971; Rogers, 1957; Truax & Carkhuff, 1967), a number of empathy training programs have been developed during the past 10 years to teach empathic responding to individuals working closely with others. Often, these programs consist of expensive, time-consuming, marathon weekends with rather vague presentations of empathic responding. Milnes and Bertcher believe their tape-assisted learning program is an inexpensive, effective, and readily available alternative learning device for those interested in developing such interpersonal skills.

The Communicating Empathy program is designed for group presentation, and the training groups should not be mixed with people of different experience levels. The program consists of nine different exercises, each lasting between 15 and 45 minutes. The basic format is clear and straightforward with numerous examples provided in each exercise. This type of training program could be useful for a number of different groups. One obvious group of individuals would be those in training as psychological counselors, therapists, social workers, or other mental health positions. Other occupational groups would be those who work in personnel positions.

In a very real sense, the main advantage of the Communicating Empathy program is also its chief disadvantage. That is, the tape-assisted, readily available program can be used in many different settings but must rely upon a group facilitator. The effectiveness of the training program hinges on the effectiveness of the group facilitator. Yet the quality of the group facilitator, the most integral component of this training program, cannot be assessed in this review because the facilitator does not come "prepackaged" as do the tapes and manual. As a consequence, the quality of the program will be lowered severely if the facilitator is not properly trained. Unfortunately, there are no training provisions of facilitators in the training manual or on the tapes. This does not mean the program will not be useful. It simply means that the program's quality cannot be properly assessed in this review because of this crucial missing ingredient.

Technical Aspects

There really is no adequate evidence that Communicating Empathy actually makes participants more effective in their empathic responding. As far as this reviewer has been able to ascertain, no research has been conducted on the final version of the training program to determine its effectiveness. This failure to document the effectiveness of the training program is a problem that needs to be rectified. At this point, potential users can only rely upon the developers' assurances that the program is effective. Much more is needed in the way of documentary evidence.

Critique

Some researchers and theorists have considered empathy to be a cognitive phenomenon and have focused on such intellectual processes as accurate perceptions of others (e.g., Dymond, 1949; Kerr & Speroff, 1954). Other researchers have stressed the emotional aspects of empathy and have studied topics such as helping behavior (e.g., Mehrabian & Epstein, 1972; Stotland, Mathews, Sherman, Hansson, & Richardson, 1978). Starting in the mid-1970s there has been a movement to integrate these two separate research traditions and approaches (Deutsch & Madle, 1975). In the research literature today, empathy is considered to be a multidimensional construct.

The two most commonly discussed and investigated aspects of empathy reflect the past research traditions. The cognitive component of empathy is commonly referred to as *perspective-taking,* and the affective component is referred to as *empathic concern* (e.g., Coke, Batson, & McDavis, 1978; Davis, 1980; Davis, 1983). Perspective-taking is defined as adopting the psychological view of others, and empathic concern is defined as experiencing "other-oriented" feelings of sympathy and concern for unfortunate others.

In their manual, Milnes and Bertcher define empathy in such a way that it is not inconsistent with the current literature; however, they do not clearly delineate for the reader the distinctions between these two types of empathy. Further, in the training exercises, no attempt is made to separate these two aspects of empathy from one another. It is the opinion of this reviewer that, although this does not create a serious impediment to participants in their understanding and use of empathic responding, it would improve the quality of the training program if the tapes and manual were updated to take recent research and theoretical advances into account.

A further weakness of the present program is that no systematic research has been conducted to determine whether the training actually results in participants responding more empathically. Although the exercises clearly describe empathic responding and appear to do an adequate job of providing necessary examples to better understand the empathic process, it still is not known whether Communicating Empathy effectively assesses and develops empathy.

Finally, even if the tape-assisted training program does have the capacity to do an adequate job in teaching people to better respond in an empathic manner, there is still the problem of facilitator quality. That is, the adequacy of the program hinges upon the effectiveness of the group facilitator in leading participants through the exercises and communicating what is involved in empathic responding. Thus, the present training program's quality in any given situation will be largely determined by the type of facilitator administering it—something that cannot be controlled by the developers or critiqued by this reviewer.

References

Bergin, A. E., & Garfield, S. L. (Eds.) (1971). *Handbook of psychotherapy and behavior change: An empirical analysis.* New York: John Wiley & Sons.

Coke, J., Batson, C., & McDavis, K. (1978). Empathic mediation of helping: A two-stage model. *Journal of Personality and Social Psychology, 36,* 752-766.

Davis, M. H. (1980). A multidimensional approach to individual differences in empathy. *JSAS Catalog of Selected Documents in Psychology, 10,* 85.

Davis, M. H. (1983). Measuring individual differences in empathy: Evidence for a multidimensional approach. *Journal of Personality and Social Psychology, 44,* 113-126.

Deutsch, F., & Madle, R. (1975). Empathy: Historic and current conceptualizations, measurement, and a cognitive theoretical perspective. *Human Development, 18,* 267-287.

Dymond, R. F. (1949). A scale for the measurement of empathic ability. *Journal of Consulting and Clinical Psychology, 13,* 127-133.

Kerr, W. A., & Speroff, B. G. (1954). Validation and evaluation of the empathy test. *Journal of General Psychology, 50,* 369-376.

Kozma, R. B., & Bertcher, H. (1974, April). *Evaluation of a self-instructional mini-course on empathic responding.* Paper presented at the annual meeting of the American Educational Research Association, Chicago.

Mehrabian, A., & Epstein, N. (1972). A measure of emotional empathy. *Journal of Personality, 40*, 525-543.

Rogers, C. R. (1957). The necessary and sufficient conditions of therapeutic personality change. *Journal of Consulting Psychology, 21*, 95-103.

Stotland, S., Mathews, K., Sherman, S., Hansson, R., & Richardson, B. (1978). *Empathy, fantasy, and helping.* Beverly Hills, CA: Sage.

Truax, C. B., & Carkhuff, R. R. (1967). *Toward effective counseling and psychotherapy: Training and practice.* Chicago: Aldine.

Allan L. LaVoie, Ph.D.

Professor of Psychology, Davis & Elkins College, Elkins, West Virginia.

DEFENSE MECHANISMS INVENTORY

Goldine C. Gleser. Cincinnati, Ohio: DMI Associates.

Introduction

The Defense Mechanisms Inventory (DMI) first appeared in print for research use nearly 20 years ago (Gleser & Ihilevich, 1969). Recently, a fine, comprehensive manual has become available (Ihilevich & Gleser, 1986). As one may judge from the title, the DMI measures a person's preferred defense mechanisms. The test's format presents a story describing a frustrating event on each page. Five general frustrating themes are represented in the stories. The first is called situational frustration and is exemplified by the irritating accidents of everyday life. The other four themes are competition, authority, independence, and sexual identity. Groups of answers assess defense preferences at four levels: behavioral, fantasy, cognitive, and affective. Each group of answers has five alternatives that correspond to each of five defense mechanism clusters. The subject selects the most and the least representative alternatives, which leads to summary scores on the five clusters of defense mechanisms: Turning Against Object (TAO), Turning Against Self (TAS), Principalization, or intellectualization (PRN), Projection (PRO), and Reversal (REV).

The clusters do not closely match any obvious grouping of classical mechanisms. For example, REV includes denial, repression, and reaction-formation. The defenses are grouped to match a series of coping mechanisms (Ihilevich & Gleser, 1986, p. 15); however, the match seems unconvincing. For example, "finding the silver lining" is described as a coping mechanism similar to the defense mechanism of denial, although it would seem to be closer to rationalization. There is ample evidence that the five clusters do not represent any final solution to the question of how to group defenses (e.g., Blodgett, 1985; Bond & Vaillant, 1986; Frank, McLaughlin, & Crusco, 1984; Juni, 1982; and Vickers & Hervig, 1981).

The DMI represents about 25 years of work by Ihilevich, whose master's degree involved work on defense mechanisms (Ihilevich, 1963, cited in Ihilevich & Gleser, 1986, p. 16); Gleser was one of the supervisors. The DMI manual (Ihilevich & Gleser, 1986) contains over 500 references, many of which are empirical reports of the DMI. Clearly, the test has become popular, at least for research purposes, since its original publication.

The currently available test stories have been professionally printed on heavy paper that will stand up well to repeated use. Directions are clear, and reading level appears to be at a high school level. Scoring templates fit the separate answer sheet well. The profile sheet is easy to use but aids little in understanding the scores, beyond the fact that the scores are profiled in standard T-scores; means and standard deviations are also printed at the bottom of the sheet for handy reference. Separate booklets and profile sheets are used for males and females. Recently, new tests have become available for adolescents and for the elderly; their use has been restricted to nonclinical applications until more data have been accumulated about their reliability and validity.

Normative data have been accumulated adequately for comparison with college students and normal adults. The norms appear to be based on incidental samples. Means and standard deviations are reported in the manual for some specialized groups such as the overweight and ulcer patients, but, generally, the numbers of subjects are small. Data on a variety of psychiatric patients also are available in the manual. The authors report that foreign language versions of the DMI are available, but no data are reported on foreign samples and the languages themselves are not identified.

Practical Applications/Uses

Numerous applications for the Defense Mechanisms Inventory come to mind as one considers the meaning of the individual scores. The test could be used in a battery of other tests to help guide treatment in clinics, in prisons, or in counseling centers. DMI scores should be useful in interpreting other test scores (e.g., a subject who extensively uses REV may be in a poor position to report on his or her own behavior accurately). For that reason, this reviewer includes it with a battery of other tests in studying sex offenders (who are notoriously reluctant to "own" their crimes). Scores on the DMI can provide a measure of the insight a client has achieved during the course of psychotherapy. Other applications can be imagined in educational settings (e.g., to complement the results of testing for learning disabilities or behavioral and emotional disorders) or forensic settings (e.g., to assist in judging the likelihood of dissimulation).

In addition to applied uses, the DMI has a history of use in several research areas, including the study of psychopathology, field independence-dependence, brain laterality and preferred defenses, neurochemical correlates of defenses, relationships between defenses and other personality characteristics, the adaptive function of various defenses, change in defenses as a function of age, sex differences in defenses, the relationship of physical illness and recovery to typical defenses, the relation of obesity and other addictive problems to defenses, and the relationship of locus of control to preferred defenses.

The foregoing uses rely on individual scores. Potential exists for even more utility from the DMI when profile summaries are available and have been empirically related to various behaviors. It would be very useful to know that someone whose major defenses are TAO and PRO can be counted on to respond in one way to frustration and that someone with a TAO and PRN pattern can be counted on to behave quite differently. If the test authors elect to focus their applied research on the question of the dimensionality of the DMI, the profile possibilities would probably be reduced in number but expanded in validity.

Administration requires no special skills, as the tests are self-explanatory for most populations. Subjects at the college level will finish in 20 to 30 minutes. Scoring by template takes about 5 minutes, and can be done with high accuracy by untrained clerks. As the marginals must be equal to one another, errors are easily detected.

The authors advise that interpretation depends on a knowledge of psychoanalysis but that a knowledge of the contents of the manual and of other recent research with the DMI is more important. Straightforward interpretations may be performed using the definitions contained in the manual. Nonetheless, in clinical settings, the interpretations should be performed by professional psychologists.

Technical Aspects

The Defense Mechanisms Inventory generally has adequate or good internal consistency, averaging about .72 for the five scales. Scores also seem to be stable over time, averaging about .73 for intervals up to 1 month. Another measurement of consistency has been provided by Blacha and Fancher (1977) who conducted a content analysis of the items. They found adequate interrater agreement with the item scoring keys for three scales: PRN (71%), TAS (72%), and REV (72%), but PRO (29%) and TAO (39%) had agreement coefficients that fall short of minimum standards. Very similar results had been obtained by the test authors (see Ihilevich & Gleser, 1986, pp. 51-53). In summary, the items for a particular scale seem to be used somewhat consistently by subjects and yield scores that are stable over at least short periods of time. But the items do not fit into the neat categories the authors hoped, at least as judged by trained graduate students. The relatively small coefficients for internal consistency probably reflect the tendency of some of the test stories to call out particular responses. Some stories consistently call out TAO responses, and others call out REV responses. Some subjects respond differently from one story to another. At the same time, the stability coefficients make it clear that subjects are responding consistently from one occasion to another.

Reliability estimates of the levels of response indicate that responses on the feeling level are more consistent than on the actual behavior level (.55 vs. .34), but the overall pattern reveals too little consistency to be useful on an individual score basis.

Validity, as usual, requires the long-term accretion of data in a very wide array of settings. The results to date for the DMI show much promise, but in no way do we have a definitive picture of validity. The questions of major concern include whether the five scales measure differentially, whether they intercorrelate as predicted, whether they can predict significant human behaviors, and whether they can advance our understanding of the functioning of defense mechanisms.

Some of these questions cannot be answered easily. For example, we know that the scales are more highly intercorrelated than is desirable. We find in the manual (Ihilevich & Gleser, 1986, pp. 66, 70, & 115) a set of correlations showing a strong tendency for PRO to be positively correlated with TAO. This relationship is predictable, for we would expect that people who aggress against others must sometimes justify it through projective defenses. Should we also expect people who use TAO to score higher on TAS, reasoning that if they aggress against others they will also aggress against self? The answer from page 66 and 70 is no, but from page 115, where the scores are based on Likert scaling, the answer is yes. Which is the correct prediction? Which is the correct answer? Unfortunately, this section of the manual has the flavor of old-time psychoanalysis, with no unambiguous predictions and never a result that one cannot rationalize.

Despite the important, unresolved issues, there exist some encouraging data to lead us to believe that the DMI has validity. In an extensive review of the literature, Ihilevich and Gleser (1986) examine the correlates of each of the scales. They report that males typically score higher on PRO and TAO than females, as would be expected from our knowledge of the greater aggressiveness of males. And, in a complementary result, females score higher on TAS as would be predicted from our knowledge of their greater tendency to depression. To summarize their extensive discussion, it would be fair to say that the data for the individual scales show great promise for future validity. Of course, the data at the moment are not completely clear. To take a couple of recent examples, Greenspan and Kulish (1985) found no relationship between the defensive style of therapists and the rate of premature therapy termination, and Moses and Reyher (1985) found DMI scores to be unrelated to visual imagery, although the images were easily classified in terms of the defenses the subjects used to produce them.

Critique

Ihilevich and Gleser have developed a potentially very useful test. Reliability and validity are entirely adequate for this stage of test refinement. Several avenues of study might profitably be pursued by them to raise the Defense Mechanisms Inventory to even higher standards. The literature makes it clear that the DMI probably consists of two dimensions, one perhaps having to do with ego strength, and the other having to do with intropunitiveness. Resolution of this question would clear the way for subsequent studies.

Another issue that might be examined in more detail concerns the situational specificity of the responses. Again it is clear from the manual (Ihilevich & Gleser, 1986, p. 43) that the stories can significantly influence the responses a subject makes. A similar conclusion was reached by Klusman (1982). If the specific story or setting makes a difference to the subject, then considerable diagnostic material could be discovered by the tendency of subjects to adapt their responses to one kind of situation but not another. Adding more stories would make the test more time-consuming to take and to score, but perhaps a more efficient response format could compensate for the increased time. And if the changes improved the quality of information, most users would consider that adequate compensation.

An additional issue has been explored by Ihilevich and Gleser (1986, pp. 112-119)— should the DMI give up its forced-choice format in favor of a free-response format? They concluded not to change, but it may be that they overlooked some advantages, including the potential savings in time. A free-response format (e.g., ratings) would allow them to obtain more data from more stories by giving up the levels approach without sacrificing reliability. In this reviewer's experience, forced-choice questionnaires have a somewhat higher respondent error rate than true/false measures or rating scales. In any case, some researchers are using the free-response format, so the authors must explore the relationships between forced-choice and free-choice formats so that users will be able to better interpret their results.

In summary, the DMI offers much to the clinician and to the researcher. It provides reliable and objective estimates of the tendencies of people to use particular clusters of defenses. With continued efforts to improve and develop the test, the

DMI could join the small group of widely used personality scales such as the MMPI.

References

Blacha, M.D., & Fancher, R.E. (1977). A content validity study of the Defense Mechanism Inventory. *Journal of Personality Assessment, 41,* 402-404.

Blodgett, C. (1985). Renal failure adjustment and coping style. *Journal of Personality Assessment, 49,* 271-272.

Bond, M.P., & Vaillant, J.S. (1986). An empirical study of the relationship between diagnosis and defense style. *Archives of General Psychiatry, 43,* 285-288.

Frank, S.J., McLaughlin, A.M., & Crusco, A. (1984). Sex role attributes, symptom distress, and defensive style among college men and women. *Journal of Personality and Social Psychology, 47,* 182-192.

Gleser, G.C., & Ihilevich, D. (1969). An objective instrument for measuring defense mechanisms. *Journal of Consulting and Clinical Psychology, 33,* 51-60.

Greenspan, M., & Kulish, N.M. (1985). Factors in premature termination in long-term psychotherapy. *Psychotherapy, 22,* 75-82.

Ihilevich, D., & Gleser, G.C. (1986). *Defense mechanisms: Their classification, correlates and measurement with the Defense Mechanisms Inventory.* Owosso, MI: DMI Associates.

Juni, S. (1982). The composite measure of the Defense Mechanisms Inventory. *Journal of Research in Personality, 16,* 193-200.

Klusman, L.E. (1982). The Defense Mechanisms Inventory as a predictor of affective response to threat. *Journal of Clinical Psychology, 38,* 151-155.

Moses, I., & Reyher, J. (1985). Spontaneous and directed visual imagery: Image failure and image substitution. *Journal of Personality and Social Psychology, 48,* 233-242.

Vickers, R.R., Jr., & Hervig, L.K. (1981). Comparison of three psychological defense mechanism questionnaires. *Journal of Personality Assessment, 46,* 630-638.

Howard Stoker, Ph.D.
Professor and Head, Instructional Development and Evaluation, Department of Education, Graduate School of Medical Sciences, University of Tennessee, Memphis.

DEGREES OF READING POWER

Touchstone Applied Science Associates. New York, New York: The College Board.

Introduction

The Degrees of Reading Power (DRP) program is, first and foremost, a unique method for measuring reading comprehension. It is designed to measure a student's ability to understand English prose. In addition, it 1) provides interpretive material that allows a teacher to determine the most difficult prose text a student can read, 2) matches materials with student ability, and 3) provides a means by which growth in reading comprehension can be documented. Two forms are available at each of five grade-equivalent equivalent levels: 3-5, 5-7, 7-9, 9-12, and 12-14. Tables are available for converting raw scores to DRP scores; norm-referenced score conversions are also possible.

The DRP was developed by the New York State Department of Education, Touchstone Applied Science Associates, and The College Board. It is relatively new, having been available for about 10 years.

Most tests of reading comprehension consist of short passages followed by one or more multiple-choice items. Such tests have been subject to criticism for years. Shell (1984), in a review of the Comprehensive Tests of Basic Skills, cited the imbalance in subject matter of the reading passages as a threat to the content validity. He reported that some items were "amazingly easy to answer without reading the passage." Filby (1978), in a review of the Iowa Silent Reading Test, cites the importance of the reader's ability to grasp key points and general ideas in the measurement of comprehension and indicates that the ISRT falls short in such measurement. The modified cloze procedures used to develop the DRP were employed in an attempt to counter some of the criticisms of reading comprehension tests.

In a typical cloze test, one deletes every n^{th} word in a reading passage. The task for the reader is to fill in the missing word, based on contextual clues. Bormuth (1962) judged cloze tests to be valid measures of comprehension ability and "efficient" where the purpose of measurement was readability. Much of the early research on the cloze technique was reviewed by Potter (1968). Panackal and Heft (1978) compared the cloze technique with a multiple-choice cloze technique. They constructed two multiple-choice versions of two cloze tests and administered them to college students. They concluded that the mutliple-choice forms were slightly less reliable, but appeared to be more valid than the cloze tests.

The modified cloze technique led to tests that have the following properties, as reported in the *DRP Handbook* (College Entrance Examination Board, 1986, p.3):

1. DRP test items are designed so that the passage in which they are embedded must be read and understood in order for the student to answer correctly.
2. All of the content information that is needed to select the correct response is contained within the DRP passage.
3. Regardless of the difficulty of the prose passage, all response options are common words.
4. DRP passages are designed to reduce the likelihood that guessing strategies, associative processes, and other non-reading activities can be used to respond to items correctly.
5. Item difficulty is linked to text difficulty.

These properties led to the production of a set of tests that resemble cloze tests, but which are not, really, cloze tests as they are described in the literature.

In his review of the DRP in the *Ninth Mental Measurements Yearbook*, Roger Bruning (1985)says,

Perhaps the most novel feature of the test, however, is that scores are scaled to the readability of text materials, rather than to grade equivalents or other more common indices. DRP scores of test materials are obtained through a standard readability analysis. A student's scores, given in the same DRP units, are interpreted as the readability of prose . . . that can be read with varying levels of comprehension.

Considerable research has been done on measuring and estimating the readability of textual material. Several reading formulas exist, including those of Fry (1968), Spache (1953), and Dale and Chall (1948a, 1948b).

DRP prose passages were scaled using the mean cloze formula developed by Bormuth (1969). A complete rule book for coding passages was developed by The College Board, which invoked the variables used by Bormuth. Readability (R), in Bormuth's formula is really a cloze score forecast of the proportion of correct restorations to be expected. DRP units are expressed as $(1-R) \times 100$. Hence, the larger the DRP score, the more difficult the passage.

DRP passages and items are produced according to precise specifications. They are, of course, reviewed for bias by gender, age, and ethnicity/race. Then they are introduced into a Rasch calibration design so that the item difficulties will be on a common scale.

Each DRP test consists of a set of passages, written specifically for the test, on topics drawn from the *Encyclopaedia Britannica*. The passages contain about 325 words each and are arranged in order of increasing difficulty. In each passage, seven sentences contain a blank space, indicating a missing word. For each blank, five single-word response options are provided. All the options are common words and all make sense within the sentence if it is taken out of the context of the passage, but only one is correct for the sentence as it appears in the passage. The choices are designed such that the student must comprehend the prose in order to select and mark the only appropriate word for the sentence.

Alternate forms are provided at each of four overlapping levels, which run from Grade 3 to Grade 12 (PA and PB series). The CP series, with two forms, is designed for use in college placement programs.

Three types of answer documents are available. The typical single-sided document is designed for one-time use and can be ordered as single sheets or in continuous form for the preprinting of students' names, classrooms, and so on. Also available is a four-sided answer folder designed for use in pretest/posttest situations. All of the answer sheets can be machine scored, and all can be hand scored.

The type styles used in the DRP are pleasing to the eye. All levels are printed in the same style and color, but the levels are easily distinguished by the coding on the covers.

A full range of services is available (for a price, of course). A Basic Service must be ordered before one can avail themselves of other services. One basic package includes four reports: an alphabetic roster, a rank order grade roster, a class summary statistical report, and a school and/or district summary statistical report. DRP text-referenced scores and national norms appear on the rank order roster, and percentile ranks and NCEs appear on the summaries. The other basic package includes all DRP text-referenced scores with no normative data. Optional services include an Individual Report, a Student Label, a Rank Order Class Roster, a Raw Score Summary Report, an Item Response Report, and Pre-Post Analyses. Research tapes are also available.

The Individual Report, which appears to be of considerable value, includes the student's scores in DRP units and a graphic representation of those scores. DRP scores are reported at three levels: Independent, Instructional, and Frustration. Descriptions of these reading-material levels appear on the report, along with a graphic representation of typical magazines and associated DRP scale values. As reported on the Individual Report,

> The Individual Level identifies the difficulty of materials which this student can read and understand without help. This student should be able to read materials of this difficulty, and easier materials, on his or her own. The Instructional Level identifies the difficulty of materials which this student can read and understand with help. Help could include teacher assistance, classroom discussions, dictionaries, and other instructional aids. Materials of about this difficulty could be used for classroom instruction. The Frustration Level identifies the difficulty of materials which this student would find hard and frustrating to read. Such materials may be too difficult to use for effective instruction.

In the *DRP Handbook* example, Johnny's scores in the three levels are reported as 40, 51, and 62. They imply that he could read few children's magazines without help, that he could read magazines such as *People* with more help, but that he would experience frustration with "teen magazines." Because Johnny took form PA-6, one could guess that he is about 12 years old and is, perhaps, more interested in comics and local sports than in reading English prose.

The *DRP Handbook*, listed as an accessory, appears to be a must for users of the DRP. A comprehensive description of the tests and the concept of readability, as represented by the DRP, are presented. A concise description of the technical characteristics of the tests appears, along with chapters on administering the tests and reporting and using the test results (how to and how not to use them). Norms, change-score norms, and conversion tables are also included. Of particular interest to many readers of the *DRP Handbook* will be the bibliography, which contains about 70 references reflecting the philosophical and instructional foundations of the DRP program. The references span 70 years, beginning with an article by E. L. Thorndike that was published in 1917.

Another valuable accessory is the Readability Report. This document lists the difficulties of published materials in DRP units. The seventh edition contains over 3,300 titles; the 1986-87 supplement adds another 300 titles. The Readability Report should facilitate the selection of texts and collateral reading materials.

Apple and IBM PC versions of MicRa, a DRP software package, are available under site license agreements. This software would allow a local school district to derive DRP values for instructional materials that have not been scaled by the publishers.

Practical Applications/Uses

Several years ago, this reviewer was asked to be a measurement "expert" on a team that was to review items for a state-wide reading assessment. The reading "experts" argued that the items, as written, did not constitute a "reading test." To resolve the issue, the title was changed to indicate that the test was an assessment of "skills related to reading." In contrast, the DRP tests look like tests of reading comprehension. One *does not* have a list of objectives (e.g., recognize the main idea, etc.). One *does* have a test that contains several prose passages. Students are asked to read the prose and select the words which will complete the prose passage. The tests appear to provide teachers with two kinds of information: 1) how well students comprehend prose at defined levels of readability and 2) the kinds of textual materials a student can read independently or with some help. As Kibby (1981) says, "Reading people will look at this test and say, 'Why didn't I think of this?'. . . the DRP is an excellent testing methodology."

Working with a class summary, the teacher can quickly determine the distribution of instructional level DRP scores. That information, coupled with the Rank Order Roster, should make the selection of appropriate texts and collateral reading materials much easier. Grade and/or school summaries appear to have the potential for helping teachers and administrators make better decisions regarding instructional materials. Text ordering decisions, based on grade-level summaries of DRP instructional level scores, should lead to the purchase of texts that would facilitate learning rather than frustrate it. This reviewer has long argued for instructional decisions based on available data. The class and grade summaries provided by the DRP are representative of data that can be used to make informed instructional decisions.

Technical Aspects

The *DRP Handbook* treats the calibration of DRP passages and items, validity, reliability, fairness, and bias in one chapter and discusses norm development in another. The DRP tests are designed as homogeneous measures of prose comprehension. The reported Kuder-Richardson (KR-20) coefficients are all about .95. Resulting standard errors are small enough to allow for adequate score interpretations (3.0-3.6).

The validity of cloze tests as measures of reading comprehension has been challenged in the literature. The *DRP Handbook* addresses the construct, content, and criterion-related validity issues of the DRP tests. It states, "DRP tests can be conceived as objective referenced tests with the single outcome objective—comprehension of expository English text. It is important to emphasize that individual DRP test items are *not* designed to sample the specific objectives often listed in the manuals of traditional reading tests." Rankin (1983) discussed the "ecological validity" of the DRP, which had been questioned by Guthrie (1981). The issue was whether the test required a response from the student similar to that which would be required in the real world. Rankin's claim was that the DRP yields results that predict the most difficult prose the student could read in the real world. Kibby (1981), referring to the content validity of the DRP, said, "The reading passages are not only clearly typical of required reading for the school populations, but also appear to be interesting." Carver (1985) claimed that a close examination of the data reported by The College Board revealed a validity problem. He states, ". . . instead of obtaining more appropriate instructional materials, DRP test users seem to be more likely to obtain less appropriate instructional materials that will stifle learning." Rankin (1986) responded to Carver's critique of the DRP, arguing that Carver ". . . failed to understand several unique characteristics of the DRP . . . ." Other articles dealing with the validity of the DRP include those by Linn (1981) and Koslin and Ivens (1986).

Questions have been raised about the adequacy of the norming sample (Kibby, 1981; Bruning, 1985). The *DRP Handbook* states, "It was assumed, and later validated, that high-quality norms could be obtained by emulating National distributions on certain demographics in the National population, using schools drawn from one very large state." Hence, a random stratified sample of New York schools, which were using the DRP tests, was drawn. Approximately 34,000 students in Grades 4 through 12 took the 10 calibration forms used in the DRP norming study. DRP norms have been compared to the 1977 California Achievement Test norms (Weiner & Kippel, 1984). The *DRP Handbook* reports that DRP norms are supported by National Assessment of Educational Progress data.

The norms tables differ from those found with typical achievement tests. Raw scores are converted to DRP units in terms of levels of comprehension. For PB-4, for example, a raw score of 50 would convert to a DRP score of 45 at the Independent Level ($p = .90$), to 56 at the Instructional Level ($p = .75$), and a 67 at the Frustration Level ($p = .50$). The prediction is that the student whose raw score is 50 can independently read material that has a readability level of 45, that the student will have some difficulty with material rated 56 on the DRP scale, and that he or she will be frustrated by any material with a readability of 67 (actually, the student would probably be frustrated by material rated much higher than 60). Reading these tables requires a different orientation from that developed for reading typical norms tables.

Critique

The Degrees of Reading Power methodology seems to make sense. At a time when schools, districts, and states seem consumed with measuring minimum competency for hundreds of minute objectives, measuring the comprehension of prose, as the DRP does, is refreshing.

The DRP tests are attractive to those who are interested in "how well messages within text are understood" (College Entrance Examination Board, 1986, p.1). The continuous scale of scores and the levels of DRP scores with associated probabilities present an interesting type of interpretation. Referencing those scores to published materials is something that is long overdue.

The DRP has its problems, as do all tests. Random samples *may* be representative of the population from which they were drawn, but there is no guarantee. When one knows the characteristics of the population, representativeness can be increased through appropriate sampling techniques. Drawing a sample from one state to represent a national sample will never be satisfactory for some users and reviewers. However, if one is interested in using the DRP tests to improve instruction, and if one utilizes appropriate sampling techniques, then representativeness of the norms sample is of lower importance.

Until reading specialists conduct more research on the DRP method for measuring comprehension of prose, debate on the validity of the DRP will continue. Given the diversity of philosophies regarding the measurement of reading comprehension, it is unlikely that the debate will be short-lived.

Not all measurement specialists accept Rasch model scaling, which underlies the DRP scales. Neither do all reading specialists accept Bormuth's method for analyzing text difficulty. Doubts and differences in methodologies should lead to numerous comparisons of DRP tests with other tests.

As is the case with all tests, DRP test scores can be misinterpreted and misused. Cautions appear in the *DRP Handbook* (e.g., "Although the Readability Report lists DRP values for thousands of widely used textbooks . . . it cannot begin to cover the wide range of materials available for use in the classroom" [p. 31], and "it should be noted that students performing 'above the norm' on DRP tests may still not be able to read . . . materials used in their classes" [p. 39]). Test publishers know well that many customers do not read test manuals and that some who do may not comprehend what they read. Perhaps The College Board should require an Independent Level DRP score on a sample of text in the *DRP Handbook* prior to the sale of the tests.

References

Bormuth, J. R. (1962). *Cloze tests as a measure of readability and comprehension ability.* Unpublished doctoral dissertation, Indiana University, Bloomington.

Bormuth, J. R. (1969). *Development of readability analyses* (Contract No. GEG-3-7-070052-0326). Washington, DC: U. S. Department of Health, Education and Welfare, Office of Education.

Bruning, R. (1985). Degrees of Reading Power. In J. V. Mitchell, Jr. (Ed.), *The ninth mental measurements yearbook* (pp. 443-444). Lincoln, NE: Buros Institute of Mental Measurements.

College Entrance Examination Board. (1986). *DRP Handbook.* New York: The College Board.

Dale, E., & Chall, J. (1948a). A formula for predicting readability. *Educational Research Bulletin, 27,* (1), 11-20.

Dale, E., & Chall, J. (1948b). A formula for predicting readability. *Educational Research Bulletin, 27*(2), 11-28.

Filby, N. N. (1978). Iowa Silent Reading Test. In O. K. Buros (Ed.), *The eighth mental measurements yearbook* (pp. 730-732). Highland Park, NJ: The Gryphon Press.

Fry, E. B. (1968). A readability formula that saves time. *Journal of Reading, 11,* 513-516, 575-578.

Kibby, M. W. (1981). The Degrees of Reading Power. *Journal of Reading, 24,* 416-427.

Koslin, B. L., & Ivens, S. H. (1986). *Validity revisited: A rejoinder.* Unpublished manuscript, The College Board, New York.

Linn, R. L. (1981). Issues of validity for criterion-referenced measures. *Applied Psychological Measurement, 4*(4).

Panackal, A. A., & Heft, C. S. (1978). Cloze technique and multiple-choice technique: Reliability and validity. *Educational and Psychological Measurement, 38,* 917-932.

Schell, L. M. (1984). Test review: Comprehensive Tests of Basic Skills (CTBS, Form U, Levels A-J). *Journal of Reading,* 586-589.

Spache, G. (1953). A new readability formula for primary grade reading materials. *Elementary School Journal, 53,* 410-413.

Weiner, M., & Kippel, G. (1984). The relationship of the California Achievement Test to the DRP Tests. *Educational and Pyschological Measurement, 44,* 497-500.

Colleen B. Jamison, Ed.D.

Professor of Special Education, California State University-Los Angeles, Los Angeles, California.

DEVELOPMENTAL ASSESSMENT OF LIFE EXPERIENCES

Gertrude A. Barber, John P. Mannino, and Robert J. Will. Erie, Pennsylvania: The Barber Center Press.

Introduction

The Developmental Assessment of Life Experiences (DALE) is an inventory intended to assess the skill development of individuals who have been isolated from society for long periods of time at institutions that treat developmental or physical handicaps. The DALE was developed and compiled by Gertrude A. Barber, John P. Mannino, and Robert J. Will at The Dr. Gertrude A. Barber Center, a community-based agency for mentally ill persons, in Erie, Pennsylvania, and was first printed in February 1974 in response to the need to assess the success of individuals being released from institutions to community agencies. The Pennsylvania State Mental Retardation Center used the 1974 edition, but as the need to assess individuals with greater handicaps grew, the development of a system that assessed the self-help skills of the profoundly to severely mentally retarded became necessary. Raters who were most familiar with the clients and responsible for their programming began to feel that they were not doing their job properly when they had to mark so many items low for their severely handicapped clients. Subsequent development of the inventory and revisions undertaken by the staff at the Barber Center resulted in two inventories. The items that assessed higher functioning community living skills became Level II. The items assessing more basic skills were placed in Level I. The revisions were published in September 1975; January 1978; September 1981; and in February 1986.

The DALE includes separate consumable booklets for Level I and Level II and the *Developmental Assessment of Life Experiences: Manual and Guidelines for Implementation* (5th ed.), which explains the system. Information from the therapist's handbook, which is no longer available, has been incorporated into the manual.

Level I is intended for individuals whose mental functional abilities fall within the profound to severe range; Level II lists "behaviors which represent higher functioning levels" (Barber, Mannino, & Will, 1986a, p. 11). The level to be used "should be determined by responsible agency personnel" (Barber et al., 1986a, p. 11). After an individual's appropriate level is identified, the corresponding level booklet is used to record each of the rating sessions for the individual.

The checklist approach of the system is reflected in the item booklet, which allows space for recording behavior observations for up to four rating periods. Each booklet also contains a chart for graphing the individual's progress. The graphs, explained in greater detail later in this review, have space for the observer to record progress for four sessions. Space for baseline ratings is provided as well. Users may choose to rule in another column on the inventories in order to parallel the recording space allotted for the graphs. Spaces at the end of each subcategory allow the user to record additional items. A page corresponding to the items in the facing inventory page is used for recording "specific strengths" relating to a particular task or skill area as they are observed and identified. The manual refers to this page as the "strength/need page," so, presumably, the page may be used to record areas of need as well as strong points. A page entitled "General Strengths" is provided at the end of each booklet for summarizing major categories of strength and need. The page is labeled with these headings: "What Resident Likes to Do," "What Resident is Good at," and "People Willing to be Involved in Implementation."

Items are expressed in positive terminology and are arranged to approximate sequential order. Level I items are categorized under the headings sensory motor, language, self-help, cognition skills, and socialization. Level II items are categorized under the headings personal hygiene, personal management, communications, residence/home management, and community access. Each category for each level includes subcategories containing from 5 to 38 items each. Each subcategory allows space for the addition of items that address individual competencies or needs. Items on the two levels are not mutually exclusive; they may appear on both inventories. The categories in each level as well as the subcategories and number of items in each follow.

LEVEL I (258 items):

Sensory Motor (43 items): Gross Motor I (5), Gross Motor II (12), Visual Perception (5), Auditory Perception (6), and Fine Motor (15).

Language (32 items): Receptive Language (7), Expressive Language (15), and Hearing Skills (10).

Self-Help (122 items): Feeding (17), Undressing (9), Dressing (17), Drinking (10), Toileting (13), Bathing (38), Nasal Hygiene (7), and Toothbrushing (11).

Cognition Skills (30 items): Identification and Discrimination (11), Attention Span (10), and Time and Number Concepts (9).

Socialization (31 items): Social Relationships (14) and Self-Concept (17).

LEVEL II (501 items):

Personal Hygiene (86 items): Toileting (6), Bathing (16), Oral Hygiene (11), Nasal Hygiene (6), Menstrual Cycle (8), Shaving Skills (15), Grooming (17), and Foot and Nail Care (7).

Personal Management (114 items): Dining Habits (22), Mealtime Tasks (19), Clothing Care (12), Laundry Skills (19), Personal Scheduling (17), Undressing (9), and Dressing (16).

Communication Skills (91 items): Verbal Skills (17), Hearing Skills (7), Telephone Skills (18), Writing Skills (13), Reading Skills (14), and Numerical Concepts (22).

Residence/Home Maintenance (77 items): General Housekeeping (18), Maintenance of Sleeping Area (13), Kitchen Maintenance (13), Maintenance of Bath Area (11), Outdoor Maintenance of Residential Area (15), and Basement/Attic (7).

Community Access (133 items): Mobility (15), Use of Public Transportation (12), Shopping Skills (20), Meal Preparation (18), Money Concepts (11), Banking Concepts (11), Supplemental Income (6), Medical (12), Sexual Awareness (12), and Social Skills (16).

Practical Applications/Uses

The DALE system requires five processes throughout habilitation: assessment, using the inventory; discussion and planning based on the results of assessment; program development to establish goals and objectives, as well as an agreement for follow through; plan implementation; and evaluation. Although direct observation is recommended, the manual states that "the procedure of using a third party informant is acceptable as long as subsequent assessments are based upon actual observation to the greatest extent possible" (Barber et al., 1986a, p. 33). The test developers recommend that the DALE be completed semiannually; accomplishments occurring between the formal assessments should be recorded as they occur. Repeating the DALE at regular intervals facilitates the measurement of an individual's progress and is the basis of the DALE system.

The DALE would be used appropriately by direct-care personnel in programs serving adults with seriously handicapping conditions; however, the system is suitable for use with school-age children as well. Although in this reviewer's opinion, additional scales should be included before using the system in educational or vocational domains, the skills assessed by the present system represent, to a large extent, those necessary for daytime and residential living.

The manual describes a goal planning package that illustrates the way in which the inventory can be used by individuals in the human service system for planning and implementing a program appropriate to the needs of each individual client. The package description includes four forms: the General Strengths Form, the Goal Plan Form, the Participation Agreement, and the Goal Plan Implementation Schedules.

Form 1, the General Strength Form, includes an area for recording "what client is good at," abstracted from the general strengths page included in the inventory booklet. The form also provides space for listing "people willing to be involved," but no area for recording the situations in which they are willing to become involved. Finally, the form allows "reinforcement preferences" to be listed. This information is taken from the column on the inventory booklet's general strengths page, which lists what the person likes to do, and from suggestions made by the client. To this reviewer, the need for two similar forms, the General Strengths page in the inventory booklet and the General Strengths Form in the manual's goal planning package, is unclear.

Form 2, the Goal Plan Form, offers space for creating objectives; developing procedures, including reinforcers and criteria; and recording dates. The term "objective," in this instance, means a subgoal used to reach the final objective or to pass the specific skill item. Presumably, the specific items chosen are determined by staff discussion. The example given in the manual is "shampooing hair." The sample form presents testing, rather than teaching, procedures. One must assume that the sample is hypothetical, given that 1) the two listings of the "final objective" are not the same, 2) the first objective is at a lower level than the baseline, 3) the criterion does not necessarily agree with the objective, and 4) the order of objectives does not seem reasonable; that is, gestural prompts are apparently given independent of physical and verbal prompts, but would be ineffective if the client had closed his or her eyes in order to shampoo. Physical prompts appear to be given independently if verbal and gestural cues are not effective. The only reinforcer, apparently, is the phrase, "Good shampooing, Dale," which is given if the client thoroughly shampoos all three areas of the hair (middle, then right, then left), which exceeds the final objective. There is no mention of shampooing the front and back of the hair. The verbal prompts, if they primarily concern the area to lather, include very sophisticated cognitive loadings, which probably is not the best use of language in this situation.

Form 3 is the Participation Agreement. It follows Form 4 in the manual sample. The target behavior, or final objective, again, is not the same as the final goal, which also is written. Target dates also are listed. All the information included on the form duplicates the content of Form 2. The sole difference is the statement, "I understand the above agreement and agree to comply to its terms," followed by the signatures of the resident and the staff. Because there are no "terms" on the form, the signatures easily could be affixed to Form 2, eliminating the need for Form 3.

Form 4, the Goal Plan Implementation Schedule, apparently lists each observation. The date, objective, and prompts needed are written in and initialled by staff members. The implementation schedule, then, is apparently a schedule of tests or trials rather than an instructional program.

Scoring is based on observation and knowledge of the individual client. Values clarification seminars are recommended for users to help minimize biasing, and regular training is encouraged for consistency in assessment. In addition, users are directed to note the basis of the scores they record; that is, observation, impression, certain number of trials, report of a third-party informant, and so on.

The DALE rating system is described on pages 27-30 of the manual as well as in the inventory booklet for each level. Two types of scores are used to rate every item on the inventory, quantitative (how often) and qualitative (how well). Speed is not considered.

The quantitative score, indicating approximately how often the skill is exhibited, implies that the scoring is more precise than it actually is. Each rating is defined by the number of correct trials out of 10, implying that 10 trials are performed for each item, an amount that is neither practical nor necessary, with the exception of certain particular items that are being monitored closely. Each item can be assigned one of six quantitative ratings per rating session:

1. 1—demonstrates the skill less than half the time (less than 5 times out of 10).
2. 2—demonstrates the skill approximately half the time (5-8 times out of 10).
3. 3—demonstrates the behavior almost all the time (9-10 times out of 10).
4. N/A—does not apply to this individual (no qualitative rating).
5. N/O—never observed (no qualitative rating).
6. INCAP.—incapable of responding (no qualitative rating).

The qualitative score indicates how well the desired skill is approximated, and the reminders needed, compared with the performance of most other people in the community. If the individual does not demonstrate the behavior, no qualitative score is recorded, of course. Although the qualitative score is necessarily subjective, it is doubtful that unreliability is a concern because each of the scores is broad in its interpretation. It is recommended that when the individual's competency is questionable, the lower rating be used. This allows time for incorporation of the skill into the repertoire. One of three qualitative ratings is assigned for each item every rating period:

1. A—poor or inappropriate approximation of the desired skill.
2. B—fair approximation of desired skill or good approximation but with frequent reminders.
3. C—excellent approximation of desired skill; few, if any, reminders required.

A quantitative rating of "3" and a qualitative rating of "C" are necessary to indicate competency for any item.

Graphing is used to provide a visual picture that summarizes progress. First, 3C ratings are plotted on a separate graph for each subscale (called "area" on the participation agreement form); that is, 21 subscale graphs are prepared for Level I or 37 subscale graphs are prepared for Level II. Then, 3C ratings are plotted on graphs for each priority area or category; that is, five more graphs for Level I or five more graphs for Level II. Finally, 3C ratings are plotted on a full-scale graph for the appropriate level to reflect the total number of 3C ratings.

Although the type of graphing suggested is probably easier to complete than percentages, the graphs are somewhat misleading, making their interpretation difficult. The items are by no means of equal difficulty in that some include several substeps; nevertheless, each item is afforded equal space on the graphs. The diversity in number of items per subscale is not considered in graphing; all subscale graphs are the same size. Spaces on the graphs are not equivalent; one space on a subscale graph may represent from one to four competencies. In addition, the graph is not the true visual representation of mastery it is intended to be. Because the total number of items in a subscale may be located from halfway up the vertical axis to the top, the vertical representation of the number of "3C" ratings does not indicate the extent to which that area has been mastered.

The system, as presented in the four forms, does not include the instructional component (plan implementation). Instead, it focuses on the plan for evaluation of the skill. The intervention is implied, and although space for commitments to it are included, space is not provided to describe the intervention plan; that is, assessment and programming are not distinct. References are made to "individual habilitation plans" and the "individual treatment plan." The manual lists, among 10 components of such a plan, "specific treatment procedures and designated staff responsibilities (including names)" (Barber et al., 1986a, page unnumbered; p. 40 assumed); however, space for this information does not appear on the forms presented.

The sample plan presented might have been better chosen. A 35-year-old man is being admitted because he cannot remain at home due to his behavior problems and the illness of his father, yet the plan does not recommend behavior goals. Although the individual is being admitted for pre-vocational service, the plan does not recommend pre-vocational goals, nor does it include community/based goals.

As presented, documentation of observations is unnecessarily cumbersome.

Some forms do not include a space for the date. Because several forms overlap, information could be contained in fewer pages. This is particularly problematic considering that the use of the entire set of forms is suggested for each specific skill item to be included in the program. Agencies electing to use the system likely will choose to tailor the forms to specific agency situations.

Technical Aspects

The manual does not present evidence supporting the validity or the reliability of the inventory. The choice of items and grouping of items, as well as the assignment of items to levels, do not seem as adequate as they might be. Some items are very specific; others are complex. Some basic items are followed or preceded by items representing quite advanced skills. Furthermore, some daily tasks one would expect to find are not included (chews with mouth closed, makes bed). Many items that are included typically are performed infrequently (cleans walls and ceilings as needed, replaces lightbulbs), and some seem unnecessary (maintains age-appropriate hair style, names items that belong in a home first-aid kit). Still other items seem unreasonable considering the intended population (provides general background information and personal medical history, maintains a record of correspondence related to benefits). Although hearing skills are assessed, including "cleans ears," visual skills are not measured. Except for the item "gestures to communicate," expressive language skills depend on the individual's hearing and speech abilities. Many items listed under the self-concept category in Level I require higher-level language and cognition (expresses feelings to others).

Items in the subcategories are not parallel; some items appear to be the result of task analysis, and others appear to be discrete items presented in order of difficulty. In addition, items are not of comparable worth nor of equivalent difficulty (in equal intervals). Subcategory items considered as a cluster do not add up to equal satisfactory performance in the area; that is, the subcategory does not contain all the items necessary for success in the area.

Several series of items are included on both Level I and Level II. Some items that one would expect to find on both levels, in that they are borderline or critical skills, appear on only one level. Some items requiring advanced skills appear on Level I, and items assessing basic skills appear on Level II. As a result, identifying an individual's appropriate level for assessment and programming is difficult indeed. An examiner might, then, either use only one level, to the exclusion of some very appropriate objectives, or both levels. For example, Level I does not include conventional grooming items except for "applies deodorant to appropriate areas," but Level II includes 47 items dealing with the menstrual cycle, shaving skills, grooming, and foot and nail care. Items related to allowing staff to provide care (shave, comb and wash hair, trim nails, etc.) are included in Level II but not in Level I. Although Level II contains six items concerning toileting, it does not contain the item "maintains bladder and bowel control night and day," which is on Level I. The gaps in the inventories and the lack of item balance place a very heavy responsibility on the staff to identify priority items that may not be included on the selected level and to disregard lower priority items that the individual might not pass.

Critique

The technical flaws in the inventories and the system as a whole greatly limit their usefulness. The booklets contain many typographical errors, for example, "feld" for "felt," "socre" for "score," "pesonal" for "personal" and grammatical errors ("less" for "fewer," "completed" for "met," "cleans self thoroughly following elimination with toilet tissue"). These are distracting to the user and devalue the system. The two instrument levels are considered to be separate, yet many individuals score on both scales, making the selection of a level difficult.

The terms used in the manual are confusing; for example, "separate charts be kept for each rating period" (Barber et al., 1986a, p. 13) refers to ratings being recorded in separate columns on the inventory; "both rating and charting be completed on a semi-annual basis" (Barber et al., 1986a, p. 13) refers to rating and graphing, although "chart" is the term used for the totality of the inventory marked with ratings for each item. The manual refers to quantitative and qualitative scores used to rate items, but the inventory booklets refer to quantitative and qualitative ratings under rating systems, under rating scales.

Many of the headings used on the forms and in the manual are not consistent and do not seem to describe the intended use of the forms. The strength/need page in the inventory booklet is headed "Specific Strengths," and a "general strength/ need list . . . to summarize major categories of strength and need" (Barber et al., 1986a, p. 12) is entitled "General Strengths," implying that it is used for recording strengths only. The same page contains a column headed "People Willing to be Involved in Implementation," but does not indicate what is to be implemented. The term "skill," rather than the manual's use of "skill level" seems to better describe items because "skill level" is related more to "how well" and does not seem to connote "how often."

The suggested implementation guidelines presented in the manual should be regarded more as suggestions and less as guidelines. In the discussion of rewards, for example, the manual states that "once a reward item is adopted, it should not vary (Barber et al., 1986a, p. 15). In other words, "That's good, Dale," should not be substituted for "Very good, Dale." Strict adherence to these guidelines, then, does not allow one to change tone of voice, facial expression, or gesture, as these changes are at least as significant as changes in wording. In reality, the constellation of social reinforcement probably is preferable because it is much more representative of the community-based reinforcement system.

The manual's bibliography is extensive; however, the most recent entry is dated 1977, 9 years prior to publication, and many entries do not include the date.

Although the inventory suggests areas of importance in programming for the developmentally delayed, technical changes should be made prior to its adoption.

References

Barber, G., Mannino, J., & Will, R. (1986a). *Developmental Assessment of Life Experiences: Instruction Manual and Guidelines for Implementation* (5th ed.). Erie, PA: Barber Center Press.

Barber, G., Mannino, J., & Will, R. (1986b). *Developmental Assessment of Life Experiences: Level I* (5th ed.). Erie, PA: Barber Center Press.

Barber, G., Mannino, J., & Will, R. (1986c). *Developmental Assessment of Life Experiences: Level II* (5th ed.). Erie, PA: Barber Center Press.

Janet A. Norris, Ph.D.
Assistant Professor of Communication Disorders, Louisiana State University, Baton Rouge, Louisiana.

DIAGNOSTIC ANALYSIS OF READING ERRORS

Jacquelyn Gillespie and Jacqueline Shohet. Wilmington, Delaware: Jastak Associates, Inc.

Introduction

The Diagnostic Analysis of Reading Errors (DARE) is intended to measure an individual's ability to transcode auditory sequences (vocal or subvocal speech sounds) into visual-motor correlates (written words). The measure of transcoding used by the DARE is the identification of a correctly spelled word from four alternatives. The manual (Gillespie & Shohet, 1979) states that error patterns in transcoding on the DARE can be used for identifying language-related learning disabilities, providing indications of the nature of the disability, and providing individual diagnostic information.

The authors of the test, Jacquelyn Gillespie, Ph.D. and Jacqueline Shohet, Ph.D. are licensed educational psychologists. Both have served as school psychologists and participated in research with the learning disabled. As a faculty member at the California Graduate Institute, Dr. Gillespie was active in research and dissertation development. Dr. Shohet has a background in test development and research with the Chicago Civil Service Commission, the New York Board of Education, and the Institute for Developmental Studies in New York City.

The authors' work on the DARE was influenced and guided by Drs. Joseph and Sarah Jastak. Jastak and Jastak (1978) defined reading and spelling as a transcoding process between visual-kinesthetic symbols and correlated sound sequences. Disabilities in reading and spelling were defined as impairments in the transcoding process, primarily between the visual and auditory modalities. This definition implicates dysfunction in the transcoding process as central to learning disabilities. The DARE was designed, therefore, to identify dysfunctions in transcoding.

The original research on the DARE was reported by Gillespie, Hays, Retzlaff, and Shohet in 1972. Initially named the Diagnostic Spelling Test, the instrument established significant differences among groups of educationally handicapped, educable mentally retarded, and general populations of high school students. Following the original research, the test was revised in format and intended populations were extended to include average and some problem readers, including junior high school students, high school students, community college students, limited-English-speaking populations, and severely aphasic and neurologically handicapped groups. No specific mention of learning disabled readers is made in the manual. The test in its current format was published in 1979.

The original analysis of the DARE was performed using groups of Southern Cal-

135

ifornia suburban high school students who displayed somewhat above-average overall achievement. A relatively small (18%) minority population was represented. Next, performance data were obtained on similar populations of upper-elementary and junior high students. Attempts to obtain generalized norms involved extension to four California high schools that drew students from a variety of multicultural inner-city and suburban areas, a "continuation" high school for students who were unsuccessful in the regular high school, and a multicultural junior high school. Norms were established only on students in regular classrooms; no data representative of learning disabled children or other poor readers was formulated. In addition, no rural populations or populations outside of Southern California are represented. Although the DARE had been administered with adequate validity to 5th- and 6th-grade students during its development, the word list was judged to be difficult for this group, resulting in potential motivation and attention problems affecting test outcome.Revisions of the DARE apparently were minimal during its development, involving primarily an adaptation to a machine-scorable form.

Stimuli for the DARE is comprised of the 46-word spelling list from the Level II Spelling subtest of the Wide Range Achievement Test (Jastak & Jastak, 1978). This spelling list contains relatively difficult words, appropriate to individuals of junior high ability or above. The one-page response form follows a multiple-choice format. The examinee is presented with four alternative spelling choices for each target word. Each of the alternative spellings for a target word is numbered and enclosed within a square, clearly distinguishing it from the other choices. The words are arranged according to increasing levels of difficulty, from three-letter, phonetically regular spellings to multisyllabic, orthographically complex words. None of the foils represent appropriate phonetic spellings of a given word in that all contain patterns that are characterized as either sound substitutions, sound omissions, or letter sequence reversals.

Four scores, including a total Correct score and three error scores, Sound Substitution, Omissions, and Reversals, are obtained for each examinee. Normative information includes standard scores and percentiles normed according to both age and grade level. Additionally, an individual profile may be developed for examining the four scores in an interpretive relationship.

Practical Applications/Uses

According to Gillespie and Shohet, the DARE systematically identifies learning disabled individuals in Grades 6-12, as well as community college. Because it is simple to administer, can be group administered, and is machine scorable, it is a relatively efficient method for screening groups of students for potential learning disabilities. In addition to identifying individuals with learning disabilities, the authors state that the patterns of errors can provide diagnostic information and suggest remedial strategies. Furthermore, as stated previously, performance on the DARE can be compared to performance on the Spelling subtest of the WRAT, purportedly differentiating specific difficulties with the transcoding process from more generalized difficulties that tend to be measured by the WRAT. The authors also state that the DARE can be used to assess minority students for learning difficulties because the word identification task is free from cultural bias.

According to the manual, the test is designed for use by classroom teachers, reading specialists, psychologists, or others interested in identifying learning disabilities. Reportedly, the test can be used efficiently to identify learning disabled individuals or to quickly yield information about the incidence of reading and spelling disabilities within a population. It also is claimed to be useful as part of a diagnostic battery. Because the two tests use the same spelling lists, the DARE yields direct comparisons of transcoding abilities with actual spelling proficiency when used in tandum with the WRAT. Because the Correct score provides a measure of overall transcoding proficiency and the three error scores provide a measure of difficulty with different aspects of the auditory-visual integration process, test results reportedly provide insights into an individual's deficits and yield information useful to instructional planning.

The examiner's manual accompanying the DARE is complete and clearly written. Information is provided on its theoretical basis, test construction and development, test administration and scoring, and implications for remediation. Although case studies are presented to exemplify profiles, they are minimally useful because 7 of the 10 represent non-native English speakers and an eighth profiles a hearing-impaired subject. Instructions for administration and scoring are straightforward and simple. Administration of the test is nonrestricted, although the authors caution that the examiner should exhibit standard English speech patterns. The test may be administered individually or in a group testing situation such as a classroom. The examiner explains the multiple-choice format of the test, refers those taking the test to the example provided on the response form, and then administers the test by presenting each target word in isolation and within the context of a sentence. Using a number 2 pencil, the examinee fills in the circle next to the perceived correct spelling. The test is untimed, generally requiring 20-30 minutees for administration, and stimulus words may be repeated. Scoring, using a template, can be completed in a few minutes. The test also can be machine scored. In short, the test can be administered and scored appropriately by anyone generally familiar with principles of testing.

As stated previously, the test yields a total correct score and sound substitution, omission, and reversal error scores. For interpretive purposes, the manual states that the Correct score reflects overall task efficiency, and the norms providing comparative data are based on this score (Gillespie & Shohet, 1979, p. 6). The manual further states that normal transcoding ability is reflected in Correct scores greater than 40; disabling difficulties are reflected in Correct scores below 30 (Gillespie & Shohet, 1979, p. 6). The derivation of these particular cutoff points is not clear. The authors note that although the norms demonstrate an increase in performance with age, the transcoding process measured by the DARE appears to be well established by adolescence, with many junior high examinees receiving perfect scores. The examiner, therefore, is directed to consider the Correct score relative to the norms and to the mastery level of 40 correct responses.

Further interpretation is recommended using the three error scores. Error scores in a normal profile should be small in number and randomly distributed across the three error categories. Deviations from this profile represent weaknesses in transcoding abilities. The manual states that a pattern of sound substitution errors reflects difficulty in making judgments concerning English sound patterns or lack of awareness of the sounds in spoken language (Gillespie & Shohet, 1979, pp. 6-7). A pattern of omissions is associated with a fundamental confusion in perceiving and manipulating sound and written symbols and is claimed to be most closely associated with learning disabilities affecting reading and writing. This error category is reportedly low among adequately functioning readers and reliable among disabled readers, although evidence in support of these differences is not clear from the manual. Reversal errors are said to be most common among severely learning disabled individuals and typically are found in conjunction with a high level of omission errors. Based on the assumption that differences in patterns are exhibited between normal and disabled readers, the manual claims that individual profiles can yield valuable diagnostic information (Gillespie & Shohet, 1979, pp. 6-8).

In summary, raw scores for correct overall responses are converted to standard scores and percentiles. The number of errors within the substitution, omission, and reversal categories yields further diagnostic information. *If* the examiner agrees with the theoretical principles on which the test is based, interpretation of the results is clear from the instructions provided by the manual. Interpretations include an analysis of error patterns, with errors of omission and/or reversal indicating confusion or disorganization of sound sequences typical of learning disabilities. Furthermore, the standard scores used by the DARE are comparable to those used by other instruments, including the WRAT and the Wechsler intelligence tests. Comparison of relative performance across these measures, therefore, can be made in an effort to determine strengths and weaknesses or to establish learning style. The interpretations of results as discussed by the manual, however, are not exhaustive and can be questioned if they are viewed from an alternative theoretical perspective.

Technical Aspects

Although grade-level norms for the DARE are available for Grades 7-12 and community-college students and age-level norms range from 12-18 years, the characteristics of the normative population are presented ambiguously in a variety of places within the test manual. All normative data was obtained from a small number of schools in Southern California. Only individuals from suburban and inner-city schools were tested. Rural areas are not represented. Because only individuals participating in classroom situations were included, special populations such as the learning disabled (whom the test is purported to identify) are not included systematically in the normative sample. No demarcation of subject characteristics is provided, although tables indicate that over 1,583 students between the ages of 12 and 18 were tested, as well as an additional 290 community-college students. A greater number of subjects were included in the 12-14-year age range than in other groups; however, at least 100 subjects were represented in each of the 15-18-year age levels. The ratio of males to females tested is approximately equal and norms are provided separately. The manual mentions the inclusion of a large population of Mexican-American students, but does not specify the proportion of other minority groups that were tested.

The DARE purports to be culturally fair because it is based on sound patterns and their written equivalents and is not dependent on knowing word meaning or on interpretation of information gained from experience. Seperate norms reflecting the socioeconomic status of participants are not provided, although a table comparing performance by suburban and inner-city groups is given. Although special populations are not included in the normative sample, studies reportedly have been conducted using the DARE with a variety of special groups. Specific information concerning these studies is not provided, however. The authors recommend caution in using general norms when the test is administered to special populations and suggest that local norms be formulated.

Some limited measures of reliability are reported for the DARE, showing adequate results only when the test is group administered. Split-half reliability was established on 39 high school seniors by comparing correct responses on odd-numbered items with those of even-numbered items, resulting in a total test reliability coefficient of .67 (.6 is considered satisfactory for group means, .9 is required for individually administered tests). Test-retest reliability was measured using three groups of 10th- and 11th-grade remedial English classes, a group selected due to its likeliness to present greater variability in performance than a group of high achievers. Under conditions of both immediate retest and retest after a 3-week interval, the correlations were satisfactory for group means (.78 and .81, respectively). No reliability studies were reported for individual administration, which is important given the test's diagnostic claims. Furthermore, the reliability data presented was based on total correct scores; sketchy information was provided about the reliability within error categories on which diagnostic claims are based. The manual provides group means related to an item analysis conducted on the responses of the 39 high school seniors. The analysis indicated that errors were randomly distributed across the three error types (i.e., substitution, omission, and reversal) and that specific words generally elicited the same type of error across subjects; the reliability of these findings was not reported, however.

Validity of the DARE was also established on group, rather than individual, data. A measure of differential prediction, or the test's ability to discriminate among groups, was obtained by comparing test performances of average, learning disabled and/or behaviorally disordered, and educable mentally retarded students in Grades 9-12. Mean differences between the groups were significant at the .01 level of confidence. Similarly, performance on the DARE differentiated between community-college students who were enrolled in college-level English classes, basic English classes, speed-reading courses, and remedial reading courses. The DARE claims construct validity in that there is a relatively even increase in Correct scores from grade level to grade level in Grades 5-12, as well as a generally random and even distribution of error scores among the three categories. This even increment is consistent with a gradual increase in academic proficiency and reading achievement expected with successive years in school. Additional evidence of construct validity was claimed on the basis of statistically significant differences reported between suburban and inner-city groups of junior high school subjects, differences that were consistent with achievement discrepancies associated with these two groups.

Correlational studies also were conducted to establish criterion- related validity. High correlations, ranging from .72 to .85, between the DARE and WRAT Spelling subtest were obtained in an analysis of 1,100 cases, including groups from junior high, high school, and community-college populations. Among community-college subjects, additional comparisons were made with Paragraph Comprehension and Vocabulary scores on the McGraw-Hill Basic Skills System tests (Raygor, 1970). Consistently low correlations were obtained. The authors claim a support of validity in that the DARE measures transcoding, a function that is different from comprehension and vocabulary knowledge.

Critique

The Diagnostic Analysis of Reading Errors defines itself as a test of reading, a procedure for identifying learning disabilities, and an instrument with diagnostic value. All of these claims are overstated grossly and are unsupported by information provided within the manual. The DARE initially was titled the Diagnostic Spelling Test, and the manual fails to explain any rationale for its renaming. Even by the authors' own, narrow definition of reading as "the process of transcoding a series of visual-kinesthetic symbols into vocal or subvocal sound sequences" (Jastak & Jastak, 1978, p. 65), the DARE is not a reading test. The respondant is never required to "read" even words in isolation; the words are read by the examiner. Even if transcoding is a legitimate construct, response to words spoken by someone else is a different process than word recognition. More to the point, by its definition, the test limits and oversimplifies the notion of reading. Reading is a communication process whereby meaning is extracted from print. It involves processes far more complex than translating sounds to visual symbols. Furthermore, it is invalid to assume that "transcoding" can be isolated as a discrete process, separate from context and meaning (Vellutino, 1982). Additional contradiction to the notion that the DARE is an assessment of reading is provided by the manual, for example, the low correlations of the DARE with reading measures on the McGraw-Hill Skills System tests (Raygor, 1970) and unpredictable correlation with the WRAT reading subtest (Gillespie & Shohet, 1979).

As an instrument for identifying learning disabilities, the DARE falls similarly short of its claims. The manual presents different patterns of errors, such as omissions and reversals, as being characteristic of individuals with learning disabilities but cites no support in the literature nor in the normative data generated by the test. Although the test purports to be designed to identify individuals with language-related learning disabilities, this population was eliminated from the normative sample. No data are provided to suggest that learning disabled individuals do, in fact, perform as the authors suggest. The study that was cited to show that the DARE differentiated among different populations merely compared group means for statistical differences. A more convincing proof would have been presented if a discriminant function analysis had been used to demonstrate that individual subjects were categorized correctly on the basis of their performance on the DARE. Furthermore, the study did not demonstrate that the profiles of the learning disabled subjects were different from other special populations, such as the educable mentally retarded, or that the learning disabled subjects actually produced a significantly higher percentage of omission and reversal errors. The manual also states that the DARE can be used for screening groups of individuals for learning disabilities, but no studies were reported to support the validity of this contention. Similarly, there is no evidence to support the stability of the performance of learning disabled students, either on the total correct score or on the profile of errors.

Problems with the test construction are not limited to issues dealing with the learning disabled. There is no reliability or validity established for the test's use as an individually administered instrument, although the manual frequently discusses comparing an individual's performance on the DARE to WRAT, WISC, and other scores. The reliability coefficients reported for group means are not sufficiently high for the standards of an individually administered test. No data support the correct placement of an individual into a specific educational group such as gifted or mentally retarded on the basis of performance on the DARE. The generalizability of the norms to any group outside of high achieving, middle class, southern California students is highly questionable because the normative population is not representative of most geographic, socioeconomic, cultural, ethnic, or ability groups. The limited data provided on special populations does not describe any of the general characteristics of the individuals, such as IQ or reading level. Contradictory statements are made relative to cultural fairness. The manual states that the test is culturally fair because transcoding is a skill that can be tested noncontextually and, therefore, responses are not dependent on experience and language differences (Gillespie & Shohet, 1979, pp. 14-15). However, the manual later provides studies demonstrating differences in performance between urban and inner-city students, indicating that influences such as culture, experience, and language do affect task performance (Gillespie & Shohet, 1979, pp. 30-31).

Even if one overlooks some of the test construction problems with the DARE, the Test Interpretation and Case Studies section provides additional reason for concern. The error profiles are treated as if they represent meaningful and important data upon which remedial intervention can be based. Even if these error categories were shown to be reliable and valid, they would still only represent tasks that correlate with reading, and correlations do not equate to a causal relationship. Recommendations directed at teaching to these "strengths and weaknesses" as a remedial strategy is not supported theoretically or by any data provided in the manual. The Case Studies section appears to have little to do with the stated purpose of the test in identifying and planning remedial instructions for learning disabilities, in that 7 of the 10 case studies are based on non-native speakers of English and an eighth profiles a hearing-impaired subject. The recommendations that are made seem to lack systematicity or a coherent theoretical foundation, resulting in general suggestions and a splinter-skill approach to teaching reading.

In summary, the DARE is an instrument that makes diagnostic claims that are unsupported, evidenced by the inherent problems with its theoretical premises, test construction, and supporting research. It is more representative of a spelling test than a reading test and is unproven in its stated ability to identify individuals with learning disabilities. Therefore, it provides little assessment information that could not be obtained better by using other instruments or procedures.

References

Gillespie, J., Hays, D.G., Retzlaff, W.F., & Shohet, J. (1972). The Diagnostic Spelling Test: A modification of the Wide Range Achievement Test, Spelling (Level 4). *California School Psychology, 19,* 26-31.

Gillespie, J., & Shohet, J. (1979). *Diagnostic Analysis of Reading Errors manual.* Wilmington, DE: Jastak Associates.

Jastak, J.F., & Jastak, S. (1978). *The Wide Range Achievement Test—Manual (rev. ed.).* Wilmington, DE: Jastak Associates.

Raygor, A.L. (1970). *Examiner's Manual, McGraw-Hill Basic Skills System.* Monterey, CA: CTB/McGraw-Hill.

Vellutino, F.R. (1982). Theoretical issues in the study of word recognition: The unit of perception controversy reexamined. S. Rosenberg (Ed.), *Handbook of applied psycholinguistics: Major thrusts of research and theory* (pp. 33-197). Hillsdale, NJ: Lawrence Erlbaum Associates.

Patrick Groff, Ed.D.

Professor of Education, San Diego State University, San Diego, California.

DIAGNOSTIC AND ACHIEVEMENT READING TESTS

Harvey Alpert and Roger Trent. Cleveland, Ohio: Modern Curriculum Press.

Introduction

The publisher of the Diagnostic and Achievement Reading Tests (DART), Modern Curriculum Press (MCP), claims the tests are designed to assess children's word-attack skills, skills that are essential for independence in attacking, recognizing, and pronouncing words. The DART is made up of 66 subtests, or "sections," presented in 10 different test booklets. There are four booklets (27 subtests) in Series A; three booklets (17 subtests) in Series B; and three booklets (22 subtests) in Series C. The subtests range from asking children to identify, from among an array, the picture that corresponds to a stimulus picture in Subtest 1 to asking children to identify from an array the dictionary syllabication of a stimulus word in Subtest 66. Other subtests assess children's knowledge of words, letters, and phonemes and phoneme-letter correspondences, as well as their ability to read silently and comprehend sentences, contractions, synonyms, antomyms, and homonyms.

Published in 1977, the DART was created by the test and research bureau of Modern Curriculum Press. It is claimed that during its creation many (unnamed) experts in the fields of reading and testing were consulted. Harvey Alpert, Ed.D., Professor of Reading and Coordinator of Diagnostic Services, Hofstra University, and Roger Trent, Ph.D., Test Development Coordinator, Ohio State Department of Education made substantial contributions to the DART.

The publishers of the DART do not explain the particulars of its developmental history. It appears, however, that the DART underwent no pretesting prior to its publication, indicating that no analysis of the items' relative difficulty to each other or to the reading ability of children was conducted. There is no explanation as to whether attempts were made to discover how discriminatory its various items were. One does not know, then, the relationship of the various items to the total scores of children on this test.

One must presume, therefore, that this edition of the DART is the first of its versions. Nothing indicates it has been published before 1977. The MCP 1987 catalog does not mention that any other forms of the test are available, for example, for the visually or hearing-impaired, for speakers of foreign languages or nonstandard English, or for slow learners.

Each of the 10 test booklets begins with an introductory section that provides information about the purpose of the test and the audience for whom it is intended.

143

This section serves as the manual for the administration of the DART. Specific directions describing how the subtests should be administered and scored and how the scores should be interpreted also are included. The introduction is followed by the subtests, each of which is printed legibly on a single-sided page. An answer key for the subtests in each booklet is printed at the end of the booklet.

Each of the 66 subtests has 10 items. Almost all these 660 items follow this format: a stimulus is presented to the left of a horizontal array of pictures, geometric forms, words, or sentences; the subject examines the stimulus and, following a given set of directions, chooses one item from the array. For example, for one item, the administrator says, "Draw a ring around the other picture that is exactly like the first picture." The test authors claim that the items in the DART ascend gradually in order of difficulty. In practice, however, many of the items in Series B of the DART are easier for a child to complete successfully than those in Series A, which require the ability to read silently and comprehend sentences.

The directions for administering the DART indicate to the administrator that it can be used with any children who have received instruction in whatever skills objectives the teacher selects to be tested. For example, after teaching children to match pictures or to syllabicate words according to the dictionary, a teacher supposedly could use the DART to determine how well the children had learned that particular skill. Usually the age range between children learning to match pictures and those being taught to syllabicate words is wide. The age group for which the DART is intended, therefore, might well include individuals in Kindergarten through Grade 6. Such a large grade span indicates that the difficulty range of the items is equally as large. It obviously is far more difficult, for example, for children to read silently and comprehend a sentence than it is to match pictures.

Practical Applications/Uses

The DART is intended to give teachers the opportunity to test certain word recognition skills that they have taught their students. It purports to inform teachers whether the pupils need additional instruction, or reinforcement, in these skills or have mastered them. It is ambiguous, however, whether the DART is designed exclusively for use by teachers *after* they have tried to teach certain reading skills. If the DART can be effectively used to diagnose skill deficiencies, as its authors claim, it could be employed to determine whether children need to be taught certain skills at all.

If the DART accomplishes what its authors purport, its application, as stated previously, would be appropriate for a wide range of students in Kindergarten through Grade 6. The DART may be administered to individual children or to small groups of children. The test booklets provide subtest forms and record sheets for only one subject, however. These forms are perforated, so they can be easily removed from the test booklet. Thus, they could be readily duplicated. However, the publisher of the DART indicates that any duplication of the test may be done only with the written permission of the publisher.

It is apparent from the administration directions provided for the DART that it could be administered successfully by classroom teachers. No specially trained psychometrist is needed. These instructions tell the teacher exactly what to say to subjects to prepare them to take the test. Before a child begins a subtest, he or she is given a sample item on which to practice. The teacher also is advised to review with subjects at that time the basic aspects (format, procedures, and special vocabulary) of the subtest.

According to the test publisher, the sequence of the 66 subtests is not intended to be adhered to rigidly. Teachers are directed simply to choose those objectives and accompanying sections that are, in their judgment, appropriate for individual children. However, the DART does not explain which subtest to administer when more than one series of subtests is appropriate for assessing a particular skill.

The numbering of the subtests, unfortunately, is unnecessarily cumbersome. Instead of numbering the subtests 1 through 66, the DART arranges them in three series, A, B, and C, each containing either 3 or 4 booklets consisting of various numbers of subtests. The first three subtests of test unit A-1 are labeled as follows: Test A-1 Section A, Test A-1 Section B, Test A-1 Section C. Confused? An administrator of the DART would be as well. This overly elaborate numbering system makes it exceedingly difficult to be sure which subtest one is administering.

The scoring of the DART is very simple. A subject answers each of the 10 items per subtest either correctly or incorrectly by marking one of its multiple-choice alternatives. One point is earned for each correct answer, for a maximum score of 10 points per subtest. Because there are only 10 items per subtest, the time needed for scoring is minimal. The DART is scored by hand by the test administrator, who records the subject's score at the bottom of each subtest. These scores then are transferred to a cumulative record sheet provided in the test booklet. Machine and computer scoring do not appear to be compatible with the DART's format.

Correct scores ranging from 0-6 on a scale of 10 are said to be *weak*, meaning that the child should be given additional instruction in the given skill. Scores of 7-8 are *marginal*, indicating that the child should be given opportunities for reinforcement work. Scores of 9-10 are considered *proficient*, meaning that the child has undoubtedly mastered the skill. Administrators of the DART are advised that there is no total test score, and no attempt should be made to compute one.

Because the DART is not norm-referenced, there is no true way to know whether a subtest score is one that many, most, or few children of a given age group would make. The interpretations the authors assign to scores are wholly subjective and arbitrary. Teachers are directed to accept without question the judgments of the DART's authors that a score of 6 on a subtest means the child's performance is weak, that a score of 7 means it is marginal, and that a score of 9 means it is proficient. Within a range of only 4 points, then, a child's performance may range from weak to proficient.

Technical Aspects

The publishers of the DART do not indicate that any validity or reliability studies were performed on its subtests. The skills measured by the DART were chosen for inclusion, it is said, because they are those considered by most authorities to be essential word-attack skills. The publishers do not explain, however, how they calculated that "most" reading experts agree with the contents of the DART. They do not name the many experts in the fields of reading and testing who were consulted in this regard.

As noted, the DART is not norm-referenced. According to its authors, it is "criterion-referenced." They assume that a child must reach a 90% accuracy level in a subtest to meet the test's criterion of success. Unless a child scores 90% on a subtest (9 out of 10 items correct), he or she is not ready for further instruction in new skills. No examples of these "new skills" are offered. The last subtest of Series A (28th in the group of 66) requires the child to read silently and to comprehend sentences. What "new skill" should the child who meets the criterion of 90% accuracy on this subtest be taught? Sentence reading and comprehension appears to be the most difficult of all the tasks in the DART.

Other questions arise at this point. Where is the material for the additional instruction that is needed to meet the 90% criterion to be found? How long should this additional instruction continue? Should not a second different subtest of silent reading and comprehension be administered when this additional instruction ends? Will not the reliability of the first subtest have been compromised if a child is tested with it a second time?

Critique

One of a teacher's most pressing problems in phonics teaching today is how he or she can know for sure whether pupils have learned certain bits of phonics information sufficiently well that they then can be taught the next aspect of decoding. There are several weaknesses in the DART, beyond those already alluded to, that prevent it from being the solution to this vexing question.

The empirical evidence on the relationships between the test items presented in the DART and children's word recognition abilities suggest that some of the 66 subtests of the DART, and/or parts of some of them, are not closely related to the reader's independence in attacking, recognizing, and pronouncing words. These experimental findings do not support spending the money, taking the time, or making the effort necessary to test certain of the skills that are measured by the DART.

For example, there appears to be no convincing evidence that one needs to teach or test for children's abilities to match pictures and geometric forms (Mason, 1984; Groff & Seymour, 1987). It is now apparent that instruction in word and letter recognition tasks are better preparation for children learning to read than is instruction in matching pictures and geometric forms. Users of the DART are advised to avoid the items of the DART that measure these abilities.

The DART directs children to listen to the initial or final speech sounds of words and then to indicate which of these words has either the initial or final sounds that correspond to the initial or final phonemes of a stimulus word. A more important and fundamental phonics skill, however, and one that is not tested in the DART, is the ability to segment spoken words into phonemes and then to blend these isolated phonemes so that recognizable spoken words are produced (Weaver, 1978; Johnson & Baumann, 1984). This phonics skill has more to do with learning to decode written words than does an awareness that some words begin and end with the same phonemes.

The DART sometimes makes an issue of testing children's understanding of the nomenclature of phonics. For example, it asks children to name which word in a set of words has the "short" sound of *e* in it. Good phonics programs deemphasize the teaching of phonics nomenclature. Instead, they teach childen that *net* and *neat* and similarly spelled words have different vowel sounds and that the spellings given these vowel sounds can be used to predict how the words will be pronounced. The DART does better in this respect when its instructions direct the administrator to say "coat" to children and then ask them to decide whether "cat," "coat," "cot," or "cut" should be circled.

The DART offers no clear-cut rationale for the arrangement of the word-attack skills that it presents to be tested. This shortcoming helps explain why the 66 subtests in the DART do not have an understandable hierarchy; that is, why they do not ascend steadily in difficulty from Series A to Series B to Series C. Subtest 7 requires children to match letters. In Subtest 28, children read silently and comprehend sentences. But Subtest 29 switches back to determining the ability to match letters. In addition, tests of children's abilities to distinguish vowel phonemes are found in all three series of subtests.

The DART claims to be a test of word-attack skills, yet it includes measures of children's abilities to read and comprehend sentences. Who were the "many experts in the field of reading" who advised the publisher of the DART that there is no significant distinction between decoding individual written words and comprehending written sentences? This reviewer knows of no reading experts who hold that view.

The DART includes five subtests that assess children's knowledge of dictionary syllabication. There is no evidence in linguistics or in educational research, however, that supports the idea that teaching children dictionary syllabication will especially help them learn to decode written words (Groff, 1981). On the contrary, research suggests that teaching dictionary syllabication is a relatively ineffective use of the time available for instructing children to decode words.

The best that the application of phonics rules can do for the reader is provide an approximate pronunciation of words (Anderson et al., 1985). The DART should, but does not, test children's ability to infer the correct pronunciation of written words after they have applied phonics rules to them in an effort to decode them. Children's ability to infer the correct pronunciation of a word after hearing an approximate pronunciation of it is an important phonics skill that the DART does not measure.

Above all, the DART should, but does not, make clear why it is superior to teacher-constructed tests of the skills it presents. There is no concrete evidence indicating that the DART's subtests are more accurate than the examinations that the teacher can find or deduce from his or her basal reader program.

In summary, the DART is a 10-year-old test that needs revisions or updating in order to bring it in line with the progress and maturing that has taken place in phonics teaching over the past 10 years (Groff, 1986). Phonics instruction is a critical aspect of teaching children to learn to read. The empirical evidence that supports this conclusion has grown stronger over the years (Chall, 1967/1983). An up-to-date, valid, and reliable test of children's phonics knowledge is notably missing from the test market. The DART does not appear to fill this void, however, for the reasons given in this review.

References

This list includes text citations and suggested additional reading.

Anderson, R. C., & committee. (1985). *Becoming a nation of readers*. Washington, DC: U. S. Department of Education.

Aukerman, R. C. (1984). *Approaches to beginning reading*. New York, NY: John Wiley & Sons.

Chall, J. S. (1967/1983). *Learning to read: The great debate*. New York, NY: McGraw-Hill.

Finn, C. E., & others. (1986). *What works*. Washington, DC: U. S. Department of Education.

Groff, P. (1981). Teaching reading by syllables. *Reading Teacher, 34,* 659-664.

Groff, P. (1983). A test of the utility of phonics rules. *Reading Psychology, 4,* 217-225.

Groff, P. (1986). The maturing of phonics instruction. *Reading Teacher, 39,* 919-923.

Groff, P. (1987). *Myths of reading instruction*. Portland, OR: National Book.

Groff, P., & Seymour, D. Z. (1987). *Word recognition: The why and the how.* Springfield, IL: C. C. Thomas.

Henderson, L. (1982). *Orthography and word recognition in reading*. New York, NY: Academic Press.

Johnson, D. D., & Baumann, J. F. (1984). Word identification. In P. D. Pearson (Ed.), *Handbook of reading research*. New York, NY: Longman.

Mason, J. M. (1984). Early reading from a developmental perspective. In P. D. Pearson (Ed.), *Handbook of reading research*. New York, NY: Longman.

Perfetti, C. A. (1985). *Reading ability.* New York, NY: Oxford.

Resnick, L. B., & Weaver, P. A. (Eds.). (1979). *Theory and practice of early reading (Vols. 1-3).* Hillsdale, NJ: Lawrence Erlbaum.

Singer, M. H. (Ed.). (1982). *Competent reader, disabled reader: Research and application*. Hillsdale, NJ: Lawrence Erlbaum.

Weaver, P. (1978) . *Research within reach*. Washington, DC: U. S. Department of Education.

Delwyn L. Harnisch, Ph.D.
Associate Professor of Educational Psychology, Institute for Child Behavior and Development, University of Illinois at Urbana-Champaign, Champaign, Illinois.

DMI MATHEMATICS SYSTEMS INSTRUCTIONAL OBJECTIVES INVENTORY

John Gessel. Monterey, California: CTB/McGraw-Hill.

Introduction

The DMI Mathematics Systems Instructional Objectives Inventory (DMI/MSI) is a set of criterion-referenced tests designed for use with students in kindergarten through eighth grade (K.6 to 8.9). The DMI/MS is designed to identify students' mathematical strengths and specific instructional needs. The results may be used to prescribe appropriate learning activities and materials. The instrument is available on seven levels, A-G.

To develop the DMI/MS, John Gessel, the test author, reviewed the contents of 14 basal math series and 18 curriculum guides, resulting in 250 objectives. Twelve reviewers noted their choices of the 50 most important objectives in the list and indicated grades appropriate for testing, which resulted in 86 tryout objectives (and assigned levels). By analyzing common errors in the parent instrument, the Diagnostic Mathematics Inventory, as well as the Comprehensive Tests of Basic Skills (CTBS), a list of common errors was compounded for use as distractors in the tryout items. The vocabulary level of compiled items was controlled on the basis of *The Living Word Vocabulary* by Dale & O'Rourke. A set of 14 tryout items and 1 sample item was produced for each level of each tryout objective. The items were administered to students in 1982. Each of the seven levels was given to three grades (the target grade, the grade above, and the grade below) except for Level G, which was given to Grades 6-10. Percentage of correct responses, point-biserials to total objective score, and item-response theory fit statistics that were used to determine items that should be deleted or switched to another level. From the remaining items in the tryout pool, two locator tests were assembled that were to be used for student placement into the appropriate test level.

The DMI/MS covers 82 instructional objectives organized into four strands of content, which in turn are divided into categories. The content strands, their categories, and number of objectives are as follows:

I. *Whole Numbers:* Numbers and operations (8), Whole number addition (4), Whole number subtraction (3), Whole number multiplication (3), Whole number division (3), and Estimation in computations (2).

II. *Fractions and Decimals:* Fraction concepts (3), Computations with fractions (3), Decimal concepts (3), Decimal addition/subtraction (2), Decimal multiplication/division (4), Estimation in computations (2), and Computations with integers (2).

III. *Measurement and Geometry:* Standard measurement (4), Measurement of length (2), Measurement of liquid capacity (2), Measurement of mass and weight (2), Estimation in measurement (1), Geometric concepts (4), and Metric geometry (3).

IV. *Problem Solving and Special Topics:* Problem solving with whole numbers (3), Problem solving with rational numbers (3), Problem solving with denominate numbers (3), Problem solving processes (4), Estimation in problem solving (1), Graphs (1), Patterns & coordinate geometry (3), Pre-algebra (2), and Statistics and probability (2).

Some objectives appear in multiple levels but with differing degrees of difficulty. Each instructional objective is tested with four items. Based on responses to the items, a student is classified as "mastered" (3-4 correct), "needs review" (2 correct), or "nonmastered" (0-1 correct).

Two locator tests (Grades K-4 and Grades 4-8) that match the student with the test level (A, K.6-1.5; B, 1.6-2.5; C, 2.6-3.5; D, 3.6-4.5; E, 4.6-5.5; F, 5.6-6.5; G, 6.6-8.9+) capable of providing the most reliable scores are available. These tests are short (Test 1 contains 23 items; Test 2 contains 30 items) and hand-scorable. A table translates the locator test raw score to the appropriate Instructional Objectives Inventory (IOI) level. The student is assigned to the level at which he or she is likely to achieve a proportion-correct score of between .3 and .8. There are separate practice exercises for developing test-taking strategies such as following directions, locating items, and marking answers. It is suggested that the student complete these exercises shortly before the actual test day.

The DMI Mathematics Systems instructional materials are organized into two systems, System 1 and System 2. System 1 contains seven kits, one for each level (A-G). Each kit covers all four strands. System 2 contains four kits, one for each strand. Each kit covers all seven levels. Each kit (from either system) contains a Systems Overview Chart, a Teacher's Management Guide, a Teacher Resource Guide, Tutor Activities, Common Error Worksheets, a Guide to Correcting Common Errors, Mastery Tests, and a Progress Monitoring Log.

Practical Applications/Uses

The DMI/MS examiner's manual gives specific instructions for establishing appropriate testing conditions. The test is group administered. Levels A through C are answered directly in the test booklet. Levels E through G use separate answer sheets. Level D can be answered either in the booklet or on separate answer sheets. The tests are hand or machine scored.

Examiners should have read through the examiner's manual and actually taken the test previous to administration. This will aid in understanding student questions and problems. None of the tests are timed, and all students should be given enough time to finish. The examiner should use the time estimates given in the examiner's manual to estimate time needed for testing. The examiner should attempt to create a relaxing environment, and administration should not follow strenuous activity. The room should be arranged to avoid cheating. Wall displays that might contain answers should be removed. The examiner should also eliminate distractions, including the use of a "Do Not Disturb" sign; schedule breaks between test sessions; schedule the test sessions far enough apart to avoid student fatigue; and try to schedule tests for Tuesdays, Wednesdays, or Thursdays, and not immediately before or after holidays or important school functions. In addition, testing should not span a weekend. Ample time for questions, and distribution and collection of test materials should be allowed.

The DMI/MS identifies those students who have mastered or failed to master specific instructional objectives in the mathematical curriculum and detects common error patterns in each student. The Objectives Mastery Report (OMR) yields the mastery or nonmastery level for each student for each objective, as well as classroom and grade summaries. The Class Grouping Report (CGR) lists, per class, students who need review of each objective and those who show nonmastery of each objective. A total score and estimated norms derived from scaling the test to the CTBS, Form U, are given.

The DMI/MS also diagnoses wrong-answer choices leading to 53 common errors (over 13 errors per level). There are at least four items containing distractors for each common error. A student is identified as making a common error if the common error distractors are chosen between 40% and 60% of the time. The classification of common errors resulted from a detailed analysis of the distractors. For example, distractors that consistently were chosen and showed a positive correlation with other distractors were included in the common error report. A scoring procedure was developed to maximize the reporting of common errors for students who were in fact making the same error consistently and to minimize the reporting of errors chosen in a random fashion. The Individual Diagnostic Report (IDR) contains student mastery scores and common errors. The Common Error Report (CER) lists all the common errors made by at least one student in a given classroom and the student(s) making those errors.

Technical Aspects

There is a very detailed description of both objective and item development in the technical manual. Several current and widely used basal math series, as well as curriculum guides, were reviewed. Educators ranked, rated, and determined the appropriate grade level for each of the objectives. "There appears to have been a sound integration of content validity criteria with a variety of empirical data bases for item selection" (Hanna, 1985). There were 2,870 trial items administered to at least 890 students.

Content validity was established using IRT statistics (fit and discrimination) and *p*-values for each item. No tests for item bias are reported. The identification of common error patterns "is based on a variety of developmental analyses that are well described and seem well founded" (Hanna, 1985, p. 245).

There seems to be a lack of validity for the establishment of the "mastery," "review," and "nonmastery" cutoff scores. No estimates of hits or misses is given. There is no reporting of attempts to relate scores to an outside criterion such as teacher grades or scores on other criterion-related tests.

The technical manual does not provide data for the reliability of the norm-referenced total test scores, diagnostic scores for each instructional objective, common error scores, or locator test assignments. Hanna (1985) states that the DMI/MS fails "to adhere to contemporary standards concerning reporting reliability data for each score provided in the basic scoring report" (p. 246). Hanna was able to obtain a report from the publisher of the test. It contained K-R 20 internal consistency statistics in two school districts for scores on each objective and for the total test. Although the total test statistics showed "very acceptable reliability" (Hanna, 1985, p. 245), the reliability for each objective averaged slightly above .5.

The Estimated Norms Class Record Sheet provides estimated CTBS U and V scores for each student, based on total scores. Students participating in the item trials took a CTBS test. The CTBS items and IOI were analyzed using the computer program LOGIST. The results were used to place the IOI items on the CTBS scale. Then the CTBS norms were reported.

Critique

Item selection is one of the strong points of the DMI/MS. Another is the strong quantitative basis for the selection of the 53 common errors. The three-way mastery decision is more appropriate than the more common dichotomous rating (mastery/ nonmastery). Confidence in the test is not helped by the fact that no rationale is given for the establishment of the mastery and review cutoff scores and that there is a lack of reliability information of the mastery and common error decisions. Other questions need to be answered. Does skipping the practice exercises change the validity of the test? And, since infinite time is often not practical nor possible in a real classroom situation, what minimum times should be given to keep the current validity of the test? The DMI/MS has some real strengths. Hopefully, the author will answer the reliability and validity questions in the near future.

References

Gessel, J. (1983). *DMI/Mathematics Systems: Preliminary technical report.* Monterey, CA: CTB/ McGraw-Hill.

Gessel, J. (1983). *DMI/Mathematics Systems Instructional Objectives Inventory: Examiner's manual, system 1, level A.* Monterey, CA: CTB/McGraw-Hill.

Hanna, G.S. (1985). Review of DMI Mathematics Systems. Journal of Educational Measurement, 22(3), 244-246.

Kenneth J. Smith, Ph.D.

Professor and Director, Interdisciplinary Educational Clinic, College of Education, The University of Arizona, Tucson, Arizona.

Darrell L. Sabers, Ph.D.

Professor of Educational Psychology, The University of Arizona, Tucson, Arizona.

DOREN DIAGNOSTIC READING TEST OF WORD RECOGNITION SKILLS

Margaret Doren. Circle Pines, Minnesota: American Guidance Service.

Introduction

The title of the Doren Diagnostic Reading Test of Word Recognition Skills is quite consistent with its stated purposes: to provide, through a group-administered test, diagnostic information that "otherwise could be obtained only by an individual diagnostic test" (Doren, 1973, p. 5). In a summary of purposes for the Doren test, the manual describes it as a measure that is inclusive of the complex elements of reading skill development and that conforms to good teaching practices in the details of its design (Doren, 1973, p. 6).

The Doren test is divided into 12 different "skill areas" that the manual states were selected after examining five basic reading programs. These "programs" are not identified, nor is the developmental research documented, but it may be worthy of note that the current edition of the test was published in 1973, with an original publication date of 1959, particularly in view of the claim that the test can be used "regardless of the basic reading program used in teaching" (Doren, 1973, p. 5). The manual suggests that the Doren test be administered at the end of second grade or at the beginning of third grade.

The test's 12 skill areas are further subdivided into "component parts." Descriptions of these skill areas (subtests) and their components follow:

1) *Letter Recognition.* Part A of this subtest is not really letter recognition, but rather a sort of visual discrimination exercise, requiring the student to identify one or more printed letters as being the same as another. Although not an entirely inappropriate area, virtually any child who could see and understand the directions could perform the task. Thus, Part A probably would present more a test of the ability to follow directions than anything else. This may provide some correlations (see discussion of Technical Aspects), but it presents a serious validity problem for a supposedly diagnostic test. Part B of Letter Recognition requires the student to recognize upper- and lowercase letter forms as "having the same name." Part C provides the uppercase printed form of certain letters in the first column, followed by upper- and lowercase printed and cursive forms, requiring the student to identify which have the "same name" as the first. Though not irrelevant, the task seems

rather trivial and the directions are more complex than those noted previously, resulting in the same validity problem.

2) *Beginning Sounds,* 5) *Speech Consonants,* and 6) *Ending Sounds.* These three phonics subtests will be discussed together because they have similar purposes and similar problems. All assume that the ability to go from speech to print (encoding) is the same as the ability to go from print to speech (decoding). That will come as a shock to all good readers who are poor spellers! All require the student to read a sentence before choosing a word that supposedly invokes a particular "skill," assuming that the student can read the rest of the sentence. For example, in the Ending Sounds subtest, the examinee must read a sentence containing the word *giraffe* in order to select *tall* or *taller* as the correct response, supposedly using the "ending sound" in order to make that selection. This same subtest requires the student to use the "ending sound" to identify *sheep* as the plural of *sheep* and *geese* as the plural of *goose.* There is such a confusion of knowledge and abilities required for these subtests that one would have little idea what was actually being tested. Scholars in phonetics do not have a term called "speech consonant," and they certainly will not know what is intended by that term through studying subtest #5. This reviewer considers these subtests useless; their potential for invalid information may well result in inappropriate instructional decisions.

3) *Whole Word Recognition.* This subtest contains two very different parts. Part A is a visual discrimination task in which given five words, the student is to circle the two that are the same. There is no need to recognize a whole word to perform this task (although doing so might facilitate its completion). Rather, the examinee can obtain the correct response in each case via letter-by-letter comparison. Part B of this subtest is actually a listening task, as the process moves from speech to print. The teacher reads a word aloud and students circle it from among two alternatives.

4) *Words Within Words.* This subtest assumes that structural, or morphemic, analysis consists of finding little words in big ones (e.g., sandbox, chalkboard). Part A involves drawing a line between two "little words." Part B requires students to find a "little word" in a big one, draw a box around it, and then write it, after being directed that the little word must be "one you can hear"; this means that the child must already know both words! (Why are we doing this?) Part C provides the box, but students must determine whether the boxed-in little word is one that they can "hear" in the big one, again requiring the identification of both in advance. Unless the "big word" is already familiar to the reader, thus making the entire exercise pointless, teaching children to look for little words in big ones can result in misperceptions such as analyzing *father* to find *fat, her, at, the,* and *he.* There are some limited uses for morphemic analysis, but they are not tested here.

7) *Blending.* This subtest uses a kind of maze procedure that requires students to fill in blanks in sentences (which ostensibly they can read) by choosing from provided alternatives, in this case, three. As each of the choices begins with the same "blend" (cluster), the examinee does not use the one supposedly being tested to differentiate among the words provided; thus, the ability to use the blend is not tested.

8) *Rhyming.* Part A of this subtest consists of having students determine whether two words pronounced by the teacher rhyme. Parts B, C, and D present the words in print and ask students to determine whether they rhyme.

9) *Vowels.* Subtest parts A and B require students to identify vowels in words read aloud by the teacher, a spelling activity. Parts C, D, and E test for "long" and "short" vowel identification; for parts C and E, the student must already know the word. Part F inaccurately indicates that there are two discrete vowels (i.e., sounds) in each word (actually a vowel digraph), and again the task requires that the student already know the word. Parts G, H, and I are similar, except that in some cases part H involves diphthongs. The items in this subtest do not test the use of vowels for word identification, but test instead, rather inaccurately, the identification of vowels in words already known.

10) *Discriminate Guessing.* This subtest includes several "riddles" that require filling in a blank, a 5-item cloze section, and a 5-item homograph section. The tasks would seem to be testing the ability to use context and, in the homograph section, spelling. However, when testing two things at once (spelling and context), one doesn't always know what the results mean.

11) *Spelling.* In this measure, the teacher reads a list of words to be spelled by the students. The list is divided into "phonetic" and "non-phonetic" words. No other indication is given of how the words were chosen. The role of this subtest in a diagnostic reading test is not clear, nor does it seem a very useful spelling test.

12) *Sight Words.* This subtest presents a list of words that, according to the manual, "must be memorized because their spelling is irregular." Students are required to select from among three alternatives for each word a respelling "that is the right sound for each word."

Test booklets, an examiner's manual, and an overlay key are provided. An Individual Skill Profile is included on the front of the booklet. The manual includes sections entitled "Introduction," "Design of the Test," "General Directions," "Guide for Administering and Scoring," "Summary of the Skills Measured," "Remedial Activities," and "Technical Data."

The examiner must give examinees directions for each part of the Doren test and, in some cases, read individual items. Most spaces on the answer forms present a multiple-choice format, but some subtests require students to write in their answers. The overlay scoring keys are very simple to use, enabling a non-professional with almost no training to score the tests accurately. After scoring, the examiner transfers the resulting data to the Individual Skills Profile and to a Class Composite. The profile includes a shaded area to indicate which skills need instructional attention.

Practical Applications/Uses

The Doren test is not restricted to any age group, but the "skills" included are commonly included in Grade 1. The manual indicates that this test is intended to provide diagnostic information in a group setting and discusses the identification of "difficulties." Thus, although the "skills" tapped are rather common to first-grade instruction, the instructional section is entitled "Remedial Activities" and it appears the test is largely intended for diagnosis leading to information for remediation. In that sense, it might be appropriate then for first-grade teachers or remedial reading teachers.

Technical Aspects

The Doren manual devotes a page and a half to "technical data" for the test. These data appear intended to support only the total score, however, which is of limited use on a diagnostic test. The data presented are derived from four school districts across four grades. It is surprising that although the test is intended for administration to entire classes, there are only 165 students in the entire sample (about 40 per grade).

The supporting data are presented in four tables, the first of which presents correlations between an unnamed achievement test and Doren total scores. The second table presents correlations of each subtest with the total score (spurious correlations resulting from combining students from all four grades). There are no data to support the diagnostic use of the subtests, as all correlations support only the reliability (or internal consistency) of the total score. Interestingly, in a case where lower correlations among subtests (and with the total score) are preferable, to avoid duplication of information among subtests, the author explains the lowest subtest-total correlation as though it were not desirable (Doren, 1973, p. 39). The third and fourth tables present pictures of skill development and growth at different grade levels, but it is not clear whether these pictures result from speculation or data. The discussion of growth in first grade is based on the assumption (apparently untested) that first-graders at the beginning of the year would score zero on this test. A table showing correlations among all the subtests is needed to document the diagnostic use of this test.

Critique

The last paragraph of the Doren test's technical data section presents a case against norms for diagnostic tests. Regardless of the merit of this argument, there is no reason for the lack of technical information supporting a diagnostic use of subtest profiles resulting from this isntrument. In summary, the Doren Diagnostic Reading Test of Word Recognition Skills currently has unacceptable content validity and inadequate technical data to support its intended application. It is, in the judgment of these reviewers, a test in which the directions are more demanding than the items, the linguistic content is seriously inaccurate, and the skills actually tested are not in each case the ones targeted for testing; therefore, except to obtain data for revision, these reviewers cannot recommend its use.

References

This list includes text citations and suggested additional reading.

Bannatyne, A. (1974). Review of the Doren Diagnostic Reading Test of Word Recognition Skills. *Journal of Learning Disabilities, 7*(9), 535-538.
Doren, M. (1973). *Doren Diagnostic Reading Test of Word Recognition Skills.* Circle Pines, MN: American Guidance Service.

Feldmann, S. (1974). Review of the Doren Diagnostic Reading Test of Word Recognition Skills. *The Reading Teacher, 28*(1), 93.

Otto, W. (1975). Review of the Doren Diagnostic Reading Test of Word Recognition Skills. *Journal of School Psychology, 13*(3), 276-277.

Schreiner, R. L. (1978). Review of Doren Diagnostic Reading Test of Word Recognition Skills. In O.K. Buros (Ed.), *The eighth mental measurements yearbook* (pp. 1247-1248). Highland Park, NJ: The Gryphon Press.

Lucille B. Strain, Ph.D.
Professor of Education, Bowie State College, Bowie, Maryland.

DURRELL ANALYSIS OF READING DIFFICULTY: THIRD EDITION

Donald D. Durrell and Jane H. Catterson. San Antonio, Texas: The Psychological Corporation.

Introduction

The Durrell Analysis of Reading Difficulty: Third Edition (DARD) is an individualized, standardized test for diagnosing strengths and weaknesses in reading ability. Although designed primarily for use with children who are beginning to read through those in the sixth grade, the DARD can be adapted for use with other individuals whose reading difficulties approximate those of children in the elementary schools. The DARD consists of a series of subtests designed for observation and evaluation of several reading abilities: 1) Oral Reading, 2) Silent Reading, 3) Listening Comprehension, 4) Word Recognition/Word Analysis, 5) Listening Vocabulary, 6) Pronunciation of Word Elements, 7) Spelling, 8) Visual Memory of Words, 9) Auditory Analysis of Words and Word Elements, and 10) Prereading Phonics Abilities. Additionally, two components are available to aid in providing an even more complete picture of an examinee's abilities, 1) Supplementary Paragraphs and 2) Suggestions for Supplementary Tests and Observations.

The DARD first was used in the Boston University Educational Clinic in 1933 and published for general use in 1937. It was revised in 1955 and again in 1987. The original test and its subsequent revisions were published by Harcourt, Brace, Jovanovich. The 1987 edition of the test is available from The Psychological Corporation, a subsidiary of Harcourt, Brace, Jovanovich.

The two authors of the DARD, Dr. Donald D. Durrell and Dr. Jane H. Catterson, have been prominent in reading and reading education for many years. Both are well known through their publications, research, and careers in prominent universities and reading clinics.

Dr. Durrell was a professor at Boston University for 35 years. During that time, he served as a dean of the School of Education and founded the New England Reading Association. Currently, he is a professor emeritus there. During his long career, Dr. Durrell served on the Board of Advisors of *My Weekly Reader,* a well-known children's news publication widely used throughout the schools. Dr. Durrell also served on the Editorial Board of the *Journal of Education.* In recognition of his national leadership in reading research in instruction, he was elected to the Reading Hall of Fame in 1973.

Widely used materials designed by Dr. Durrell for use in reading instruction include *Speech-to-Print Phonics, Phonics Practice Program, Plays for Echo Reading, Vocabulary Improvement Practice, Sound Start Reading Program,* and *System 80 Phonics*

Programs. Besides the DARD, Dr. Durrell is responsible for two other assessment instruments, the Durrell Listening-Reading Series and the Murphy-Durrell Reading Readiness Analysis.

Dr. Catterson is a professor in the Reading Department of the Faculty of Education at the University of British Columbia. She has served as director of reading clinics at the University of Saskatchewan, the University of Calgary, and the University of British Columbia. Dr. Catterson is the author of several articles in the field of reading and co-author of *Word Analysis Practice*, a curriculum material. She is the editor of *Children and Literature* (Catterson, 1970). Dr. Catterson has authored *A Readiness Program* and *Grouping for Progress in Skills* for the government in Quebec.

The test authors credit changes in the third edition of the DARD to several sources. Two hundred professors who use the DARD as an integral part of their college courses offered suggestions related to the revision of the test. Their names appear in the Acknowledgement Section of the test manual. Other changes stemmed from critical reviews of the DARD over the years, evaluations of it in textbooks and in research articles, reports of its usage in clinical reports, and insights stemming from results of competing instruments and changing attitudes about the nature of reading. Many of the changes in the new edition are also a result of the research and clinical services of the authors.

Primary changes in the third edition were made in response to the need for 1) new normative data, 2) an updating of paragraph content, and 3) clarification that would allow more effective use of some of the subtests. A new Listening Vocabulary subtest, new measures of prereading abilities, and some new inventories in word analysis have been added, although the structure of the test remains basically unchanged from that of the previous edition. Most of the subtests included previously are retained in the third edition.

Materials for administering, scoring, and interpreting the DARD are simple, compact, and easily handled. These include 1) a reusable, spiral-bound booklet containing passages for oral reading, silent reading, listening comprehension; supplementary paragraphs; and tests measuring prereading abilities, 2) a tachistoscope (a quick exposure device) with accompanying cards for testing word recognition and word analysis, 3) an individual record booklet, and 4) the *Manual of Directions*. Materials are portable and easily filed for convenient use. A stopwatch, which is not provided, is recommended for use with several of the subtests.

The spiral-bound booklet, which measures approximately 6" ×8½," contains cardboard pages. Colors are used to separate major sections of the booklet. For example, passages for oral reading and listening comprehension and tests for prereading abilities are presented on blue pages, and passages for silent reading and supplementary reading are printed on white pages. Except for the passages for beginning levels, which are indicated as 1A and 1B and appear together on the first page of each of the relevant sections, each successive passage for oral, silent, and supplementary reading is contained on a single page. Eight paragraphs (five primary and three intermediate) are provided for assessing oral reading ability, silent reading ability, and supplementary, respectively. Six graded paragraphs comprise the section on listening comprehension. The final section contains tests designed to evaluate the prereading phonics abilities considered essential to success in learning to read. Specifically, these tests relate to syntax matching, identification of phonemes (speech sounds) in spoken words, naming lowercase letters, and writing letters (of the alphabet) from dictation.

The tachistoscope, provided as part of the DARD materials, is a device essentially consisting of a sleeve, a shutter, and a series of graded word lists. It permits the controlled short-time exposure of words (one at a time) necessary to test an individual's word-perception abilities.

The individual record booklet provides the examiner with an opportunity for recording all of an examinee's responses. A form for a profile chart is provided on the front cover of the booklet. The profile chart permits the examiner to record information related to oral reading, silent reading, listening comprehension, listening vocabulary, word recognition, word analysis, and spelling according to whether an individual measures high, medium, or low in a particular category. Other major forms provided in the record booklet include a checklist for instructional needs and a General History Data form, which is intended for recording significant preschool problems, school and medical records, psychological factors, and home history.

All the tests presented in the booklet are reproduced in the individual record booklet. Questions to be asked by the examiner in order to assess comprehension and a checklist for recording observed difficulties, as well as applicable normative data, follow each test passage.

The *Manual of Directions* contains all information necessary for administering, scoring, and interpreting test results. Descriptions of standardization procedures and data relating to validity and reliability also are presented in the manual.

Practical Applications/Uses

Like the previous versions, the purpose of the third edition of the DARD is to provide reading tasks through which the strengths and weaknesses of an individual's reading ability can be observed. Through the checklists, an analytic record of the difficulties displayed by an individual can be identified, thus permitting direct and specific remediation. The DARD also estimates general reading ability and, through the norms accompanying most of the subtests, an examinee's progress in relation to that of other individuals with similar characteristics can be ascertained.

Through the DARD, it is possible to estimate an individual's instructional, independent, and capacity reading levels. The instructional level, the level at which an individual can read a selection under the guidance of a teacher, is determined as a result of the oral reading passages. The independent level, the level at which an individual can read effectively without assistance, is assessed on the basis of performance on the silent reading passages. Capacity level, the level at which an individual can understand a passage when it is read to him or her, is shown by responses to the listening tests. The assumption underlying the capacity level is that if an individual can use oral language for communication, he or she can read at the same level if the decoding aspects of reading are accomplished.

In recognition of imagery's impact on comprehension, some provisions are made in the DARD for its measurement. The examiner is given instructions regarding the appropriate paragraph to use as well as suggestions for the types of open-ended questions that should be asked. In the manual, the authors of the test call attention to some of the classic studies supporting the importance of imagery (Galton, 1883; Halleck, 1897) and to their own studies reported in the *Journal of Education* (Durrell & Catterson, 1963).

The DARD is suitable for use in settings that permit thorough and lengthy individual testing and close observation of an examinee's response and general behavior. Reading specialists who are called upon for a more thorough diagnosis of an individual's reading difficulty than is feasible for a regular classroom teacher will find the DARD useful. Clinicians working in reading clinics who are interested in diagnosis and remediation of especially troublesome cases of reading difficulties and their impact on the social and emotional adjustment of an individual will find the DARD a serviceable instrument. The DARD also can be useful in literacy education. Individuals who are well past the ages for which the test is recommended but who are having reading difficulties similar to those encountered in school-age children can be diagnosed using the appropriate DARD subtests.

History of the DARD indicates that it has proven itself valuable as an integral part of college courses designed to prepare personnel working or planning to work in clinical and other positions involving the development of reading ability. Because of its comprehensive approach to the specific, separate factors believed to comprise reading ability, it is useful for training observers in what to look for when an individual is experiencing reading difficulty. For the same reason, classroom and special education teachers can improve their understanding of reading and observational techniques by studying the DARD.

As mentioned earlier, the DARD is designed primarily for children in Grades 1-6. This does not negate its usefulness with readers beyond elementary school age, however. When an individual's reading characteristics approximate those of children with reading difficulties, several portions of the test can be adapted to meet testing needs. For example, despite the inappropriateness of the norms for adults who are learning to read, the checklists can provide insight into the nature of specific difficulties. The DARD would appear to justify its adaptations for use with special populations such as the blind or non-English-speaking individuals who are learning to read.

Administration of the DARD should take place on a one-to-one basis between the examiner and the examinee in a casual, friendly, and pleasant setting. Good rapport between the examiner and the examinee is crucial to the success of the testing and the examiner is advised to avoid testing during times such as recess, after school, or especially enjoyable classroom activities. Testing should also be avoided when the examinee is in less than the best physical and emotional condition.

Instructions in the test manual are explicit regarding the conditions of the room and lighting in which the test is administered. The room should be quiet with good lighting. Care should be taken to make sure that shadows or bright light do not fall directly on the material the examinee is expected to read.

The examiner can select any order for giving the separate tests in the DARD, although the authors recommend that administering the Oral Reading and Word Recognition/Word Analysis Tests early in the testing can produce results useful in determining subsequent tests that should be given. In order to improve and maintain the examinee's motivation to take the test, it is suggested that he or she be kept informed of the purposes and results of the separate tests as the testing proceeds. The authors of the test maintain that "the child is more concerned than anyone about his or her difficulties and responds quickly when there is hope of improvement" (Durrell & Catterson, 1987, p. 13).

Accurate timing is important in that the norms for the Oral and Silent Reading Tests are based on the time required for reading each passage. Although a stopwatch is recommended, a watch with a sweep hand can be substituted if the examiner carefully notes the time at the beginning and at the end of each test. The time should be recorded in hours, minutes, and seconds. Time required for administration of the entire test will vary according to the testing situation. An estimation of the approximate administration of time required for the reading section is from 40-60 minutes (Wilson, 1981, p. 432).

The DARD is designed for use by experienced persons familiar with behaviors associated with reading ability. Although specific formal training in testing is not required for its administration, the successful administrator will be thoroughly familiar with all aspects of the test. The authors suggest that administration of the DARD is best learned under the direction of a person who has had experience in the analysis and correction of reading difficulties (Durrell & Catterson, 1987, p. 12).

Because of the variety of subtests comprising the DARD, scoring procedures are varied. For this reason, it is necessary to follow directions in the manual for each subtest under consideration. While scoring is not complicated, it is specific in terms of the behaviors being observed. Provisions for scoring and the behaviors to be observed are immediately at hand in the individual record book as well as in the manual. Interpretation of DARD results is based on both norms and examiner judgment. Objective test scores, for example, are available as results of Oral Reading, Silent Reading, Listening Comprehension, Word Recognition, Word Analysis, Listening Vocabulary, and Spelling. Check lists accompanying all subtests require judgment by the examiner. Interpretation of the results of the check lists requires knowledge of the nature and quality of behaviors required in effective reading at various levels of development. Scoring and interpretation of the test results can be accomplished efficiently and effectively by an examiner qualified to administer the test.

Technical Aspects

As a standardized test, the DARD provides norms as a means for estimating an examinee's proficiency in reading in relation to other individuals of the same age and grade for some of its subtests. The manual reports that norms were obtained by testing a relatively small, but adequate, number of children. The manual calls attention to some of the difficulties encountered in testing large numbers of children with an individually administered test. For standardization of the DARD, testing 200 children per grade in each of Grades 1-6 was deemed statistically sound and economically feasible. The regional distribution of the standardization group included the New England, Central, South, Southwest, and Pacific regions, including the states of Massachusetts, Illinois, Ohio, North Carolina, Texas, and California. A total of 1,224 children were involved in the standardization process.

Validity and reliability data for the entire battery and for several specific subtests also are reported in the test manual. A claim for content validity is based on the test's potential for yielding a clear, accurate, and complete description of abilities essential for planning effective remediation for an individual. In this way, the DARD fulfills its major purpose. Further evidence of validity is based on the test's performance during the years since its introduction in the early 1930s and the recent critical assessments made by almost 200 college reading teachers.

The test authors assert that the vocabularies of the paragraphs assessing oral reading, silent reading, and listening comprehension are designed to be representative of reading content at the specified grade levels. It is claimed that vocabularies for these tests were screened both by standard word lists and by careful field testing. The words used in the Listening Vocabulary Test are reportedly selected from *Roget's Thesaurus'* word classes, screened by standard word lists and by item analysis during field testing (Durrell & Catterson, 1987, p. 57). Among the word classes in *Roget's Thesaurus* (Morris et al., 1966) are abstract relations, space, matter, intellect, volition, and affections.

Predictive validity of several of the subtests at the prereading level was assessed by correlating the results of measures administered to first graders in September with reading achievement the following June. The correlations reported ranged between .55 and .65.

Because the DARD consists of a series of short tests assessing different aspects of reading at different levels, a variety of treatments were used to estimate reliability. Reliability of the use of rate of reading (time) as the basis for assigning grade levels in oral and silent reading was determined on a randomly chosen population of 200 children drawn from Grades 2-6. Each child read the two adjacent paragraphs deemed appropriate, and the times required for reading were correlated, resulting in correlations of .85 between paragraphs assessing oral reading and of .80 between those measuring silent reading. Reliabilities of other subtests were determined by Kuder-Richardson Formula 21 and involved at least 200 children taking each test regardless of their grade level. All the resulting correlations are reported in the manual. Relationships among selected tests and with grade-equivalent scores of the Metropolitan Reading Tests also are presented in the manual. Reliabilities of other subtests were determined by use of the Kuder-Richardson formula 21 and involved at least 200 children taking each test regardless of their grade level. The resulting coefficients are reported in the manual (p. 58) as follow:

Test	Grade Range	$r21$
Word Recognition (Primary)	1	.81
Word Analysis (Primary)	1	.83
Word Recognition	2-6	.89
Word Analysis	2-6	.91
Listening Vocabulary	1-6	.79
Spelling (Primary)	1-3	.70
Spelling (Intermediate)	4-6	.97
Phonic Spelling	4-6	.73
Visual Memory of Words (Primary)	1-3	.63
Visual Memory of Words (Intermed.)	4-6	.73
Hearing Sounds in Words	1-3	.81
Giving Sounds of Blends	1-3	.88
Giving Sounds of Phonograms	1-3	.91

All norms essential for interpreting the results of the DARD are provided in the individual record booklet: grade norms for reading time, recall, listening, word recognition and word analysis, sounds in isolation, and phonic spelling of words. For the Prereading Phonics Abilities inventories, norms are provided for syntax matching, identifying letter names in spoken words, identifying phonemes in spoken words, naming lower case letters, and writing letters from dictation. The remaining prereading phonics abilities tests are intended primarily as records of progress for use only with children having very severe reading difficulties.

Critique

In the third edition of the DARD, the test authors have given attention to improvement of several weaknesses cited by users and reviewers of the previous edition. In the current revision, the DARD seems to have reasonable potential for providing test results capable of forming the basis for a remedial-reading plan at various levels. This view was not shared, for instance, by Ekwall and Shanker regarding certain tests in the previous edition. These authors expressed concern about the prereading phonics tests of the previous edition, claiming that the test lacked enough "depth in phonics testing to plan prescriptive instruction" (Ekwall & Shanker, 1983, p. 126). These authors cited the subtests "Identifying Sounds in Words" and "Sounds in Isolation" as focusing on low-level phonics' skills and, then, only superficially. Two new measures of early phonics abilities are included in the current edition of the DARD. Measures to assess an examinee's awareness of separate words in spoken sentences and ability to hear letter names in spoken words are included in the new battery. Newly designed tests to provide for retention of the examinee's interest and attention are other improvements. Although the tests are relatively brief, their focus on skills primarily needed to understand relationships between oral and written language provide evidence crucial in determining an examinee's readiness for formal reading instruction.

Previous editions of the DARD were frequently cited for their lack of normative data. Among others, Gabriel Della-Piana claimed that "the Durrell Battery does not even present oral reading norms for various error-types," going on to cite the test for failure to defend the practice of administering the subtests in any order (Della-Piana, 1969, p. 75). Both of these weaknesses are addressed in the third edition of the DARD. Normative data are provided for the objective tests included in the battery. Norms for oral reading are presented for determination of the level at which an examinee can read with instructional assistance. Norms for silent reading are presented for determination of an examinee's independent reading level. Although no definite order is yet demanded for administration of the entire test, recommendation is made for beginning testing with certain components and a rationale is given for the order recommended. The self-contained nature of the subtests, however, clearly indicates that an examiner can determine the order in which the tests are administered with no loss in effectiveness. This is especially appropriate when the examiner is already aware of the nature of many reading behaviors shown by a given reader.

The DARD's comprehensiveness in terms of the specific reading behaviors assessed negates the need to use several different instruments for acquiring a total picture of an examinee's reading behavior. At the same time, the DARD is comprehensive in terms of the grade levels to which it applies: its range of evaluation includes the non-reader and all grade levels up to and through Grade 6.

In comparisons with other individual, standardized reading tests, the DARD has, over the years, maintained its competitiveness in terms of the number of abilities measured (Trela, 1966; Wilson, 1981). Its compactness clearly marks it as superior in portability to the use of a variety of instruments to achieve similar purposes. In comparison with group tests designed to indicate the nature of instruction needed by an individual, the DARD provides a more detailed picture.

The time allotment estimated for administration of the DARD makes it somewhat less than practical for use by classroom teachers, whose responsibilities include more than making precise diagnoses of individuals' reading behaviors. Few classroom teachers can spare 45 to 60 minutes to test one individual. This limitation becomes even more pronounced if one takes the test manual's stipulation literally that testing should not occur during recess or after school. The time allocations required for complete administration of the total DARD limit its use by regular classroom teachers.

Although directions for administration, scoring, and interpretation of DARD results are immediately clear to persons experienced both in testing and in observation of reading behaviors, they become somewhat complex when attempted by a less-experienced examiner. The DARD can be used best by specialists in testing and in reading instruction.

Provision of a ready-made tachistoscope with accompanying cards for testing skills with letters and words is a time-saver for the examiner and assures that testing of these behaviors will be done appropriately. The cardboard pages in the spiral-bound student test booklet assure long-term reusability as well as ease of use. The individual record booklet in which all student responses are recorded facilitates the recording of responses and their interpretation. The materials provided for administration of the DARD have many positive features.

The technical aspects of the DARD indicate its capability of producing valid and reliable results. The authors have given careful attention to content and predictive validity. Reliability coefficients, although not uniformly high in all instances, are adequate in light of the short lengths of many of the subtests involved.

The DARD provides an organized set of diagnostic materials that can be used to provide a thorough diagnosis of an examinee's reading behaviors. For a specialist, the DARD can be considered efficient in the use of effort and materials, highly recommended as part of the basic equipment essential to the diagnosis and remediation of reading difficulties.

References

Catterson, J. H. (1970). *Children and literature.* Newark, DE: International Reading Association.

Della-Piana, G. (1968). *Reading diagnosis and prescription.* New York: Holt, Rinehart and Winston.

Durrell, D. D. (1936). Individual differences and language learning objectives. *Childhood Education, 12,* 149-151.

Durrell, D. D., & Catterson, J. H. (1963). Boston University studies in elementary school reading. *Journal of Education, 146,* 44, 50.

Durrell, D. D., & Catterson, J. H. (1987). *Durrell Analysis of Reading Difficulty: Manual of directions.* San Antonio, TX: The Psychological Corporation.

Ekwall, E., and Shanker, J. L. (1983). *Diagnosis and remediation of the disabled reader.* Boston: Allyn & Bacon.

Galton, F. (1883). *Inquiries into the human faculty and its development.* London: Macmillan.

Halleck, R. P. (1897). *Education of the central nervous system.* New York: Macmillan.

Morris, W., Chadsey, C. P., & Wentworth, H. (Eds.). (1966). *The Grosset dictionary and thesaurus* (rev. ed.). New York: Grosset & Dunlap.

Trela, T. M. (1966). What do diagnostic reading tests diagnose? *Elementary English, 43,* 370-372.

Wilson, R. M. (1981). *Diagnostic and remedial reading* (4th ed.). Columbus, OH: Charles E. Merrill.

David J. Hansen, Ph.D.
Assistant Professor of Psychology, West Virginia University,
Morgantown, West Virginia.

DYADIC PARENT-CHILD INTERACTION CODING SYSTEM

Sheila M. Eyberg and Elizabeth A. Robinson. San Rafael,
California: Social and Behavioral Sciences Documents.

Introduction

The Dyadic Parent-Child Interaction Coding System (DPICS) is a direct observation procedure for assessing the quality of interactions between parents and children ranging in age from 2 to 10 years. The DPICS, which is to be used in a clinic or laboratory setting, was designed to 1) provide a direct observational measure of parent and child behaviors that may serve as an adjunct to full psychological evaluation of childhood behavior problems and parenting skills, 2) serve as a pretreatment assessment of parent-child interactions, 3) provide a measure of progress for therapy that focuses on changing parent-child interaction patterns, and 4) serve as a behavioral observation measure of treatment outcome (Eyberg & Robinson, 1981; Eyberg, 1985). Frequency of a variety of discrete positive and negative parent and child behaviors is recorded.

Sheila Eyberg, the orginator of the DPICS, is currently a professor of clinical health psychology and pediatrics at the University of Florida. Elizabeth Robinson, co-author of the DPICS, is an affiliate associate professor of psychology at the University of Washington. The test authors acknowledge the contributions of their students and research assistants at the University of Oregon-Eugene, Oregon Health Sciences University, and University of Washington, who served as research coders and trainers, provided coding examples, and identified problem areas (Eyberg & Robinson, 1983).

The developers of the DPICS intended to provide a comprehensive, manageable coding system that could be practical in a therapy hour. Their intent was to construct an observational system that would provide detailed information without requiring elaborate or expensive recording equipment, home visits, lengthy observation periods, or additional coders (Eyberg & Robinson, 1981, 1983).

The first version of the DPICS was developed by Eyberg in 1974, and the manual was later revised by Eyberg, Robinson, Kniskern, and O'Brian (1978). The current, third edition of the manual (Eyberg & Robinson, 1981) contains the same behavioral categories as the original (Eyberg, 1974). The revisions have improved and refined the guidelines for distinguishing between code categories and refined and expanded decision rules for coding.

Several of the parent behaviors were derived from definitions in direct observation coding manuals by Hanf (1968; cited in Eyberg & Robinson, 1981) and Patter-

167

son, Ray, Shaw, and Cobb (1969). Other categories were added to allow coding of every parent verbalization. Parent behaviors, which are coded "exhaustively," are 1) direct command, 2) indirect command, 3) descriptive statement, 4) reflective statement, 5) descriptive/reflective question, 6) acknowledgement, 7) irrelevant verbalization, 8) unlabeled praise, 9) labeled praise, 10) physical positive, 11) physical negative, 12) critical statement, 13) responds to deviant behavior, and 14) ignores deviant behavior.

Because distinguishing praise and critical statements reliably from descriptive statements was one of the most difficult coding problems, an empirical survey of 13 clinical psychologists was conducted (Eyberg & Robinson, 1981). Sentences for which 90% or more of the panel agreed were included in the praise or critical statement section of the coding manual. All other sentences in the survey were included in the descriptive statement category (Eyberg & Robinson, 1981).

Most of the child behaviors were selected from a list of empirically derived child deviant behaviors (Adkins & Johnson, 1972, cited in Eyberg & Robinson, 1981). The following child behaviors are coded: 1) cry, 2) yell, 3) whine, 4) smart talk, 5) destructive, 6) physical negative, 7) change activity, 8) compliance with command, 9) noncompliance with command, and 10) no opportunity for compliance with command.

An "Other" category is provided on the Data Recording Sheet for clinicians or researchers assessing a relatively infrequent or unique child or parent behavior or sequence. For example, the DPICS authors have used the "Other" category to measure "child appropriate talk" (Eyberg, 1985; Eyberg & Robinson, 1983).

A variety of summary variables of parent behavior have been reported in the literature: 1) total praise (labeled praise + unlabeled praise) (Eyberg & Matarazzo, 1980; Eyberg & Robinson, 1982; Packard, Robinson, & Grove, 1983; Robinson & Eyberg, 1981; Webster-Stratton, 1984, 1985a, 1985b, 1985c), 2) total commands (direct commands + indirect commands) (Robinson & Eyberg, 1981; Webster-Stratton, 1984, 1985a, 1985b), 3) direct command ratio (direct commands/total commands) (Eyberg & Robinson, 1982; Robinson & Eyberg, 1981; Webster-Stratton, 1984), 4) no opportunity ratio (no opportunity/total commands) (Robinson & Eyberg, 1981), 5) total negative behavior (direct command + indirect command + critical statement + physical negative + descriptive/reflective question) (Packard et al., 1983), 6) total nontargeted behaviors (irrelevant statement + acknowledgement + reflective statement) (Packard et al., 1983), 7) rate of reinforcement (physical positive + unlabeled praise + labeled praise) (Zangwill & Kniskern, 1982), and 8) rate of punishment (physical negative + critical statements) (Zangwill & Kniskern, 1982).

Several child summary variables also have been reported in the literature: 1) total deviant (whine + cry + physical negative + smart talk + yell + destructive) (Eyberg & Robinson, 1982; Robinson & Eyberg, 1981; Kniskern, Robinson, & Mitchell, 1983; Webster-Stratton, 1984, 1985a, 1985b, 1985c, 1985d; Zangwill & Kniskern, 1982), 2) compliance ratio (compliance/total commands) (Eyberg & Robinson, 1982; Robinson & Eyberg, 1981), 3) noncompliance ratio (noncompliance/ total commands) (Eyberg & Robinson, 1982; Robinson & Eyberg, 1981; Webster-Stratton, 1984), and 4) compliance ratio to direct and indirect commands [compliance/(compliance + noncompliance)] (Zangwill & Kniskern, 1982).

The 88-page DPICS manual (Eyberg & Robinson, 1981) includes 1) a description of the development of the coding system; 2) information on reliability, validity, and normative data; 3) a data recording sheet; and 4) definitions, examples, guidelines, and decision rules for each behavior category. This publication is available for $16.50 as Ms. No. 2582 from Social and Behavioral Sciences Documents (previously the APA's Psychological Documents), Select Press, P.O. Box 9838, San Rafael, CA 94912.

Practical Applications/Uses

The DPICS was designed to measure parent-child interactions thoroughly. The developers have emphasized the practicality and ease of use of the system (Eyberg & Robinson, 1981, 1983). The DPICS may be helpful in an evaluation of a family experiencing a parent-child or child conduct problem or in the evaluation of parent training progress or outcome. The wide range of parent and child behaviors coded, as well as the numerous summary variables that can be calculated, make the procedure very useful.

A direct observation system such as the DPICS has uses in a variety of applied and research settings. The DPICS appears to have utility for clinical, counseling, and health-related professionals whose goals are evaluation or improvement of parent-child interaction. The DPICS clearly has been used in research settings but the extent of its use by nonresearch clinicians is unknown. Trained observers may be a luxury that often is not available in applied settings, yet observations may be conducted by the clinician.

Eyberg and Robinson (1981) recommend that observations be conducted with one parent-child dyad at a time in a playroom equipped with a sound system, two-way mirror, table by the mirror, three or four chairs, and several toys. The observer(s) is located in an observation room behind the mirror, and the child is not informed that the interaction is being observed (Eyberg & Robinson, 1981).

Eyberg and Robinson (1981) suggest toys that allow creative, relatively quiet play, such as toy buildings with toy people and animals, building materials (e.g., blocks or Legos), or drawing materials (e.g., crayons and paper). Inappropriate toys include those that 1) elicit aggressive or highly active behavior (e.g., punching bags, hammers, etc.), 2) are potentially difficult to cleanup (e.g., finger paints), 3) elicit stereotyped responses (e.g., toy telephones), or 4) have pre-set rules (e.g., games such as checkers) (Eyberg & Robinson, 1981). A standard set of three toys is recommended for pre-post treatment comparison research.

The structure of the DPICS observation in brief, dyadic, standard laboratory situations was derived from the work of Hanf (1972; cited in Eyberg & Robinson, 1981). The parent and child are observed in three standard situations, and directions are provided to the parent by the observer, usually through a bug-in-the-ear microphone (Eyberg & Robinson, 1981). In the first situation, labeled Child-Directed Interaction (CDI), the parent is instructed to allow the child to choose any activity and to play along with the child according to the child's rules. In the second situation, labeled Parent-Directed Interaction (PDI), the parent is instructed to choose an activity and keep the child playing according to the parent's rules. In the third situation, Clean-up, the parent is instructed to get the child to put away all the toys without assistance. The structure of the situations allows the parent and the child to interact naturally with increasing requirements for parental control. The three situations always are observed in this same sequence, and each situation is coded for 5 minutes.

Each of the 24 standard behaviors is coded continuously during each situation (i.e., CDI, PDI, clean-up), and the total frequency of each behavior per 5-minute interval is obtained. Behavior is coded by making a tally mark in the appropriate space on the Data Recording Sheet each time the behavior occurs. The manual defines each category by 1) a general definition, 2) a series of examples, 3) specific guidelines to aid discrimination between behaviors, and 4) decision rules to provide prioritizing when there is uncertainty between behaviors (Eyberg & Robinson, 1981).

Coding procedures follow several basic coding rules: 1) each clearly demarcated sentence defines one verbal behavior (i.e., a "one-sentence rule"), 2) each time a behavior stops for 2 seconds and then continues, the continuation after the pause is coded as a new behavior (i.e., a "2-second rule"), 3) a "5-second rule" is applied to nondiscrete, continuous behavior (e.g., crying) where each new 5-second interval is coded as one occurrence of the behavior, and 4) each discrete behavior is coded into only one category.

The total frequency of individual behaviors or summary categories within each of the 5-minute intervals can be examined. The frequencies can be compared to available normative data or used for evaluation of interventions for parent-child interactions (e.g., baseline vs. posttreatment frequencies).

The DPICS has been used with a variety of populations. For instance, the primary standardization and validation study for the coding system was conducted on parent-child dyads with conduct-problem or normal children between 2 and 7 years of age, with siblings (who also were observed) between the ages of 2 and 10 (Robinson & Eyberg, 1981). This study included both mothers and fathers and sons and daughters.

Other studies have included a variety of subjects: 1) mother-child dyads of neglectful mothers, mothers of behavior problem children, and mothers of nonproblem children between the ages of 2.3 and 9.4 (Aragona & Eyberg, 1981); 2) mother-child dyads of speech- and language-delayed children between the ages of 4.0 and 9.0; 3) mother-child dyads of behavior problem children between the ages of 2.0 and 7.0, with siblings between 2.0 and 10.0 who also were observed (Eyberg & Robinson, 1982); 4) mother-child dyads recruited from the community with children between the ages of 2 and 8 (Kniskern et al., 1983); 5) mother-child pairs recruited from nursery schools and day-care centers with an average child age of 39.7 months (Packard et al., 1983); 6) mother-child dyads with conduct problem children with a mean age of 4.7 years (Webster-Stratton, 1984); 7) abusive and nonabusive mother-child dyads with children with mean ages of abused children at 55.9 months ($SD = 19.0$) and nonabused children at 58.9 months ($SD = 16.1$) (Webster-Stratton, 1985a); 8) mother-child dyads with conduct problem children with a mean age of 58.1 months ($SD = 17.9$) for father-present families and 62.2 months ($SD = 19.3$) for father-absent families (Webster-Stratton, 1985b); 9) mother-child dyads with conduct problem children with a mean age of 57.45 months ($SD = 13.39$) and nonclinic children with a mean age of 47.53 ($SD = 9.51$) (Webster-Stratton, 1985c); 10) mother-child dyads with conduct problem children between the ages of 3 and 8 years of age (Webster-Stratton, 1985d); 11) mother-child dyads with conduct problem children between the ages of 2 and 8 (Zangwill & Kniskern, 1982). Although the majority of data appear to concern mother-son dyads, other dyads (mother-daughter, father-son, father-daughter) have been assessed in the literature. The sex of the subject occasionally is not reported.

Technical Aspects

Researchers have evaluated interrater reliabilities obtained when using the DPICS with a variety of populations and various ages of children. Robinson & Eyberg (1981), in a standardization and validation study of the DPICS, compared fathers, mothers, siblings, and target children in 42 behavior problem and normal families. Coders were trained to initial 90% reliability and received additional reliability feedback throughout the study. Two coders were present for every observation when scheduling permitted (244 of 276 5-minute observations). The mean correlation between raters for parent behaviors was .91 (range = .67 to 1.0) and for child behaviors was .92 (range = .76 to 1.0).

Several other reports of adequate to excellent interrater reliabilities can be found in studies that utilized the DPICS as a dependent measure. Eyberg and Matarazzo (1980) obtained Pearson product-moment correlations between independent raters ranging from .735 to .999 for each code category. Aragona and Eyberg (1981) obtained Pearson product-moment correlations between observers for each behavioral category ranging from .65 to 1.0, with only Acknowledgement during PDI being below .85. Kniskern et al. (1983) reported a mean interrater correlation of .84 with a range of .47 to .96. Just two behaviors had correlations below .72, irrelevant verbalization (.47) and labeled praise (.57). Zangwill and Kniskern (1982) reported mean interrater agreement of 68% in the home and 69% in the clinic. The Pearson prduct-moment correlation median was .87, with a range of .42 to 1.0. Packard et al. (1983) reported product-moment interrater reliabilities ranging from .78 to 1.0 for several DPICS behaviors.

Webster-Stratton conducted a series of studies in which the DPICS was one of the major dependent measures. For a variety of summary and singular DPICS behaviors, she reported obtaining generally excellent interrater reliability with mean interrater agreements of 78.6 and Pearson product-moment correlations that ranged from .61 to .99 (Webster-Stratton, 1984, 1985a, 1985b, 1985c, 1985d).

Although the primary purpose of the Webster-Stratton studies was not to establish the interrater reliability of the DPICS, the studies clearly provide significant data to support the ability to achieve adequate to excellent interrater reliability with a variety of subjects, a variety of observers and researchers, and in both clinic and home settings.

Much research has addressed the validity of the DPICS. Research with the DPICS suggests that observation in a laboratory or clinic setting is a valid predictor of behavior in the natural setting when the analog activities and individuals are similar to the natural situation. Zangwill and Kniskern (1982) compared the behavior of 15 families with conduct problem children both in the home and in the clinic. Spearman correlation coefficients indicated that there were moderate to high correlations between home and clinic for three of four behavior categories tested: mother's rate of reinforcement (.53), mother's rate of punishment (.61), and child's rate of compliance (.89). Only child deviant behavior did not receive a significant correlation. Repeated measures analyses of variance, however, indicated that there were significant differences in rate of behavior between settings: in the clinic, mother's reinforcement and punishment was significantly higher, children's compliance was significantly lower, and deviance was significantly more.

Kniskern et al. (1983) examined the effects of four environmental variables on the interactions of 40 mother-child dyads: location, presence of a sibling, task structure, and day of observation. The results of the DPICS, which was the primary measure, along with parent-report measures of child behavior indicated that task structure (CDI, PDI, Clean-up) and presence of a sibling had large and systematic effects but that location (home, laboratory) had minimal impact on behavior. There were no significant differences between occasions (2 consecutive days).

These studies support the use of the DPICS for direct observation in the clinic. The clinic analog is hypothesized to be valid because both the parent's and the child's "behavior are free to vary, and, thus, the interactive nature of the problem is captured" (Eyberg, 1985, p. 133).

One method of assessing the validity of a measure is to demonstrate its ability to discriminate between groups that would be expected to differ on the measure. In the DPICS standardization and validation study (Robinson & Eyberg, 1981), the discriminant function analysis of DPICS observations correctly classified 100% of the 22 normal families, 85% of the 20 conduct-problem families, and 94% of all families. The conduct-problem children displayed higher rates of noncompliance than normal children, and their parents gave more critical statements, more direct commands, and fewer descriptive questions than normal parents. The conduct-problem children demonstrated more whining, yelling, noncompliance, and total deviance than normal children. In conduct-problem families, the referred child and his or her sibling exhibited behavior problems, but the referred child demonstrated deviant behavior in a greater variety of situations than the sibling. Multiple linear regression showed that the DPICS predicted 61% of the variance in parent report of home behavior problems as measured by the Eyberg Child Behavior Inventory (Robinson, Eyberg, & Ross, 1980).

Aragona & Eyberg (1981) found significant differences in verbal behavior in neglectful mothers, behavior problem mothers, and normal mothers using the DPICS. Compared to control dyads, the interactions of neglectful mothers were most negative and controlling during CDI. Behavior-problem mothers were most negative and as controlling as neglect mothers during PDI.

Webster-Stratton (1985a) used the DPICS in a comparison of abusive and non-abusive families with conduct-disordered children. There were significantly more instances of physical negatives, total commands, and criticisms for abusive mothers than for nonabusive mothers. There were no significant differences in child misbehavior between abusive and nonabusive families. Webster-Stratton (1985c) also used the DPICS in a comparison of the clinic-referred and nonclinic mothers' perceptions and mother-child interactions. Clinic mothers gave significantly more praise, commands, and criticism.

Another, related, method of establishing validity is to demonstrate a measure's sensitivity to pretreatment and posttreatment differences. Eyberg and Matarazzo (1980) found significant changes in the behavior of parents of speech- and language-disordered children following five sessions of parent-child interaction training. During CDI, parents showed a substantial decrease in direct and indirect commands, questions, and critical statements and a substantial increase in descriptive statements and labeled praise. In PDI, these parents decreased indirect commands and critical statements and increased labeled praise. Children showed a substantial decrease in noncompliance.

Eyberg & Robinson (1982) treated seven families with conduct-problem children for an average of nine parent-child interaction training sessions. DPICS' observational data indicated significant improvement in a variety of the mothers', fathers', childrens', and siblings' behavior in CDI and PDI.

Webster-Stratton (1984) used the DPICS and other measures in an evaluation of two parent-training programs (individual vs. videotape modeling with group discussion) for families with conduct-disordered children. Both treatments showed immediate and long-term (1 year) improvements in mother behaviors and decreases in child deviance and noncompliance. No significant differences between treatments were found with the DPICS.

Webster-Stratton (1985b) used the DPICS and other measures in an evaluation of the effects of father involvement in parent training for conduct-problem children. Following treatment, both the father-involved and father-absent families showed significant increases in mother praise and reductions in mother negative behaviors, child noncompliance, and deviance. Webster-Stratton (1985d) also used the DPICS in the evaluation of a behavioral parent-training program.

Packard et al. (1983) used the DPICS and parent report to evaluate the effects of training on the maintenance of parental relationship building skills. Laboratory observations during child-directed play 11 weeks after training showed that mothers who were coached via bug-in-the-ear following self-instruction continued to make gains in nondirective skills after training was discontinued and issued more descriptive statements, fewer indirect commands, and less negative behavior at follow-up than placebo-control mothers.

A common procedure with many measures that is less common with direct observation coding systems, however, is collection of normative data. Robinson and Eyberg (1981) collected normative data on a sample of 22 families recruited from the community who had at least two children, one between the ages of 2.0 and 7.0 years and another between the ages of 2.0 and 10.0 years and no history of psychotherapy or treatment for behavior problems. The families were observed in CDI and PDI on each of two observation periods, 7 days apart. The normative data are reported in Eyberg and Robinson (1981) and in Robinson and Eyberg (1981). Data for boys and girls and for mothers and fathers are not presented separately.

Several other studies previously described (e.g., Aragona & Eyberg, 1981; Webster-Stratton, 1985a, 1985c) have reported normative or descriptive data for small samples of clinical and nonclinical populations of parents and children. Descriptive data for retarded mothers have also been reported (Peterson, Robinson, & Littman, 1983).

Critique

Because the DPICS has been developed carefully and researched extensively, there is considerable evidence to support its use. The behaviors were chosen and described intelligently; however, a code category for physical positive behavior by children is notably absent. The adequate to excellent interrater reliability that can be obtained has been well documented. The validity of the DPICS has been shown in several ways, including correspondence with observations in the home, discrimination between various groups of clinic and nonclinic subjects, correspondence with parent reports of home behavior problems, and sensitivity to the effects of treatment. The small amount of normative data that is available may be useful for comparing family functioning or for evaluating treatment outcome.

The generalizability of a direct observation system is important (cf. Cronbach, Gleser, Nanda, & Rajaratnam, 1972). Although many studies already have been conducted, further assessment of the generalizability of the DPICS across subjects (e.g., older children, children or parents with other clinical problems), across settings (e.g., to the home), and across methods (e.g., with parent report) seems necessary. Generalizability across observers has been addressed considerably in analyses of interrater agreement and reliability.

Generalizability across time needs to be assessed to determine whether the results can be assumed to be accurate and representative of other time periods. One investigation showed a high relationship between data collected on 2 consecutive days (Kniskern et al., 1983). Use of traditional methods of assessing internal consistency, such as split-half reliability (e.g., odd vs. even days of baseline) and presentation of stable baseline data (Cone & Foster, 1982), may help. However, this does not determine directly whether data observed at one time are representative of data that would be obtained at times that have not been assessed.

The structure of the DPICS procedures may be responsible for its effectiveness and apparent ease of use (Eyberg & Robinson, 1981). For instance, use of continuous recording contributes to validity and utility by providing a thorough account of behavior and permitting data to be collected in less time than usually required of interval sampling methods (Eyberg & Robinson, 1981). In addition, the structure of the three situations allows both the parent and the child to proceed naturally under varying degress of parental control, which may increase the likelihood of parents and children exhibiting both typical and problem behaviors. The brevity and simplicity of the system also permits use of periodic assessments to determine treatment course and progress. Eyberg and Robinson (1981) note that the DPICS has considerable potential for facilitating therapist accountability by making reliable and valid direct observation data easily obtainable within a clinical setting. While the developers emphasize the practicality of the DPICS and its utility for clinicians, the playroom, observation room with two-way mirror, intercom, and bug-in-the-ear radio transmitter that they recommended are not available in many clinical settings.

Evaluation of the psychometric properties of the system when video recording is used seems worthwhile because it could increase the cost effectiveness and convenience of the system for clinical and research purposes. As with any instrument, even small changes may affect the DPICS' psychometric properties; therefore, such changes in procedures should be used cautiously.

The DPICS uses a frequency, or event, recording system of the simplest type. Every behavior for which there is a code is recorded every time it occurs. Event recording is relatively straightforward and often is used by clinical researchers. Event recording, however, has several limitations, including 1) the time at which behaviors occur is not recorded and sequences or temporal patterns cannot be determined from observer records, 2) less reliability than other procedures when the initiation and termination of a behavior are difficult to discriminate, 3) when low interrater reliability occurs it is difficult to determine which occurrence of the behavior caused the disagreement, and 4) observers can end up with the same frequencies although they recorded different behaviors during the observation (Cone & Foster, 1982). These problems could be reduced by using an interval system that divides the observation period into many equal intervals and recording the occurrence of the behavior within the interval. Although interval recording with short intervals is better for a more thorough analysis of the data, it is more cumbersome because a signal device is needed, which is not efficient for collecting data on low frequency behavior or for long periods of time (Cone & Foster, 1982). The use of total frequency of DPICS behaviors during the 5-minute interaction does seem to be appropriate for visual or statistical analysis, and the unit of analysis is an appropriate unit for assessment of reliability (Hartmann, 1977).

Duration of behavior may often be a characteristic of interest, and the DPICS generally considers duration in the assessment of continuous behaviors. For instance, crying is coded at the beginning of each 5-second period during the duration of the crying. For example, continuous crying would be coded when it began and at 6, 11, 16, and 21 seconds, and so on. This appears to necessitate use of a stopwatch or some type of signal or timing device. The DPICS' assessment of duration behavior may not be adequate for some research questions but should be adequate for clinical assessment.

The generally thorough and well-prepared DPICS manual deserves mention. Explicit rules and instructions are critical for observation systems to be reliable and remain as accurate as possible, and the clear strength of the manual is in the description and definition of code categories. The definitions, examples, guidelines, and decision rules are well written and refined and should serve as an example to individuals preparing coding definitions. Of notable absence in the manual are guidelines for training observers. Given the extensive research with the DPICS by Eyberg, Robinson, and others, at least a general procedure, if not a specific one, that has been successful surely could be described. Inclusion of training guidelines would seem to ensure satisfactory interrater reliabilities by users of the system. Another related addition that could add significantly to the system would be a videotape, available from the developers of the DPICS, of parent-child interaction segments that have been scored by the author and that could be used for training and criterion assessment. Because the manual was prepared prior to publication of many of the studies discussed in this review, the discussions of reliability, validity, and normative data are relatively brief and only moderately convincing. Given the bulk of recent research, the quality and utility of the system seems much more impressive.

References

Aragona, J. A., & Eyberg, S. M. (1981). Neglected children: Mothers' report of child behavior problems and observed verbal behavior. *Child Development, 52,* 596-602.

Cone, J. D., & Foster, S. L. (1982). Direct observation in clinical psychology. In P. C. Kendall & J. N. Butcher (Eds.), *Handbook of research methods in clinical psychology* (pp. 311-354). New York: John Wiley & Sons.

176 Dyadic Parent-Child Interaction Coding System

Cronbach, L. J., Gleser, G. C., Nanda, H., & Rajaratnam, N. (1972). *The dependability of behavioral measures*. New York: John Wiley & Sons.

Eyberg, S. M. (1974). *Manual for coding dyadic parent-child interactions*. Unpublished manuscript, Oregon Health Sciences University, Department of Medical Psychology, Portland.

Eyberg, S. M. (1985). Behavioral assessment: Advancing methodology in pediatric psychology. *Journal of Pediatric Psychology, 10*, 123-139.

Eyberg, S. M., & Matarazzo, R. G. (1980). Training parents as therapists: A comparison between individual parent-child interaction training and parent group didactic training. *Journal of Clinical Psychology, 36*, 492-499.

Eyberg, S. M., & Robinson, E. A. (1981). *Dyadic Parent-Child Coding System* (Ms. No. 2582). San Rafael, CA: Social and Behavioral Sciences Documents, Select Press.

Eyberg, S. M., & Robinson, E. A. (1982). Parent-child interaction training: Effects on family functioning. *Journal of Clinical Child Psychology, 11*, 130-137.

Eyberg, S. M., & Robinson, E. A. (1983). Dyadic Parent-Child Interaction Coding System: A manual. *Psychological Documents, 13*, 24.

Eyberg, S. M., Robinson, E. A., Kniskern, J., & O'Brian, P. (1978). *Dyadic parent-child interaction coding system: A manual (revised)*. Unpublished manuscript, Oregon Health Sciences University, Department of Medical Psychology, Portland, Oregon.

Hartmann, D. P. (1977). Considerations in the choice of interobserver reliability estimates. *Journal of Applied Behavior Analysis, 10*, 103-116.

Kniskern, J. R., Robinson, E. A., & Mitchell, S. K. (1983). Mother-child interaction in home and laboratory settings. *Child Study Journal, 13*, 23-39.

Packard, T., Robinson, E. A., & Grove, D. C. (1983). The effect of training procedures on the maintenance of parental relationship building skills. *Journal of Clinical Child Psychology, 12*, 181-186.

Patterson, G. R., Ray, R. S., Shaw, D. A., & Cobb, J. A. (1969). *A manual for coding family interactions* (Document No. D1234). New York: ASIS National Auxiliary Publications Service.

Peterson, S. L., Robinson, E. A., & Littman, I. (1983). Parent-child interaction training for parents with a history of mental retardation. *Journal of Applied Research in Mental Retardation, 4*, 329-342.

Robinson, E. A., & Eyberg, S. M. (1981). The Dyadic Parent-Child Interaction Coding System: Standardization and validation. *Journal of Consulting and Clinical Psychology, 49*, 245-250.

Robinson, E. A., Eyberg, S. M., & Ross, A. W. (1980). The standardization of an inventory of child conduct problem behaviors. *Journal of Clinical Child Psychology, 9*, 22-28.

Webster-Stratton, C. (1984). Randomized trial of two parent-training programs for families with conduct-disordered children. *Journal of Consulting and Clinical Psychology, 52*, 666-678.

Webster-Stratton, C. (1985a). Comparison of abusive and nonabusive families with conduct-disordered children. *American Journal of Orthopsychiatry, 55*, 59-69.

Webster-Stratton, C. (1985b). The effects of father involvement in parent training for conduct problem children. *Journal of Child Psychology and Psychiatry, 26*, 801-810.

Webster-Stratton, C. (1985c). Mother perceptions and mother-child interactions: Comparison of a clinic-referred and a nonclinic group. *Journal of Clinical Child Psychology, 14*, 334-339.

Webster-Stratton, C. (1985d). Predictors of treatment outcome in parent training for conduct disordered children. *Behavior Therapy, 16*, 223-243.

Zangwill, W. M., & Kniskern, J. R. (1982). Comparison of problem families in the clinic and at home. *Behavior Therapy, 13*, 145-152.

Michele Paludi, Ph.D.

Women's Studies Program, Hunter College, The City University of New York, New York, New York.

EATING DISORDERS INVENTORY

David M. Garner, Marion P. Olmsted, and Janet Polivy. Odessa, Florida: Psychological Assessment Resources.

Introduction

The Eating Disorder Inventory (EDI) is a 64-item, 6-point forced-choice inventory assessing several behavioral and psychological traits common in two eating disorders, bulimia and anorexia nervosa. Specifically, the EDI consists of eight subscales measuring traits commonly associated with these disorders:

Drive for Thinness: excessive concern with dieting and weight gain.

Bulimia: tendency toward episodes of binging and purging.

Body Satisfaction: specific body parts are perceived as too large.

Ineffectiveness: feelings of general inadequacy and not being in control of one's life.

Perfectionism: excessive personal expectations of superior achievement.

Interpersonal Distrust: sense of alienation.

Interoceptive Awareness: lack of confidence in recognizing and identifying emotions or hunger and satiety.

Maturity Fears: wish to return to security of preadolescent years.

The EDI, a self-report measure, may be utilized as a screening device, outcome measure, or part of typological research. It is not purported to be a diagnostic test for anorexia nervosa or bulimia.

The EDI was developed in 1983 by David M. Garner, Ph.D., Marion P. Olmsted, M.A., and Janet Polivy, Ph.D. in order to assess objectively attitudes or behaviors found in anorexia nervosa, an eating disorder involving a preoccupation with thinness that includes 1) an overall sense of personal ineffectiveness or helplessness, 2) disturbances in internal perceptions, and 3) disturbances in body image (Bruch, 1973). The American Psychiatric Association (1980) outlines the following diagnostic criteria for anorexia nervosa:

    A. Intense fear of becoming obese, which does not diminish as weight loss progresses.

    B. Disturbance of body image, e.g., claiming to "feel fat" even when emaciated.

    C. Weight loss of at least 25% of original body weight or, if under 18 years of age, weight loss from original body weight plus projected weight gain expected from growth charts may be combined to make the 25%.

    D. Refusal to maintain body weight over a minimal normal weight for age and height.

    E. No known physical illness that would account for the weight loss. (p. 69)

Garner et al. asked clinicians who were familiar with the research and treatment of anorexia and bulimia to generate items designed to measure psychological traits that have been found to be fundamental in the development of these eating disorders. Garfinkel and Garner (1982) proposed a modified version of the criteria employed for diagnosing patients with anorexia nervosa:

1) No restrictions on age of onset;
2) No loss of appetite required; the term *anorexia* is actually misleading. Weight loss of 25% or more of original body weight is not strictly required; if an individual was relatively thin at the onset or still growing and lost only 15-20%, it should not preclude a positive diagnosis;
3) A distorted, implacable attitude towards eating, food, or weight that overrides hunger, admonitions, reassurance, and threats;
4) No known medical illness that could account for the weight loss;
5) No other known psychiatric disorder with particular reference to primary affective disorders, schizophrenia, obsessive-compulsive neurosis, and phobias;
6) At least two of the following manifestations: amenorrhea, lanugo, bradycardia, periods of overactivity, episodes of bulimia, and/or vomiting (may be self-induced). (Garner & Olmsted, 1984, p. 1)

The authors of the EDI wanted to distinguish anorexics who restrict their dietary intake from bulimic anorexics, who eat with rapid ingestion of large quantities of food followed by self-induced purging.

Recognizing that anorexia nervosa is a multidimensional disorder, Garner et al. devised the EDI to assess the cognitive and behavioral characteristics clinically observed in anorexia nervosa. A pool of 146 items was initially generated by clinicians, and the eight previously mentioned constructs met the authors' final psychometric requirements for the scale.

Two groups of respondents participated in the validation of the EDI. The criterion group ($n = 129$) was composed of three subsamples of women, mostly anorexics, who were seen in consultation at the Clarke Institute of Psychiatry. These women averaged 20% below the expected weight for their height and age according to norms from Health and Welfare Canada. In this sample of 129 women, 56 were classified as "restricters" (anorexics who restrict their dietary intake) and 73 were diagnosed as "bulimic." Although no significant differences in age or duration of the eating disorder were obtained between these groups, significant differences were found in mean percentage of average weight, with the bulimics weighing more. An additional 155 anorexic women were tested, and subsequent psychometric analyses are based on these women.

The comparison group ($n = 770$) consisted of three samples of female university students who were enrolled in introductory and upper-level psychology classes. These volunteers were administered the EDI in their classes.

Garner et al. used the anorexic and college women to select items from the original item pool. Retained items were those that 1) significantly differentiated the two groups and 2) were highly correlated with only the subscale to which they were intended to belong.

Following the administration of the EDI to these initial samples of women, additional questions were developed for the Interoceptive Awareness and Maturity

Fears subscales. These new items were subsequently evaluated on the second and third samples of anorexic and college women. Garner et al. found no significant mean subscale score differences within the criterion groups across the three validation samples.

The EDI has been used for research purposes as well as in clinical settings.

Practical Applications/Uses

The Eating Disorder Inventory measures behavioral and psychological traits common in bulimia and anorexia nervosa. The EDI profile may thus provide a therapist with important clinical information related to treatment. Respondents may retake the EDI in order for the therapist to assess change as a result of treatment. Additionally, the EDI may be useful in identifying subtypes of anorexia nervosa, providing information that will help with the treatment of the eating disorder.

The EDI may be used as a screening device to indicate which individuals are weight-preoccupied. However, several limitations of the EDI in nonclinical populations should be noted:

1) The EDI is sensitive to response style bias and inaccurate reporting by the respondent.

2) The EDI may lack external validity. Its development was based upon its ability to differentiate between a criterion group and nonclinical samples. Elevated subscale scores among nonclinical populations should not be interpreted as indicating the identical pathology inferred for a clinical group.

3) The EDI does not assess all the psychopathological characteristics of anorexia nervosa.

4) The EDI, by itself, should never be used as the basis for screening for or diagnosing anorexia nervosa. Clinical diagnoses must confirm the presence of this eating disorder.

The EDI is purported to be used both in individual and group settings. The test developers have used the EDI with people as young as 12 years, with the assistance of an examiner to answer questions the young adolescents had. The presence of an examiner is recommended for older respondents as well. The examiner should be able to clarify items because the EDI does not measure idiosyncratic item interpretations.

Respondents are asked to read sixty-four 6-point, forced-choice items and rate whether each applies always, usually, often, sometimes, rarely, or never. Respondents must be told to answer all of the questions. Subscale scores should not be computed when more than one of the items in the subscale have been omitted.

Respondents should be encouraged to offer comments on the back of the test booklet. There is room on the cover page of the EDI test booklet for demographic information (e.g., name, age, marital status, occupation, parents' occupation) as well as weight history.

Respondents are given no time limit for completion of the EDI. Twenty minutes is the average amount of time required.

Administrators of the EDI do not have to be trained in clinical psychology. They must, however, be knowledgable about the purpose and application of the EDI's subscales and be able to provide clarification of the items to respondents.

The EDI comes with a set of hand-scoring keys, one key for each subscale. The most extreme anorexic response ("always" or "never," depending on the keyed direction) earns a score of 3. The consecutively adjacent responses receive a 2 and a 1. The three choices opposite to the most anorexic response receive 0. The subscale scores are the summation of all item scores for that particular subscale. These scores are recorded on the EDI booklet. Subscale scores are then plotted on profile forms, which allow comparison with subscale scores for the anorexic and female college student groups.

On the profile forms, the shaded areas indicate the 99% confidence intervals for the groups. Patients' scores that fall in these shaded areas are not significantly different ($p < .01$) from the normative sample means. Patients' individual subscale scores need to be graphed. This same procedure should be followed for the standard error for each subscale. Percentile scores may be reported as well and are based on the appropriate comparison group.

The EDI kit includes the manual, scoring keys, 25 test booklets, and 25 profile forms. Separate scoring keys, profile forms (package of 25), and test booklets (25 or 50) may also be obtained.

Test interpretation should be done by an individual with psychometric assessment skills. In clinical use, a patient's EDI score is compared with normative data for anorexia nervosa patients so as to generate hypotheses to be verified clinically. For nonclinical use, the EDI may be used as a screening device to locate people who are weight-preoccupied or who have "ego deficits." The EDI developers have identified college women as weight preoccupied if they score at or above the mean score for anorexia nervosa patients on Drive for Thinness. People who have high scores on all eight subscales may be at high risk for anorexia nervosa.

Technical Aspects

The average item total correlation of the eight subscales of the Eating Disorder Inventory was .63 ($SD = .13$). Reliability information was based on 271 college women on whom completed information on all subscales was obtained. Reliability coefficients (standardized Chronbach's alphas) for the anorexia nervosa group ranged from .83 (Interoceptive Awareness) to .93 (Ineffectiveness). Reliability coefficients for the female college studeuts ranged from .72 (Maturity Fears) to .92 (Body Dissatisfaction).

Criterion-related validity studies were performed by comparing the EDI patient profiles with the judgments of clinicians familiar with the patient's psychological presentation. A subgroup of 49 of the anorexia nervosa patients who had completed the EDI was assigned two raters: a psychologist and psychiatrist who were familiar with the patients, being their primary therapist or consultant. The raters were instructed to "rate the relevancy of each of these traits or characteristics for this patient compared to other anorexics that you have treated" (Garner, Olmsted, & Polivy, 1983) on an analogue scale divided into 10 centile intervals. The raters were provided with the description of the subscale content and with the patients' total score percentile rank within the entire anorexic sample. All interrater correlations were significant at the $p < .001$ level and ranged from .43 (Maturity Fears) to .68 (Ineffectiveness).

In addition, criterion-related validity was demonstrated in that the bulimic and restricter anorexia groups scored in the theoretically expected manner on specific subscales. A discriminant function analysis correctly classified 85% of anorexia nervosa subjects into restricter and bulimia subtypes based on their Bulimia subscale score. Convergent and discriminant validity was determined for the subsamples of anorexia nervosa patients.

Critique

The Eating Disorder Inventory is a valuable psychological tool that, when used in conjunction with clinical diagnoses, helps to delineate subtypes of anorexia nervosa. However, this reviewer is concerned with the wording of many items in the test because many adolescents (especially younger ones) may not understand the questions and therefore answer them inaccurately. Secondly, several items deal with issues that are developmentally appropriate for adolescents to be concerned or preoccupied with (e.g., body image, feeling part of the "in group"). Causal relationships linking anorexia with high scores on these items are dangerous. Thirdly, and most importantly, there is a growing scientific literature addressing the importance of gender role development and cultural prescriptions of ideal feminine weight. None of these issues are discussed or made part of the items of the EDI. This reviewer considers this absence a significant drawback to this instrument.

References

These lists include text citations and suggested additional reading.

Feminist Issues in Eating Disorders

Chernin, K. (1981). *The obsession: Reflections on the tyranny of slenderness.* New York: Hayser Colophon Books.

Boskind-Lodahl, M. (1976). Cinderella's stepsisters: A feminist perspective on anorexia nervosa and bulimia. *Signs, 2,* 120-146.

Mintz, L.B., & Betz, N.E. (1986). Sex differences in the nature, realism, and correlates of body image. *Sex Roles, 15,* 185-195.

Response-Style Bias

Vandereycken, W., & Vanderlinden, J. (1983). Denial of illness and the use of self-reporting measures in anorexia nervosa patients. *International Journal of Eating Disorders, 2,* 101-107.

Personality Features not Tapped by the EDI

Stonehill, E., & Crisp, A.H. (1977). Psychoneurotic characteristics of patients with anorexia nervosa before and after treatment and at follow-up 4-7 years later. *Journal of Psychosomatic Research, 21,* 187-193.

Strober, M. (1980). Personality and symptomatological features in young, nonchronic anorexia nervosa patients. *Journal of Psychosomatic Research, 24,* 353-359.

Research Using the EDI

Cooper Z., Cooper, P.J., & Fairborn, C.G. (1985). The specificity of the Eating Disorder Inventory. *British Journal of Clinical Psychology, 24,* 129-130.

Garfinkel, P.E., & Garner, D.M. (1982). *Anorexia nervosa: A multidimensional perspective.* New York: Brunner/Mazel.

Garner, D.M. (1985). Iatrogenesis in anorexia nervosa and bulimia nervosa. *International Journal of Eating Disorders, 4,* 701-726.

Garner, D.M., Olmsted, M.P., & Garfinkel, P.E. (1985). Similarities among bulimic groups selected by weight and weight history. *Journal of Psychiatric Research, 19,* 129-134.

Garner, D.M., Olmsted, M.P., & Polivy, J. (1983). Development and validation of a multidimensional eating disorder inventory for anorexia nervosa and bulimia. *International Journal of Eating Disorders, 2,* 15-34.

Olmsted, M.P., & Garner, D.M. (1986). The significance of self-induced vomiting as a weight control method among nonclinical samples. *International Journal of Eating Disorders, 5,* 683-700.

Criteria for Diagnosis of Anorexia

American Psychiatric Association. (1980). *Diagnostic and statistical manual of mental disorders* (3rd ed.). Washington, DC: Author.

Bruch, H. (1973). *Eating disorders.* New York: Basic Books.

Gurmal Rattan, Ph.D.
Associate Professor of Educational Psychology, Indiana University of Pennsylvania, Indiana, Pennsylvania.

Edward M. Levinson, Ph.D.
Assistant Professor of Educational Psychology, Indiana University of Pennsylvania, Indiana, Pennsylvania.

FLORIDA INTERNATIONAL DIAGNOSTIC-PRESCRIPTIVE VOCATIONAL COMPETENCY PROFILE

Howard Rosenberg and Dennis G. Tesolowski. Chicago, Illinois: Stoelting Company.

Introduction

Handicapped and disabled individuals historically have been underrepresented among our country's employed. In an attempt to address this problem, recent federal legislation (e.g., PL 94-142) has focused on providing these individuals with appropriate vocational training necessary to become productive members of society. In this regard, the Florida International Diagnostic-Prescriptive Vocational Competency Profile (FIDPVCP; Rosenberg & Tesolowski, 1979) was developed to evaluate an individual's vocational functioning.

The FIDPVCP was developed by Howard Rosenberg, Ed.D. and Dennis G. Tesolowski, Ed.D., co-directors of the Vocational Education for the Handicapped certificate program at the Florida International University in Miami, Florida.

The FIDPVCP was designed to assist professionals in secondary schools (including special education and vocational education classes) and post-secondary settings (e.g., vocational schools, vocational rehabilitation facilities, adult education classes, and sheltered workshops) in response to a paucity of behaviorally oriented measures. According to the authors, an instrument that was ". . . behaviorally based; required sustained behavioral observation; included elements of behavioral, psychometric, and situational evaluation strategies; and easily translated into an individualized vocational prescription" (Rosenberg & Tesolowski, 1979, p.2) was unavailable prior to development of the FIDPVCP.

The FIDPVCP consists of 70 items distributed within the following six domains: vocational self-help skills (14 items), social-emotional adjustment (14 items), work attitudes-responsibility (7 items), cognitive-learning ability (14 items), perceptual-motor skills (12 items), and general work habits (9 items). Each item contains five behaviorally based statements reflecting varying degrees of vocational competency ranging from Level 1 (unable to perform the task) to Level 5 (ready for competitive employment). Rosenberg & Tesolowski (1979) state that content validity for the

FIDPVCP was established vis à vis a literature search, on-site visits to vocational training programs, and interviews with trainers. From this procedure, 70 items were gleaned from a pool of several hundred. However, it is not clear what criteria were used in item selection and if other allied professionals participated in screening items. No additional information is provided to ensure that items comprising the FIDPVCP constitute a representative sample of questions from the domain of vocational competency.

The complete FIDPVCP includes a manual, a set of rating forms, and a set of Individualized Vocational Prescription (IVP) forms. These IVP forms are similar in structure to the Individual Education Plans (IEPs) developed for many handicapped students in compliance with federal law. The IVP includes sections that describe the individual's strengths and weaknesses, the selection of training programs as a function of the individual's interests and abilities, the development of annual and short-term goals, and ancillary services that may be required (e.g., speech therapy, counseling). Additionally, the IVP provides an individual weekly schedule along with signatures of committee members who are authorizing the service delivery.

The examiner's record form contains information relating to the student's biographical and academic background, an abbreviated description of the rating scale, and a listing of the 70 items across the six domains. The last page of the record form includes a scoring table and chart that can be used to profile the individual's strengths and weaknesses. However, the rating scale is not sufficiently detailed, requiring the examiner to refer back to the manual.

Practical Applications/Uses

General uses of the FIDPVCP are purported to include assessing an individual's current level of vocational competency, identifying strengths and weaknesses for use in vocational programming, selecting individuals for training in specific vocational programs; monitoring growth and development of an individual's vocational competency, and determining the differential effects of various training programs as a function of different populations. The authors do not state the age ranges or types of populations with whom the test may be used appropriately. However, standardization information is provided on five client populations (economically disadvantaged, educable mentally retarded, trainable mentally retarded, seriously emotionally disturbed, and specific learning disabled) with ages ranging from 13 years, 3 months to 64 years, 7 months.

Rosenberg & Tesolowski (1979) suggest that "valid and reliable observation data is contingent upon the length of service of the client/student within the training program" (p. 3). However, no constraints are placed on the length of examiner observation time prior to rating an individual's behavior. Such an unstructured approach may adversely affect the meaningfulness of obtained scores. Additionally, the authors do not indicate the length of time necessary to administer and score the instrument and do not clearly define the "professional staff" who are qualified users of this measure. Hence, the lack of standardized procedures and other descriptive chararacteristics of this measure limit its potential usefulness.

The rating of FIDPVCP behaviors are completed by assigning a numerical value reflecting vocational competency. The ratings are judged on a 5-point scale: a 1 is assigned if the individual is "essentially unable to perform the task"; a 2 indicates "below average performance in training programs"; a 3 suggests "average performance in training programs"; a 4 is assigned to "above average performance"; and a 5 predicts "ready for placement in competitive employment." Composite scores can be computed for each domain and converted to percentages. Similarly, a total score reflecting an individual's vocational functioning can be calculated and converted to a percentage score. The resultant scores can be plotted to amplify an individual's strengths and weaknesses. A scoring example is provided in the manual to assist users.

The hand-scoring procedure for the FIDPVCP is simple and straightforward. The instructions provided in the manual are presented clearly and minimize errors in converting raw scores to percentages. Although scoring is "user friendly," no provisions are made to handle missing data for responses that may not be a part of the training program (e.g., use of power tools). This problem will need to be addressed to ensure that the FIDPVCP is applicable to most vocational settings.

The behavioral criteria used to assign ratings for each item are generally well written, specific, and observable. However, only an abbreviated rating scale is listed on the protocol. This may have an adverse effect by encouraging the examiner to make a subjective judgment in rating behaviors. Such potential misuse could have been remedied by including the behavioral descriptions for each item on the record form.

Rosenberg & Tesolowski (1979) do not indicate the method by which they determined that a certain level of behavior reflected "below average performance in a training program" or a readiness "for placement in competitive employment." For example, can we be sure that reading "at [a] third to sixth grade level" reflects "average performance in [a] training program"? A description of the means by which such "average" performance was determined would have increased the confidence raters could place in the interpretations of their ratings. Additionally, the behavioral criteria associated with certain items appear to be related only marginally to the trait allegedly measured by the item. For example, "reaction to praise," "perseverance," and "concentration" are all measured by production rate, which may be affected only partially by the construct underlying each of these traits.

Technical Aspects

The FIDPVCP was "standardized" on a sample of 100 adolescents and adults, 50 of which were attending a secondary educational facility and ranged in age from 13 years, 3 months to 20 years, 1 month. The balance were attending a post-secondary facility and ranged in age from 19 years, 8 months to 64 years, 7 months. Ten subjects participated in each of five distinct diagnostic groups at both the secondary and post-secondary levels. As noted earlier, the groups consisted of educable mentally retarded, trainable mentally retarded, specific learning disabled, seriously emotionally disturbed, and economically disadvantaged. Normative data for each group are presented in the manual.

Although descriptive characteristics of each group are presented, no objective criteria are offered on how individuals were placed into diagnostic groups. This is especially evident for the learning disabled (LD) and emotionally disturbed (ED) groups. For example, the deficits of LD participants consisted of ". . . the person's imperfect ability to listen, think, read, write, spell or to perform mathematical calculations . . ." (Rosenberg & Tesolowski, 1979, p. 21), but no objective criteria are documented to establish "imperfect ability." Similarly, participants with ". . . marked deviation from age-appropriate emotional and/or social behavior that significantly interfered with their own development, the lives of others or both" (p. 21) are included in the emotionally disturbed group without objective or psychometric criteria for inclusion in this group. Moreover, the questionable qualifications of social workers, physicians, and individuals in related disciplines to diagnose emotional disturbance adds further to the tenuousness of this diagnostic category. Additionally, a rationale for including individuals with organic brain syndrome in the LD group is not clear. Such individuals often present with emotional and personality disorders (American Psychiatric Association, 1980) and may be better placed in the ED group.

The lack of demarcations between the LD and ED groups can be further evidenced by the normative data presented for the Socio-Emotional Adjustment scale. The LD group at the secondary level achieved a mean score of 43.95 in contrast to 47 for the ED group. The low score for the LD group indicates a greater deficit in socioemotional adjustment than the ED group, a result contrary to expectations. Although this relationship is reversed at the post-secondary level, the results suggest that these groups are heterogeneous and may not represent meaningful diagnostic groups for purposes of vocational planning.

Attempts to demonstrate psychometric credibility for the FIDPVCP consist of estimates of interrater reliability and concurrent validity. Two evaluators administered the FIDPVCP to each participant in the 10 groups. Coefficients ranged from .74 for the seriously emotionally disturbed group (secondary level) to .90 for the trainable mentally retarded group (secondary level). Most of the remaining coefficients ranged from .80 to .90 indicating that scoring criteria are sufficiently objective for examiners to obtain a similar score for a given individual. Although establishing interrater reliability is an important first step, it is not sufficient to address the stability of this measure. A test-retest procedure would have been more appropriate in order to provide evidence that the construct being measured is consistent over time.

Evidence of concurrent validity was established by having two evaluators rank-order each participant's vocational competency within his or her respective group. The same evaluators then administered the FIDPVCP and computed the amount of agreement between their independent ratings and subject performance on this instrument. The resulting Spearman rank-order correlation coefficients ranged from .76 to .96.

Although the magnitude of these correlations meet psychometric standards, the meaningfulness of these data is suspect. A major procedural flaw in establishing concurrent validity entailed having the same evaluators perform both the independent ratings and the administration the FIDPVCP. The high validity coefficients may well have resulted from the evaluators' prior knowledge of each participant's performance. It would have been more meaningful to have different evalutors perform each segment of the evaluation. More importantly, however, an alternate and perhaps more significant evidence of validity would have entailed a comparison of FIDPVCP performance with other "known" measures of vocational competency (e.g., the MDC Behavior Identification Format; Materials Development Center, 1974).

Overall, the psychometric properties of the FIDPVCP are weak. Its utility therefore is affected, and the uses outlined in the manual need to be interpreted with caution. This is especially evident when developing an Individualized Vocational Program based on individual strengths and weaknesses. In view of the fact that performance levels between diagnostic groups were not evaluated for statistical significance, these reviewers question the justification for differential vocational programming. Moreover, the relatively flat profiles for several groups at both the secondary and post-secondary levels (e.g., economically disadvantaged, educable mentally retarded, and trainable mentally retarded) may not warrant differential intervention based on reported "strengths and weaknesses." Notwithstanding the small sample size, however, the FIDPVCP does provide an approximation of an individual's current vocational functioning.

Critique

The FIDPVCP was designed to assess an individual's vocational competency, primarily for program planning and placement. In view of the limited sample size, lack of clear demarcations between diagnostic groups, and inadequate procedures used to establish reliability and validity, the efficacy of the FIDPVCP has not been fully demonstrated. Instead, the FIDPVCP may be considered at an experimental stage of development and the results therefore interpreted accordingly. Further research and refinement will be required in order to realize the clinical utilty of this instrument.

References

American Psychiatric Association. (1980). *Diagnostic and statistical manual of mental disorders* (3rd ed.). Washington, DC: Author.

Materials Development Center (1974). *MDC Behavior Identification Format.* Menonomie, WI: University of Wisconsin-Stout.

Rosenberg, H., & Tesolowski, D. G. (1979). *The Florida International Diagnostic-Prescriptive Vocational Competency Profile.* Chicago, IL: Stoelting Co.

Kenneth J. Smith, Ph.D.
Professor and Director, Interdisciplinary Educational Clinic, College of Education, The University of Arizona, Tucson, Arizona.

Darrell L. Sabers, Ph.D.
Professor of Educational Psychology, The University of Arizona, Tucson, Arizona.

GILLINGHAM-CHILDS PHONICS PROFICIENCY SCALES, SERIES I AND II

Anna Gillingham, Bessie W. Stillman, Sally B. Childs, and Ralph de S. Childs. Cambridge, Massachusetts: Educators Publishing Service, Inc.

Introduction

The Gillingham-Childs Phonics Proficiency Scales are designed to evaluate an individual's progress in mastering the sequential steps of the coding, or phonic, method of teaching initial written language skills. Two levels are available. Series I, the basic level, is intended to assess basic reading and spelling skills; Series II measures only advanced reading skills. Although the authors themselves do not use the term to describe the tests, the Gillingham-Childs Phonics Proficiency Scales are clearly criterion-referenced tests intended to assess the phonic skills the authors consider essential.

The tests are designed to be consistent with the Gillingham method of "teaching coding." It is one of several visual-auditory-kinesthetic-tactile (VAKT) methods that utilizes a combination of synthetic phonics, writing, and tracing in order to make use of multiple senses. It also emphasizes spelling as part of reading instruction. Gillingham's procedures are very structured, with teachers instructed to follow a very precise set of steps.

Anna Gillingham developed the method to be consistent with the theories of Dr. Samuel Orton, with whom she worked in the Language Research Project in the New York Neurological Institute from 1932-36. She and Bessie Stillman, a friend and remedial reading teacher, devised a manual for the procedures, *Remedial Training for Children with Specific Disability in Reading, Spelling, and Penmanship* (Gillingham & Stillman, 1960); thus, the approach is sometimes referred to as the Gillingham-Stillman method.

Although no age groups are identified for the series, Series I clearly is intended for beginning readers, and Series II is designed for more advanced students. The phonics content of Series II is similar to that found in phonics programs through Grade 2. However, some of the vocabulary would be unfamiliar to most second-grade students (e.g., injection, impede, conclusion, and deltoid).

The 12 primary subtests, called "scales," for Series I are Letter-Sound Connections—Reading, Sound-Letter Connections—Spelling, 3-Letter Regular Words, 3-Letter Nonce Words, Consonant Digraphs and Blends—Regular Words, Consonant Digraphs and Blends—Nonce Words, Monosyllables Ending in F, L, S, Vowel-Consonant-E—Regular Words, Vowel-Consonant-E—Nonce Words, Syllable Division between Two Consonants—Regular Words, Syllable Division between Two Consonants—Nonce Words, and Doubling Final Consonants in Monosyllables. A separate section, "Extra Skills," includes the following four additional subtests: Knowing the Alphabet, RED WORDS for Reading and Spelling—First Level, Basic Plural Rule, and Vowel-Consonant-E—Adding Suffixes. These identifications are probably familiar to most readers of this review. "RED WORDS" are defined by the authors as "irregular" words, apparently meaning those that do not follow common spelling patterns.

Series II has 13 primary scales and 5 "extra" scales. Six of the 13 scales test vowel rules, and 2 scales test hard and soft "c" and "g" and a combination of some inflectional endings and final consonants. The five "extra" scales are X-1 Prefixes and Suffixes, X-2 Words Irregular for Reading—Level 2, X-3 Words Irregular for Reading—Level 3, X-4 Reading Pronunciations, and X-5 Syllable Division—Rule 2.

Each level includes a "Directions for Use" manual, a record booklet (two separate ones for reading and spelling in Series I), and a booklet containing the items. For each reading scale, or subtest, the student "reads" each item orally while the teacher checks the responses for accuracy. Spelling words, including "nonce," or "nonsense," words, are dictated.

The answer forms for reading in both Series I and Series II reproduce the test items, allowing the teacher to mark incorrect items directly on the answer sheet. A place is provided at the bottom of the answer sheet for each subtest for entering the date and the number of correct responses. A summary sheet is included on the last page. The answer booklet for the spelling test in Series I merely provides blanks for the child to use in attempting to spell each word and also a summary page. Each record booklet provides a glossary of terms.

The "Directions for Use" manuals are grossly inadequate for a commercial product. It is probably not necessary that they include data on prior administrations (standardization sample) for this type of instrument; however, there is no section on interpretation of results. In addition, they lack information regarding the technical aspects of the test.

Practical Applications/Uses

In order for a teacher to find the Gillingham-Childs Phonics Proficiency Scales useful, he or she must agree at least with the test authors' choices of important phonic skills. No evidence is presented to indicate how those choices were made. Further, one would have to agree with the directions the authors provide for determining errors and scoring in the "Directions for Use" manual for Series I:

> Every sound must be clearly and correctly made, and any ambiguity, addition, omission, substitution, reversal or self-correction is an error. The goal is perfect performance so that any error indicates the need for further practice.

Furthermore, one must accept that "pure" sounds are essential and that regional variations in the pronunciation of the short sounds of *a* and *o* are permitted. The authors do not indicate whether social dialectal and "bilingual" pronunciations are permitted. Moreover, the authors state that if a child must sound out a word, even silently, the child is not reading, which will offend many phonics advocates. However, it is clear that the authors believe that the ability to produce the appropriate "pure" sounds is reading.

The testing procedure that is followed is generally appropriate. It is true, as the authors indicate, that it is virtually impossible to test phonic abilities in a group situation because using phonics in reading requires proceeding from print to speech, not from speech to print. Thus, the student must pronounce words and may not, as in group test, mark written choices in a booklet after hearing a teacher pronounce words or merely use words that he or she already knows as "sight" words. Also, the use of "nonce" words has value, as students may recognize real words from sight.

Test administration time cannot be estimated because not all subtests need to be administered. Examiner qualifications need not be great if one accepts all of the definitions and explanations provided by the test authors; however, in order to filter out the errors, examiner qualifications would have to be substantial.

Although the administration procedures are fitting for a test of this nature, the following examples illustrate that the instrument itself is replete with linguistic errors. First, the directions for Series I instruct the teacher to test the "blending" of "sounds" into words, which, given the authors' admonition against "sounding out" words, is technically and practically impossible; it certainly appears to be a contradiction in the directions. Second, the mixture of (consonant) digraphs and blends (clusters) in the same subtest (Consonant Digraphs and Blends, Series I), with a single total score, is confusing because digraphs are simple spellings that consist of two letters associated with one phoneme, such as *th-*, *sh-*, *-ng*, and so forth, and "blends" refer to closely associated phones, or speech sounds, such as *sp-* in *spin*, *cr-* in *crack*, *fl* in *flat*, and *str-* in *string*. A total score of the mixture would be essentially meaningless. In Scale 5, Vowel-Consonant-E (Series I), one is instructed that long *u* is /yo͞o/, not /o͞o/ and that *duke* is /do͞ok/, not /do͞ok/. /yo͞o/ is of course two phonemes, and /o͞o/ is one of them. Further, these reviewers do not pronounce *duke* as /do͞ok/, indicating the authors' lack of awareness of dialectal variation. Finally, including base words and suffixes in a phonics test is inappropriate because it confuses morphemic (meaning units), or structural, analysis with phonics.

As mentioned previously, the manuals do not address interpretation of test results. It seems unforgivable that the test authors make no provision for the user to obtain the background necessary to interpret the test results in the way prescribed by the authors. Given that no interpretation section is included, these reviewers cannot suggest the ways in which the results can be used or misused.

Technical Aspects

To the knowledge of these test reviewers, no technical data exist for the Gillingham-Childs Phonics Proficiency Scales. It seems, however, that some data on technical aspects of the tests could be produced easily, even if the tests are intended to produce criterion-referenced scores. For example, some reliability or precision data could be provided, especially because subtests provide total scores (even though providing total scores conflicts somewhat with the test's apparently criterion-referenced intentions). More importantly, the manual should include an extensive section describing why the test authors chose the content and skills that they did.

Critique

The Gillingham-Childs Phonics Proficiency Tests are better than most phonics tests. They appropriately proceed from print to speech, and the skills tested are reasonable, although no justification is given for choosing them. However, given the number of practically important linguistic errors, the failure to recognize dialectal and "bilingual" differences in pronunciation, and the lack of reliability or item precision data, these reviewers cannot recommend the use of this test to anyone in its present form except for purposes of obtaining data necessary for a revision. Because too many aspects of the tests must be revised to order to make the tests acceptable, it is not suggested that the development of technical data be an immediate goal. However, it would not be difficult to improve these tests and obtain data that might support their use.

References

This list includes text citations and suggested additional reading.

Aukerman, R. C. (1981). *Approaches to beginning reading instruction* (2nd ed.). New York: John Wiley & Sons.

Childs, S. B., & Childs, R. (1973). *Gillingham-Childs Phonics Proficiency Scales: Series 2— Advanced Reading.* Cambridge, MA: Educators Publishing Service.

Gillingham, A., & Stillman, B. W. (1960). *Remedial training for children with specific disability in reading, spelling and penmanship* (7th ed.). Cambridge, MA: Educators Publishing Service.

Gillingham, A., Stillman, B. W., & Childs, S. B. (1970). *Gillingham-Childs Phonics Proficiency Scales: Series 1—Basic Reading and Spelling.* Cambridge, MA: Educators Publishing Service.

Orton, S. (1937). *Reading, writing, and speech problems in children.* New York: W. W. Norton.

Tierney, R. J., et al. (1980). *Reading strategies and practices: A guide for improving instruction.* Newton, MA: Allyn & Bacon.

Robert L. Heilbronner, Ph.D.
Clinical Psychologist, Section of Neuropsychology, HCA-Presbyterian Hospital, Oklahoma City, Oklahoma.

Michael Ayers, Ph.D.
Clinical Psychologist, Section of Neuropsychology, HCA-Presbyterian Hospital, Oklahoma City, Oklahoma.

THE GRADED NAMING TEST

Pat McKenna and Elizabeth K. Warrington. Windsor, England: NFER-Nelson Publishing Company Ltd.

Introduction

The Graded Naming Test (GNT) was developed by Pat McKenna and Elizabeth K. Warrington in 1983 to screen for naming deficits and to estimate an individual's premorbid level of intelligence. The test authors assumed that naming skills, like general verbal abilities, varied across the general population. However, there was "no test of naming ability appropriate to a clinical population that is graded in difficulty to take into account an individual's premorbid naming vocabulary" (McKenna & Warrington, 1983, p. 3).

Until the GNT was developed, naming tests tended to rely on very common objects in order to ensure a familiar vocabulary for all subjects. Patients typically were assessed by reference to known groups of dysphasics (persons with acquired language deficits) rather than by reference to their own premorbid naming ability level. The difficulty with this approach is that less frequent items are more vulnerable to word retrieval difficulties than the more commonly used, well-practiced items (Rochford & Williams, 1965; Newcombe, Oldfield, & Wingfield, 1965). An individual with a good naming vocabulary who develops an insidious word finding difficulty following brain insult may continue to function in the normal range when presented with items that are high in frequency and are used more commonly. Therefore, the GNT was developed in order to "sample the more vulnerable items on the boundary of the individual's vocabulary" (McKenna & Warrington, 1983, p. 3).

Normative standards for the GNT are based on the performance of 100 subjects within the age range of 21 to 76. Of these subjects, 72 were inpatients at the National Hospital in England and had known extracerebral neurological disorders. The remainder of the sample consisted of nonhospitalized volunteers. Only subjects who were educated in the normal English system were included in the study. The sample was divided into six (20-30, 31-40, 41-50, 51-60, 61-70, and 71 and older) age groups that were represented approximately equally in number; however, there were only two subjects in the 70 and older age group. The participants included 69 males and 31 females, but the manual does not provide information relative to the distribution of males and females per age category. In addition, the nature of the subjects' extracerebral disorders is not known.

Subjects were tested in one session. Four tests of intelligence were selected as baseline measures. The Vocabulary and Picture Completion subtests of the Wechsler Adult Intelligence Scale (WAIS; Wechsler, 1955) were chosen because they are the individual Verbal and Performance subtests that correlate most highly with the WAIS Full Scale IQ. The National Adult Reading Test and the Schonell Graded Word Reading Test also were chosen because they were felt to provide a stable measure of intellectual level relatively unaffected by normal aging or brain damage. The two WAIS subtests were administered, followed by the two Reading tests. Finally, a "naming test," consisting of a pool of 61 items, was administered. These items were selected from a variety of sources, including illustrated dictionaries, newspapers, and magazines. Four main criteria were used for selecting the 61 items: 1) the pictorial representation of the name had to be distinctive and not perceptually or verbally confusable with other stimulus items, 2) the pictorial representation had to be unambiguous such that only a single response could be elicited by the item, 3) items requiring a "specialist" knowledge were excluded, and 4) items in very common usage were excluded. Thirty object names were selected from the 61-item pool. Items that were too easy or too difficult to produce good discrimination between subjects were eliminated as were items that gave rise to ambiguous responses.

The results of the standardization study revealed high correlations between vocabulary and reading ability. High correlations also were demonstrated between object naming, spoken vocabulary, and reading vocabulary. Age did not correlate significantly with reading vocabulary, and it only weakly correlated with naming ability. The nonverbal IQ measure, Picture Completion, correlated least highly with the 30 items. Because the obtained mean age-corrected scaled scores on the WAIS Vocabulary subtest were higher than the WAIS standardization group mean scores, the GNT authors felt it inappropriate to standardize the scores for object naming on the same scale. However, the manual does provide tables and conversion equations for the object naming scores in order to determine comparable baseline equivalents on the WAIS Vocabulary subtest, the NART, and the GWRT. The mean score for the object naming test was 22.5 correct responses.

The GNT consists of an object picture book, a test manual, and a record form for recording an examiner's responses. The object picture book consists of 30 black-and-white line drawings in a spiral-bound format for easy page turning. The manual describes the development of the test and results from a validation study and presents administration, scoring, and interpretation guidelines. Tables and appendices describing relevant standardization sample data and a conversion table for converting scores on the GNT to equivalent scales from other tests also are included. The record sheet contains spaces for demographic information (name, age, patient number, and date), and the 30 item names are listed with space beside them for recording an examinee's verbatim responses.

Practical Applications/Uses

The GNT was designed to measure naming deficits in brain-damaged patients. Naming impairments are not always demonstrated in the majority of neuropsychological evaluations. Rarely are they the primary presenting problem of a person undergoing neuropsychological testing. A failure to name objects, however, can be quite debilitating and could negatively impact an individual's ability to function effectively in his or her environment. Indeed, impaired naming abilities can manifest within the context of other langauge-related deficits and one could easily mistake this type of deficit for a memory problem. Perhaps the most useful application of the GNT would be within a battery of neuropsychological tests, especially those designed to qualitatively assess language abilities. If a patient showed particular difficulty on the Verbal subtest of the WAIS-R or on other tests purported to measure the integrity of the dominant cerebral hemisphere, the GNT could assist in identifying the nature of the impairments and, quite possibly, have some utility in localizing a suspected lesion. The results of testing with the GNT should be compared to scores on other language-dependent tests. Consequently, analyses of the test results would be understood maximally by clinical neuropsychologists or speech and language pathologists trained and practiced in the administration and interpretation of various neuropsychological and language-dependent tests, respectively.

The GNT consists of 30 black-and-white line drawings that are presented to the examinee according to a prearranged order of difficulty. The examiner points to a drawing and asks "What is this?" and then records the examinee's answer verbatim. The test is untimed. The authors suggest that it is necessary to question the examinee's responses when, for example, individuals misperceive items (e.g., "flowers" instead of shuttle cock) or give generic responses (e.g., "building" instead of pagoda). The test manual includes the following principles to guide the examiner as to when to question a response:

1. *Pointing.* The examiner directs the examinee's attention to the salient feature of the item by pointing.
2. *Perceptual reorientation.* When the subject completely misperceives the stimulus, the examiner provides another cue, such as "No, it is something else."
3. *Semantic reorientation.* When the subject's response is insufficient or imprecise, the examiner provides another cue, such as "What is another name?" or "What else could it be?"

The examinee's responses are scored on a record sheet. The correct name of each object appears on the left side of the page. The examiner records the subject's verbatim response in a space to the right. If an object is named correctly, that item is given a score of 1; if not, no point is awarded. The authors suggest that the obtained raw score reflects the subject's present naming vocabulary and does not measure anomia or word finding difficulties directly. An object naming score of 25 or greater is labeled superior; scores between 19 and 24, bright normal; between 11 and 18, average; and of 10 or below, dull normal. Appendix 3 in the manual lists the 30 graded naming items in order of difficulty and the percentage of subjects per item that gave a correct response. An appendix that gives equivalent scores on existing clinical tests, including the WAIS Vocabulary subtest, the National Adult Reading Test (NART; Nelson, 1977), and the Schonell Graded Word Reading Test (GWRT; Schonell, 1942) is provided.

Technical Aspects

No reliability information is provided in the test manual for the Graded Naming Test. Predictive validity is the primary source of validity for the GNT. Predictive, or criterion, validity reflects the extent to which a test adequately samples the subject matter that is the domain of the test. In terms of the GNT, predictive validity is concerned with which individuals do and do not have deficits in naming and, ultimately perhaps, left-cerebral hemisphere dysfunction at the time the test is administered.

The original validation study of the GNT sought to establish a naming test that would detect naming difficulties in patients with left-hemisphere lesions. The experimental group consisted of 46 patients, 29 males and 17 females, with radiologically documented unilateral left-hemisphere lesions. Thirty-two patients had well-localized, space-occupying lesions; 14 had left-hemisphere vascular lesions. Importantly, there is no information provided in the manual as to the precise anatomic locus of the lesions. The patients ranged in age from 20 to 67 years (mean age was 46.2 years). As in the case of the normative sample, patients not educated in the English school system were excluded. Results of the validation study revealed highly significant differences ($p > .001$) between the performance of the experimental group and that of the normative sample on all three covariates (Vocabulary, NART, GWRT). With the exception of the three most difficult items, the failure rate for the experimental condition was 30% greater than that of the normative sample and 10% to 35% greater over all the items.

Construct validity reflects the extent to which a test measures a psychological trait or construct. To the extent that the GNT measures the construct being investigated (naming abilities) or correlates with one or more observable variables or tests shown to be sensitive to this domain, construct validity is assumed. Likewise, concurrent ability is assumed to the degree that the GNT correlates with other tests assessing object naming. However, because the GNT has not been compared to other, more well-established tests measuring verbal naming abilities, construct and concurrent validity cannot be measured adequately. As a test requiring the subject to name objects, however, the Graded Naming Test appears to have face validity.

Critique

The test authors are to be credited for attempting to refine naming procedures. The Graded Naming Test would be particularly useful as an easy-to-use resource for objects to be named. It provides the clinician with an available set of items, thereby eliminating the need to search the office, hospital room, or other setting for increasingly difficult objects to be named. However, the Graded Naming Test does have a number of problems in terms of test development, standardization sample, and diagnostic validity. As a face-valid test of naming abilities, the GNT lacks adequate reliability and construct validity. Furthermore, the manual provides no information on concurrent or criterion validity. One cannot be completely sure whether a deficient performance on the GNT reflects impaired verbal naming abilities, faulty word retrieval, perceptual deficits, or some other ability that has been compromised by brain damage. Unlike the more widely used Boston Naming Test

(Kaplan, Goodglass, & Weintraub, 1983), the GNT does not include stimulus cues or phonemic cues that help maximize a patient's ability to name an item, even if a spontaneous response is not forthcoming. Ideally, the GNT performance of normals and left-hemisphere brain-damaged subjects should be correlated with scores from the more empirically validated Boston Naming Test.

In terms of its content, the GNT employs some items that are essentially indigenous to England. Many of these objects would not be recognizable to even the most well-educated American, which certainly limits its use within the context of a battery of neuropsychological tests in the United States. The drawings themselves are rather ill-defined and ambiguous. With the exception of one item (the periscope on a submarine), the subject is not likely to know whether the item to be named is one feature of the drawing or the whole object itself. Furthermore, some of the items easily can be mistaken for others: the picture of the anteater looks more like a tapir than anything else. In terms of scoring, it is unknown why the mean score from the standardization sample (22.5) is considered as being in the bright normal range and not in the average range.

As a test reportedly developed to consider individual differences when taking into account an individual's premorbid naming vocabulary, the GNT does not consider the role that age and education may play in test performance. The standardization sample does not include an adequate number of individuals in the 70 and over age group. In order to be most useful, the Graded Naming Test needs norms and cutoff scores for different age groups and years of education. Establishment of norms for different age groups potentially could demonstrate if and how naming abilities are likely to change with advancing age, which could be especially useful in differentiating a progressive dementia from depression or normal aging.

References

Kaplan, E., Goodglass, H., & Weintraub, S. (1983). *The Boston Naming Test.* Philadelphia: Lea and Febiger.

McKenna, P., & Warrington, E. K. (1983). *Manual for the Graded Naming Test.* Windsor, Berkshire, England: NFER-NELSON.

Nelson, H. E. (1977). *National Adult Reading Test (NART)* (test manual). Windsor, Berkshire, England: NFER-NELSON.

Newcombe, F., Oldfield, R. C., & Wingfield, A. (1965). Object naming by dysphasic patients. *Nature, 207,* 1217-1218.

Rochford, G., & Williams, M. (1965). Studies in the development and breakdown of the use of names: Pt. IV. The effects of word frequency. *Journal of Neurology, Neurosurgery and Psychiatry, 28,* 407-413.

Schonell, F. J. (1942). *Backwardness in the basic subjects.* Edinburgh, Scotland: Oliver and Boyd.

Wechsler, D. (1955). *Manual for the Wechsler Adult Intelligence Scale.* New York: Psychological Corporation.

Francis X. Archambault, Jr., Ph.D.

Professor and Department Head, Department of Educational Psychology, University of Connecticut, Storrs, Connecticut.

GRADUATE AND MANAGERIAL ASSESSMENT

Psychometric Research Unit, The Hatfield Polytechnic. Windsor, England: NFER-Nelson Publishing Company, Ltd.

Introduction

The Graduate and Managerial Assessment (GMA) is a series of three tests, each with a distinct rationale and item layout, measuring numerical, verbal, and abstract abilities. The GMA is intended to be used in the recruitment and selection of upper-level employees for industry, commerce, and public service. It also is designed to identify individuals with management and promotion potential in these same settings. The GMA is aimed at the upper 12.5% of the British population. Personnel of the Psychometric Research Unit of the Hatfield Polytechnic in Hertfordshire, England began to develop the GMA in 1983. Validating studies reported in the test manual were conducted in 1984. Additional studies are now under way.

Numerical Test (GMA-N). The GMA-N was designed to provide an assessment of reasoning in a numerical context. The test publishers suggest that the test has particular value in the recruitment of prospective managers into "finance-related occupations, and more generally in assuring the employer of a minimum level of competence in those numerical skills which are increasingly required of the modern manager" (Hatfield Polytechnic, 1985, p. 3). The GMA-N was designed to emphasize problem solving strategies rather than computational skills. Consequently, the arithmetic involved is fairly simple (i.e., about algebra level). The test also was intended to ensure that very little time would be spent assimilating information from the item stem, thereby affording larger portions of time for actual problem solving.

The GMA-N has two forms, A and B, each containing 33 items equally distributed across 11 item stems. The three items per stem are supposed to measure mathematics reasoning at different levels of difficulty. The publishers offer the following comments on this design feature:

> The attempt to cross content area with item difficulty affords the user two kinds of information. If a subject is able to answer the first question within each group of three, but tends to get the second, and especially, the third wrong across content areas, this points to a reasonable knowledge of mathematics at this level, but possibly weak reasoning skills. On the other hand, if a subject gets all three questions within some groups of three correct, but other groups of three completely wrong, this may indicate sound reasoning ability but poor mathematical knowledge. (Hatfield Polytechnic, 1985, p. 4)

197

Subjects respond to the GMA-N by selecting their intended response from among 16 possible options per stem. This format was chosen so that scores could not be inflated by guessing the correct response, a shortcoming the publishers went to great lengths to avoid.

Verbal Test (GMA-V). The verbal reasoning measure purportedly assesses verbal comprehension and critical thinking skills. It descends in part from the Watson-Glaser Critical Thinking Appraisal, but, according to the test publishers, the immediate forebears of the test are verbal critical reasoning tests written at the Hatfield Polytechnic. The guiding principle in the GMA-V is that "the evaluation and interpretation of written materials is an integral part of many, if not all, high-level jobs, and appropriate levels of skill in this area may be crucial to successful appointments" (Hatfield Polytechnic, 1985, p. 7).

The GMA-V, like many other verbal reasoning measures, presents information in a short prose passage and then asks questions requiring examinees to use that information. The authors indicate that the four statements relating to each of the 15 passages (i.e., a total of 60 items) require the examinees to detach themselves from their own beliefs and prejudices and use only the newly absorbed information in formulating their responses (i.e., "true" based on information provided, "false" based on information provided, or "can't tell" from information provided). No technical expertise or specialized knowledge is required of the examinees, who supposedly could include high school dropouts as well as senior executives. The suggested testing time is 30 minutes, which the authors claim makes the GMA-V a power test rather than a speed test.

Abstract Test (GMA-A). The GMA-A is a measure of an individual's ability to think flexibly, recognize order in the midst of apparent chaos, and focus on certain aspects of a task while ignoring irrelevant details. In short, it attempts to measure what has been called "divergent thinking," "fluid intelligence," or "flexibility." It is intended to be useful in the selection of higher-level managers whose positions often require the capacity to perceive new patterns, devise new methods, and operate effectively at different levels of analysis. It is also useful in providing an assessment, through difference scores, of the extent to which performance on the GMA-N and the GMA-V represents "flexible intellectual resources in contrast to the results of effective drilling in specific skills" (Hatfield Polytechnic, 1985, p. 10).

The GMA-A was designed to rely on educational attainment as little as possible and to emphasize the stages of thinking leading to insight into the nature of a solution, rather than the implementation of the solution once the principle has been discovered. In practice, examinees are presented with 23 different item stems, each consisting of two groups of four linked patterns. The four patterns within each group are held together by a common rule or concept, which also differentiates them from the patterns in the other group. Examinees must determine whether five patterns associated with each stem belong to one or the other of the two groups of patterns, or to neither group. Students are given 30 minutes to complete this supposed power test.

Although the GMA-A appears similar to other high-level measures of ability using diagrammatic item content (e.g., Raven's Advanced Progressive Matrices), it is intended to be dissimilar from them:

GMA-A was designed explicitly to avoid certain features of existing high-level measures of ability which use not dissimilar diagrammatic, low attainment-loaded item content. Taking as a starting point Spearman's celebrated definition of intelligence as consisting of the apprehension of experience and eduction of relations and correlates (Spearman, 1927), it may be noted that most frequently it is the eduction of correlates, that is the deductive element, the working out of a principle once discovered, which dominates the difficulty gradient in tests such as Raven's Advanced Progressive Matrices, or SHL Diagrammatic Reasoning (DT8). There is, in these tests, an emphasis on sequential transformations, and sometimes multiple simultaneous sequential transformations, that even when an item has been solved in principle and the transformation identified, the correct solution may require a capacity to handle transformational complexity far beyond the inductive capacity required to reach insight into the nature of the item. (Hatfield Polytechnic, 1985, p. 10)

The GMA-A, on the other hand, is said to focus more on induction and on the premise that inductive problem solving primarily requires the generation of an appropriate description of the problem material. The authors perceive this as a divergent thinking task: to generate as many different descriptions of the item content as possible. However, according to the manual, "convergent processes are also involved: in almost every item there are a few descriptors which very nearly satisfy the conditions, as candidates who do not temper divergent exuberance with careful checking will go astray" (Hatfield Polytechnic, 1985, p. 11).

Practical Applications/Uses

The three tests comprising the Graduate and Managerial Assessment may be administered individually or as a series. Each test requires 30 minutes of actual testing time and about 10 minutes of preparation time (i.e., establishing rapport, explaining purpose of the tests, checking to see that each examinee has appropriate test materials, reading directions, going over examples, etc.). Consequently, it will take 2-2 ¼ hours to complete the series. The tests can be administered in any order, but usually the Numerical test is administered first, followed by the Verbal test and the Abstract test, respectively. No special training is required to administer the GMA. However, administrators should familiarize themselves thoroughly with the procedures described in the *Manual and User's Guide*. Directions for administering each of the subtests are printed on color-coded cards as well as in the manual. The procedures sections of both the manual and the cards are well organized, thorough, clear, and easy to use.

Examinees record their answers on answer sheets designed in such a way that they can be used with both Form A and Form B. Scoring is straightforward and is accomplished manually. Tables allowing conversions to percentiles, T-scores, and Z-scores are provided. However, in regard to these conversions, the authors caution that because the GMA target group is individuals in the upper end of the ability range, there is no reason to expect that scores will be distributed normally. Consequently, the authors suggest that test users interested in standard scores (i.e., T-scores or Z-scores) should first convert to percentiles and then, using the tables provided, convert to standard scores, which will have the effect of forcing a normal distribution of the standardization data.

In addition to these "basic scores," as they are referred to by the authors, "alternative scores" also may be calculated for the Numerical and Abstract tests. For the GMA-N, alternative scores are obtained by counting the number of items correct for each of the three levels of item difficulty. For the GMA-A, the alternative score is the number of item stems for which all five choices are correctly classified. An understanding of test norms is required to interpret these results, but no license or certificate is necessary to interpret successfully their meaning.

Technical Aspects

Test standardization data, as well as reliability and validity information, are presented clearly in the manual. Traditional approaches to test construction (Nunnally, 1978) were followed in the standardization and validation process. Norms are based on 596 students (302 males, 294 females) in institutions of higher education in Great Britain. Approximately two thirds of the students in the norming group were enrolled in universities, and the remainder were enrolled in polytechnics and colleges of higher education. A total of 336 students were in their second year of study, 175 were in their third year, 60 were in their fourth year, and 25 were either first-year postgraduate students or "other." Of these, 293 students were liberal arts majors and 302 were science majors (one student did not report either major). Nonetheless, "the most able students are somewhat under-represented in the sample" (Hatfield Polytechnic, 1985, p. 18), which is surprising given the test's stated goal of discriminating among the top 12.5% of the British population. Because each student took one form of each test, form norms are based on a sample size of about 300. Correlations between tests are based on samples of about 150 people, and correlations among the three tests are based on samples of about 75 people. Criterion-related validity data are based on even smaller samples.

As noted above, the authors suggest that both basic and alternative scores should be calculated for the Numerical and Abstract tests. For the former, norms are presented for the basic score only, a shortcoming of the instrument in its present form. For the latter, norms are available for both basic and alternative scores, but the authors prefer the alternative or "harsh" scoring method, a decision based on what appears to be more stable correlations with other tests across samples, as well as the suspicion that the basic score may be susceptible to guessing. Correlations between basic and alternative scores ranged from .62 to .89 for the small samples providing such data.

Evidence supporting the parallelism of forms A and B of the three tests include the comparibility of test format, item performance data (i.e., item difficulty and item total correlations), and total test performance (i.e., means and standard deviations). These data are derived from small samples of subjects, but they do suggest comparability, as does an inspection of the tests themselves. Alternate form correlations are somewhat less encouraging, however. For the GMA-N, the measure of equivalence was a reasonably high .75, but for the GMA-V and the GMA-A the correlations were a low .57 and .47, respectively. If the forms were truly parallel and if what they are measuring were stable across the interval between testing, then higher correlations would be expected.

In addition to alternate form reliabilities, the authors also present alpha reliabilities for each subtest by the total group, sex, and college major. For the GMA-N, these reliabilities ranged from .70 to .85 with a standard error of measurement (SEM) ranging from 2.11 to 2.23. For the GMA-V the corresponding figures are .61 to .79 with standard errors between 3.12 and 3.39. For the GMA-A (lenient scoring), the figures are .83 to .92 with standard errors of 4.07 to 4.42; and for GMA-A (harsh scoring) the estimated reliability lies between .80 and .90 with an estimated standard error between 1.0 and 1.5 raw score points. The reliability of the GMA-A for both scoring methods approaches the stringent criterion of .90 suggested by Nunnally (1978) for tests designed for use in applied settings. The reliabilities of the GMA-N and the GMA-V, on the other hand, are somewhat lower than this criterion. One way of interpreting these reliabilities is to consider what a change in raw score points of one SEM would mean in percentiles. For the GMA-N, a change of 2.24 raw score points at the mean of 14 (a mean of 13.95 is reported) would result in a percentile shift from about 55 to about 75. For the GMA-V, a change of 3.26 for a score of 37 (again, the mean) would result in a percentile shift from about 60 to about 85. For the GMA-A (harsh scoring), a change of 1.83 for a score of 8 would translate into a percentile shift from about 55 to about 75. Because changes at the middle of a distribution result in larger percentile shifts than those at the extremes, the effect of less-than-perfect reliabilities is somewhat dramatized by the above examples. The point remains, however, that extreme caution must be exercised in making far-reaching decisions on the basis of the results of these tests.

Validity data, or perhaps more appropriately the lack of certain types of validity data, also would lead one to this same conclusion. The authors of the GMA series state:

> The authors main concern was to ensure adequate support for the construct validity of the GMA series prior to publication. Concurrent and predictive validity studies were begun when the final published versions had been settled. . . . It is intended to continue with criterion-related validity studies throughout the life of the series and to provide up-to-date information with the results of such studies as they become available. (Hatfield Polytechnic, 1985, p. 25)

In this reviewer's opinion, such validity data should have been available before the test was published.

Claims for the construct validity of the test series are based on zero-order correlations of GMA scores with other tests, including Raven's Advanced Progressive Matrices; the Abstract Reasoning and Space Relations subtest of the Differential Aptitude Test (DAT), the Spatial Relations subtest of the Primary Mental Abilities test, Wide Range Vocabulary, the Watson-Glaser Critical Thinking Appraisal, and the Thurstone Test of Mental Alertness. Many of these correlations follow from theory. For example, the GMA-V scores correlate higher with the Vocabulary test than either the GMA-N or the GMA-A scores, and the GMA-A scores correlate higher with Raven's Advanced Progressive Matrices than any other criterion measure. However, some relationships are less predictable. For example, it is unclear why the GMA-N correlates more highly with the Watson-Glaser Critical Thinking Appraisal than the GMA-V, or more highly with the Advanced Progressive Matrices than the GMA-A. Perhaps these unexpected results are due in part to the size of the samples (*N*s = 36-69) that produced them. Regardless of the explanation, however, more evidence of construct validity is needed before these tests can be recommended for widespread use. More evidence of criterion-related validity also is required. Because this test is intended for use in the recruitment and selection of upper-level employees and in the identification of individuals with management and promotion potential, the paucity of evidence concerning the tests' ability to support such decisions is particularly problematic.

Critique

The three tests in the Graduate and Managerial Assessment series rest on a sound theoretical foundation and are well constructed using widely accepted principles of classical test design. The tests are easy to administer, reasonably short, and, from the examiner's perspective, generally easy to understand. The manual is clear, scoring is straightforward and painless, and the test fills a real need. In short, the GMA has a number of appealing features. But it also has a number of shortcomings that must be addressed before the instrument is considered for widespread adoption. Clearly, additional reliability and validity data are required, particularly data supporting the instrument's ability to predict who will be be productive employees or successful managers. This reviewer also would like to see additional data about the effect of the 30-minute time limit, more evidence regarding the parallelism of the test forms, further explanations of the concurrent validity findings, additional research on the meaning of test scores, and norms based on larger and more representative samples. In particular, this reviewer would like to see more research on the Verbal test. On that test, examinees are required to discount what they know from other sources in determining whether a statement is made in, implied by, or logically follows from a passage. It is relatively easy to determine whether a passage contains a statement. It is more difficult, and perhaps artificial and counterproductive, to require examinees to discount the knowledge they have acquired from other sources while also determining whether a statement is implied by or logically follows from a passage.

The test manual reveals that some of the previously stated concerns are being addressed already, which is encouraging because the GMA does appear to have some merit, particularly for use in Great Britain. If the authors and publishers intend to distribute the test in the United States, however, some item revision will be required to eliminate language differences (e.g., "labour" should be changed to "labor," "defences" to "defenses," "colour" to "color," etc.) as well as other problems (e.g., the numerical test requires familiarity with British currency). However, it appears that these revisions could be made quite easily.

References

Atwell, C. R., & Wells, F. L. (1945). *Wide Range Vocabulary Test*. New York: The Psychological Corporation.

Bennett, G. K., Seashore, H. G., & Wesman, A. G. (1972). *Differential Aptitude Test*. New York: The Psychological Corporation.

Hatfield Polytechnic. (1985). *Graduate and Managerial Assessment: Manual and users guide.* Windsor, England: NFER-Nelson.

Nunnally, J. C. (1978). *Psychometric theory* (2nd ed.). New York: McGraw-Hill.

Raven, J. C. (1972). *Advanced Progressive Matrices.* London: H. K. Lewis & Co., Ltd.

Thurstone, L., & Thurstone, T. (1962). *Primary Mental Abilities Test.* Chicago: Science Research Associates.

Thurstone, T., & Thurstone, L. (1968). *Thurstone Test of Mental Alertness.* Chicago: Science Research Associates.

Watson, G., & Glaser, G. M. (1964). *Watson-Glaser Critical Thinking Appraisal.* New York: The Psychological Corporation.

Michael D. Franzen, Ph.D.

Director of Neuropsychology, Associate Professor of Behavioral Medicine and Psychiatry, West Virginia University Medical Center, Morgantown, West Virginia.

GRASSI BLOCK SUBSTITUTION TEST

Joseph R. Grassi. Lisse, The Netherlands: SWETS and Zeitlinger B.V.

Introduction

The Grassi Block Substitution Test (GBST) is designed to measure abstract functioning in the adolescent or adult subject. However, because it requires visuospatial and visuo-constructive skills for its successful completion, it can also be said to be a measure of these functions. This test is simple to administer and can be completed by most subjects in about 20 minutes.

The GBST was developed in 1947 by Joseph Grassi, a psychologist at the Child Development Center at the University of Miami. During the early stages of development of the GBST, Dr. Grassi was associated with Fairfield State Hospital in Newton, Connecticut. Dr. Grassi was interested in constructing a test that was sensitive to reductions in intellectual efficiency. The GBST was based on the Kohs Blocks, the Block Design subtest of the Wechsler-Bellevue Intelligence Scale, and the Goldstein-Sheerer Cube Test. Rather than constructing a broad screening device, Grassi intended to devise an instrument that was sensitive to only one aspect of mental functioning (Grassi, 1970). The test was also intended to show impairment in both the concrete and the abstract degrees of behavior. In addition, each task of the GBST was designed to be administered in both an easy and a difficult level. As an improvement over the Goldstein-Sheerer test, the results of the GBST were intended to be amenable to quantitative measurement.

The original standardization sample for the GBST was composed of inpatients from the Fairfield State Hospital in Newton, Connecticut. Later, further standardization (Anderson, 1951) was conducted, resulting in a sample of 86 normals, 86 schizophrenics, 72 organics, and 30 postlobotomy patients. The diagnostic system by which these categories were composed is not specified. Because of the age of the test, it is possible that these diagnoses are no longer applicable under today's concepts of diagnostic nosology. In particular, the manner in which subjects were classified as organic (EEG neurological exam, neurosurgical investigation) and the types of subjects included in the category (e.g., strokes, head injuries, congenitally mentally retarded) are not specified.

The GBST consists of four cubes and five blocks, the blocks being the size of four of the cubes joined in a solid square. The sides of each of the four cubes display either a solid color or a color combination: red, blue, yellow, white, half blue and half yellow, half red and half white. The five blocks resemble combinations of the four cubes.

204

Practical Applications/Uses

In the manual, Grassi (1970) states that because the Grassi Block Substitution Test can be used to demonstrate an inability to shift from a concrete to an abstract attitude, it is sensitive to early, minimal changes in organic integrity. Throughout the manual, reference is made to organicity, despite the author's earlier statements describing the GBST as a test of a specific function rather than as a screening device. From the manual's discussion of Goldstein's concept of abstract behavior, it appears that abstraction skills are the area that the GBST is intended to measure. However, as mentioned earlier, successful completion of the test also requires intact visuo-spatial and visuo-constructive skills.

The test requires the subject to reproduce the designs of the blocks using the cubes. First, the subject is asked to reproduce only the top part of the design. Next, the subject is asked to reproduce the top part but to substitute different colors than are used in the stimulus block. Third, the subject is asked to reproduce both the top and the side parts of the stimulus block. Finally, the subject is asked to reproduce the top and side parts of the design but to substitute different colors for those used in the stimulus. In each case, the examiner shows the block to the subject and provides the instructions.

Scores are assigned on the basis of a numerical system that is described on each score record sheet. The subject is given one point for each correct solution and an additional half-point if the differences between the time to solve different steps for each block is less than 10 seconds. One-half point is subtracted if the subject requires greater than 120 seconds for any solution. Two points are deducted if the subject fails step four for each block, and one point is deducted if there are three or more failures on all five blocks.

In addition, one point is subtracted for each manifestation of a behavioral sign designated as "Level I," and two points are deducted for each manifestation of a behavioral sign designated as "Level II." Level I behavioral signs include the categories "Trial and Error" and "Reassurance." Level II behavioral signs include the categories "Spatial Disorientation" and "Correction." Although the inclusion of the behavioral signs is a well-intentioned attempt to increase the amount of information derived from the test, the categories are not always defined well enough to facilitate their scoring. For example, "Reassurance" is scored when frequent reassuring remarks are sought by the subject. Unfortunately, the quantification of "frequent" is not specified. Other behavioral signs have incompletely defined characteristics.

Grassi (1970) suggests that interpretation be based on three aspects of test performance: behavior, test score, and intellectual level. He further states that the differentiation between functional impairment and organic impairment is conducted by examination for the presence of behavioral signs (p. 25). The presence of impairment is conducted by comparison of the score to cutoff points. Finally, the presence of new damage is determined by comparison to indications of premorbid intellectual functioning. There are no objective rules provided for the comparison of premorbid functioning to test score.

Grassi states that severe deterioration is evidenced by scores between 0 and 16 points; moderate deterioration is evidenced by scores between 16 and 20; and scores of 20 or greater are evidence for the absence of deterioration. The norms are said to be derived from the results of the original study; however, the rules or methods by which these cutoff points were determined is not stated. The manual contains 15 sample cases that can be helpful in interpretation.

Technical Aspects

Grassi states that the test-retest reliability of the Grassi Block Substitution Test was determined by administering it to a group of normal, schizophrenic, organic, and lobotomized patients. The reliability coefficient was reported to be .85; however, the length of the retest interval was not specified and neither was the test-retest reliability of the behavioral signs. But most importantly, the interrater reliability of the behavioral signs was not reported either.

In order to validate the GBST, data were collected from a sample of 350 individuals. This sample included 97 individuals with an organic condition, 68 schizophrenics, 55 normals, 37 nondeteriorated alcoholics, 26 deteriorated alcoholics, and 67 neurotics. There were no significant age or IQ differences among the groups. However, there were no statistical analyses conducted on the actual test data. Inspection of the groups' means indicates that performance on the test tends to decline from normals to neurotics, schizophrenics, alcoholics, and, finally, organics. In addition to the need for information regarding the accuracy of the GBST in detecting different types of neurological impairment, information is needed supporting the construct validity of the test.

The manual cites results from two studies that the author states support the validity of the test. The first study (Hayashi, 1968) was reported in the *Bulletin of Kyoto University of Education* (the full reference is not given). Hayashi administered the test to 20 "feebleminded" children, 50 schizophrenics, and 10 organic patients. In general, the results seemed to indicate that these individuals scored worse than the normals in the initial standardization sample. There were no control comparison groups or statistical analyses reported. The second study was a dissertation in which the GBST, the Hunt-Minnesota Test for Organic Brain Damage, and the Bender Visual Motor Gestalt Test were administered to seven brain-damaged individuals, seven schizophrenics, and seven normal controls. Although there were significant differences between the groups, there was also overlap in the distributions of the scores, indicating that application to individual cases may be problematic. The scores of these 21 individuals were combined with the scores of 54 additional subjects who were again equally divided among the three diagnostic groups. The GBST was reported to be the most accurate of the three tests used in classifying the subjects.

Critique

The Grassi Block Substitution Test is a nice attempt to refine previous tests of abstraction by defining quantitative scoring. The four-step administration procedure appears to be conducive to separating some of the components of performance on the test. The GBST is short and easy to administer, having potential as a quick test of abstraction functions. However, the norms need to be updated, and more information is needed regarding the reliability of the test. Finally, future research should be directed at uncovering the limits of validity for different problems in different populations.

References

Anderson, A. L. (1951). The effect of laterality localization of focal brain lesions on the Wechsler-Bellevue subtests. *Journal of Clinical Psychology, 7,* 149-153.

Bender, L. (1938). A visual motor gestalt test and its clinical use. *American Orthopsychiatric Association Research Monographs* (No. 3.)

Goldstein, K., & Scheerer, M. (1941). Abstract and concrete behavior: An experimental study with special tests. *Psychological Monographs, 53,* (2, Whole No. 239).

Grassi, J. R. (1970). *The Grassi Block Substitution Test for measuring organic brain pathology.* Springfield, IL: Charles C. Thomas.

Hunt, H. F. (1943). A practical clinical test for organic brain damage. *Journal of Applied Psychology, 27,* 375-386.

Wechsler, D. (1944). *The Measurement of Adult Intelligence* (3rd ed.). Baltimore: Williams and Wilkins.

Wyman E. Fischer, Ph.D.
*Professor of Psychology and Chair, Department of Educational
Psychology, Ball State University, Muncie, Indiana.*

HALSTEAD CATEGORY TEST

*Michael Hill and Norman McLeod, based on the original by Ward
C. Halstead. Jacksonville, Florida: Precision People, Inc.*

Introduction

The Halstead Category Test (HCT), on which the Halstead Category Test: A
Computerized Version (HCT-CV) is based, was one of 27 behavior indicators of
biological intelligence developed by Ward C. Halstead under the patronage of the
Otho S. A. Sprague Memorial Institute and the Division of Psychiatry in the University of Chicago's Department of Medicine. Studies involving these behavioral
indicators began in 1935 and continued for approximately 12 years. A monograph
reporting results of these studies, relating primarily to the behavioral effects of
brain lesions in man, was published in 1947 (Halstead, 1947).

The HCT, briefly described by Halstead (1947, p. 39) as "a nonverbal test involving visually induced abstractions of size, shape, color, etc., presented by means of a
multiple choice projection apparatus," was subsequently included in a subbattery
of 13 tests selected from the original 27 behavioral indicators. The criteria for inclusion in the subbattery were ease of objective scoring to facilitate statistical treatment and ability to reflect some component of Halstead's conception of biological
intelligence (Halstead, 1947). These 13 behavioral measures, including the HCT,
were subjected to a number of factor analyses from which four basic factors
emerged. These factors were identified by Halstead as "C" (the integrative field
factor), "A" (the factor of abstraction), "P" (the power factor), and "D" (the directional factor). Because the HCT loaded heavily on factors C and A, Halstead concluded that the test was a measure of the central integrative process or, stated in
another way, a measure of the ability to make appropriate mental/cognitive shifts as
a situation demanded it. Similarities between this conception and the fluid intelligence notion proposed by Horn and Cattell (1966) are apparent. In addition,
Halstead viewed the HCT as a measure of abstraction. He stated that

> For successful performance on the test the subject is thus forced to recognize
> recurrent similarities in the presence of marked dissimilarities and vice versa
> and to make appropriately differential responses. The ability to carry out such
> tasks has long been known in psychology as the ability for abstraction. Since
> this ability is a prime requisite for successful performance on Test 2 [Category
> Test], it would appear justifiable to identify it as a "factor of abstraction", or the
> A factor . . . (Halstead, 1947, p. 59)

In further refinement of his composite measure of biological intelligence,
Halstead selected the 10 behavioral indicators that demonstrated the greatest

degree of differentiation between brain-injured and non-brain-injured individuals and combined them, on the basis of cutoff scores, into a single score, which he termed the "impairment index." The HCT was included among the 10 measures composing this index. Although the impairment index has been modified by Reitan (1969), a former student of Halstead, the HCT has been retained.

In its original form, the HCT was composed of 336 items or stimulus figures arranged into nine subtests on the basis of specific organizing principles required for correct responses (e.g., matching 1) Arabic numerals with Roman numerals, 2) the number of objects displayed on a screen with the appropriate numbered key, and 3) missing figures of a quadrant with response keys). Reitan (1969) subsequently reduced the number of subtests from nine to seven and the number of items from 336 to 208. It is this form of the HCT, presently included in the adult Halstead-Reitan Neuropsychological Battery (HRNB), that was computerized by Michael Hill and Norman McLeod and is the subject of this test review. Dr. Hill, the senior author of the HCT-CV, presently is employed as the chief psychologist for the North County Education Services in Berlin, New Hampshire, and also maintains a private practice in psychology. Norman McLeod was the computer programmer for the project.

Like the original HCT projection box apparatus developed by Halstead and still used in most clinics utilizing the HRNB, the HCT-CV requires the examinee to match a visually presented stimulus with a number from one through four for all the items on the seven subtests.

The HCT-CV is packaged in a black, plastic loose-leaf binder. Dependent on the user's request, the kit contains either an Apple or IBM compatible 5-inch program disk and a typed manual. Instructions for producing a backup copy of the program disk are provided. The cost of the total unit at the time this review was written was $199.00

The manual for the HCT-CV is brief and committed almost entirely to the operation of the Apple and IBM computer programs. The step-by-step instructions should prove particularly helpful to examiners unsophisticated in computer operation. A brief introduction to the Halstead Category Test is presented in two short paragraphs, and two pages are devoted to textual and tabular reporting of a validity study intended to justify the use of the HCT-CV as a substitute for the examiner-administered HCT projection box version. The section of the manual labeled "Interpretation" covers less than one-half page and identifies age-adjusted adult cutoff scores between normals and brain-impaired individuals.

Practical Applications/Uses

The Halstead Category Test is among the better known of the subtests comprising the Halstead-Reitan Neuropsychological Battery and is one of the original measures that has survived the test of time. The HCT still is considered by many (Dean, 1985a) to be the mainstay of the HRNB, though a few (Swiercinsky, 1978) may question its significance. Its sensitivity to deficits in nonverbal reasoning (Russell, 1982) and its ability to detect subtle cognitive defects among individuals suspected of neurological impairments (Cullum, Steinmam, & Bigler, 1984; Kilpatrick, 1970) has sustained its popularity among clinical neuropsychologists.

At the same time, however, the sheer cumbersomeness, lack of portability, high cost, and extensive examiner/examinee time required with the traditional projection box version has prompted the development of a number of alternative forms of the test. Some of these are merely shortened forms still utilyzing the projection box apparatus (Calsyn, O'Leary, & Chaney, 1980; Gregory, Paul, & Morrison, 1979), while others are either card form, paper-and-pencil, or booklet versions not requiring a mechanical apparatus (Kimura, 1981; Adams & Trenton, 1981; & DeFillippis, McCampbell, & Rogers, 1979). The validity of these alternative forms, both shortened and modified, generally has been substantiated (Kupke, 1983; Boyle, 1986; Byrd & Warner, 1986; Sherrill, 1985).

Although personal computer versions of the HCT, other than the one currently under review, have not been reported in the literature, Beaumont (1975) demonstrated that the HCT administered by on-line computer was a viable alternative to the projection box apparatus and, in fact, had definite advantages because of its 1) ability to maintain numerous subject records and summary scores, 2) total elimination of scoring errors, and 3) control for interexaminer variability in administration. Neuropsychological clinics interested in maintaining administration consistency and reducing both administration and scoring time will agree with Beaumont's assessment that the computer provides a superior means of administering and scoring the HCT subtests.

In the first subtest (8 items), the examinee must pair Roman numerals projected on the computer screen with Arabic numbers on the computer keyboard. In Subtest 2 (20 items), the subject must press the numerical key that is equivalent to the number of items on the screen in order to receive credit. In Subtest 3 (40 items), the subject must identify the position of the figure that is at variance with the rest of the figures presented. Subtest 4 (40 items) involves locating the correct quadrant in which a part of the figure is missing or to which attention is directed. A single principle operating for Subtest 5 (40 items) and Subtest 6 (40 items) requires the examinee to determine the proportion of the figure or figures drawn in solid versus dotted patterns. Subtest 7 (20 items) supposedly contains items randomly selected from the previous six subtests. However, because several of the items in Subtest 7 did not appear previously, more than short term memory may be required for success.

Unlike the HCT projection box apparatus, which dictates examiner involvement in reading directions, projecting the test figures on the screen, and recording responses, the HCT-CV is totally self-administered and automatically scored. Directions for entering and exiting the program and for examining previous test results for a maximum of 50 subjects are stated clearly on the main program menu for both Apple and IBM personal computers. Because all subject data is stored on the program disk, only a single disk drive is needed. Performance records for individual subjects are stored in the form of raw error scores for each individual subtest and a composite error score for the total test. The program does provide for viewing of individual subject records, but the examiner must view them in the order in which they were stored as there is no provision in the program for requesting a specific record. General directions regarding the nature of the test and specific directions relating to individual subtests appear on the computer screen at the appropriate time in the administration. Except for minor variations, these directions are similar to those in the HRNB examiner's manual (Reitan, 1969). The omissions and/or additional prompts in the HCT-CV program are not likely to affect the subject's performance.

Because the authors of the HCT-CV were interested in developing a version that could be used with both color and monochrome computer monitors, they found it necessary to substitute texture for color on those items in Subtest 3 and Subtest 7 for which color was either the total or partial differentiator. A small sample study conducted by Hill & McLeod (1984) indicated that this modification did not alter the results of the test significantly when compared with the HCT. Examiners with color monitors may choose the color option, in which case all items on the test are essentially identical to those used with the HCT projection box.

The authors recommended a cutoff error score of 50 to differentiate those with suspected brain impairment from neurologically unimpaired groups. They further stated that in age groups above 60, the actual chronological age of the subject should become the cutoff score (Hill & McLeod, 1984). This recommendation probably was made in recognition of the general finding that age is related significantly to neuropsychological test results (Hesselbrock, Weidenman, & Reed, 1985; Loo, 1978; Bak & Greene, 1980).

The Halstead-Reitan Neuropsychological Test Battery, which includes the HCT as one of its subtests, is a highly sophisticated instrument requiring extensive training in neuropsychological theory and assessment for proper interpretation (Dean, 1985b). Because the HCT-CV makes administration and scoring of the subtest relatively simple, professionals and consumers alike should be aware of the dangers of an oversimplified interpretation made by an improperly trained examiner.

Technical Aspects

The HCT, as modified by Reitan and used in numerous neuropsychological clinics in the United States and several other countries, has been found to be the most sensitive test in the Halstead-Reitan Neuropsychological Battery for detecting brain impairment (Golden, 1979) and has achieved up to 90% accuracy in differentiating normals from the brain injured (Wheeler, Burke, & Reitan, 1963). Though its validity is well established, the nature of the test as a complex measure (Royce, Yeudall, & Bock, 1976), including an element of learning (Fischer & Dean, 1987), restricts the use of traditional methods of reliability. Marked intrasubject changes in score upon readministration are expected as a function of experience with the test (Reitan & Davison, 1974).

Although statistical studies and references to computerized versions of the HCT are sparse, the one study that does exist seems to imply that validity attributed to the HCT projection box version may apply to computerized versions as well. Hill and McLeod (1984) compared the HCT-CV (monochrome monitor option) with the HCT projection box version as modified by Reitan. Ten neurologically normal adults and 10 adult subjects with verified brain lesions were administered both versions in counterbalanced order with a 2-week time period between administrations. Both samples were drawn from an outpatient population of a medical clinic. Each of the versions significantly differentiated the two groups, but no significance was found between scores on the two tests. Though not a statistical study comparing versions of the HCT, Beaumont (1975) reported that an on-line computer administration of the HCT presented no problems for either brain-damaged subjects or hospitalized controls and found this method of administration completely satisfactory.

Critique

Numerous studies cited in previous sections of this review support the notion that both the length and method of presentation of the Halstead Category Test may be modified without significantly affecting its validity. In light of this, it seems desirable to employ the technological advantages of a computer to eliminate, or at least minimize, some of the disadvantages inherent in the HCT projection box apparatus. It appears that the HCT-CV has accomplished some of this. The HCT projection box is unwieldy (practically immobile), mechanically fragile, and very expensive. The HCT-CV, compatible with both IBM and Apple II series computers, is much more compact and, without the computer, available for a fraction of the cost. The HCT projection box requires a trained technician for administration and scoring. The HCT-CV scores responses automatically and is basically self-administering. Except for occasional monitoring and note-taking of significant happenings, the examiner is free to engage in other activities while the HCT-CV is being administered. This, together with the fact that the subject has greater control over the speed of the administration, should reduce the total amount of time involved in administering and scoring the total Halstead-Reitan Neuropsychological Battery.

There are, however, a number of concerns that relate specifically to the HCT-CV, including its validity, the examiner instructions, record keeping, and the comprehensiveness of the HCT-CV manual.

One study (Hill & McLeod, 1984) involving a small sample size seems insufficient to establish the comparability of the HCT and the HCT-CV (monochrome monitor option). Additional studies comparing both color and monochrome options of the HCT-CV with the HCT are needed.

The examiner instructions to the subject as they appear on the computer screen vary slightly from those appearing in the Reitan (1969) manual. Although it is unlikely that this variance would have any appreciable affect on subject performance, it is not clear why the authors of the HCT-CV did not attempt an exact duplication of the HCT's examiner instructions and specific help statements.

The scoring of responses, general record keeping, and summarization of subject data require modification for maximum usefulness. The HCT-CV, in its present form, maintains a record of subject error scores on each of the seven subtests and also calculates a composite error score for the entire test. However, a record of specific responses to individual items on the test, similar to that of the HCT projection box version, is not maintained. As Stephaniv (1985) suggested, the pattern (perseveration, etc.) of individual responses may have both clinical and research significance and could increase the usefulness of the measure, particularly in cases involving frontal lobe lesions.

The main menu of the HCT-CV program provides an option for screen display in tabular form of individual subject score profiles. A maximum of 50 such records may be stored on the program disk and are available for retrieval in the order in which they were stored. It would be helpful if the program contained a search feature allowing for the display of a specifically selected record from those presently in storage. It also would be beneficial if the examiner could select specific records for deletion from storage. The HCT-CV, in its present form, allows only mass deletion.

The HCT-CV program has no printout option. Consequently, the examiner must copy scores by hand from the screen display. A print option would increase the practicality of the program, particularly if it were modified to provide a protocol including subject responses to all test items. Research interests also would be well served if the program calculated simple descriptive statistics (e.g., means, standard deviations, etc.) relating to the cases in storage and printed them in tabular form as well.

A recent study by Rattan, Dean, and Fischer (1985) indicated that a measure of response time or latency may enhance the usefulness of the HCT as a measure of neuropsychological functioning. Such data was collected by the on-line computer administration described by Beaumont (1975) and would be a worthwhile feature to consider for the HCT-CV as well. It is possible that some of of the programming modifications suggested above may not be possible or practical for personal computer operation. They should, nevertheless, be given careful consideration in subsequent revisions of the HCT-CV.

The manual of the HCT-CV, though clearly written and geared specifically toward the novice computer operator, might contain a more comprehensive introduction to the Halstead-Reitan Neuropsychological Test Battery and the specific contribution that might be expected from the HCT-CV. The manual should also explicitly state that the administration should be supervised by either a neuropsychologist or a trained technician and that the interpretation of the test results should be attempted only by professionals with appropriate neuropsychological training. Presently, the manual contains no such admonitions. It is also of concern that the manual contains no section on administration procedures for subjects with reading impairments, either long standing or the result of neurological problems. Because the examiner instructions appear in written form on the computer screen, it must be assumed that the test is self-administered. It cannot be taken for granted that all subjects, particularly those with minimal educational backgrounds, left hemisphere lesions, or progressive dementia, would not be penalized by this deviation from the HCT projection box version, where all instructions are given verbally by the examiner. This is a serious concern that requires attention in future revisions of the HCT-CV.

In summary, the HCT-CV represents a good beginning in the application of computer technology to enhance the efficiency and utility of a well-known and heavily documented measure of neuropsychological impairment. The program is easy to use and the graphics (computer reproduced test items) are well done and closely approximate those of the HCT projection box apparatus. Although the HCT-CV package in its present form has weaknesses, they are not beyond correction in future revisions.

References

Adams, R. L., & Trenton, S. L. (1981). Development of a paper-and-pen form of the Halstead Category Test. *Journal of Consulting and Clinical Psychology, 49,* 298-299.

Bak, J. S., & Greene, R. L. (1980). Changes in neuropsychological functioning in an aging population. *Journal of Consulting and Clinical Psychology, 48,* 395-399.

Beaumont, J. G. (1975). The validity of the Category Test administered by on-line computers. *Journal of Clinical Psychology, 31,* 458-462.

Boyle, G. J. (1986). Clinical neuropsychological assessment: Abbreviating the Halstead Category Test of brain dysfunction. *Journal of Clinical Psychology, 42,* 615-625.

Byrd, P. B., & Warner, P. D. (1986). Development of a booklet version of the Halstead Category Test for Children age nine through fourteen years: Preliminary validation with normal and learning disabled subjects. *International Journal of Clinical Neuropsychology, 8,* 80-82.

Calsyn, D. A., O'Leary, M. R., & Chaney, E. F. (1980). Shortening the Category Test. *Journal of Consulting and Clinical Psychology, 48,* 788-789.

Cullum, C. M., Steinman, D. R., & Bigler, E. D. (1984). Relationship between fluid and crystallized cognitive functions using Category Test and WAIS scores. *International Journal of Clinical Neuropsychology, 6,* 172-174.

Dean, R. S. (1985a). Neuropsychological assessment. In J. D. Cavenar & S. B. Guze (Eds.)., *Psychiatry* (Vol. 1, pp. 1-17). Philadelphia, PA: J. B. Lippincott.

Dean, R. S. (1985b). Review of Halstead-Reitan Neuropsychological Test Battery. In J. V. Mitchell, Jr. (Ed.), *The ninth mental measurements yearbook* (pp. 644-646). Lincoln, NE: The Buros Institute of Mental Measurements.

DeFillippis, N. A., McCampbell, E., & Rogers, P. (1979). Development of a booklet form of the Category Test: Normative and validity data. *Journal of Clinical Neuropsychology, 1,* 339-342

Fischer, W. E., Dean, R. S. (1987, October). *The multidimensional nature of the Halstead Category Test.* Paper presented at the National Academy of Neuropsychology Conference, Chicago.

Golden, C. J. (1979). *Clinical interpretation of objective psychological tests.* New York: Grune & Stratton.

Gregory, R. J., Paul, J. J., & Morrison, M. W. (1979). A short form of the Category Test for adults. *Journal of Clinical Psychology, 35,* 795-798.

Halstead, W. C. (1947). *Brain and intelligence: A quantitative study of the frontal lobes.* Chicago: The University of Chicago Press.

Hesselbrock, M. E., Weidenman, M. A., & Reed, H. B. (1985). Effects of age, sex, drinking history and antisocial personality on neuropsychology of alcoholics. *Journal of Studies on Alcohol, 46,* 313-320.

Hill, M., & McLeod, N. (1984). *Halstead Category Test: A computer version.* Jacksonville, FL: Precision People.

Horn, J. K., & Cattell, R. B. (1966). Refinement and test of of the theory of fluid and crystallized intelligence. *Journal of Educational Psychology, 57,* 253-270.

Kilpatrick, D. G. (1970). The Halstead Category Test of brain dysfunction: Feasibility of a short form. *Perceptual and Motor Skills, 30,* 577-578.

Kimura, S. D. (1981). A card form of the Reitan-Modified Halstead Category Test. *Journal of Consulting and Clinical Psychology, 49,* 145-146.

Kupke, T. (1983). Effects of subject sex, examiner sex, and test apparatus on Halstead Category and Tactual Performance Tests. *Journal of Consulting and Clinical Psychology, 51,* 624-626.

Lezak, M. D. (1983). *Neuropsychological assessment* (2nd ed.). New York: Oxford University Press.

Loo, R. (1978). Relationship of field dependence, MMPI scales, age, and education to performance on the Halstead Category Test. *Psycholooy, 15,* 23-26.

Rattan, G., Dean, R. S., & Fischer, W. E. (1986). The role of reaction time in assessing Category Test performance. *Archives of Clinical Neuropsychology, 1,* 68.

Reitan, R. M. (1969). *Manual for administration of neuropsychological test batteries for adults and children.* Indianapolis: Author.

Reitan, R. M., & Davison, L. A. (Eds.). (1974). *Clinical Neuropsychology: Current status and application.* New York: John Wiley & Sons.

Royce, J. R., Yeudall, L. T., & Bock, C. (1976). Factor analytic studies of human brain damage: I. First and second-order factors and their brain correlates. *Multivariate Behavioral Research, 11,* 381-418.

Russell, E. W. (1982). Factor analysis of the revised Wechsler Memory Scale tests in a neuropsychological battery. *Perceptual and Motor Skills, 54,* 971-974.

Sherrill, R. E. (1985). Comparison of three short forms of the Category Test. *Journal of Clinical and Experimental Neuropsychology, 7,* 231-238.

Stephaniv, W. M. (1985). *Verification of the Narratives Test with perseveration responses on the Wisconsin Card Sorting and Category Tests.* Unpublished doctoral dissertation, Ball State University, Muncie, IN.

Swiercinsky, D. P. (1978). *Manual for the adult neuropsychological evaluation.* Springfield, IL: C.C. Thomas.

Wheeler, L., Burke, C. H., & Reitan, R. M. (1963). An application of discriminant functions to the problems of predicting brain damage using behavioral variables. *Perceptual and Motor Skills Monograph Supplement 3-V16, 16,* 417-440.

Peter D. Lifton, Ph.D.
Management Consultant, Bethesda, Maryland.

Ellen D. Nannis, Ph.D.
Assistant Professor of Psychology, University of Maryland, Catonsville, Maryland.

HOGAN PERSONALITY INVENTORY

Robert Hogan. Minneapolis, Minnesota: NCS Professional Assessment Services.

HOGAN PERSONNEL SELECTION SERIES

Robert Hogan and Joyce Hogan. Minneapolis, Minnesota: NCS Professional Assessment Services.

Introduction

The Hogan Personality Inventory (HPI) and the Hogan Personnel Selection Series (HPSS) are two newly developed self-report inventories. The HPI assesses "six dimensions of broad, general importance for personal and social effectiveness" (Hogan, 1986, p. 5); the HPSS "measures constructs fundamental for successful performance in virtually any job" (Hogan & Hogan, 1986, p. 2).

The HPI and HPSS are published and marketed as separate inventories; that is, National Computer Systems (NCS) markets separately the HPI and the four individual inventories that compose the HPSS. However, all five inventories are based on the same pool of 310 items, all of which are contained within the HPI. A person's responses to these 310 items yield scores for *all* HPI scales and *all* scales on the four HPSS inventories; consequently, this review examines the HPI and HPSS concurrently. However, the HPSS gives interpretive statements not available in the HPI and requires that the user be qualified at the "C" level. The HPI requires an "A-1" or "B-1" level of qualification.

The HPI and HPSS were developed by Robert Hogan, Ph.D. and Joyce Hogan, Ph.D. Dr. Robert Hogan is a McFarlin professor and Chairperson of the Department of Psychology, University of Tulsa, Oklahoma. Some of his work on the HPI and HPSS was conducted while he was Professor of Psychology at the Johns Hopkins University in Baltimore. Dr. Hogan is a leading figure in the field of personality theory, personality assessment, and trait psychology. More recently, he has applied his knowledge in these areas to organizational psychology, specifically the measurement of personality dimensions for predicting successful performance in various occupational classes. Dr. Joyce Hogan is an associate professor in the psychology department at the University of Tulsa. Her areas of expertise and research include industrial/organizational psychology, personnel selection, and human factors.

216

The development of the HPI was influenced conjointly by the research of Harrison G. Gough, creator of the California Psychological Inventory, and that of factor analytic psychologists, particularly Raymond Cattell and Hans Eysenck. The HPSS reflects exactly the same influences and theoretical rationales as the HPI because it comprises "a set of empirically keyed measures of occupational performance derived from the HPI" (Hogan & Hogan, 1986, p. 1). The Gough influence on the HPI is reflected by the assumption that a broad range of normal (non-pathological) everyday life behavior can be described and predicted from underlying personality dimensions. The factor analytic influence is seen in the assumption that there is a finite number of personality dimensions affecting virtually all behavior.

In addition to these influences, the HPI reflects Hogan's own socioanalytic theory of personality psychology (Hogan, 1976; Hogan, Johnson, & Emler, 1978; Hogan, 1982; Hogan, Cheek, & Jones, 1985; Hogan & Briggs, 1986). For over a decade, Hogan and his colleagues have published papers arguing that responses to items contained within any self-report personality inventory constitute a form of self-presentation virtually identical to self-presentations in everyday social interactions. People present images of themselves to others for the purpose of influencing how others will perceive them. Self-presentations ("public selves") reflect self-images ("private selves"), and people behave in a manner that maximizes the consistency between their public self-presentations and their private self-images. Therefore, according to Hogan (1986, p. 2), "responses to personality inventories are *not* self-reports. Rather, the responses reveal how the person wants to be regarded." Consequently, a person's response to the content of a particular item in a self-report personality inventory does not necessarily reflect with any accuracy actual past behavioral tendencies. Instead, the response reflects the behavior that a person wants other persons to believe he or she engages in, thereby influencing how he or she is perceived by other persons. Furthermore, the aggregation of a person's responses to items in a self-report personality inventory (i.e., his or her scale scores) do not necessarily reflect personality traits or behavioral tendencies; rather, scale scores reflect a person's characteristic interpersonal style.

Hogan (1986, p. 5) describes the six primary HPI personality scales and the validity scale as follows:

Intellectance (INT) 33 items: measures the degree to which a person is perceived as bright, creative, and interested in cultural and educational matters;

Adjustment (ADJ) 46 items: measures such basic themes as self-esteem and self-confidence as well as freedom from anxiety, depression, guilt, and somatic complaints;

Prudence (PRU) 54 items: measures conscientiousness, responsibility, dependability, and planfulness or, conversely, risk-taking, thrill-seeking, and impulsivity;

Ambition (AMB) 28 items: measures energy level, initiative, leadership potential, and achievement strivings;

Sociability (SOC) 23 items: measures extraverted and affiliative tendencies or, conversely, shy and introverted tendencies;

Likeability (LIK) 28 items: measures the degree to which a person is seen as cordial, even tempered, and cooperative;

Validity (VAL) 16 items: measures random or careless response patterns.

The HPI's primary personality scales are composed of Homogeneous Item Composites (HICs), which are "small, very homogeneous clusters of items . . . that reflect facets or aspects of the primary scales" (Hogan, 1986, p. 5). The HICs range in length from three to seven items, and the number of HICs per scale ranges from five to nine. For example, five HICs make up Sociability: Entertaining, Exhibitionistic, Likes Crowds, Likes Parties, and Expressive. The HPI manual (Hogan, 1986, pp. 9-13) contains a complete listing of each scale's HICs.

The items that compose an individual HIC are highly intercorrelated; HICs that compose an individual scale also are highly intercorrelated. However, HICs from one scale are correlated minimally with HICs from another. Hogan (1986, p. 5) argues that "HICs from various scales can be recombined to form new empirical scales to predict behaviors that a researcher may deem important." The HPI contains six such scales developed by the process of empirical-recombination of HICs. These six secondary HPI scales define six aspects of occupational performance:

Service Orientation (SOI) 87 items: identifies persons who are courteous, tactful, helpful, and pleasant;

Reliability (REL) 69 items: identifies persons who are conscientious, rule abiding, and dependable;

Stress Tolerance (STR) 55 items: identifies persons who rarely miss work because of illness and who rarely suffer on-the-job accidents;

Clerical Potential (CLE) 25 items: identifies persons who follow directions carefully, attend to details, and communicate accurately;

Sales Potential (SAL) 24 items: predicts success in occupations that require initiative, persistence, social skill, and persuasiveness;

Managerial Potential (MAN) 57 items: predicts success in occupations that require leadership ability, planning, and decision-making skills (Hogan, 1986, p. 5).

As noted previously, the HPSS comprises four inventories. The first, the Prospective Employee Potential Inventory (PEPI), is composed of the HPI scales SOI, REL, STR, and VAL. These scales individually total 227 items; item overlap reduces the PEPI to 198 items. The second, the Clerical Potential Inventory, is composed of the PEPI and the HPI scale CLE. These scales individually total 252 items; item overlap reduces them to 215 items. The third, the Sales Potential Inventory, is composed of the PEPI and the HPI scale SAL. These scales individually total 251 items; item overlap reduces them to 216 items. Finally, the Managerial Potential Inventory is composed of the PEPI and the HPI scale MAN. These scales individually total 284 items; item overlap reduces them to 223 items.

There is no item overlap among the six primary personality scales plus the validity scale. In fact, these seven scales require only 228 of the 310 items in the HPI. Hogan (1986, p. 5) states that "a small amount of item overlap does occur between the six primary scales and the six occupational scales . . . they are not entirely independent." However, only 82 items (310 minus 228) are unique to the occupational scales, which suggests more than just a "small amount of overlap" between the two sets of scales.

Practical Applications/Uses

The HPI is a self-administered test and the 310 item statements are presented in a reusable test booklet. Using computer-scored answer sheets, subjects respond "True" if they agree with the statement or "False" if they do not. The content of all the statements is innocuous (e.g., "It is easy for me to talk to strangers"; "I am a relaxed, easygoing person"; and "I would enjoy skydiving"); persons responding to the HPI should not regard the instrument as an invasion of privacy or feel the statements are personally offensive (unlike other personality inventories, particularly the Minnesota Multiphasic Personality Inventory). Examinees are encouraged to respond to all statements and to work as quickly as possible; most complete the inventory in 25 to 35 minutes. A fourth-grade reading level is required to comprehend the statements. However, Hogan assumes that personality does not become consistent before early adolescence and advises that the HPI not be used with persons less than 14 years of age.

At the time of this writing, NCS charges $12.00 for an HPI specimen set, which includes a manual, a test booklet, a prepaid computer-scored answer sheet that may be completed and returned to NCS for an actual computer-generated report, and a sample computer-generated report on a fictitious subject. Purchased separately, prices range currently from $8.50 for a test manual, $12.00 to $14.00 for a package of 25 reusable test booklets (price varies depending on quantity ordered), and $6.00 to $7.00 for one computer scorable answer sheet, which includes the cost of scoring and deriving a computer-generated report. NCS promises 24-hour turn-around time for prepaid computer scorable answer sheets. Also available are computer-scorable answer sheets for either Arion II teleprocessing or Microtest computer software, which cost $5.40 to $7.00 per administration depending on quantity ordered and type of computer service. NCS does *not* market hand-scorable answer sheets.

Development and use of the HPSS presumes that the attitudes, values, and motives that form people's public and private selves are associated with competent performance in virtually any job, in virtually any organization. These attitudes, values, and motives, derived empirically, are grouped into three general constellations—Service Orientation, Reliability, and Stress Tolerance—as well as three specific occupational performance scales—the 215-item Clerical Potential Inventory, the 216-item Sales Potential Inventory, and the 233-item Managerial Potential Inventory. The three general constellations, along with the HPI validity scale, combine, as noted, to form the 198-item Prospective Employee Potential Inventory (PEPI). Each of these three inventories also yield scores on the same three constellations that compose the PEPI as well as on the HPI validity scale. As there is no HPSS inventory *per se*, Hogan and Hogan's statement (1986, p. 2) that the "HPSS takes 20-25 minutes to complete" is puzzling.

NCS currently charges $11.00 per specimen set for each of the four HPSS inventories, which includes a combined test booklet/answer sheet for the HPSS inventory of choice (PEPI, Clerical Potential, etc.). The test may be completed and returned to NCS for a computer-generated report. Each set also contains the same HPSS manual; therefore, for users who wish to review each of the four inventories, NCS recommends the purchase of one specimen set plus the individual inventories for the remaining three at $5.95 apiece. The HPSS manual, which costs $8.50 when purchased individually from NCS, is redundant with the HPI manual, because the HPSS overlaps the HPI. Computer-scorable answer sheets range from $4.50 to $9.25 per answer sheet, including scoring and a computer-generated report, depending on the quantity ordered. Because these tests are used for employee selection, it is rather unlikely that one individual would take all four tests. However, in order to evaluate an examinee's potential for clerical, sales, and managerial positions, the qualified professional could opt instead to administer the HPI and thereby derive all the occupational potential scores.

The remainder of this review will focus on the HPI only, as it contains all the HPSS scales as well as unique personality scales.

Technical Aspects

Selection of Scales, HICs, and Items. The HPI was developed initially in 1979 using the rational approach for personality inventory construction. Hogan, working in conjunction with his colleagues and graduate students at the Johns Hopkins University, reasoned that six primary trait dimensions were sufficient to explain the breadth of human personality and behavior: Intellectance, Adjustment, Prudence, Ambition, Sociability, and Likeability. Hogan (1986, p. 9) concedes that the issue of adequate sampling of domain "is an empirical question to be answered by other researchers." Hogan et al. next created items or statements that described each aspect of these six trait dimensions. This initial item pool of 425 items was reduced between 1979 and 1984 into the HPI's present form of 310 items, 43 HICs, and six scales.

This reduction was conducted empirically and with two goals in mind: 1) no item overlap among the six primary scales nor among the 43 HICs; and 2) high internal consistency among items on the same scale and items on the same HIC. The approximate "rule of thumb" seemed to be eliminating items that reduced a scale's coefficient alpha below .80 and/or an HIC's coefficient alpha below .50.

The sample populations used for the reduction and organization of items totalled approximately 1,700 persons. These samples are noteworthy for their heterogeneous nature. For example, about 80% of the persons were non-college educated. (Hogan, unlike many academic researchers, did not rely exclusively on college sophomores as the basis for his test development population.) The samples also contained a reasonable proportion of women and minorities. Overall, the HPI was developed on samples that included undergraduate students, military personnel, blue collar workers, police officers, felons, and professional and technical workers.

The Validity scale, the seventh HPI scale, was developed based on endorsement frequency. Sixteen items were found to have endorsement frequencies of 92% or greater. The content of these items was adjusted to provide a balanced scoring direction (8 "true" items, 8 "false" items). Hogan claims an accuracy rate of 96% for identifying invalid (i.e., random response) profiles.

The six occupational scales were developed by using subsamples culled from the overall HPI sample population. These occupational scales were created by combining HICs found to correlate significantly with an external criterion that operationally defined an important occupational dimension or a general occupational class. The SOI scale was developed by correlating HIC scores for 101 nurses against ratings by their supervisors on the service orientation of each nurse. The REL scale was developed by comparing HIC scores for a sample of 80 delinquents against a sample of 156 nondelinquents. The STR scale was developed by correlating HIC scores for 56 truck drivers against their medical absence record. Hogan also selected some HICs for this scale on a purely rational basis. The CLE scale was developed by correlating HIC scores for 107 clerical workers with ratings by their supervisors on the overall job performance of each worker. The SAL scale was developed by correlating HIC scores for 127 sales representatives with ratings by various levels of managers familiar with the overall job performance of each salesperson. Finally, the MAN scale was developed by correlating HIC scores with the level of achieved organizational status for 372 trucking firm employees. For each scale, virtually no attempt was made to adjust for low internal consistency among the HICs that correlated with a particular criterion, unlike the six primary HPI scales where items were eliminated that lowered internal consistency.

Reliability of Scales and HICs. Hogan assessed the internal consistency reliability of the HPI personality scales and HICs using samples of approximately 800 adults across a variety of occupational settings. The test-retest reliability was based on 90 college undergraduates tested over a 4-week interval. The reliabilities of the HPI personality scales are extremely high, with alpha coefficients ranging from .76 to .89 and test-retest correlations ranging from .74 to .99. The reliabilities of the HICs are lower, with alpha coefficients ranging from .39 to .83 and test-retest correlations ranging from .38 to .99. The HPI personality scales show remarkable consistency; the HICs are less consistent but still adequately reliable, given how few items define each HIC. Given the high degree of internal consistency, the HPI would seem a likely candidate for computer adaptive testing, thereby reducing the administration time to perhaps as little as 10 to 15 minutes.

The internal consistency and test-retest reliabilities of the HPI occupational scales are relatively low, ranging from .19 to .63 and .42 to .76, respectively. Hogan attributes the low alpha coefficients to the inherent heteronomous nature of the scales, because each scale combines HICs from several different basic personality dimensions. Although this explanation seems appropriate to explain the lowered internal consistency of the scales, it does not explain the lowered reliability across time.

Content Validity of Scales and HICs. The HPI personality and occupational scales, HICs, and items were developed rationally, at least in their inception. In part, the goal of this rational approach was to identify a range of qualities associated with each broad dimension defined by each HPI scale. It is always difficult, if not impossible, to ensure adequate coverage of the content domain of any personality or occupational scale. The HPI, however, has applied a creative solution to this problem through the use of HICs, which provide the user with a clear indication of the personological and behavioral qualities sampled by each scale. In short, the HICs are an innovative and highly useful means for breaking down a personality or occupational scale into its components. HICs allow for an easier determination of the adequacy of content sampling and coverage. Perusal of the HICs that define each of the 12 HPI scales suggests an adequate coverage of the content of each scale.

One can argue, however, that the HPI's reliance on just six personality dimensions does not provide an adequate sampling of the entire range of personality and behavior. In fact, given Hogan's theoretical rationale, his six dimensions sample only the interpersonal domain of personality and behavior. For example, the first reviewer has found from experience that the HPI loses its predictive and descriptive utility with samples of persons who are highly effective in interpersonal situations (e.g., high-powered salespersons). The HPI profiles among such persons are essentially undifferentiated from one another (high scores on all HPI personality scales except PRU, which is comparatively lower). The HPI falters in its ability to identify individual differences among these persons because the HPI assesses their interpersonal style to the exclusion of other differentiating qualities. Such samples, however, are rare.

Criterion-Related (Concurrent and Predictive) Validity of Scales and HICs. The criterion-related validity of the HPI personality and occupational scales and HICs seems well documented and surprisingly strong for an instrument as new as the HPI. Hogan (1986, pp. 18-19) provides research findings that indicate adequate concurrent and predictive validity for the HPI. For example, the HPI was able to identify truck drivers who violated traffic laws on a regular basis, workers who filed grievances against their companies, workers who received commendations for superior performance, and salespersons who generated superior sales revenues. The HPI also predicted the success of persons participating in Navy diver training courses and military bomb disposal training school. These findings are particularly noteworthy because of the diversity of the sample populations and the objectivity of the criterion with which the HPI was related.

Construct Validity of Scales and HICs. There appears to be strong evidence to support the construct validity of the HPI personality and occupational scales and HICs. Hogan (1986) and Hogan and Hogan (1986) present innumerable sets of data in support of the HPI's construct validity. (In fact, their presentation in the test manuals should be a model for other test authors to follow.) The HPI scales and HICs correlate positively with scales measuring similar constructs (convergent validity) and correlate negatively or insignificantly with dissimilar or irrelevant constructs (divergent validity).

For example, the HPI personality scales were correlated with those of the California Psychological Inventory (CPI) for a sample of 125 military personnel. The highest correlate for each HPI scale was its counterpart on the CPI. INT correlated .44 with Intellectual Efficiency, ADJ correlated .57 with Sense of Well-Being, PRU correlated .46 with Socialization, AMB correlated .45 with Dominance, SOC correlated .48 with Empathy and .46 with Social Presence, and LIK correlated .50 with Sociability. Similarly, in a correlation with the Minnesota Multiphasic Personality Inventory for a sample of 109 police cadet applicants, the HPI scales correlated negatively and significantly with virtually all MMPI clinical scales.

The data supporting the HPI's construct validity generally satisfy only the "multitrait" portion of the Campbell and Fiske (1959) multitrait-multimethod approach to determining construct validity. Most of the data presented by Hogan et al. compared the HPI, a self-report inventory, with other self-report inventories; consequently, the magnitude of the correlations reflect in part a similarity in method. Although there is no reason to believe the HPI would not withstand multimethod comparisons, Hogan et al. need to pursue more construct validation research that emphasizes this multimethod approach.

Critique

The Hogan Personality Inventory is an innovative and important contribution to the fields of industrial/organizational and personality psychology. The HPI is the first major attempt to develop a new self-report personality inventory specifically for use in occupational settings since the Strong Vocational Interest Blank and Holland's Vocational Preference Inventory. Consequently, the HPI has direct, practical application for professionals involved in actual personnel selection as well as professionals involved in research issues related to personnel selection. Further, the HPI seems to satisfy both the *Uniform Guidelines on Employee Selection Procedures* (U.S. Civil Service Commission, 1978) in terms of validity and the various Equal Employment Opportunity guidelines regarding sex or racial differences. Although the HPI should never be used as the sole criterion for personnel selection, it is a valuable supplement to more standard screening procedures (i.e., job interview, matching an applicant's SKAs with the SKAs required for a job).

The HPI, however, is not without its problems and drawbacks. First, the inventory is overly dependent on the NCS computer-generated report system for profile interpretation. The use of such systems is quite controversial and in fact may detract from the validity of the personality inventory on which the report is based (Matarazzo, 1986). Both the American Psychological Association and the U.S. Office of Personnel Management are clear in their positions that the validity of the computer-generated report must be established by a test author and publisher, not just the validity of the test. There is little if any evidence to support the validity of the NCS computer-generated report for the HPI (or the HPSS). Furthermore, these reviewers found the report more "flashy" than substantive, creating the impression of providing more information than in fact it conveys and of being more scientific. Hogan (1986) and Hogan and Hogan (1986) provide a detailed presentation for interpreting the HPI without reliance on the computer-generated report that these reviewers strongly recommend.

Second, the HPI may reflect situation-specific personality characteristics rather than more stable, core personality characteristics. Given that the HPI is intended for use in job application settings, the probability increases that a person's self-presentation in the situational role of "job applicant" may not be consistent with that same person's self-presentation in the situational role of "on-the-job employee." The degree of consistency between the HPI personality profile of a person applying for a job and the same person as an employee needs to be researched more thoroughly.

Third, the HPI Validity scale measures only the response set of randomness. In job application settings, the more common and problematic response set is social desirability, that is, "faking good." Hogan and Hogan (1986, p. 20) argue that "frequency of faking in the actual employment process is rare and infrequent" and that even if faking occurred, it is an indication of social skills (i.e., a positive attribute). These reviewers disagree with both assertions. Based on the first reviewer's extensive experience using the California Psychological Inventory in personnel selection settings, elevated scores on Good Impression and lowered scores on Communality, both measuring the social desirability response set, are far more common under these circumstances than when the CPI is administered in research settings. Furthermore, rather than an indicator of social skills, faking responses to a personality inventory more often is a sign of insecurity and unwillingness to portray oneself in terms of strengths *and weaknesses*. Utility of the HPI thus would improve if its Validity scale provided "faked good" responses in addition to random ones.

Fourth, there appears to be overlap between the samples used to develop the initial HPI personality scales and HICs and those used to validate the HPI occupational scales and HICs. For example, it appears that the samples of hospital personnel, truck drivers, military personnel, sales representatives, and school administrators were used to develop the HPI personality scales and HICs and then again to validate the HPI occupational scales and HICs. As the occupational scales are based on the HICs that define the personality scales, this "contamination" of samples is problematic.

Why use the HPI rather than another personality inventory, particularly the California Psychological Inventory (CPI)? This question takes on added importance given the recent revision of the CPI (Gough, 1987) for the specific purpose of increasing its applicability to occupational settings. For example, the revised CPI contains Work Orientation and Managerial Potential scales. Both of these scales correlate highly with the HPI scale of Adjustment, .54 and .53, respectively, which is higher than any of the correlations with the HPI occupational scales. The revised CPI also provides four general themes reflecting work style as well as a worker's general level of competence. The question of HPI versus CPI is more easily raised than it is answerable. Some differences between the two instruments are practical—the HPI takes about 15 to 20 minutes less to administer. Some differences are objective—the CPI has accumulated over 30 years of published research compared to less than 3 for the HPI—and some will be subjective—psychologists may prefer whichever of the two instruments is more familiar to them and whichever seems better suited to their specific requirements. Ultimately, Hogan et al. will need to determine the specific and unique utility of the HPI as compared with its nearest, more established counterpart, the CPI.

In summary, the Hogan Pesonality Inventory is a reliable and valid instrument for assessing general personality structure and general occupational potential. For such a new instrument, the HPI already has established a surprisingly strong "track record." It may become the future instrument of choice in terms of personality assessment for the purpose of occupational selection.

References

Campbell, D., & Fiske, D. (1959). Convergent and discriminant validity by the multitrait-multimethod matrix. *Psychological Bulletin, 56,* 81-105.

Gough, H. (1987). *Administrator's guide to the revised California Psychological Inventory.* Palo Alto, CA: Consulting Psychologists Press.

Hogan, R. (1976). *Personality theory: The personological tradition.* Englewood Cliffs, NJ: Prentice Hall.

Hogan, R. (1982). A socioanalytic theory of personality. In M. Page (Ed.), *Nebraska symposium on motivation* (Vol. 29, pp. 55-89). Lincoln, NE: University of Nebraska Press.

Hogan, R. (1986). *Manual for the Hogan Personality Inventory.* Minneapolis, MN: National Computer Systems.

Hogan, R., & Briggs, S. (1986). A socioanalytic interpretation of the public and the private selves. In R. Baumeister (Ed.), *Public and private self* (pp. 179-188). New York: Springer-Verlag.

Hogan, R., Johnson, J., & Emler, N. (1978). A socioanalytic theory of moral development. In W. Damon (Ed.), *New directions in child development* (Vol. 2, pp. 1-18). San Francisco: Jossey-Bass.

Hogan, J., & Hogan, R. (1986). *Manual for the Hogan Personnel Selection Series*. Minneapolis, MN: National Computer Systems.

Hogan, R., Cheek, J., & Jones, W. (1985). Socioanalytic theory: An alternative to armadillo psychology. In B. Schlenker (Ed.), *The self and social life* (pp. 175-198). New York: McGraw-Hill.

Matarazzo, J. (1986). Computerized clinical psychological test interpretation. *American Psychologist, 41,* 14-24.

U.S. Government Civil Service Commission. (1978). *The uniform guidelines on employee selection procedures*. Washington, DC: U.S. Government Printing Office.

Gregory H. Dobbins, Ph.D.

Associate Professor of Management, Program in Industrial and Organizational Psychology, University of Tennessee, Knoxville, Tennessee.

HOW SUPERVISE?

Q. W. File and H. H. Remmers. San Antonio, Texas: The Psychological Corporation.

Introduction

How Supervise? is a 70-item instrument designed to assess respondents' knowledge of supervisory principles. It is based upon the assumption that supervisory skills generalize across different types of work settings, organizations, and industries. The instrument emphasizes the importance of human relations skills for achieving high levels of organizational effectiveness.

Q. W. File (1945) recognized the importance of effective supervision. As part of his dissertation at Purdue University, he developed How Supervise? to assess understanding of supervisory principles. He defined a supervisor as "an individual who actually directs productive processes at the scene of operation" (p. 323). Content areas of How Supervise? were identified based upon training manuals for industrial supervision and human relations skills, suggestions from actual supervisors, and labor leaders. Using this information, 204 items were generated and submitted to 8 "experts" in supervision (individuals writing books and articles on industrial supervision) and 37 government supervisors in the Division of Vocation Training for War Production Workers. Experts identified ambiguous items and rated the desirability of each item. Their modal response to each item was used to construct a scoring key.

Final selection of the items was accomplished by administering them to 577 industrial supervisors. The effectiveness of each supervisor in the sample also was rated by superiors. Analyses indicated that 23 of the items significantly discriminated between the top and bottom 27% of the sample. The 140 items that were most discriminating were retained, and the other 164 were deleted. Seventy of the discriminating items were assigned to Form A, and 70 were assigned to Form B. A third form, Form M, was constructed by selecting 100 of the items that appeared most relevant for middle- and upper-level management. Data evaluating the appropriateness of items selected for Form M are not available (File & Remmers, 1971).

How Supervise? was developed in 1945 and, unfortunately, has not been revised. The test first was marketed by The Psychological Corporation in 1948. The manual that accompanies the test was updated in 1971. The revised manual provides norms from 12 fairly large samples of supervisors employed in electronics, chemicals, and heavy manufacturing. The median score in each sample varies dramatically (e.g.,

$M = 49$ at an electronics plant and $M = 31$ at a metal fabricating company). Thus, type of industry appears to affect the distribution of scores. Furthermore, normative data are not available for service-oriented industries (e.g., banking, retailing, etc.). In addition, the norms reported in the manual may be obsolete; the most recent sample included was collected in 1970. Given the dramatic changes that have occurred in the U.S. economy and supervisory training, it seems problematic to interpret How Supervise? scores based upon the normative data included in the manual. If companies wish to use How Supervise?, they must be willing to develop local norms. Hopefully, more recent and comprehensive normative data will be presented in upcoming revisions of the instrument.

All forms of How Supervise? are divided into three sections. The first, Supervisory Practices, asks respondents to indicate whether various supervisory practices are desirable or undesirable. The second section, Company Policies, asks respondents to review different company policies concerning promotions, labor unions, and disciplinary actions and to indicate the desirability of each. The final section, Supervisor Opinions, requires respondents to indicate whether they agree or disagree with various opinions that may be held by supervisors. This final section focuses on whether respondents believe that workers are consistent with McGregor's (1960) Theory X (i.e., workers must be controlled and directed through external punishments and rewards) or Theory Y (i.e., workers are motivated to seek out responsibility without external rewards and punishers). Somewhat surprising is that although the three subscales are scored individually, they are not interpreted individually. Instead, they are summed to form a single index.

The manual that accompanies How Supervise? provides an extremely clear description of the procedures used to administer it. How Supervise? can be administered either individually or in groups. Prior to testing, all respondents should be presented with a straightforward explanation of the company's intended use of the results. Respondents are given as much time as needed for completing How Supervise?, but are encouraged to work "as rapidly as possible without being careless."

The "Company Policies" and "Supervisory Practices" sections of How Supervise? require respondents to read a series of statements describing common supervisory practices and company policies and circle "D" if the statement is desirable, "U" if the statement is undesirable, or "?" if they are uncertain about the desirability of the statement. The "Supervisor Opinions" section of How Supervise? presents respondents with opinions that are commonly held by supervisors and asks them to circle "A" if they agree with the opinion, "DA" if they disagree with the opinion, or "?" if they are uncertain about the opinion.

All three sections of How Supervise? are scored by comparing individuals' responses to a scoring key. Respondents receive 1 point for each question answered correctly and lose 1 point for each question answered incorrectly. Questions on which respondents answered "?" are not graded. Scores on the three sections of How Supervise? are summed, and the raw score is transferred to a percentile score based on norm tables included in the manual. Unfortunately, as the norms are based on data collected in the 1940s and the distribution of scores varies as a function industry, users of How Supervise? should develop local norms to interpret the instrument. Each instrument requires approximately 10 minutes to score. Forms A, B, and M are administered and scored following identical procedures.

Practical Applications/Uses

How Supervise? has several applications in organizations. First, because the instrument assesses knowledge of supervisory skills, it could be used as part of a selection program to identify individual employees with supervisory potential. Such an instrument is needed because employees with excellent technical skills do not always develop into effective supervisors. Individuals could be screened based upon How Supervise?, and those who perform well on the instrument could be selected and trained for supervisory positions. Unfortunately, past reseach with the instrument does not support such use (Bass, 1957; Parry, 1968; Weitz & Nuckols, 1953).

How Supervise? also is used commonly to document the effectiveness of supervisory training. In essence, one version of the instrument is given prior to training, and a second version is administered following training. Increases in How Supervise? scores are interpreted as evidence of training effectiveness. Two points concerning the test's use to demonstrate the effectiveness of supervisory training should be emphasized. First, a control group must be included if organizations wish to demonstrate unambiguously that training affects How Supervise? scores. Without a control group, numerous other interpretations of the change in scores are plausible, including history, maturation, testing, instrumentation, and statistical regression (Cook & Campbell, 1979, pp. 51-55). A second limitation of using How Supervise? to evaluate training effectiveness is that the instrument does not measure actual behavior, but instead assesses a respondent's knowledge of generally accepted supervisory practices. In other words, the test can measure the cognitive change produced by a training program but not the behavioral change. In most settings, it is more important to demonstrate that criteria external to the supervisor (e.g., profitability, subordinate satisfaction, absenteeism, turnover) have been affected by training. If organizations wish to assess the effects of training on criteria external to the supervisor, then How Supervise? is not an appropriate instrument to use.

How Supervise? also can be used to identify areas in which supervisors possess inadequate understanding of supervisory principles. By tabulating supervisors' responses to each item, areas of weakness can be identified and training programs targeted at them. In addition, feedback of test results to individual supervisors may facilitate discussion concerning appropriate supervisory styles and behaviors. Such discussion possibly may result in positive behavioral change.

Technical Aspects

The reliability of How Supervise? has been reported in several studies. The alternate-form reliability coefficients are about .80, a value that is clearly sufficient for experimental use but somewhat below the levels desired if the scale is to be used as a selection device. When Form A and Form B are added together and used as one scale, the reliability of the instrument increases to .86. Limited data assessing the reliability of Form M are available. Split-half reliability was .87 in one sample of high-level supervisors.

Given the length of How Supervise?, it is somewhat surprising that its reliability is not higher. One possibility is that the three subscales (Supervisory Practices, Company Policies, and Supervisor Opinions) measure slightly different constructs and should be scored and interpreted separately. Unfortunately, past research has not reported the correlations between the subscales or the reliability of each subscale. In addition, it would be useful to examine intercorrelations of items and the correlation between each item and total scores. Several of the items probably are correlated negatively with the others and should be eliminated. Furthermore, coefficient alpha and test-retest reliabilities should be reported.

Three types of validity studies have been conducted with How Supervise?. The first type has compared the scores obtained by different levels of supervisory employees. For example, File and Remmers (1946) administered How Supervise? to 828 supervisors working in the rubber industry. They found that higher-level supervisors obtained significantly higher mean scores than lower-level supervisors. Such a finding is consistent with the notion that How Supervise? is assessing supervisory knowledge because high-level supervisors should have more supervisory knowledge than low-level supervisors. The results also could have been produced by any factor that is confounded with level of supervision, including age, socioeconomic status, and intelligence.

A second type of validity study has examined the relationship between How Supervise? scores and ratings of on-the-job performance. Unfortunately, most of this research has been conducted with concurrent designs. For example, Carter (1952) correlated How Supervise? with peer ratings of supervisory performance and found a significant relationship in two independent samples ($r = .22$ and $r = .33$). Several other studies report similar findings (e.g., Dooher & Marting, 1957; Holmes, 1950).

Other studies have failed to find significant correlations between How Supervise? and ratings of supervisory performance. For example, Decker (1956) found that How Supervise? correlated only .108 with superiors' ratings of supervisory effectiveness. Meyer (1956) used a predictive validity design and found that How Supervise? was not correlated significantly with superiors' ratings of "overall success as a supervisor" or union representatives' ratings of "human relations skills." Meyer did find a curvilinear relationship between How Supervise? and job performance, with respondents performing in the middle half of the distribution on How Supervise? being evaluated as more effective than those that scored either high or low on the instrument. In a second predictive validity study, Dicken and Black (1965) found that How Supervise? scores did not predict superiors' ratings of leadership or management potential at a follow-up period of 3.5 years.

A final group of studies has investigated the convergent and discriminant validity of How Supervise?. The instrument's scores are correlated fairly highly with adaptability ($r = .62$ and .71) and general intelligence ($r = .38$, .67, and .33). More systematic multitrait-multimethod studies need to be conducted following the rationale of Campbell and Fiske (1959). In addition, the relationship between How Supervise? and other leadership scales needs to be determined.

It is interesting to note that several critical areas have been neglected in past psychometric evalautions of How Supervise?. First, investigators have not examined the relationship between the three subscales of How Supervise? and supervisory performance. It may be that one subscale (e.g., Company Policies) is related highly to performance and that summing the three scales dilutes this effect. Similarly, the extent to which How Supervise? adds to the prediction of supervisory performance above and beyond more cognitively oriented predictors needs to be determined. Third, researchers should investigate the relationship between How Supervise? scores and actual supervisory behavior. For example, what percentage of supervisors that indicate that they obtain subordinate input prior to making important decisions actually engage in such a practice? Fourth, the possibility of differential validity needs to be examined; that is, the relationship between How Supervise? and supervisory performance needs to be considered separately for majority and minority groups. Finally, and perhaps most importantly, published validation studies of How Supervise? are outdated. The validity of the instrument needs to be established in the workplace of the 1980s and in a variety of industries.

Critique

The importance of an instrument like How Supervise? cannot be over-emphasized. Identification of supervisory talent is critical for the survival of any organization. File's proposal that supervisory selection programs should consider skills other than technical competence has been supported strongly (Bass, 1981). Thus, How Supervise? could be, in theory, an extremely useful instrument.

Unfortunately, several major problems limit its usefulness. First, and perhaps most serious, How Supervise? assesses only knowledge of supervision. It is based upon the assumption that supervisors who do not understand the principles of supervision cannot implement appropriate behaviors in their day-to-day interactions with subordinates. Unfortunately, the converse of this assumption is not true. In fact, many supervisors who have knowledge of supervisory principles are unable to implement these principles in their interactions with subordinates. Although a supervisor may have a good understanding of principles of supervision, he or she may lack the human relations skills that are necessary for implementing the principles. It is for this reason that assessment-center methodologies and managerial work samples have been so effective in identifying managerial personnel (Byham, 1986), especially leaderless group discussions (Bass, 1954) and business games (Wolfe, 1976).

Despite this limitation, however, a revised version of How Supervise? could assess respondents' cognitive understanding of supervision satisfactorily. Several major revisions need to be incorporated into the instrument. First, each question should present the respondent with more information about the situation. Research in leadership has demonstrated that there is not one best supervisory style. Instead, there are several effective styles of behavior depending upon the situation (see Bass, 1981, for a review). How Supervise? forces respondents to indicate whether an action is desirable or undesirable without any reference to situational factors.

The revised version of How Supervise? also needs to be constructed more systematically than the original version. More care needs to be given in identifying the content of supervisory knowledge and writing questions that accurately capture this content. Extensive statistical analyses should be performed on each item and special attention should be focused on inter-item and item-total correlations. Furthermore, factor analyses should be conducted to identify the major dimensions that are assessed with the instrument.

A revised How Supervise? would enable organizations to determine those areas in which their supervisors lack knowledge, and training programs then could be focused on those areas. Such training programs should employ techniques such as role-playing, simulation, and case studies to insure that supervisors develop both an understanding of supervision and the skills necessary to transfer this knowledge to supervisor-subordinate interactions. This reviewer strongly encourages The Psychological Corporation to engage in such a revision and possibly develop a supporting supervisory training program. Without such work, How Supervise? has questionable value and will provide organizations with little, if any, useful information.

References

Bass, B. (1954). The leaderless group discussion. *Psychological Bulletin, 51,* 465-492.

Bass, B. (1957). Validity information exchange, No. 10-25. *Personnel Psychology, 10,* 343-344.

Bass, B. (1981). *Stogdill's handbook of leadership: A survey of theory and research.* New York: Free Press.

Byham, W.C. (1986). *Assessment centers: More than just predicting management potential.* Pittsburg, PA: Development Dimensions International.

Campbell, D.T., & Fiske, D.W. (1959). Convergent and discriminant validity by the multitrait-multimethod matrix. *Psychological Bulletin, 56,* 81-105.

Carter, G.C. (1952). Measurement of supervisory ability. *Journal of Applied Psychology, 36,* 393-395.

Cook, T.D., & Campbell, D.T. (1979). *Quasi-experimentation: Design and analysis issues for field settings.* Chicago: Rand-McNally.

Decker, R.L. (1956). Item analysis of *How Supervise?* using both internal and external criteria. *Journal of Applied Psychology, 36,* 406-411.

Dicken, C.F., & Black, J.D. (1965). Predictive validity of psychometric evaluations of supervisors. *Journal of Applied Psychology, 40,* 406-411.

Dooher, M.J., & Marting, E. (1957). *Selection of management personnel.* New York: American Management Association.

File, Q.W., & Remmers, H.H. (1946). Studies in supervisory evaluation. *Journal of Applied Psychology, 30,* 421-425.

File, Q.W., & Remmers. H.H. (1971). *Manual: How Supervise?* Cleveland, OH: The Psychological Corporation.

File, Q.W. (1945). The measurement of supervisory quality in industry. *Journal of Applied Psychology, 29,* 323-337.

Holmes, F.J. (1950). Validity of tests for insurance office personnel. *Personnel Psychology, 3,* 57-69.

McGregor, D. (1960). *The human side of enterprise.* New York: McGraw-Hill.

Meyer, H.H. (1956). An evaluation of a supervisory selection program. *Personnel Psychology, 9,* 499-513.

Parry, M.E. (1968). Ability of psychologists to estimate validities of personnel tests. *Personnel Psychology, 21,* 139-147.

Weitz, J., & Nuckols, R.C. (1953). A validation study of *How Supervise?. Journal of Applied Psychology, 37,* 7-8.

Wolfe, J. (1976). The effects and effectiveness of simulations in business policy teaching applications. *Academy of Management Review, 1,* 47-56.

Julian Fabry, Ph.D.

Counseling Psychologist, Meyer Children's Rehabilitation Institute, Omaha, Nebraska.

HUMAN FIGURES DRAWING TEST

Eloy Gonzales. Austin, Texas: PRO-ED.

Introduction

The Human Figures Drawing Test (HFDT; Gonzales, 1986) is a revision of the Goodenough-Harris Draw-a-Person (DAP) test. The current 1986 revision is intended to be a nonverbal measure of cognitive maturation for children ages 5-11. Children's level of maturity or intelligence is derived from the human figures they draw. Influenced by, and building on the investigations of Koppitz (1968), the author of the HFDT, Eloy Gonzales, Ph.D., believes that the drawings yield an understanding of children's perceptions of their environment as well as of themselves. In turn, these environmental and self-perceptions reflect a level of general intelligence and gross- and fine-motor coordination maturation. According to Gonzales, the HFDT is also a measure of visual perception. Furthermore, he appears to believe that children's drawings, unlike those of adults, are not projective assessments of personality traits (e.g., impulsivity) or states (e.g., anxiousness).

As a special educator at the University of New Mexico, Dr. Gonzales has been interested in determining a nonverbal and culture-free assessment of cognitive abilities for children aged 5-10 (Gonzales, 1982). These earlier efforts, which demonstrated that differences exist between various subcultures (e.g., Navajo, Mexican-American, black, etc.) in the frequency of drawing features or scoring criteria endorsed on the DAP, seem to have impelled the author's current undertaking.

In the opening chapter of the HFDT manual, Gonzales, in tracing the historical perspective that has given shape to the current 1986 revision, recognizes the original work of Goodenough (Draw-a-Person; 1926), the revision by Harris (Draw-a-Person; 1963), and the efforts of Koppitz (Human Figure Drawing; 1968). The original work assumed that children's drawings were valid and reliable measures of intelligence. They also were considered to be nonverbal because little understanding of language was required of the examinee.

Goodenough (1927) reviewed a variety of studies accomplished up to the time she presented the DAP. She appeared to be influenced by the work of Claparede (1907) and Ivanoff (1909). Claparede proposed that developmental stages in drawing existed in addition to the relationship between drawing aptitude and general intellectual ability as measured by school achievement. Ivanoff developed scoring criteria for drawings and compared these with teacher ratings of abilities, achievement in various school subjects as well as certain moral and social traits. Goodenough surmised that a child's drawing was dependent on the concept of the object rather than the imagery or artistic appreciation. She concluded that as developmental changes take place at a conceptual level, so do the features represented in children's drawings.

The first revision, undertaken by Harris in 1963, changed the original scoring criteria of 51 items to 73 items in an attempt to make the test more objective and empirical in nature. Harris also converted the original ratio IQ to a deviation IQ. The 1963 revision introduced a parallel, Draw-A-Woman form. In addition, Harris restandardized the original standardization sample, which consisted of children ranging in age from 4-10 drawn mainly from New Jersey and New York City, to represent the occupational distribution of the United States in 1950. His sample was derived from four geographical areas and composed of children ranging in age from 5-15. Harris' conversion of the original ratio IQ to a deviation IQ has been kept in the current revision. The parallel form (i.e., the Draw-a-Woman) that was introduced in Harris' revision remains in the 1986 version to the extent that the child is asked to draw a figure of the opposite sex after the initial drawing.

In the current 1986 revision, Gonzales has reduced the 73 items of the 1963 version to 38. The items were reduced to account for the developmental criteria which allowed for proportional increases in the occurrence of a drawing feature over the age groups being studied. It is difficult at this time to determine how the reduction of items has affected the objectivity and the empiricism of the HFDT because few studies external to those accomplished by Dr. Gonzales have been undertaken. From the information provided in the manual, there does not appear to be a significant effect at this point in time.

Koppitz (1968) created a scoring system based on developmental criteria which accounted for the occurrence of various expected and exceptional drawing features over 5- to 12-year-olds. Gonzales (1982) utilized her developmental criteria to study five ethnic groups in the Southwest and these results impelled him to modify them for the HFDT.

The sample for the 1986 revision consisted of public-school children from 20 states representing various geographic locations (e.g., Maine, Iowa, Oregon, Georgia, and Illinois). Gonzales attempted to consider the general population figures as reported in the 1985 *Statistical Abstract of the United States*, which takes into account the distribution of children by their gender, place of residence (i.e., urban/rural), and race or ethnicity.

Practical Applications/Uses

The HFDT was updated to provide a general and global assessment of intelligence and cognitive maturity in young children. It requires relatively little time to administer and score, which should appeal to individuals seeking a quick, short, screening instrument that yields a single index of intellectual potential. Because the test has a minimal language requirement, it is considered relatively nonverbal in nature. It can be used as one index of cognitive maturity with the hearing-impaired and children with strong subcultural influences. The nonthreatening drawing format of the HFDT relieves children of their fears of being tested, which has been thought to hinder the performance of some youngsters. As is true of past versions, the HFDT is designed to help the examiner determine which child may need further and more extensive individually administered assessments in order to insure the child's proper evaluation, correct placement within the curriculum offered, or informed referral to another agency, with the ultimate goal of optimizing the child's potential to learn.

In the 1963 version, the DAP was used with children ranging in age from 3 to 15, but the current revision is intended to be used with 5- to 11-year-olds. Anastasi (1972) has remarked that efforts to extend the DAP to adolescents has not proved successful. This lack of success may be due, in part, to the theoretical notion postulated by Harris (1963) that children cease to show increments of cognitive maturity as they move from concrete to more formal operations, which occurs around 10 or 11 years of age. The HFDT also seems to be limited in its ability to test children with a cognitive maturity below the base of 55 standard score points and above the ceiling of 145, established by Gonzales as the range of the HFDT's utility.

The manual provides instructions for both individual and group administration. Essentially, each child is given two 8½"x11" sheets of blank paper and asked to draw a picture of himself or herself. Upon completion of the drawing, they are instructed to draw someone of the opposite sex on the remaining sheet. Unidentifiable parts on each of the drawings are located by the examiner, who queries the child for help in labeling them.

This assessment can be used by clinical, counseling, and school psychologists, in addition to special educators, pediatricians, and family physicians. Although anyone can ask a child to draw a picture, Gonzales stresses that examiners should have some formal training in assessment. He suggests some education in statistics and test administration, as well as some knowledge about the evaluation of mental abilities.

Administration of the HFDT takes approximately 5-20 minutes; about the same amount of time is needed for scoring. Because there are no restrictions as to time or place of administration, it appears that these are left to the discretion of the examiner.

The actual scoring consists of comparing the child's drawings with the 38 criteria (e.g., whether the drawing has a head) proposed by the author to measure cognitive maturity. Each feature is scored as present by assigning it a value of 1; if it is considered absent, it is scored zero. The raw score represents the summation of all the features identified on the child's drawing. This raw score then is translated to a standard score by comparing the results with other children within the same age range. Each range represents 1 year (e.g., 5.0 to 5.11). The standard score, or quotient, that is derived is equivalent to a deviation intelligence quotient, which can be equated to other measures of intellectual potential, although the interpreter should bear in mind that the HFDT score appears to be an unrefined global assessment.

The results are scored and profiled on a form that can be obtained along with the manual from the publisher. The form consists of six parts. In the first part, the examiner provides information such as the child's name, address, date of birth, date the test is given, and the examiner's name and title. The next part, section I, is labeled for the HFDT scores and accommodates the raw and standard scores (quotients), as well as the percentile ranks. Two tables are provided in the appendices of the manual. Table A converts raw scores to standard scores, and Table B converts quotients (standard scores) to percentile ranks. Section II accommodates other testing information (i.e., other test scores), and the next part, section III, can be used to identify various testing conditions that may affect the child's performance. Section IV is for the examiner's general comments or recommendations, and the final section contains the actual scoring criteria used by the examiner to check the child's drawing.

The manual contains a chapter devoted to the interpretation of results. It addresses issues related to equating the results with other tests and conditions that may affect the child's performance. Specifically, the author demonstrates equating the results of several standardized tests (e.g., the Detroit Tests of Learning Aptitude-Primary Tests-Revised) by converting the findings to standard scores (HFDT Equivalent = (15/SD)(X-M)+100). He further contends that noises, interruptions, and other conditions may affect a child's performance, as well as the results, and therefore should be noted. Gonzales further explains how to analyze test scores in light of deviation standard scores and bands of performance (e.g., average, below average). A standard error of measurement, which is available (Table 5) for each age group, can assist the interpreter in assessing the degree of reliability obtained from the results. Again, some knowledge of statistics, functions of measurements, and general testing principles would be helpful in making interpretations.

Technical Aspects

Building upon his previous work of investigating the specific item endorsements of various cultural groups, Gonzales approached the restandardization process for the HFDT in a manner that would be representative of children in the United States. The current sample (N = 2,400) accounts for the distribution of public school children according to geographic region, sex, race or ethnicity, and place of residence (i.e., urban or rural).

Various aspects of reliability have been addressed for this particular assessment procedure. Anastasi (1972) indicated that the interscorer reliability (.9) determined for the 1963 test reflected the "fullness of the scoring instructions" (p.670) and the explicitness of the scoring criteria. It appears that this notion has been fulfilled in the current 1986 revision. Gonzales taught three graduate students how to score the HFDT in two 1-hour training sessions. The graduate students then were given 10 randomly selected profiles from the normative sample of 5-, 8-, and 10-year-old children. Their score correlations yielded an average coefficient of .97. Previous internal consistency reliability coefficients ranged from .7 to .8 utilizing the split-half method. Using a coefficient alpha, Gonzales calculated correlations ranging from .73 to .85. The test-retest correlations on the 1963 version ranged from .6 to .7. For the present revision, Gonzales sampled 50 kindergartners, as well as third- and fifth-graders from two schools in New Mexico. He determined that the 2-week test-retest reliability coefficients were .87 for the kindergartners, .91 for the third-graders, and .89 for the fifth-graders. A standard error of measurement calculated for each of the age groups ranged from 8 points for the 5-year-old children to 6 points for the 6-, 9-, and 10-year-olds. These results appear to be somewhat broad when compared to the standard error of measurement on the revised edition (1974) of the Wechsler Intelligence Scale for Children (WISC-R), yet they represent a more global and unrefined assessment.

Apparently, previous validity studies between the DAP and the Stanford-Binet, Form L-M, as well as the WISC, have resulted in substantial and significant correlations (Anastasi, 1972). In another commentary, Dunn (1972) stated that only one concurrent validity study was accomplished between the Harris (1963) revision and the original. The study, performed using a group of Canadian-Indian children, yielded validity coefficients ranging from .91 to .98 between the 1926 and the 1963 scales. He further commented that no relationships between either version of the DAP and achievement were investigated. Gonzales has sought to establish the relationship between the HFDT and achievement, in addition to substantiating the assessment's validity.

The developmental criteria for various age groups proposed by Koppitz (1968) has been posited by Gonzales as evidence for content validity. In the manual, he has presented the percentage of children in each age group of the normative sample (Table 2) who can accomplish each criteria or drawing feature, thereby demonstrating the level of item difficulty.

To establish concurrent validity, Gonzales correlated the results of the HFDT with scores derived from the Harris scoring criteria for a group of 30 students (10 each from the first, third, and fifth grades). A relationship of .66 was calculated. In another study, Gonzales correlated the HFDT with the K-ABC for 60 students. The results yielded coefficients of .42 for the Sequential composite, .57 for the Simultaneous composite and .52 for the Total score. He also correlated the HFDT with the WISC-R for 30 students. A correlation of .53 was found between the HFDT and the Verbal scale, a .31 relationship was determined for the Performance scale, and a .5 for the Full scale. Gonzales believes that the correlations determined for the HFDT represent a moderate degree of validity.

To demonstrate construct validity, Gonzales compared the HFDT with achievement scores and indicated that the test differentiated between various ages and groups. He specifically correlated the HFDT with the Woodcock-Johnson Psychoeducational Battery (WJPEB) and found coefficients of .20 with Reading, .52 for Math, and .42 for Written Language. Gonzales further maintains that the higher scores obtained with older children are evidence for the relationship between the HFDT and age. He also studied two markedly different groups to ascertain the assessment's utility and construct validity. From a sample of 30 children enrolled in the gifted programs of the Albuquerque, New Mexico, public schools, he calculated a mean HFDT Quotient of 128. He then investigated 30 mentally handicapped children and determined a mean HFDT Quotient of 66. He concluded that the test does discriminate between intellectually exceptional children.

Critique

Individual intelligence tests were promoted primarily as aids for determining the capabilities of people to function in society. If, as a result of this testing, examinees were determined to have a verbal deficit, then some program of treatment or remediation could be made available to them to rectify the problem rather than to restrict the person. Because our primary means of interpersonal communication is through spoken and written language, assessments that cannot ascertain verbal capabilities or difficulties are of limited value.

The HFDT appears to be a global, unrefined assessment of general intellectual potential. Although the author has made a painstaking effort to ensure the assessment's psychometric properties of reliability and validity, the test lacks the diagnostic specificity of lengthier individual tests such as the WISC-R.

Because the HFDT appears to correlate as well with verbal tasks such as the WISC-R Verbal scale ($r = .53$) as it does with the Performance scale ($r = .31$), the question of whether it is a nonverbal assessment is moot. Moreover, it is unclear whether the drawings actually represent the efforts of a child's right or left hemisphere of the brain or both. For the majority of right-handed children it would appear that the drawings are largely a right-hemisphere function, although some language is needed to understand the directions and to identify verbally the various features of the drawing for the examiner. The children also may identify drawing features subvocally or through inner speech developed within the context of their respective environments.

Gonzales has demonstrated that children from various subcultural groups have different perceptions of the important features of human figure drawings, thereby implying a distinctness of cognitive potential developed within their social milieu. However, it remains to be investigated what these special qualities or features mean or how they can be channeled into developing the child's potential.

The HFDT does have a lengthy history of use, although limitedly for the information obtained regarding intellectual potential. It appears that in recent years more examiners have been concerned with its projective potential based on the belief that art is a reflection of the artist's emotionality and cognition of the world. It is, therefore, seen as a representation of personality. Handler (1985) contends that the DAP is the most sensitive assessment of pathology available, provided that the feature or sign is found commonly within a pathological group and that the context, as well as other possible meanings, have been considered. Handler's comments suggest that a consistent, infallible meaning associated with a drawing feature is lacking, although interpretation may focus on the possibility of selected behavior occurring in the future, if it is not being manifested overtly already.

In conclusion, the HFDT provides the user with a broad, unrefined, general, and global representation of a child's emotions and cognitions developed within an historical context and social milieu that can pave the way for more specific assessments or inquiries.

References

This list includes text citations and suggested additional reading.

Anastasi, A. (1972). Review of the Draw a Person. In O.K. Buros (Ed.), *The seventh mental measurements yearbook* (pp. 669-671). Highland Park, NJ: The Gryphon Press.

Claparede, E. (1907). Plan d'experiences collectives sur le dessin des enfants. *Archives de Psychologic, 6,* 276-278.

Dunn, J. (1972). A Review of the Draw a Person. In O.K. Buros (Ed.), *The seventh mental measurements yearbook* (pp. 671-672). Highland Park, NJ: The Gryphon Press.

Gonzales, E. (1982). A cross-cultural comparison of the developmental items of five ethnic groups in the Southwest. *Journal of Personality Assessment, 46,* 26-31.

Gonzales, E. (1986). *Human Figures Drawing Test.* Austin, TX: Pro-Ed.

Goodenough, F.L. (1926). *Measurements of intelligence by drawings.* New York: Harcourt, Brace & World.

Goodenough, F.L. (1975). *Measurements of intelligence by drawings.* New York: Arno Press.

Hammer, E. (Ed.). (1958). *The clinical application of projective drawings.* Springfield, IL: Charles C. Thomas.

Handler, L. (1985). The clinical use of the Draw-A-Person test (DAP). In Charles S. Newmark (Ed.), *Major psychological assessment instruments.* Newton, MA: Allyn & Bacon.

Harris, D.B. (1963). *Children's drawings as measures of maturity.* New York: Harcourt, Brace & World.

Ivanoff, E. (1909). Recherches experimentales sur le dessin des ecoliers de la suisse romande: correlation entre l'aptitude au dessin et les autres aptitudes. *Archives de Psychologic, 8,* 97-156.

Koppitz, E.M. (1967). Expected and exceptional items of human figure drawings and IQ scores of children age 5-12. *Journal of Clinical Psychology, 23,* 81-83.

Koppitz, E.M. (1968). *Psychological evaluation of children's human figure drawings.* New York: Grune & Stratton.

Machover, K. (1949). *Personality projection in the drawing of the human figure.* Springfield, IL: Charles C. Thomas.

Reisman, J.M., & Yamoskoski, T. (1973). Can intelligence be estimated from drawings of a man? *Journal of School Psychology, 11,* 239-244.

Wechsler, D. (1974). *Manual for the Wechsler Intelligence Scale for Children—Revised.* New York: Psychological Corporation.

Kenneth W. Wegner, Ed.D.
Director, Division of Counseling Psychology, Boston College, Chestnut Hill, Massachusetts.

INDIVIDUAL CAREER EXPLORATION

Anna Miller-Tiedeman, in consultation with Anne Roe.
Bensenville, Illinois: Scholastic Testing Service, Inc.

Introduction

The Individual Career Exploration (ICE) is a self-administered, self-scored inventory developed to help individuals focus on future occupations as they relate to individuals, current interests, abilities, experiences, and ambitions. It has two forms, the Picture Form for Grades 3-7 and special education classes, and the Verbal Form for Grades 8-12. The Picture Form is designed to help younger students gain an awareness of the world of work. The Verbal Form is designed to help older students understand their interests, abilities, and experience and to narrow their occupational choices. The exercises are structured according to the author's theoretical position on career development and decision-making and Anne Roe's Classification of Occupations (1976).

The author of the ICE, Anna Miller-Tiedeman, has collaborated with David Tiedeman (1984) for many years in the area of career decision-making theory and research. She is currently the president of the Life Career Foundation in Los Angeles. The structure of the ICE is based on Roe's (1956) occupational classification system, and the theoretical foundations are based on David Tiedeman's work in career decision making (Tiedeman & O'Hara, 1963). The ICE was developed in the early 1970s and field-tested in Illinois schools. The Picture Form (Grades 3-7) focuses on career exploration as an early aspect of career development. The Verbal Form (Grades 8-12) assists in career decision making, a later aspect of the career-development process. The current Verbal Form was published in 1976, the Picture Form in 1977. There are no separate forms for non-English-speaking or visually handicapped students. No information on normative populations is provided in the specimen set or manual.

The materials needed for the Picture Form include a manual, picture form booklet, student record sheet, booklet describing job trends from 1960-70, and *Classification of Occupations by Group and Level* (Roe, 1976). Almost every page of the Picture Form booklet is illustrated with drawings of people and objects. The instructions and other narrative material must be read to the examinees. The introductory section defines the term "career," the Roe occupational classification system, the change process in personal career development and the job market, and sex stereotyping. The first section of the booklet, Things You Might Like To Do, contains pictures of tasks people do. Examinees respond by marking "yes" or "no" to indicate whether they like performing each task. The 24 items are arranged by eight occupational groups and three levels within groups. The next three sections of the

239

booklet—Places You Would Like to Work, Things You Have Pretended Doing or Things You Have Done, and Tools Would Like to Use—are identical in format. The final section, Jobs You May Want to Work, follows the same format as the three previous sections, except the items are presented in words rather than as pictures. Generally, when the drawings contain people, the sex of the person is not distinguishable. Some drawings are shaded to represent different racial or ethnic backgrounds.The booklet is buff-colored and printed in black.

The Picture Form is designed for students ages 8-13 in Grades 3-7. The material appears to be at a reading level appropriate for this age group. However, it is recommended that the examiner be present at all times to read the material to younger students, help with unfamiliar vocabulary, and complete the record forms. Examinees mark answers directly on the booklet, then add the number of "yes" marks by group and translate these scores to the record sheet. The scores for each of the subtests described previously are totaled for each of eight occupational groups: 1) service, 2) business contact, 3) organization, 4) technology, 5) outdoor, 6) science, 7) general culture, and 8) arts and entertainment. The two highest group scores are listed at the bottom of the record sheet. The manual then suggests that examinees look up jobs in those groups in *Classification of Occupations by Group and Level* (Roe, 1976), which lists 600 occupations by the eight groups and indicates for each a 1) type (subspecialty) or location of employment, 2) *Dictionary of Occupational Titles* (1977) code, and 3) decision level. There are six decision levels ranging from unskilled labor through professional and managerial occupations. After listing these occupations on a separate sheet, it is suggested that the *Job Trends 1960-1970* booklet and other secondary sources be used for further exploration of particular jobs. This booklet, which is divided into the eight occupational groups, lists census data on the number of males and females employed in various jobs (about 88% of labor market) in 1960 and 1970 and the percentage increase or decrease of employment for each occupation over that decade.

The materials needed for the Verbal Form include a manual, student inventory booklet, job information checklist, and *Classification of Occupation by Group and Level* (Roe, 1976). The student inventory booklet, which is buff-colored and printed in red, contains some stick-person illustrations. The instructions and narrative are printed in the booklet for the examinees to read. The introductory material consists primarily of definitions and descriptions of the eight occupational groups and six decision levels outlined earlier in this review. The first section, Interests, contains a list of 14 activities for each of the eight groups. Examinees rate each item on a 4-point scale ranging from "no interest" (0) to "strong interest" (3). When completed with the section, the examinee totals the scores for each group and lists the two highest groups on the record sheet at the back of the booklet. In section two, Experience, 14 activities for each of the eight groups are rated on a 4-point scale ranging from "never done this activity" (0) to "often do this activity; very good at it" (3). Scoring and recording are completed as in section one. Section three, Occupations, lists 11 occupations for each of the eight groups. Each item from each group is rated from "no interest" (0) through "strong interest" (3). Scoring and recording are performed as in the previous two sections. The fourth section, Skills, contains very brief descriptions of skills required in the eight groups (e.g., Service: Helping People). The groups are ranked in terms of examinees' estimates of their best skills.

Then, the two highest-ranking groups are entered on the record sheet. Section five, Abilities, requires examinees to list the grades they received in three specified courses related to each of the eight groups and then indicate the two groups in which they received the best grades on the record sheet. The sixth section, Decision Level, requests that examinees use a scale ranging from low (1) through high (3) to rank brief, single descriptions of the six occupational levels in terms of income level, hard work, exertion, learning motivation, finishing tasks, and endurance. These ratings are totaled, converted into decision levels, and entered on the record sheet. The final section, Values, involves a simple ranking of single value statements (e.g., helping others) related to the eight occupational groups. The top two groups are then transferred to the record sheet. The record sheet also contains space to enter the interest inventory scores from the American College Testing Program, if taken.

Next, the information on the record sheet is summarized to determine examinees' most frequently occurring top two group scores. These two scores then are used to seek out applicable careers in the *Classification of Occupations by Group and Level* (Roe, 1976). Next, examinees are referred to the *Occupational Outlook Handbook* (1986) or a computerized career information system to further explore various aspects of the careers listed for their groups. A job information checklist is provided to assist in comparing suggested careers on 23 aspects, including values, interests, job and training requirements, employment outlook, and salaries. One checklist is needed for each career explored.

The examiner's main role in testing individuals and groups is to introduce the material and be available to answer questions about instructions. The age range for the ICE Verbal Form is 14 through 18, although adults could use it. The booklet is self-contained in that responses are made on the booklet and final results translated to the record sheet that serves as a summary profile.

Practical Applications/Uses

The two levels of the ICE were designed to assist third- through twelfth-grade students in exploring their interests, experiences, abilities, and ambitions as they relate to careers. The format permits the summarization of responses to develop a relationship to the most likely job areas for exploration among the eight groups. Although designed for use with school children, it could be used with adults and in other career counseling settings. Because the intent of the inventory is to measure the examinee's current career orientations, presumably, one form or the other could be used at various levels of an individual's career development. The development of a Spanish version for use in bilingual situations would be helpful.

Judging from the information presented in the manual and other available data, the only subjects recommended for this inventory are English-speaking typical elementary or secondary school students. However, either level of the test could be read to the blind. The administration instructions imply that providing assistance in understanding items would not invalidate its use with other special populations. Foreign-language translations also would be possible. However, the content is designed to fit the American culture and labor market, which might invalidate its use in other countries.

The ICE is relatively easy to administer in either individual or group settings. The inventory materials, a pencil, and a table or desk on which to work are all that is required. Examiner qualifications are not specified in the manual; however, a counselor or teacher with some special training in occupational classification systems, the labor market, and career information resources is necessary for effective use. The Picture Form must be read to the examinee, but the upper-level Verbal Form could probably be self-administered by the average high school student or adult. A resource person is necessary when the inventory is completed to assist in using the results in the labor market information search and other suggested activities. The administration procedures in the manual are adequate. Administration is simple and flexible. Sections could be independently administered in any order if necessary, and under any schedule, ranging from one section per session to all sections in one session. The manual does not provide specific instructional statements to be read to examinees.

Both levels of the ICE can be self-scored by section and the results translated to a summary record sheet. The examiner must provide instructions for scoring of the Picture Form and should be available to assist on the Verbal Form. Scoring involves such simple activities as counting "yes" marks, adding single-digit numbers, or recording rank orders in subsections. Scores are entered on the record sheet as the first or second highest in each section, and there is no "total" score for the inventory. Scoring directions are relatively specific in the Picture Form manual and more general in the Verbal Form manual. No scoring instructions appear in the Picture Form itself, but they are quite specific in the Verbal Form booklet. Since the ICE is self-scored, total time required could vary with age-level but would probably average about 10-15 minutes of the recommended 2-hour administration time. Younger examinees might experience difficulties or errors in scoring and translating scores to the record sheets. It is probably useful to check the self-scoring results to reduce the error rate. The ICE can be scored only by hand. A machine- or computer-scoring answer sheet could be devised for the Verbal Form, but these scoring methods would not provide the immediate feedback of results and follow-up activities designed for the inventory.

Interpretation of the ICE is based on objective scores and rankings translated to a summary record form. The information on the summary record form is then used in the self- and world-of-work exploration exercises suggested in the manual. Thus, the ICE is essentially self-interpreted with the aid of the examiner and career-related resource materials. The examiner needs some general training in career and decision-making theory and specific and current background in career resource materials and systems (e.g., a computerized career information system, if available) in order for the interpretation and exploration process to be optimal. No normative data are provided for interpretation.

Technical Aspects

There are no validity or reliability studies or data contained in the specimen set. Correspondence with both the test author and the test publisher, as well as a reference search, did not produce an extensive bibliography of such information for the ICE. The rationale for the instrument is briefly described in the manual. The author also has summarized the theoretical rationale for the ICE in "Career Decision-making: An Individualistic Perspective" (Tiedeman & Miller-Tiedeman, 1984). The validity of the ICE relies on the validity of the research on Roe's (1956) occupational classification.

Critique

The ICE appears to have a sound theoretical basis for its development and structure. It resembles very closely the approaches used by Holland (1985) in The Self-Directed Search (SDS) and Harrington and O'Shea (1982) in the Harrington-O'Shea Career Decision-Making System (CDMS). Although the structure of the ICE is based on Roe's (1956) eight occupational groups, the summary data also can be related to the Holland hexagonal structure used in the SDS and CDMS. All three inventories are intended to assist examinees in exploring themselves and the relationship of their characteristics to the world of work. The major advantage of the ICE is that the Picture Form can be used with younger populations. From a technical point of view, the ICE suffers from having only a loose "validity generalization"-type relationship to studies conducted on other instruments, and reliability data do not appear to be readily available. In contrast, a wealth of such data have been published for the SDS and CDMS over the past decade. The SDS and CDMS appear to have had much more widespread use for the same purposes. These instruments also have undergone more recent updating, whereas the only available ICE editions and supplementary reference materials were published in the mid 1970s. In summary, from technical and user points of view, the ICE would not appear to have any advantage over other similar inventories, except for its availability in picture form for younger populations.

References

Harrington, T.F., & O'Shea, A.J. (1982). *The Harrington-O'Shea Career Decision-Making System*. Circle Pines, MN: American Guidance Service.

Holland, J.L. (1985). *The Self-Directed Search professional manual*. Odessa, FL: Psychological Assessment Resources, Inc.

Roe, A. (1956). *The psychology of occupations*. New York: John Wiley & Sons.

Roe, A. (1976). *Classification of occupations by group and level*. Bensenville, IL: Scholastic Testing Service, Inc.

Tiedeman, D.V., & Miller-Tiedeman, A. (1984). Career decision-making: An individualistic perspective. In D. Brown & L. Brooks, (Eds.), *Career choice and development* (pp. 281-310). San Francisco: Jossey-Bass.

Tiedman, D.V., & O'Hara, R.P. (1963). *Career development: Choice and adjustment*. New York: College Entrance Examination Board.

U.S. Department of Labor. (1977). *Dictionary of occupational titles* (4th ed.). Washington, DC: U.S. Government Printing Office.

U.S. Department of Labor. (1986). *Occupational outlook handbook*. Washington, DC: U.S. Government Printing Office.

Delwyn L. Harnisch, Ph.D.
Associate Professor of Educational Psychology, Institute for Child Behavior and Development, University of Illinois at Urbana-Champaign, Champaign, Illinois.

INDIVIDUAL CRITERION REFERENCED TESTING MATH BASICS+

Educational Progress Corporation. Tulsa, Oklahoma: Educational Development Corporation.

Introduction

Individual Criterion Referenced Testing (ICRT) Math Basics+ was designed as an individualized criterion-referenced testing program to measure student performance relative to a specific set of math objectives for Grades 1-8. This testing program differs in several important ways from common norm-referenced testing instruments. First, administration of the test is untimed and informal. Teachers may vary the directions and provide examinees with additional examples. Second, multiple test levels are provided to evaluate students at their individual functional levels. Thus, it is probable that students in a single class would take different levels of the test at the same time. Finally, a primary function of criterion-referenced testing is diagnosis of individual progress relative to specific objectives. In this case, teaching for the test is desirable and is facilitated by a listing of objectives correlated to materials available to the instructor. Teaching for the test would invalidate the results of most norm-referenced tests. The teacher's guide (Teichert & Daniel, 1980) suggests that the testing program 1) provides diagnostic and prescriptive information for successful instructional planning; 2) customizes the test to the approximate working ability of students through the use of 51 test booklets; 3) identifies skills students have mastered, need to review, and need to learn; 4) focuses instruction upon specific unlearned objectives; 5) assists in classroom grouping and management of individual student progress; and 6) links diagnostic information directly to identified existing materials in the classroom.

The ICRT Math Basics+ was developed by Educational Progress Corporation (EPC), a division of the Educational Development Corporation in Tulsa, Oklahoma. Initial field testing involving over 80,000 students took place during the 1972-73 school year. The teacher's guide was prepared in 1980 by Phala Daniel and Marlene Teichert. Development of the technical manual was supervised by Bill Hill, president of EPC, and Ted Bussman, the director of testing at EPC, with the assistance of

The reviewer wishes to acknowledge the contributions of Debi Switzer and Michael Connell, doctoral students in the Department of Educational Psychology who assisted in the writing of this review.

Ronald K. Hambleton, H. Swaminathan, Janice Gifford, and Craig Mills, all of the University of Massachusetts at Amherst. Subsequently, over 3,000,000 students have taken some form of the ICRT.

The ICRT technical manual normative data were gathered in the spring of 1979 and the fall of 1979 when 42,000 students (between 777 and 1,419 students for each grade level) were given both the ICRT and the 1978 Metropolitan Achievement Test (MAT). Schools were selected by EPC salespeople in each region based on their geographical, ethnic, and public/private characteristics. The ICRT norm group was adjusted by differential weighting of test data from different schools to closely match the samples used in the MAT (Educational Progress Corporation, 1981, p. 35). This adjusted norm group was comprised of only 32,000 students organized into four geographic regions. When compared to the 1973 United States census data, the adjusted norm group underrepresented the Northeast (14% of the norm group vs. 30% of U.S. population) and the Southeast (25% of norm group vs. 31% of U.S. population). Also, compared to the U.S. 1976 National Center for Education statistics, whites were underrepresented in the adjusted norm group (66% of norm group vs. 75% of U.S. population), and Hispanics were overrepresented (10% of norm group vs. 4% of U.S. population).

There are six color-coded levels consisting of nonconsumable test booklets ranging in length from two to six pages. For a given testing session, the student takes five sequential tests, determined by performance on an initial placement test. Each test booklet contains 16 items measuring eight objectives. (Each objective is tested with 2 items for a total of 40 objectives measured by the five tests.) The objectives are grouped according to six strands. These six strands and the objectives they cover are 1) whole number operations (concepts, numeration, operations, and problem solving); 2) fractional operations (concepts, numeration, operations, and problem solving); 3) measurement (size, length, area, volume, and weight); 4) geometry (concepts, angles, lines, shapes, and solids); 5) decimal/percent (concepts, numeration, operations, and problem solving); and 6) special topics (graphs, tables, integers, money, ratio, roman numerals, scientific notation, statistics, temperature, and time). For each objective, the student is classified as mastered (two items out of two answered correctly), needs review (one out of two correct), or nonmastered (zero out of two correct).

In addition to the test booklets, the ICRT Math Basics + contains machine-scorable student answer sheets for recording student responses to the five test booklets used in each testing session; class, school, and transmittal forms for processing the completed tests; a list containing over 50 programs and materials correlated to the ICRT Math Basics + objectives; placement tests; parallel forms A and B; practice answer sheets; special answer sheets for younger students; and hand-scoring templates for use with the practice answer sheets and starting booklets.

Five reports are available: 1) the Student Summary, which provides mastery decision for each of the 40 objectives; the Instructional Grouping by Objective Report, which lists, for each objective, the students needing review or instruction; the Instructional Grouping by Approximate Working Level Report, which groups students according to the objective they should start learning; the Administrative Report, which contains summaries by class, building, and district for each objective; and the Program Evaluation Report, which contains norm-referenced information (scale scores, grade equivalents, percentile ranks, and normal curve equivalents). Optional materials described in the teacher's guide include a student profile folder for use in recording student progress and an objective-by-objective testing kit, Benchmarks: Math Basics+, containing validated test items for each of the ICRT Math Basics+ objectives for interim testing and for monitoring progress.

Practical Applications/Uses

ICRT Math Basics+ is designed to strengthen the relationship between teaching and learning by improving the quality of information about the instructional needs of students from Grades 1-8. The information provided by the program would be useful in diagnosing specific procedural difficulties and of special benefit for systems following an outcome-based curriculum because it provides, at an objective level, information concerning student understanding.

To administer the ICRT Math Basics+, all that is required is a normal classroom environment. Desks and tables should be cleared of all distractions and positioned as far apart as possible. Because the test is untimed, it is suggested that students who are not involved directly with testing continue with their normal activities. Because the assignment of tests to individual students will vary, it is helpful to sort the testing booklets prior to the beginning of the testing session. Prior to administration, student answer sheets should be gridded with the student's name, school district, classroom teacher, subject, grade level, and other optional information. The first time that students are tested, the examiner must determine the proper placement tests to administer to each student based upon standardized mathematics scores, the most recent mathematics textbook the student has completed, grade equivalent, or review of the ICRT Math Basics+ objectives. Student responses to the placement tests are scored using a hand-scoring template made of durable plastic, and the number of correct responses are recorded on a class list or class placement table. Students then are assigned five testing booklets based upon the level of the placement test used and their performance on the placement tests. The assigned booklet numbers then are recorded at the head of five columns on the student answer sheet. The examiner reads the instructions for completing the test booklets, cautioning students that answers should be placed in the correct columns on the answer sheet. The sequence of the tests taken and the time spent completing the tests will vary according to the individual student. Although there is no time limit, the test authors suggest that testing sessions be divided into three blocks of 45-60 minutes each, with orientation, initial testing, and completion of testing each taking place in a separate block.

The ICRT Math Basics+ is scored by optical scanners that read the answer sheets. The examiner is responsible for ensuring that all information from the students' answer sheets has been recorded correctly and that any stray marks have been erased. A completed submission packet consists of the transmittal forms, the school form, and forms for the individual classes within the school, and the student answer sheets.

Interpretation of the results is aided if a listing of objectives correlated to available instructional resources is requested. Teachers then would be directed to specific lessons and materials corresponding to the objectives measured by the test. If such a listing is not requested, the instructor must select the appropriate instructional strategy and materials.

Technical Aspects

Content validity for the ICRT Math Basics+ was determined in the fall of 1979, when five specialists in mathematics reported that, on the average, 83.1% of the mathematics objectives were highly appropriate, 91.3% of the objectives were written clearly, 98.4% of the items measured the intended objectives, and 95.9% of the items were rated as technically sound (Educational Progress Corporation, 1981). There is no mention of changes, if any, that were made to the test as a result of these findings. For example, although 16.9% of the objectives were viewed as inappropriate, the technical manual does not mention whether any were modified.

Concurrent validity was tested by correlating the 1979 norming sample's ICRT scores to their scores on the Metropolitan Achievement Test, which also was administered at the time of the ICRT testing. The scores were correlated for each grade level and for the fall and spring administrations. In mathematics, the correlations ranged from .67 to .82 with a median value of .77.

During the 1978-79 school year, 400,000 students were given the ICRT. Averaging over all the objectives, 98% of the estimates of decision consistency exceeded .80. The Subkoviak (1976) method of split-half reliability estimation was used; however, this method tends to produce estimates of decision consistency that are substantial overestimates of those that would be obtained from test-retest or parallel-form reliability methods when used with short tests (Educational Progress Corporation, 1981, p. 15). Because the ICRT tests are extremely short (only two items per objective tested), these numbers do not produce confidence in the reliability of the mastery decision. In addition, this reliability estimate is based on mastery (two items out of two correct) versus nonmastery (zero or one out of two correct), although the actual results are reported as mastery, needs review, and nonmastery. Although parallel forms A and B exist, no parallel forms reliability has been reported and no reliability information is available for Form B.

Each grade level of the norm group took the same five test booklets. By using the Rasch model, all the ICRT tests were linked to form a continuous scale. By taking any combination of five test booklets, the student is given national norms for his grade level.

Critique

A major strength of ICRT Math Basics+ lies in providing information concerning both teaching and learning in the classroom relative to the ICRT objectives. This program could be beneficial to systems beginning an outcome-based program by providing an existing framework of carefully chosen objectives. As the strength of the program ultimately rests on the appropriateness of these objectives, it is important that the user select test booklets that accurately reflect the local mathematics program. Provided this is done, in this reviewer's opinion, reports would be of great benefit to administrators and teachers. The list of materials available in the classroom provided in the Student Summary and Instructional Grouping by

Objective reports would be extremely timesaving for a teacher interested in individualizing assignments or in quickly locating supplementary material for additional review and practice. For administrative purposes, the information contained in the Administrative and Program Evaluation reports provides an excellent overview to student progress at class, building, and district level and useful information in meeting federal reporting regulations.

Another strong point of ICRT Math Basics + is the use of a three-way mastery decision: mastery, needs review, or nonmastery, which could supply important information for planning effective instruction, provided the decision is accurate. However, in a single administration of the test, only two items measure each objective, resulting in a situation in which guessing correctly could play an important role in determining the mastery decision. It would be extremely important to see reliability measures on this seemingly delicate three-way classification process. However, the reported reliability measures dealt only with a dichotomous, not a three-way, decision, which does not create for this reviewer a high level of confidence in the format.

Additional concerns arise when one considers the issues of test sequencing and timing. Although the test booklets can be given in any order, no mention of how this flexibility affects reliability was made. This is an important concern as the booklet sequence will vary widely depending upon the student's functional level and the instructional program used. Because students are allowed an unlimited amount of time to complete the test, several separate blocks of time can be used at the teacher's option. As a result, administration and test format vary from administration to administration. Despite the fact that this flexibility is encouraged in the teacher's guide, possible performance variations due to differences in administration (for example, giving the test in one block or over as many as 5 days) are not mentioned in the manuals.

Before using this test for diagnostic purposes, the prospective user should consider carefully the information concerning the norm group, item and objective selection, content validity decisions, time considerations, and the reliability of the three-way mastery decisions.

References

This list includes text citations and suggested additional reading.

Educational Progress Corporation. (1981). *Individualized Criterion Reference Test: Technical manual.* Tulsa, OK: Educational Development Corporation.

Gronlund, N. E. (1985). *Measurement and evaluation in teaching* (5th ed.). New York: Macmillan.

Huynh, H. (1976). On consistency of decisions in criterion-referenced testing. *Journal of Educational Measurement, 13(4),* 253-264.

Lord, F. (1977). Practical applications of item characteristic curve theory. *Journal of Educational Measurement, 14(2),* 117-138.

Subkoviak, M. J. (1976). Estimating reliability from a single administration of mastery test. *Journal of Educational Measurement, 13(4),* 265-276.

Teichert, M., & Daniel, P. (1980). *Individualized Criterion Reference Test: Teacher's Guide.* Tulsa, OK: Educational Development Corporation.

Wright, B. D. (1977). Solving measurement problems with the Rasch model. *Journal of Educational Measurement, 14(2),* 97-116.

Susan Homan, Ph.D.

Associate Professor of Reading, University of South Florida, Tampa, Florida.

INFORMAL READING COMPREHENSION PLACEMENT TEST

Ann Edson and Eunice Insel. New York, New York: Educational Activities, Inc.

Introduction

The Informal Reading Comprehension Placement Test (IRPT) is designed to be used by classroom teachers and reading specialists to assess a student's instructional and independent reading levels. The instructional reading level is recognized as the level at which students will make optimum progress in reading instruction.

The instrument's two authors are Ann Edson, M.Ed., and Eunice Insel, M.S. Ann Edson is the principal of Plaza Elementary School in New York as well as the district reading correlator for the Baldwin, New York public schools. Mrs. Edson also has served as a curriculum consultant for the New York State Education Department. In addition, she was the project director for an instructional resource and reading program that was cited for excellence in the federal government publication *222 Elementary Programs*. Eunice Insel is the director and coordinator of reading and media at the Plaza Elementary School Learning Center in Baldwin, New York. During her 23 years as an educator, she has taught at various levels. Ms. Insel, who has published over 50 projects in reading, also has been an associate in education in reading at Hofstra University. Both authors have been involved in the reading education of children for a significant number of years. They developed this informal testing instrument to meet the following three objectives:

1. provide quick instructional, independent, and frustration reading placement levels for students.

2. allow classroom, content area, and resource teachers and curriculum specialists to attain a quick placement for students using a minimum amount of teacher time.

3. determine a quick placement for new entrants.

The IRPT, an informal assessment instrument, has not been normed; however, over a 10-year period, it has been administered to over 3,000 students in Grades 1-6 attending Baldwin, New York public schools.

The IRPT consists of a plastic-covered three-ring binder and an Apple computer disk. The 14-page binder contains information describing the content of the test and explaining administration and scoring procedures. The student takes the test while seated at the computer, allowing the teacher freedom to teach other students while the test is being taken.

The IRPT is designed for use with students in Grades 1-6 developmentally and Grades 7-12 remedially. It also may be used with special education students.

The IRPT results yield both a word comprehension and a passage comprehension score. The Word Comprehension test contains 64 items assessing the student's word meaning knowledge. Eight sets of eight words are presented in an analogy format. The vocabulary was controlled using both EDL and Dolch word lists. The Passage Comprehension test contains eight reading selections covering prereading through eighth-grade levels. Passage readability was assessed using the Spache, Fry, and Dale Chall readability formulas. Each passage is followed by four comprehensive questions. One question measures the main idea, one inference, one detail, and one vocabulary from context. The computer program scores the analogies and passages and reports an independent and frustration level. Students must answer correctly 75%, or three out of four, of the questions at each passage level to continue the test at each successive level.

Practical Applications/Uses

An accurate IRPT score would provide classroom teachers with the information necessary for placing students at the reading level at which they would make optimum progress in reading instruction. This information is of value to classroom teachers, as well as to clinicians, because proper book placement is of major importance to the classroom teacher as well as to the diagnostician.

Although the manual suggests that the IRPT is suitable for students in first grade, the analogy portion of the test might be difficult for many first-graders. However, the direct-choice (two-option) format would help to alleviate some of the confusion for younger children. Because a computer is used for administration, the IRPT also would be appropriate for hearing-impaired students. In addition, teachers of remedial students may find that the computerized aspect of the test rates high in terms of sustaining student interest and of minimal use of teacher time.

Because the IRPT is self-administered through a computer, it frees the teacher to work with other students. A teacher or an aide can set the computer program in place and explain the testing procedures to the students taking the test. The computer program itself also provides some directions to the student, including the direction to call the teacher for help or when testing is completed. Testing time is approximately 20 minutes for Part I, Word Comprehension, and 30 minutes for Part II, Passage Comprehension. However, these times may vary greatly due to the fact that a student must answer 75% or more of the questions correctly in order to continue with the remaining passages.

The computer program scores responses as the student progresses. In Part I, each of the eight analogy questions is progressively more difficult that the previous level. If a student correctly completes at least six of the eight questions (75%), the next group of analogies appears. When the student fails to correctly answer the appropriate number of questions correctly, the next set is not presented.

In Part II, Passage Comprehension, each of the eight passages is followed by four questions. The student must complete successfully three of the four questions in order for the next passage to appear. Again, the passages are presented in order of increasing difficulty.

The scoring component of the instrument appears to be simple, well explained, and useful. The teacher may consult a scoring chart provided in the manual for further direction. For example, based on this chart, a teacher could determine that a student who had answered two or more questions incorrectly on Level 4 of the Passage Comprehension test would be placed at a third-grade reading level, information that is valuable for a teacher seeking to place students at appropriate instructional levels quickly. Likewise, the computer program provides this information for the teacher when the individual has completed the test. A Total Instructional Level (TIL) based on the student's scores from both Parts I and II can appear on the screen or be printed out.

Interpretation is based strictly on objective scores. As explained previously, to pass any section of Part I or Part II, a student must answer at least 75% of the questions in that section correctly. Any teacher or clinician trying to determine a student's instructional level would be able to interpret the scores and levels as presented by the IRPT program.

Technical Aspects

The IRPT has not been normed. As mentioned previously, the manual states that the IRPT has been administered to over 3,000 students in Grades 1-6 over the past 10 years. No further normative information is available.

Critique

Due to the ease with which the IRPT can be administered on the computer, the instrument would be an asset to almost any classroom. After using the instrument, a teacher should be aware of the appropriate reading level for every child in the classroom. In addition, the IRPT provides a way to determine instructional level while leaving the teacher free to be with the rest of the class.

However, because no validation studies have been completed on the IRPT, its reliability and validity are questionable. In terms of face validity, the samples presented in the manual appear to be appropriate for measuring word and passage comprehension. Studies of the reliability of the IRPT for determining instructional level placement should be completed. If the results demonstrate good reliability and validity, then teachers and clinicians could utilize this placement tool with confidence.

References

Dolch, E. W. (1960). *Teaching primary reading* (3rd ed.). Champaign, IL: Garrard.
EDL core vocabularies in reading, mathematics, science, and social studies. (1979). New York: McGraw-Hill.
Edson, A., & Insel, E. (1983). *Informal Reading Comprehension Test* (manual). New York: Educational Activities.

Jane L. Swanson, Ph.D.

Professor of Psychology, Southern Illinois University, Carbondale, Illinois.

INTEREST DETERMINATION, EXPLORATION AND ASSESSMENT SYSTEM

Charles B. Johansson. Minneapolis, Minnesota: NCS Professional Assessment Services.

Introduction

The Interest Determination, Exploration and Assessment System (IDEAS; Johansson, 1983) is a 112-item measure of vocational interests designed for use with students in junior high through early high school. Using the data base from the Career Assessment Inventory (CAI), IDEAS was designed to provide a quick, easy-to-score inventory for use in tandem with career planning programs and curricula.

IDEAS was developed by Charles B. Johansson, Ph.D., whose credits include the development of the CAI as well as a number of other inventories published by National Computer Systems (NCS) that are designed for use in career planning in conjunction with the CAI or IDEAS: 1) the Self-Description Inventory (SDI), designed to measure personality traits, 2) the Temperament and Values Inventory (TVI), designed to measure work-related values and personality dimensions, and 3) the Word and Number Assessment Inventory (WNAI), designed to measure verbal and numerical aptitudes. Johansson received his Ph.D. in counseling psychology from the University of Minnesota in 1970; as a graduate student, he contributed to revisions of the women's and men's forms of the Strong Vocational Interest Blank (SVIB).

According to Johansson (1980), his goal in developing IDEAS was to construct a short, easily administered interest inventory, with a "modest number" of scales (characterized by validity, reliability, ease of interpretation, and comprehensiveness), that would be equally useful for females and males. An additional goal was to balance psychometric concerns with the desire for a simple, understandable inventory. IDEAS was published in 1977, shortly after the release of the first edition of the CAI, and its development relied heavily upon data collected for the CAI. In fact, IDEAS has been described as simply a shortened version of the CAI (Weeks, 1985). Since its initial publication, revisions of IDEAS have focused on updating the accompanying interpretive materials to reflect changes in the *Occupational Outlook Handbook* (U.S. Department of Labor, 1984).

Test materials are packaged in a self-contained booklet that includes the test items, answer sheet, profile sheet, and interpretive materials. The reading level is reported to be at the sixth grade. The test itself consists of 112 items representing "activities related to various careers" (Johansson, 1983) to which students respond on a 5-point scale ("like very much," "like somewhat," "indifferent," "dislike somewhat," "dislike very much"). Items for IDEAS, which were selected from the item pool of the CAI, are contained in 14 scales presented in six groupings corresponding to John Holland's theory of vocational choice (Holland, 1985). The 14 scales are Mechanical/Fixing, Electronics, Nature/Outdoors (Realistic); Science, Numbers (Investigative); Writing, Arts/Crafts (Artistic); Social Service, Child Care, Medical Service (Social); Business, Sales (Enterprising); Office Practices and Food Service (Conventional).

The answer sheet and profile sheet are arranged in the test booklet so that self-scoring is possible. The profile is flanked by brief descriptions of each of the 14 scales, which include a definition of the interest area and related activities and occupations. An accompanying two-page "additional references" section lists related courses, occupational examples, and page numbers for the *Occupational Outlook Handbook* and the *Dictionary of Occupational Titles* (U.S. Department of Labor, 1984) for each scale.

Practical Applications/Uses

IDEAS was designed to provide an expedient measure of vocational/career interests for use with junior high and early high school students. The test publisher, National Computer Systems, suggests in a 1986 product sheet that IDEAS can be used in "career programs, guidance units and social studies units to help students explore career paths that match their interests and aptitudes, plan their high school coursework, explore higher education alternatives, [and] consider post-high school career entry."

The format of IDEAS allows considerable flexibility in test administration. A trained test administrator is not required. The test can be administered easily to a large group (e.g., to a classroom of students) or to individuals; it also can be given to students to complete outside of class. Total completion time, including the scoring and plotting of results, is estimated at 20-30 minutes.

Instructions for scoring the inventory are written for the student and include pictorial examples. As a student circles the appropriate letter (L=Like very much, l=Like somewhat, I=Indifferent, d=Dislike somewhat, D=Dislike very much) on the answer sheet to indicate his or her response to each item, the marks are transferred via carbon to corresponding numbers (L=4, l=3, I=2, d=1, D=0) on the next page. The student is then instructed to score his or her responses as follows: 1) each circled number is copied onto an adjacent line, 2) eight responses in each of 14 columns are summed to arrive at a score for each of the 14 interest area scales, 3) these 14 scores are written on a profile sheet on the next page, 4) each score is plotted on a conversion chart to produce a standard score, and 5) the standard scores are transferred to and plotted on the student's own profile sheet. Two separate profile sheets are available, one for students in Grades 6-8 and one for those in Grades 9-12.

Although the series of scoring steps listed above seems relatively complex, it apparently does not cause difficulties for students: In a survey undertaken by the test publisher (Johansson, 1987), 91% of IDEAS users indicated that it was either "very easy" or "somewhat easy" for their students to complete.

True to its nature, the IDEAS inventory is not only self-administered and self--

scored but also self-interpreted. The descriptions of the scales included on the student's profile provide concise definitions of each of the interest areas measured by the scales. Students are instructed to interpret their own standard scores relative to an average score of 50, with most students scoring between 40 and 60 on each scale. A summary section reminds students that IDEAS measures interests, not abilities, and that low scores may result from lack of information or experience in a particular interest area. The accompanying two-page "additional references" section directs students to further information about the activities and related occupations for each scale.

Technical Aspects

To develop the 14 IDEAS scales, all 305 CAI items were first intercorrelated using three samples: 1) a general sample of adults, apparently the norm group for the CAI (n=750 females and 750 males), 2) the 6th-8th-grade students used in the IDEAS norming sample (n=306 females and 292 males), and 3) the 9th-12th-grade students used in the IDEAS norming sample (n=1,681 females and 1,755 males). These item intercorrelations then were subjected to a cluster analysis; within the adult sample, the "best" eight items were selected from each of 14 clusters, and each of the eight items was correlated with the total scale. This initial item selection then was verified in the 6th-8th-grade and 9th-12th-grade samples. Selection of items for the final version of the scales was "based on a review of all the analyses to guarantee that a high degree of internal consistency was evinced for all the samples" (Johansson, 1980, p. 2).

The IDEAS scales were standardized by scoring them for subjects (the two student samples used in the cluster analyses for scale development) who had been administered the CAI. The two separate student samples were chosen for the normative studies because interests tend to become more crystallized with age (Johansson, 1980). The raw score means and standard deviations for these samples were converted so that the standard score mean is 50 and the standard deviation is 10 for each scale.

The student's profile reflects norms based on a composite sample of females and males; however, norms developed separately for females and males are presented in the manual. The rationale for using combined-sex norms is to avoid students' "undue concern . . . if their scores are not within the average boundaries for their sex" (Johansson, 1980, p. 9). An examination of the separate-sex norms reveals considerable sex differences for nine of the 14 scales. In some cases (e.g., Mechanical/ Fixing, Electronics, and Child Care), there is no overlap between the bars representing the middle 50% range of scores for each sex. The manual encourages counselors to use the separate-sex norms if a student has significantly higher scores than other individuals of his or her sex; however, because these norms are not available on the profile, their use at appropriate times seems unlikely.

Evidence for content, construct, and concurrent validity of the 14 IDEAS scales is presented in the manual. Content validity was addressed during the test's development. In constructing each scale, items that met the statistical criteria were excluded if they detracted from the scale's "purity." Internal consistency coefficients were reported to be "all in the high .80s and low .90s" (Johansson, 1980, p. 3). Construct validity of the scales was addressed by computing correlations between IDEAS scales and three other interest inventories: the Career Assessment Inventory (CAI), the Strong-Campbell Interest Inventory (SCII), and the Minnesota Vocational Interest Inventory (MVII; Clark & Campbell, 1965). The correlations between the IDEAS scales and the CAI scales (from which the IDEAS scales were derived) were reported as .91 and higher (Johansson, 1980, p. 4). The correlations between IDEAS and the SCII and MVII were described as being "quite high—.80s and above" (Johansson, 1980, p. 4).

Evidence for concurrent validity was determined by considering CAI scores "from over 100 samples of students in career programs and adults in occupations" (Johansson, 1980, p. 4). Johansson summarized these data as follows:

> "Overall, good concurrent validity was exhibited for the fourteen scales on the IDEAS inventory . . . Generally, the differentiation between the highest and lowest scoring occupation on each scale was about two standard deviations. (Johansson, 1980, p. 4)

For example, electricians received higher scores on the Electronics scale than people in other occupations.

Test-retest reliability coefficients for 1-week, 2-week, and 30-day intervals were reported as being "in the high .80s and low .90s" (Johansson, 1980, p. 4).

Overall, it is difficult for the reader to evaluate the validity and reliability evidence for the IDEAS inventory primarily because it is presented in the manual in such a brief manner that crucial information is omitted. In all cases, the actual data are not presented but rather are summarized as ranges of coefficients. Characteristics of the samples—size, gender and racial/ethnic composition, age, and other relevant information—are not provided. Details regarding the method or procedures also are inadequate; for example, it is not clear whether the sample of subjects used to compute the IDEAS-CAI correlations were administered only the CAI (from which the IDEAS scales were then extracted, resulting in part-whole correlations) or actually were administered both instruments. Additionally, the evidence for validity often is based on data from the CAI, which cannot be substituted for data based on the IDEAS inventory because even though the IDEAS items are a subset of the CAI items, they form a separate instrument that must be evaluated as such.

Although Johansson cited evidence regarding concurrent validity, the issue of predictive validity for IDEAS is as yet unaddressed. Both types of criterion-related validity are important for career inventories: concurrent validity refers to the power of the inventory to differentiate between people currently belonging to different groups of some sort (such as in different occupations); predictive validity refers to the power of the inventory to differentiate between people who will eventually belong to different groups (such as entering different occupations). Parenthetically, the lack of predictive validity evidence has been cited as a concern in evaluations of the CAI as well (McCabe, 1985).

It appears that one cannot, and should not, describe the IDEAS inventory without reference to the CAI, particularly regarding item selection, scale construction, reliability and validity, as discussed previously. However, in spite of the common genesis of the two instruments, there are several important differences that should be mentioned here. The most notable differences, which are interrelated, are the type of scales, method of scale construction, intended audience, and end result—the profile that the respondent receives. Johansson developed the CAI in 1975 to complement the SCII; it was intended to cover occupations and careers ranging from those requiring no postsecondary education to those requiring a college degree. The version of the SCII available at that time focused almost exclusively on college-degreed occupations. The CAI resembles the SCII in many ways. It consists of 1) rationally constructed theme scales measuring John Holland's theoretical types, 2) homogeneously constructed basic interest scales measuring relatively pure areas of interests, and 3) empirically constructed occupational scales measuring the similarity of the respondent's interests to incumbents of various occupations. Because of the number and complexity of scales, the CAI must be computer-scored. The IDEAS inventory, on the other hand, consists of 14 homogeneously constructed scales measuring specific areas of interest that can be hand-scored easily by the examinee.

One of the disadvantages of considering IDEAS as a shortened version of the CAI is that the latter instrument was designed as a "blue collar" version of the SCII (Bodden, 1978), and, therefore, it has been concluded, perhaps inappropriately, that the use of IDEAS is limited to students who are not college bound (Weeks, 1985). Because the CAI includes scales measuring interest in specific occupations, its use is limited to people who are interested in vocations not requiring extensive postsecondary education; however, because IDEAS does not include such scales, its use is not so limited. At the risk of oversimplification, the CAI tends to focus on *occupations*; IDEAS tends to focus on areas of *interest*. This divergence in focus seems to stem from the stated purpose and intended audience of each instrument: the CAI is used most often with students who are making at least tentative occupational or educational choices; IDEAS is designed for use with students who are at a much earlier stage in the career exploration process. Contrary to Weeks' (1985) statement, both the item pool and occupational alternatives of IDEAS appear to represent a broad range of occupational levels.

Critique

The IDEAS inventory could fulfill an important role as an early career exploration tool. Its major strengths include the following:

1. It offers a quick assessment of vocational interests for students in Grades 6-12.

2. It was developed with a solid empirical data base by an author with considerable technical expertise.

3. The ease and flexibility of administration make it suitable for classroom or individual use.

4. It "demystifies the testing process because students score their own inventory" (Johansson, 1987, p. 3) and receive immediate feedback.

5. The interpretive materials present a broad range of educational and occupational alternatives.

However, several shortcomings in the instrument and the manual warrant further attention: 1) inadequacy of the scale development and technical information presented in the manual, 2) representativeness of the scales, 3) potential for scoring errors, 4) lack of theoretical discussion, and 5) issues of sex fairness.

First, although a previous reviewer praised the manual as "extremely well done for such a brief and simple inventory" (Weeks, 1985, p. 698), the present reviewer must disagree. As Johansson stated, the manual "presents a cursory summary of the extensive research done" (Johansson, 1980, p. 1). The information presented does indeed seem impressive, but it lacks the detail necessary to fully understand and evaluate the inventory. Consulting the CAI manual provides some supplementary information, but this information should be available in the IDEAS manual itself. Reading the IDEAS manual tends to leave one with more questions than answers. For example, regarding scale development, How many items were in the initial pool from the CAI?, Was the sample used to determine interitem correlations the same sample used to norm the CAI?, What were the specific statistical criteria used to determine which items to retain?, How was the decision made to drop items in order to preserve scale "purity"?, Why were eight items chosen for each scale?, and Why are there 14 scales? Questions regarding the reliability and validity information also arise: What were the actual coefficients summarized in the manual?, What samples were used to compare the CAI, IDEAS, and the MVII?, and What evidence exists for the predictive validity of the instrument?

The second concern involves the representativeness or comprehensiveness of scales on the IDEAS inventory. A comparison of the IDEAS and CAI manuals raises questions regarding the scales on each inventory. IDEAS consists of 14 scales, whereas the analogous section of the CAI consists of 22 scales. No mention is made of why only 14 of the 22 CAI scales were developed for IDEAS nor why the particular 14 were chosen. Scales that were not carried over from the CAI to IDEAS are Carpentry, Manual/Skill Trades, Agriculture, and Animal Service in the Realistic area; Performing/Entertaining in the Artistic area; Teaching and Religious Activities in the Social area; and Clerical/Clerking in the Conventional area. The omission of these scales seems to create a gap in the instrument, particularly if one considers whether the remaining scales are a representative sample of the content domain of vocational interests. For example, because high school students, particularly girls, often express an interest in teaching occupations, omitting the Teaching scale seems to be a serious oversight.

The third concern relates to potential errors in the self-scoring of IDEAS profiles. Scoring errors are possible in each of the following operations: 1) copying the circled number to a line (which is actually a box), which creates the potential for writing an incorrect number and for writing it in the incorrect box; 2) adding numbers down columns, with the potential for both arithmetic errors and the transposing of columns; 3) transferring summed numbers across a page break, with potential for error and transposition; 4) choosing the appropriate profile sheet, either Grades 6-8 or Grades 9-12 (a relatively easy task, but one that at least a few students will miss); 5) marking scores on a standard score conversion chart, and 6) transferring the standard scores for the student's personal copy of the profile sheet. Given that self-scoring errors have been documented (Gelso, Collins, Williams, & Sedlacek, 1973; Prince, 1984) for the Self-Directed Search (SDS; Holland, 1985), which requires fewer computations, and given that errors could occur at 126 possible points (nine types of errors multiplied by 14 scales), errors in final IDEAS scores seem hopelessly inevitable. The instrument probably should not be administered without close supervision by a teacher or counselor. A discussion of possible scoring errors and corrections in the test booklet instructions is desirable. Relatedly, research needs to be conducted regarding the frequency and severity of scoring errors.

Fourth, there is a glaring lack of reference to John Holland's theory in any of the test materials, other than use of the six occupational types described by his theory. Because Holland's theory has been so well integrated into the career counseling profession, it may not seem necessary to mention it by name. Nevertheless, it is an unfortunate omission. At the very least, citing Holland seems to be professional courtesy. More importantly, when one uses a theoretical framework in test development, one ought to describe the theory and provide a justification for its inclusion. In fairness to Johansson, it apparently was not he, but rather the test publisher, who added the six Holland types to the revision.

Fifth, the issue of sex fairness is not addressed, nor is an adequate justification given for the choice of combined-sex norm groups. The debate regarding the relative merits of combined-sex norms, separate-sex norms, or raw scores for achieving sex fairness has at times been heated and will not be reviewed here. However, the National Institute of Education's *Guidelines for Assessment of Sex Bias and Sex Fairness in Career Interest Inventories* (Diamond, 1985) provides clear recommendations for test authors and publishers, including guidelines regarding information to be furnished in the test manual.

Most of the aforementioned shortcomings or concerns could be rectified by expanding the manual; the publisher presently is considering a revision of the inventory and manual that would address these issues. However, several concerns, such as errors in self-scoring and evidence for predictive validity, require further research not only by the test publisher but also by independent investigators.

In a previous review, Weeks (1985) summarized his reactions regarding the usefulness of IDEAS: "As a counseling tool it is, at best, a preliminary indicator of general directions of vocational interests . . . IDEAS has limited utility as a predictor for vocational choice and/or future vocational success" (Weeks, 1985, p. 698). He recommended that IDEAS can be useful as a career exploration tool used *in conjunction* with other structured programs or activities, such as career units in social studies classes, "provided its users are cautioned not to allow their IDEAS profiles to weigh too heavily on vocational choices" (Weeks, 1985, p. 698). His comments suggest that the use of IDEAS might be most appropriate with the lower portion of the age range for which it is intended; that is, with junior high school students. At this stage, vocational interests are not crystallized fully yet and, therefore, career and educational choices must be tentative. In the senior high age range, however, it may be more appropriate to use an interest inventory that can provide precision in differentiating between a greater number of interest areas, with demonstrated predictive power for future choices.

In summary, the initial development of the IDEAS inventory proceeded from a strong technical and empirical background. With further attention to the issues discussed in this review, IDEAS has potential as a useful career exploration instrument.

References

Bodden, J. L. (1978). Review of the Career Assessment Inventory. In O. K. Buros (Ed.), *The eighth mental measurements yearbook*. Highland Park, NJ: Gryphon Press.

Clark, K. E., & Campbell, D. P. (1965). *Manual for the Minnesota Vocational Interest Inventory.* Minneapolis: University of Minnesota Press.

Diamond, E. E. (Ed.). (1975). *Issues of sex bias and sex fairness in career interest measurement.* Washington, DC: Department of Health, Education, and Welfare, National Institute of Education, Career Education Program.

Gelso, C. J., Collins, A. M., Williams, R. O., & Sedlacek, W. E. (1973). The accuracy of self-administration and scoring on Holland's Self-Directed Search. *Journal of Vocational Behavior, 3,* 375-382.

Holland, J. L. (1985). *The Self-Directed Search.* Odessa, FL: Psychological Assessment Resources.

Holland, J. L. (1985). *Making vocational choices: A theory of careers* (2nd ed.). Englewood Cliffs, NJ: Prentice-Hall.

Johansson, C. B. (1980). *Manual for IDEAS: Interest Determination, Exploration and Assessment System.* Minneapolis, MN: National Computer Systems.

Johansson, C. B. (1982). *Manual for the Career Assessment Inventory.* Minneapolis, MN: National Computer Systems.

Johansson, C. B. (1983). *IDEAS: Interest Determination, Exploration and Assessment System.* Minneapolis, MN: National Computer Systems.

Johansson, C. B. (1987, April). *Customer survey results: IDEAS Inventory.* Paper presented at the annual meeting of the American Association for Counseling and Development, New Orleans.

McCabe, S. P. (1985). The Career Assessment Inventory. In D. J. Keyser & R. C. Sweetland (Eds.). *Test critiques* (Vol. II, pp. 128-137). Kansas City, MO: Test Corporation.

Prince, J. P. (1984). *Investigating college student attrition: An application of Holland's theory of vocational choice.* Unpublished doctoral dissertation, Minneapolis, MN: University of Minnesota.

U.S. Department of Labor. (1984). *Dictionary of Occupational Titles* (4th ed.). Washington, DC: U.S. Government Printing Office.

U.S. Department of Labor. (1984). *Occupational Outlook Handbook, 1984-85.* (Bureau of Labor Statistics Bulletin No. 2205). Washington, DC: U.S. Government Printing Office.

Weeks, M. O. (1985). Review of the Interest Determination, Exploration and Assessment System. In J. V. Mitchell, Jr. (Ed.). *The ninth mental measurements yearbook.* Lincoln, NE: Buros Institute of Mental Measurements.

Loy O. Bascue, Ph.D.

Independent Practice, Devon, Pennsylvania.

INTERPERSONAL STYLE INVENTORY

Maurice Lorr and Richard P. Youniss. Los Angeles, California: Western Psychological Services.

Introduction

The Interpersonal Style Inventory (ISI) is a multidimensional measure of "an individual's characteristic ways of relating to other people" (Lorr, 1986, p. 1). The ISI contains 300 true/false statements that assess 15 primary dimensions of personality. The 15 scales are each empirically assigned to one of five interpersonal areas:

Interpersonal Involvement: Sociable (SOC), Help Seeking (HLP), Nurturant (NUR), and Sensitive (SEN).

Socialization: Conscientious (CON), Trusting (TRS), and Tolerent (TOL).

Autonomy: Directive (DIR), Independent (IND), and Rule Free (RLF).

Self-Control: Deliberate (DEL), Orderly (ORD), and Persistent (PRS).

Stability: Stable (STA) and Approval Seeking (APP).

The ISI must be machine scored, and the current process for doing so requires that the completed answer sheet along with a transmittal form for authorization be mailed to the test publisher. In turn, the publisher returns to the examiner a computer printout of about 15 pages that includes the results and an individualized interpretation.

The principal author of the ISI is Maurice Lorr, a psychologist and professor emeritus of psychology at Catholic University of America, Washington, D.C. The development of the ISI seems to have grown out of his interest in the interpersonal domain of behavior, an interest that has spanned at least the last 20 years (Lorr & McNair, 1965; Lorr, Bishop, & McNair, 1965). Dr. Lorr has conducted extensive research related to the development of the ISI and is also the author of the inventory's manual (Lorr, 1986), which describes the development, administration, and psychometric properties of the inventory.

The current version of the ISI is identified as Form E to acknowledge that the inventory was refined from four earlier versions. The initial data collection in the development of the ISI (Form A) consisted of the administration of 340 items to 457 university students (Lorr & Youniss, 1973). The results were factor analyzed into 14 bipolar scales. A second version (Form B) was then administered to 370 men and women, representing both university students and employed adults. This investigation confirmed the 14 factors identified by the previous version and also identified a social desirability dimension. Further refinements followed the administration of Form C to another sample of 216 male and female college students and 327 high school boys (Lorr, O'Connor, & Seifert, 1977) and Form D to a sample of 648 high school boys and girls as part of a longitudinal study of personality development (Lorr & Manning, 1978b).

The final version of the ISI (Form E) is made up of 15 scales or first-order factors comprising 20 items each, keyed half true and half false. As one aspect of the development of the ISI, the author determined that the 15 scales intercorrelate into five higher-order dimensions of personality (Lorr & DeJong, 1984). These higher-order factors are asserted to represent general, broader dimensions of personality, each made up of the selected primary factors that the 15 ISI scales measure.

Raw scores for the 15 scales are converted into T-scores and presented on a profile form that is part of the interpretive report. The conversion of raw scores to T-scores is based upon sex and one of two sets of normative data. As justification for using sex as a variable for conversion, the author cites three studies in the manual that reflect some sex differences with regard to the ISI. One set of normative data came from 411 women and 354 men attending colleges and universities in "12 diverse settings" (Lorr, 1986, p. 44). The second set was collected from five Baltimore, Maryland, high schools and includes 423 girls and 225 boys. The manual states that the high school students represented a wide range of socioeconomic levels and ranged in age from 15 to 19 years. The manual asserts that if people who take the ISI are not clearly like the normative sample applied, T-scores should be interpreted with caution.

Practical Applications/Uses

In contrast to instruments that focus on the assessment of individual personality characteristics, the Interpersonal Style Inventory is of particular value because it appraises the domain of interpersonal style. It can be used with people aged 14 and older and can be administered individually or in groups. Instructions for administering the inventory are presented clearly in the manual. It seems to take about 45 minutes to complete.

The manual is well organized and provides information on the development, reliability, and validity of the instrument. According to the manual, the ISI can be used for self-understanding, planning psychotherapeutic treatment, research, and personnel guidance. It is in the area of personnel and vocational guidance that the instrument can be especially useful by providing data related to the interpersonal behavior of candidates for employment, training, or promotion. The ISI also has value for planning and evaluating psychotherapeutic treatment, and it is likely to have particular appeal for therapists with an interpersonal orientation, perhaps especially for those who want to evaluate the interpersonal style of married or engaged couples.

Chapter 3 of the manual concerns the scoring and interpretation of the ISI, and the material in the chapter brings into focus two problems related to interpretation of the inventory results. First, the 10 sample cases presented to aid in interpretation are each only a few paragraphs long, with typically only one paragraph devoted to discussing the results of the ISI. In contrast to the interpretative report, which is 15 pages long, these brief case studies can leave an examiner wondering whether the case examples failed to incorporate all the useful data from the report or, alternatively, whether the report contains a considerable amount of information that is not helpful. The second problem evident in chapter 3 is the absence of objective guidelines for interpretation of the test results. An example of the problem can be seen in cases C, D, and E on page 29 of the manual. In case C, the female client obtained a T-score of 64 on the scale of Rule Free (RLF) and was described as "not needing to comply with authority." In case D, another female client obtained a T-score of 60 on RLF and was described as "prone to follow her own way of doing things and resist conventional rules and norms." In case E, a male client achieved a T-score of 63 on the same scale, which was interpreted to suggest "a rebellious way of relating to authority." It might be that the relationship of RLF to other scales influenced these interpretative statements, but the difference in language and intensity of discription among cases with similar T-scores points out the absence of firm criteria or detailed guidelines for the interpretation of the results.

Further, since the ISI must be computer scored and interpreted, it is impossible to separate the utility of the instrument from the printed report. The manual cautions that the computer report is a professional-to-professional communication and not directly interpretable without reading the manual. Although the manual provides a sample report that is interpreted and explained, understanding the report requires examiners to be familiar with probability theory. Such familiarity is needed because the analysis of data upon which a report is written includes several statistical tests of which the examiner should have knowledge. For example, response patterns are analyzed to determine the probability of an invalid protocol by checking the frequency of shifts between true and false responses and the number of unanswered responses. The scale scores within each of the five higher-order factors are examined to determine if the scales within a given factor are different from one another. Further, the scales within each higher-order factor are compared to the scales in all other higher-order factors to determine if the factors themselves differ one from another. Scale scores are also tested to discover if they differ from one another and if unusual pairs of scores exist. Finally, scores are compared to six interpersonal types identified by the author and his associates (Lorr & Youniss, 1973) to determine the probability of membership in each of these six groups. The groups are 1) disagreeable, assertive extraverts; 2) agreeable, assertive extraverts; 3) stable, agreeable extraverts; 4) disagreeable, controlled introverts; 5) agreeable, compliant introverts; and 6) neurotic, compliant introverts.

Technical Aspects

Data on the reliability and validity of the Interpersonal Style Inventory are presented in chapter 5 of the manual. Coefficients of internal consistency for the individual scales, based upon the scores of 554 male and 411 female college students, ranged from .72 to .89. Test-retest reliability coefficients for the scales were computed from a sample of 60 college women with a 2-week interval between testing and ranged from .81 to .95. The manual also contains information on the test-retest reliability of three samples of high school students and one sample of student nurses with a 2- to 3-year span between testing, but the data were collected on Form D of the inventory.

The manual also provides evidence for both construct and criterion-related validity. In terms of construct validity, the manual contains summaries of six studies in which the ISI was administered along with other inventories that measure the same interpersonal constructs. Although the results of these comparisons provide support for the validity of the ISI, five of the studies were done with the early forms of the inventory, and the extent to which these findings can be generalized to Form E is not clear. The sixth study included a comparison of Form E with three other personality inventories, but no statistical data concerning the relationship among the inventories is cited, although it is reported that two psychologists reviewed the contents of the inventories and concluded that the ISI measures similar constructs to those of the other instruments.

Lorr has conducted research to determine the existence of the five higher-order factors of Form E. His justification for researching these higher-order factors is because they are broader, more inclusive personality constructs that might have more predictive value than primary scales (Lorr & DeJong, 1984). In one study (Lorr & Manning, 1978a) that sampled 225 male and 423 female high school students and college freshmen in Baltimore and 220 high school boys in Washington, D.C., the analysis of the results identified five factors that accounted for between 45% to 48% of the variance, depending upon the sex of the sample. A second study (Lorr & DeJong, 1984) of 144 high school boys and 173 girls and a third study (Lorr & Nerviano, 1985) involving 50 hospitalized alcoholics also identified the same five factors. Throughout his research, Lorr has pointed out that the five higher-order factors of the ISI are similar to the factors measured by other personality inventories, and he asserts that the ISI along with more "well-established inventories measure a very similar set of higher order dimensions" (Lorr & DeJong, 1984, p. 1381).

Criterion-related validity is demonstrated in the manual by reporting the results of several studies that distinguish between groups in the predicted direction with a wide variety of subjects. However, only one of the studies actually used Form E of the ISI, and the remainder either used earlier versions of the ISI or the version used was not identified. The study using Form E involved two samples of college students' self-ratings and peer ratings that correlated with the ISI scores, with a resulting median correlation of .57 for self-ratings and .45 for peer ratings.

Critique

Because the Interpersonal Style Inventory permits the specific assessment of an individual's interpersonal style, it is a useful addition to the field of instruments available for the evaluation of personality. Lorr and his associates are to be given credit for their efforts over the last 15 years to develop and refine the final version of the ISI. Because most of the validity studies of the instrument seem to have used earlier versions, it is particularly important for the authors, and others as well, to continue investigating the validity of Form E. The ISI will be strengthened if the 15 scales of Form E are found to correlate with other instruments that measure similiar facets of personality, and especially if the scales are found to correlate with specific external behavioral criteria that one would rationally expect to be associated with those interpersonal scales.

A major consideration in a decision to use the ISI concerns the ability of an examiner to understand the scores and interpretative material that make up the computer printout. With the exception of a discussion in the manual concerning the interpretation of T-scores, guidelines and examples for interpretation seem limited to the 10 brief case studies in the manual. Although the results for one person are interpreted in the computer printout, caution is needed for any examinee that does not fit one of the norm groups used for generating interpretative statements. Moreover, determining if a person fits one of the norm groups is difficult because relatively little descriptive information is given in the manual about the specific characteristics of each norm group. Clearly, additional work needs to be done both to specify the interpretive rules that examiners can use in understanding the relationship among the 15 scales of the ISI and also in identifying the characteristics of the norm groups that examiners use for comparison.

Overall, the ISI appears to have been developed in a detailed and disciplined manner, so that what results is a strong statistical foundation, especially for an instrument that is new. If used and interpreted with caution, the ISI has value for professionals interested in the specific assessment of interpersonal style, and that value is likely to increase with further research and manual development.

References

Lorr, M. (1986). *Interpersonal Style Inventory (ISI) manual.* Los Angeles: Western Psychological Services.

Lorr, M., Bishop, P., & McNair, D. (1965). Interpersonal style among psychiatric patients. *Journal of Abnormal Psychology, 70,* 468-472.

Lorr, M., & DeJong, J. (1984). Second-order factors defined by the ISI. *Journal of Clinical Psychology, 34,* 1378-1381.

Lorr, M., & Manning, T. (1978). Higher-order personality factors of the ISI. *Multivariate Behavioral Research, 13,* 3-7.

Lorr, M., & Manning, T. (1978b). Personality correlates of the sex role types. *Journal of Clinical Psychology, 34,* 884-888.

Lorr, M., & McNair, D. (1965). Expansion of the interpersonal circle. *Journal of Personality and Social Psychology, 2,* 823-830.

Lorr, M., & Nerviano, V. (1985). Factors common to the ISI and the 16PF Inventories. *Journal of Clinical Psychology, 41,* 773-777.

Lorr, M., O'Connor, J., & Seifert, R. (1977). A comparison of four personality inventories. *Journal of Personality Assessment, 41,* 520-526.

Lorr, M., & Youniss, R. (1973). An inventory of interpersonal style. *Journal of Personality Assessment, 37,* 165-173.

G. Cynthia Fekken, Ph.D.
Queen's University, Kingston, Ontario, Canada

THE INWALD PERSONALITY INVENTORY

Robin E. Inwald. New York, NY: Hilson Research, Inc.

Introduction

The Inwald Personality Inventory (IPI) is a structured measure of various personality characteristics and behavioral patterns relevant to the "psychological fitness" of candidates applying for law enforcement positions. The IPI is intended to be used as one component of a more comprehensive assessment battery designed to identify applicants who are emotionally or psychologically unqualified for training as police officers, security guards, or corrections officers.

Dr. Robin E. Inwald of Hilson Research Incorporated developed the Inwald Personality Inventory to meet the particular needs of the law enforcement selection process. First, individuals must be selected who possess personal characteristics that will not impair effectiveness in handling the high stress, often hostile, situations that law enforcement officers inevitably face. Specifically, emotional unsuitability, a history of psychoses, or drug or alcohol abuse would interfere seriously with an officer's ability to cope on the job. Second, a mechanism for documenting the specific bases (i.e., candidates' behavioral characteristics) for hiring decisions is required by selection boards should these decisions be legally contested. Third, the procedure for collecting personality and behavioral data on candidates must be time- and cost-efficient. Although the Minnesota Multiphasic Personality Inventory (MMPI) was used widely to meet these requirements, Dr. Inwald (1982) noted that, in fact, few published research studies evaluated the validity of the MMPI for screening applicants to law enforcement and corrections jobs. Moreover, Dr. Inwald's own research (1982) on 2,500 candidates pretested with the MMPI indicated that it did not always detect difficulties in job adjustment and behavioral problems subsequently uncovered during clinical interviews.

Noticeably absent from the IPI manual, and from the published literature, are descriptons of how the IPI was conceptualized, how the test specifications were derived, and how items were selected. A manuscript by Shusman (in press) offers a superficial description of how the rational approach (Burisch, 1984) was employed to construct the IPI. According to Shusman, the first stage of development was to outline three overall categories of behavior and personality characteristics: 1) "acting out" behavior measures, 2) "internalized" conflict measures, and 3) interpersonal conflict measures. Within each of these categories, constructs were defined (e.g., Job Difficulties, Illness Concerns, Family Conflicts) that formed the basis for the 25 clinical scales in the test's final form. In addition, a validity scale designed to detect "guardedness" was developed. The 310 IPI items were derived from preemployment interviews with over 2,500 law enforcement candidates and were written within the context of the construct definitions. Items were designed

265

especially to detect high stress reactivity in a law enforcement context as well as deviant patterns of behavior. In fact, many items are based on actual behaviors (e.g., "I have to admit, I once took money from an employer") to minimize the number of inferences required to relate item responses to the construct. The 25 clinical scales also reflect an emphasis on behavior, that is, they are intended explicitly to distinguish between the tendency to express socially unusual or undesirable attitudes and the tendency to behave in a socially unusual or undesirable fashion. Dr. Inwald argues that this distinction is essential given the high proportion of lower SES males and minority-group members in the applicant pool who may show unusual attitudes but not necessarily unusual behaviors.

The IPI is scored on 26 distinct scales organized into three clinical groupings and a validity measure. The first clinical group, "Acting Out" Behavior Measures (11 scales), is further subdivided into Specific "External" Behavior (6 scales) and Attitude and Temperament (5 scales) to reflect the behavior-attitude distinction espoused by Inwald in the underlying conceptualization of the IPI. The scales of the Specific "External" Behavior subdivision are 1) Alcohol (13 items), 2) Drugs (13 items), 3) Driving Violations (6 items), 4) Job Difficulties (22 items), 5) Trouble with the Law and Society (21 items), and 6) Absence Abuse (19 items). The Attitudes and Temperament facet includes 1) Substance Abuse (20 items), 2) Antisocial Attitudes (27 items), 3) Hyperactivity (42 items), 4) Rigid Type (19 items), and 5) Type "A" (21 items). The second clinical grouping is designated "Internalized" Conflict Measures and contains 7 scales assessing various dimensions of psychological or psychiatric dysfunctioning: 1) Illness Concerns (14 items), 2) Anxiety (15 items), 3) Phobic Personality (34 items), 4) Obsessive Personality (13 items), 5) Depression (27 items), 6) Loner Type (17 items), and 7) Unusual Experiences/Thoughts (26 items). An eighth scale in this grouping, Treatment Programs (3 items), identifies individuals who have previously participated in a substance abuse treatment program, psychotherapy, or a psychotropic drug regimen. The third clinical grouping, Interpersonal Conflict Measures, assesses the quality of interpersonal interactions. The six scales are 1) Lack of Assertiveness (14 items), 2) Family conflicts (23 items), 3) Spouse/Mate Conflicts (8 items), 4) Interpersonal Difficulties (27 items), 5) Undue Suspiciousness (22 items), and 6) Sexual Concerns (5 items). Finally, the 19-item validity scale, Guardedness, identifies individuals who are defensive or lack insight and who respond in a socially desirable fashion. Definitions of all 26 scales are provided in the IPI manual along with descriptions of what high and low scale scores may indicate.

Many of the IPI items are scored on more than one scale, resulting in both conceptual and empirical confounding. Shusman and Inwald (1987) report on a reorganization of items into a 12 scale version of the IPI that was initiated to remove item overlap and to increase scale homogeneity. Internal consistencies and the validity of the inventory for the 12-scale version were not seen as improved sufficiently to warrant its adoption over the 26-scale version.

Extensive norms are available for the IPI (Inwald, Knatz, & Shusman, 1982.) Over the 3-year period from 1979 to 1982, the IPI was administered as one part of a larger test battery to all candidates applying to two large, urban law enforcement agencies in New York state. Norms were developed for two distinct groups of male ($n = 1884$; $n = 2007$) and female ($n = 523$; $n = 431$) corrections officer candidates and for male (*N*=2,397) and female (*N*=147) police officer candidates. Subsequently, norms for hired male (*n*=748) and female (*n*=157) corrections officers and for hired male (*n*=329) and female (*n*=15) police officers were derived. Furthermore, the larger norming group of corrections officer candidates was subdivided by sex and race (i.e., white, black and Hispanic) into groups ranging in size from 60 to 748 individuals, providing adequate bases for comparison. In conjunction with Hilson Research Incorporated, local, agency, or regional norms may be prepared and used when scoring.

The IPI test materials are similar to those of most typical personality tests. The test booklet is comprised of a cover page and one page of general instructions followed by the 310 items. Test items are organized spaciously and printed in large type to enhance readability. The computer-scored answer sheet (by Scan-Tron) contains an area in which the candidate prints demographic data and a section in which the test administrator codes background data. The main body of the answer sheet contains seven columns of numbered pairs of boxes marked with a "T" and an "F." The respondent indicates his or her response by darkening the box reflecting a true or false response to the corresponding test item.

The role of the test administrator is minimal. General administration instructions provided in the IPI manual describe how respondents should complete the demographic data on the answer sheet and how these data, as well as departmental identification data, should be coded on the answer sheet. Administrators also are requested to read test instructions to candidates to ensure that they understand what is expected of them, especially with regard to using the computer)scored answer sheets.

Feedback on the IPI scales is presented in the form of a computerized report that includes a written evaluation, a listing of critical items, a printout of the individual's item responses and how they were scored to obtain the 26 scale scores, and a personality profile group comparing the individual's scores to a normative group. The written evaluation which runs three to four pages long, begins with an introduction to the purpose of the IPI that warns that the test should not serve as the sole basis for a hiring decision. The body of the narrative is organized around the validity scale and the clinical groupings. Within these groupings, the respondent's relative standing on each scale is outlined using a small number of descriptive statements that reflect not only actual test responses, but also the test developer's clinical experiences with similar test scores. Potential problems areas (i.e., elevated scale scores) are highlighted throughout the narrative by asterisks. Concluding the written evaluation is a paragraph summarizing the overall similarity of the respondent's scores to the normative group scores and giving the probable job success of the respondent.

The written evaluation is followed by the IPI Critical Item List, which consists of endorsed items requiring external verification or follow-up discussion. For example, items may have been answered incorrectly or may have acquired a different connotation when the response was given a context by the candidate. This component of the feedback report is not discussed in the IPI manual. Hence, the selection of critical items is obscure. It is clear, however, that the Critical Item List does not constitute a full list of endorsed items. Furthermore, critical items are grouped into categories, which also lack a clear origin. For example, some categories (e.g.,

Depression) are identical to scale names, other categories may represent a combination of scales (e.g., the Alcohol and/or Drug Use category presumably represents the Alcohol and the Drugs scales). Still other categories appear to be independent of IPI scales (e.g., the Phobic/Stress/Physical Symptoms category).

The Critical Item List may have been developed as an addendum to the user "unfriendly" IPI Critical Item Printout, which is a listing of the raw respondent data as it is input into the computer, as well as a list of the items that contributed to the scores for each of the 26 scales. Scales and items are referred to by mnemonics that are probably quite meaningless to individuals unfamiliar with the IPI. To illustrate, for a sample test protocol, a score of 3 was obtained on the AL scale as follows: 5T SIXPK *157T SIXPKEZ 183T SOCDRINK. Users of the IPI are likely to learn the 26 scale abbreviations, but learning the short forms for 310 items is an improbable and unnecessary task. If this is truly central information, a complete listing of the items contributing to each scale should be generated.

Normative feedback on the 26 IPI scales is presented in the form of T-scores depicted on a profile graph. T-scores ranging from 0 to 100 are demarcated along the ordinate of the graph. The abscissa is subdivided into sections for the validity scale, for each of the Actions and Attitudes scale groups within the "Acting Out" Behavior Measures, for the "Internalized" Conflict Measures, and for the Interpersonal Conflict Measures. Within each section, scales are labelled using the two-letter abbreviation given in the manual. Actual T-scores and raw scores are printed directly below the relevant scale label. In distinction to common practice, T-scores are represented on the profile graph as vertical bars that are parallel to the ordinate and extend from the zero point to the exact T-score. Normative comparisons routinely are performed separately for males and females; however, as previously noted, more specialized norms may be available. General information on how to interpret elevated T-scores is presented adequately in the manual.

To the credit of Dr. Inwald and Hilson Research, the feedback report seems to have changed over time. That is, the sample narrative report included in the 1982 manual provides less information than that provided in more recent versions (1984, 1987). However, the manual should be updated or an addendum provided to ensure that test users have the background required to evaluate the test protocols. For example, a discussion of IPI Critical Items for Follow-up Evaluation is entirely lacking in the manual. A page entitled "IPI prediction of law enforcement performance rating provided by psychologist" is included in the more recent narrative report. The description in the 1984 report is simply inadequate; in the 1987 report, the description is better but, not surprisingly, still too cursory to permit an informed evaluation of this new index.

Practical Applications/Uses

According to Dr. Inwald (Knatz & Inwald, 1983), the role of preemployment testing is to screen out "poor performers" rather than to predict "good performers." In keeping with this provision, the purpose of the Inwald Personality Inventory is to identify candidates who are psychologically unfit for law enforcement work. In addition, the IPI can serve as a record of specific problematic behaviors that may be used in court to help justify hiring decisions. Even so, the IPI was designed expressly to form only one component of the law enforcement selection process. In both the IPI manual and various publications, the collection of additional data through biographical questionnaires and clinical interviews is encouraged strongly (Inwald, 1984, 1985). Paper-and-pencil tests such as the IPI may be susceptible to guardedness and defensive or desirable responding. Hence, data collected through other means, especially direct behavioral observation in the interview setting, are necessary to substantiate findings on the IPI.

The IPI assesses a wide range of maladaptive behaviors, including alcohol and drug abuse, traffic violations, legal problems, job difficulties, antisocial attitudes, hyperactivity, tendency to be habitually late or absent from work, hypochondriasis, anxiety, depression, rigidity, lack of assertiveness, workaholism, interpersonal conflicts, paranoia, and psychotic tendencies. It is intended for use with candidates applying for such law enforcement positions as police officer, corrections officer, transit officer, court officer, security guard, and so forth. The IPI is not appropriate for screening candidates to positions outside of law enforcement unless a job analysis and legal justification would deem otherwise. Currently the IPI is available only in English; however, a Spanish translation is underway. As always, a translated test will require careful empirical evaluation.

Administration of the IPI is quite straightforward. Administrators simply provide candidates with the test materials and read the general instructions. The administration directions provided in the manual are generally clear with one notable exception: the instructions printed in the manual do not match those found in the test booklet exactly. An administrator who directs candidates to read along in the test booklet as he or she presents the instructions from the manual would engender confusion. The content of the two sets of instructions is quite similar, however. The obvious solution is for the administrator to read the directions aloud from a test booklet rather than the manual. Once the instructions have been read, candidates should be able to respond to the items independently.

The reading level demanded by the IPI is fairly high. The manual does not recommend explicitly a minimum educational qualification for candidates, but it does state that the IPI was developed using individuals with at least a high school education. Adopting such an educational criterion should ensure that candidates have the minimum verbal ability to respond to the IPI.

The total time required to administer the IPI should not exceed 1 hour. The IPI may be administered conveniently in a group setting, although individual administration is equally appropriate. The test administrator should possess some type of professional status. A mental health professional, staffing consultant, or personnel officer would be qualified to administer the test, whereas clerks and secretaries would not. The professional qualifications of the administrator should enhance the cooperation and minimize the defensiveness of candidates in light of the sensitive information being requested.

Scoring keys are not provided routinely in the manual. Rather, answer sheets may be sent to the test publisher, Hilson Research Incorporated, for computerized processing. As previously described, the full IPI computerized narrative report contains a three- or four-page written evaluation, two lists of critical items, and a personality profile graph. Tests may be scored using individual agency, local, or regional norms in addition to the normative bases available through Hilson. For test users requiring immediate scoring, Hilson has developed a remote user system that scores tests via teleprocessing and allows users to print out the results on site within minutes. In order to use this facility, certain equipment (e.g., a microcomputer with various peripherals and a printer) must be available. Finally, agencies may request that results be returned in the form of an ASCII data file that contains test information for each candidate who was tested. The data file includes demographic variables, raw scale scores, T-scores, and the results of several discriminant prediction equations developed by Inwald and her colleagues.

Interpretation of the IPI is based on objective scores rather than on clinical judgment. The full IPI narrative report (1987 version) provides five sources of information pertaining to the interpretation of IPI test results. The most elaborate source is the written evaluation, which offers a clear and direct verbal explanation of the candidate's 26 scale scores. Also given is a brief summary of the candidate's potential for performing law enforcement work and an outline of the limitations of IPI data. The personality profile graph and, to a lesser degree, the two critical item lists, serve to substantiate the narrative. The personality profile graph indicates the degree to which the scale scores are elevated; the item lists may illustrate how score elevations were obtained by the candidate. The graph is interpreted readily by experts in the selection process but may cause non-experts somewhat more difficulty, despite the description in the manual and the concomitant warnings regarding the need to take normative comparisons into account. Critical item lists engender the risk for expert and nonexpert alike of overemphasizing individual items. The IPI-generated predictions of future law enforcement performance constitute additional sources of objective data that bear directly on the probability of job performance problems. Because this aspect of the report involves reference to technical analyses, these important data may be misinterpreted by a nonexpert. Overall, interpretation of IPI results, designed to be done in the context of a comprehensive assessment battery, is best performed by an expert in preemployment psychological testing or a staffing or personnel officer with considerable relevant training.

Technical Aspects

Both the test-retest reliability and the internal consistency of the IPI have been studied using large samples of law enforcement officer candidates. Because only one form of the IPI has been developed, no data exist on parallel forms reliability. The 26 IPI scales demonstrate acceptable levels of test-retest reliability. Based on a 6-8-week interval, retest reliabilities ranged from .58 to .87 and from .60 to .79 for male ($n=321$) and female ($n=171$) correction officer candidates, respectively. The internal consistency of IPI scales is modest for most scales and unacceptable for others. Coefficients alpha varying from .16 to .82 and from .28 to .80 are reported in the manual for male ($n=534$) and female ($n=518$) corrections officer candidates, respectively. Similar findings were obtained for large samples of police officer candidates (Inwald, Knatz, Shusman, 1982; Shusman & Inwald, 1987). Not surprisingly, the scales with the fewest items (e.g., Treatment Programs, 3 items; Sexual Concerns, 5 items) consistently exhibit low internal consistency. However, some longer scales (e.g., Rigid Type, 19 items; Unusual Experiences, 26 items) also generally have coefficients alpha below .60 across the various subsamples. These scales do not appear to access a homogeneous construct, which may be a function of an ill-defined or multifaceted domain or of inadequate item selection procedures. Partly out of recognition of the instrument's weak internal consistency, Shusman and Inwald (1987) evaluated a 12-scale version of the IPI. The original 310 items were allotted to 12 non-overlapping scales, mainly according to judgmental criteria. Coefficients alpha were generally higher for the revised scales (comprised of 19 to 39 items) than for the original scales, ranging from .61 to .81 for 1,593 male and from .57 to .82 for 687 female police officer candidates.

The dimensionality of the IPI was evaluated through a series of factor analyses conducted for large samples of male and female police officer and correction officer candidates (Inwald, Knatz, & Shushman, 1982; Inwald & Shushman, 1984b). Unfortunately, the results of these analyses have limited utility. First, IPI scales show substantial item overlap, which violates one of the basic assumptions of factor analysis regarding independent data. At the very least, this problem must be addressed when interpreting factors. Second, a description of the factor analytic technique employed is entirely lacking. Inwald and her coauthors (1982) reference the factor analysis subprogram of a specific computing package. Is the reader to assume that the analysis was run with all the default options? No information is given on what was placed in the diagonal of the correlation matrix, what criterion was used for factor extraction, or how a factor loading was defined. Third, the factor analyses are described inaccurately. For each of the four analyses, the loadings of 11 to 13 scales on each of three "varimax rotated factors" are presented along with related eigenvalues and the percentage of variance for which the factor accounted. The latter two pieces of data are clearly in error—squaring and summing the factor loading obviously yields a different estimate of the eigenvalues and, hence, variance associated with factor. It is quite likely that the eigenvalues are, in fact, correctly associated with the unrotated solution for the full 26 scales. If this were the case, the size of the first factor (always accounting for well over 60% of the common variance) would argue strongly for a one-factor solution. A single-factor solution would suggest that IPI scales all are related to one dimension, such as general psychopathology, or to a pervasive response style, such as social desirability. In view of the scale dependence and the thoroughly inadequate description of the factor analytic procedures, the factor interpretations and anecdotal factor comparisons of Inwald and her colleagues (1982; 1984b) are difficult to evaluate. An unpublished factor analysis of IPI scale scores for 169 male candidates applying to the New York State Department of Correctional Services (Malin & Morgenbesser, n.d.) unfortunately is flawed in much the same way. Again, a basic description of the factor analytic technique employed is lacking. Seven factors were retained, the first of which accounted for approximately 50% of the variance. The authors' interpretation and comparison of their solution to the ones reported in the IPI manual (Inwald et al., 1982) is superficial and hard to judge in the absence of further information.

Given the stated purpose of the IPI, its validity for identifying poor performance among law enforcement officers is of central importance. Inwald and her associates have examined the predictive validity of the IPI both for police officers (Inwald et al., 1982; Inwald & Shusman, 1984a; Shusman, Inwald, & Knatz, 1984) and for corrections officers (Inwald, et al., 1982; Inwald & Shusman, 1984b; Shusman, Inwald, & Landa, 1984). Essentially, linear combinations of scale scores were derived empirically to classify officers' performance with regard to various (generally dichotomous) criteria, such as retention versus termination from the force, or the number of absences, late arrivals, or disciplinary interviews. Across the various studies, discriminant functions derived from the IPI generally classify 65-75% of officers correctly. For example, Shusman, Inwald, and Landa (1984) report that on the four respective criteria outlined above, the IPI accurately identified 73, 69, 67, and 67% of 716 corrections officer recruits. In related analyses, the retained officers were divided into an original ($n=400$) and a cross-validation ($n=265$) sample. A linear combination of IPI scores correctly predicted absences, lateness, and disciplinary interviews for at least 67% of the officers in the original sample and at least 58% in the cross-validation sample. A similar study with police recruits (Inwald & Shusman, 1984a; Inwald, Shusman, & Knatz, 1982) demonstrated that the discriminant function classification accuracy of the IPI for nine distinct performance criteria ranged from 60-82%.

Classification accuracy may be increased by conducting analyses separately by sex (Inwald & Shusman, 1984b) or by sex and race (Inwald, Knatz, & Shusman, 1982). For example, the accuracy rates for predicting termination versus retention on the force for 596 males was 72%; for 143 females, the accuracy rate was 83% (Inwald & Shusman, 1984a). Likewise, the IPI manual reports separate discriminant function analyses by race for male corrections officers in an attempt to increase predictive accuracy. However, such studies must be evaluated in the context of the overall base rate for the behavior to be predicted (i.e., termination). Given that 86% of males and 95% of females in the Inwald and Shusman (1984b) study were retained, use of the IPI linear equations actually results in worse overall classification than simply hiring everyone using the selection procedure currently in place. The fairest evaluation of the IPI would involve evaluating, classifying, and hiring *all* candidates. The predictive validity of the IPI then would be examined using the termination/retention criterion for all candidates. This, of course, is impractical, if not unethical. The accuracy of the IPI relative to the base rates cited above may be justifiable if the IPI significantly identifies false positives, perhaps at the expense of false negatives. From the point of view of the law enforcement organization, hiring an incompetent police officer presumably is perceived to have more serious consequences than failing to hire a candidate who would perform well. Shusman and Inwald (1987) recognize this in an unpublished manuscript in which they report that 17-41% more of the false positives (defined by various criteria) were identified by the IPI relative to chance predictions. Such findings support the validity of the IPI for identifying unfit candidates.

The MMPI has been found to be the single, most highly used preemployment psychological screening test in law enforcement selection (cf. Inwald, 1982). How does the IPI compare? This has been one of Dr. Inwald's longstanding concerns, and many of her research studies involve a comparison of the two inventories. The linear prediction equations and the multiple correlations developed on the IPI scales marginally outperform those developed on the MMPI clinical scales (possibly along with the McAndrew Alcoholism Scale) across various studies (e.g., Inwald, Shusman, & Knatz, 1982; Inwald & Shusman, 1984a, 1984b; Shusman,

Inwald, & Landa, 1984). Together, the discriminant functions developed on both sets of scales generally predict slightly better than the IPI alone. To illustrate, in the study conducted by Shusman, Inwald, & Landa (1984) the termination criterion was predicted correctly for 73, 63, and 73% of officers using the IPI alone, the MMPI alone, and the IPI and MMPI together, respectively. For the absence criteria, the IPI, MMPI, and combined IPI and MMPI showed 65, 62, and 67% accuracy, respectively. A redundancy analysis (Shusman, in press) indicates that the IPI and MMPI share approximately one sixth overlapping variance. Therefore, the unique variance assessed by each measure may be expected to add to the prediction independently. Shusman (in press) hypothesizes that the pathology measured by the MMPI has some relevance for predicting failure among law enforcement officers but that the behaviorally oriented scales of the IPI are sensitive to more central variables. In particular, the acting out and antisocial scales for males and the scales tapping fearfulness and withdrawal for females appear related to poor job performance (Inwald & Shusman, 1984b).

Evidence for the construct validity of the IPI is considerably weaker than for predictive validity. The manual reports that the Phobic scale was able to differentiate significantly between clinically diagnosed phobics and law enforcement officers. In addition, a linear function based on IPI scale scores was able to discriminate among three categories of phobics. Whether the hit rates for correct classification were above chance and could be cross-validated is not reported. More importantly, it is unclear how successful discrimination among groupings of clinically phobic patients relates to the construct validity of a multiscale inventory designed to reflect psychological unfitness for law enforcement positions. Another investigation of construct validity (Inwald et al., 1982) related the IPI scores of 115 to 166 candidates to five interview criteria based on the pooled judgments of three experienced corrections officer raters. Unfortunately, the criteria are only described using vague titles, such as "attitudes" and "response to background questions." Thirteen IPI scales correlated with at least one criterion, but in the absence of a well-articulated set of predictions, it is not clear how these findings demonstrate construct validity. A similar complaint may be voiced with regard to the corelations reported in the manual between IPI scales and biographical data. No explicit predictions regarding the expected pattern of correlations is offered. Various intuitively meaningful relationships do appear, however. For example, the Driving Violations scale correlates .67 with the number of moving violations ($n = 2,259$). If biographical criteria such as the number of moving violations, the length of the longest job held, and marital status are expected to be important predictors of success as a law enforcement officer, one wonders if a biographical questionnaire is not to be preferred over the IPI. Again, the relative merits of various types of data from the point of view of construct validity are left unexplored.

Two unpublished studies also provide equivocal support for the construct validity of the IPI. Ostrov (1985) reported a relationship between the high scores on the Drug scale and positive urine drug test results. Although the study appears to be methodologically shaky, it does appear that the Drug scale may elicit admissions of drug use that occurred too far in the past to be detected by standard urine tests. A second study (Brobst & Brock, in press) related IPI scales to candidates' behavioral self-admissions and indices of deception collected using a polygraph. That is, IPI scales were correlated with the number of admissions 86 security officer candidates made to 14 categories of questions while hooked up to a polygraph. Six IPI scales showed significant correlations with content relevant polygraph scales (e.g., the number of admissions to drug-related questions correlated .36 with Drug scale scores). However, IPI scales such as Alcohol Abuse, Driving Violations, and Trouble with the Law failed to relate to presumably relevant polygraph scales such as Alcohol, Driving Record, and Serious Crime. Overall, Brobst and Brock's findings provide only weak support for construct validity.

In addition to demonstration of appropriate convergence, construct validity demands an evaluation of the ability of scales to show discrimination for irrelevant scales. Nowhere in the manual is discriminant validity expressly addressed. Particularly worrisome is the lack of apparent concern with demonstrating that the IPI is unconfounded by stylistic variance, such as that associated with acquiescence or social desirability. Predictive validity likely is to be attenuated by stylistic variance; construct validity is certain to be affected. In order to demonstrate that a theoretically circumscribed construct is being measured, the scale must show no correlations with irrelevant scales. The evidence that is available would suggest that IPI scales lack discriminant validity. For example, Phobics' *mean* score on the Type A scale was elevated relative to the critical scores reported for the various normative groups (Inwald et al., 1982). There is no clear theoretical basis for predicting a relationship between the phobia and Type A scales. As another example, consider the relationship of IPI scales to the State-Trait Anxiety Inventory (STAI). State Anxiety and Trait Anxiety scores correlate with the IPI Anxiety scale at .40 and .35, respectively, providing some evidence for the IPI Anxiety scale's convergent validity. However, the STAI correlates more highly with the IPI Depression ($r = .46$) and Interpersonal Difficulties ($r = .48$) scales. Furthermore, 39 of the 52 correlations between State Anxiety and Trait Anxiety and the 26 IPI scales are significant. A similar pattern of general intercorrelations is reported between the MMPI and the IPI scales (Inwald et al., 1982). For other multiscale inventories, IPI scales all tend to correlate with a particular scale (e.g. California Personality Inventory Sense of Well-being scale) or they do not (e.g., Edwards Personal Preference Schedule Achievement scale). Taken together, the pattern of correlations between IPI scales and other scales is indicative of a single, pervasive, underlying construct, which is likely to be general psychopathology or social desirability.

Critique

The Inwald Personality Inventory is used appropriately as one component in the process of screening candidates for positions as law enforcement officers. The greatest strength of the IPI is its demonstrated ability to predict correctly the status of approximately two thirds of candidates on such job-relevant criteria as termination from the force, excessive absence from the job, habitual lateness, and disciplinary actions. Moreover, the IPI predicts these criteria above chance for those candidates who eventually prove to be unacceptable employees, albeit at the cost of overidentifying candidates who turn out to be acceptable. In general, the IPI does a credible job of meeting its major goal to flag for further evaluation those candidates who may be psychologically unfit for law enforcement work.

A second, explicitly stated purpose of the IPI is to document the bases for hiring decisions. That is, when a selection board is required to justify its decisions in court, the IPI should serve as a record of "the specific behavioral *characteristics* [italics added] of candidates" (Inwald et al., 1982, p. 1). Meaningful interpretation of individual differences in test scores requires more than face validity: evidence for the construct validity of the test is mandatory. The Inwald Personality Inventory appears to lack construct validity. First, it is unclear whether the constructs underlying the IPI scales were defined explicitly and, further, how items subsequently were selected to ensure accurate construct measurement. Additionally, assessment of distinct constructs is additionally compromised by the item overlap on the IPI scales. Dr. Inwald does stress one aspect of the IPI conceptualization, namely, the difference between the expression of socially deviant attitudes and the expression of socially deviant behaviors (Inwald, Knatz, & Shusman, 1982, p. 2). Whether this distinction is operationalized adequately in the IPI is never evaluated empirically; neither is its predictive merit. Second, the construct validity of the IPI is compromised by uncontrolled stylistic variance. The test lacks a balance of true- and false-keyed items to mitigate acquiescence. Social desirabiliity is unevaluated, despite the fact that the nature of the items, the empirical relationships among IPI scales, and the correlation of IPI scales with other personality scales all suggest that social desirability is a confound. The Guardedness scale is inadequate for addressing the desirability issue, given its heterogeneous description as a measure of denial, defensiveness, and lack of insight, as well as social desirability. Finally, the failure of the IPI to demonstrate discriminant validity attests to its deficient construct validity. As detailed above, IPI scales correlate appropriately with some relevant criteria. However, they also correlate with theoretically irrelevant criteria, indicating that they cannot be interpreted conclusively in terms of one or the other underlying construct. Thus, whereas IPI test scores may be useful for predicting candidates' future performance on certain criteria, these same test scores are not amenable to meaningful interpretation in terms of particular underlying constructs or characteristics. Arguments dependent upon explicating candidates' scale scores (i.e., position on a construct or behavioral characteristic) are likely to meet with failure both inside and outside the courtroom.

References

Brobst, K.E., & Brock, R.D. (in press). A study of the relationship of the polygraph with the MMPI, the IPI and the Stanton Survey Phase II when used with security officer applicants. *Security Management*.

Burisch, M. (1984). Approaches to personality inventory construction: A comparison of merits. *American Psychologist, 39*, 214-227.

Inwald, R.E. (1982). Psych screening test helps police corrections save money, avoid legal battles. *Criminal Justice Digest, 1*, 1-4.

Inwald, R.E. (1984, Dec.). Law enforcement officer screening: A description of one pre-employment psychological testing program. In J. T. Reese and H. A. Goldstein (Chairs), *National Symposium on police psychological services*. FBI Academy, Quantico, VA.

Inwald, R.E. (1985). Proposed guidelines for conducting pre-employment psychological screening programs. *Crime Control Digest, 19*, 1-6.

Inwald, R.E., & Shusman, E.J. (1984a). The IPI and MMPI as predictors of academy performance for police recruits. *Journal of Police Science and Administration, 12,* 1-11.

Inwald, R.E., & Shusman, E.J. (1984b). Personality and performance sex differences of law enforcement officer recruits. *Journal of Police Science and Administration, 12,* 339-347.

Inwald, R.E., Knatz, H.F., & Shusman, E.J. (1982). *Inwald Personality Inventory Manual.* New York: Hilson Research.

Knatz, H.F., & Inwald, R.E. (1983). A process for screening out law enforcement candidates who might break under stress. *Criminal Justice Journal, 2,* 1-4.

Malin, S.Z., & Morgenbesser, L.I. (n.d.). *New York State Department of Correctional Services Pre-Employment Psychological Screening Program.* Unpublished manuscript.

Ostrov, E. (1985). *Validation of police officer recruit candidates' self-reported drug use on the IPI Drug Scale.* Unpublished manuscript.

Shusman, E.J. (in press). A redundancy analysis for the IPI and MMPI. *Journal of Personality Assessment.*

Shusman, E.J., & Inwald, R.E. (1987). *Internal consistency and predictive validity of the IPI.* Unpublished manuscript.

Shusman, E.J., Inwald, R.E., & Knatz, H.F. (1984, August). *A cross-validation of police recruit performance as predicted by the IPI and MMPI.* Paper presented at the Annual Convention of the American Psychological Association, Toronto, Canada.

Shusman, E.J., Inwald, R.E., & Landa, B. (1984). Correction officer job performance as predicated by the IPI and MMPI: A validation and cross-validation study. *Criminal Justice and Behavior, 11,* 309-329.

Ronald K. Hambleton, Ph.D.
Professor of Education and Psychology, School of Education, University of Massachusetts at Amherst, Amherst, Massachusetts.

IOWA TESTS OF BASIC SKILLS, FORMS G AND H

A.N. Hieronymous and H.D. Hoover. Chicago, Illinois: The Riverside Publishing Company.

Introduction

The Iowa Tests of Basic Skills (ITBS) are achievement tests intended to measure fundamental skills in the areas of word analysis, vocabulary, reading, language, work-study, and mathematics for students in Grades K to 9. In addition, there are supplemental subtests for measuring listening, social studies, science, and writing skills. All of the ITBS use a multiple-choice item format except the writing skills subtest, which employs a writing sample instead.

Broadly speaking, the tests that make up the core of the ITBS can be used to assess students' current skills and growth in the areas of reading, language, and mathematics. The ITBS are intended to provide scores applicable for individualizing instruction, providing guidance, assessing group (class, school, or district) performance, and evaluating school programs. Scores for individuals and groups may be interpreted from a criterion-referenced as well as a norm-referenced perspective.

The documents reviewed by this writer provided almost no discussion of the history of the Iowa Testing Program or the test authors. This omission is unfortunate, because both the program and the authors have outstanding reputations in the testing field. The earliest edition of the ITBS was published in 1935, when the Iowa Testing Program was under the direction of E. F. Lindquist, one of the major contributors to the field of psychometrics in the 1930s, '40s, '50s, and '60s. Over the more than 50 years of its history, the ITBS have become widely regarded as among the very best educational achievement tests in the testing field. Major authors of the current edition of the ITBS, Al Hieronymous and H. D. Hoover, not only have directed work on the ITBS through several highly rated previous editions, but they also have contributed important papers to the field of testing in the areas of test development, item bias, test score scaling, norming, and test score equating. Both authors are Professors of Education at the University of Iowa and have outstanding reputations in the psychometrics field.

None of the ITBS documents that the publisher provided this reviewer addressed the main changes in the 1985 edition from the previous 1978 edition (Forms 7 and 8). In addition to the obvious updates associated with any new achievement test battery, it would appear, however, that new tests in Social Studies and Science for Grades 1 to 9 were added to the ITBS along with Listening and Writing subtests for

Grades 3 to 9. Since 1963, Riverside Publishing Company has also published the Cognitive Abilities Test. This set of tests for Grades K to 12 measures verbal, quantitative, and non-reasoning abilities and was normed on the same student population as the ITBS. With the availability of both achievement and ability test results on students, the use of the test data for diagnosing student strengths and weaknesses and prescribing instruction should be enhanced.

Forms G and H of the ITBS consist of 10 levels, organized into three batteries:

Battery	Level	Average Age	Development Level Grade
Early Primary	5	5	K.1-1.5
	6	6	K.8-1.9
Primary	7	7	1.7-2.6
	8	8	2.7-3.5
Multilevel	9	9	3
	10	10	4
	11	11	5
	12	12	6
	13	13	7
	14	14	8-9

Users of the ITBS undoubtedly will appreciate its comprehensiveness and flexibility. However, with these features go a very large number of products that can be confusing. This reviewer counted well over 100 different products in the material he was sent to review.

What follows is a list of most of the main documents associated with the ITBS:
Early Primary Battery, Levels 5-6
1) one form, denoted G;
2) two levels, denoted Levels 5 and 6;
3) two editions, machine scorable and hand scorable
   a) for the machine-scorable edition, two types of answer sheets are possible: MRC, which are scored by the Riverside Scoring Services, and NCS, which are scored by schools with National Computer System equipment;
4) Level 5, which measures Listening, Word Analysis, Vocabulary, Language, and Mathematics;
5) Level 6, which adds a Reading subtest;
6) a teacher's guide, with complete information about test purposes, test preparation, test administration directions, scoring directions, test score interpretations, uses of test results, and norms tables; tests are untimed (except for Reading) and orally administered;
7) scoring keys for the hand-scorable booklets;
8) NCS directions for administration;
9) practice tests;
10) parent/teacher report forms; and
11) special norms booklets.
Primary Battery, Levels 7-8

1) all of the information previously described for the Early Primary Battery for two forms, denoted G and H;
2) Basic and Complete Battery, and Complete Battery plus Social Studies and Science supplements (Form G only)
   a) the Basic Battery includes tests to measure Word Analysis, Vocabulary, Reading Comprehension, Spelling, and Mathematics Concepts, Problems, and Computation;
   b) the Complete Battery includes the Basic Battery plus tests to measure Listening, Capitalization, Punctuation and Usage and Expression in the language arts area, and Work-Study Skills;
   c) all tests except Vocabulary, Reading, Capitalization and Punctuation are untimed and orally administered;
3) four available editions: Complete, machine scorable; Complete plus Social Studies and Science, machine scorable; Basic, machine scorable; and Basic, hand scorable.

Multilevel Battery, Levels 9-14

1) two forms, denoted G and H;
2) all of the information previously described for the Early Primary Battery;
3) Basic and Complete Batteries with Social Studies, Science, Listening, and Writing supplements
   a) the Basic Battery has six tests to measure Vocabulary, Reading Comprehension, Spelling, and Mathematics Concepts, Problem-Solving, and Computation;
   b) the Complete Battery includes the Basic Battery plus five tests to measure Capitalization, Punctuation, Usage and Expression, and Work-Study (visual and reference materials);
4) spiral-bound versions of the Basic and Complete Multilevel Batteries; spiral-bound versions for each level within the Basic and Complete Multilevel Batteries; Social Studies and Science, Listening, and Writing supplements are packaged separately (only Form G is available for the supplements); and
5) five different types of answer sheets.

Basically, the foregoing descriptions of these batteries are not totally accurate or complete because of many additional variations, options, exceptions, and features that are omitted. Readers will need to refer to the complete set of ITBS documents for the detailed information. Table 1.8 in the *Manual for School Administrators* will be helpful to users interested in a summary. Table 1 below provides an overview of the content of the ITBS at each level and the number of test items in each subtest.

Practical Applications/Uses

According to the authors, the three ITBS batteries were designed to address nine main uses:

1) to determine the developmental level of students in order to better adapt materials and instructional procedures to individual needs and abilities;
2) to diagnose specific qualitative strengths and weaknesses in students' educational development;
3) to indicate the extent to which individual students have the specific readiness skills and abilities needed to begin instruction or to proceed to the next step in a planned instructional sequence;

Table 1[1]

Content for the Iowa Tests of Basic Skills, Forms G and H

Tests		*Number of Items*									
	Level	5	6	7	8	9	10	11	12	13	14
Practice Page		10	10								
Li: Listening		31	31	32	32						
WA: Word Analysis		35	36	47	50						
V: Vocabulary		29	29	30	30	30	36	39	41	41	41
R: Reading/Reading Comprehension			58	56	61	44	49	54	56	57	58
Words			13								
Word Attack			7								
Pictures			13	23	23						
Sentences			13	14	14						
Picture Stories/ Stories			12	19	24						
L: Language		29	29								
L1: Spelling				27	29	30	36	40	41	41	41
L2: Capitalization				60	66	28	29	30	30	31	32
L3: Punctuation				46	60	28	29	30	30	31	32
L4: Usage and Expression				27	27	33	36	38	40	43	43
W: Work Study											
W1: Visual Materials				28	29	33	36	43	45	49	51
W2: Reference Materials				30	33	33	39	40	41	42	43
M: Mathematics		33	33								
M1: Mathematics Concepts				33	36	28	32	35	39	41	42
M2: Mathematics Problem Solving				22	28	24	26	27	29	30	32
M3: Mathematics Computation				27	32	34	37	39	41	42	43
Social Studies				39	39	38	40	42	43	45	45
Science				35	35	38	40	42	43	45	45
Listening Supplement						31	33	34	36	38	40
Writing Supplement						5*	5*	5*	5*	5*	5*

[1]This table is adapted from Table A in the 1987 edition of the Riverside Publishing Company *Test Resource Catalog* (p. 5).

*The optional writing test has a choice of five different prompts (directions to the student) for Levels 9-11 and the same number for Levels 12-14.

4) to provide information that is useful in making administrative decisions in grouping or programming to better provide for individual differences;

5) to diagnose strengths and weaknesses in group performance (class, building, or system) that have implications for changes in curriculum, instructional procedures, or emphasis;

6) to determine the relative effectiveness of alternate methods of instruction and the conditions that determine the effectiveness of the various procedures;

7) to assess the effects of experimentation and innovation;

8) to provide a behavioral model to show what is expected of each student and to provide feedback that will indicate progress toward suitable individual goals; and

9) to report performance in the basic skills to parents, students, and the general public in objective, meaningful terms.

All nine uses are generally of interest to school districts, though uses 1, 2, 4, 5, and 9 are probably the most popular. Validity data to support the recommended uses of the ITBS will be discussed later in this review.

The ITBS seem well suited for regular students in Grades K through 9 in the United States. In addition, the publishers report that braille and large-print editions can be prepared. They further note that revised directions may be used to administer the ITBS (or parts thereof) to students with severe reading disabilities and other handicaps.

Apart from these special administrations, the ITBS are intended to be administered by teachers (with the help of proctors, if possible) to students in their classrooms. Groups of 25 to 30 students, with one or two proctors or aides, and a room with good lighting and desk space are the physical arrangements needed for a good test administration. The *Teacher's Guides* are clear. However, test administrators should be very familiar with the directions and tests before they begin. Advanced planning is essential. A teacher's checklist for testing is available and would be invaluable.

The time required to administer an ITBS battery depends upon the level and the choice of options. The Basic Battery administration time is approximately 2 hours. The Complete Battery administration time is about 4 hours. Roughly speaking, the supplemental tests in social studies, science, listening, and writing require another 40 minutes each. Practice tests are also available, and these tests are strongly recommended for students who have little experience in taking tests.

Scoring options and methods are well documented in the various teacher's guides. Hand scoring is possible but very tedious, and many of the group score summaries so important to accomplishing most of the main uses are not feasible with hand scoring. The hand-scoring option seems most useful when only a few students are involved in the testing and only individual reports (needed quickly) are of interest. In addition to hand scoring by the school district and machine scoring by the test publisher, local scoring with an optical scanner is also possible and feasible. With the correct equipment, personnel, and planning, test scoring could be done very quickly at the local level with fast returns of individual and group reports. Cost savings could also be realized in this way. However, ensuring quality control can be a problem with local scoring.

A complete list of score reports and scoring options is described in Riverside

Publishing Company's 1987 *Test Resource Catalog*. A school district would be hard pressed to come up with a score-reporting request that is not included among the 26 main options (which the publisher refers to as services).

A listing and brief description of the main services follows:

1) Student criterion-referenced skills analysis
   —permits criterion-referenced interpretations of student performance on the skills measured in the battery
   —up to 85 skill scores may be reported on a typical student report
   —normative basic skill scores also appear on the report
   —areas of relative strength and weakness are highlighted

2) Pupil profile narrative
   —profiles student performance on the basic skills
   —both student and parent reports are available

3) Building criterion-referenced skills analysis
   —for each grade in each building, group performance on the skills and tests in the battery are reported along with district and national averages

4) Group narrative report
   —highlights group performance for all tests and all grades in each building and in the district via the use of graphic profiles and verbal descriptions

5) Class, building, and system summaries
   —reports various summary derived score statistics for each grade in a building (reporting each class average and the building average) and for each grade in the system (reporting each building average and the system average)

6) Group item analysis report
   —for items and skills, percent correct is provided for each class for each grade in each building, and for each grade for the district as a whole

7) Test results by class section
   —reports student performance in configurations of interest (and different from classrooms)
   —are of interest when students have different teachers for different subjects

8) Student press-on labels
   —includes the normative scores for a student requested in Service 9
   —on the right of the label is a portion of the label that can be forwarded to parents on one of several available reports for parents

9) List reports of student scores
   —are available by homeroom, in alphabetical order, or in ranked order
   —many options are available for organizing and reporting student normative scores from the test administration
   —district has some choice of normative scores to be reported

10) Frequency distributions
   —provides easy to read information about score distributions and descriptive statistics on the basic skills tests
   —report serves to highlight the diversity of skills in grades and schools of a school district

11) Pre/post reporting
   —provides pre- and posttest scores for students

—data are useful, for example, in Chapter 1 reporting or program evaluation studies

12) Individual performance profile
    —provides skills information for each student (and national norms) along with a graphical representation of the student skill performance data
    —provides GEs and percentile ranks on the basic skills
    —provides similar information to Service 1 but with the aid of graphics

13) Individual item analysis
    —main part of report provides student percent correct and national pecent correct for each test and skill, and the student's response to each item
    —normative scores on the basic skills test also are printed

14) Class/building/system diagnostic report
    —is similar to the Student Criterion-Referenced Skills Analysis but instead of one page per student, the skills for all students in a class (classes in a school, or schools in a district, depending upon the district's preference) are presented side by side.

In addition, each of the available services is cross-classified with the audience (or audiences) that would be interested in the service. Seven audiences are used in cross-classifying the services: students, parents, classroom teachers, curriculum coordinators, building or district administrators, counselors, and researchers. Overall, the set of reports appears to be clearly presented, comprehensive, and responsive to the various information needs of school districts. In fact, the score reporting capabilities are nothing short of outstanding.

National percentile norms tables at both the student and school level are available for fall, midyear, and spring administrations. Large-city school norms, Catholic school norms, international school norms, and low and high socioeconomic norms are also available. Local school system norms make up another one of the options provided by the test publisher. In addition, raw score to grade-equivalent score, grade-equivalent to percentile rank and developmental standard score, and percentile rank to stanine and normal-curve equivalent score conversion tables are available.

Those who wish can obtain an analysis of student performance on the skills measured in the ITBS. Using norm-referenced achievement tests (NRT) to provide test data at the skill level for absolute score interpretations is not without problems, however. For example, the specifications of the skills measured by the ITBS are not as complete as they are with high-quality criterion-referenced tests (CRTs; see, for example, Berk, 1984; Popham, 1978). Without more content specifications, there is also the danger that schools may do a skills match using the skills names only. This could result in erroneous matches and ultimate misuse of the CRT results obtained from the ITBS. Further, as item selection criteria for CRTs are very different from those used with NRTs, the representativeness of the ITBS items as measures of the skills of interest is unknown. With these cautions in mind, the available criterion-referenced ITBS information can be used to identify student, classroom, building, and district strengths and weaknesses. At the student level it is noted that the number of items per skill is typically low, and information on the reliability of these skill scores is not reported in the teacher's guides, though confidence bands (reflecting reliabilities) do appear in the Individual Performance Profile Reports. Users should proceed with caution until they have established the reliability and validity of these scores for themselves.

Technical Aspects

The *Manual for School Administrators* and the accompanying *Teacher's Guides* are clearly written, and the technical discussions are excellent. The authors offer rationales for important decisions that were made in the test development process (e.g., decisions about test content), they offer complete details about their actual developmental procedures (e.g., content reviews, item writing and reviews, item bias studies), and they were willing to draw attention to possible weak areas in the ITBS (e.g., content match to particular school curricula; various ITBS derived scores).

The *Manual for School Administrators* and the *Teacher's Guides* are among the best in the testing field. In fact, the manual's "Part 6—Technical and Other Considerations" should be required reading for students studying educational tests and measurements. The authors provide an eminently sensible statement about the roles of the test authors, test publishers, and potential users of the ITBS in test validation, as well as some guidelines for conducting validity investigations and interpreting the results. Ultimately, as they note, the final determination of test validity is a judgment based on a review of technical evidence compiled by the test authors, other considerations such as the norms, and factors unique to the school district, such as their informational needs and curricula.

Among the types of technical evidence reported in the administrator's manual are development steps for each subtest, test and subtest reliabilities, item statistics, item bias results, and correlations between test scores and a variety of criterion measures (e.g., school grades).

Sufficient information about test reliability has been reported to keep even the severest critic happy. Estimates of equivalent form and internal consistency reliability (along with other descriptive statistics) are reported for each form of each subtest at each level in both the fall and spring of the grades in which the forms are scheduled for administration. A number of coefficients representing the stability of test scores over the period of a year are also reported. However, the value of this information, as noted by the authors, is very limited for achievement tests; the information seems to say more about the nature of instruction than the tests themselves.

One special feature of the reliability reporting is the inclusion of information pertaining to the precision of score estimation at various ability levels. Such data are becoming available for testing programs built within an item response theory framework, but they are seldom, if ever, reported when test reporting is done within a classical measurement framework. Test reliabilities for the subtests tend to be in the .80s and .90s, though they are somewhat lower for the Level 5 and 6 subtests and for the Listening subtest at all four levels in which it is included.

Predictive validity evidence for the new forms is confined to six studies carried out on earlier editions. Fortunately, 1) the usefulness of predictive validity evidence is limited for achievement tests, and 2) the studies reported were based on large samples and were supportive of the ITBS. Among the validity results reported are those that show strong relationships between 1) fall ITBS scores for kindergarten and first-grade students and their grades at the end of the school year, and 2) ITBS scores in Grades 4, 6, 8 and high school and first-year college GPAs. Also, data addressing ceiling and floor effects, test completion rates, readability levels, relationships between achievement and ability, sex differences, and socioeconomic, sex, and cultural bias provide additional evidence for the validity of scores obtained from the ITBS.

The writing assessment seems to be one part of the ITBS especially in need of additional technical support. Questions about the reliability of scoring across judges as well as about the consistency of student performance over various writing prompts are especially interesting and need to be fully answered before the writing scores can be used with confidence. Steps were taken by the authors to standardize the administration and scoring. Interrater reliabilities are high, though reliability of writing scores over similar prompts are only modest, ranging from about .39 to .70.

In view of the long history of the ITBS and its careful development, one would expect more validity evidence to support the various recommended uses of the ITBS. Certainly content validity evidence for achievement tests is important, and this evidence is well documented by the authors and highly supportive of the ITBS. However, additional evidence to show that the ITBS scores, for example, are useful for "grouping . . . students to better provide for individual differences" and other recommended uses would be helpful. Each application of the ITBS scores is associated with a different type of inference, and therefore evidence for the validity of each recommended use should be reported in the technical documentation.

Test norming for the ITBS was carried out in the fall of 1984 and the spring of 1985. Three stratifying variables were used to classify public school districts and played a key role in the school sample selection: geographic region, school district size, and socioeconomic status of the community. The main goal was to draw a sample that could represent the national population with respect to both ability and school achievement; the use of previous years' achievement data was helpful in achieving the goal. The number of students per grade was never less than 10,000 for the fall administration and, except for kindergarten, was never less than 12,800. In total, over 126,000 participated in the fall norming. About a third of the sample also participated in the spring 1985 norming. Weights were used to adjust the actual data for minor departures from the desired national demographic statistics.

Baglin (1981) was one of the first researchers to draw attention to the possible bias in norms tables due to the non-participation of selected school districts, as well as to the prevalence of users of the test battery in the norming group. Unfortunately, information with which to address these two possible sources of bias was not reported in the technical documentation. It is true, however, that the authors did replace non-participating school districts with similar school districts that were willing to participate.

Critique

The ITBS present an outstanding example of a nationally normed achievement test battery. In every respect, these tests appear to be built upon modern measurement principles and practices. In some areas, such as the assessment of item bias and reliability and the scale development, the authors actually developed the methodologies they used, and these methodologies influence the measurement field more broadly than through the ITBS only. The concurrent availability of the Tests of Achievement and Proficiency (for Grades 9 to 12, or Levels 15 to 18) and the Cognitive Abilities Test (for Grades K to 12), along with the ITBS measures of science, social studies, listening, and writing, provides school districts with a comprehensive Grades K to 12 achievement and ability testing program.

About the only criticism this reviewer has of the ITBS concerns its packaging. Accompanying the very large number of options in test booklets, supplemental tests, levels, forms, answer sheets, score reporting, and so on are complexity in packaging and possible user confusion about what one actually needs. The extensive options obviously have been made available to meet broad user needs as well as to provide flexibility to schools in designing sensible school testing programs; therefore, perhaps the time has arrived for the authors to prepare a detailed flowchart representing the desired sequence of discussions and decisions for working through the myriad documents, options, and considerations encountered in building a school standardized achievement testing program. Possibly such a system could even be prepared for microcomputers, allowing users to work interactively.

More validity evidence addressing the authors' recommendations for possible uses of the ITBS would also be desirable. Certainly the recommended uses seem reasonable, but evidence to support them would further enhance an outstanding testing package.

References

Baglin, R.F. (1981). Does "nationally" normed really mean nationally? *Journal of Educational Measurement, 18,* 97-107.

Berk, R. A. (Ed.), (1984). *A guide to criterion-referenced test construction.* Baltimore, MD: Johns Hopkins University Press.

Popham, W. J. (1978). *Criterion-referenced measurement.* Englewood Cliffs, NJ: Prentice-Hall.

Andres Barona, Ph.D.
Assistant Professor of School Psychology, Arizona State University,
Tempe, Arizona.

KINDERGARTEN LANGUAGE SCREENING TEST

Sharon V. Gauthier and Charles L. Madison. Austin, Texas: PRO-ED.

Introduction

The Kindergarten Language Screening Test (KLST) is a language screening device designed to identify the developmental receptive and expressive verbal language skills of four- and five-year-old children entering kindergarten. The test is not intended to specify the exact nature of a language problem but rather to

> differentiate between those children who use language as an effective means of communication and those who have areas of language deficit as measured by more intensive language testing, take a minimum of time per child, and be easily administered. (Gauthier & Madison 1978, p. 1)

The KLST was developed by Sharon Gauthier, a communications disorders specialist in Ferndale, Washington, and Charles Madison, coordinator of the Communication Disorders Clinic at Washington State University. Based on the premise that defective language skills should be developed before they affect academic achievement, the authors designed the test as a general verbal screening instrument that would allow age and grade comparisons for children at the kindergarten entry level.

The KLST first was published in 1978 and a revised edition was issued in 1983. The 1983 edition does not provide information regarding the differences between the two editions. Test items were selected on the basis of developmental information obtained from a survey of the literature: selected items included aspects of both receptive and expressive language generally present in the normal kindergarten child.

The test authors report that following item selection, construct validity was established in 1973 by administering the KLST, the Utah Test of Language Development (Mecham, Jex, & Jones, 1967), and portions of the Illinois Test of Psycholinguistic Abilities (Kirk, McCarthy, & Kirk, 1968) to 41 kindergartners. Administration of the KLST to a subsequent sample of 113 kindergartners resulted in the deletion of items found to have low discrimination. Separate test-retest reliability and predictive validity studies were conducted later.

Normative data in the form of means and standard deviations are provided separately for a sample of 485 children and another sample of 141 Headstart children. No further description of the population regarding such factors as ethnicity, geographic location, or sex is provided. There is no data on the intelligence level of the standardization group nor on how it was selected. Percentile rankings by age in months also are provided, although it is not clear which sample these data are based upon.

The KLST is an individually administered test consisting of eight items and taking approximately 5 minutes to administer. Both receptive and expressive language skills are examined through the child's responses to both oral and visual stimuli. The child is asked to provide his or her name and age, identify colors and body parts, and demonstrate knowledge of number concepts. The child's ability to repeat sentences and follow a three-part command also are evaluated, and a sample of spontaneous language is obtained.

No special training is required to administer or score the KLST. The examiner's primary role is to build sufficient rapport with the child in order to obtain results that are representative of the child's ability to communicate verbally. In addition, the examiner must be sure to have the child's full attention before presenting an item.

The individual screening test form is designed simply and clearly. The child's name and identifying information appear on the front page along with a checklist to indicate whether further testing is recommended. The remainder of the four-page form lists each test item, with examiner queries and prompts printed in bold relief. No other instructions regarding test administration are provided on the form. Space is provided on the side of each page for scoring the individual items.

Practical Applications/Uses

The KLST is a screening instrument designed to compare a kindergartner's language abilities with the language abilities of similar-aged children. Both receptive and expressive aspects of language are examined. Because the KLST does not pinpoint areas of language deficit, its usefulness is limited diagnostically. Because administration is brief and easy and the test is relatively nonthreatening, either a teacher, clinician, or paraprofessional may use the instrument to determine gross-language abilities for children upon school entry. The obtained results can provide teachers with ideas for instruction and enrichment activities as well as suggest the need for further evaluation.

Because only the test portfolio, a pencil, and the test scoring booklet are required, the test may be administered in a classroom, office, or small test room. Ideally, no extraneous stimuli should be present to distract the child from attending to the tasks presented; therefore, every effort should be made to test the child in a relatively isolated setting.

The KLST appears appropriate for use with children about to enter kindergarten, currently attending kindergarten, or about to graduate from kindergarten. Comparing a child's performance at entry and exit points may give the school professional both quantitative and qualitative information regarding language development.

Instructions for test administration are presented clearly in the manual, as well as suggestions for working with children who appear to be having difficulty with test items. The test is administered by asking the child to respond to a series of verbal and visual prompts. Credit is given for a response only if the child responds correctly and verbally. The individual screening test form is easy to use.

Procedures for scoring generally are simple and clear, although scoring instructions for item 5 appear to be somewhat confusing. Test responses are recorded in a test booklet and are scored as correct (+) or incorrect (-) according to specified criteria. For example, to receive credit on the "colors" item, the child must name accurately and verbally the colors presented. A total score is obtained by tallying all correct (+) responses in the test booklet. This score then may be compared against normative data in the form of means, standard deviations, and percentile rankings for children between 48 and 83 months of age. It is suggested that a total score of 19 or below indicates the potential for later school problems and the need for further diagnostic testing.

Although they do not provide specific information as to whether items can be presented in a modified sequence, the test authors do stress that the child, rather than the presentation of the instrument, is of primary importance. Thus, this reviewer assumes that item sequence may be changed or items presented again if the clinician feels a more accurate sample of language ability may be obtained.

Technical Aspects

The content validity of the Kindergarten Language Screening Test reportedly was determined by the research on which each item was based. The work of Gesell et al. (1940) and Berry (1969) heavily influenced the types of items included in the test. In addition, several items were taken from those included in the Houston Test for Language Development (Crabtree, 1958) and the Utah Test of Language Development.

To establish construct validity, the KLST, the Utah Test of Language Development (UtLD), and three subtests of the Illinois Test of Psycholinguistic Abilities (ITPA; Kirk, McCarthy, & Kirk, 1968) were administered to 41 kindergartners (Gauthier, 1973). Although the ethnicity of this sample was reported (20 Caucasian, 21 Nez Perce Indian), no additional information on the population's characteristics was reported. Significant positive correlations of .60 and .51 were found between the KLST and UTLD and ITPA subtest sums, respectively.

In a study designed to evaluate the predictive value of low KLST scores (Gauthier, 1975), the KLST was administered to 233 kindergartners. Two and one-half years later 82% of the 30 children who had received scores of 19 or below were found to be functioning below their academic grade level and had been either retained or placed in some form of special education services. Gauthier and Madison concluded that low KLST scores are good predictors of poor receptive and expressive syntax as well as of concept development and academic achievement. Again, however, information regarding the sample's characteristics (e.g., intelligence, socioeconomic status) was not provided. In addition, it should be noted that this study did not provide information on the number of valid negatives, false negatives, or false positives for the remainder ($n=203$) of the sample. Thus, it is not known whether children who actually required further language assessment and possibly intervention were overlooked by the screening.

Research involving the KLST has not been extensive. In one study (Illerbrun,

Haines, & Greenough, 1985) that compared a number of language screening tests, the KLST was found to be less effective in correctly identifying kindergarten children with language problems than the Language Identification Screening Test for Kindergarten (Illerbrun, McLeod, Greenough, & Haines, 1984), the Bankson Language Screening Test (Bankson, 1977), the Clinical Evaluation of Language Functions—Elementary Screening Tests (Semel & Wiig, 1980), and the Fluharty Preschool Speech and Language Screening Test (Fluharty, 1978). Of these, the KLST was found to be the poorest predictor of language difficulties, with the lowest number of accurate identifications.

Test-retest reliability was examined in a study of five-year-old Headstart children (Madison & Garner, 1974). After initially screening 88 Headstart children with the KLST, 22 children from the original sample were selected randomly and retested by the same examiner within a 4-week period. A Pearson r technique of the test-retest data yielded a significant correlation of .87 and indicated an acceptable level of stability.

Critique

With its simple construction, the KLST appears to be both a quick and effective screening device that can provide useful information for the school professional. However, several issues should be considered before a final evaluation of the test is made. There is sparse information on the standardization sample and, therefore, the generalizability of the norms is questionable. For example, without specific information on such characteristics as level of intelligence, socioeconomic status, geographic region, and ethnicity, a decision regarding the comparability of test results cannot be made. Similarly, the characteristics of the populations used in the reliability and validity studies were described in scant detail. Finally, the results of at least one externally conducted study (Illerbrun et al., 1985) has indicated the KLST to misclassify more students than a number of other language screening instruments.

It appears that the KLST is not a poor screening instrument but perhaps one that needs additional refinement. It is obvious that much thought and effort went into the selection of the test items. At this point, however, it appears that more in-depth research, and possibly a renorming of the test with a better delineated sample, would be useful in more adequately establishing the test's psychometric properties. A larger, more diverse, and more clearly described sample would be beneficial. Although the paucity of research involving the KLST precludes a final evaluation of the test's usefulness, it seems that, at the present time, the KLST is not the most useful kindergarten screening instrument available.

References

Bankson, N.W. (1977). *Bankson Language Screening Test*. Baltimore: University Park Press.

Berry, M.F. (1969). *Language disorders of children*. New York: Appleton-Century-Crofts.

Crabtree, M. (1958). *The Houston Test for Language Development*. Houston, TX: Margaret Crabtree.

Fluharty, N.B. (1978). *Fluharty Preschool Speech and Screening Test*. New York: Teaching Resources.

Gauthier, S.V. (1973). *Development of a language screening device.* Master's project, Washington State University, Pullman, WA.

Gauthier, S.V. (1975). *A predictive validity study of the Kindergarten Language Screening Test.* Unpublished manuscript.

Gauthier, S.V., & Madison, C.L. (1983). *Kindergarten Language Screening Test.* Tigaard, OR: C.C. Publications.

Gesell, A.L., Halverson, H.M., Ilg, F., Thompson, H., Castner, B.M., Ames, L.B., & Amatruda, C.S. (1940). *The first five years of life.* New York: Harper.

Illerbrun, D., McLeod, J., Greenough, P., & Haines, L. (1984). *Language Identification Screening Test for Kindergarten.* Saskatoon: University of Saskatchewan, EDEXC Occasional Publications.

Illerbrun, D., Haines, L., & Greenough, P. (1985). Language Identification Screening Test for Kindergarten: A comparison with four screening and three diagnostic language tests. *Language, Speech, and Hearing Services in Schools, 16,* 280-291.

Kirk, S., McCarthy, J., & Kirk, W. (1968). *The Illinois Test of Psycholinguistic Abilities* (rev. ed.). Urbana: University of Illinois Press.

Madison, C.L., & Garner, C.K. (1974). *Test-retest reliability of the Kindergarten Language Screening Test.* Unpublished manuscript.

Mecham, M., Jex, J.L., & Jones, J.D. (1967) *Utah Test of Language Development.* Salt Lake City, UT: Communication Research Associates.

Semel, E.M., & Wiig, E.H. (1980). *Clinical Evaluation of Language Functions-Elementary Screening Tests.* Columbus, OH: Charles E. Merrill.

Rik Carl D'Amato, Ed.D.
Assistant Professor and Co-Director, School Psychology
Program, Department of Educational Psychology, Mississippi State
University, Mississippi State, Mississippi.

KNOX'S CUBE TEST

Mark Stone and Benjamin Wright. Chicago, Illinois: Stoelting Company.

Introduction

Knox's Cube Test (KCT) originally was designed as a brief index of mental ability (Knox, 1914) but more recently has been viewed as an important component of overall mental functioning (i.e., attention span and short-term memory). Indeed, the importance of attention span and short-term memory functions cannot be negated. For learning to take place, clearly some degree of attention is needed. Based on this reasoning, the test has been viewed specifically as a measure of visual short-term memory for spatial location (Dean, 1985a). Thus, the KCT would seem to have potential as a valuable instrument for evaluating an important part of the learning process.

The KCT consists of a line of four 1-inch black wooden blocks, each glued 2 inches apart on a 10.5-inch wooden strip. The examiner uses a similar companion block to tap out sequences, which the examinee attempts to imitate.

Not surprisingly, the KCT has a rich and rather robust history. The test initially was developed by Knox (1914) as a measure of mental impairment for immigrants at Ellis Island. Soon after this introduction, Pintner included the measure in A Scale of Performance Tests (Pintner & Paterson, 1923). Numerous other authors have utilized versions of the instrument to evaluate various populations, including United States army personnel (Yerkes, 1921), deaf and hard-of-hearing children (Drever & Collins, 1928), Ontario school children (Amoss, 1936), and preschool children (Goodenough, Maurer, & Van Wagener, 1940). Another version of the KCT was included in each of the revisions of the once popular Arthur Point Scale of Performance Tests (Arthur, 1925, 1943, 1947). More recently, Babcock (1965) developed a KCT version that was standardized on a sample of 3,000 persons aged 7-50. The most recent revision of the KCT has been offered by Stone and Wright (1980). Obviously, early on the measure had achieved a degree of popularity that has not been rivaled since that time. Although some of these versions used slightly different tapping sequences, Stone and Wright (1980) have argued that the differences have not effected performance and, thus, they view all the past versions as comparable and compatible with their revision.

The reviewer would like to thank Joe Khatena for his constructive comments regarding this review.

Practical Applications/Uses

The test is purported to be appropriate for all subjects because it reportedly does *not* depend on language, cultural, or numerical abilities. In fact, the test authors have claimed that the test is useful for evaluating the "mental status of deaf, nonverbal and foreign speaking children and adults" (Stone & Wright, 1980, p. 1). However, after analyzing the testing task, it is clear that subjects need to display a certain degree of fine-motor development in order to manipulate and tap the 1-inch cube. Moreover, although adequate hearing is certainly not a requirement, it would seem that subjects may benefit concomitantly from the auditory presentation. So too, subjects with adequate or better vision may profit from the use of such abilities. Indeed, it has been established fairly well that subjects often process information using a number of styles (Dean, 1985b; Hartlage & Telzrow, 1983).

The nonverbal format of the block imitation test makes it simple to administer. Two test forms, corresponding to different age groups, are available. The Junior Form (16 items) is used for children aged 2-8. The Senior Form (22 items) is suggested for individuals ranging in age from 9 through adult. The items on the Junior Form begin with a 2-tap series and end with a 6-tap series; the items on the Senior Form begin with a 3-tap series and end with an 8-tap series. The two forms contain 12 identical, overlapping items. The items on both are arranged in order of increasing complexity.

To begin the test, the examinee must complete two practice items correctly. The test ends when the examinee responds incorrectly to five successive items. If young subjects correctly answer all items on the Junior Form, it is suggested that the examiner continue testing using the Senior Form. The manual suggests repeating the two practice items until they are mastered *completely*, because directions are not repeated and no further help is provided to the examinee after the items are completed.

To administer the test, the examiner uses the companion cube to tap out the sequences recorded on the report form at the rate of one tap per second. In what seemed to be an effort to establish a high degree of administration standardization, a figure is presented to designate the right to left 1-2-3-4 suggested cube numbering/tapping system. After tapping a sequence, the cube then is deposited midway between the examinee and the row of blocks, and the examiner tells the examinee to "do what I did." The manual also specifies that pantomimes can, if needed, be used in place of oral directions. Although the manual lists testing time as "brief," Dean (1985a) has indicated that complete administration can be performed in 15 minutes or less. By way of contrast, Lezak (1983) has maintained that it can be administered in 2 to 5 minutes; however, based on the procedures involved, Dean's (1985a) claim appears more realistic. In general, administration procedures in the test manuel are written clearly and easy to understand.

It is interesting to note that although the manual indicates that the test can be useful for a wide variety of psychological and educational evaluations, it does not delineate the qualifications of the test administrator. One could surmise that training in psychology would be required if the measure was to be viewed as a test of mental ability. However, educational specialists could argue that they are trained appropriately to administer educational measures of short-term memory and attention. Indeed, it becomes a quandary to decide who can administer this instrument appropriately.

Although scoring for both forms is rather basic, scoring instructions are presented in a complex and confusing manner. In essence, the total number of correct responses are summed and can be converted to an age-in-years score. Additional information, such as "median taps," "median reverses," "median distances," and "error boundaries" are presented but seem to have little relevance from an *actuarial* perspective. In fact, it would seem that these clinical keystones of information should be viewed only as informal indicators. Although four cases are presented in the manual, they provide little direction in helping to clarify scoring procedures.

A number of difficulties arise during score interpretation. Most importantly, it is common practice to consider the subject's scores in concert with the composition of the norming sample. In other words, examiners may want to consider their subjects in light of the subjects that compose the norming group. For example, if the subject correctly responds to seven of the tapping sequences, after consulting the norms one might be tempted to propose that since the subject achieved a suggested age-in-years of six, the subject would be functioning (i.e., in tapping skills) at an average 6-year-old level. However, for two reasons, this instrument cannot be interpreted in the standard fashion. First, rather unusual circumstances, which will be discussed in detail later in this review, surround the norming of the test. Second, the use of age-in-years scores creates inherent difficulties. In fact, the significant problems associated with such a developmentally based scoring system (e.g., inaccurate performance levels, use of interpolation and extrapolation, and typological thinking) have been frequently discussed elsewhere (Salvia & Ysseldyke, 1981; Sattler, 1982). It is definitely disappointing to find that the KCT manual presents *only* age-in-years scores.

A closely related issue that further clouds interpretation is the ceiling effect that occurs for the senior form at age 18 (or with 16 correct items), which significantly restricts the interpretation of scores for adolescents and adults. This ceiling effect falls in opposition to numerous studies that have argued that development cannot be viewed as static when evaluating children and adults (e.g., D'Amato, in press; D'Amato, Gray, & Dean, in press; Dean, 1985b).

Overall, if the KCT is used as part of a complete evaluation, comparison between measures becomes problematic, as the test does not present any type of standard score equivalents (e.g., X = 100, *SD* = 15). Indeed, after using the measure, the examiner is left in an actuarial abyss that leaves interpretation uncertain.

Technical Aspects

The technical aspects of the KCT should determine whether it is to be used. Unfortunately, considerations like ease and speed of administration and ease and speed of scoring often take precedent over technical aspects, although clearly they should not (Anastasi, 1982).

As previously detailed, standardization was not completed in the usual fashion of norming the test on a predetermined group of subjects. Instead, Stone and Wright (1980) used Rasch's (1960) measurement procedures to sophisticatedly integrate and calibrate previously collected norms (e.g., Arthur, 1947; Babcock, 1965) from studies using different versions of the instrument. To this end, information one might want to consider concerning the sample (e.g., sex and racial characteristics) is not available. Also of interest is that although previous studies have covered more than a 60-year period, age norms have changed little since that time. It is necessary to accept this assumption if one is to utilize the Stone and Wright (1980) version because some of the *integrated* norms were 65 years old. Dean (1985a) has asserted that if one is to believe that the norms have not changed since that time, then *why not* utilize the older, more traditionally normed, versions (e.g., Babcock, 1965; Levinson, 1956).

No information about the tests validity and reliability is presented in the manual. This is of great concern and unfortunate because a review of the literature suggests that some studies have evaluated both the reliability and the validity of the instrument (e.g., see Lezak, 1983).

Critique

The KCT appears to be a potentially useful and easy-to-administer measure of short-term visual memory for spatial location. In fact, in view of the current focus on neuropsychological functioning (D'Amato & Dean, 1987) and the emphasis on processing styles (e.g., simultaneous-sequential; Dean, 1984; Hartlage & Telzrow, 1983), it would seem that the KCT may be useful in the evaluation of sequential, time-dependent functions (Lezak, 1983; Horan, Ashton, & Minto, 1980). From a historical perspective, Stone and Wright have done an excellent job of presenting detailed information about past revisions of the test. However, the technical information presented in the manual about the current version is sadly lacking. Sattler (1985) has argued that "it is inexcusable for both the authors and publishers to allow this potentially useful test to be published without supplying reliability and validity coefficients by age groups" (p. 794). Certainly, major norming revisions would need to be undertaken before the KCT could be recommended for general use.

References

Amoss, H. (1936). Ontario School Ability Examination. Toronto: Ryerson.

Anastasi, A. (1982). *Psychologial testing* (5th Ed.). New York: Macmillan

Arthur, G. (1925). A new point performance scale. *Journal of Applied Psychology, 9,* 390-416.

Arthur, G. (1943). *Arthur Point Scale of Performance Tests.* Chicago: Stoelting.

Arthur, G. (1947). *Arthur Point Scale of Performance Tests* (revised form II). New York: The Psychological Corporation.

Babcock, H. C. (1965). The Babcock Test of Mental Deficiency. Beverly Hills, CA: Western Psychological Services.

D'Amato, R. C. (in press). Subtyping children's learning disorders with neuropsychological, intellectual and achievement measures (Doctoral dissertation, Ball State University, 1987). *Dissertation Abstracts International.*

D'Amato, R. C., & Dean, R. S. (1987). Neuropsychology. In C. R. Reynolds and L. Mann (Eds.), *Encyclopedia of Special Education: A reference for the education of the handicapped and other exceptional children and adults* (pp. 1099-1100). New York: John Wiley & Sons.

D'Amato, R. C., Gray, J. W., & Dean, R. S. (in press). A comparison between intelligence and neuropsychological functioning. *Journal of School Psychology.*

Dean, R. S. (1984). Functional lateralization of the brain. *Journal of Special Education, 18,* 239-256.

Dean, R. S. (1985a). Review of Knox's Cube Test. In J. V. Mitchell, Jr. (Ed.), *The ninth mental measurements yearbook* (pp. 793-794). Lincoln, NE: Buros Institute of Mental Measurements.

Dean, R. S. (1985b). Perspectives on the future of neuropsychological assessment. In B. S. Plake and J. C. Witt (Eds.), *Buros Nebraska series on measurement and testing services: Future directions for testing* (pp. 203-244). Hillsdale, NJ: Erlbaum.

Drever, J., & Collins, M. (1928). *Performance tests of intelligence.* Edinburgh: Oliver and Boyd.

Goodenough, F., Maurer, K., & Van Wagener, M. (1940). *Minnesota Preschool Scale.* Circle Pines, MN: American Guidance.

Hartlage, L. C., & Telzrow, C. F. (1983). The neuropsychological basis of educational intervention. *Journal of Learning Disabilities, 16,* 521-528.

Horan, M., Ashton, R., & Minto, J. (1980). Using ECT to study hemispheric specialization for sequential processes. *British Journal of Psychiatry, 137,* 119-125.

Levinson, B. M. (1956). The Knox cube backward as a performance test of general intelligence. *Journal of Clinical Psychology, 12,* 185-187.

Knox, H. A. (1914). A scale based on the work at Ellis Island for estimating mental defect. *Journal of American Medical Association, 62,* 741-747.

Lezak, M. D. (1983). *Neuropsychological assessment* (2nd ed.). New York: Oxford.

Pintner, R., & Paterson, D. (1923). *A Scale of Performance Tests.* New York: Appleton.

Rasch, G. (1960). *Probablistic models for some intelligence and attainment tests.* Copenhagen, Denmark: Danmarks Paedogogiske Institut.

Salvia, J., & Ysseldyke, J. E. (1981). *Assessment in special and remedial education* (2nd ed.). Boston: Houghton Mifflin.

Sattler, J. M. (1982). *Assessment of children's intelligence and special abilities* (2nd ed.). Boston: Allyn & Bacon.

Sattler J. M. (1985). Review of Knox's Cube Test. In J. V. Mitchell, Jr. (Ed.), *The ninth mental measurements yearbook* (pp. 794-795). Lincoln, NE: Buros Institute of Mental Measurements.

Stone, M. H., & Wright, B. D. (1980). *Knox's Cube Test.* Chicago: Stoelting.

Yerkes, R. M. (Ed.) (1921). *Memoirs of National Academy of Sciences: Vol. 15. Psychological examining in the U. S. army.* Washington, DC: U.S. Government Printing Office.

John A. Zarske, Ed.D.
Director, Northern Arizona Psychological Services, P.C., and
Chairperson, Department of Neuropsychology, Flagstaff Medical Center.

LATERALITY PREFERENCE SCHEDULE

Raymond S. Dean, Ball State University, Muncie, Indiana.

Introduction

The Laterality Preference Schedule (LPS; Dean, 1978a)) is a relatively new multi-factorial measure of lateral preference patterns. The LPS is derived from tasks included in most neuropsychological test batteries that allow comparison of the patient's right-side performance with that of their left side. The LPS differs from standard clinical comparisons of the two sides of the body (i.e., strength of grip, hand preference, finger recognition) by allowing the examiner to assess the degree of lateralization for various functions, including the hands, arm, eyes, ears, and feet. In this reviewer's opinion, the LPS, in its current form, represents the most thorough and broad-based self-report measure of laterality preference available to clinical neuropsychologists and other neuroscientists. It has become increasingly popular among researchers interested in learning problems, brain damage, and lateralization of function.

The author of the LPS, Raymond S. Dean, Ph.D., has been interested in the psychometric properties and development of the LPS for most of his professional career. He has written many articles dealing with the topic of cerebral lateralization. Although the LPS is used in a number of clinical settings, Dr. Dean seeks no financial gain from the measure. His primary interest relates to his research on cortical lateralization of functions. In addition to numerous studies employing the LPS, he currently is developing a technical manual to accompany the test.

The LPS was normed initially on a sample of 125 males and 131 females with a mean chronological age of 23.4 and 22.4 years, respectively. Additionally, Dr. Dean and others (Dean, Schwartz, & Smith, 1981; Oldfield, 1971) have published numerous studies regarding the downward extension of the test to much younger age levels. Currently, ample research support exists to warrant using the LPS with individuals ranging in age from early childhood (age 8) to late adulthood. Because the test requires a minimum level of reading proficiency (approximately fifth grade), one would anticipate that the items could be read to young patients.

According to the LPS literature provided by Dr. Dean, two forms of the test have been developed, a 33-item, self-report form (used in the early phases of test development) and a 49-item self-report inventory. The current form, comprised of 49 items in a self-report format, queries the respondent's laterality preferences for a number of tasks involving the hands, legs, feet, ears, and eyes. For each task, the subject indicates his or her laterality preference on a 5-point Likert-type scale. Factor analytic studies performed by Dr. Dean have isolated six salient dimensions of the measure: 1) General Laterality, 2) Visually Guided Activity, 3) Visual, 4) Auditory, 5) Strength, and 6) Foot Use.

The Laterality Preference Schedule consists of a 5-page test form. Additionally, the author has reported that a scoring template and a scoring-interpretive manual will be available for test users. The record booklet contains clearly written instructions and is easy to hand score. The patient responds directly on the record booklet.

The examiner is provided with clearly labeled boxes for totaling subtest scores. The form could be improved by providing a summary table wherein scores for all subtests could be recorded, rather than locating them on separate pages.

The historical antecedents for the development of the LPS stem from early 19th-century studies that began to link complex psychological functions to specific areas of the brain. Although efforts in the localization of specific cerebral functions to structures of the brain have not been very productive, over the last century and particularly over the last several decades, basic principles of brain-behavioral organization have been delineated (Dean, 1985a). It generally is established now that a patient's early medical and developmental history, as well as individual differences in brain chemistry and structure, make the entire enterprise of specific structural localization of functions a questionable endeavor. Thus, although differences in cortical processing between the two hemispheres is acknowledged, highly specified localization of functional activities is considered less likely.

In reviewing the literature on cerebral lateralization, one generally comes to the conclusion that the left hemisphere of the brain has been associated more intimately with the processing of speech, language, and calculation (Sperry, 1969; Reitan, 1955) than the right. It has been hypothesized that the left hemisphere is more equipped to process information that requires an analytical, logical, or sequential approach. Conversely, the right hemisphere has been linked more closely to visual spatial processes, and recent research indicates that the right hemisphere more efficiently processes tasks requiring holistic or simultaneous processing of nonverbal gestalten and complex transformations of visual patterns (Dean, 1985b; Milner, 1962).

The performance of unimanual activities on one side of the body is served by the contralateral brain hemisphere. Consistent with the now outdated idea of cerebral dominance, it long has been hypothesized that lateral preference may be a behavioral expression of the left hemisphere's degree of functional specialization for language and other functions. As early as 1937, Orton hypothesized a corollary to the notion of hemisphere dominance by hypothesizing that anomalous preference patterns (e.g., failure to establish proficiency with at least one side of the body) may underlie functional disorders. The relationship between atypical patterns of lateral preference and cortical functioning remains one of the most studied and controversial issues in the neurosciences (Dean, 1985a). Of course, the implicit assumption is that observable patterns of preference would reflect functional lateralization of the cortical hemispheres. As Dean has pointed out in numerous publications, most research regarding this topic has concentrated on hand preference (Dean, 1983).

Dr. Dean has expressed concern that the focus on hand preference has developed both because of its deceptive ease in assessment and because of various reports of a higher incidence rate of mixed hand preference for individuals with a number of expressive and receptive language disorders (Orton, 1937). The catalyst for the development of the LPS derives from the inconsistent findings that have been demonstrated in the study of lateral preference. Dean has criticized the literature for using simple hand preference as the predominant and sometimes only measure of laterality. Furthermore, he has argued that some of the confusion in past research on lateral preference may be related to the specific index of preference employed (Dean, 1982). Additionally, recent research demonstrates that the relationship between hand preference and cerebral lateralization is less than robust (Dean, 1979, 1982). Although it generally is agreed that in normal subjects language is served by the left hemisphere and the right side of the body is preferred for peripheral activities, a corollary notion suggests that both language and peripheral preference are continuous variables that can be lateralized to varying degrees (Whittaker & Ojemann, 1977). As such, though a patient's preference for peripheral activities may reflect aspects of cortical organization, the relationship is not a simplistic one and has led Dean to express concerns that early theoretical notions that offer hand preference alone as the definitive indicator of cerebal organization are outdated. In addition, there is evidence that discrepancies of eye/hand (Dunlop, Dunlop, & Fenelon, 1973) and ear/hand preference (Bryden, 1978) actually may be more sensitive measures of underlying cerebral confusion. Because studies tend to suggest that cerebral confusion or competition between the hemispheres (for execution of various functions) underlies many forms of learning disabilities and language disorders, the investigation of degree of lateralization across various multiple bodily functions becomes an obviously important issue. In this reviewer's opinion, the LPS represents a state of the art measure that takes into account these complex research and theoretical issues in the study and assessment of cerebral dominance and lateralization of function. The reference section of this review contains readings that provide detailed discussions of the research and clinical issues leading to the development of the LPS.

Practical Applications/Uses

Obviously, the Laterality Preference Schedule has been used primarily in research settings. Although the test has obvious theoretical and practical applications for the neuroscientist, clinical neuropsychologists in private practice will find the test valuable as a measure of degree of lateralization of function across various modalities, which would be important for diagnosing patients referred due to brain impairment, language disorders, or suspected learning disabilities. From a clinical point of view, the psychologist may administer the LPS in order to compare patient ratings against standard clinical procedures employed in the assessment of laterality preference, such as the Aphasia Screening Test (Reitan, 1984) and the Sensory Perceptual Examination (Reitan, 1984).

In its present form, the LPS could be used with most psychiatric and rehabilitation inpatient and outpatient referrals, as well as with patients ranging in age from early childhood to late adulthood who may be referred to the private psychological practitioner. By having the examiner simply read the items and ask the patient for a verbal or motor response, the test can be adapted easily for brain-damaged subjects who may be unable to read.

A doctoral level education is not required of the test administrator, but considering the lack of interpretive guidelines, an experienced clinician should be consulted regarding interpretation of the results. When using the LPS for most standard clinical purposes involving adolescents and adults, the test could be completed by the patient and scored within 10-15 minutes.

The LPS instructions ask the respondent to use a 5-point Likert-type scale (LA, left always; LM, left mostly; E, left and right equally; RM, right mostly; and RA, right always) to rank the degree to which he or she would use one side of the body or the other to accomplish the task presented in each of the 49 statements.

The respondent circles his or her choice for each of the items. The 49 items on the test form are presented according to the following categories: Visual Activities, Auditory Activities, Foot Use, Strength, General Laterality, and Visually Guided Activity. An additional 10 items that focus on Maternal and Paternal laterality preferences are included. The complete test, then, is comprised of 59 items. Scoring is a simple additive operative whereby, for each category, the examiner sums the frequency of circled responses in each of the five rating categories. Because the technical manual currently is only under development, no specific scoring procedures or interpretive guidelines are offered. However, the experienced psychologist or neuroscientist should have little difficulty discerning the patient's relative laterality preferences across the various bodily functions measured.

As indicated previously, interpretation is not based upon objective scores. The clinical acumen and judgment of the psychologist are the primary sources of interpretation at this time, underscoring the need for a technical manual. A basic training in the neurosciences, neuropsychology, and physiological psychology would represent an appropriate level of sophistication and training required for the examiner to interpret the test adequately and properly in its present form.

Technical Aspects

Dr. Dean provided this reviewer with several articles discussing the multiple validity and reliability studies conducted on the Laterality Preference Schedule and the use of the test with various populations. Studies by Dean (1978b, 1982) establish that the LPS is a reliable estimate of children's lateral preference patterns and has satisfactory predictive validity of actual manual patterns of lateral preference. In one study (Dean, 1978b), 50 children (25 males, 25 females) with a mean age of 10.9 years (SD = 1.57) were administered the LPS on two occasions, 4 weeks apart. Standard scoring yielded data to be correlated with scores from the first administration. The correlation of scores of the scale and those after a 4-week delay was .91 ($p < .001$). Additionally, scores on the scale were correlated with clinicians' ratings of actual, observed manual performance, yielding an r of .83 ($p < .001$). This value was not significantly different ($p > .05$) from the .89 reported by Dean (1978b) in the standardization sample of 200 undergraduates. The LPS appears to be a fairly straightforward measure of an individual's degree of lateral preference. Factor analytic studies that rendered the aforementioned six-factor solutions that form the test's categories further attest to the validity of the constructs under study (Dean, Schwartz, & Smith, 1981).

Among the reliability information provided to this reviewer was a table of reliability coefficients based upon various studies undertaken by Dr. Dean and his colleagues. These studies suggest reliability coefficients for adults and children ranging from .62 to .98. Reliability studies were conducted on relatively large samples of children (n = 579), although the range was restricted to children in Grades 4-6. General alpha coefficients for this sample fell in the range of .97. The test-retest stability for children, reported at .91, was discerned from a sample of 50 male and 50 females (mean age = 10.9 years) after a 4-week delay. For adults, a standardization sample of 1,000 males and females with a mean age of 20.1 years yielded a general alpha coefficient of .98. A 100-subject sample of 50 males and 50 females retested after a 1-week delay resulted in a test-retest stability coefficient of .88. Such reliability coefficients suggest that the LPS is a consistent measure across time.

Of course, the author will improve the technical information regarding the LPS by providing a technical manual that summarizes, in one location, the numerous studies that have been conducted regarding the test's reliability for different populations and factor analytic studies addressing validity.

Critique

The LPS represents a broad-based, multivariate measure of laterality preference. In this reviewer's opinion, Dr. Dean has contributed significantly to the field of neuropsychological assessment by providing a self-report instrument that is sensitive to the complex research and theoretical issues underlying the assessment of cerebral lateralization and peripheral preference patterns. The test shows much promise as a research tool as well as a clinical asessment device for psychologists treating patients who experience learning disabilities, language disorders, and brain impairment.

References

Bryden, M.P. (1970). Laterality effects in dichotic listening: relations with handedness and reading ability in children. *Neuropsychologia, 8*, 443-450.

Dean, R.S. (1979, September). *Lateral preference in reading comprehension.* Paper presented at the annual meeting of the American Psychological Association, New York.

Dean, R.S. (1978a). *Laterality Preference Schedule.* Tempe, AZ: Arizona State University.

Dean, R.S. (1978b). Reliability and predictive validity of the Dean Laterality Preference Schedule with preadolescents. *Perceptual and Motor Skills, 47*, 1345-1346.

Dean, R.S. (1982). Assessing patterns of lateral preference. *Journal of Clinical Neuropsychology, 4*, 124-128.

Dean, R.S. (1983, February). *Dual processing of prose and cerebral laterality.* Paper presented at the annual meeting of the International Neuropsychological Society, Mexico City, Mexico.

Dean, R.S. (1985a). Foundation and rationale for neuropsychological bases of individual differences. In L. C. Hartledge & C. F. Telzrow (Eds.), *The neuropsychology of individual differences: A developmental prospective.* New York: Plenum.

Dean, R.S. (1985b). Neuropsychological assessment. In J.D. Cavenar, R. Michaels, H.K.H. Brodie, A.M. Cooper, S.B. Guze, L.L. Judd, G.L. Klerman, & A.J. Solnit (Eds.), *Psychiatry.* Philadelphia: Lippincott.

Dean, R. S., Schwartz, N. H., & Smith, L. S. (1981). Lateral preference patterns as a discrimination of learning difficulties. *Journal of Consulting and Clinical Psychology, 49*(2), 1227-1235.

Dunlop, D.B., Dunlop, P., & Fenelon, B. (1973). Vision laterality analysis in children with reading disabilities: The results of new techniques of examination. *Cortex, 9*, 227-237.

Milner, B. (1962). Laterality effects in audition. In V.B. Mountcastle (Ed.), *Interhemispheric relations in cerebral dominance*. Baltimore: Johns Hopkins University Press.

Oldfield, R. C. (1974). The assessment and analysis of handedness: The Edinburgh Inventory. *Neuropsychologia, 9*, 97-113.

Orton, S.T. (1937). Specific reading disability—strephosymbolia. *Journal of American Medical Association, 90*, 1095-1099.

Reitan, R.M. (1955). Certain differential effects of left and right cerebral lesions in human adults. *Journal of Comparative and Physiological Psychology, 48*, 474-477.

Reitan, R. M. (1984). *Aphasia and sensori-perceptual deficits in adults*. Tucson, AZ: Neuropsychology Press.

Sperry, R.W. (1969). A modified concept of consciousness. *Psychological Review, 76*, 532-536.

Whitaker, H.A., and Ojemann, G.A. (1977). Lateralization of higher cortical functions: A critique. In S.J. Dimond and D.A. Blizard (Eds.), *Evolution and lateralization of the brain*. New York: New York Academic Sciences.

Joyce A. Eckart, Ed.D.
Assistant Professor of Education, Department of Curriculum,
Instruction and Leadership, Oakland University, Rochester, Michigan.

LEADERSHIP SKILLS INVENTORY
Frances A. Karnes and Jane C. Chauvin. East Aurora, New York: D.O.K. Publishers.

Introduction

The Leadership Skills Inventory (LSI) is a 125-item, self-administered, Likert-type assessment designed to assist individuals at the upper-elementary, secondary, and post-secondary levels in analyzing the strength of their leadership skills. The authors of the LSI are Frances A. Karnes, Ph.D., and Jane C. Chauvin, Ph.D. Dr. Karnes is the director of the Center for Gifted Studies and a professor of special education at the University of Southern Mississippi, Hattiesburg. She is the past president of The Association for the Gifted and has co-authored numerous journal articles and four books about gifted education. Dr. Chauvin is an associate professor in the Department of Education at Loyola University in New Orleans. She received her doctorate at the University of Southern Mississippi in 1982.

Karnes and Chauvin were influenced strongly by R. M. Stodgill, who attempted to factor analyze leadership traits. After a thorough review of the literature (Karnes & Chauvin, 1985, 1986), the test authors determined that the skills necessary to the development of leaders fell into nine categories: 1) fundamentals of leadership, 2) written communication, 3) oral communication, 4) group dynamics, 5) problem-solving, 6) personal development, 7) values clarification, 8) decision-making, and 9) planning. Item statements were written for each of the categories. No information is provided about the number of items that made up the original item pool. Two panels then were asked to review the items for content validity. One panel consisted of adults working in developing leadership in youth, and the other panel drew its members from elementary and secondary students in leadership roles in scouting, gifted education programs, or student councils. The suggestions of the panel members were incorporated into the final inventory.

The Leadership Skills Inventory, published in 1985, is the first part of a three-part Leadership Skills Development Program (Karnes & Chauvin, 1986) designed to assess and develop the leadership potential of upper-elementary and secondary school students. Only one form of the inventory has been developed.

Part II of the program is comprised of activities designed to practice leadership skills. Karnes and Chauvin have produced a *Leadership Skills Activities Handbook*. For each test statement in the Leadership Skills Inventory, there is an activity described in the handbook. By working through the suggested activities, one might grow in strength or frequency of performance in any skill area. A list of

303

materials needed for each activity is also provided. The appendix includes supplemental activities and resources that students and group facilitators could use for leadership development. This reviewer found the activities well matched to the assessment statement performances.

Part III represents the authors suggestions for a structured *Plan for Leadership*. The components of this plan include reflective thinking time to summarize what has been learned about leadership and formulation of a scheme to incorporate this knowledge into daily actions. Strategies for practice include setting up goals, objectives, and time lines (Karnes & Chauvin, 1986).

The LSI is well organized and the 8½" × 11" test booklet is attractive, with wide margins, statements arranged uniformly at the left side of each page, clear and readable print, boldface subskills titles, and heavy paper stock. Each examinee receives an 11-page booklet with space on the cover for entering his or her name, grade, age, sex, and school. There is also a place to mark whether the instrument is being used for pre- or postassessment. The purpose of the LSI is stated in nonthreatening terms and directions are explained in terms appropriate for upperelementary, secondary, or postsecondary students.

The LSI is divided into nine categories consisting of clusters of statements focused on traits commonly possessed by leaders. The administration manual offers the following examples of the skills surveyed by the nine categories:

1) *Fundamentals of Leadership* (FL; 9 items)—defining terms and identifying various leadership roles;

2) *Written Communication Skills* (WCS; 12 items)—outlining, writing a speech, and research reports;

3) *Speech Communication Skills* (SCS; 14 items)—defining one's viewpoint on issues, delivering a speech, and offering constructive criticism;

4) *Value Clarification* (VC; 17 items)—understanding the importance of free choice, identifying things that one values and prizes, and affirming one's choices;

5) *Decision Making Skills* (DMS; 10 items)—gathering facts, analyzing the consequence of certain decisions, and reaching logical conclusions;

6) *Group Dynamic Skills* (GDS; 19 items)—serving as a group facilitator, effecting compromise, and achieving consensus;

7) *Problem Solving Skills* (PSS; 6 items)—identifying problems, revising strategies for problem solving, and accepting unpopular decisions;

8) *Personal Development Skills* (PDS; 21 items)—self-confidence, sensitivity, and personal grooming;

9) *Planning Skills* (PS; 17 items)—setting goals, developing time lines, and formulating evaluation strategies (Karnes & Chauvin, 1985, p. 7).

Each item, for the most part, is written in language that affirms the particular skill to be assessed by the item. The examinee, using the 4-point Likert-type scale, circles in the booklet the response that corresponds to the level to which he or she possesses the skill. The examinee is responsible for totaling the scores for each subskill area and entering them on the profile sheet. Although the examiner does not appear to play a specific role in test administration, the manual suggests that the items can be read to examinees and responses recorded by the examiner. The four possible item responses are "almost always" (3 points), "on many occasions" (2 points), "once in a while" (1 point), and "almost never" (0 points).

Practical Applications/Uses

The LSI was designed to help examinees analyze the strengths and weaknesses of their leadership characteristics. As mentioned previously, the inventory currently is being used as the first step in a Leadership Skills Development Program (Karnes & Chauvin, 1986). Those who train youth for leadership roles both in the school and in extracurricular settings could use the LSI and the accompanying activities manual to plan for growth and development of leadership skills. The inventory is appropriate for upper-elementary and secondary students and might be useful for some postsecondary groups. With little modification, the inventory can be administered to the visually or physically handicapped and the hearing impaired.

The LSI is written clearly, the reading level is appropriate for the intended groups, and the instructions are easy to follow. The inventory is quite flexible. One could use each of the nine skill categories independently and offer workshops for each of the skills or administer the entire inventory at one sitting and use the resulting profile to select growth-promoting exercises. The LSI can be given in a group setting or individually. Although it is not timed, the instrument takes approximately 45 minutes to complete. The role of the examiner for a normal group is minimal; however, younger students might need to be reassured that there are no correct answers on a self-report instrument.

The last page of the inventory booklet contains a profile. The examinee locates his or her raw score for each of the nine categories on the grid, circles it, and connects the circles with straight lines. T-scores are reported for the norm group along the margins of the profile. The result of this charting is a graphic representation of the examinee's standing with respect to others from the norm group. Although the scoring procedure is straightforward, the clarity of score interpretation appears to depend on the examinees' knowledge of T-score concepts or on the explanation of the score interpreter. This is especially true for younger examinees.

Technical Aspects

The manual reports the procedure used for norming the LSI. A total of 452 students in eight samples from seven states—California, Illinois, Kansas, Louisiana (two groups), Massachusetts, Mississippi, and Nebraska—comprise the population. However, no criteria for selection of these geographic locations is included. Gender and age means and ranges are reported for each of the groups. Of the 452 students in the group, 422 reported gender (207 male and 215 female). Age ranges from 9 to 18 with a mean age of 14.6 years was reported for the group. In reporting ages within the sample, it should be noted that only one sample contained subjects in the 9-11-year-old range, and the number of those students was not reported. Other than gender and age, consistent information is not included on the normalizing group.

Means and standard deviations for each of the nine categories are reported on each of the eight samples. However, because more information has not been given about these groups, statistical comparisons are inappropriate.

Little is reported on the validity of the instrument. The authors state that the suggestions of two panels of youths in leadership positions and professionals who train youth were incorporated into the instrument.

Split-half and Spearman-Brown coefficients of reliability are reported for each category. They are all above .78, indicating that the instrument appears to be quite consistent. Kuder-Richardson coefficients for internal consistency of items is above .62 for each of the categories reporting.

One of the two samples in Louisiana reported test-retest coefficients of stability after 4 weeks. These coefficients indicate that self-reporting on Speech Communication remained the most stable (.62) and that Fundamentals of Leadership scores proved the least stable (.30).

Critique

The Leadership Skills Inventory (LSI) contains several flaws. In general, no reference has been made to the growing body of leadership literature from business and industry. Moreover, no subjects in the norming population were from business and industry or from postsecondary units, yet the instrument is reported to be appropriate for personnel development specialists in business and industry and persons involved in leadership in training and management. Rather than declare the instrument useful to personnel development specialists, it would be more effective to limit it to upper-elementary and secondary school students and to youth organizations. However, the instrument could become stronger if several of the concepts of effective leadership used in industry could be adopted into the item pool. Also, one must be aware that the LSI is a self-report inventory rather than a performance-based assessment. It should not be used to select individuals for leadership positions. Possessing knowledge of the components of leadership and practicing effective leadership techniques are two completely different things. Moreover, results of self-report devices can be misleading, and the objectivity of such instruments cannot be insured.

A more specific weakness of the LSI is its statistical analysis. First, a better validation process should be established. Second, if the results of the instrument are to be generalized to upper-elementary, secondary, and postsecondary subjects, proper sampling must be insured. Third, the selection criteria for subjects to be used in the normalization procedure must be stated. Fourth, there are too few items in several of the categories to even weakly represent the domains of leadership knowledge assessed. Not only might this underrepresentation of items lead to inaccurate conclusions, but the use of split-half reliability with a 6-item section (e.g., Problem Solving Skills) seems improper.

Within the instrument itself, the Fundamentals of Leadership section should be lengthened and strengthened. Nine items are too few to begin to assess knowledge in the domain of leadership, which is the focus of the inventory. More items should be added to better represent the construct of leadership. The same is true of the 6-item Problem Solving Skills section.

In conclusion, the Leadership Skills Inventory contains several weaknesses in statistical norms that cannot be overlooked. However, an examination of the literature by this reviewer has not found another instrument that serves a similar purpose. If the LSI is used as a guide in selecting activities (Karnes & Chauvin, 1985) for leadership growth, and if the scores are not explained in terms of norms, its use becomes more suitable. If the instrument is used as part of a Leadership Development Program, it can be useful for promoting the development of leadership characteristics.

References

Karnes, F.A., & Chauvin, J.C. (1985). *Leadership Skills Inventory: Administration manual and manual of leadership activities.* East Aurora, NY: D.O.K. Publishers.

Karnes, F.A., & Chauvin, J.C. (1986). The leadership skills: Fostering the forgotten dimensions of giftedness. *G/C/T, 9*(3), 22-23.

NOTE: B. A. Kerr and S. W. Lee also have reviewed this instrument in an update of Mitchell's *Ninth Mental Measurements Yearbook,* accessible on-line via BRS (#1012-129) at the time of this writing.

Robert J. Drummond, Ed.D.
Program Leader, Counselor Education, and Interim Chairperson,
Division of Educational Services and Research, University of North
Florida, Jacksonville, Florida.

LEARNING STYLE INVENTORY

Rita Dunn, Kenneth Dunn, and Gary E. Price. Lawrence,
Kansas: Price Systems, Inc.

Introduction

The Learning Style Inventory (LSI; Dunn, Dunn, & Price, 1987) is a 104-item paper-and-pencil Likert-type instrument designed to assess the learning styles of students in Grades 3-12 by identifying the conditions under which they prefer to learn. The inventory measures preferences in 22 different areas and is based on the premise that each individual's learning style is founded on a complex set of reactions to varied stimuli, feelings, and previously established patterns that tend to be repeated in learning contexts (Dunn, Dunn, & Price, 1987, p. 5).

Rita Dunn, Ed.D., is director of the Center for the Study of Learning and Teaching Styles at St. John's University in Jamaica, New York. Dr. Dunn's work in developing a graduate teacher education program that focused on helping underachievers perform better in reading and mathematics, conducted through the auspices of the State Education Department in New York, provided the impetus for the inventory's development. In the ensuing years, she identified 21 characteristics that appeared to influence how individuals learn. The Center has conducted numerous research studies on learning style, and its investigators have received numerous awards. Dr. Kenneth Dunn is Professor and Coordinator of Administration, as well as Supervisor, Department of Educational and Community Programs, at Queens College of the City University of New York. He has written extensively on translating the findings of individual styles into schools and industrial management strategies and has applied research on individual styles in both educational and industrial settings. Dr. Gary Price is a professor in the department of counseling psychology and a counselor in the university counseling center at the University of Kansas. He has written about as well as given workshops on individual learning styles. The authors of the LSI have published and presented extensively on the topic and are recognized nationally and internationally as experts in the field of learning styles.

The LSI, first published in 1975, was developed from a content and item analysis of the pool of 223 items on the Learning Style Questionnaire, originally published by Dunn and Dunn in 1972. Factor analytic studies conducted on the 1975 version led to a revision of some of the scales, resulting in the publication of the 1978 instrument. The 1984, 1985, and 1986 versions eliminated items that were confusing, could be interpreted in different ways, or were not clear in their assessment of the defined areas (Dunn, Dunn, & Price, 1987, p. 29). In addition, the administration format was simplified to include the questions printed on the answer sheet. The response format for the form for Grades 5-12 was changed to include a 5-point Likert scale, and for Grades 3-4, a 3-point Likert scale. These revisions led to increased reliability and a sharpening of the factors measured. The Learning Style Inventory—Primary Version is available for Grades 1-2 (Perrin, 1982), and the Productivity Environmental Preference Survey (1981) is available for adults. Research editions have been translated into Spanish, French, and Hindu.

There are two forms of the LSI. The version for third- and fourth-graders consists of 104 items printed in blue on the front and back of the answer sheet. Using a 3-point scale, students mark whether they disagree, are uncertain, or agree with each statement presented on the answer sheet.

The version for Grades 5-12 contains 104 items printed in red. The questions are of this type:

I like to sit in a recliner when I study.
I like to eat a snack as I study.

Students indicate on a 5-point Likert scale the extent to which they would agree or disagree with each statement if they had to learn something new or difficult.

The 22 scales included on the LSI are 1) Noise Level—Quiet or Sound, 2) Light—Low or Bright, 3) Temperature—Cool or Warm, 4) Design—Informal or Formal, 5) Unmotivated/Motivated, 6) Impersistent/Persistent, 7) Irresponsible/Responsible, 8) Structure, 9) Learning Alone/Peer Oriented Learner, 10) Authority Figures Present, 11) Prefers Learning in Several Ways, 12) Auditory Preferences, 13) Visual Preferences, 14) Tactile Preferences, 15) Kinesthetic Preferences, 16) Requires Intake, 17) Functions Best in Evening/Morning, 18) Functions Best in Late Morning, 19) Functions Best in Afternoon, 20) Mobility, 21) Parent Figure Motivated, and 22) Teacher Motivated.

The LSI yields an individual profile that plots the individual's scores in one of three divisions: 1) low (below -1 SD), 2) high (above +1 SD), and 3) average (between -1 and +1 SD). The LSI also can provide group summaries of individuals with standard scores of 60 or higher and standard scores of 40 or lower in any of the areas. The scoring service also provides a subscale summary that indicates the number and percentage of the total group falling within the high or low categories.

Practical Applications/Uses

The LSI was designed as the "first step toward identifying the conditions under which an individual is most likely to learn, remember, and achieve." (Dunn, Dunn, & Price, 1987, p. 5). The information from the test can help teachers, counselors, or psychologists 1) identify how students prefer to learn, 2) describe students' preferred learning style, 3) provide indicators of how students learn best, 4) yield information for developing strategies for instructional and environmental alternatives that complement students' revealed learning style, and 5) identify strategies for facilitating students' involvement in their unique learning prescriptions. The LSI would be useful to teachers, counselors, and psychologists who are working with individuals for which learning style information might be useful. The LSI is designed to provide information about the environmental, emotional, sociological, and physical preferences of students in Grades 3-12 and how these preferences relate to students' functioning, learning, concentration, and performance during educational activities (Dunn, Dunn, & Price, 1987, p. 6). The results can be used by teachers to match selected learning style characteristics with instructional methods. The inventory does not measure the underlying psychological dynamics of students' behavior, nor does it measure the cognitive skills related to learning.

The LSI has been used widely in both educational and research settings. It has been administered to normal, learning disabled, and gifted students. It has been used for placement purposes and for assigning students to counseling groups on the basis of the students' learning styles. Unless administered orally, the test is not appropriate for students with reading problems.

Procedures for administering the LSI are discussed in the manual. The LSI may be self-administered, but it also can be administered "in writing, by computer, on tape, orally, or through a combination of these" (Dunn, Dunn, & Price, 1987, p. 10). The examiner can administer the test to an individual or to large groups. The examiner's primary role is to monitor the testing and answer any questions students may have about the procedure. The examiner should stress that examinees should "give immediate reactions to each question on a feeling basis" (Dunn, Dunn, & Price, 1987, p. 10). The items on the test are stated directly and clearly, but the authors do not provide any information about the reading level required for the two forms. A computerized self-administered form is available. No audio- or videotape of the test is available from the publisher. Most students complete the LSI in 20-30 minutes.

A computer scoring program can be purchased from the publisher, from whom computerized scoring and interpretative services are available as well. No instructions for handscoring are given in the 1987 manual. A research report (Price, Dunn, & Dunn, 1977) does present the item composition of the scales. The scoring service provides the user with raw scores for each of the 22 areas, standard scores with a mean of 50 and a standard deviation of 10, and a graph of the relative location of each person's standard scores in each of the 22 areas. The standard scores are based upon the scores of more than 500,000 students who have taken the LSI. With the computerized version, students can receive immediate results, either on the screen or from the printer, of their test profile.

The test authors present interpretation suggestions for each of the 22 dimensions on the LSI for students with standard scores of 40 or less or 60 or above. They also guide the user to be alert to certain types of relationships between such dimensions as motivation and persistence; motivation and responsibility; motivation and structure; and learning alone, with peers, with authorities, or in several ways; and auditory, visual, tactile, and kinesthetic preferences. Part VII of the LSI manual presents suggested guidelines for matching selected learning style characteristics with instructional methods. The suggested procedures are based upon a wide spectrum of research focusing on these dimensions. The user must be familiar with the conceptual model of learning styles defined by the authors as well as have a basic background in measurement theory and practice. Workshops are available to train individuals in the proper use and interpretation of the results.

Technical Aspects

The primary effort of the test authors has been to establish the construct validity of the LSI through the use of factor analytic analysis. The validity of the items was studied extensively and item revisions were made. The intercorrelations between the 22 scales are not presented in the 1987 manual. Concurrent validation studies have been reported indicating that students in various categories, such as learning disabled, normal, and culturally disadvantaged, differ significantly on a number of the variables measured by the LSI. Studies have shown that students performing in their preferred learning styles or environments score higher than students who are not. Studies comparing the scales on the LSI with achievement, attitude, and personality measures also have been conducted. The 1987 manual, however, does not report specific validity coefficients.

The LSI reports (Dunn, Dunn, & Price, 1987, pp. 99-101) Hoyt's reliability coefficients for the 22 scales based upon a sample of 1982 students in Grades 5-12. The coefficients range from a low of .40 on Functions Best in Late Morning to a high of .84 on Learning Alone/Peer Oriented Learner and Intake. The median coefficient was .68. Sixteen of the 22 coefficients were .60 or higher. The coefficients reported for Grades 3-4 based upon 770 students using Hoyt's method ranged from .35 on Teacher Motivated to .84 on Intake. Fourteen of the 22 scales had coefficients of .60 or higher. The test-retest reliabilities also were presented for the 22 scales based upon 100 students and ranged from .00 on Impersistent/Persistent to .74 on Visual Preferences. Studies by Virostko (1983) and Copenhaver (1979-1980) provide reliability information. The authors report the standard error of measurement for each of the scales.

Critique

The LSI has been one of the most widely used instruments both in research and in practice. There are several reasons for its widespread popularity and use. First, a Learning Styles Network (Professor Rita Dunn, Center for the Study of Learning and Teaching Styles, St. John's University, Grand Central and Utopia Parkway, Jamaica, New York 11439), co-sponsored by the National Association of Secondary School Principals and St. John's University, supports a Learning Styles Network Hotline (718/990-6161, Ext. 6412) and publishes newsletters and annotated bibliographies. Second, applied resource books and materials to help teachers translate information on learning styles into instructional strategies are available (Dunn & Dunn, 1978; Carbo, Dunn, & Dunn, 1986). Third, the LSI has face validity. It has been based upon observations and research.

Although the authors have worked to improve the validity and reliability of the LSI since it was first published in 1975, there are still areas in which they could improve. They fail to report in the manual much of the information that the *Standards for Educational and Psychological Testing* (American Educational Research Association, American Psychological Association, National Council on Measurement in Education, 1985) call for and which is needed in order to make a comprehensive evaluation of the LSI. For example, the 1987 manual does not present an adequate description of the norming group, scoring procedures, items included on each scale, intercorrelations of the scales, means, standard deviations for each scale by grade and by sex, and standard errors of measurement. The section on reliability and validity focuses on reliability. No evidence of the criterion-referenced or construct validity of the test is given. The reliabilities reported were based upon a heterogeneous grouping, Grades 3-4, and Grades 5-12; no reliability information is provided for each grade. The factor analytic studies used to guide the revisions of the scale are mentioned, but data are not provided in the manual.

The LSI is user friendly. Computer scoring services and a microcomputer version are available. For scoring, users, in general, would have to rely on the scoring service provided for the publisher or buy the microcomputer version, which provides for administration, scoring, and interpretation of the scale. It would be useful to most practitioners if scoring keys and directions for hand scoring were available, although the use of a 3- or 5-point Likert scale might make hand scoring complicated.

The LSI presents one conceptual model of learning styles. Although there are an increasing number of competing models and systems being published, the LSI still remains a classic in this area. This reviewer has found that students and teachers find the information it yields useful, understandable, and translatable into interventions and instructional and learning strategies.

Users need to exercise caution in their use of the instrument and recognize the problems inherent in the type of approach it uses to measure learning styles. The authors of the LSI should continue to refine and develop the instrument so that it conforms to the criteria set forth by the *Standards for Educational and Psychological Testing*. The research base does indicate the validity of the instrument and preliminary evidence indicates that the constructs can be measured with some degree of reliability.

References

American Educational Research Association, American Psychological Association, & National Council on Measurement in Education. (1985). *Standards for educational and psychological testing*. Washington, DC: American Psychological Association.

Carbo, M., Dunn, R., & Dunn, K. (1986). *Teaching students to read through their individual learning styles*. Englewood Cliffs, NJ: Prentice-Hall, Inc.

Copenhaver, R. W. (1979-1980). The consistency of learning styles. *The Teacher Educator, 15*(3), 2-6.

Dunn, R. (1981). *A learning styles primer.* Arlington, VA: National Association of Elementary Principals.

Dunn, R., & Dunn, K. (1978). *Teaching students through individual learning styles*. Reston, VA: Reston Publishing Co., Inc.

Dunn, R., Dunn, K., & Price, G. E. (1981). *Productivity Environmental Preference Survey.* Lawrence, KS: Price Systems.

Dunn, R., Dunn, K., & Price, G. E. (1987). *Manual for the Learning Styles Inventory (LSI).* Lawrence, KS: Price Systems, Inc.

Perrin, J. (1982). *Perrin Learning Style Inventory (Primary Version).* Jamaica, NY: St. John's University, Learning Styles Network.

Price, G. E., Dunn, R., & Dunn, K. (1977). *Learning Style Inventory: Research report.* Lawrence, KS: Price Systems.

Virostko, J. (1983). *An analysis of the relationships among academic achievement in mathematics and reading, assigned instructional schedules, and learning style time preference of third, fourth, fifth, and sixth grade students.* Unpublished doctoral dissertation, St. John's University, Jamaica, New York.

Louis M. Hsu, Ph.D.

Professor of Psychology, Fairleigh Dickinson University, Teaneck, New Jersey.

THE LEEDS SCALES FOR THE SELF-ASSESSMENT OF ANXIETY AND DEPRESSION

R. P. Snaith, G. W. K. Bridge, and Max Hamilton. High Wycombe, England: The Test Agency Ltd.

Introduction

The Leeds Scales for the Self-Assessment of Anxiety and Depression (Leeds Scales) consist of four 6-item self-assessment scales of state anxiety and depression. These scales are of two types: Specific (one Anxiety and one Depression), which were developed to measure the severity of disorder in patients with primary diagnoses of affective illness, and General (one Anxiety and one Depression), designed to measure the severity of depressive and anxiety symptoms in patients with a primary diagnosis other than affective disorder (e.g., obsessional, phobic, or depressive neuroses, schizophrenia, alcoholism, psychosomatic disorders, etc.). A fifth scale, the "Diagnostic" scale, is defined in terms of two of the basic scales and was developed to measure "homogeneity of symptom composition" (p. 12)* in patients with primary affective disorder diagnoses.

The Leeds Scales evolved from the Wakefield Self-Assessment of Depression Inventory (SAD; Snaith, Ahmed, Mehta, & Hamilton, 1971). Items included in all four Leeds scales were drawn from a pool of 22 symptom description statements (e.g., "I feel miserable and sad"). Twelve of these 22 items consisted of the SAD items, which in turn were derived from the 10 items most frequently endorsed by depressed patients on the Zung Self-Rating Depression Scale (see Duckitt, 1985). The remaining 10 items were included in the original 22-item pool to more nearly cover the range of common symptoms of depressive illness and anxiety states than did the original 12 items of the SAD. Two of these 10 items were drawn from the Anxiety Scale of Kellner and Sheffield's (1973) Symptom Rating Test (misspelled "Kalin" and Sheffield in the Leeds Scales manual [1976a]). The source of the remaining 8 items included in the original 22-item pool is not specified in the manual. Items related to weight loss, loss of libido, diverse phobic and obsessive behaviors, and somatic symptoms were not included in the item pool.

The process of constructing the Leeds Scales from the original 22-item pool involved tne use of both exclusion and inclusion criteria. Items that were found to correlate significantly with either age or sex in a mixed group of 137 patients (Snaith, Bridge, & Hamilton, 1976b) were excluded. This criterion resulted in the

---

*Page references in this review refer to the Leeds Scales manual (Snaith et al., 1976a) unless otherwise indicated.

elimination of three items (concerned with weeping, tiredness, and headaches). The authors' justification for adoption of this criterion is that their scales were designed to "measure illness in either sex at any age" (p. 2).

The criteria for inclusion of any of the remaining 19 items in the Specific and General scales involved item validity statistics, which were calculated from data of various subsamples of a sample of 137 patients suffering from a variety of psychiatric disorders (diagnosis of type of disorder was made by psychiatrists from previous knowledge of the patients or from preliminary interviews).

More specifically, an item was included in the Specific Anxiety scale if it met two criteria: 1) its correlation with psychiatrists' global ratings of severity of illness among those diagnosed as Anxiety Neurotics had to be at least .58, and 2) its mean score for moderately or severely ill Anxiety Neurotics had to be higher than its mean score for moderately or severely ill Endogenous Depressives. Similarly, an item was included in the Specific Depression scale if it correlated at least .58 with psychiatrists' ratings of severity of illness among those diagnosed as Endogenous Depressives and if its mean score for moderately to severely ill Endogenous Depressives was higher than its mean score for equally ill Anxiety Neurotics. These criteria reflected the authors' belief that an item in a Specific scale should show a high correlation with an independent measure of severity of the illness and that if the scales were to be used to distinguish endogenous depression from anxiety neurosis, then items in a scale should show higher scores for patients in the relevant diagnostic group (p. 7). Application of these inclusion criteria resulted in the 6-item Specific Anxiety and Specific Depression scales. The two Specific scales were used to define what the authors describe as the Diagnostic scale. The "diagnostic score" is the difference between a patient's Specific Depression scale score and his Specific Anxiety scale score. This is the only one of the Leeds Scales that can be viewed as ipsative.

The criteria for inclusion of an item in either the General Anxiety or General Depression scales involved correlations of the item with psychiatrists' anxiety and depression ratings of the patients. These ratings were generated from subsets of items from the Hamilton Anxiety Rating Scale (Hamilton, 1959) and the Hamilton Depression Rating Scale (Hamilton, 1967). More specifically, for an item to be included in the General Anxiety scale, it had to correlate at least .52 with psychiatrists' ratings of anxiety, and it had to account for a greater proportion of the variance of psychiatrists' ratings of anxiety than of the variance of their ratings of depression for the same patients. Similarly, for an item to be included in the General Depression scale, it had to correlate at least .52 with psychiatrists' ratings of depression, and it had to account for a greater proportion of the variance of psychiatrists' ratings of depression than of the variance of their ratings of anxiety for the same patients. For both of these scales, the difference in the proportions of variances accounted for by the item had to be at least .05. Application of these criteria resulted in the 6-item General Anxiety and General Depression scales.

Application of the exclusion and inclusion criteria described above resulted in the inclusion of only 15 of the original 22 items in the Leeds Scales. Thus, the answer sheet, on which are typed the items used to determine Leeds Scales scores, includes only 15 items. Instructions typed on the Leeds Answer Sheet read: "Please indicate how you are feeling now, or how you have been feeling in the last day day [*sic*] or two, by UNDERLINING the correct response to each of the following items." Four response options are listed below each item: a) Yes definitely, b) Yes sometimes, c) No not much, and d) No not at all.

Norms are not provided in the manual identified as such. However, the authors include frequency distributions of General scale scores in their description of a cross-validation study (Snaith et al., 1976b) involving 50 normals, 9 mildly ill patients, and 34 moderately to severely ill patients. They also provide a histogram representing the distributions of the "diagnostic scores" of two cross-validation groups: a group of 18 Endogenous Depressives and a group of 16 Anxiety Neurotics. The authors' interpretive comments, which are based mostly on the data from these cross-validation groups, suggest that these data are to be viewed as the norms.

Practical Applications/Uses

The authors describe the Leeds Scales for the Self-Assessment of Anxiety and Depression as self-assessment instruments designed for the measurement of severity of illness (p. 21). They further indicate that the Specific scales measure severity of diagnosed affective illness and that the General scales measure the severity of depressive symptoms and of anxiety symptoms in patients who have not received a primary diagnosis of affective illness but of some other psychiatric disorder (p. 22). The authors indicate in the manual that the scales may be used in research (p. 22), but they do not suggest that the scales should not be used for diagnostic purposes in clinical settings, given the present stage of their development. Furthermore, they note that the scales were "designed to detect or measure illness in either sex at any age" (p. 2). These statements suggest that the authors view their scales as having a very broad range of applicability. Use of these scales for other than research and scale development purposes, however, may be premature given the 1) total absence of reliability information, 2) extremely limited norms, and 3) very limited evidence of validity and generalizability.

It should be noted that the publisher of the Leeds Scales includes with the test materials a sheet containing the following warning: "The [Leeds Scales] must be administered and interpreted only by professional psychologists with post-graduate clinical qualifications such as would be accepted by the British National Health Service or in the context of research by professionals with equivalent and appropriate qualifications." However, in line with information in the manual noted above, the publisher does not suggest that these scales be used exclusively in research and not for diagnostic purposes in clinical settings.

Instructions for administration of the Leeds Scales (pp. 22-24) are clearly presented. Part of the instructions suggests that the authors are concerned about the possibility of faking: "When asking the patient to complete the scale, do nothing to give him the impression that a major decision, e.g. admission to or discharge from hospital, will depend on the way he completes it" (p. 23). It is doubtful that the administrator's not drawing the patient's attention to the possible use of the scales in such decisions would prevent the typical patient from being aware of this fact. This may be a problem with the Leeds Scales because of the transparency of the symptom-descriptive items and because no attempt has been made either to detect faking or to adjust Leeds Scales scores for faking.

The scoring instructions, which call for the placement of a transparent stencil over the answer sheet, are clear, concise, and easy to follow. Learning to score the scales should take no more than a few minutes, and the scoring time, once instructions have been mastered, should be under 1 minute per answer sheet. There is a minor problem of misalignment of the stencil and answer sheet that this reviewer received from the publisher; when item #1 on the stencil was lined up with item #1 on the answer sheet, the last item (#15) was off by one line. The response options for each item are scored one way (e.g., 3 points for "Yes definitely") on a 4-point scale (3, 2, 1, 0) for 12 of the 15 items, and the other way (e.g., 0 for "Yes definitely") for the remaining 3 items.

No information is provided in the manual or by the publisher about availability of machine scoring for the Leeds Scales. However, it may be noted that microcomputerized administration, scoring, and normative interpretation would be very simple to develop.

The authors provide very limited guidelines for interpreting Leeds Scales scores. These guidelines are primarily based on results of the cross-validation sample reported in Snaith et al. (1976b). Because these results are most directly related to the validity of the Leeds Scales, description of these limited guidelines will be presented in the context of an examination of the results of the cross-validation study in the Technical Aspects section of this review.

The authors are clearly aware of the possible influence of response sets in self-rating scales: "Self-rating scales in psychiatric practice . . . possess certain inherent drawbacks: . . . they are liable to be influenced by the patient's wish to present himself in a certain light. Other shortcomings [include] 'overall agreement set,' 'social desirability,' 'end-users versus middle users,' and 'position bias'" (p. 1). However, their attempts to control response sets in the Leeds Scales appear to be very limited. Specifically, they reversed the polarity of only 4 of the 22 original (and 3 of the 15 final) items, presumably to control for such response sets as 'overall agreement' and/or 'positional bias.' It is perhaps questionable whether reversing the polarity of 20% or less of the items can control the relevant response sets, especially if one notes that this polarity reversal did not affect any of the items in either of the two Leeds Anxiety scales. Apparently, no systematic attempts have been made to control any of the other response sets. It might be noted that several other authors of anxiety and depression scales have taken significant steps to control effects of response sets in their scales (e.g., Costello & Comrey, 1967; Dobson, 1985a).

Technical Aspects

No evidence is presented in the manual concerning any aspects of the reliability of the Leeds Scales for the Self-Assessment of Anxiety and Depression. That is, there are no internal consistency statistics, no test-retest reliability statistics, and no interrater reliability statistics for these scales. Two interrater reliability coefficients are mentioned (p. 4), but these are for a set of 20 joint ratings for the Hamilton rating scales and thus have no direct bearing on the reliability of the Leeds Scales.

Several factors are related, directly or indirectly, to the validity of the Leeds Scales. The following factors will be covered in this section (while others will be covered in the Critique section): 1) item selection strategies, 2) misclassification rates in the cross-validation groups, 3) the correlation of the Leeds Scales with psychiatrists' ratings, and 4) the correlation of the Leeds Scales with age.

Some of the details of the authors' strategies for the inclusion and exclusion of items were described in the Introduction. An evaluation of these strategies is provided here.

With respect to the construction of valid anxiety and depression scales, the authors' exclusion of items that correlated with sex would appear to make sense only if sex differences in scores on these items did not reflect differences in actual levels of anxiety or depression experienced by males and females. Three items were eliminated using this exclusion criterion, but no evidence is presented in the manual that these items only reflected differences in scores and not in the levels of experienced anxiety and depression by the two genders. Thus, some perfectly valid items may have been eliminated by this exclusion criterion. The authors' use of the differences between means of the two diagnostic groups on an item because these differences were considered to measure the item's separation of groups ignores the effects of item variances on group separability. More specifically, items with the largest mean differences would not necessarily be the items that best separated the groups. Also, the authors' criterion of large within-group correlations for inclusion of an item in a Specific scale could be expected to produce relatively invalid items from the perspective of group separation because of the dependence of correlations on ranges (variabilities). That is, because correlations tend to increase with variances, it could be expected that items with large correlations would tend to have large variances. Because separability of groups, for any fixed difference between group means, is negatively related to within group variances, it would be expected that items with the larger correlations would also be the ones that discriminated more poorly between the groups, given any fixed difference between means. Finally, the authors' criteria for exclusion and inclusion of items did not include any measures of inter-item correlations. Darlington and Bishop's (1966) work suggests that item selection strategies that ignore item intercorrelations consistently yield scales that perform worse on cross-validation than scales constructed using methods that take these intercorrelations into account. Failure to utilize item intercorrelations can also result in the omission of good suppressor items. These items can potentially contribute more to the validity of a scale than an equal number of other items whose individual item validity statistics are much more impressive than those of the suppressors.

Concerning misclassification rates in the cross-validation groups, the authors note that performance of the cross-validation sample on the General scales indicates that a cutoff between scores of 6 and 7 provides the most satisfactory division between healthy and sick populations, for both the Depression and Anxiety scales. They note that using this cutoff for the General Depression scale results in misclassification rates of 6% for normals, 33% for mildly ill, and 3% for moderately or severely ill. For the General Anxiety Scale, the respective figures are 6%, 22%, and 11%. Two points should be noted in evaluating these statements. First, the cutoff was apparently selected from the distributions of what the authors describe as their "cross-validation" populations, for the purpose of minimizing misclassification rates. Estimates of misclassification rates obtained from the distributions of the groups that were used to select the optimal cut scores can be expected to be biased. A more conventional "cross-validation" would require evaluation of misclassification rates in samples other than the one used to generate the cutoff scores. Second, it should be recalled that the cross-validation samples (described as populations by the authors) consisted of only 50 normals, 9 mildly ill, and 34 moderately to severely ill patients, so that the stability of the estimated misclassification rates would be questionable, even if these estimated rates were not biased.

The authors indicate that validity coefficients of .87 and .72 were obtained for psychiatrists' global ratings and the Specific Depression and Anxiety scales, respectively, in the cross-validation samples (p. 13). Sample sizes associated with these correlations were 32 and 20, respectively. Correlations of observer-rated depression with General Depression scores and observer-rated anxiety with General Anxiety scores were .85 and .83, respectively, for a cross-validation sample of 31. In general, the sizes of these validity coefficients are quite respectable for anxiety and depression scales. However, the sample sizes are small. Two additional cautions appear to be in order. First, very little information is available from the manual concerning the characteristics of the six psychiatrists whose ratings were used in the cross-validation study. In fact, no information is available about four of the six. Second, the manual contains virtually no information about the characteristics of the cross-validation patients, which would be of importance with respect to generalizability. Therefore, there is insufficient information in the manual to permit plausible inferences about generalizability of the validity findings of the cross-validation study to other psychiatrists or patient samples.

Although the authors attempted to include items in their scales producing scores that would not be related to age, data resulting from their cross-validation samples indicate that their anxiety scales, both Specific and General, *are* related to age. The authors therefore suggest specific point "adjustments" to raw scores "to nullify the relation between age and anxiety" (p. 14). However, they do not indicate either how these adjustments were determined or the conditions under which they should be applied. Certainly, one would expect that if actual levels of anxiety do decrease with increasing age, a valid scale of anxiety should reflect this decrease and not be "adjusted" for it. If there are, in the authors' view, conditions when this adjustment is appropriate, a description of these conditions in the manual would be very desirable.

Critique

The manual does not link the development or validation of the Leeds Scales for the Self-Assessment of Anxiety and Depression to any explicitly defined model of anxiety and depression, and it provides no information on the relation of these constructs to other variables. One of the principal consequences of this is that it is not clear what correlates of Leeds Scales scores should have been partialled out in the process of constructing and validating the scales—or should be partialled out in the measurement of anxiety and depression with these scales.

Thus, for example, the authors' exclusion criterion required elimination of items correlated with either age or sex. But no rationale or theory was presented that would require the adoption of this criterion. Similarly, having found evidence that, in spite of the exclusion criterion, the Anxiety scales correlated with age, the authors provided a recommendation for making point adjustments to the scale scores without specifying any relevant context and/or theory calling for such adjustments.

The authors also note (p. 13) that age has been found to be related to the Diagnostic scores, and they indicate the need to partial out the effects of age in evaluating the efficacy of these scores at separating the depressive and anxious groups. They accomplish this by performing an analysis of covariance of the groups' Diagnostic scores (with age viewed as a covariate). The resulting ANCOVA F ratio of 6.26 ($p < .025$) was less than one half the size of the F ratio of 16.22 ($p < .005$) of the corresponding analysis of variance. They do not comment about the loss of discriminative ability of the Diagnostic score after partialling out effects of age, but they instead note that the difference between the two groups is still significant. Two related points might be made with respect to these analyses. If age needs to be partialled out, as suggested by the authors (p. 13), then 1) the drop in F from 16.22 to 6.26 (with negligible changes in degrees of freedom) suggests a very sharp drop in discriminative ability of the Diagnostic scores, and, therefore, 2) the cross-validation histogram showing the distributions of unadjusted Diagnostic scores of the two groups (p. 11), together with the authors' comments related to this histogram, are highly misleading. That is, if age adjustment is needed as indicated on page 13 of the manual, then the authors should have illustrated the discriminability of groups with histograms of covariate-adjusted scores and not with histograms of unadjusted scores.

The authors recommend partialling out the effects of another variable, physical illness. They describe evidence of relation of physical illness to depression and anxiety (p. 18) and suggest that "if the Leeds scales are to be used in studies where a large proportion of patients are physically sick, then the effect of the variable should first be determined . . . and . . . removed by appropriate statistical techniques" (p. 18). Although there may very well be contexts or theories in which following these recommendations makes sense, these contexts or theories are not presented in the manual.

Thus, an important principle guiding the construction, cross-validation, and interpretation of the Leeds Scales appears to have been that the effects of variables (viz., sex, age, and physical illness) associated with performance on depression and anxiety items and scales should be controlled or partialled out. However, it would seem that, in contexts where these variables affect not only performance on the items and scales but also levels of experienced anxiety and depression, application of such a principle could decrease instead of increase the validity of the resulting scales. Perhaps this is what happened in the case of the authors' Diagnostic scale, for which the F ratio for group discriminability dropped considerably when age was partialled out. An earlier reviewer (Duckitt, 1985) focused on the smaller ANCOVA F ratio and concluded that the discriminant validity of the Specific scales was rather dubious. It would seem, however, that if age is associated not only with performance on the scales but also with experienced anxiety, then it is the larger of the two F ratios—the ANOVA F—that better reflects the separability of the anxious and depressed groups.

One of the authors' objectives in developing the Leeds Scales was to create two different types of anxiety and depression scales. One type (Specific) could be used to measure anxiety and depression, both normatively and ipsatively, in patients with primary diagnoses of affective disorders. The second (General) could be used to measure severity of depressive and anxiety symptoms in patients suffering from a variety of disorders. Because the criteria for inclusion of items in scales differed for the two types of scales, there was the empirical possibility that the Specific and General scales would have few or no items in common. However, this did not happen; the Specific and General Anxiety scales share four of six items, and the Specific and General Depression scales share five of six items. Duckitt (1985) notes that the authors' cross-validation does not demonstrate the utility of the two types of scales and that, in fact, no evidence is presented that the Specific scales predict severity of ratings of Anxiety Neurotics or Endogenous Depression better than the General scales. Duckitt also notes that a study by Forrest and Berg (1982) indicates a correlation of .70 between Specific scales. Duckitt concludes that "instead of indexing specific syndromes both scales might be primarily reflecting generalized psychiatric impairment or distress" (1985, p. 846).

In defense of the Leeds Scales, it might be noted that, even if the correlation between the Specific Anxiety and Depression scales had been much higher than .70, this would not necessarily imply that these scales measure only one construct (any more than a high correlation of height and weight scales imply that they tap only one construct). Additionally, obtaining such a correlation would not rule out the possibility that a linear combination of these scales—such as the authors' Diagnostic scale—could discriminate between Anxiety Neurotics and Endogenous Depressives much better than either scale alone. A search for the best linear combination of Anxiety and Depression scores for the discrimination of anxious from depressed groups might involve the use of discriminant function analysis, a technique that was not used by the authors. It should be noted in this context that other investigators (e.g., Prusoff & Klerman, 1974) have used discriminant function analysis for this purpose, with anxiety and depression scales that were developed from the results of factor analyses.

In conclusion, the Leeds Scales are short, easily administered, and easily scored scales of anxiety and depression. The extensive item overlap of corresponding Specific and General scales suggests that the two types of scales provide redundant information. The Diagnostic scale may prove to be useful as an ipsative measure of anxiety and depression, although at this time relevant validity evidence is very limited.

Ideally, anxiety and depression scales should be developed in the context of clearly formulated theories, using both large and well-defined populations in item selection and item selection methods that take iteritem correlations into account (e.g., multiple regression and factor analytic methods). They should also be carefully evaluated for dimensionality, internal consistency, and test-retest reliability, in addition to being systematically evaluated for concurrent, predictive, and construct validity—again, with large, well-defined subject populations. The Leeds Scales were found wanting in all of the above areas. Depression and/or anxiety scales having most of the foregoing characteristics include scales developed by Dobson (1985a, 1985b), Prusoff and Klerman (1974), Costello and Comrey (1967),

Spielberger, Gorsuch, and Lushene (1970), and Beck, Ward, Mendelson, Mock, and Erbaugh (1961).

References

Beck, A. T., Ward, C. M., Mendelson, M., Mock, J., & Erbaugh, J. (1961). An inventory for measuring depression. *Archives of General Psychiatry, 4,* 561-571.

Costello, C. G., & Comrey, A. L. (1967). Scale for measuring depression and anxiety. *Journal of Psychology, 66,* 303-313.

Darlington, R. B., & Bishop, C. H. (1966). Increasing test validity by considering interitem correlations. *Journal of Applied Psychology, 50,* 322-330.

Dobson, K. (1985a). Defining an interactional approach to anxiety and depression. *The Psychological Record, 35,* 471-489.

Dobson, K. (1985b). The relationship between anxiety and depression. *Clinical Psychology Review, 5,* 307-324.

Duckitt, J. (1985). Review of Leeds Scales for the Self-Assessment of Anxiety and Depression. In J. V. Mitchell, Jr. (Ed.), *The ninth mental measurements yearbook* (pp. 845-846). Lincoln, NE: Buros Institute of Mental Measurements.

Forrest, G., & Berg, I. (1982). Correspondence: Leeds Scales and the G.H.Q. in women who had recently lost a baby. *British Journal of Psychiatry, 141,* 429-430.

Hamilton, M. (1959). The assessment of anxiety states by rating. *British Journal of Medical Psychology, 32,* 50-55.

Hamilton, M. (1967). Development of a rating scale for primary depressive illness. *British Journal of Social and Clinical Psychology, 6,* 278-296.

Kellner, R., & Sheffield, B. F. (1973). A self-rating scale for distress. *Psychological Medicine, 3,* 88-101.

Prusoff, B., & Klerman, G. (1974). Differentiating depressed from anxious neurotic outpatients. *Archives of General Psychiatry, 30,* 302-309.

Snaith, R. P., Ahmed, S. N., Mehta, S., & Hamilton, M. (1971). The assessment of the severity of primary depressive illness. *Psychological Medicine, 1,* 143-149.

Snaith, R. P., Bridge, G. W. K., & Hamilton, M. (1976a). [Manual for the] *Leeds Scales for the Self-Assessment of Anxiety and Depression.* London: J. Bellers Press.

Snaith, R. P., Bridge, G. W. K., & Hamilton, M. (1976b). The Leeds Scales for the Self-Assessment of Anxiety and Depression. *British Journal of Psychiatry, 128,* 156-165.

Spielberger, C. D., Gorsuch, R. L., & Lushene, R. E. (1970). *STAI—Manual for the State-Trait Anxiety Inventory.* Palo Alto, CA: Consulting Psychologists Press.

Jennifer Ryan Hsu, Ph.D.
Associate Professor and Chairperson, Department of Communication Disorders, William Paterson College, Wayne, New Jersey.

"LET'S TALK" INVENTORY FOR ADOLESCENTS

Elizabeth H. Wiig. San Antonio, Texas: The Psychological Corporation.

Introduction

The Let's Talk Inventory for Adolescents (LTIA) is an assessment procedure that assesses the ability to express and interpret speech acts representing the four communication functions: ritualizing, informing, controlling, and feeling. There are two types of items for each function. One requires "speech act formulation" in which the examinee is requested to describe what an individual would say given a description of the speaker's intention, as well as the situational and audience context. Each description is accompanied by a pictorial illustration of the context. This first type of item probes the ability to express the appropriate intent and degree of formality (termed "social register" by the test author) for selected speech acts, first, in a peer context and, then, in an adult/authority context. The second type, termed "speech act association items," is a picture selection task that probes comprehension of speech acts and the ability to associate them with appropriate contexts depicted in illustrations.

The inventory was developed by Elizabeth H. Wiig, a retired professor of communication disorders at Boston University. Professor Wiig has published extensively in the area of communication disorders. In addition, she has developed a comprehensive assessment instrument (Semel & Wiig, 1980), an intervention program (Semel & Wiig, 1982), and guides for observing and training communication functions and intents in children and adolescents (Wiig, 1982a; Wiig 1982c; Wiig & Bray, 1983; Wiig & Bray, 1984). At the time that the LTIA was developed, there were no standardized procedures for assessing the ability to produce and understand speech acts. Thus, the inventory was developed to provide a uniform and standardized method of eliciting and probing selected communication functions and speech acts.

The items included in the inventory were designed to feature four communication functions as well as speech acts identified by Wells (1973, cited in Wiig 1982c) and Wood (1981). The four communication functions—ritualizing, informing, controlling, and feeling—were selected based, in part, on the ease of developing pictures and narratives expressing intent. The particular speech acts included in the inventory were considered to 1) range from simple to complex, 2) be acquired relatively late, and 3) be germane to everyday interpersonal interactions in real-life and educational settings. Speech acts that have been found to be delayed in special populations also were included. Wiig (1982c) reports that the selected speech acts are mastered between the ages of 11 and 13. Items that discriminate on the basis of sex were excluded.

322

The contexts for each speech act were developed on the basis of recommendations from a group of teenagers who identified situations and interactions in which the speech acts would naturally occur. The situational contexts for the peer interactions were selected to include a variety of settings, such as school, classroom, home, and the outside environment. The situational contexts for the adult/authority interactions included the classroom, other school settings, and professional offices. Later, some items were modified based on either preliminary or subsequent field testing.

The manual provides normative data (i.e., means and standard deviations) for four age groups: 7-8-, 9-10-, 11-12- and 13-14-year-olds. The developmental data were obtained from 108 children, 55 males and 53 females, with normal language and academic achievement. Overall, the children ranged from 7.0-13.10 years of age. There were 22 children in the 7-8-year-old age group, 29 in the 9-10-year-old group, 28 in the 11-12-year-old group, and 29 in the 13-14-year-old group. The author reports that all children included in the developmental sample came from urban-suburban educational settings with lower- to upper-middle-class socioeconomic backgrounds. All were from standard English language backgrounds and, with the exception of two children, were Caucasion. Every child in the sample was reported to have normal vision and hearing.

The children in the developmental sample represented several geographic regions. The largest group attended public schools in a middle-income suburb of a major midwestern city. The second largest group was from urban and suburban middle-class homes and attended a parochial school in an urban northeastern setting. Smaller samples of children were from suburban settings on the West Coast and in the Northeast, Midwest, and South.

The test author states (Wiig, 1982c) that the size of the sample used to establish developmental data will be expanded for the younger age groups and that revised central tendency measures will appear in future printings and updates. At the present time, these data are not available in the manual and there have been no revisions of the test.

The LTIA is available in a Spanish translation. According to the manual, although the translation has not been field tested, individuals interested in using it may obtain assistance from the author for developing scoring criteria and developmental data. The manual also states that a copy of the Spanish translation may be obtained.

The LTIA contains a manual for administration and scoring of the test, a picture manual containing the test stimuli, and record forms. The inventory consists of four sections that probe 40 speech acts representing the communication functions of ritualizing (8 speech acts), informing (10 speech acts), controlling (11 speech acts), and feeling (11 speech acts). There are two speech act formulation items (designated Segment A and Segment B) and one speech act association item (designated Segment C) associated with each speech act. Thus, there are a total of 120 items, 80 speech act formulation items, and 40 speech act association items.

Segment A speech act formulation items involve a child interacting in a peer context, and Segment B items involve a child interacting in an adult/authority context. The peer context is administered first and the adult/authority context second. Each segment is presented with a picture and a narrative describing the interaction and the intent, followed by an instruction for the formulation of a speech act appropriate to the context. For example, the examiner will say, "What do you think _____ said to _____?". Both the peer and adult/authority contexts probe the same communication function and intent, although two segments are used in order to determine whether examinees can select appropriate informal forms for peer contexts and appropriate formal (i.e., polite) forms for adult/authority contexts.

If inappropriate responses are given to both segments, then a speech act association item, Segment C, is administered. These items involve a statement of the communicative intent and a set of three separate pictures displayed on one page in a diagonal line. One picture is located on the bottom left-hand corner, a second in the middle, and the third on the top right-hand corner. The examinee is told, "Here are some more people. Who do you think said: ' < *statement expressing a speech act* > '." The examinee then selects the picture depicting a situational and audience context that is believed to best portray the intent and audience of the speech act. Each of the speech act association items features a speech act appropriate for either a peer or an adult/authority context. The intent of each item parallels the intent of the two formulation items that precede it. A summary of the test's organization follows.

Communication Functions	Speech Act Formulation Items			Speech Act Association Items	Grand Total
	A	B	Total	C	
1. Ritualizing	8	8	16	8	24
2. Informing	10	10	20	10	30
3. Controlling	11	11	22	11	33
4. Feeling	11	11	22	11	33
Total	40	40	80	40	120

The LTIA is intended to be used with individual's ranging from age 9 to adulthood. The level of difficulty for language-disordered populations will depend on the nature and severity of the problem. Administration of the test requires the examiner to present each item to the child. Responses should be audiotaped for later transcription on the record forms. These forms provide space for writing the response for each formulation item and circling the picture that was selected in the association task. Space is provided on the form for recording the number of morphemes for each utterance and the score for each segment. A summary section that indicates the number correct, percent correct, and rank for the formulation items is included. There is no summary section for the morpheme counts or the picture association task.

Practical Applications/Uses

The purpose of the LTIA is to identify, in a standardized format, an individual's strengths and/or weaknesses in 1) expressing speech acts related to the four communication functions (ritualizing, informing, controlling, and feeling), 2) selecting appropriate formal or informal forms (i.e., using "social register") depending on the audience and context, and 3) associating speech acts with appropriate contexts and intents. Thus, the inventory probes the knowledge and use of two aspects of the pragmatic component of language: speech acts and social register. The test author clearly states that the inventory was not designed to identify the nature and bases of inadequate social-verbal language and communication skills, nor was it designed to determine educational placement (Wiig, 1982c). The author also states that the inventory was designed to probe speech acts germane to a standard American English language community. It was not designed to probe the adequacy of using nonstandard, racio-ethnic or dialectical variations of general American English (Wiig, 1982c).

The items in the inventory represent a sample of the range of possible speech acts and contexts that require different forms. Weaknesses observed on the items would suggest the need to probe additional speech acts and contexts. Analysis of the content of responses to segments A and B may provide information concerning an individual's ability to take into account listener needs. However, further probing would be required in this area as well. Although discourse functions are not assessed by the items included in the inventory, some information regarding these functions can be obtained in the recorded language sample.

Because the inventory is not intended to be a diagnostic instrument, it should be used with individuals known to have language impairments. Furthermore, it should be used with individuals who have pragmatic problems that involve expressing speech-act functions and selecting forms appropriate to peer or adult/ authority contexts. It is important to note that the latter problem may be related to deficits in the syntactic component as well. Relevant populations would include learning disabled, hearing-impaired, mentally retarded, and autistic individuals, as well as neurologically or perceptually impaired individuals who have pragmatic deficits. The inventory also would be relevant for individuals with limited English proficiency. The goal for the individuals in the latter population group would be to assess their knowledge of speech acts in standard American English.

Information obtained from the inventory can be used to guide further testing of an individual's ability to express speech act functions and to select appropriate formal forms (i.e., social register). In addition, information obtained from the inventory, in conjunction with observation and other test results can be used to determine appropriate short-term and long-term goals for an intervention program concerned with pragmatic skills. The standardized format facilitates comparison of performance over time. However, because there are no alternate forms, carry over may influence retest responses.

Professionals likely to use the inventory include speech and language pathologists, special educators, bilingual and ESL teachers, and other educational specialists. The special educators include a large group of professionals such as teachers of hearing-impaired, mentally retarded, autistic, or learning disabled individuals. In addition, psychologists concerned with the pragmatic skills of their clients may find the test useful.

The LTIA is administered individually. According to the test author, it may be administered by an examiner without general training if he or she has experience in giving individually administered educational, clinical, or psychological tests. The author also states that the examiner should administer, score, and interpret a minimum of five inventories under supervision before the results can be considered valid. Because it takes 2-3 hours to administer, transcribe, and score one protocol, such training would require 10-15 hours. The manual does not indicate who should supervise the training. Presumably, it would be someone with experience in administering and scoring the test and with a background in the area of language disorders and language assessment instruments. Anyone using the test should have a good understanding of the pragmatic component of language.

The manual clearly describes the procedures for administering the LTIA. The picture manual is placed in an upright position between the examinee and the examiner. The examiner presents each item by showing a picture and reading a narrative printed on the page facing him or her. Each stimulus picture and its corresponding narrative is keyed according to its function, item number, and segment (Segment A = peer context and Segment B = adult/authority context). For example, the first item testing the ritualizing function in an adult/authority context would be keyed "R-1B." As stated previously in this review, if the examinee fails to respond correctly to both speech formulation items (Segments A and B), then the speech act association item (Segment C) is presented. Each association item follows segments A and B and the manner in which it is presented is the same as that described previously except that three pictorial choices, rather than one, face the examinee. The examiner records the picture selected by the examinee by circling "a", "b" or "c" on the record form. If appropriate responses are given to either Segment A or Segment B, then segment C is omitted. Responses to segments A and B may be recorded during administration or audiotaped for later transcription.

There are some difficulties in recording responses to Segment C, the picture association task. It is not readily clear from the manual which of the three pictures correspond to the three choices, a, b, and c. Although the test author states in the manual (Wiig, 1982c, p. 19) that pictures a, b, and c are presented from left to right in the picture manual, it is not clear whether "left to right" refers to the examiner's or examinee's perspective. A key to all pictures is provided on pages 42 and 43 of the manual. However, administration and recording of responses would be facilitated greatly if the pictures were clearly labeled on the page facing the examinee as well as on the page facing the examiner. Furthermore, the examinees could then respond by naming their choice (i.e., saying "picture a", etc.). In the present format, the examinee will typically point to a picture and the examiner must determine where the examinee is pointing and then figure out whether the response corresponds to picture a, b, or c. The suggested alternative format would greatly facilitate administration of this section of the test.

The items in the inventory may be administered in their entirety or by category. When administered by category, all items within a particular communication function should be given in one session. In Appendix E, the manual identifies a subset of items that may be used for screening purposes. It is suggested that if the quality of the responses to the screening items appear immature or delayed, all items of the inventory should be administered (Wiig, 1982c, p. 46).

The author states that the inventory can be administered within a 30-45-minute time period. This estimate assumes that an examinee will respond to each item within 30 seconds and that items in Segment C are not administered. Inclusion of items from Segment C increases the administration time by as much as 10-15 minutes. The time estimate also assumes that the examinee does not engage in extraneous conversation. With the exception of recording responses to Segment C, administration of the inventory is easy and straightforward.

The speech act formulation responses to segments A and B are scored in terms of 1) the number of morphemes and 2) the extent to which the response expresses the intent and uses the appropriate social register (i.e., required polite or formal forms). Rules for identifying morphemes are summarized in two separate tables in the manual. The rules are similar to those outlined by Brown (1973), except that morphemes are counted for all sentences given in response to an item and the mean length of utterance is not calculated.

Expression of intent and use of social register are scored by assigning numbers on a scale of 0 to 2. Responses to Segment A, the peer context, are scored only for adequacy of intent. Thus, only scores of 0 or 2 may be assigned. Responses to Segment B, the adult/authority context, are scored for both intent and social register. Thus, scores of 0, 1 (correct intent) or 2 (correct intent and appropriate formal forms) are assigned.

The rules for assigning the scores are outlined in three separate tables and summarized on pages 25 and 43 of the manual. Table 5.4 and Table 5.5 include examples and corresponding scores for responses to each item. The tables also include a general scoring rule for each item. In addition, items that have been found difficult to score are discussed. For most items, scoring intent and social register is straightforward. The discussion of difficult-to-score items is helpful in resolving doubtful responses. However, only a limited number of items are discussed. Because there tends to be a small percentage of responses that are difficult to score despite the examples and discussion provided in the manual, an expansion of the examples of responses and corresponding scores would further facilitate scoring. For some items, the general rule that is provided includes terminology that is not clear (e.g., "non-involvement" in I-7A), for other items the general rule specifies general requirements for scoring a 1 but not those for scoring a 2 (e.g., C-7B, C-10B, and F-5B), and for one item (viz., R-8B) the general rule seems to contradict the examples.

The scores for Segment C also range from 0 to 2. A key for scoring each response is provided on the record form. For most items either a 0, 1, or 2 is assigned. Scoring is unambiguous and extremely easy.

All scoring must be done by hand. The rules for counting morphemes and assigning scores to each formulation item are easy to learn if the manual is studied and two to three protocols are scored. All scoring can be completed within 30-60 minutes.

The time-consuming aspect of the procedure is transcribing the responses for segments A and B. The recommended procedure for administration requires audiotaping, which allows interaction between the examiner and examinee as well as quick administration of the items. However, audiotaping requires later transcription, which involves reviewing the tape and writing the responses, a process that adds at least 1-2 hours to the procedure. As a result, the entire process of administering, transcribing, and scoring is time-consuming. There does not seem to be a solution to this problem. Transcription during administration undoubtedly would increase the time required to give the test. In this case, administration time likely would exceed the attention span of many language-disordered individuals.

The front page of the record form provides a summary section. The scores for segments A and B of each communication function are summed to obtain an overall raw score. The raw scores are recorded in the summary section and converted to percentages. These data are used to rank the communication functions in terms of order of difficulty. A column for ranking is included in the summary section. The term "rank" is explained only in the manual. Because the word has several possible interpretations, it would be helpful if the term also were explained on the record form. The manual provides cutoff scores for age levels, which can be used to determine whether a score falls below -1 or -2 standard deviations from the mean for a particular age group. In addition, means, standard deviations, and standard error of the mean for segments A and B of the four communication functions are provided for each age level.

The record form does not include a space for summarizing in one section the morpheme counts or the scores on the speech act association items. However, the manual provides means and standard deviations for the morpheme counts for each item by age level. Means and standard deviations for scores on the association items are provided by age level for each communication function.

The test author recommends that the scores be interpreted both qualitatively and quantitatively. With respect to the quantitative interpretation, Wiig suggests that the scores for segments A and B of each communication function be considered separately. The scores should be compared to the measures of central tendency for the appropriate age level. In addition, the raw scores may be compared to the cutoff scores to determine whether they fall either -1 or -2 standard deviations below the mean. Because the inventory is not to be used to diagnose a delay or to determine educational placement, these comparisons would be used only to identify areas of weakness and possible targets for intervention. The comparisons, in conjunction with the rankings, would help identify those functions in which an individual has difficulty expressing the appropriate intents or polite forms.

Morpheme length for any speech act may be compared to central tendency measures. Cutoff scores are not provided but must be calculated. If an individual's score is above or below 1 or 2 standard deviations from the mean, then a qualitative analysis is recommended. Excessively long utterances may reflect word retrieval problems or verbosity. Excessively short utterances may reflect lack of formality, syntactic problems, or lack of intent. Because the scoring of the formulation responses would detect the pragmatic problems and other tests should be used to detect syntactic or word retrieval problems, it is not clear that such an analysis is worth the time and effort.

A qualitative interpretation may evaluate several aspects of the responses to segments A and B, including analyzing individual responses to determine if there is an awareness of the perspective of others in interactive situations. Several examples of such an analysis are given. In general, responses across several items must be analyzed and compared. Responses on segments A and B also may be compared. Expression of the majority of intents on Segment A but consistent use of the direct and informal forms on Segment B (typically indicated by a score of 1) would indicate difficulty in the use of polite forms. In addition, responses may be analyzed for the use of reason, logical sequencing of speech act components, and relevance of the response. Finally, each response may be analyzed in terms of the reasons for awarding a 0 or a 1, revealing specific problems associated with particular speech acts, such as failure to use reciprocal introductions, express disapproval, use specific types of reminders, and so on. Examples of this type of analysis are provided on page 69 of the manual. Once the error analysis is completed, consistent patterns may be identified.

Not only is a qualitative analysis of the responses time-consuming, but the manual provides only general guidelines for conducting it. It would be extremely helpful if the record form identified the name of each speech act as the manual does in Tables 5.4 and 5.5 and Appendix B. Furthermore, including a checklist of possible errors could help guide the qualitative analysis and reduce the time required for interpretation. It also would facilitate formulation of specific objectives for an intervention program. In the present form, the qualitative analysis is difficult. It requires clinical judgments that should be based on knowledge of the nature of the pragmatic component of language and how that knowledge interacts with the ability to take the perspective of others in an interactive situation.

The section on the interpretation of performance does not discuss the speech act association items. Central tendency measures are provided that are based on performance of a subset of 64 children from the original developmental sample of 108. This information would be useful only if all the items are administered. However, administration of all the items is unlikely, given the directions stating that Segment C items should be administered only when responses to items from both Segment A and Segment B are inappropriate. Thus, the responses on Segment C are used primarily to evaluate the ability to recognize an intent and associate it with the proper context when there is difficulty in expression of the intent. Performance on Segment C will be helpful in determining the ordering of goals for specific speech acts. Thus, it would be useful if the record form included a summary of performance on items in Segment C as compared to those in segments A and B.

In general, the interpretation of performance on the LTIA is the most difficult aspect of the procedure. Most individuals administering the inventory should be able to evaluate relative difficulty in expressing each of the communicative functions and the degree to which polite forms are used in adult/authority contexts. However, many individuals are likely to need help in performing a valid qualitative analysis of the responses and in determining appropriate goals for intervention. The difficulty of interpretation could be reduced by including on the record form the name of each speech act, an error analysis for each item as well as groups of items, and a summary of performance on Segment C as compared to segments A and B.

Technical Aspects

Several reliability measures for the LTIA are reported in the manual: test-retest (stability of performance), split-half (internal consistency) and interrater (examiner agreement). The data were obtained from subsets of children included in the original developmental study.

The test-retest data were based on responses to eight speech acts with two examples from each communication function. Responses for both the peer (Segment A) and adult (Segment B) contexts were included, resulting in a total of 16 items. The items were administered to 52 children, with roughly equal distribution across age levels, on two separate occasions within a 1-2-week interval. Scores were obtained for each speech act for Segment A and Segment B, individually, as well as segments A and B combined. The Pearson product moment correlation coefficients were .59, .59, and .77 for Segment A, Segment B, and the combined segment scores, respectively. All were significant at a $p < .01$ level. A similar study (Wiig, 1982c; Wiig, Bray, & Colquhoun et al., 1983) was conducted with 13 adolescents and young adults (termed EMR group) with developmental delays and language retardation. A Pearson product moment correlation coefficient for the test-retest scores was .68 ($p < .01$). The test-retest reliabilities reported for both studies are lower than those recommended for clinical tests (Kaplan & Saccuzzo, 1982). These authors recommend that tests used for diagnostic purposes should have reliabilities of at least .90. In view of this, Wiig's caution that the test should not be used to diagnose a delay or determine educational placement must be taken seriously.

Internal consistency was evaluated in two ways. The first evaluation involved the internal consistency among sections. Scores obtained on the ritualizing and controlling functions were combined, as were the scores on the informing and feeling functions. Data were obtained on a total of 66 children with roughly equal distribution of subjects across age levels. With respect to Segment A, Pearson product-moment correlations for 7-8-, 9-10- and 11-12-year-olds were .44, .39, and .40, respectively. For Segment B, the correlations were .85, .56, and .61 for the respective age groups. The correlation of .44 was significant at $p < .05$. All the correlations for Segment B were significant at $p < .01$.

A split-half procedure also was used to evaluate internal consistency. Response scores for every other item in Segment A and in Segment B were summed for the 7-8- and 9-10-year-olds. Pearson product moment correlations for Segment A were .49 and .46 for the 7-8- and 9-10-year-olds, respectively. These correlations were significant at $p < .05$. The correlations for Segment B were .76 and .81 for the 7-8- and 9-10-year-olds, respectively. These correlations were significant at $p < .01$.

The correlation coefficients for the internal consistency studies are relatively low. As the author suggests, the inventory evaluates communication functions and speech acts that are not necessarily homogeneous in nature. The author also suggests that the somewhat higher correlations obtained for Segment B as compared to Segment A may reflect the fact that the scores are based on both expression of intent and the ability to use polite forms. The latter ability appears to be scored consistently across items.

Interrater reliability was determined by comparing the scores assigned independently to 26 protocols by the author and a linguistically trained doctoral student. The protocols were obtained from children who were roughly equally distributed across the three age groups identified in the normative study. The guidelines for scoring were based on general guidelines outlined in Table 5.3 of the manual. The examiners did not have the specific rules and examples that are now included in the manual as Tables 5.1, 5.4, and 5.5.

The percentage of agreement between the examiners scores for morpheme counts was determined for each communication function and for each age group. Agreement ranged from 99-100%. The percentage of agreement in scores for speech act intent and quality, which presumably means the use of polite forms, was also determined for each communication function and age group. The percentages ranged from 91.4% to 99%. Finally, Pearson product moment correlation coefficients were determined within each communication function for the two examiners' total scores, their scores on Segment A, and their scores on Segment B. Twelve coefficients ranging from .83 to 1.00 are reported. Eight of the 12 were above .90. All correlations were significant at or beyond $p < .001$. Although interrater reliability seems to be high (.90 or greater for most comparisons), the correlations are based on only two raters who seem to have extensive experience with the test. More meaningful measures of interrater reliability would be those obtained from practitioners who use the test in various clinical settings.

It also should be noted that although almost all of the reliability coefficients reported by Wiig are statistically significant at conventional levels, such significance is considered a necessary, but by no means sufficient, condition for acceptable reliability in applied settings.

The manual discusses content and diagnostic validity. The former refers to whether the items represent an adequate sample of speech acts and the latter can be viewed as a form of criterion-related validity (*viz.*, concurrent validity), which refers to the effectiveness of a test in predicting an individual's behavior in specified situations. The manual does not discuss predictive or construct validity, at least not under separate headings. With respect to content validity, Wiig states that the communication functions and speech acts evaluated in the inventory are based upon those outlined in Well's parametric model and that they "must be considered to be representative of the universe of speech acts which may be expressed in an interactive social context" (Wiig, 1982c, p. 61). It is not clear that this is necessarily true. Bach and Harnish (1982) and Levinson (1983) note that there are a number of systems for classifying speech acts. The systems do not identify the same general categories of communicative functions nor do they identify the same set of speech acts. Furthermore, the models differ in terms of the classification of particular speech acts according to type of communicative function (Levinson, 1983). The four main communication functions identified in Well's model parallel functions identified by Bach and Harnish and other models as well. However, Well's model does not include declaratives that are included in Searle's model (Levinson, 1983). A comparison of the speech acts identified by the Bach and Harnish model with those included in the LTIA reveals that a large number of acts are identified in the former but not in the latter. Examples include thanking, predicting, appraising, verifying, withdrawing, excusing, volunteering, and so on. As outlined previously in this review, item selection was based on considerations of complexity, age of acquisition, and relevance to everyday interactions. These considerations would not result necessarily in a representative sample of the universe of speech acts. A more important consideration is whether the items sample a relevant universe of speech acts. That is, the items should sample speech acts used by adolescents and young adults. In general, more information is needed to establish the content validity of the inventory.

Wiig (1982c) states that diagnostic validity was evaluated by administering the test to two clinical groups. One group consisted of five 7-9-year-old children who were diagnosed as language disordered and enrolled in language therapy in a suburban public school in the Northeast. Two of the five language disordered children manifested scores that fell below -2 standard deviations below the mean on several communication functions, and two additional children manifested scores that fell below -1 standard deviations below the mean. A speech pathologist independently ranked the five children in terms of the severity of their language disorders. The author reports that this ranking paralleled the ranking obtained from results on the inventory. She suggests that the results support the diagnostic validity of the test. In view of the small sample size, the results constitute an extremely limited validation study. In the second study, the inventory was administered to a sample of 13 EMR individuals who ranged in age from 15 years, 3 months to 25 years and attended a nongraded vocational high school. The scores for segments A and B of each communication function were compared to the central tendency measures for normal 13-14-year-olds. Between three to nine individuals fell below -2 standard deviations from the mean on specific communication functions and segments. Wiig, Bray, Colquhoun, et al. (1983) also report a comparison in the performance of the EMR subjects and normal children 7-8- and 9-10-years-of-age. Speech act scores differed significantly for the three groups on all four communication functions ($F(2) = 4.58$, $p < .05$ for ritualizing; $F(2) = 3.72$, $p < .05$ for informing; $F(2) = 11.02$, $p < .001$ for controlling; and $F(2) = 5.64$, $p < .01$ for feeling). The mean scores for all the 9-10-year-old normal children were higher than those of the EMR group. Wiig, Bray, Colquhoun, et al. (1983) also report results of a third study in which the inventory was administered to 11 learning disabled children ranging in age from 9 years, 4 months to 14 years, 2 months. Two to seven children manifested scores on segments A and B that fell below -1 standard deviation from the mean of normal children with equivalent ages. The results of these studies suggest that the inventory will identify deficits in the expression of speech acts among a variety of language-disordered populations. Thus, the inventory appears to have criterion-related validity in that it detects deficits in children who are known to have a language impairment. However, none of the studies described above indicated that the children were known to have specific pragmatic deficits. Thus, the results indicate only that the inventory will identify pragmatic problems in individuals who are likely to have them. To establish more specific criterion-related group validity, the inventory should be administered to children who are known to have deficits in speech act formulation, and additional measures that specifically assess this ability should be compared to the inventory.

Feudo and Wiig (1987) report results that are relevant to establishing concurrent validity of the inventory. They compared the number of speech acts elicited by 11 of the inventory items that measure the controlling function and a planned conversation also designed to elicit the same controlling speech acts. The results indicate that both methods elicit speech acts. However, the structured format of the inventory resulted in either an equal number or significantly more of the targeted speech acts. The inventory required significantly less time to complete and it resulted in language that was similar in other respects to that obtained in the planned conversation as well as a spontaneous language sample. The results suggest that the inventory may be a more efficient method of evaluating knowledge and use of speech acts than planned conversations.

Although the manual does not explicitly address the issue of construct validity, some of the results reported in the manual and elsewhere provide some limited data. Both the manual and Wiig, Bray, Colquhoun, et al. (1983) report significant differences in speech act scores across age groups of normal children. Thus, the skills assessed by the inventory undergo developmental changes as would be expected of an aspect of linguistic competence. Furthermore, the diagnostic studies reported above indicate that the skills assessed by the inventory are related to general linguistic competence. Individuals who manifest deficits in language are likely to manifest deficits in pragmatic skills as well.

In general, validity studies on the inventory are limited. They involve extremely small sample sizes, and only one type of validity, namely, diagnostic or criterion-related, was investigated empirically. Further research is needed to establish the validity of the inventory.

Critique

The LTIA seems to be the first standardized, commercially available procedure for evaluating aspects of the pragmatic component of language. As such, it represents a significant contribution to the area of language assessment. The test successfully elicits the expression of targeted speech acts and communication functions. Furthermore, it appears to be more efficient than planned conversations for eliciting specific speech acts. It is likely that the inventory will prove to be more efficient than spontaneous language samples as well. One of the major problems of naturalistic methods is the limited range of forms that may occur and the associated ambiguity in determining whether omitted forms are due to lack of knowledge or the fact that the context failed to require their use. The standardized procedures used in the LTIA guarantee that a range of speech acts are assessed. In addition, the speech act association task included in the LTIA represents a novel method for assessing an aspect of the pragmatic component of language. Although pragmatics is generally considered to involve the use of language in social situations, there do appear to be aspects of pragmatic knowledge that can be assessed via a comprehension task. However, it is important to note that speech acts and social register constitute only one aspect of pragmatic knowledge. In cases where a general assessment of pragmatic functions is required, the LTIA must be supplemented with other procedures.

It is clear that the LTIA should not be used to diagnose a delay or to determine educational placement. It is primarily useful in providing information on strengths and weaknesses and in determining short- and long-term goals with respect to the specific speech acts and communication functions included in the inventory. Further probing will be required for speech acts and functions that are not included in the inventory. A major advantage of the LTIA is that a set of objectives and therapy materials for developing the functions assessed by the inventory have been developed by the author (Wiig & Bray, 1984). Thus, practitioners will find objectives, specific remediation procedures, and therapy materials for each speech act included in the inventory. Sequences for objectives and activities also are provided. The connection between assessment, objectives, and remediation procedures should be extremely helpful to practitioners. A checklist for monitoring progress also is included in the package.

In general, the LTIA includes interpretable pictures and appropriate narratives. Most of the targeted speech acts are elicited by the stimuli. The ambiguous items that did occur tended to be stimuli used in the picture association task. The narrative for R-7C describes a situation that is not clearly depicted in the picture, and the statement associated with F-8C does not seem sufficiently formal for an adult context. In other examples, such as I-4C, although formal forms may be required for authority contexts, they are not necessarily inappropriate for peer contexts. In these examples, the lower score does not seem warranted. With respect to segments A and B, most of the narratives clearly indicated the desired speech act. However, the narrative for R-4A does not make clear that the speaker's name is unknown. Nevertheless, the scoring rule requires mention of the name for a score of 2. For most of the items in the inventory, segments A and B represent parallel situations differing only with respect to an adult/authority context in Segment B. In most cases, Segment B elicits more polite or formal forms. One item, F-1, may be an exception. Endearment is elicited first in the context of talking to an animal and then in the context of talking to a cousin. Although segments A and B require different forms, it is not clear that Segment B necessarily requires a formal form.

A major advantage of the LTIA is that it is easy to administer and score. There is likely to be a small percentage of responses on segments A and B that are difficult to score. Additional examples and discussion would probably solve this problem. For some items, the rules for scoring either a 1 or a 2 seem rather arbitrary and result in penalizing what appear to be normal responses. This seems to be the case for items I-2A and F-10B. Scoring Segment C is extremely easy. As noted previously, the major problem with this segment is recording the responses, a difficulty that could be solved easily by indicating numbers or letters both on the pictures and on the page facing the examiner.

A major disadvantage of the LTIA is that the entire process of administration, transcription, scoring, and interpretation is time-consuming. Many practitioners do not have the time to devote 2-3 hours to the evaluation of such a limited aspect of language for a single client. In view of this, the inventory is likely to be primarily useful in cases where the problems are known to involve the specific aspects of language that are assessed by the inventory.

As stated earlier in this review, most practitioners should be able to identify from individual protocols whether there are difficulties in 1) expressing the four communication functions, 2) expressing specific speech acts, 3) using formal forms, and 4) understanding the intent and contexts of specific speech acts. Any additional interpretation of the responses will be difficult for most practitioners unless there is a specific guide for analyzing errors on each item and identifying patterns on groups of items. In any case, interpretation will be facilitated if the record form is revised to include the names of each speech act and summary sections for responses to Segment C as compared to Segment A and Segment B. It is not clear that the morpheme counts provide any useful information beyond that provided by the scores for intent and social register. Eliminating this step will reduce the time required for analyzing and interpreting responses.

There is limited information on the reliability and validity of the LTIA. The studies that have been conducted reveal low test-retest reliability and low internal consistency. The latter finding suggests that the speech acts and communication functions included in the inventory are not homogeneous. Interrater reliability needs to be established for practitioners using the LTIA in clinical situations. With respect to content validity, the relevant population of speech acts needs to be defined. Some socially important acts, such as thanking, have been omitted whereas others that may not be important for adolescents or young adults have been included. For example, the inclusion of challenging/bragging is somewhat surprising to this reviewer. This speech act, which was difficult for the examinee who participated in the present evaluation of the LTIA, may be one that parents discourage. More research is needed to establish the criterion-related validity of the inventory. In particular, the inventory needs to be compared with other measures that evaluate speech act functions, and it needs to be administered to groups with known difficulties in these areas.

References*

Bach, K., & Harnish, R.M. (1982). *Linguistic communication and speech acts.* Cambridge, MA: The MIT Press.

Brown, R. (1973). *A first language: The early stage.* Cambridge, MA: Harvard University Press.

Feudo, D., & Wiig, E.H. (1987). *A comparison of speech act elicitation procedures in mentally retarded adults.* Manuscript submitted for publication.

Kaplan, R.M., & Saccuzzo, D.P. (1982). *Psychological testing: Principles, applications and issues.* Monterey, CA: Brooks/Cole Publishing Company.

Levinson, S.C. (1983). *Pragmatics.* New York: Cambridge University Press.

Semel, E.M., & Wiig, E.H. (1980). *Clinical evaluation of language functions.* Columbus, OH: Charles E. Merrill.

Semel, E.M., & Wiig, E.H. (1982). *Clinical language intervention program.* Columbus, OH: Charles E. Merrill.

Wiig, E.H. (1982a). *Let's Talk: Developing prosocial communication skills.* Columbus, OH: Charles E. Merrill.

Wiig, E.H. (1982b). *Let's Talk Inventory For Adolescents.* Columbus, OH: Charles E. Merrill.

Wiig, E.H. (1982c). *Let's Talk Inventory for Adolescents: Manual.* Columbus, OH: Charles E. Merrill.

Wiig, E.H., & Bray, C.M. (1983). *Let's Talk for Children.* Columbus, OH: Charles E. Merrill.

Wiig, E.H., & Bray, C.M. (1984). *Let's Talk: Intermediate level.* Columbus, OH: Charles E. Merrill.

Wiig, E.H., Bray, C.M., Colquhoun, A., Posnick, B., Vines, S., & Watkins, A. (1983, November). *Elicited speech acts: Developmental and diagnostic patterns.* Miniseminar presented at the annual convention of the American Speech-Language-Hearing Association, Cincinnati, OH.

Wiig, E.H., & Semel, E.M. (1984). *Language assessment and intervention for the learning disabled.* Columbus, OH: Charles E. Merrill.

Wood, B.S. (1981). *Children and communication: Verbal and nonverbal language development* (2nd ed.). Englewood Cliffs, NJ: Prentice-Hall.

*Tests formerly published by Charles E. Merrill are now published by The Psychological Corporation, San Antonio, Texas.

Ann Robinson, Ph.D.

Associate Professor, Department of Teacher Education, University of Arkansas-Little Rock, Little Rock, Arkansas.

MATRIX ANALOGIES TEST

Jack A. Naglieri. San Antonio, Texas: The Psychological Corporation.

Introduction

The Matrix Analogies Test (MAT; Naglieri, 1985) is a measure of nonverbal reasoning constructed from factors similar to those identified for Raven's Coloured Progressive Matrices, such as pattern completion and reasoning by analogy (Corman & Budoff, 1974; Carlson & Jensen, 1980). The MAT, therefore, has its psychometric roots in Spearman's theory of general intelligence. The MAT measures a person's ability to reason by analogy from pictorial, or nonverbal, forms. The test includes four kinds of items: 1) pattern completion, 2) reasoning by analogy, 3) serial reasoning, and 4) spatial visualization. It is designed for use with persons ranging in age from 5 to 17. Two forms of the MAT are available. The expanded form, MAT-EF (64 items), is administered only to individuals. The short form, MAT-SF (34 items), is a screening instrument designed for group administration.

The test author, Jack A. Naglieri, is currently a member of the faculty of The Ohio State University, where he teaches intellectual and academic assessment in the school psychology program. His publications are primarily in the area of validity, reliability, and interpretation of intelligence and academic tests. He also has experience as a practicing school psychologist.

According to the test manual, the MAT was developed for two major reasons. First, nonverbal tests provide information about an important area of intellectual functioning, particularly for individuals with limited language proficiency. Second, the MAT is described as overcoming some of the limitations of existing nonverbal measures embedded in the major intelligence test batteries, for example, the Wechsler Intelligence Scale for Children—Revised Intelligence Scale (WISC-R) and the Stanford-Binet Intelligence Scale. In particular, the Performance scale of the WISC-R requires comprehension of verbal directions and is timed stringently, two factors that may affect the performance of children who have difficulty understanding verbal directions or who have motor difficulties. In addition, the MAT is a nonverbal measure with up-to-date U.S. norms, the key factor in limiting the usefulness of Raven's Progressive Matrices in the United States.

The initial national try-out of the MAT was undertaken in 1984 using an 88-item version. The standardization versions of the test, which were administered later in 1984, included the MAT-EF, an expanded form containing 64 items and the MAT-SF, a 34-item short form. Standardization was conducted in two phases. The MAT-SF was administered to classroom-sized groups (Group sample), and the MAT-EF was administered only to individuals during a second standardization sample (Individual sample).

336

Both the Group (N = 4,468) and the Individual (N = 1,250) standardization samples consisted of individuals in *each* of the 1-year intervals ranging from ages 5 to 17 years, 11 months, 30 days that represented the various geographic regions, socioeconomic groups, and races of the U.S. Efforts were made to approximate the distributions provided by the 1980 census. Information about test development and standardization is reported in the manuals at a sophisticated technical level.

Test forms for Spanish-speaking or visually impaired subjects have not been developed. However, because the test format includes very simple verbal directions that permit the examinee to point to or say the number of the response, the test could be used with individuals of limited English proficiency or who have language, auditory, or physical disabilities such as those associated with cerebral palsy.

Materials for the MAT-EF include a spiral-bound stimulus manual consisting of 64 plates printed in black, yellow, and blue on glossy paper and divided into four sections corresponding to the four item types (pattern completion, reasoning by analogy, serial reasoning, and spatial visualization). Each of the four sections is separated by a cover page for ease in administration. Items within each of the four sections are arranged in order of increasing difficulty. Each item consists of a series of figures at the top of the page and six response figures at the bottom of the page (1 correct, 5 distractors). A cardboard fold-out easel allows the examiner to place the stimulus materials in an upright position and to screen the answer sheet from the examinee during test administration.

In addition to the stimulus manual, the MAT-EF test materials include an answer sheet and the examiner's manual, which contains the norms data, technical information on the statistical properties of the test, and guidelines for administration scoring, and interpretation. The front of the single-page answer sheet contains a total test scoring summary that is used for entering the student's raw score, standard score, percentile rank, age equivalent, and ability classification (e.g., very superior, superior, etc.). The standard score, percentile rank, and age equivalent are recorded with error bands and confidence levels. The back of the sheet contains the answers for the 64 items divided into four columns corresponding to the four item groups, with 16 items per column. For convenience, the correct answers are printed in blue to contrast with the black answer sheet. Answer sheets can be purchased in packets of 50.

Materials for the MAT-SF version include an examiner's manual, answer sheets, and individual, colored test booklets. The MAT-SF examiner's manual is not as extensive as that of the MAT-EF; however, it contains information on test development, standardization, and statistical properties of the instrument, as well as guidelines for test administration, scoring, and interpretation. The MAT-SF provides raw scores, percentile ranks, stanines, and age equivalents. It does not provide a standard score.

The examiner's role for the two versions of the test differs. The MAT-SF, a 25-minute group-administered test, requires only that standardized directions be given orally and the test booklets collected and scored. The examiner's manual states that the test may be given by a number of educational and clinical personnel, including teachers familiar with group ability tests. In contrast, the MAT-EF is individually administered. The examiner gives verbal directions, manipulates the stimulus manual, and records the responses on the answer sheet. The manual recommends that individuals familiar with individual assessment and with training in the interpretation of ability tests administer the MAT-EF.

Practical Applications/Uses

Because of their nonverbal nature, the MAT-EF and the MAT-SF are useful for testing persons with communication disorders, limited language development, or non-English language backgrounds. In addition, both versions of the MAT could be useful in research settings as relatively quick measures of ability. For example, the group administration of the MAT-SF makes it particularly appealing for data collection. With minimal training, a graduate research assistant could administer the MAT-SF in the 25-minute suggested time limit.

The MAT-SF might have potential as a screening device for gifted programs. Although Raven's Progressive Matrices have been used in this capacity, no studies as yet exist in the literature to support the MAT-SF's use in such situations. The manual states that the MAT has advantages over Raven's Progressive Matrices because they lack up-to-date U.S. norms; however, the tests are in the process of restandardization, thus eliminating one of the reasons for selecting the MAT over the Raven's more thoroughly researched instruments.

The MAT-EF is easy to administer and score. Students are allotted 12 minutes to respond to the questions in each of the four item groups in the expanded form. The total administration time for the MAT-EF, then, is approximately 50 minutes to 1 hour. Items should be administered in order. Testing is discontinued when the examinee fails to answer four consecutive items correctly *within* the item group. Students follow the same directions for answering questions in each of the four item groups.

Three sets of examiner directions are provided for the MAT-SF. One set is for children in Kindergarten-Grade 3 who are instructed to mark their answers directly in the test booklet. A second set of directions is provided for children in Grades 4-6 who use an answer sheet to complete the test. The third set permits the examiner to administer the test individually. The answer sheets are pressure sensitive. The child marks his or her answers on the top sheet by darkening the circle that corresponds to the response. When the child has completed the test, the examiner tears off the top sheet, and the bottom sheet contains the child's responses with the correct choices keyed over by a darker circle. Instructions for scoring and the test score summary are located at the bottom of the answer sheet. The design of the answer sheet is efficient for use with groups of children.

Raw scores for the MAT-EF are converted to standard scores by means of an appended table. The standard scores then can be converted to percentile ranks and stanines using a second conversion table. A third conversion table allows the examiner to convert raw scores to age equivalents. Likewise, the MAT-SF tables convert raw scores into percentile ranks, stanines, and age equivalents. All the tables are easy to use and are located in the appendices of the manuals.

Interpretation of the MAT-EF and the MAT-SF is based on objective scores. The MAT-EF manual includes the descriptive ability classifications (e.g., very superior, superior, etc.) common to individual intelligence tests. The MAT-SF manual also provides descriptive classifications associated with the stanines. For example, a child whose MAT-SF score is in the first stanine is described as "at risk of academic failure."

A careful examiner should not have difficulty interpreting the two versions of the MAT. However, one caution is in order for the expanded form. This version provides separate raw scores for each of the four item groups (pattern completion, reasoning by analogy, serial reasoning, and spatial visualization) and a standard score conversion table for item groups. In addition, the answer sheet contains space for recording the group scores separately. Although Naglieri states that the item groups should not be considered subtests and that the item group scores have been included only to encourage research, they could be generalized unwisely by an unwary practitioner in a diagnostic or clinical setting.

Technical Aspects

The MAT is a new test and, therefore, has not appeared extensively in the literature yet. However, Naglieri conducted several analyses in conjunction with the test's development and standardization. These are reported in the examiner's manuals, particularly in the manual for the MAT-EF (Naglieri, 1985).

Reliability was assessed through internal consistency and test-retest reliability. Cronbach's alpha internal consistency coefficients range from .88 to .95 for the MAT-EF and from .63 to .89 for the MAT-SF. The test-retest reliabilities for the MAT-EF were calculated with a fifth-grade sample ($n = 65$). The total score test-retest reliability for the MAT-EF was .77. For the four item groups, test-retest coefficients ranged from .40 to .67 for the fifth-grade sample. Test-retest reliabilities for the MAT-SF ranged from .51 to .91 for age level groups. The overall test-retest reliability coefficient for the MAT-SF was .94.

Several validity analyses were undertaken as a part of test development. The MAT-EF was examined in relationship to other measures of nonverbal ability, such as the WISC-R Performance scale and Raven's Progressive Matrices. In an analysis of 82 normal subjects, the correlation between the WISC-R Performance scale and the MAT-EF was .41. In a sample of 200 first- and second-grade children, the MAT-EF correlated .71 with Raven's Coloured Progressive Matrices. The MAT-SF was used with a sample of 70 deaf students in order to examine its relationship with other measures of nonverbal ability, primarily the WISC-R Performance scale (54 cases). The .72 correlation between the MAT-SF and other measures of intelligence (i.e., the WAIS-R, 5 cases; the Kaufman Assessment Battery for Children, 1 case; the Leiter International Performance Scale, 8 cases; and the Nonverbal Test of Cognitive Skills [Johnson & Boyd, 1981], 2 cases) indicated that the MAT-SF yields similar results for hearing and hearing-impaired individuals.

Analyses of sex and racial differences indicate that there are few variations in test results due to sex or race. Girls and boys perform similarly, as do blacks and whites. The MAT-EF manual reports these analyses in detail (Naglieri, 1985).

The MAT-EF manual also reports the factor analyses extensively. Naglieri cautions the reader that the MAT-EF is designed to measure a *single* construct of nonverbal reasoning ability. This is a caution well heeded because the item groupings (pattern completion, reasoning by analogy, serial reasoning, and spatial visualization) should be researched further. Upon visual inspection, a few items appear to be appropriate for more than one grouping.

Critique

The Matrix Analogies Test—Expanded Form and the Matrix Analogies Test—Short Form are new and promising measures of nonverbal ability. They are easy and efficient to administer and score. As such, they should be popular research instruments. The short form may be useful as a screening device although the forthcoming restandardization of Raven's Progressive Matrices may make this use less likely (Raven, 1987). In light of reports of the differences found between blacks and whites on other nonverbal, "culture fair" measures of ability (Sattler, 1974), the test author's analyses reporting few differences due to sex and racial characteristics should be ripe for further investigation.

The test materials are packaged attractively and of good quality. The level of detail in the manuals is particularly rewarding. However, one or two colored plates may test too fine a discrimination of size and proportion. They remind one of the picture puzzles in the Sunday newspaper that ask the reader to detect at least six differences between two very similar cartoon frames.

Further research on the MAT-EF and MAT-SF should assist in establishing their usefulness in clinical settings and subsequently their educational applications.

References*

Carlson, J. S., & Jensen, C. M. (1980). The factorial structure of the Raven Coloured Progressive Matrices Test: A reanalysis. *Educational and Psychological Measurement, 40,* 1111-1116.

Corman, I., & Budoff, M. (1974). Factor structures of retarded and nonretarded children on the Raven's Progressive Matrices. *Educational and Psychological Measurement, 34,* 407-412.

Johnson, G.O., & Boyd, H.F. (1981). *Nonverbal Test of Cognitive Skills.* Columbus, OH: Charles E. Merrill.

Naglieri, J. A. (1985). *Matrix Analogies Test—Expanded Form, examiner's manual.* Columbus, OH: Charles E. Merrill.

Raven, J. (1987, August). The 1985 U.S. norming of the Raven Progressive Matrices in an international context. Paper presented at the Seventh World Conference on Gifted and Talented Children, Salt Lake City, UT.

Sattler, J.M. (1974). *Assessment of children's intelligence.* Philadelphia, PA: W.B. Saunders.

*Tests formerly published by Charles E. Merrill are now published by The Psychological Corporation, San Antonio, Texas.

Selma Hughes, Ph.D.
Associate Professor of Special Education, East Texas State University, Commerce, Texas.

METROPOLITAN READINESS TESTS: 1986 EDITION

Joanne R. Nurss and Mary E. McGauvran. San Antonio, Texas: The Psychological Corporation.

Introduction

The Metropolitan Readiness Tests: 1986 Edition (MRT) are norm-referenced tests designed to assess the school readiness of children beginning school. The tests are skills-based and measure the underlying competencies needed for success in beginning reading-, mathematics-, and language-related activities. Primarily, they are group tests designed for administration to a class as a whole in order to determine the strengths and weaknesses of kindergarten and first-grade children; they may be given on an individual basis, if needed. The MRT form the core of a comprehensive assessment program that includes an early-school inventory of developmental skills (both cognitive and noncognitive), an early-school inventory assessing preliteracy, a resource book of ideas and activities for the development of these skills, and a form for reporting the test results to the parents.

The 1986 revision of the Metropolitan Readiness Tests is the fifth in a series of revisions that began in 1933. The program subsequently was revised in 1949, 1964, and 1976. The tests have over half a century of background experience and use that has been brought to bear on the construction of the present tests. Nurss and McGauvran (1976) describe the historical background of the tests and summarize the changes that have taken place from 1933 to 1976. The MRT were devised by Gertrude H. Hildreth and Nellie Griffiths and published by the World Book Company of Yonkers, New York, in 1933. The tests were among the first to be developed, and because they included both reading and number readiness material, they were more comprehensive than the Lee-Clark Readiness Test (1931), which predated them. The 1933 MRT consisted of one form only and was designed for use with beginning first-grade students. It consisted of six subtests and one optional subtest. The number of items correct on each subtest constituted the raw score, which could be converted to percentile ranks, and critical scores for determining readiness were provided.

The 1949 revision included two forms, R and S. Each form contained six subtests, which had been revised in content, and one optional subtest. In addition to percentile ranks and total scores for the six subtests, five letter levels, together with a reading readiness score and a number readiness score, were available.

The third edition, published in 1964, was available in forms A and B and, like the two earlier editions, was designed for first-grade entrants. In addition to percentile ranks, the six subtests and one optional copying test yielded five letter levels with stanines for the total scores and quartiles for each subtest. The authors of the third edition were Hildreth and Griffiths, the original authors, and Mary McGauvran, one of the authors of the present 1986 edition of the MRT.

The 1976 edition by Joanne Nurss and Mary McGauvran was a radical revision and included, for the first time, two levels of the tests. Level I tests were designed for administration to pupils early in the kindergarten year and for older students for whom the Level II tests were too difficult or inappropriate. Level II tests were designed for use with pupils at the end of the kindergarten year or the beginning of first-grade and were comparable to earlier, single-level editions of the test (Nurss & McGauvran, 1976). The number of subtests at Level II was changed from six to eight and one optional subtest of copying. The Level II MRT yielded raw scores, stanines, and performance ratings (low, average, high) for each skill area and a composite Pre-Reading Skills score. The new Level I of the MRT consisted of six subtests and one optional copying test and yielded similar information with separate ratings for Visual Skills and Language Skills.

The 1986 revision reflects current thinking and research related to readiness, which, as a construct, is controversial and has undergone considerable change over the years. Historically, the construct came into being in the 1920s and 1930s and reflected the views of child development that prevailed at the time. Readiness was described in terms of levels of maturation reached with the passing of time. Durkin (1970) gives a succinct summary of the development of the concept and how it was influenced by the writings of Arnold Gesell and his predecessor G. Stanley Hall. Both of these theorists linked readiness to factors such as "intrinsic growth" and "neural ripening."

The ability to learn to read became associated with a particular stage of development, and it was logical to assume that a child's problem with beginning reading was a sign that he or she had not reached the stage of development necessary for formal reading instruction and was, therefore, "unready" for reading. Tests were developed to measure a child's "readiness to learn to read" and readiness programs came in to being to help the child prepare for learning to read.

This concept of readiness was criticized soundly by the reading specialist Arthur Gates in the 1930s. Gates' many studies supported the conclusion that readiness to read was not entirely dependent on the nature of the child but was, in large measure, determined by the nature of the reading program. In other words, readiness was a function of the methods and the material (Durkin, 1970). The current view of readiness supports Gates' viewpoint that readiness is very much affected by the kind and quality of instruction offered to the child.

According to Salvia and Ysseldyke (1985), two dilemmas become apparent when assessing readiness. The first is that the performance of preschool children is so variable that readiness tests provide little in terms of long-term prediction, particularly the tests that are short and quickly administered. The second problem is that readiness tests may be used as measures of current functioning and current achievement.

In response to the first criticism, it should be pointed out that the MRT are quite comprehensive in scope and assess a broad variety of skills related to beginning school learning. The MRT are not short and quick; rather, they are quite thorough and have demonstrated predictive validity. The content of the tests has changed over the years to include skills more directly related to beginning reading, which improves their predictive validity. For example, the 1976 revision included the additional subtest of auditory skills, which are an essential component of linguistic awareness. In a recent article, Nurss (1980), one of the authors of the MRT, discusses in-depth the importance of linguistic awareness and learning to read.

In response to the second criticism, it is true that readiness tests may be used like achievement tests (measures of current functioning), and there is a need to question whether there is an appropriate linkage between the curriculum and the content of the readiness tests. However the fault lies not in the readiness tests themselves, but rather in how they are used. Durkin (1987) discusses the uses and abuses of readiness tests in a cogent article on the subject.

Readiness tests, according to Durkin, can provide information about what children know in relation to the content of the curriculum. Depending on the results, testing should lead to the development of a program altered to suit the needs of children requiring it. In practice, however, Durkin found no evidence that reading programs (i.e., instruction) were changed to meet individual needs. In fact, Durkin reported that a teacher in her study said that what was especially difficult about kindergarten teaching was that "there is great variation in what comes to you, but by the end of the year they are more levelled out" (Durkin, 1987).

The present 1986 edition of the MRT contains test items drawn from the pool of items from previous editions of the test and new items in areas where changes were considered necessary. An extensive review of the literature on child development, reading, and early-school learning led to the formulation of objectives and test specifications on which the present test is based. The authors describe the four-stage editorial process that followed the development of the new items, including two national item analysis programs (Nurss & McGauvran, 1986b).

Level I of the 1986 edition includes six subtests and one optional subtest. It is intended for use with four-year-olds in preschool settings, prior to entry into kindergarten, and during the kindergarten year. Level I tests also may be used with groups judged to be at relatively low levels of skill development in the first few weeks of Grade 1.

The six subtests that constitute Level I are 1) Auditory Memory, 2) Beginning Consonants, 3) Letter Recognition, 4) Visual Matching, 5) School Language, and 6) Listening and Quantitative Language. The second subtest, Beginning Consonants, replaces the Rhyming subtest, which was a part of the 1976 edition. Level I also includes an optional Copying Sheet on which the child is required to copy his or her own first name from a model provided by the teacher. As with earlier editions, an MRT practice booklet is available to help pupils prepare to take the tests. By completing the items in the practice booklet, children learn skills such as working from left to right across the page and down the page, as well as how to indicate the correct answer by drawing a line across the chosen option. Teachers may provide as much help as the pupils need on the first six items of the practice booklet but not on the last four items, which must be answered independently.

The Auditory Memory Subtest on Level I requires the ability to remember two, three, or four word sequences and match them with the appropriate visual symbols. The Beginning Consonants subtest requires the child to match a given sound to one of four objects shown in pictures. Letter Recognition requires the ability to discriminate the graphemic representation of sounds from four given choices. Visual Matching requires the child to match a series of printed shapes and upper- and lowercase letter combinations to samples. In School Language and Listening, the child marks the picture that shows the actions or objects that have been described. The Quantitative Language subtest measures understanding of comparison of adjectives and concepts such as one-to-one correspondence, as well as matching numerals to the number of objects designated.

In order not to tire the children who are taking the test, it is recommended that the Level I subtests be given at different sittings that vary in terms of administration time. The overall length of time required for the tests is about 95 minutes although there are no time limitations.

At Level II, the same practice booklet is used to familiarize students with the testing procedures. There are eight subtests at Level II. In Beginning Consonants, the child identifies the sound at the beginning of two words, which are given orally, and matches it with the beginning sound of words previously identified. Sound-to-Letter Correspondence requires the child to match sounds to letters or letter combinations. Visual Matching requires the ability to discriminate between visual symbols and match them to a sample. The visual symbols include meaningless symbols, upper- and lowercase letter combinations, and combinations of numerals. The Finding Patterns subtest requires the ability to find one letter or a number of letters embedded in longer letter combinations or words. School Language assesses receptive language skills by requiring students to mark the picture that illustrates the concept described. The Listening subtest is similar to the School Language subtest, but it requires reasoning ability because the pupil must arrive inductively at the correct answer before selecting it from a choice of four visual symbols. The subtest of Quantitative Concepts requires spatial reasoning, counting, and one-to-one correspondence skills.

The optional subtest at both levels of the MRT is a copying subtest. At Level 1, the copying subtest assesses the students' skills in reproducing their own names or in reproducing a sentence printed on the sheet. The copying sheets are evaluated by comparing them to samples rated on a scale from one to five. The percent of cases attaining various ratings at midyear of kindergarten and in the fall of Grade 1 is shown for comparison purposes.

The Early School Inventory (ESI)—Preliteracy may be administered with either level of the tests. It is individually administered and provides information about the child's understanding of concepts about print and familiarity with the written language code. The performance ratings are "proficient" (+), "acquiring the skills" (/), or "needing help" (-). The cumulative percentages of children ages 5-7 are shown for the raw scores in each of the four areas assessed by the ESI.

The Early School Inventory (ESI)—Developmental is an enumeration of 20 skills under each of four areas: physical, language, cognitive, and social-emotional. Behaviors to assess each skill are reported as either "observed" or "not observed" by the parents and teacher. Activities that parents could use to help develop the skills not observed are suggested.

The test booklets at each level consist of over 20 pages of large-size black-and-white line drawings and large-size print symbols enclosed in clearly defined colored boxes. The material provides good cues to the line of boxes to be studied through the provision of solid pictures at the beginning of each line. These enable the teacher to indicate the line of options. For example, "Put your finger on the leaf—heart—moon—flowers—etc." The pupil is required to shade a small circle under the box that indicates the correct answer. The answer forms are pleasing in format and not obstructed by unnecessary detail. They are most suitable for the age level for which the MRT are intended.

Some of the oral directions could be confusing initially although the practice booklet is intended to make the nature of the tasks clearer. For example, for the Beginning Consonants subtest on Level I, the pupils listen to two words beginning with the same sound. The next statement says "I am going to say some more words. Tell me if they begin with the same sound." Pupils may be listening for the same sound as the first pair of words instead of listening for the similarity of sound at the beginning of the two words presented. The nature of the task would become clearer by the second item, and one could argue that adding to the directions would make them longer and more difficult for the younger students to understand.

Practical Applications/Uses

The purpose of readiness testing should be to determine the additional teaching or alternative instruction that is needed to facilitate reading development. The MRT are an appropriate means of screening children to find out areas in which more instruction is needed. The MRT may be used to screen a small number of children from a larger group to determine whether additional help is needed to prevent failure in academic learning. The MRT are also suitable for use by classroom teachers to determine whether a particular method of instruction is appropriate for use with particular children or whether another method of instruction should be utilized.

Nurss and McGauvran (1981) acknowledge that, in the past, measures of a child's overall level of functioning were frequently used for placement decisions. The dichotomous classification of "ready" and "not ready" derived from the total score on the readiness tests may have been used to justify retention in kindergarten or refusal of entry to first grade in schools that did not have kindergarten.

The way in which the tests were used was clearly different from that which was intended by the test authors. Nurss and McGauvran (1981) point out that from the beginning, the instructional use of the MRT has been stressed. That is to say, the tests were and are intended to show a child's relative strengths and weaknesses in the various instructional areas so that appropriate intervention may be taken. The manuals of the first four editions of the test stress skill development in areas assessed by the subtests, and the fifth edition of the MRT incorporates an activities handbook as part of the total readiness program.

Because the national standardization sample included both white and black children, as well as children of Spanish and other unspecified ethnic origins, the MRT are suitable for use with children of all ethnic groups. The National Standardization Program for both levels of the MRT utilized a stratified sample that took into account factors such as public and nonpublic schools, size of the school enrollment, geographic location, and so on. The tests are, therefore, suitable for use in a wide variety of school situations and with children of the diverse ethnic groups included in the sample.

The demographic characteristics of the children participating in the standardization sample corresponded to the total school enrollment of the United States in terms of socioeconomic status. There is no specific mention of the developmental status of the children used in the sample; that is, whether children with physical or mental handicaps were included. Nevertheless, the MRT could be used with exceptional children in kindergarten and preschool to assess readiness skills for beginning instruction in academic areas. Although not normed on handicapped children, Lerner, Mardell-Czudnowski, and Goldberg (1987) include the MRT program as an example of tests that could be used by special educators during the early childhood years.

The MRT could be used on an individual basis with children with some handicapping conditions (e.g., physical handicaps). Because minimum movement is required to fill in the circle in the test booklet, it would be simple to adapt the response. However, the tests place great emphasis on verbal instructions, so they may be inappropriate for use with children who have severe auditory impairments or receptive language problems. The visual aspect of the tests make them inappropriate for use with children with severe visual or perceptual problems.

No special training is required to administer the MRT, and no specialized personnel are needed. It is suggested that better results are obtained when the pupil's classroom teacher administers the tests because rapport already exists between the pupil and the teacher. No special equipment is needed for test administration other than the test materials and a chalkboard to demonstrate the marking system. Comprehensive directions are given in the manual; a test schedule showing the order in which the tests should be administered is provided. Options for administering the tests in up to seven sittings are discussed, which should minimize fatigue for younger children. The estimated time for each sitting is about 15 minutes.

It is recommended that students in Kindergarten and Grade 1 be tested in groups of no more than 15, and prekindergartners in groups no larger than 10. More than one sitting may be held in one day provided that a change of activity follows the first administration of the test. The practice test should be administered a day or so prior to the first sitting.

The tests may be machine- or hand-scored. The norms booklet provides directions for hand-scoring the tests. It also gives very clear directions for converting the raw scores for each test to a performance rating and for converting the Skills Area scores to norm-referenced scores. It does this by indicating the tables to be used with the exact page numbers. The procedures to be followed are outlined clearly.

Class record sheets and class analysis charts provide a systematic summary of the tests' results and directions are given (with a completed model) on how to complete them. No special keys, codes, or templates are needed for the hand-scoring; only the teacher's ability to follow directions is required. The interpretation of the test results is straightforward and based on objective criteria; it does not require clinical judgment on the teacher's part. Minimal training or sophistication is required to interpret the test results, which are user friendly.

Technical Aspects

Two aspects of validity relevant to the MRT are content validity and predictive validity. Content validity seeks to determine whether the content of the test is related to the skills of beginning reading and mathematics. The authors of the test point out that an extensive review of the literature was conducted prior to the revision of the MRT to identify the specific skills that are important to early learning (Nurss & McGauvran, 1986b). The MRT is considered to have adequate content validity.

Predictive validity measures the extent to which some criterion scores may be predicted from scores on another test. Predictive validity on the MRT seeks to show how reading and mathematics achievement scores may be predicted from the readiness scores on the MRT. Two studies of predictive validity were conducted during the 1985-86 school year for Level II of the MRT. First-grade children were tested with Level II of the MRT in the fall of 1985 and then with the Primary Level of either Form L of the Metropolitan Achievement Tests, 6th Edition (MAT-6) or Form E of the Stanford Achievement Tests, 7th Edition (SAT-7) in the spring of 1986. The number of children tested was 600 and the process by which the MRT and the achievement tests was equated (using the Rasch model) is available from the test publishers.

The correlations quoted between the MRT and the achievement tests vary from .34 to .65 for the MAT-6 and from .47 to .83 for the SAT-7. There is a higher correlation between the Prereading Composite score and the Total Reading score than between the Skill Areas scores and the Total Reading score, which is to be expected. The Prereading Composite score on the MRT shows .77 correlation with the Total Reading score on the SAT-7 and .83 correlation with the Total Complete Battery, indicating that the child's score on the MRT is a good predictor of the reading score on the SAT-7 and an even better predictor of the overall score on the SAT-7 test battery.

The predictive validity of the MRT to the Metropolitan Achievement Tests is not quite as high as for the Stanford Achievement Tests nor is the predictive validity of the skill areas as great as the predictive validity of the Prereading Composite score. Nevertheless, the correlations quoted are quite impressive and more than adequate to justify the use of these scores.

The Quantitative Skill Area score on the MRT correlates well with the Total Mathematics score on both the MAT-6 (.58) and the SAT-7 (.74). The MRT do, therefore, measure what they purport to measure and the tests are good predictors of later scores in reading and mathematics achievement.

The 1986 edition of the MRT also gives the intercorrelations between each of the skill areas of the MRT. This data is provided for both levels of the test. An analysis of the scores of over 4,000 children at each level was made and provides useful data to test users. For example, there is a strong correlation (greater than .90) between the Auditory Skills subtest and the Beginning Consonants subtest, which indicates that any intervention resulting in improved recognition of beginning consonants may also result in improved performance on auditory skills.

A test can never be valid unless it is at the same time reliable. Two kinds of reliability data are provided for the MRT, internal consistency data and test-retest data.

The Kuder-Richardson Formula 20 (KR-20) was used to show the lack of error due to internal inconsistency, and the coefficients that were computed range from .66 to .88 for the skill areas to .93 for the Prereading Composite. The MRT demonstrates high internal consistency.

Test-retest reliability provides evidence of the stability of test scores over time. High scorers on the test on one administration should be high scorers on a later administration of the test. The number of children who were tested twice within a 2-week period with Level I of the MRT is not stated, but the reliability coefficients obtained varied from .64 to .87 for the Skill Areas to .92 for the Prereading Composite. For Level II, the test-retest coefficients ranged from .62 to .88 for the Skill Areas to .88 for the Pre-Reading Composite. The MRT, therefore, demonstrates high test-retest reliability.

In addition to the above data, the norms booklet provides the standard error of measurement (SEM), an indication of how close the child's true score is to the obtained score, for each level of the test for each skills area. Using what is known about the SEM, it is possible to estimate each child's true score to a predetermined level of confidence. Extensive data on the technical background of the test is available in the norms booklet, which does show satisfactorily that the MRT is technically sound and well constructed.

Critique

The Metropolitan Readiness Tests are designed specifically to determine children's levels of cognitive development as a prerequisite for beginning reading and mathematics instruction. They form the core of a readiness assessment program that also appraises developmental skills and preliteracy skills and identifies the particular instructional needs of each child. A handbook of skill development activities contains small group and individual activities that can be used to remediate weak areas identified by the tests. The MRT are well designed and highly suitable for use by classroom teachers. They are easy to administer, and comprehensive instructions are provided to aid in interpreting the results. The format of the tests is suitable for the younger age range for which the tests are intended.

The major limitation of the MRT is the relatively limited behavior sample (Salvia & Ysseldyke, 1985). However, this is precisely what gives the MRT its good predictive validity. It makes use of a sampling of specific skills that are strongly predictive of success in beginning reading and mathematics. The predictor measure, the MRT, incorporates many of the skills assessed in standardized achievement tests. The sixth edition of the Metropolitan Readiness Test is a valuable instrument for groups or for individualized use as needed.

References

Durkin, D. (1970). *Teaching them to read.* Boston: Allyn and Bacon.

Durkin, D. (1987). Testing in kindergarten. *The Reading Teacher, 40,* 766-770.

Hildreth, G. H., Griffiths, N. L., & McGauvran, M. (1965). Metropolitan Readiness Tests. Cleveland, OH: The Psychological Corporation.

Lee, J. M., & Clark, W. (1931). *Lee-Clark Readiness Test.* Monterey, CA: CTB/McGraw-Hill.

Lerner, J., Mardell-Czudnowski, & Goldberg, D. (1987). *Special education for the early childhood years.* Englewood Cliffs, NJ: Prentice Hall.

Nurss, J. (1980). Linguistic awareness and learning to read. *Young Children, 35,* 57-66.

Nurss, J., & McGauvran, M. (1976). *Metropolitan Readiness Tests.* Cleveland, OH: The Psychological Corporation.

Nurss, J., & McGauvran, M. (1981). Readiness and its testing (Metropolitan Readiness Tests Technical Report No. 1). New York: The Psychological Corporation.

Nurss, J., & McGauvran, M. (1986a). *Metropolitan readiness assessment program.* San Antonio, Texas: The Psychological Corporation.

Nurss, J. & McGauvran, M. (1986b). *Metropolitan Readiness Tests level 1 and 2 norms booklet.* San Antonio, Texas: The Psychological Corporation.

Salvia, J., and Ysseldyke, J. (1985). *Assessment in special and remedial education.* Boston: Houghton Mifflin.

Gordon R. Simerson, Ph.D.

Assistant Professor, Industrial/Organizational Psychology, University of New Haven, West Haven, Connecticut.

MINNESOTA JOB DESCRIPTION QUESTIONNAIRE

Work Adjustment Project. Minneapolis, Minnesota: Vocational Psychology Research, University of Minnesota.

Introduction

The Minnesota Job Description Questionnaire (MJDQ) is designed to measure the working environment of a job in terms of a profile of reinforcer (need-satisfier) characteristics. This profile, called an Occupational Reinforcer Pattern or "ORP," describes the stimulus conditions in a job that are believed to contribute to the worker's job satisfaction. Typically, employees, their supervisors, and job analysts are given the MJDQ to rate a job along a fixed set of dimensions. An ORP is constructed by combining all of these ratings. It then can be used as a standard against which to compare an individual's profile of needs along the same set of dimensions. If the individual's need profile matches the ORP for a certain occupation, it is predicted that the individual will experience higher job satisfaction and be better adjusted in that occupation than individuals whose ORPs differ from their need profiles.

The MJDQ is one of a battery of instruments developed by the Work Adjustment Project at the University of Minnesota to test the propositions of the Minnesota Theory of Work Adjustment (MTWA). Other related instruments include the Minnesota Importance Questionnaire (MIQ), the Minnesota Satisfaction Questionnaire (MSQ), the Minnesota Satisfactoriness Scales (MSS), and the Minnesota Job Requirements Questionnaire (MJRQ), some of which have been reviewed in earlier volumes of *Test Critiques* (Bolton, 1985, 1986; Benson, 1985). The principal authors of the Minnesota Theory of Work Adjustment, Rene Dawis, Lloyd Lofquist, and David Weiss, are psychology professors at the University of Minnesota. Many graduate students in industrial/organizational and counseling psychology have contributed to the Work Adjustment Project since its inception in 1957. The first account of the MTWA appeared in 1964 (Dawis, Lofquist, & England), but the theory has been updated and expanded more recently (Dawis, Lofquist, & Weiss, 1968; Dawis & Lofquist, 1984; Dawis, 1986).

According to the theory, an employee's vocational adjustment is defined as his satisfactoriness to his employer and his satisfaction with his job. Satisfactoriness is postulated to be a function of the correspondence of the employee's job-relevant abilities with the ability requirements of the job. Similarly, job satisfaction is postu-

This reviewer gratefully acknowledges the assistance of Noah L. Durkin of the University of New Haven and of Drs. Allen Due and David Weiss of the University of Minnesota.

lated to be a function of the match between the employee's profile of work-relevant needs and the profile of reinforcer (need-satisfying) characteristics of the job. The theory originally focused on vocational rehabilitation and disabilities, and its application has been primarily in the area of assisting counselees with occupational choices. Many of the instruments developed in connection with the MTWA, however, have been used widely in a variety of research projects not related directly to the Work Adjustment Project.

The underlying rationale—that one's adjustment to a job depends upon the appropriateness of the person-job match—is not new. The classic work by Parsons (1909) introduced the man-job matching model and has influenced other well-known authors such as Strong, Super, Roe, and Holland. Although several systems were developed in the ensuing years to describe work environments, Schaffer (1953) and Darley and Hagenah (1955) lamented the absence of a standardized system for measuring directly the extent to which families of jobs could satisfy a broad set of work-related needs. The MJDQ introduced in 1967, was developed in response to this challenge. The broad set of reinforcers that the instrument measures are adapted directly from the MIQ and MSQ dimensions, thus allowing direct comparison with those measures.

The MJDQ is available in two forms: Form S is used with job analysts and supervisors; Form E is for employees currently in the target job. Form E includes the 20-item short form of the MSQ and asks for slightly different information about the respondent than Form S (e.g., "How long have you been employed in that job?" as opposed to "How long have you been a supervisor?"). Otherwise, the two forms are essentially the same. Both forms were revised in 1979 to eliminate gender-specific language but otherwise are unchanged since their introduction.

The 21 reinforcer dimensions measured by the MJDQ are 1) ability utilization, 2) achievement, 3) activity, 4) advancement, 5) authority, 6) company policies and practices, 7) compensation, 8) co-workers, 9) creativity, 10) independence, 11) moral values, 12) recognition, 13) responsibility, 14) security, 15) social service, 16) social status, 17) supervision-human relations, 18) supervision-technical, 19) variety, 20) working conditions, and 21) autonomy. The first 20 dimensions are the same as those measured by the 20-item MIQ and MSQ, but the items are reworded slightly in order to tap what workers can *expect* to experience along these dimensions rather than how *important* (MIQ) or *satisfying* (MSQ) those experiences are.

A methodological peculiarity necessitated the inclusion of an additional dimension, autonomy. The MJDQ is constructed using a multiple-rank-order method. Each of the 21 reinforcer dimensions is represented by one item. Twenty-one blocks of items are presented, five items to a block, and the respondent is instructed to rank the items in each block from "1" to "5" in terms of their descriptiveness of the job. Each statement appears in five different blocks, each time with a different set of statements and in a different position within the block. Balance in the exhaustive juxtaposition of items cannot be achieved with 20 items, but it can be achieved using 21. The "autonomy" dimension was suggested by the factor analyses performed during the development of the MSQ although it is not included in that instrument.

The multiple-rank-order system allows for assessing consistency of responses in a way comparable to the more simple paired-comparison format. Although the former method requires more complex judgments, it requires only about half as many item presentations and significantly less administration time.

A second shorter section of the MJDQ requires examinees to make an absolute judgment for each of the same 21 statements. The respondent is instructed to indicate ("yes" or "no") whether the statement is generally descriptive of the job.

The test booklets are 8½" × 5½". Form S devotes one page to instructions, the presentation of a sample item, a code number for the respondent, and a gaurantee of confidentiality. Five pages contain the ranked blocks of items, one page presents the absolute judgment items, and the last page asks for information about the respondent. Form E devotes an additional two pages to the short, 20-item form of the MSQ. The booklets, available only from Vocational Psychology Research (VPR) at the University of Minnesota, cost from $0.53 to $0.60 each (plus postage), depending on quantity. The technical manual is free to purchasers of the questionnaires, but copies of the catalogue of ORPs already computed by the publisher cost extra.

Although hand scoring of the MJDQ is theoretically possible, it is so complex, time-consuming, and fraught with error that the computer-scoring services offered by the test publisher are a practical necessity. This service costs about $1.25 per booklet (scoring the MSQ section costs extra), and the publisher will arrange the booklets into groups if requested. The University of Minnesota will provide a computer-generated ORP for each job and may assist in its interpretation.

The reading level required to complete the MJDQ is estimated to be about fifth or sixth grade. An intimate familiarity with the job is a more relevant and stringent requirement for the respondents, however. Because the questionnaire essentially is self-administered, no participation by an examiner is involved. Competent interpretation of the ORPs, though, requires an advanced and thorough knowledge of psychometrics and a general familiarity with theory on work adjustment.

Practical Applications/Uses

It is difficult to separate the use of the MJDQ from the wholesale application of the Minnesota Theory of Work Adjustment. The MJDQ is designed specifically to provide raw data from which ORPs of selected jobs are constructed. These ORPs, in turn, provide reference points that facilitate the interpretation of MIQ and MSQ scores. Although the MSQ and MIQ have each been used widely in research studies not associated with the Work Adjustment Project, the MJDQ would seem to be less flexible, and its independent use by the average user is unlikely.

Use of the MJDQ by sophisticated users is another matter, however. The fact that the instrument is dedicated to the MTWA is a possible advantage because the theory, taken as a whole, is widely understood and well documented. The uniqueness of the MJDQ is that it is the basis for a job taxonomy based upon dimensions that are *psychologically meaningful* to job incumbents. Thus, when the MJDQ is used as a supplement to job analysis systems that focus on behaviors or tasks, the analyst's understanding of the jobs is potentially broader and richer.

With this idea in mind, two alternative uses besides vocational counseling are noted. The Minnesota Occupational Classification System (MOCS; Dawis, Lofquist, Henly, & Rounds, 1979/1982; Dawis, Dohm, Lofquist, Chartrand, & Due, 1987) is based upon data from the MJDQ, the MIQ, and the General Aptitude Test

Battery. This system generates "clusters" of jobs, taking into account both the abilities required of the incumbent and the reinforcer characteristics preferred by the incumbent. Second, the MJDQ and MIQ might contribute to a broad-brush system for personnel selection purposes when used in concert with job-oriented systems for job analysis. It probably would be advisable, however, to rely on local norms when using the MJDQ this way rather than to assume that the ORPs catalogued by Vocational Psychology Research are accurate. The reason for this caveat is discussed in the next section. The rest of this review focuses only on ORPs' intended use for vocational counseling.

The preferred respondents for the MJDQ are the supervisors of the jobs being investigated, although the publishers invite its use as well with job analysts and incumbents. Although normally self-administered, individual administration for those with physical disabilities is possible and should not affect the validity of the results. The most important and exclusive criterion for subject selection is that the respondents be intimately familiar with the job being rated.

The MJDQ is most demanding in terms of scoring and interpretation. Whether one uses the catalogue of ORPs already available from Vocational Psychology Research or asks that new ORPs be constructed, interpretation of the computer-generated results calls for advanced knowledge of psychometrics and familiarity with theory on vocational adjustment. Graduate-level training in both areas is recommended.

The use of the ORPs for vocational counseling is made easier and somewhat more mechanical, however, by relying on additional information provided in the ORP catalogue (Stewart et al., 1986). The profiles for the jobs are presented along with a list of the most descriptive of the 21 dimensions based on VPR's statistical interpretation of the profiles. Corresponding codes from the *Dictionary of Occupational Titles* (U.S. Department of Labor, 1982) and the job's Occupational Aptitude Pattern (U.S. Department of Labor, 1979) are provided for comparison purposes, as are names of other occupations falling into the same empirically derived ORP cluster. Factor analyses performed by VPR on MIQ data resulted in six value dimensions: achievement, comfort, status, altruism, safety, and autonomy. These have been incorporated into the display of the profiles in the ORP catalogue to facilitate the matching of an individual's values detected by the MIQ to the dimensions of jobs appearing in the ORPs most relevant to those particular values.

Although only 81 occupations were included in the original catalogue of ORPs (Borgen, Weiss, Tinsley, Dawis, & Lofquist, 1968a), data continue to accumulate allowing VPR periodically to expand the coverage, particularly among professional and managerial occupations previously underrepresented. The number of occupations covered by 1972 was 145 (Borgen, Weiss, Tinsley, Dawis, & Lofquist, 1972; Rosen, Hendel, Weiss, Dawis, & Lofquist, 1972), and by 1986 the number covered had climbed to 185 (Stewart et al., 1986). The norms for occupations included in earlier editions are updated in newer editions.

Technical Aspects

The most comprehensive technical account of the MJDQ and the resultant ORPs is found in the manual (Borgen, Weiss, Tinsley, Dawis, & Lofquist, 1968b). Because the development of the MJDQ was approached from a "content validity" orientation using MIQ dimensions, one must consult the technical manual of the MIQ (Gay, Weiss, Hendel, Dawis, & Lofquist, 1971) and critiques of that instrument (e.g., Benson, 1985; Zedeck, 1978a; Albright, 1978) to fully assess the conceptual content of the MJDQ.

A distinction must be drawn between the technical properties of the MJDQ and the ORPs it generates. Because the MJDQ is not used to assess an attribute of an individual respondent vis-à-vis a normative distribution, but rather to accumulate perceptions of many respondents in order to assess an environment, evaluation of the MJDQ as an instrument is not as productive as evaluating the ORPs.

Several methodological precautions have been taken to insure the reliability of the data on which the ORPs are based. First, before constructing the ORP for a given occupation, at least 20 supervisors and/or workers must have contributed data on the MJDQ. Although most occupations are well represented, some others are either comparatively rare or typically are not supervised (e.g., an optometrist in private practice), making larger samples hard to find. Supervisors are the preferred respondents for the MJDQ, but evidence exists that employees and their supervisors generally agree on MJDQ ratings, and the level of agreement is better for higher-level and more sophisticated jobs (Tinsley & Weiss, 1971). Setting the minimum sample size cutoff at 20 and using employees' ratings seems adequate in light of that finding, at least for reporting the preliminary results for those jobs about which supervisors and workers could be expected to agree.

Second, the multiple-rank-order method used to design the MJDQ allows for monitoring the consistency of responses by individuals. Because each item is compared to every other item at least once in the ranking blocks portion, a check on transitivity is possible. The rationality of an individual's data could be suspect if a pattern emerged where, for example, item A was ranked higher than item B, and item B higher than item C, but item C later was ranked higher than item A. A few such circular triads are to be expected, particularly among items whose true rank positions are not extremely high or low. A total circular triads (TCT) score can be computed for each respondent. Further, the expected distribution of TCTs for entirely random responding is known. An apparently conservative maximum TCT score cutoff has been set at a point three standard deviations below the mean of this distribution. Less than 1% of the original data were rejected using this cutoff. This is more noteworthy in light of the test publisher's practice of allowing missing data for up to six items to be replaced by randomly generated digits.

A third control relies on the absolute judgment section of the MJDQ. Merely rank ordering a set of stimuli yields no information about their strength in an absolute sense. By also providing a dichotomous, "yes/no" response to each item, its position relative to a psychological neutral point can be estimated. The neutral point then can be treated statistically as an additional stimulus but with a known absolute level, allowing for more accurate calibration of stimulus judgments made by multiple respondents and more reliable scaling of the stimuli (in an absolute sense) based on aggregated data.

A reliability study of the ORPs is reported in the manual (Borgen et al., 1968b). A measure of internal consistency was calculated for each occupation by splitting the derivation sample in half and computing the correlations between the profiles generated by each half. The uncorrected "split-half" coefficients ranged from .78 to .98 with a median of .91; the number of subjects ranged from 11 to 48. By comparison, the correlations between profiles for different occupations ranged from about -.08 to about .97 with a median of .55. In order for an ORP to be included in the catalogue (Stewart et al., 1986), its split-half reliability must exceed .85.

Zedeck (1978b) noted that many of the supervisors who contributed the raw data were recruited from the same organization or from a small number of organizations. The samples of supervisors may or may not be responding representatively to the full range of reinforcer conditions characteristic of those occupations. Therefore, the high levels of agreement among those raters evidenced by the high split-half coefficients might only reflect the homogeneity of their environments and experience. It is for this reason that the earlier caveat was raised about using existing ORPs without first establishing that the derivation sample is representative of one's own organization.

Other reliability evidence has accumulated regarding the stability of ORPs and group differences. Flint (1980) measured the "test-retest" reliability of ORPs over an 11-year period. For two occupations, professional nurse and real estate agent, the stability of profiles (.88 and .93, respectively) compared favorably to their internal consistency. The same study established the high similarity of MJDQ ratings made by male and female respondents. Another study by Jenson (1975) found no differences in ratings as a function of the level of professional development respondents had attained within the occupation of school psychologist.

Establishing the validity of the MJDQ requires information focusing on two levels of analysis. First, it must be shown that MJDQ responses can be relied upon to derive ORPs; and second, it must be shown that ORPs provide meaningful information about occupations that conforms to theory.

Much of the information needed to address the first level of analysis has been presented. Convergent validity evidence is found in intrarater agreement (raters respond rationally and consistently) and in intraoccupation/interrater agreement (multiple raters respond to the same occupation similarly). No evidence is reported in the manual regarding the stability of a given rater's responses over time. However, the low incidence of rejections based on TCT scores constitutes supportive evidence for intrarater agreement. Evidence for intraoccupation/interrater agreement is not as clear. Although the split-half coefficients are high, the variance in the ratings for particular reinforcers in given occupations is also high. Given that the most extreme scale values in the ORPs (those reflecting that certain reinforcers are very characteristic or very uncharacteristic of an occupation) are given the most weight in interpretation and that the greatest variance usually appears in those dimensions falling around the neutral point, the positive split-half reliability evidence should be considered more compelling.

Discriminant validity is evidenced in interoccupation disagreement—the ability of the MJDQ data to differentiate among occupations. As noted above, the ORPs' internal consistency tend to be higher than the average correlation among ORPs for different occupations. Taken as a whole, positive evidence exists that MJDQ data can be relied upon to derive ORPs.

The most compelling evidence for the validity of ORPs would be found in 1) meaningful group differences based on ORPs, 2) meaningful convergence with other taxonomies of occupations, and 3) meaningful relationships between ORPs and other variables that conform to theory.

Stepwise cluster analysis performed on the original 81 ORPs resulted in nine interpretable clusters accounting for 59 of the occupations. The clustering seemed to depict an occupational hierarchy. Holland (1973) noted that these same clusters seemed to align conceptually with the six dimensions measured in his own Environmental Assessment Technique. All of the ORPs catalogued by 1986 are accounted for by a set of six occupational clusters based on the Minnesota Occupational Classification System-II (Dawis, Lofquist, Henly, & Rounds, 1982).

Less forceful evidence concerns the relative abilities of the 21 individual MJDQ scales to discriminate among occupations. One-way analyses of variance were conducted to test how well each MJDQ scale could discriminate among the original 81 occupations (Borgen, Weiss, Tinsley, Dawis, & Lofquist, 1968b). Omega-squared ranged from .07 to .37 with a median value of .16. Although all 21 Fs were statistically significant ($p < .00001$, with 20 and 2895 degrees of freedom for each analysis), some scales, such as Creativity, Responsibility, Compensation, Ability Utilization, and Social Service, were clearly better able to provide information about occupational differences than others, such as Moral Values, Recognition, and Supervision-Human Relations.

Because the MJDQ has been used so seldom in research conducted outside of the Work Adjustment Project, evidence relating ORPs to other variables is accumulating slowly (e.g., Betz, 1969; Dunn & Allen, 1973; Elizur & Tziner, 1977). However, these studies provide at least tentative support that, in a concurrent design, MIQ-ORP correspondence is predictive of job satisfaction. This approach to measuring need-reinforcer correspondence has been adapted successfully by researchers for purposes other than vocational counseling (e.g., Seiler & Lacey, 1973), thus attesting to its conceptual appeal, rigor, and flexibility.

Critique

Whether the VPR instruments are used widely, the MTWA is a solid, mainstream theory of work adjustment with abundant conceptual appeal. Since its inception, the Work Adjustment Project has maintained a reputation for thorough methodology. A difficulty surrounding the MJDQ is that its design, and, hence, its validity is largely dependent on that of the MIQ, on which it was based. Neither the MIQ manual nor the MJDQ manual describe in sufficient detail the item selection process for these instruments. A thorough technical review of the history of the Work Adjustment Project requires examination of many of the more than 30 technical monographs in the *Minnesota Studies in Vocational Rehabilitation* series.

The ORPs developed from MJDQ data appear psychometrically sound, particularly with respect to reliability. Their validity is somewhat less clear due mainly to the lack of information, as noted above. More information on the social desirability of MJDQ items and the representativeness of the samples of respondents would be helpful. The most serious vulnerability of the ORPs is that MJDQ data from raters in a limited array of work environments would both inflate the apparent level of inter-rater agreement and create a bias in the ORP obscured by that high level of reliability (cf. Zedeck, 1978b).

Practitioners applying the ORP methodology would be interested in more information on the influence of moderator variables such as race, tenure, and educational level, as well as the disability and job history information presented in the ORP catalogue. Researchers familiar with the intrinsic/extrinsic structure of the MSQ would be assisted by information about whether the measures of occupational reinforcers share that structure.

The application of the VPR system and measures, both for vocational guidance and basic industrial research, requires well-developed interpretive skills. However, the system is "user-friendly"—at least on par with most other vocational guidance systems—for individuals possessing such a background.

The MJDQ form itself requires patience and concentration on the part of the respondent. The repetitiveness of the items may become boring, yet the ranking judgments are complex and demanding. This apparently is the price to be paid for the psychometric integrity of the ORPs, which is due in large part to the tedious multiple-rank-order method.

Overall, the MJDQ reflects the conscientious efforts of the Work Adjustment Project. Provided that potential users have familiarized themselves assiduously with the role of the MJDQ in the Minnesota Theory of Work Adjustment, its use is recommended.

References

Albright, L.E. (1978). Review of The Minnesota Importance Questionnaire. In O.K. Buros (Ed.), *The eighth mental measurements yearbook* (pp. 1671-1673). Highland Park, NJ: The Gryphon Press.

Benson, P.G. (1985). Minnesota Importance Questionnaire. In D.J. Keyser & R.C. Sweetland (Eds.), *Test critiques* (Vol. II, pp. 255-265). Kansas City, MO: Test Corporation.

Betz, E. (1969). Need-reinforcer correspondence as a predictor of job satisfaction. *Personnel and Guidance Journal, 47,* 878-883.

Bolton, B. (1985). Minnesota Satisfactoriness Scales. In D.J. Keyser & R.C. Sweetland (Eds.), *Test Critiques* (Vol. IV, pp. 434-439). Kansas City, MO: Test Corporation.

Bolton, B. (1986). Minnesota Satisfaction Questionnaire. In D.J. Keyser & R.C. Sweetland (Eds.), *Test Critiques* (Vol. V, pp. 255-265). Kansas City, MO: Test Corporation.

Borgen, F.H., Weiss, D.J., Tinsley, H.E.A., Dawis, R.V., & Lofquist, L.H. (1968a). *Occupational reinforcer patterns: I.* Minneapolis: University of Minnesota, Vocational Psychology Research.

Borgen, F.H., Weiss, D.J., Tinsley, H.E.A., Dawis, R.V., & Lofquist, L.H. (1968b). *The measurement of occupational reinforcer patterns* (Minnesota Studies in Vocational Rehabilitation No. 25). Minneapolis: University of Minnesota, Work Adjustment Project.

Borgen, F.H., Weiss, D.J., Tinsley, H.E.A., Dawis, R.V., & Lofquist, L.H. (1972). *Occupational reinforcer patterns: I.* Minneapolis: University of Minnesota, Vocational Psychology Research.

Darley, J.G., & Hagenah, T. (1955). *Vocational interest measurement: Theory and practice.* Minneapolis: University of Minnesota.

Dawis, R.V. (1986). The Minnesota Theory of Work Adjustment. In B. Bolton (Ed.), *Handbook of measurement and evaluation in rehabilitation* (2nd ed., pp. 203-217). Baltimore: Paul Brookes.

Dawis, R.V., Dohm, T.E., Lofquist, L.H., Chartrand, J.M., & Due, A.M. (1987). *Minnesota occupational classification system-III.* Minnesota: University of Minnesota, Vocational Psychology Research, Work Adjustment Project.

Dawis, R.V., & Lofquist, L.H. (1984). *A psychological theory of work adjustment: An individual differences model and its applications.* Minneapolis: University of Minnesota.

Dawis, R.V., Lofquist, L.H., & England, G.W. (1964). *A theory of work adjustment. (Minnesota Studies in Vocational Rehabilitation No. 15).* Minneapolis: University of Minnesota, Work Adjustment Project.

Dawis, R.V., Lofquist, L.H., Henly, G.A., & Rounds, J.B., Jr. (1979/1982). *Minnesota occupational classification system-II.* Minneapolis: University of Minnesota, Vocational Psychology Research, Work Adjustment Project.

Dawis, R.V., Lofquist, L.H., & Weiss, D.J. (1968). *A theory of work adjustment (a revision) (Minnesota Studies in Vocational Rehabilitation No. 23).* Minneapolis: University of Minnesota, Work Adjustment Project.

Desmond, R.E., & Weiss, D.J. (1970). Measurement of ability requirements of occupations *(Minnesota Studies in Vocational Rehabilitation No. 34).* Minneapolis: University of Minnesota, Work Adjustment Project.

Dunn, D.J., & Allen, T. (1973). Vocational needs and occupational reinforcers of vocational counselors. *Vocational Evaluation and Work Adjustment Bulletin, 6*(4), 22-28.

Elizur, D., & Tziner, A. (1977). Vocational needs, job rewards, and satisfaction: A canonical analysis. *Journal of Vocational Behavior, 10,* 205-211.

Flint, P.C. (1980). *Sex differences in perceptions of occupational reinforcers.* Unpublished doctoral dissertation, University of Minnesota, Minneapolis.

Gay, E.G., Weiss, D.J., Hendel, D.D., Dawis, R.V., & Lofquist, L.H. (1971). Manual for the Minnesota Importance Questionnaire *(Minnesota Studies in Vocational Rehabilitation No. 28).* Minneapolis: University of Minnesota, Work Adjustment Project.

Holland, J.L. (1973). *Making Vocational Choices: A theory of careers.* Englewood Cliffs, NJ: Prentice-Hall.

Jenson, G., III. (1975). *An application of a theory of work adjustment to selection for graduate training in school psychology.* Unpublished doctoral dissertation, University of Minnesota, Minneapolis.

Parsons, F. (1909). *Choosing a vocation.* Boston: Houghton Mifflin.

Rosen, S.D., Hendel, D.D., Weiss, D.J., Dawis, R.V., & Lofquist, L.H. (1972). *Occupational reinforcer patterns: II.* Minneapolis: University of Minnesota, Vocational Psychology Research.

Schaffer, R.H. (1953). Job satisfaction as related to need satisfaction in work. *Psychological Monographs, 67*(14, Serial No. 364).

Seiler, D.A., & Lacey, D.W. (1973). Adapting the work adjustment theory for assessing technical-professional utilization. *Journal of Vocational Behavior, 3*(4), 443-451.

Stewart, E.S., Greenstein, S.M., Holt, N.C., Henly, G.A., Engdahl, B., Dawis, R.V., Lofquist, L.H., & Weiss, D.J. (1986). *Occupational reinforcer patterns.* Minneapolis: University of Minnesota, Vocational Psychology Research, Work Adjustment Project.

Tinsley, H.E.A., & Weiss, D.J. (1971). A multitrait-multimethod comparison of job reinforcer ratings of supervisors and supervisees. *Journal of Vocational Behavior, 1*(3), 287-299.

U.S. Department of Labor, U.S. Employment Service. (1979). *General Aptitude Test Battery, B-1002, section II: Norms, occupational aptitude pattern structure.* Washington, DC: U.S. Government Printing Office.

U.S. Department of Labor, U.S. Employment Service. (1982). *Dictionary of Occupational Titles.* Washington, DC: U.S. Government Printing Office.

Zedeck, S. (1978a). Minnesota Importance Questionnaire. In O.K. Buros (Ed.), *The eighth mental measurements yearbook* (pp. 1673-1675). Highland Park, NJ: The Gryphon Press.

Zedeck, S. (1978b). Minnesota Job Description Questionnaire. In O.K. Buros (Ed.), *The eighth mental measurements yearbook* (pp. 1675-1677). Highland Park, NJ: The Gryphon Press.

Brian Bolton, Ph.D.
*Professor, Research and Training Center in Vocational
Rehabilitation, University of Arkansas, Fayetteville, Arkansas.*

MOTIVATION ANALYSIS TEST

*Arthur B. Sweney, Raymond B. Cattell, John L. Horn, and IPAT
Staff. Champaign, Illinois: Institute for Personality and Ability
Testing.*

Introduction

In contrast to standard personality questionnaires and interest inventories,
which measure relatively stable characteristics of human personality based on the
respondent's conscious and subjective statements of preference, the Motivation
Analysis Test (MAT; Institute for Personality and Ability Testing, 1975) uses a series
of *objective* tests of ergic strength to measure the dynamic or motivational trait structure of the examinee. The tests are objective in the sense that the respondent is not
aware of the nature of the trait being measured and, thus, cannot systematically
answer the items in such a way as to present a desired outcome. The Motivation
Analysis Test, then, measures factorially derived trait-dimensions that account for
individual differences in motivated human behavior.

The architect of the MAT, Raymond B. Cattell, initiated programmatic research
on human motivation in the late 1940s. Working with graduate students and
research fellows in psychology at the University of Illinois during the 1950s and
1960s, Professor Cattell mapped scientifically the domain of dynamic functioning
and developed practical devices for the clinical assessment of basic human motives.
The coauthors of the MAT, John L. Horn, Arthur B. Sweney, and John A. Radcliffe,
are now professors of psychology at the University of Southern California, Wichita
State University, and the University of Melbourne (Australia), respectively. Since
1973, Professor Cattell has continued his research and writing at the Hawaii School
of Professional Psychology.

The Motivation Analysis Test is the product of a program of scientific research
that produced a model of human motivation known as the dynamic calculus (Cattell, 1985), an interaction model in which behavior is postulated to be the result of
a) stimuli that characterize a particular situation and b) unique trait characteristics
of the individual. Further, this model recognizes that the individual's existing drive
levels are the product of previous learning and physiological predispositions and
that it is the entire drive complex that influences and determines motivated behavior. Two systems of factors comprise the basic referents of the dynamic calculus
model: two motivational components and 10 dynamic structure factors.

The two motivational components are the avenues or modes of expression
through which any motive may be expressed. The Integrated dimension of
expression is the cognitively controlled component that represents the integration
achieved through ego function and the impulse control established through cul-

tural values internalized in the superego. In contrast, the *U*nintegrated dimension of expression is the unrestrained, spontaneous component that is bound neither by reality considerations nor by the learning implicit in the superego and is thought to be in part unconscious or preconscious. In brief, the *I* component is the reality-oriented, culturally channeled component of goal-directed activity, and the *U* component represents the impulsive, undifferentiated aspect of human motivation.

The 10 dynamic structure factors are divided into two main classes of influence that determine the development and courses of action of human motives: ergs and sentiments. The ergs, or primary drives, represent the basic physiological dispositions, and the sentiments, or secondary drives, are patterns that reflect learned combinations of ergic expression associated with particular objects and institutions. The MAT measures five ergs and five sentiments, each in the *U*nintegrated and *I*ntegrated modes of expression.

The five ergs assessed by the MAT and their defining attitudes are described briefly in the following paragraphs. It should be emphasized that the attitudes are *not* MAT items, but rather served as conceptual guides to the content and focus of the 208 items comprising the instrument.

Mating (Ma): Strength of the normal, heterosexual or mating drive. Attitudes measured are falling in love and satisfying sexual needs.

Assertiveness (As): Strength of the drive to self-assertion, mastery, and achievement. Attitudes measured are dressing smartly and increasing personal status.

Fear (Fr): Level of alertness to external dangers. Attitudes measured are protection from attack and avoidance of disease and injury.

Narcism-Comfort (Na): Level of drive to sensuous, self-indulgent satisfactions. Attitudes measured are enjoyment of life and seeking relaxation and ease.

Pugnacity-Sadism (Pg): Strength of destructive, hostile impulses. Attitudes measured are destruction of enemies and desire to view violent displays.

The five sentiments assessed by the MAT and their subsidiating attitudes are described as follows:

Self-Concept (SS): Level of concern about the self-concept, social repute, and more remote rewards. Attitudes measured include maintenance of reputation and self-respect, control of impulses, and desire for self-knowledge.

Superego (SE): Strength of development of conscience. Attitudes measured include satisfaction of sense of duty and avoidance of impropriety.

Career (Ca): Extent of development of interests in a career. Attitudes measured are learning occupational skills and acquiring career tenure.

Sweetheart-Spouse (Sw): Strength of attachment to spouse or sweetheart. Attitudes measured are spending time with and giving gifts to one's sweetheart or spouse.

Home-Parental (Ho): Strength of attitudes attaching to the parental home. Attitudes measured are pride in parental home and maintenance of an affectionate relationship with one's parents.

The concept of motive in the dynamic calculus model is analogous to the concept of energy in physics in that the MAT assesses motive strength or ergic investment by calculating expenditure of motivation in various possible outlets. The MAT consists of 208 standardized stimuli (items) for gathering controlled responses indicative of the respondent's pattern of dynamic expression. Each item requires the respondent to choose from among either two or four alternatives. Although the MAT looks similar to dozens of other paper-and-pencil inventories, it was constructed on a new scientific foundation that is generally accepted as unexcelled in motivation measurement.

Each of the 28 attitudes measured by the 208 items is assessed via four objective devices: Paired Words (or Association), Information (or Knowledge), Uses (or Projection), and Estimates (or Autism). The Integrated dimension of motivational expression derives from Association and Knowledge items. Association items require a choice between two words, while Knowledge items assess specific information elicited directly from the respondent. The logic is that persons expressing various motives in the *I* mode will associate thematically relevant words and be better informed in areas of interest in which they are invested more. Research has demonstrated that Association and Knowledge devices do in fact measure respondents' realistic interests.

The Unintegrated mode of expression is quantified using items from the Projection and Autism devices. Projection items ask respondents how they would prefer to spend their time and resources. The underlying principle is that persons will project their salient motives, of which they may not be consciously aware, in selecting goals to which they would allocate personal resources. Items using the Autism vehicle request that respondents offer opinions about a variety of open-ended social and political issues and judge the correctness of various statistically oriented statements. In contrast to the Knowledge items, which assess Integrated motives, Autism items do not have factually correct answers. Hence, Autism items allow respondents to generate speculations that reflect their wishes and needs. In other words, Autism items assess distortion or "wishful thinking" in dynamically satisfying directions. Research has confirmed that Projection and Autism devices measure unconscious expression, reflecting the operation of the general pleasure principle rather than conformance to social reality.

In the dynamic calculus system, *conflict* or frustration is defined operationally by the difference between the *I* and *U* modes of expression for each of the ergs and sentiments individually, as well as across the entire domain of dynamic structure factors. Total conflict is the summary expression of overall frustration, whereas total motivation is indicated by the sum of *I* and *U* components for each of the ergs and sentiments and summed for all 10 factors to give a Total Personal Interest score.

Practical Applications/Uses

The MAT can be used to assess the motive structure of adults in educational, industrial, and clinical settings. Although the *Handbook for the Motivation Analysis Test* (Cattell, Horn, Sweney, & Radcliffe, 1964) does not specify a minimum educational level, it may be reasonable, based on the overall level of item sophistication, to restrict MAT applications to individuals with the equivalent of a high school education. Even the device with the simplest format, Association, involves a fairly high level of abstraction in order to comprehend the choices available.

The MAT is not a time-limited assessment procedure. Usually, it can be administered individually or to small groups in about 1 hour. Because the *Handbook* provides detailed directions for test administration, the task can be readily accomplished by a psychometric assistant. Scoring is somewhat more complicated than most paper-and-pencil instruments, but instructions for hand scoring are presented clearly and an example illustrates the process. A computerized scoring service for large-volume users is available through the publisher.

The MAT protocol yields 45 scores, all of which are reported on the easily understood Sten (*Standard ten*) scale. Consistent with the previous description, a set of four profiles is generated across the 10 dynamic structure factors (i.e., the five ergs and five sentiments). The *Integrated* and *Unintegrated* modes constitute the primary profiles, whereas total motivation and conflict profiles are derived by summing the *I* and *U* expressions and taking the discrepancies of the *I* from the *U* expressions, respectively.

The most broadly based of the five global scores is Total Personal Interest, which is an average of the total motivation scores over the 10 dynamic areas. Total Conflict is an analogous condensation of the conflict profile that indicates extent of internal strife and frustration. Total Integration is an index of reality contact. General Information is a rough measure of verbal intelligence that derives from the Knowledge items. General Autism-Optimism provides an index of the favorableness of the respondent's overall life outlook.

All MAT scores are calculated with reference to a standardization group composed of 1,094 males and 753 females that is reasonably representative of the young adult population of the United States. The group, whose mean age is 21 years, includes 866 undergraduate college students, 156 Air Force enlistees, and 825 persons employed in a variety of semiskilled and skilled occupations.

The appropriate clinical application of the MAT is premised on an understanding of the dynamic calculus model of motivation and familiarity with the basic interpretive principles that have emerged from 25 years of research. Overviews written at a level suitable for practitioners are provided in chapters by Dielman and Krug (1977), Horn and Sweney (1970), and Sweney, Anton, and Cattell (1986). Two excellent sources for clinical interpretation of the MAT are the manual by Sweney (1969) and a chapter by H. Cattell (1986).

Motivated human behavior is postulated to occur through the mechanism of the *subsidiation chain*, which is a series of linkages that connect the basic ergic drives, expressed through the developed sentiment structures and goal-interests, to the pursuit of particular courses of action. The fundamental principle, then, is that all behavior, represented initially as choices of activities, must ultimately satisfy ergic drives that have their origins in the biological nature of the organism. In other words, the ergs are the energy sources that propel behavior.

Because human behavior is complexly determined and influenced by numerous constitutional and cultural factors, the subsidiation chains for several courses of action usually are interconnected and depend upon the simultaneous attainment of multiple subgoals and the satisfaction of several ergic drives. The *dynamic lattice* is the network that traces the connections among the courses of action and their ergic determinants, thereby outlining the individual's dynamic life patterns. The structures that integrate the subsidiation chains in order to reduce ergic conflict are the sentiments, which are culturally shaped and learned, but derive from the ergs. The function of the sentiments, then, is to channel ergic energy into socially acceptable behavior while minimizing intrapsychic conflict.

It follows from their intermediate role in the subsidiation process that motivation resulting from the sentiments is typically more stable and dependable than that produced directly by the ergs. This is so because ergic strength fluctuates from day to day depending on the degree of stimulation and satisfaction occurring, whereas sentiment strength is relatively more constant, reflecting the moderating influence of learning and past experience. It also has been determined that the dynamic traits are essentially independent of temperamental characteristics, with the exception of the two largest sentiment structures, the self-sentiment and the superego sentiment, which are related closely to their corresponding personality traits. The difference between these two classes of human traits is that the dynamic characteristics are goal directed, while the temperamental traits are more general-orienting or predisposing structures.

Because of the central organizing role it plays in human personality functioning, the self-sentiment is of special concern to clinicians. The self-sentiment is concerned with the maintenance of self-respect, social reputation, self-control, and personal health and well-being. The self-sentiment not only exercises primary control over the expression of the ergs, but it also organizes the less dominant sentiments into an integrated personality system, which is the hallmark of the well-adjusted person whose goals and courses of action are consistent and unified.

Assisting the self-sentiment structure in maintaining the regulatory process are the superego sentiment and the ego. The superego sentiment reflects the individual's ethical values and moral standards that provide the guidelines for his or her behavior. Ego strength includes the capacity to defer immediate gratification of impulses and to compromise among competing ergic drives, thus reducing psychological conflict and enhancing ultimate drive satisfaction. The interrelationships among the ego, self-sentiment, and superego are discussed in depth by Gorsuch and Cattell (1977) and Cattell (1980, chapter 6).

Technical Aspects

The MAT is the product of more than 20 years of psychometric research on the nature of human dynamic functioning. The first stage of the research program addressed the question "How are motives expressed?" To locate the major avenues of motivational expression, replicated factor analyses of 68 measurement principles or devices were carried out for samples of adults and children. Examples of the measurement devices analyzed are autism, reasoning distortion, fantasy choice, defensive fluency, superego projection, perceptual integration, perseveration, impulsiveness, decision strength, reminiscence, and reflex inhibition.

Seven motivational components or avenues of dynamic expression were identified at the primary or first-order level: *Alpha* or conscious id satisfaction, *Beta* or realized ego expression, *Gamma* or superego restraints, *Delta* or physiological needs, *Epsilon* or unconscious memories, *Zeta* or impulsiveness, and *Eta* or resistance to distraction. When the interrelationships among the seven motivational components were examined by factor-analytic procedures, two second-order modes of motivational expression were discovered. The Beta, Gamma, and Eta components defined the *Integrated* dimension of expression, and Alpha, Delta, Epsilon, and Zeta comprised the *Unintegrated* dimension.

The second phase of research on the dynamic calculus model addressed the question "What are the major human motives?" In contrast to the motivational component investigations, which focused on the ways that motives are expressed, research on dynamic structure was concerned with identifying the types of motives that are operative in members of a given society; that is, the fundamental motives that determine observed behavior in all situations.

In order to address the question of motive structure, it was first necessary to operationalize the unit of observation. The key concept in dynamic structure research is *attitude*, which embodies the central idea of an *incentive* to which the respondent reacts (i.e., it operates in the stimulus-response framework). Examples of attitudes that were mentioned previously are satisfaction of sexual needs, avoidance of disease and injury, control of impulses, and pride in parental home.

Attitudes are measured by test items constructed according to a standard formulation. The prototypic strategy or measurement paradigm is "In these circumstances I want so much to do this with that." This formulation encompasses a) a stimulus situation, b) a particular course of action as a response, c) a goal connected with the course of action, and d) an intensity with which the course of action is desired.

The objective in motivation measurement is to hold constant elements a), b), and c) in a defined condition that fixes the attitude to be measured, and then to discover devices or vehicles that will quantify element d), the *intensity* with which the course of action is pursued. The first three conditions are built into the item when it is constructed, whereas intensity of expression is indicated by means of the response elicited by the item.

In a series of factor-analytic investigations of adults and children that typically included 40 to 50 attitudes, each measured by at least a half dozen items with two or more devices, the major dynamic structure factors or motives were identified. Two classes of motives were clearly evident, ergs and sentiments. The 10 largest unitary motive structures that emerged from this research, the five ergs and five sentiments discussed in the Introduction, were selected for inclusion in the MAT.

The psychometric characteristics of the MAT reported in the *Handbook* (Cattell et al., 1964), including internal consistency, reliability, stability, and concept validity, are modest in their scope and support for clinical applications. Two types of internal consistency data are presented, alpha coefficients and split-half coefficients. Mean coefficients for the 10 ergs and sentiments are .50, with ranges from .33 to .70, for both indices of internal consistency. Raymond Cattell has long argued that because substantial breadth of concept coverage is essential to scale validity and utility, scale homogeneity should not be too high. It also should be noted that internal consistency only establishes a lower limit for scale reliability.

Hence, it should not be surprising that test-retest reliabilities for the 10 MAT scales average .66, with a range from .51 to .81. Even short-term stability coefficients over a 5-week interval averaged .51, with a range from .39 to .69, for a sample of college students. It is important to emphasize, however, that these indices apply to the total scale scores; reliabilities of the *I*ntegrated and *U*nintegrated components of the ergs and sentiments, and especially the conflict scores for each MAT scale, are certainly much lower. For this reason, clinical interpretation of the dynamic profiles for *I*, *U*, and conflict should be undertaken with considerable caution.

One important property of the MAT is its *partial ipsativity*; that is, high scores on some scales must be accompanied by low scores on other scales. Ipsativity occurs when the alternative choices on an item are scored on different scales. In other words, the respondent is actually forced to choose between two ergs or two senti-ments. This circumstance is not unreasonable if it is assumed that an individual has a limited amount of expendable motive energy. Working upon this assumption, if a certain amount of energy is invested in one course of action, that energy source is not simultaneously available for expenditure in pursuing other goals.

The MAT is only partially ipsative, because while the 48 Association items are multiply scored in the Integrated mode, the 56 Knowledge items are scored on just one scale. Similarly, the 48 Projection items are multiply scored in the Unintegrated mode of expression, but the 56 Autism items are not. The net result of this partial ipsativity is that the 10 MAT dynamic scales are typically either uncorrelated or are slightly negatively correlated; that is, *r*s are typically in the .00 to -.35 range.

The only validity evidence reported in the *Handbook* is concept validity, and it pertains to the psychometric foundation of the MAT. It has no direct relevance to practical applications of the instrument. Correlations between the 10 MAT scales and their corresponding dynamic structure factors (i.e., concept validity coeffi-cients) located in developmental research averaged .65, with a range from .52 to .76. Some indication of the clinical validity of the MAT can be inferred from various investigations published in the research literature that examined relationships between the MAT scales and different types of life criteria.

The following investigations illustrate the range of validity-relevant evidence that exists for the MAT. Krug and Henry (1974) examined the relationship of the MAT to patterns of drug use in high school seniors and college freshmen. Boyle (1984) administered the MAT to college students before and after they viewed a film segment that portrayed graphic scenes of automobile accident victims. Brown (1976) studied the relationship of the MAT to the frequency of moving traffic vio-lations. Kawash and Busch (1978) analyzed the MAT protocols of married couples to identify factors in expressed marital happiness.

Singh and Vanvaria (1977) used the MAT to investigate the motive to avoid suc-cess among Indian women. Skinner (1982) administered the MAT to volunteers for painful experiments. Hinman and Bolton (1980) analyzed the motivational dynam-ics of disadvantaged women with the MAT. Kline and Grindley (1973) and Kline (1976) examined the longitudinal interrelationships between the MAT and personal diaries for college students. Birkett and Cattell (1978) illustrated how by P-technique objective motivation measures can be used to diagnose the causal factors in a patient's alcoholism.

Readers interested in pursuing technical issues concerning the dynamic calculus model and the MAT are referred to *Human Motivation and the Dynamic Calculus* (Cat-tell, 1985) for a succinct overview and to *Motivation and Dynamic Structure* (Cattell & Child, 1975) for an in-depth treatment. The following chapters present technically oriented summaries: Cattell (1958, 1959, 1973, 1979 [chapter 4], 1980 [chapter 2]), Horn (1966), and Sweney (1967a, 1967b).

Critique

The MAT is the product of an innovative, conceptually sophisticated, psycho-metrically advanced program of basic research on human motivation. Whereas instruments like the Edwards Personal Preference Schedule and the Personality Research Form were developed to assess rationally derived motives (i.e., needs selected from Murray's well-known formulation), the MAT evolved from a series of investigations designed to discover the fundamental components of dynamic functioning. Cattell's dynamic calculus model is unique among theories of human motivation for its unsurpassed psychometric foundation. No other instrument in this area even approaches the MAT in this regard.

Unfortunately, the development and refinement of the MAT ended for all practical purposes more than 20 years ago. Although the current edition was published in 1975, it was updated only cosmetically then rather than truly revised. Although sexist language was excised and many items were recast in simpler or less awkward phraseology, the item content remained unchanged. Because the modifications were considered so minor and a study found only small score differences in comparison to the 1964 edition, the 1964 normative sample was retained. However, correlations between the scale scores in the two MAT editions were not reported.

In this reviewer's opinion, the MAT needs thorough revision. The content of many items, and some of the colloquial and idiomatic expressions used, can be described most accurately as "old-fashioned," which is not surprising considering that the MAT item pool was assembled in the 1950s and early 1960s. For young adults, who are most often candidates for motivational assessment, the MAT is clearly the product of an earlier generation. Cattell has repeatedly emphasized that the dynamic domain is more culturally sensitive than other areas, and this caution certainly applies to intergenerational differences within cultures.

Of course, new norms also will have to be constructed and the *Handbook* should be revised to include accumulated information about clinical and research applications of the MAT. A systematic validation program in mental health and industrial settings would provide concrete evidence of the MAT's relevance to practical assessment decisions. Because of its exemplary developmental foundation, which resulted in the discovery and verification of the basic components of motivated behavior and the main dynamic trait structures, the MAT merits the investment of resources necessary for a complete revision.

Reviewer's Note: In light of my highly favorable evaluation of the dynamic calculus model of human motivation and my necessarily critical comments about the current (1975) edition of the MAT, it is with great relief that I can announce that a thoroughly revised MAT will be available in the fall of 1988. The new edition will adhere to the same basic design as previous versions in that it will use four measurement devices to assess 10 ergs and sentiments resulting in 45 motivation scores. However, the 1988 revision probably will be slightly longer to enhance scale reliabilities. In addition, the revised MAT will be re-normed, and the *Handbook* will be updated and expanded. A computer-generated narrative report incorporating the excellent interpretive materials for the MAT also will be available. With this new edition of the MAT, assessment specialists in educational, industrial, and clinical settings have the option of including a modern, scientifically based motivation instrument in their psychological testing batteries.

References

Birkett, H., & Cattell, R. B. (1978). Diagnosis of the dynamic roots of a clinical symptom by P-technique: A case of episodic alcoholism. *Multivariate Experimental Clinical Research, 3,* 173-194.

Boyle, G. J. (1984). Effects of viewing a road trauma film on emotional and motivational factors. *Accident Analysis & Prevention, 16,* 383-386.

Brown, T. D. (1976). Personality traits and their relationship to traffic violations. *Perceptual & Motor Skills, 42,* 467-470.

Cattell, H. B. (1986). The art of clinical assessment by the 16PF, CAQ, and MAT. In R. B. Cattell & R. C. Johnson (Eds.), *Functional psychological testing: Principles and instruments* (pp. 377-424). New York: Brunner/Mazel.

Cattell, R. B. (1958). The dynamic calculus: A system of concepts derived from objective motivation measurements. In G. Lindzey (Ed.), *The assessment of human motives* (pp. 197-238). New York: Rinehart.

Cattell, R. B. (1959). The dynamic calculus: Concepts and crucial experiments. In M. R. Jones (Ed.), *The Nebraska symposium on motivation* (pp. 84-137). Lincoln, NE: University of Nebraska Press.

Cattell, R. B. (1973). Key issues in motivation theory (with special reference to structured learning and the dynamic calculus). In J. R. Royce (Ed.), *Multivariate analysis and psychological theory* (pp. 465-499). New York: Academic Press.

Cattell, R. B. (1979). *Personality and learning theory* (Vol. 1). New York: Springer.

Cattell, R. B. (1980). *Personality and learning theory* (Vol. 2). New York: Springer.

Cattell, R. B. (1985). *Human motivation and the dynamic calculus.* New York: Praeger.

Cattell, R. B., & Child, D. (1975). *Motivation and dynamic structure.* New York: Halstead.

Cattell, R. B., Horn, J. L., Sweney, A. B., & Radcliffe, J. A. (1964). *Handbook for the Motivation Analysis Test.* Champaign, IL: Institute for Personality and Ability Testing.

Dielman, T. E., & Krug, S. E. (1977). Trait description and measurement in motivation and dynamic structure. In R. B. Cattell & R. M. Dreger (Eds.), *Handbook of modern personality theory* (pp. 117-138). New York: Halstead.

Gorsuch, R. L., & Cattell, R. B. (1977). Personality and socioethical values: The structure of self and superego. In R. B. Cattell & R. M. Dreger (Eds.), *Handbook of modern personality theory* (pp. 675-708). New York: Halstead.

Hinman, S., & Bolton, B. (1980). Motivational dynamics of disadvantaged women. *Psychology of Women Quarterly, 5,* 255-275.

Horn, J. L. (1966). Motivation and dynamic calculus concepts from multivariate experiment. In R. B. Cattell (Ed.), *Handbook of multivariate experimental psychology* (pp. 611-641). Chicago: Rand-McNally.

Horn, J. L., & Sweney, A. B. (1970). The dynamic calculus model for motivation and its use in understanding the individual case. In A. R. Mahrer (Ed.), *New approaches to personality classification* (pp. 55-97). New York: Columbia University Press.

Institute for Personality and Ability Testing (1975). *The Motivation Analysis Test (MAT).* Champaign, IL: Author.

Kawash, G., & Busch, N. (1978). Personal dynamic conflict as a predictor of expressed marital happiness. *Journal of Clinical Psychology, 34,* 171-176.

Kline, P. (1976). Personal diaries and responses to the Motivation Analysis Test (MAT): Two case studies. *British Journal of Projective Psychology & Personality Study, 21,* 29-35.

Kline, P., & Grindley, J. (1973). A 28-day case study with the MAT. *Multivariate Experimental Clinical Research, 1,* 13-22.

Krug, S. E., & Henry, T. J. (1974). Personality, motivation, and adolescent drug use patterns. *Journal of Counseling Psychology, 21,* 440-445.

Singh, S., & Vanvaria, K. (1977). Correlates of fantasy-based motive to avoid success in Indian women. *British Journal of Projective Psychology & Personality Study, 22,* 7-13.

Skinner, N. F. (1982). Personality characteristics of volunteers for painful experiments. *Bulletin of the Psychonomic Society, 20,* 299-300.

Sweney, A. B. (1967a). Designing objective motivation instruments to measure motivation components. In R. B. Cattell, F. W. Warburton, F. L. Damarin, Jr., & A. B. Sweney, *Objective personality & motivation tests: A theoretical introduction and practical compendium* (pp. 127-148). Urbana, IL: University of Illinois Press.

Sweney, A. B. (1967b). Objective measurement of strength of dynamic structure factors. In R. B. Cattell, F. W. Warburton, F. L. Damarin, Jr., & A. B. Sweney, *Objective personality & motivation tests: A theoretical introduction and practical compendium* (pp. 149-186). Urbana, IL: University of Illinois Press.

Sweney, A. B. (1969). *Descriptive manual for individual assessment by the Motivation Analysis Test.* Champaign, IL: Institute for Personality and Ability Testing.

Sweney, A. B., Anton, M. T., & Cattell, R. B. (1986). Evaluating motivation structure, conflict, and adjustment. In R. B. Cattell & R. C. Johnson (Eds.), *Functional psychological testing: Principles and instruments* (pp. 288-315). New York: Brunner/ Mazel.

Maxine B. Patterson, Ed.D.
Assistant Professor of Education, University of Tennessee,
Memphis, Tennessee.

NATIONAL ADULT READING TEST

Hazel E. Nelson. Windsor, England: NFER-Nelson Publishing Company, Ltd.

Introduction

The National Adult Reading Test (NART) is designed to facilitate the clinical assessment of intellectual impairment in adult patients suspected of suffering from dementia. Specifically, the NART is a tool for "estimating the premorbid intelligence levels" (Nelson, 1982, p.4) of dementing patients in cases where the patient history includes no measures of intellectual abilities prior to the onset of mental deterioration. As an instrument for assessing word-reading ability, the NART yields an error score that enables the clinician to predict premorbid Full Scale, Verbal, and Performance IQs on the Wechsler Adult Intelligence Scale (WAIS). Given the predicted premorbid WAIS IQ and the current IQ, the clinician then is able to compute a discrepancy score, which purportedly serves as an indicator of impairment in intellectual functioning.

Hazel E. Nelson is on the clinical staff of the National Hospital for Nervous Diseases in London, England. Her research in the area of dementia assessment has spanned more than a decade and has been the basis for articles related to the use of reading ability to estimate the premorbid intelligence levels of dementing adults.

The development of the NART had its roots in a series of studies that showed a positive correlation between word-reading ability and general intelligence in a group of 98 normal adults. The studies further indicated that word-reading ability remained relatively stable at premorbid levels in a group of dementing patients. On the basis of those findings, the author concluded that word-reading ability could provide a useful indicator of the premorbid intelligence level of the demented patient. Because existing word-reading tests were inadequate for discriminating among high levels of literacy in the adult population, the construction of a new word-recognition test for adults was undertaken.

A pool of 140 words, selected for their nonconformity to common pronunciation rules, was given to 28 nondementing subjects, who were asked to read the words. Ninety of the 140 words were eliminated on the basis of accuracy of pronunciation by all subjects, vulnerability to "successful guesswork" (Nelson, 1982, p. 7), or difficulty in scoring objectively. The remaining 80 words constitute the NART in its present form.

Normative data for the NART is based on the assessment of 120 patients hospitalized for extra-cerebral disorders at the National Hospital for Nervous Diseases in London, England. The standardization population included both males and females ranging in age from 20 to 70 years. Although the manual indicates that "an effort was made to include approximately equal numbers of subjects in all age decades" (Nelson, 1982, p. 7), it gives no indication of distribution by age. Likewise, it provides no data on distribution by gender. The mean IQ level of the sample, in which higher socioeconomic levels are overrepresented, is above average.

No translations of the test have been documented. The number of languages that would accommodate translation would be limited to those which, like English, embody numerous words that cannot be pronounced correctly simply by the application of common pronunciation rules. Translation to a "phonetic" language such as Spanish, for example, would be virtually impossible.

The NART consists of 50 words listed in order of increasing difficulty. Varying in length from 4 to 11 letters, the words are all "irregular" in the sense that their correct pronunciation cannot be determined by phonemic decoding. The test is administered individually and has no time limit. The list of words is printed on a card that the examiner gives to the subject with instructions that the words are to be read aloud. A scoring sheet, which provides space for error notations, along with a pronunciation guide for use by the examiner, is provided.

According to the manual, the NART is intended for use with subjects between the ages of 20 and 70, the age range of the standardization sample. A recent study using a sample of older patients (O'Carroll & Gilleard, 1986, pp. 157-158), however, implies that the age ceiling can be raised to 86 years.

Practical Applications/Uses

Designed as a clinical tool, the NART may be useful to psychologists and psychiatrists in the assessment of mental disorders. Clinicians also may find it helpful in assessing the effects of drugs or alcohol on a subject's level of intellectual functioning. To date, however, documented applications have been limited to research (Nebes, Martin, & Horn, 1984; O'Carroll & Gilleard, 1986), which currently appears to be the most appropriate use of the instrument. The manual includes suggestions for its use as a criterion for group matching in investigations involving alcoholics or drug addicts and as a tool for research into reading processes. Because of the over-representation of higher socioeconomic levels in the standardization sample, the NART is most appropriate for British subjects of high socioeconomic status. Its inclusion of one word that is classified as archaic in standard American English, and the further inclusion of four words whose accepted American pronunciation differs from that indicated in the pronunciation guide, render the NART unsuitable for American subjects. As stated in the manual, the NART is not appropriate for use with non-native speakers of English or with subjects identified as extremely poor readers in the premorbid state. In either case, the test will underestimate premorbid intelligence.

To administer the NART, the examiner gives the subject a card containing 50 words, with instructions that the words are to be read aloud. The subject is asked to pause after each word until the examiner gives a cue to proceed. As the subject pronounces the words, the examiner records errors on the answer/record sheet. The subject should be encouraged to attempt every word on the list and to guess when necessary. The examiner may stop the test if 14 of 15 consecutive responses are incorrect. The test should be continued, however, if any doubt exists about whether the subject's limit has been reached. To reduce test anxiety, the examiner should give the subject reassurance and reinforcement throughout the test.

The NART can be administered by any English-speaking adult with an understanding of the diacritical marks used in the pronunciation guide. To avoid misinterpretation of responses, however, the NART's author recommends administration by experienced examiners only. Although no time limit is attached, actual administration should require no more than 15-30 minutes. Because the test must be administered individually and because of the sensitive nature of subject disorders, administration of the NART probably should be limited to the clinical setting.

Although the manual provides relatively clear instructions for test administration, the explanation would be improved with a more concisely written description of the testing procedures. Strong features of the manual are its identification of problems that might be anticipated in the administration process, specific recommendations for dealing with problems that may arise during administration, and concrete suggestions for reducing the subject's test anxiety.

Basically, scoring the NART is a very simple procedure. An answer/record sheet, on which the examiner records errors, is provided. Since the examiner records the errors as the subject reads the word list, scoring is virtually complete when the subject pronounces the final word. All that remains is to sum the errors to obtain the "error score," a process that should require no more than 20 seconds. The answer/record sheet provides space for recording "the actual errors" (Nelson, 1982, p. 5). The manual suggests that the examiner may want "to look at these later" (Nelson, 1982, p. 5), but fails to define "actual errors" or to identify the reason for looking at them later.

NART scores may be interpreted easily and objectively. The error score is entered into the appropriate table in the manual to obtain the predicted premorbid IQ on WAIS Full Scale, Verbal, and Performance IQ scales. Although virtually no knowledge of statistics or psychometrics is necessary for a basic interpretation of results on the NART, an understanding of concepts related to the measurement of intelligence and to the interpretation of IQ scores is essential to any further use of the results.

Technical Aspects

Split-half reliability (Cronbach alpha = .98) of the NART is based on the responses of 120 subjects in the standardization sample. Nelson's assumption that "the nature of the words [used on the test] makes it very unlikely that they would be affected by repetition" (Nelson, 1982, p.10) is unwarranted in the absence of test-retest reliability data.

Validity data, as traditionally reported, are not provided. Construct validity possibly could be inferred, however, from research evidence that performance on the NART is insensitive to the effects of dementia. In one study, Nelson & O'Connell (1978) administered the NART to 40 elderly subjects (mean age = 48) with bilateral cortical atrophy and compared the results with those of a control group consisting of the 120 normal adults used in the standardization study. Although the mean score obtained for the atrophy group on the WAIS ($M = 92$) was lower than that of the control group (M = 109), the mean NART error scores for the two groups were close (atrophy group, 23.9; control group, 22.4). Those results were replicated in a study of an older group (mean age = 71) of 20 Alzheimer patients (Nebes, Martin, & Horn, 1984). O'Carroll and Gilleard (1986) have documented further support for the dementia insensitivity of the NART in an investigation involving 30 demented patients between the ages of 65 and 86.

A series of regression equations, derived from the standardization data, shows a positive relationship between the NART and the WAIS and, thus, provides a semblance of support for concurrent validity. The assumption that correlation with the WAIS validates the prediction of premorbid intelligence solely on the basis of word-reading performance, as indicated by NART results, is questionable, however.

Critique

Au (1985, p. 1033) states that results of the NART should be used with caution, "particularly if one is basing conclusions on predicted WAIS scores." Bradley (1985, p. 1033) concurs and suggests that "the NART also greatly needs to be validated with dementia patients already having premorbid intelligence test information. . . . Until such retro-predictive validity information is gathered, the use of the NART should probably be confined to research."

In addition to investigating the retropredictive validity of the NART, future research should include examination of the test's equivalent forms and test-retest reliability. Such studies would address potential questions generated by the tendency of the split-half technique, used in the standardization study, to exaggerate an instrument's reliability. Moreover, they would provide additional information about the test's validity. The ease of its administration, the simplicity of its format, and the nonthreatening nature of its tasks for subjects with presumed intellectual deterioration make the NART very practical for use as a research and clinical device. At the same time, the relative newness of the test and the limited amount of evidence to support its underlying assumptions warrant extreme caution in using it for clinical decision making. In any case, in its present form, the NART is inappropriate for use with speakers of American English.

References

Au, K. (1985). Review of National Adult Reading Test. In J.V. Mitchell, Jr. (Ed.), *The ninth mental measurements yearbook* (p. 1033). Lincoln, NE: The Buros Institute of Mental Measurements.

Bradley, J. (1985). Review of National Adult Reading Test. In J.V. Mitchell, Jr. (Ed.) *The ninth mental measurements yearbook* (pp. 1033-1035). Lincoln, NE: The Buros Institute of Mental Measurements.

Nebes, R., Martin, D., & Horn, L. (1984). Sparing of semantic memory in Alzheimer's disease. *Journal of Abnormal Psychology, 93*, 321-330.

Nelson, H. (1982). *National Adult Reading Test (NART) test manual.* Windsor, Berkshire, England: NFER Publishing Company Ltd.

Nelson, H., & O'Connell, A. (1978). Dementia: The estimation of premorbid intelligence levels using the new adult reading test. *Cortex, 14*, 234-244.

O'Carroll, R., & Gilleard, C. (1986). Estimation of premorbid intelligence in dementia. *British Journal of Clinical Psychology, 25*, 157-158.

Leonard J. West, Ph.D.

Professor of Education, Baruch College, The City University of New York, New York, New York.

NATIONAL BUSINESS COMPETENCY TESTS

National Business Education Association. Reston, Virginia: National Business Education Association.

Introduction

The National Business Comptency Tests (NBCT) are described by their publisher (in a 1-page cover sheet) as "a series of achievement tests measuring marketable productivity in one or more basic office areas." The tests are intended for administration to students coincident with completion of the terminal secondary or postsecondary course in the area covered by the test. Test administrators (i.e., teachers) are urged (in the various test manuals) to explain to examinees (students in their classrooms) that the purpose of the tests is "to help teachers and employers determine whether prospective employees [students] are ready for entry-level [office] jobs."

Five tests are currently available in the NBCT battery (copyright dates in parentheses): 1) Accounting Procedures Test, Trial Edition (1984), 2) Office Procedures Test (1981), 3) Secretarial Procedures Test (1983), 4) Typewriting Test (1979), and 5) Word Processing Test (1986). Because the inclusion of a stenographic test in the competency battery was still under consideration as of mid-1987, one from an older series, the National Business Entrance Tests (NBET), was submitted for review here: a 1972 revision of the Stenographic Test originally copyrighted in 1955.

The "competency" battery, the first one so named, is a replacement for the NBET, a series with a history of more than 30 years. Its subtests had been reviewed from time to time in the "Business Education" section of the various *Mental Measurements Yearbooks*, extending through the eighth yearbook (Buros, 1978). The NBCT's Word Processing Test and an optional portion of the Accounting Procedures test entitled Microcomputers in Accounting are new components of the testing program of the publisher, the National Business Education Association (NBEA). All other components of the NBCT battery, aside from the entirely new contents of all its tests, are largely indistinguishable in scope, purpose, and general characteristics from the variously titled components of the earlier NBET battery. Except for updating that reflects modern office technologies, "competency" appears to be a new label for the same sort of testing as that of the NBET.

For the earlier NBET and the current NBCT, overall supervision of test development came under the aegis of a test committee appointed by the publisher. For each test in the battery, a separate group of test consultants was responsible for test construction and its concomitants. The committee members and consultants include a few high school teachers, but the majority are college and university specialists in business-teacher education and in business subjects at the levels encompassed by the contents of the NBCT battery. None of the dozens of these individuals is primarily a specialist in tests and measurements.

The National Business Compentency Tests are primarily terminal achievement tests; that is, they test what has been taught. For example, performance items in the tests closely resemble in content and format the materials and exercises in students' textbooks and workbooks, whereas objective items in the NBCT have fewer counterparts in instructional materials for students. In any event, to the extent that what has been taught and, in turn, test content capture the job duties of beginning office employees, outcomes may be construed as measures of readiness for entry-level employment.

Each of the five NBCT tests consists of a test booklet for the examinee and a test manual for the administrator. Except for the (trial edition) Accounting Procedures Test, each 4-page manual (an 11" x 17" sheet folded in half) devotes 1-2 pages to descriptive information about the test and the remaining space to an answer key or model answers. The first 1-2 pages identify, in a series of short, separately headed sections, the members of the test construction committee and describe the content of the test, student population, standards, test administration procedures, scoring of test papers, and reporting of test results.

"Student Population" specifies secondary and postsecondary students who have completed the terminal course in the subject covered by the test. Test content, however, is generally confined to secondary school curricula and suits postsecondary students only in instances of identical curricular scope at both school levels (confined at postsecondary levels to proprietary business schools and community colleges). Also, to varying degrees the test manuals recognize that the term "terminal" lacks a uniform meaning. In some of the areas tested there are both 1- and 2-year programs. All in all, the tests are probably more suitable to high school than to postsecondary students. Data in support of that judgment reside in the only tally of examinees known to this reviewer: For an administration of the NBET typewriting test, high school examinees outnumbered postsecondary examinees in a ratio of 6.5:1 (West, 1978).

Each of the National Business Competency Tests can be administered either in one 2-hour session (presumably as a specially scheduled final examination) or in a 50-minute period on each of two days. In selected instances, however, the test booklets are not assembled and collated for efficient and secure administration on separate days: Part 2 is sometimes begun on the reverse side of the last page of Part 1.

Practical Applications/Uses

Accounting Procedures Test. Aside from an optional test on Microcomputers in Accounting (10 four-option multiple-choice questions), the trial edition of the Accounting Procedures Test consists in Part 1 (80 points) of 25 four-option multiple-choice items covering basic accounting procedures and of performance items covering payroll, bank reconciliation, and worksheet. Part 2 (100 points) consists of a journalizing and posting project for which source documents are provided. In awareness of the existence of both 1- and 2-year high school bookkeeping programs, the manual states that "it is not expected that students completing one year of accounting will finish the entire test."

The performance tests, which account for 155 of the test's 180 points, involve the maintenance of manual records. Research has shown, however, that financial recordkeeping has for years been overwhelmingly computerized and that entry-level positions are essentially clerical ones. These positions require little if any knowledge of accounting concepts and use record forms for computer input that bear little resemblance to the record forms of manual systems (West, 1973).

The multiple-choice test of basic accounting procedures mostly tests knowledge of bookkeeping and accounting terminology and covers a broad range of terms and concepts. The performance tasks range from common, everyday tasks (filling out a bank deposit slip) to tasks that form the major basis for high school bookkeeping instruction: making journal entries, posting to ledger accounts, and preparing a trial balance. The worksheet and the journalizing and posting tasks capture the important objectives of instruction and account for 120 of the 180 total test points. Of the remaining 60 points, the multiple-choice test accounts for 25, payroll records for 15, a bank reconciliation for 15, and a bank deposit slip for 5. However modest the relevance of manual record maintenance to entry-level job requirements, overall test content nicely represents the content of high school instruction.

Several reservations about internal test details deserve expression. First, the error penalty in the performance tests is typically 1 point for each error or omission in tasks for which the point values approximate the number of discrete responses the tasks require. More than a few of the total points per test item, however, apply to mere copying of information stated in the problem (such as writing a person's name in the name column of a form) and require no knowledge of bookkeeping concepts. Second, the ledger account forms on which examinees are to make entries are so reduced in size that they necessarily require minuscule and doubtfully legible responses by examinees. Third, the entries on the payroll task's model form need only be mimicked to guarantee correct responses. Again, no knowledge of bookkeeping is required. Fourth, the multiple-choice test of basic concepts includes three items for which the stems act as specific determiners that signal the correct response on the basis of ordinary vocabulary. The item that uses the word "reconciliation" in its stem has "in agreement" in the correct option; the item with "printed report" in its stem includes "printout" in the correct option; the item with "combining" in its stem includes "merging" in the correct option. Fifth, of the 10 multiple-choice items in the optional Microcomputers in Accounting test, 7 are confined to microcomputer terminology in general, without particular relevance to the maintenance of financial records. Sixth and finally, the test booklet is not printed and assembled conveniently for administration as a 2-day test; the answer page for Part 2 of the test (journalizing and posting) has to be torn from its Part 1 facing page and provides no blank for recording the examinee's name.

Office Procedures Test. The two-part Office Procedures Test is intended for administration to students in office procedures, secretarial practice, or cooperative office education, preferably the final course. Part 1 consists of 12 "jobs": 1) making handwritten responses on an application form for a job as "a general office worker," 2) comparing words and numbers, 3) proofreading, 4) editing, 5) mail services, 6) telephoning, and 7-12) filing. Part 2 consists of nine jobs involving computation and accounting services applied to payroll, bank forms, purchase orders, invoices, and computation of costs, losses, and revenues. Examinees respond on ruled forms. Of the 393 points available on Part 2, 202 (51%) are based upon correct arithmetic; the remaining points are awarded for correctly copying or transcribing pro forma information on the various forms.

Test content appears to cover that of the pertinent school courses and the job duties of several different sorts of entry-level clerks. This reviewer has reservations concerning a number of internal test details, however. Mainly, discriminating judgment has been sacrificed to ease of scoring. The point value of each task reflects the number of responses the task requires. However, the point allotments are uniform within tasks and questionable between tasks. Within tasks, a trivial response is worth as much as an important one; as many points are given to transcribing an entry from one form to another as to determining what the original entry should be. Trivial tasks are sometimes allotted relatively large numbers of points in comparison to the number allotted to important tasks. For example, filling in a bank deposit slip and then a bank check and its stub are worth 21 and 26 points respectively; a simple purchase order allots only 6 points to computation but 40 points to mere copying; an invoice allots 10 points to computation but 48 points to mere copying. In contrast, job skills about which employers persistently complain are slighted in the allotment of points (e.g., 30 points for proofreading and 15 points for editing). In other instances, 4-option multiple-choice questions are worth 1 point each, whereas 3-option questions are worth 2 points each. The test user should ignore the scoring prescriptions (unfortunately printed within the test rather than in the manual) and substitute more discriminating point allotments.

Several other features of the test are also troublesome. Although the test is timed (50 minutes for each of its two parts), the test manual suggests 25-38 minutes for Part 2 if students use calculators rather than pencil arithmetic on scrap paper. Such a provision assumes that every student has a personal calculator or that the test is administered in a room that provides a calculator for each student. Moreover, machine arithmetic is not cognitively equivalent to manual arithmetic. Different skills are being measured when different modes of doing the arithmetic are employed—a distinction that would require separate norms for the two kinds of arithmetic. Contributing to general procedural looseness is that "up to five bonus points can be given for completion before the time allowed"; as well, "an extra bonus point will be given for each completed job that is accurate and neat." How accurate and neat is accurate and neat enough? A final reservation applies to the packaging of the test booklet for 2-day administration. The test manual advises that Part 1 should be administered and collected before Part 2 is administered. Part 2, however, must be torn out of the booklet and lacks a space for the student's name.

In summary, test content is generally good, but because scoring procedures are undiscriminating and in a few instances imprecise, the potential validity of the test as a predictor of job performance is greatly reduced.

Secretarial Procedures Test. Part 1 of this test consists of 1) handwritten completion of an application form for a job as a secretary (15 points); 2) a 30-item, 4-option multiple-choice test of secretarial procedures that samples the topical areas of filing, phoning, mail, reference books, office equipment, and secretarial tact and judgment in a variety of person-to-person situations (60 points); and 3) a 25-item editing test requiring the examinee to locate and correct errors in punctuation, grammar, spelling, and word usage (25 points). Part 2 (100 points) consists of the priority scheduling of six "tasks" and, except for handwritten corrections to be made on a calendar, the typing, in priority order, of 1) the minutes of a meeting, 2) a letter and envelope, 3) an agenda, 4) a memo, and 5) a purchase order. Desirably— because more than half of inputs to employed typists are handwritten or include handwritten editing (West, 1983, p. 354)—all the materials to be typed are handwritten, and all but the purchase order are lightly edited to require corrections or changes.

For Part 2, the test manual instructs the teacher to "follow your usual practices and standards for evaluating secretarial skills and job performance" but prescribes point penalties for typographical and major form errors, incomplete tasks, and the typing of the tasks in the wrong order.

The topical content of the Secretarial Procedures Test overlaps to some extent the contents of the Office Procedures Test and the Typewriting Test, but the secretarial test usually invokes more detailed and higher-level information and skills. For the stenographic skills required of some secretaries, the separate Stenographic Test (from the earlier NBET battery) is available.

Test content is generally excellent. However, a few minor matters and one major one concern this reviewer. Among the minor matters, the first task of Part 2 is printed on the reverse side of the page containing the task in Part 1, interfering with secure and efficient 2-day test administration. In addition, the test manual provides that "up to five points can be given for completion of Part 1 before the time allowed." Part 1, however, is scored objectively, and its content makes a time-limit power test appropriate, regardless of speed. Third, although the test manual "recommend[s]" 100 points for each of the two parts of the test, the values of the components of Part 2 (inferred from the specified penalties) do not sum to 100 points. Fourth, scoring for the quality of the typed tasks of Part 2 might not be sufficiently discriminating. Are "major form" errors ones that are not correctable? The test manual does not say, nor does it mention minor (i.e., correctable) errors. Fifth, one of the Part 1 multiple-choice items has, as worded, a doubtful model answer—not all atlases include both air and highway mileage distances between cities in the United States. It may be that no one kind of reference book always supplies both kinds of information.

This reviewer's major complaint is that the test manual does not prescribe a measure of the examinee's speed at the typed tasks of Part 2 (e.g., as completion time to the nearest quarter or half or whole minute), so that an overall score incorporating both speed and quality could be generated. That deficiency, however, also pervades classroom practices; for the important aspects of typewriting and stenographic skills, time-limit tests prevail among teachers, and discriminating between more and less productive workers is rare.

Typewriting Test. Part 1 of the Typewriting Test consists of two 5-minute straight copy timings (with numbers and symbols included in the second timing), intended for administration during part of one class period. Part 2, for administration in a 50-minute period the next day, consists of five tasks: the typing of 1) personal information on a ruled form, 2) a simple business letter (with carbon copy) from clearly printed test copy, 3) a simple handwritten 3-column table, 4) a 2-page report (including two footnotes) of about 300 words from typed copy that includes handwritten corrections to be made, and 5) printed and handwritten invoice information (a printed invoice form is provided).

Correspondence, reports, tables, and forms are the major kinds of tasks performed by employed typists, and Part 2 nicely samples from those domains. The straight copy timings of Part 1 can be justified only because they are a common (and sometimes the only) component of employment testing. Not widely known either or applied by employers or teachers is that stroking skills play only a modest role in accounting for proficiency at realistic tasks, and straight copy accuracy, in particular, is only negligibly correlated with stroking accuracy in realistic tasks (West, 1983, chapter 10).

The test manual designates students in advanced typing (a 4th-semester course) as the target population, but grants that examinees with less training may not be able to complete all five of the tasks of Part 2.

A few minor internal features of the test require correction. The straight copy prose timing purports to be of average difficulty, but is in fact a little below average. The number and symbol timing is lavish with numbers but stingy with the number and variety of symbols. In Part 2, the business letter includes a reminder to type reference initials and an enclosure notation, and the letter is accompanied by a word count (the major basis for determining the margins to use for the letter). No employer, however, would or could supply such information to a typist. Including that information in the test reduces the validity of the measure. The report, although admirable in other respects, includes an uncorrected misspelling (of the name of a major publisher) in one of its footnotes. In the test manual, the model typing of page 2 of the report lacks a page number.

The test manual provides model typing of the tasks of Part 2 but, unfortunately, instructs the teacher to "follow your usual practices and standards for evaluating basic skills and job performance." The omission of explicit scoring procedures makes impossible the uniform use of the test to determine whether students can meet employment requirements and precludes the compilation of consistently interpretable norms.

Quality assessment is one requirement, speed assessment is another. Each task in Part 2 is accompanied by a cumulative count of "standard" (i.e., 5-stroke) words to permit expression of speed in words-per-minute form. However, as a time-limit test without a measure of the actual finishing time of examinees, there is no way to determine productivity. Besides, measuring speed as words per minute is predicated on the incorrect supposition that stroking speed is the major determinant of proficiency at realistic tasks; instead, planning the layout of work on the page is the dominant factor (West, 1983, chapter 10).

Word Processing Test. Part 1 of the Word Processing Test, intended for administration in a 50-minute period, consists of 1) sixty 5-option multiple-choice questions on word processing theory, principles, terminology, equipment, and procedures and 2) the processing of a handwritten business letter of approximately 120 words that requires the examinee to identify and correct "spelling, punctuation, and other irregularities."

Part 2, for administration in another 50-minute period, is entitled "Advanced Word Processing Applications" and consists of 1) a handwritten 4-column table requiring the computation of column totals in its two money columns, 2) a 2-page manuscript from a typed draft requiring the examinee to make the corrections indicated in longhand, and 3) a revision of the foregoing manuscript that requires the examinee to make the indicated corrections.

In this reviewer's opinion, test content is good, and the test manual provides explicit scoring instructions. As with several other tests in the NBCT battery, test standards are supplied for "excellent," "good," and "competent" students, for which percentage correct is roughly equivalent to letter grades of low A, low B, and low C, respectively. It goes without saying that percentile norms would be the proper basis for relative standards, but none are available.

One reservation applies to the multiple-choice test of Part 1, however. It might be superfluous. The Psychological Corporation's (1985) Word Processing Test (West, in press) is entirely a performance test; paper-and-pencil objective testing was not considered necessary. For the Word Processing Test of the NBCT battery, correlations between scores on the objective and on the performance items could suggest whether or not the objective test is a useful component of the total test.

Stenographic Test. The Stenographic Test, as previously stated, is not a "competency" test, but rather the 1972 revision of an NBET component first copyrighted in 1955. As such, a few of its features are outdated: 1) erasers have given way to speedier modes of making corrections, 2) electric and electronic typewriters largely have replaced manual ones, and 3) xerography has largely replaced carbon copies. Those minor matters, however, are easily correctable via modified instructions and some modifications in scoring by the teacher. The substance of what is meant by stenographic skills is unaffected.

The 2-hour test allows 90 minutes for transcribing on the typewriter 12 business letters and an interoffice memo dictated at 80 "standard" words (112 speech syllables) per minute. Among the letters, an attention line and a subject line occur once each, four of the letters require an enclosure notation, and six of the letters require one or two carbon copies. Dictation applies from the salutation through the closing of each letter; inside addresses are supplied in the student's test booklet. The cover page of the booklet shows the point value of each of the 13 items. As the length of each item during the dictation increases from 80 to 200 standard words, the dictation time per letter increases accordingly (1-2½ minutes), as does the point value of each item (9-22 points, totalling 180 points). A 15-second pause between letters during dictation and a 5-minute interval between the end of dictation and the beginning of typewritten transcription are provided. The materials for dictation are marked in 15-second intervals, and a detailed correction manual prescribes the penalties for various kinds of correctable and uncorrectable errors in the typewritten transcript.

The foregoing conditions commonly apply to classroom testing of stenographic dictation and transcription proficiency at high school levels; college-level instruction often extends to dictation speeds ranging up to 120 words per minute.

An additional feature of this test attempts to simulate a stenographer's actual working conditions and consists of "redictation" periods of 2 minutes following letters 4, 8, and 11 and after the memo that ends the dictation. During those intervals, students may ask for redictation of words, phrases, and so forth that they missed or wrote unintelligibly during the dictation. Although students are advised to mark trouble spots in their notes distinctively during the dictation, implementing redictation does not seem to be practicable. For a class of only 20 students, a 2-minute redictation period allows each student only 6 seconds. That is barely enough time to ask the question, let alone for the teacher to locate the materials in the dictation and answer the question. Besides, the mode of implementing redictation is a poor simulation of reality; an employed stenographer questions the dictator during or after each letter, not after a group of them.

The vocabulary and topical content of the dictated materials are reasonably broad, and the prescriptions for scoring the transcript are very detailed. However, the acceptability, without penalty, of equivalents that preserve the original meaning (in instances of missing or illegible shorthand outlines) might not have high agreement among scorers, thus reducing the reliability of scores. For example, if a student substitutes "large" or "very large" for "tremendous" in the expression "tremendous savings" in one of the dictated letters, teachers might disagree about whether the substitution is sufficiently equivalent in meaning.

This reviewer has, in addition, two major reservations and a few minor ones about the test. One major matter concerns the test's failure to measure two of the three criteria of stenographic proficiency: dictation speed and transcription speed. Surely, the faster notetaker is the more proficient stenographer, as is the faster transcriber. However, the constant dictation speed of 80 words per minute permits no discrimination, and the attempt to take transcription speed into account is worded ambiguously, wrongly implemented, and unlikely to accomplish its proper purpose. Regarding transcription speed, in addition to the 180 points for the 13 items to be transcribed in 90 minutes (probably an unreasonably high expectation), the cover page of the student's folder of supplies follows the listing of points per item with "Bonus, 1 point for each minute saved (if all transcripts are acceptable)." Probably only a modest percentage of students could complete the transcription within 90 minutes, and those few are likely to be confined to a narrow range little below 90 minutes—adding not more than a few points to a quality score with a maximum of 180 points. Furthermore, "acceptable" as a prerequisite for a speed bonus is undefined. Does it mean perfect? If not, how good is good enough? Should the fast transcriber go unrewarded for speed if, say, just one of the 13 items falls a bit short of being "acceptable"? The test pays lip service to the measurement of transcription speed, but the provision for such measurement cannot be taken seriously.

The second major matter refers to excessive test length, resulting from a number of letters probably well beyond what is sufficient for reliable measurement. Sets of three or four letters, each set at a different dictation speed, with students selecting the set they wish to transcribe, would probably provide a sufficient sample and permit a measure of differences in notetaking speed. Procedures that take all three criteria of proficiency into account have been described by West (1975).

Minor reservations about the test concern discrepancies in marking the materials for dictation in quarter-minute intervals. At 80 standard words per minute, 28 speech syllables equal 20 words, but departures from that yardstick across entire letters vary from as few as 2 to as many as 11 syllables. Syllable counting is often careless. Parallel carelessness applies to the marking of the interoffice memo for dictation. Although the dictation is marked to begin with the first word in the memo heading (TO:), the syllable count for the first quarter minute excludes the "To/From/Subject" verbiage and begins with the first word of the message. As a result, the first quarter minute contains 49, not 28, syllables—not 20 standard words but 35, not a speed of 80 words per minute for that first quarter minute but 140.

A final point does not affect the usefulness of the test but instead could correct a long-standing, pervasive fallacy. The supposition that the average word in the English language contains 1.4 syllables harks back to the 1920s and substantially underestimates the true average for the vocabulary of written business communication. Recent studies (reviewed by West, 1983, pp. 366-367), using source materials far more representative than those of the 1920s, show 1.65 to be the average number of syllables per word in a weighted-for-frequency business vocabulary. Not 28 but 33 syllables should equal 20 standard words; the traditional 80 words per minute is really 68 words per minute.

Technical Aspects

Neither norms nor statistical evidence for test validity and reliability is supplied for any of the five NBCT tests or the Stenographic Test. Instead, the manuals for the Office Procedures, Secretarial Procedures, and Typewriting tests recommend that performance norms and indices of validity and reliability resulting from local test administration by research individuals and groups be reported to NBEA or in its journal. Whether NBEA intends to supply such information for the current accounting and word processing tests at some future date is not determinable—but seems highly unlikely. There are, in fact, other instances of absence of uniformity among the five tests in the battery.

Despite the absence of norms as a sound basis for standards, some but not all of the manuals suggest how many points "superior" and "good" students should be expected to earn. Similarly, and bearing on motivational tactics, some but not all of the manuals suggest that students "might" be told that test results will count as part of their final grades. Variation among tests also applies to scoring. Some manuals assign explicit point values to each element in their tests and specify precise criteria of acceptability; in other instances, scoring is discretionary and according to local preferences. There is, in short, much looseness in important aspects of testing.

Critique

With the occasional exceptions mentioned, test content in the NBCT battery adequately samples what has been taught and what is required of entry-level employees. The test items are, in general, probably better than ones that could be constructed by the average classroom teacher. Carelessness, errors, and omissions that occur occasionally in some of the tests need to be cleaned up, however, and reformatting of some of the tests for efficient and secure 2-day administration is highly desirable.

Test content excepted, the NBCT battery makes evident little sophistication in the requirements for acceptable testing on a national scale. The NBEA relies on volunteer, unpaid services from business educators. NBEA finances and, one must suppose, a low priority for test development in relation to other organizational objectives probably also account for jamming each test into as few printed pages as possible, as well as for the total absence of norms and of statistical evidence for validity and reliability—despite the existence of some of the tests in the battery long enough to permit such information. Inconsistency among the tests on features that should have been uniform, as well as serious weaknesses in some of the tests, might to some extent be attributable to the advisory, rather than supervisory or directive, role of the test committee vis-à-vis the separate groups of test consultants.

For teachers accustomed to the casualness of typical classroom testing with respect to factors affecting validity and reliability, the NBCT battery will be acceptable. For more sophisticated teachers, however, the absence of norms, the frequent looseness or even absence of scoring prescriptions, and the occasional point allotments that do not distinguish the important from the trivial will render the NBCT battery disappointing.

References

Buros, O. K. (Ed.). *The eighth mental measurements yearbook*. Highland Park, NJ: Gryphon Press.

The Psychological Corporation. (1985). *Word Processing Test*. San Antonio, TX: Author.

West, L. J. (1973). *Survey of entry-level bookkeeping activities in relation to the high school book-keeping curriculum* (Research Report No. 73-1). New York: City University of New York, Institute for Research and Development in Occupational Education. (ERIC Document Reproduction Service No. 086 873)

West, L. J. (1975). Principles and procedures for testing of typewriting and stenographic proficiency. *Business Education Forum*, 29(5), 24-27, 30-32.

West, L. J. (1978). Typewriting Test: National Business Entrance Tests. In O. K. Buros (Ed.), *The eighth mental measurements yearbook* (pp. 329-330). Highland Park, NJ: Gryphon Press.

West, L. J. (1983). *Acquisition of typewriting skills: Methods and research in teaching typewriting and word processing* (2nd ed.). Mission Hills, CA: Glencoe/Bobbs-Merrill.

West, L. J. (in press). Word Processing Test. In D. J. Keyser & R. C. Sweetland (Eds.), *Test critiques: Volume VII*. Kansas City, MO: Test Corporation.

Leslie J. Fyans, Jr., Ph.D.
Research Design and Psychometric Testing, Department of Planning,
Research, and Evaluation, Illinois State Board of Education, Springfield,
Illinois.

NEUROPSYCHOLOGICAL SCREENING EXAMINATION

John Preston. Dallas, Texas: The Wilmington Press.

Introduction

The Neuropsychological Screening Examination (NSE) was designed by Preston and Preston to bring to bear indicants of brain/behavior dysfunction from a variety of sources and measurements. It appears to be a synthesis of Luria's (1973) theory of mental systems organization and that of Springer and Deutsch (1981), Kaufman and Kaufman (1983), and Reitan (1969). The package of the NSE is composed not only of the examination tools, but also contains an educative cassette tape describing neuropsychology and neuropsychological assessment. The discussion on the cassette functions also as a step-by-step guide through the NSE. Additional material is provided that graphically portrays certain dysfunctions, including an extensive outline of learning disabilities that may be uncovered through measurement with this examination.

The NSE is composed of six major assessment domains, with certain domains broken into even greater specificity. These six domains are measures of 1) arousal and attention, 2) lateral dominance/handedness, 3) basic perceptual ability, 4) output, 5) higher level cognitive and integrative abilities, and 6) quantitative aspects of spelling errors. Each of these will be evaluated in turn.

Arousal and Attention. In this domain, the NSE contains four indicants. It includes a vigilance test of 60 random letters, an Attention subtest (composed of the Seashore Measures of Musical Talents Rhythm test and the WISC-R Digit Span test), and two indicants based upon observation: Signs of Distractibility/Motor Arousal and Ability to Concentrate.

Lateral Dominance/Handedness. In this area, the NSE requires examinees to write their full name with both their preferred and non-preferred hand, with the time required monitored for each. Examinees are also asked to demonstrate, with each hand, how they would cut with a pair of scissors, unscrew a lid from a jar, and throw a ball.

Basic Perceptual Abilities. This domain of the NSE is subdivided into five distinct components: 1) Tactile, 2) Kinesthetic/Cerebullar, 3) Receptive Speech (auditory-verbal perception/processing), 4) Auditory (nonverbal perception/processing), and 5) Visual. Tactile ability is measured by the Fingertip Number Writing test; supplementary material links performance on this tactile task to reading performance. Kinesthetic/cerebullar ability is assessed by having examinees stand on one foot, stand on toes, and touch their nose with extended arms while their eyes are closed. Receptive speech ability is measured by components of the Peabody Picture Vocabulary Test, the Speech-Sounds Perception test of the Halstead-Reitan Neuropsychological Test Battery, and the Luria-Nebraska Neuropsychological Battery Receptive Speech scale. Auditory nonverbal perception/processing is measured by the Rhythm scale of the Luria-Nebraska and the Seashore Rhythm test. Visual ability is assessed by the Motor-Free Visual Perception Test (scored for the visual discrimination, figure-ground, visual memory, and visual closure scales and perceptual quotient), and the Trail Making Test of the Halstead-Reitan.

Output Measures. According to the NSE, this cluster of tasks is designed to tap frontal lobe, motor strip, and Broca area functions. The output measures examine motor skills and expressive speech. The tests of motor skills include free motor speed and coordination (Luria), motor shifting task (Luria), finger tapping (Reitan), and observation of pencil grasp. Expressive speech is measured by a scoring of the rate, flow, and melody of spontaneous speech, a scoring of dysarthric speech, and the vocabulary test of the Weschler.

Higher Level Cognitive & Integrative Abilities. Measurement of these abilities relies upon the construction of scores of seven factors of the Weschler, three scales of the Kaufman Assessment Battery for Children, and evaluation of the Trail Making Test of the Halstead-Reitan.

The examinee also responds to tasks involving spelling. The responses are scored in terms of 1) phonetically based errors and 2) visually based errors. The latter includes phonetically correct spelling in which errors involving reversals, omissions, and substitutions are present.

Practical Applications/Uses

The Neuropsychological Screening Examination is designed to obtain a comprehensive evaluation of neuropsychological functioning of individuals from age 9 through 14. The cassette discusses its utility for school-based assessment, especially as regards differential expectations and remedial strategies. Discussion is also given regarding the type of language that can be used in composing a report of neuropsychological dysfunction derived from the NSE. Graphs are provided that illustrate the correlation between performance on the reading ability components of the NSE and errors on a Wechsler block design task that may be expected from an individual with neuropsychological impairment. The NSE is designed to be individually administered, and it is estimated that administration will require 3 to 4 hours.

Technical Aspects

No reliability or validity information is provided for the Neuropsychological Screening Examination. Even though face or content validity may be assumed, one would expect to see evidence of interscale correlational structure. Better still would be evidence of the consistency of test results in examining the same client (Fyans, 1983). None of these conventional psychometric qualities are substantiated.

The authors do provide some information on the norms and norm development.

Preston and Preston indicate that the normative data were obtained through their own development and from collating data from studies of 10 other authors. However, exact references to the publications of these authors are not given. In most cases, the norms involve both means and standard deviations; in one case, the standard deviation is unavailable.

The NSE is purportedly sensitive to the level of impairment as reflected by the performance of the examinee. However, no information is provided to demonstrate empirically how the cutoff scores were established, nor is there information concerning the rates of false positive and false negative misclassifications.

One of the more extensive technical developments of the NSE is the computation of factor scores for six of the seven Wechsler factors used in the cognitive domain. For these computations, the NSE draws upon the research of several authors, and exact references are given (e.g., Luria, Wechsler, Halstead). However, no references or substantive backup is given for the seventh factor discussed by the NSE.

Critique

The Neuropsychological Screening Examination does a superior job of synthetically integrating a wide range of indicants of neuropsychological functioning. Based upon the extensive documentation (both written and upon the cassette) that describes all manner of neuropsychological functions, the origin and characteristics of certain neuropsychological dysfunction, and the aspects of assessments, the NSE appears to have been developed toward this systematic orientation. Thus, the NSE is embedded in and intertwined with coherent theory.

Furthermore, Preston and Preston directly discuss concepts that may prove useful in evaluations of workmen's compensation and/or disability status. The diagnosis of work-related head and brain trauma may become inextricably involved with neuropsychological issues (Fyans, 1986), and the NSE provides evaluation of the examinee's competencies in the areas of reasoning, attention, and information processing.

However, there are several serious drawbacks to the NSE. First, as indicated previously, there is a decided lack of conventional psychometric information. No reliability, validity, generalizability, or consistency data are given. This severely limits the implementation of the NSE as a clinical tool. Even though the authors extensively develop the theory surrounding the NSE, they must also include some information concerning 1) how ratings on one component should link or match ratings on any other component (profile validity); 2) how the use of the NSE improves decisions or placements from the examinee's evaluation accompanied to other neuropsychological indices (incremental validity); 3) whether the scores on the NSE taken either as a composite or as subcomponents accurately predict levels of functioning where the levels are know a priori (criteria validity); and 4) whether the NSE would be useful to distinguish among clients with different types of disabilities and dysfunctions (differential validity). Although rather extensive documentation is provided with the NSE describing the nature of various types of disabilities, the linkage of them to differential profile scores on the NSE is not provided. Until information regarding validity, correlational structure, reliability, and norm development are provided, the NSE should have tentative use only.

Other drawbacks are also apparent with the NSE. First, there is not a direct assessment of memory skills. The authors discuss their rationale for this; however, this reviewer would recommend that a direct assessment of memory be included. Second, the administration instructions lack detail and clarity. Although an individual thoroughly trained in neuropsychology may have little trouble following them, they will not be clear to clinicians studying neuropsychological assessment or those conducting measurement in the schools. Third, the title is slightly misleading. It defines the NSE as a screening tool, a first test or assessment prior to more extensive evaluaton. However, it is clear that the NSE is intended to be a comprehensive, multibattery assessment. As administration may take as long as 4 hours, the NSE apparently does not function as a pretest. Added to this is the pragmatic implication of cost to the clinician. The NSE uses subtests from other neuropsychological batteries, and the clinician will have to purchase several extensive and expensive neuropsychological tests in order to implement the NSE. Perhaps a more useful version of the NSE would be one that presented a cogent rationale for the inclusion and exclusion of certain subtests from other assessments. The provision of this explanatory material and expanded psychometric information would go a long way toward making the NSE an instrument of substantial utility for neuropsychological assessment.

References

Fyans, L.J., Jr. (Ed.). (1983). *Generalizability theory*. San Francisco: Jossey-Bass, Inc.

Fyans, L.J., Jr. (1986). *Workman's compensation & psychology: Necessary issues, necessary questions, necessary information*. Springfield, IL: Department of Central Management Services.

Kaufman, A.S., & Kaufman, N.L. (1983). *Kaufman Assessment Battery for Children*. Circle Pines, MN: American Guidance Service.

Luria, A.R. (1973). *The working brain*. New York: Basic Books.

Preston, J., & Preston, B. (1983). *Neuropsychological Screening Exam* (2nd ed.). Dallas: Wilmington Press.

Reitan, R.M. (1969). *Manual for administration of neuropsychological test battery for adults and children*. Indianapolis: Author.

Springer, E.P., & Deutsch, G. (1981). *Left brain, right brain*. San Francisco: W.H. Freeman & Co.

James E. Jirsa, Ph.D.
School Psychologist, Madison Metropolitan School District, Madison, Wisconsin.

THE OFFER PARENT-ADOLESCENT QUESTIONNAIRE

Daniel Offer, Eric Ostrov, and Kenneth I. Howard. Chicago, Illinois: Institute for Psychosomatic and Psychiatric Research andTraining, Michael Reese Hospital.

Introduction

The Offer Parent-Adolescent Questionnaire (OPAQ) is a 50-item instrument designed to measure 11 areas of parental perception regarding adolescents aged 13 to 19. The OPAQ is based on a wide range of theoretical perspectives, pilot testing, and subsequent revision.

The authors of the OPAQ have collectively published approximately 100 articles, books, chapters, and research reports. Although their research, individually and collectively, shows a commonality of interest in the area of adolescence, a diversity of topics is also evident. For example, Offer seems most interested in the general developmental period of adolescence, also focusing on specific subjects such as suicide, psychopathology, sexual behavior, and the concept of normality. Ostrov has concentrated on abused adolescents, delinquents, self-image issues, and instrument development. Howard has written extensively on psychotherapy, parenting, and clinical judgment as well as on the application of multivariate statistical techniques to research data.

Daniel Offer, M.D., received his medical degree in 1957 from the University of Chicago. He was an intern (1958) at the University of Illinois Research and Educational Hospital and has been associated, in various capacities since 1961, with the Michael Reese Hospital and Medical Center, Institute for Psychosomatic and Psychiatric Research and Training. Offer has functioned as a consultant to the Illinois Department of Mental Health and is a member of the American Society for Adolescent Psychiatry (president, 1972-73), the Illinois Psychiatric Society, and the Chicago Society for Adolescent Psychiatry. He has been a fellow at the Institute of Medicine and at the Center for Advanced Study of Education.

Eric Ostrov, Ph.D., J.D., received his doctoral degree in 1974 from the University of Chicago and his law degree from the same institution in 1980. He has held a variety of research posts, including, since 1982, serving as the director of the Police Evaluation Project at Rush Presbyterian-St. Luke's Medical Center, Chicago, Illinois, where he is also an assistant professor in the Department of Psychiatry. Ostrov is a lecturer at the Chicago School of Professional Psychology and at St. Xavier College, and functions as the Director of Forensic Psychology at Michael

Reese Hospital and Medical Center. He is a member of the American Bar Association, the American Psychological Association, and numerous local organizations.

Kenneth I. Howard, Ph.D., earned his doctorate at the University of Chicago in 1959 and is currently a professor of psychology at Northwestern University and a senior consultant at the Institute for Juvenile Research, State of Illinois, and at Michael Reese Hospital and Medical Center. Howard is a member of the American Psychological Association (fellow), the Society for Psychotherapy Research (president), and the Society for Multivariate Experimental Psychology.

The OPAQ is derived from an earlier instrument, the Offer Self-Image Questionnaire (OSIQ), and the OPAQ itself has been subsequently modified to generate the Offer Therapist-Adolescent Questionnaire (OTAQ) and the Offer Teacher-Student Questionnaire (OTSQ), which tap a therapist's or teacher's view of the adolescent as client or student, respectively.

The original OSIQ was developed in 1962 by Offer in order to provide a defensible method for selecting a representative group of adolescents from a larger group. The selected high school students were chosen on the basis of their "normal" responses on the OSIQ and were involved in Offer's longitudinal follow-up studies (1969, 1975). Since that time, the OSIQ has been extensively used in various studies, and data have been accumulated on many diverse adolescent populations, both in terms of type (e.g., males, females, normal, delinquent, disturbed, physically ill) and geographic location (e.g., the United States, Australia, Israel, Ireland, Canada, West Germany, etc.). At the present time, the OSIQ has been administered to over 20,000 adolescents.

As a result of a 1984 revision, each OPAQ item is now a direct restatement of an OSIQ item. For example, the OSIQ item "I think that I will be a source of pride to my parents in the future" is rephrased as "My daughter thinks that she will be a source of pride to her parent(s) in the future."

The OPAQ consists of 50 statements presented in a reusable paper folder accompanied by an answer sheet. Different forms are provided depending on whether the adolescent in question is a son or daughter of the respondent. Items in each form are identical except for the son/daughter wording differences. The response key is explained on the first page of the folder along with a sample question and response example. The response key is also repeated at the top of each of the three pages of statements for quick and easy reference. Each item is answered by circling a response from 1 to 6, corresponding to the following descriptions (example from "daughter" form given): 1) Describes her very well; 2) Describes her well; 3) Describes her fairly well; 4) Does not quite describe her; 5) Does not really describe her; 6) Does not describe her at all.

Item difficulty is very low and should present few if any problems for the respondents. Similarly, the process of indicating responses on the answer sheet is simple and straightforward, offering little opportunity for confusion or misunderstanding of directions.

The OPAQ is composed of 11 specific subtests under five major headings, each of which defines a separate "self" according to the authors:

1. Psychological Self—concerns, feelings, wishes, fantasies; sense of control over impulses; body conception.
   Impulse Control

Mood
Body Image
2. Social Self—perceptions of interpersonal relationship, moral attitudes, and future plans.
Social Relations
Vocational and Educational Goals
3. Sexual Self—the integration of sex drives into the adolescent's psychosocial life.
Sexual Attitudes and Behavior
4. Familial Self—attitudes toward family milieu.
Family Relations
5. Coping Self—individual strengths; psychiatric symptoms; coping strategies.
Mastery of the External World
Psychopathology
Superior Adjustment (Coping)
Idealism

The individual OPAQ items reflect a number of perspectives on adolescence, such as those postulated by Erikson (1950, 1968), Freud (1946, 1958), and Murray (1938), among others. Q-sort techniques (Engel, 1959; Marcus et al., 1966) were also used in questionnaire development, but no details on the item development process are provided.

Practical Applications/Uses

In its most straightforward application, the Offer Parent-Adolescent Questionnaire is designed to provide information that describes how a parent views his or her adolescent son or daughter. The subtests addressed by the OPAQ reflect factors that parents would probably be interested in because a successful adaptation within each of the "selves" would tend to predict intact adult functioning. It would seem then that one of the most common uses of the OPAQ would be in a clinical setting as an information-gathering instrument prior to or during individual or family therapy. Used in conjunction with the OSIQ in therapy, the OPAQ could identify a number of areas of contention between parents and adolescents, thus presenting this information clearly so that it could be approached by the participants.

Because the OPAQ is simple to administer and requires only 20 minutes for completion, it could be used by a wide range of individuals who deal with adolescents and their parents. This potential user group could include psychologists, social workers, school counselors, members of the clergy, and physicians, among others. The OPAQ would also be an appropriate instrument for use in various research projects.

Scoring of the OPAQ is handled primarily through the Michael Reese Hospital and Medical Center, which provides standard scores indicating the degree to which individuals deviate from a reference group comprised of normal adolescents. The reference group totaled 1,385 adolescents who were tested in 1979 and 1980 from high schools across the United States. Within this framework, "a standard score of 50 corresponds to a subject's having a raw score equal to the mean of the appropriate control group's mean. A standard score higher than 50 reflects better adjustment with scores lower than 50 reflecting poorer adjustment" (Offer, Ostrov, & Howard, 1984, p. 61). Interpretation then is very simple, consisting of an examination of standard scores for each questionnaire submitted.

In addition to the basic standard scores, other data analyses are available that provide raw scores and parent agreement scores, which are correlations between parent ratings of both sons and daughters. Hand scoring on the local level is possible because raw score to standard score conversions are provided in the manual. However, this would be a very time-consuming and tedious task.

Technical Aspects

In order to minimize response set, half of the Offer Parent-Adolescent Questionnaire items are stated positively and the remaining items negatively; the location of both types in the questionnaire is scrambled. The negatively worded items are handled by reflection, or subtracting the circled value from 7. For example, a 4 on a negatively worded item is a 3 following reflection (7 - 4 = 3). The raw score is the sum of the circled responses for the positive items and the reversed (reflected) negative items. In the raw score situation, lower scores imply positive adjustment, and high scores imply poor adjustment. Conversion of raw to standard scores follows commonly accepted practice, except that the standard deviation is 15 instead of 10. Raw score means and standard deviations, by scale and respondent (father/mother) and referent (son/daughter), are also available (Wisniewski, 1985).

Internal consistency data (Cronbach's alpha) have been reported for the OPAQ's 11 scales (Wisniewski, 1985). The available alphas were based on a 1984 study involving 160 fathers and 191 mothers of 16- to 19-year-old students from two midwestern high schools. In this case, alpha values ranged from a low of .30 (mothers responding for sons; Body and Self-Image Scale) to a high of .91 (fathers responding for daughters; Family Relationships Scale). Most OPAQ alphas could be considered within the moderate range.

The remaining reliability information and all validity data available appear to relate primarily to the OSIQ and not specifically to the OPAQ. The construction of the OPAQ scales was based on interitem correlations of all OSIQ items, not OPAQ items separately. Although it can be argued that the OPAQ possesses sufficient content validity to be a useful instrument in practice, definitive and specific data are not available to substantiate such a position from a psychometric point of view.

Critique

Offer and his colleagues have been conducting research and practicing in the area of adolescence for many years. The individual and group reputations of those associated with the Adolescent Research Center at Michael Reese Hospital are very solid and consistently positive. The original instrument (OSIQ) from which the Offer Parent-Adolescent Questionnaire was developed displays a number of problems with item construction and psychometric properties. In turn, the OPAQ possesses these same limitations and suffers from an additional lack of specific validity data. Such limitations contribute to major difficulties with respect to score interpretation and overall meaningfulness. Given these problems, it would be risky to rely on the OPAQ in isolation. However, if used in conjunction with other information-gathering techniques, especially interviews, its utility would be improved, and it could be expected to contribute a unique perspective to the process of understanding adolescent-parent relationships.

References

This list includes text citations and suggested additional reading.

Engel, M. (1959). The stability of the self-concept in adolescence. *Journal of Abnormal and Social Psychology, 58*, 74-83.

Erikson, E. H. (1950). *Childhood and society.* New York: W. W. Norton.

Erikson, E. H. (1968). *Identity, youth, and crisis.* New York: W. W. Norton.

Freud, A. (1946). *The ego and the mechanisms of defense.* New York: International Universities Press.

Freud, A. (1958). Adolescence. *Psychoanalytic Study of the Child, 13*, 225-278.

Howard, K. I., Orlinsky, D. E., & Perilstein, J. (1976). Contribution of therapists to patients' experiences in psychotherapy: A components of variance model for analyzing process data. *Journal of Consulting and Clinical Psychology, 44*(4), 520-526.

Marcus, D., Offer, D., Blatt, S., & Gratch, G. (1966). A clinical approach to the understanding of normal and pathological adolescence. *Archives of General Psychiatry, 15*, 569-576.

Murray, H. A. (1938). *Explorations in personality.* New York: Oxford University Press.

Offer, D. (1966). Studies of normal adolescents. *Adolescence, 1*(4), 305-320.

Offer, D. (1969). *The psychological world of the teenager: A study of normal adolescent boys.* New York: Basic Books.

Offer, D. (1973). The concept of normality. *Psychiatric Annals, 3*(5), 20-29.

Offer, D., & Offer, J. (1975). Three developmental routes through normal male adolescence. *Adolescent Psychiatry, 4*, 121-141.

Offer, D., & Offer, J. B. (1975). *From teenage to young manhood: A psychological study.* New York: Basic Books.

Offer, D., Ostrov, E., & Howard, K. I. (1981). *The adolescent: A psychological self-portrait.* New York: Basic Books.

Offer, D., Ostrov, E., & Howard, K. I. (1982). *The Offer Self-Image Questionnaire for Adolescents: A manual.* Chicago: Michael Reese Hospital.

Offer, D., Ostrov, E., & Howard, K. I. (1984). Body image, self-perception, and chronic illness in adolescence. In R. M. Blum (Ed.), *Chronic illness and disabilities in childhood and adolescence* (pp. 59-73). New York: Grune & Stratton.

Ostrov, E., & Offer, D. (1978). Loneliness and the adolescent. *Adolescent Psychiatry, 6*, 34-50.

Wisniewski, S. (Ed.). (1985, May). OPAQ developments. *OSIQ Newsletter, 1*(2), 2-4.

Gurmal Rattan, Ph.D.
Associate Professor of Educational Psychology, Indiana University of Pennsylvania, Indiana, Pennsylvania.

Robert H. Hoellin, Ph.D.
Professor of Educational Psychology, Indiana University of Pennsylvania, Indiana, Pennsylvania.

PEDIATRIC EARLY ELEMENTARY EXAMINATION

Melvin D. Levine and Leonard Rappaport. Cambridge, Massachusetts:Educators Publishing Service, Inc.

Introduction

The Pediatric Early Elementary Examination (PEEX) is an individually administered screening device offered to provide information on a child's neurodevelopmental functioning. The PEEX is designed to evaluate children between the ages of 7 and 9 years, while a companion measure, the Pediatric Examination of Educational Readiness (PEER), offers similar information for children aged 4 to 6 years. Older children from 9 to 15 are evaluated by the Pediatric Examination of Educational Readiness at Middle Childhood (PEERAMID).

The PEEX was designed as an adjunct to other psychoeducational measures to delineate a child's learning problems. Additionally, it offers a profile of the child's neurodevelopmental strengths and weaknesses for purposes of remedial planning. However, the clinical utility of this screening device is not fully detailed in the manual.

The PEEX was developed by Melvin D. Levine, M.D., F.A.A.P. and Leonard Rappaport, M.D., F.A.A.P. Dr. Levine is chief of the Division of Ambulatory Pediatrics at the Children's Hospital in Boston and Associate Professor of Pediatrics at the Harvard Medical School. Similarly, Dr. Rappaport is on faculty at the Division of Ambulatory Pediatrics at the Children's Hospital in Boston and Instructor of Pediatrics at the Harvard Medical School.

Initial development resulted from testing 148 children in the Massachusetts area during 1978 and 1979 (Levine, Meltzer, Busch, Palfrey, & Sullivan, 1983). Scoring criteria were modified subsequently and further testing on the PEEX was carried out a year later.

The five major sections of the PEEX cover 1) developmental attainment, 2) neuromaturation, 3) task analysis, 4) associated (behavioral) observations, and 5) general health assessment. The neurodevelopmental attainment section consists of 32 tasks emphasizing current functioning in six primary developmental areas: fine motor functions, visual-motor integration, visual processing, temporal-sequential organization, linguistic skills, and gross motor functions. The remaining sections augment interpretation of the Developmental Attainment score. Items for this measure were adapted from a standard pediatric neurological exam, neuropsychological measures, and developmental scales.

The theoretical underpinnings of the PEEX are predicated on the assumption that neurological deficits form the bases for children's learning disorders. Although soft neurological signs are often used as diagnostic markers and indicators of learning problems, neurological involvement only offers a partial explanation for such difficulties. From epidemiological studies, only about one half of the 15% of children identified with learning disorders showed signs of neurological deficits (Myklebust & Boshes, 1969); the remaining variability has been associated with sociological and environmental factors (Gaddes, 1985; Rattan & Dean, 1987). Although the PEEX cannot account for all sources of children's learning problems, it does present the examiner with a structured approach in assessing the neurodevelopmental integrity of the child. More specifically, detection of neurological indicators such as motoric impersistence, choreiform movements (involuntary jerky movements), dysdiadochokinesis (difficulty coordinating rapid alternating movements), dystonic posture, and dyskinesia (slow movement) is outlined.

In addition to these neurological indicators, the child's cognitive tempo or response style is used as an adjunct measure in evaluating his or her neurological status. In the manual, Levine and Rappaport (1983) argue that it is clinically meaningful to differentiate between children whose learning problems derive from a pervasive impulsive response style and those whose impulsivity is a reaction to learning problems (p. 3). From this perspective, the authors state that long response latencies provide a positive index of neurodevelopmental deficits. Although research suggests an interaction between response style and learning/behavioral problems (e.g., Messer, 1976; Rattan & Dean, 1987), it is not clear how Levine and Rappaport interpret the behavior of children who are slow or have long response latencies: are long response latencies interpreted as a function of low cognitive abilities or as indicative of a reflective (slow but accurate) response style? Kagan's (1965) work on the dimensions of cognitive tempo included children who were either impulsive or reflective. The impulsive children were characterized by fast response times, and they were noted to have learning problems. Conversely, the reflective children were characterized by long response latencies and no academic difficulties. Although slow response times may be diagnostic, the PEEX authors do not differentiate between reflective children and those who are slow and inaccurate and have numerous learning problems; this may result invariably in misinterpretation of neurodevelopmental data and thus in inappropriate remedial programs. In short, a child with long latencies but a reflective response style could be interpreted as positive for neurological delays.

Directions for the administration and scoring of the PEEX are contained in a 43-page manual. A description of the underlying construct of each task is also presented, but the rationale for its inclusion is not always well documented. Additional materials for the PEEX consist of a stimulus booklet, a response booklet, an examiner's record form, and the PEEX kit (ball, paper clip, wooden block, key, rubber bank, and eye-hand board). The only materials required other than those provided are a pencil and a stopwatch.

Practical Applications/Uses

As stated in the examiner's manual, the PEEX is designed as an ". . . evaluation tool that enables health care and other professionals to derive empirical descriptions of a school child's development and functional neurological status" (Levine & Rappaport, 1983, p. 1). As the middle level in a trilogy of pediatric examinations for assessing school-age children, the PEEX could be utilized for initial evaluation or as a follow-up assessment of development/neurological functioning progress if the child was evaluated originally with the educational readiness version (PEER, ages 4-6) of the pediatric examination.

Scoring the PEEX primarily involves rating tasks on the Developmental Attainment section. Each of the 32 tasks are rated on a 4-point scale:

Level One—Performance clearly below expectations for entire age group.
Level Two—Performance appropriate for younger children (ages 7-8) in age group.
Level Three—Performance appropriate for upper range (ages 8-9) of age group.
Level Four—Performance considered strong for anyone in age group.

The authors' emphasize the importance of utilizing attainment levels for each task rather than computing a total score. Although a composite score for the total test can be calculated, the authors argue in favor of using subtest scores to identify patterns of strengths and weaknesses. A higher frequency of Level One scores is viewed as clinically significant and may justify further diagnostic procedures.

Neuromaturational delays or soft neurological signs can be inferred from observed synkinesias or associated movements for 12 of the PEEX tasks. For example, on the Motor Speed task, the child is required to open and close the fingers of one hand quickly while keeping the other hand still. Associated movement observed in the contralateral hand or mouth is rated as 2 (absent), 1 (suggestive), or 0 (clearly present). A composite score in this area can be calculated to provide a "crude" estimate of the child's neuromaturational integrity; however, the lack of behavioral criteria in rating associated movements coupled with the lack of normative data limit the diagnostic usefulness of this information.

The Task Analysis section of the PEEX attempts to ascertain cognitive and motoric components common to specific tasks. In this way, skills underlying poorly performed tasks can be identified. For example, poor performance on a copying task may be a function of deficits in either visual perception, fine-motor coordination, or both. However, if a pattern emerges implicating problems in one specific area, this will enable the clinician to develop a remedial program to ameliorate the deficit area. Although this procedure is clearly an asset for any psychological measure, the authors unfortunately do not demonstrate the efficacy of the method through detailed instructions or a case study approach.

The test manual does not identify examiner qualifications or restrict its use to specific professionals; however, it does recommend some familiarity with normal and atypical child development. Additionally, knowledge of physical medicine would be required in order to administer and interpret the general health assessment section of the evaluation. In sum, a background in pediatric neurology would serve well for both administration and interpretation of PEEX test scores.

The administration and interpretation section of the PEEX manual describes the 32 tasks, the rationale for each item, administration instructions, and the criteria or methods for scoring. In general, instructions for administration are clearly written and include pictorial examples to amplify the directions. However, the specific directions presented to the child are embedded in the general text, making the novice examiner manual-bound when administering the instrument. Using boldface type or some other method of "setting off" oral directions and scoring methods/criteria would aid administration.

Associated with examiner efficiency is the time required to administer the PEEX. Administration time is estimated to be 45 minutes (Levine et al., 1983), based on the mean time of completing tasks for children in the pilot studies. Because the majority of the children were "normal," the time for completion of tasks may be considerably longer for children with significant developmental delays and/or neuromaturation deficits. Additionally, although the manual indicates that professionals in mental health facilities and educational settings may administer the PEEX, as noted previously administration should be limited to those with training in neurodevelopmental assessment because of the heavy clinical judgment required in scoring the child's behaviors. Moreover, the equivocal relationship between specific auditory-visual processes and academic achievement adds to the tenuousness of the current scoring procedures.

Technical Aspects

The PEEX manual does not report any normative data. However, the authors indicate that norms were developed based upon several studies: "A series of preliminary studies on normal volunteer children in several middle-class communities, as well as some pilot testing in a clinic for youngsters with learning disorders, have helped establish norms for each item of the PEEX . . ." (Levine & Rappaport, 1983, p. 2). These norms refer to criteria used to determine performance levels (One to Four) for individual tasks. This should not be confused with normative data in which standard scores are provided to ascertain an individual's performance (e.g., percentile ranks) relative to peers.

In establishing performance levels for individual tasks, the authors used a sample of 187 second-grade children from two suburban middle-class communities in Boston (Levine et al., 1983). In addition, 59 children with learning problems were included in the study. Participants were administered the California Achievement Test (CAT), Wechsler Intelligence Scale for Children—Revised (WISC-R), and the PEEX. Scores on the CAT and WISC-R indicated that the sample was negatively skewed, with performance levels being above average. In fact, overall CAT scores (Grade 3.24) were one grade level above subjects' current second-grade placement. Similarily, the average Full Scale IQ on the WISC-R was 115, one standard deviation above the mean. Although the authors strongly suggest the development of local norms, use of the current criteria to establish performance levels for individual tasks may result in false negatives or an underreferral of children with neurodevelopmental delays.

In addition to the absence of normative data, no reliability or validity studies have been conducted on the PEEX. From a psychometric perspective, this would categorically exclude it as a "standardized" instrument.

Critique

In summary, the PEEX is presented as a screening measure to detect neurological deficits that may account for children's learning problems. As an adjunct to other psychoeducational measures, the PEEX is a welcome addition. Neurodevelopmental information gleaned from the device may well add to existing clinical data in facilitating differential remediation. However, given the current development of the PEEX, it should be viewed as an experimental instrument and caution should be used in making diagnostic or placement decisions. Moreover, lack of normative data, reliability, and validity precludes its use in isolation for purposes of program development.

References

Gaddes, W. H. (1985). *Learning disabilities and brain function: A neuropsychological approach* (2nd ed.). New York: Springer.

Kagan, J. (1965). Impulsive and reflective children: Significance of conceptual tempo. In J. D. Krumkoltz (Ed.), *Learning and the educational process* (pp. 133-161). Chicago, IL: Rand McNally.

Levine, M. D., Meltzer, L. J., Busch, B., Palfrey, J., & Sullivan, M. (1983). The Pediatric Early Elementary Examination: Studies of a neurodevelopmental examination for 7-to-9 year old children. *Pediatrics, 71,* 894-903.

Levine, M.D., & Rappaport, L. (1983). *Pediatric Early Elementary Examination* (examiner's manual). Cambridge, MA: Educators Publishing Service.

Messer, S. (1976). Reflection-impulsivity: A review. *Psychological Bulletin, 83,* 1026-1052.

Myklebust, H. R., & Boshes, B. (1969). *Final report, minimal brain damage in children.* Washington, DC: Department of Health, Education, and Welfare.

Rattan, G., & Dean, R. S. (1987). The neuropsychology of children's learning disorders. In J. M. Williams & C. J. Long (Eds.), *The rehabilitation of cognitive disabilities* (pp. 173-190). New York: Plenum Press.

Shari Eubank, Ed.D.
Director of Special Education, Lancaster-DeSoto Special Education Cooperative, Lancaster, Texas.

PICTORIAL TEST OF INTELLIGENCE

Joseph L. French. Chicago, Illinois: Riverside Publishing Company.

Introduction

The Pictorial Test of Intelligence (PTI) was designed by Joseph L. French (1964) to assess the general intellectual functioning of normally developing and handicapped children aged 3 to 8 years. French developed the PTI's picture format, in part, from his direct observations while assessing young, physically handicapped students. These observations led the author to develop an alternative individually administered intelligence test minimizing the amount and complexity of verbal and motor responses.

An earlier version of the PTI was entitled the North Central Individual Test of Mental Ability (NCITMA; French, c. 1955). Originally designed for students aged 3 to 6, NCITMA items were extrapolated from existing tests and developmental scales (e.g., Stanford-Binet Intelligence Scale, Peabody Picture Vocabulary Test, Columbia Mental Maturity Scale). Only those items that could be modified and adapted to the target population were included on the NCITMA. Test items extrapolated, modified, and adapted produced a limited stimulus pool. Child behavior records and direct observations also were used for item development (French, 1964). Test items were screened further to minimize cultural biasing.

The NCITMA initially was standardized in two states using 1950 census data as the basis for stratification. Four hundred subjects, from the 3-, 4-, 5-, and 6-year-old age levels, were drawn from eastern Nebraska and northern Illinois. Each age level was stratified according to the size of the community in which the subject resided and the occupational classification of the subject's father (French, 1964). Following the initial construction and regional standardization, the target population of the NCITMA was expanded to include 7- and 8-year-olds, which led to the national standardization study. This expanded version of the NCITMA was renamed the Pictorial Intelligence Test.

National standardization, completed in 1962, appears to be representative of the 1960 census. Procedures for conducting the sampling are discussed in detail in the manual (French, 1964, pp. 10-12) and indicate efforts to gain a stratified random sampling. However, several standardization concerns that may be related, in part, to the age of the normative group and other factors are evident. One specific concern focuses on the lack of explicit information regarding ethnic and minority-group membership. French (1964) reported that "the final sample included 16 percent non-whites as compared to approximately 11.4 percent in the United States according to the 1960 Census data" (p. 12). He also reported that "two examiners in the deep South tested only Negro children" (p. 12). An obvious question arises regarding justification for differences in the data collection process for the deep South as compared to the data collection procedures used for the remaining standardization group. A further concern relates to the obvious dichotomy of "nonwhite" and "white." French does not mention specifically either Asian/Pacific Islander or Hispanics in the discussion of standardization, even though data were collected in locales that would suggest high concentrations of these groups (e.g., Honolulu, HI; Corpus Christi, TX).

The PTI is organized according to six subtests, each of which corresponds to one of three broad categories: 1) verbal comprehension, 2) perceptual organization, and 3) spatial and numerical symbolic manipulation. The six subtests are 1) Picture Vocabulary, 2) Form Discrimination, 3) Information and Comprehension, 4) Similarities, 5) Size and Number, and 6) Immediate Recall. Each subject is administered the entire test, with the possible exception of the Similarities subtest, which is discontinued if the subject fails any four of the first six test items.

As the test title implies, the PTI is a nonverbal assessment technique utilizing a picture stimulus and response format. Subjects are not required to respond either verbally or with gestures. French suggests a variety of possible responses, including eye movements if the examiner can determine the intended movement of top, bottom, left, and right (French, 1964, p. 4). Subtests have a ceiling of six consecutive failures; however, each subject begins with the first item of each subtest.

The Picture Vocabulary subtest (32 verbal stimuli and 29 response cards) measures word comprehension. Subjects respond to a spoken stimulus by selecting on the response card one of four drawings that best represents the meaning of the stimulus word. This task is comparable to other picture vocabulary tests.

The second subtest, Form Discrimination (27 stimuli cards and 24 response cards), measures the subject's ability to match forms and discriminate between similar shapes. It requires the subject to match a stimulus card drawing with one of four response card drawings, some of which have been duplicated from the common geometric symbols used in other intelligence tests (French, 1964). Most of the cards appear to be commonly known figures. The last five response cards contain four partially completed drawings. The subject responds by indicating the drawing that, when completed, matches the stimulus card.

Information and Comprehension, the third subtest, utilizes 24 response cards to obtain 29 answers. This subtest purports to measure the subject's verbal comprehension by sampling his or her range of knowledge and general understanding levels. Although the subject uses past experience to solve verbal stimuli, French (1964) attempted to develop cross-cultural items that minimize specialized knowledge (p. 4). The subjects' responses may provide relevant information regarding their environmental, educational, and cultural opportunities (p. 5).

The fourth subtest, Similarities (22 items), measures the subject's ability to generalize visual information (French, 1964). A response card showing three common elements and one unrelated drawing is presented to the subject, who is instructed to identify the one drawing that is different from the others. All items are elaborations of concepts known to most children. Some 3-year-olds may fail to understand the directions and concepts being tested. In such cases, the subtest is discontinued.

The fifth subtest, Size and Number (8 stimuli, 19 response cards), attempts to measure the subject's perception and recognition of size, number symbol recognition and comprehension, ability to count, and ability to solve simple arithmetic problems. To test number symbol comprehension, seven of eight stimuli cards are used for two response card items. Large numerals are printed on 17 response cards used for identifying and counting numbers. The subtest also includes basic fact problems and fractions. However, recognition of number symbols is not necessary for the solution of the problems (French, 1964).

The final subtest, Immediate Recall, is composed of 19 stimuli and response cards measuring the subject's ability to retain momentary perception of size, space, and form relationships (French, 1964). A stimulus card is presented for 5 seconds and then withdrawn. The examiner then immediately presents the subject with a response card from which he or she is to identify the stimulus drawing from among four drawings. This subtest is similar to Binet-type memory tests in which a picture is presented, removed, and presented again in conjunction with several distractors. With the exception of the initial five response cards, all the cards in the subtest previously were used in the Form Discrimination subtest. Stimuli are presented progressively, with each higher-level stimuli having more subtle distractors within the response field.

The PTI utilizes a total of 54 stimuli and 137 picture-response cards. Each stimulus card is a 3″ square with centered drawings and/or geometric shapes. The 137 response cards are 11″ squares with a drawn figure centered along each of the four edges (i.e., top, bottom, left, right). Subjects respond to orally presented questions by indicating which one of four alternatives is the correct answer. Both the response and the stimulus cards are presented in serial, consecutive, order. The numerical order is printed on the reverse side of each card. When a card is used for more than one response, duplicates are enclosed to facilitate continuous testing.

Visually, the PTI's record booklet is reminiscent of that of the Stanford-Binet Intelligence Scale. The descriptive information (i.e., name, address, date of testing, date of birth, etc.) and the behavior observations required during testing are similar in content and format to those of the Stanford-Binet.

Practical Applications/Uses

The Pictorial Test of Intelligence is a nonverbal test of general intelligence for children 3 to 8 years of age. Its value would seem to be in its potential ability to measure symbolic language concepts using a nonverbal format, which apparently would assist appraisal personnel in assessing subjects who may be unable to give verbal responses and subjects with unintelligible or immature verbal skills. A variety of responses (i.e., gesturing, pointing, eye movements, etc.) are acceptable. Spacing and location of the alternatives on the response cards make these various communication modes feasible. Because the record booklet is keyed to each response card, the examiner can monitor responses easily. The subtest format would appear to accommodate the short attention span of most young children and presumedly would allow the examiner to suspend the testing session when the child's attention wanes, a procedure advocated by French (French, 1964, p. 27).

A further potential value would seem to be the concepts measured by the PTI.

Each subtest appears to measure a general cognitive demand required in the school academic environment. The PTI could be helpful in determining the learning potential of high-risk students being considered for placement in special or support programs. Using a subtest analysis, a profile of learner potential strengths and weaknesses could be developed.

The PTI takes approximately 45 minutes to administer and 30 minutes to score. Examiner cues listed on the back of response cards facilitate administration. However, these are procedural cues only. Administration of the subtests requires the use of the test manual as well as the response cards because the directions are, at times, confusing and detailed.

Overall, administration procedures for the PTI are consistent with many other individually administered assessment procedures. The PTI can be given by any professional trained to administer individualized assessments. Directions for administration and scoring are relatively clear and easy to follow. The most significant demand placed on the examiner is manipulating the number of large cards required to administer the test. To proficiently administer the PTI, the examiner is highly encouraged to review and practice the manipulation of the test materials.

Data reporting for the PTI is based on raw score equivalents. Tables, easily read and interpreted, are provided for the long and short forms of the test as well as for each subtest and total score reporting metric. Each subtest score is reported as a mental age (MA) with an overall test reporting metric of derived intelligence quotient (DIQ). Also provided is a MA equivalent for the DIQ as well as percentile rank.

Interpretation of the PTI is based on the DIQ ($M = 100$, $SD = 16$) and MA equivalents. French suggests that DIQs exceeding more than one standard deviation above or below the mean be interpreted to indicate that the subject's test performance is very well or very poor in relation to his or her chronological age (p. 47). Further, the MA scores, overall and subtest, are to be interpreted as the average score earned by subjects at a given chronological age. French cautions examiners that a 2-3-year fluctuation is common among subtest MAs and that significant differences may be present when deviations exceed this 2-3-year fluctuation.

Technical Aspects

Technical data for the Pictorial Test of Intelligence are reported in the test manual and documented in the professional literature. The professional literature reviews question issues related to the test's content, criterion-related, and construct validity (Sawyer, Stanley, & Watson, 1979; Hynd, Quackenbush, Kramer, Connor, & Weed, 1979; Newland, 1972; Smith, 1972; Sawyer, 1968; Pasewark et al., 1967).

Item selection, related to reliability and validity, is a primary concern. Final item selection consisted of administering the preliminary test to 25 subjects at ages 3-6 and determining correlation coefficients for each item with the total test score. French (1964) points out that to improve the test and to extend its range through age 8, certain items were eliminated and some new items were added. New items were administered to 100 8-year-old subjects.

The resulting 160 items were analyzed using biserial r. Correlations ranging from .04 to .95, $M = .60$, are reported for the standardization sample. Examining

Appendix A, Item Analysis Data, in the test manual it is apparent that the contribution to variance for several items may be questioned, assuming better test items contribute more to the total variance than lesser contributing items (Ferguson, 1981). Kuder-Richardson Formula 20, equivalent to the alpha coefficient, indicates correlation coefficients of .90 and above for age levels 4-7, high .80s for 3- and 8-year-olds, and .86 and .88 on the short form for 3- and 4-year-olds, respectively.

Test-retest reliability coefficients present some difficulty. Reliability coefficients are reported for five groups with minimal identification of the group's characteristics. A group of 6- and 7-year-olds were tested within a 2-4-week interval. The test-retest coefficient for the 6-year-olds ($n = 30$) was at an acceptable level, $r = .90$. Twenty-five 7-year-olds tested at 2-4-week intervals were reported to have a reliability coefficient of .94. A third group of 5-year-olds ($n = 31$) was tested at 2-6-week intervals ($r = .91$). The fourth group, consisting of a mixture of 3-, 4-, and 5-year-olds ($n = 27$), was tested at 3-6-week intervals and obtained a reliability coefficient of .96. The final study reported consisted of a group ($n = 49$) originally tested as 3- and 4-year-olds on the NCITMA and retested as 8- and 9-year-olds on the PTI. The resulting coefficient ($r = .69$) was attenuated on the basis of the test's inability to discriminate when used with average or better ability subjects exceeding the chronological age of 8 (French, 1964).

Although four of the five reported test-retest coefficients equal or exceed traditional standards for test consistency (Aiken, 1979; Guilford & Fruchter, 1978; Helmstadter, 1964; Nunnally, 1978), questions regarding the size of these samples should be raised. Sawyer (1968) reported considerably lower test-retest coefficients than those reported by French (1964). Himelstein (1972) concluded that test-retest reliabilities were inadequate. Further, considerable speculation should be cast on the attenuated reliability coefficient between the NCITMA and PTI. French reported that test items were altered from the original (NCITMA) to the present test. It would appear that two different instruments were developed as the attenuated correlation accounted for less that 50% of the variance between the two measures.

Validity measures appear potentially inadequate. Content validity is ensured by the plan and procedures used in the construction of the test (Brown & Bryant, 1984). The basis for establishing content validity was based on the presumption of constructing test items similar to those existing on recognized tests of intelligence, such as the Stanford-Binet Intelligence Scale and the Wechsler Intelligence Scale for Children (Newland, 1972). The biserial r for the 160 items ranged from .04 to .95. Apparently, all items were retained in the final test version.

Criterion-related validity also seems to be highly questionable. Predictive validity data were determined from the original NCITMA and group achievement and intelligence test scores. Rank order correlations *(rho)* of .75, .82, and .74 are reported between the NCITMA and group achievement tests for three groups ($n = 11, 11,$ and 19) following a 4-year interval (French, 1964, p. 19). Two larger groups ($n = 28$, $n = 18$) are reported to have yielded rank correlations of .68 and .77 following a 3-5-year interval (French, 1964, p. 20). It is apparent that predictive validity for the PTI must be questioned seriously, a position similarly held by Himelstein (1972) and Newland (1972).

Concurrent validity, another form of criterion-related validity, appears to have a fate similar to predictive validity. A sample of 32 first graders was administered the PTI and three other individual intelligence tests, the Stanford-Binet Intelligence Scale, the Wechsler Intelligence Scale for Children (WISC), and the Columbia Mental Maturity Scale (CMMS). Intercorrelations between the PTI and these three measures were reported as .72, .65, and .53, respectively (French, 1964, pp. 20-21). However, the distribution of the 32 subjects across the three instruments must have been scant; therefore, the interpretation value of these correlations must be held in abeyance. Similar conclusions are rendered for the correlation coefficients reported between PTI subtests and the Stanford-Binet, WISC, and CMMS scores.

Two remaining studies attesting to the PTI's concurrent validity were reported for samples of 30 and 32 first-graders. The first study ($n = 30$) was conducted as a component of a test-retest reliability study. The California Test of Mental Maturity (CTMM) and the New Basic Reading Test, a reading readiness test, were adminsitered in the spring, with test-retest PTIs administered the following fall. Pearson coefficients of .62 and .51 were reported for the PTI and CTMM comparisons and .81 and .79 for the PTI and reading readiness comparisons (French, 1964, p. 22). A second study consisting of 32 first-graders was administered the Lorge-Thorndyke and PTI. A Pearson coefficient of .61 was obtained for this comparison (French, 1964, p. 22).

The manual presents support for content validity by referring the test user to discussions of the standardization and item analysis processes. The interrelationships between the six subtests suggest that discrete properties may not be measured. Approximately 30-50% of the total variance seems to be accounted for by each subtest, an overlap that raises questions regarding general versus specific test factors. Sawyer, Stanley, and Watson (1979) factor analyzed PTI scores for 90 students. Their findings indicate that examiners should be cautious in interpreting the PTI's subtests as discrete entities. However, they did support the overall contention that the PTI may assess the general intellectual levels of young children.

Critique

The obvious strength of the Pictorial Test of Intelligence is its pictorial and nonverbal response format. The item stimuli seem to be familiar to and within the backgrounds of most young children. Response cards are organized to facilitate a variety of acceptable responses. The PTI seems to be a reasonable alternative for assessing aptitude in physically handicapped and language-disordered children.

Although questions regarding its standardization sampling, item selection, reliability, and validity have arisen, the most obvious shortcoming of the PTI is its lack of currency. Brown and Bryant (1984) view test currency as "the extent to which the standardization group is representative of the most recent census report" (p. 56). The PTI's data base is over 25 years old. Thus, it would seem that any deliberation regarding the PTI's utility must be suspended in light of its lack of currency.

Regardless of previous discussions relative to the technical construction of the PTI, its most salient limitation is standardization. It is doubtful, in the opinion of this reviewer, that the PTI has any degree of representational validity after 25 years. Accurate professional judgments regarding identification, placement, and intervention plans cannot be based on a measure without some degree of currency.

References

Aiken, L. (1979). *Psychological testing and assessment.* Boston: Allyn & Bacon.

Brown, L., & Bryant, B. (1984). A consumer's guide to tests in print: The rating system. *RASE, 5,* 55-61.

Ferguson, G. (1981). *Statistical analysis in psychology and education* (5th ed.). New York: McGraw-Hill.

French, J. (1964). *Pictorial Test of Intelligence.* Lombard, IL: Riverside Publishing.

Guilford, J., & Fruchter, B. (1978). *Fundamental statistics in psychology and education.* New York: McGraw-Hill.

Helmstadter, G. (1964). *Principles of psychological measurement.* New York: Appleton-Century-Crofts.

Himelstein, P. (1972). Pictorial Test of Intelligence. In O.K. Buros (Ed.), *The seventh mental measurements yearbook* (pp. 748-749). Highland Park, NJ: The Gryphon Press.

Hynd, G., Quackenbush, R., Kramer, R., Connor, R., & Weed, W. (1979). Clinical utility of the WISC-R and the French Pictorial Test of Intelligence with native American primary grade children. *Perceptual and Motor Skills, 49,* 480-482.

Newland, T. (1972). Pictorial Test of Intelligence. In O.K. Buros (Ed.), *The seventh mental measurements yearbook* (pp. 749-750). Highland Park, NJ: The Gryphon Press.

Nunnally, J. (1978). *Psychometric theory* (4th ed.). New York: McGraw-Hill.

Pasewark, R., Sawyer, R., Smith, E., Wasserberger, M., Dell, D., Brito, H., & Lee, R. (1967). Concurrent validity of the French Pictorial Test of Intelligence. *Journal of Educational Research, 61,* 179-183.

Sawyer, R. (1968). An investigation of the reliability of the French Pictorial Test of Intelligence. *Journal of Educational Research, 61,* 211-214.

Sawyer, R., Stanley, G., & Watson, T. (1979). A factor analytic study of the construct validity of the Pictorial Test of Intelligence. *Educational and Psychological Measurement, 39,* 613-623.

Smith, T. (1972). Review of the Pictorial Test of Intelligence. *Journal of School Psychology, 10,* 213-215.

Robert H. Bauernfeind, Ph.D.

Professor of Education, Northern Illinois University, DeKalb, Illinois.

PRE-PROFESSIONAL SKILLS TEST

Educational Testing Service. Princeton, New Jersey: Educational Testing Service.

Introduction

The Pre-Professional Skills Test (PPST) was developed in 1983 to measure skills judged to be important for people planning to become elementary or high school teachers. The test publisher, Educational Testing Service (ETS), reports that some 50% of the tests are administered to students wishing to enter teacher training and the other 50% are given after teacher training, as a part of the teacher certification process. In most instances, students must "pass" the PPST before they can enroll in education courses or before they can be certified (licensed) as teachers. The pass/fail cutoff lines are established by each college or state based on local standards (Livingston & Zieky, 1982).

The PPST program was developed in the early 1980s to measure basic proficiencies in the areas of Reading, Mathematics, and Writing. The initial planning for the PPST program was conducted by a "Policy Council" composed of teachers, school administrators, deans of colleges of education, and state legislators. It was this group that established policy guidelines, including guidelines for the test specifications (ETS, 1985). After the general guidelines had been established, a new group—the "Teacher Council"—assembled to monitor the total program. This council, meeting several times each year, is composed of 20 teachers, school administrators, and state legislators.

Construction of the multiple forms of PPST is handled by ETS professionals and outside consultants. The ETS staff conducts the item-analysis, reliability, norms, and other psychometric studies of each form.

When we talk about "teacher competence" we encounter an interesting phenomenon. Our children, even our youngest children, can tell us whether a teacher is unduly grouchy or hostile. But our children, even our oldest children, may not recognize that a teacher often misreads and misquotes articles, consistently communicates with more words than necessary, or frequently makes gross errors in arithmetic estimates.

When those teachers were in college, their professors may have been caught in a strange trap. First, the professors may have awarded high grades based on very narrow criteria that did not assess "teaching competence" as we have defined it here. Later, seeing the same students commit gross errors in reading, mathematics, or writing, the professors rationalized that the students must be performing satisfactorily because they maintained grade point averages (GPAs) of 2.3, 2.6, or

---

All sample test items cited within this review are from the *Bulletin of Information* (ETS, 1986a) and are reproduced with the permission of Educational Testing Service.

higher. Some badly incompetent students are not "caught" until they enter student teaching; others are not discovered until their first year on the job; and, as we all know, some incompetent teachers are not recognized until after attaining teacher-tenure.

It seems reasonable to ask those who would teach our children to demonstrate that they *can* read well, handle arithmetic relations well, and recognize English sentences that can be improved. Thus, many of us endorse the idea of a "pre-professional" screening test for individuals planning to become teachers. Is the PPST an appropriate test for such a purpose? And, if our answer is "yes," where do we set cutoff scores? Those are the tough questions this review attempts to answer.

The PPST provides scores in the usual three areas of "basic skills"—reading, mathematics, and writing:

Reading (50 minutes): The Reading test consists of 40 questions that assess one's ability to understand, analyze, and evaluate written messages. The test contains long passages (around 200 words), short passages (around 100 words), and a few very short statements to be analyzed. Each segment is followed by one or more 5-choice multiple-choice questions.

Mathematics (50 minutes): The Mathematics test consists of 40 questions that assess one's ability to judge mathematical relations. The problems minimize computational arithmetic. Instead, they stress the conceptual skills needed by professional educators, including estimating answers, interpreting graphs, comparing fractions with percents, reading scales, interpreting formulas, and following the logic of a problem. Each problem is followed by a 5-choice multiple-choice question.

Writing (30 + 30 minutes): The Writing test is presented in two parts. First, the students complete a 45-item test of functional written English in 30 minutes. Each problem is followed by a 5-choice multiple-choice question. In the second part of the test, students are given 30 minutes in which to write an essay on an assigned topic. Scores from the two exercises are combined into a single Writing score.

All three basic skills tests are reported on a unique scale ranging from a low of 150 to a high of 190. ETS does not combine the three scores into a "total" or "composite" score. Individual programmers can do so if they wish, but ETS advises against this.

Practical Applications/Uses

The PPST is a group-administered test requiring, with breaks, a total administration time of about 3 hours. The PPST testing program is conducted approximately 10 times each year, on specified dates at ETS-approved testing centers. (These testing procedures are much the same as those of the ACT, the Law School Admissions Test, and other high-security testing programs.) The program is conducted by directors of these centers with the assistance of trained proctors. Candidates register with one of the ETS-approved testing centers and pay a fixed fee ($35.00 per examinee at the time of this writing). These fees are then mailed to ETS, and ETS later pays the administrators and proctors for their work in conducting the testing program. These procedures are detailed in the *PPST Supervisor's Manual*, published yearly by ETS.

Test results are sent to the candidate and to the institutions requested by the candidate. Turnaround time from the testing date to receipt of one's scores is typically 1 month.

The three multiple-choice tests are scored for number correct. The essay is graded "holistically" on a 0-to-6 scale by two readers. In "holistic" gradings, the readers attempt to follow the logic, or "the *flow,*" of the essay, with minimal regard for grammar, spelling, capitalization, or punctuation. "The placing, scoring, or grading occurs quickly, impressionistically, after the rater has practiced the procedure with other raters" (Cooper, 1977, p. 3). The student's essay score is the sum of the two readers' scores. If the two scores are 2 points or more apart, the essay is read a third time, and the student's score is then the sum of the two closest ratings.

Educators who are annoyed with the proliferation of measurement scales in recent years will be displeased greatly with that aspect of the PPST. The PPST scales are essentially *raw-score* scales—41 possible raw scores (0-40) yielding 41 possible scale scores (150-190). (The Writing test, 45 items and the essay score, is scaled back as if it were derived frcm 0-40 raw scores.) These are not expanded standard score scales, nor are they equal-interval scales; they are essentially raw scores, negatively skewed, derived from three 40-item tests with similar distributions of raw scores:

> Raw 0 = scale 150
> Raw 10 = scale 160
> Raw 20 = scale 170
> Raw 30 = scale 180
> Raw 40 = scale 190

Although this inventive scale seems strange and unnecessary, the fact remains that the three scale score distributions are enough similar (ETS, 1986b, p. 5) that a student's three scale scores could be cross-compared, although ETS advises against that practice (ETS, 1986a, p. 6).

Technical Aspects

In an article published more than 25 years ago, Ebel (1961) urged educators to think less in terms of predictive validity and more in terms of precision of measurement (reliability) and precision in the definition of that which is to be measured (content validity). With the PPST and similar tests, we need to concentrate on the fact that *content validity* is at issue. There is no practical criterion measure to be placed on the X-axis of a scattergram. There is no practical criterion measure that would show us which teachers make ghastly mistakes in speaking, in arithmetic, or in recognizing subtle implications in a note from a parent or a memorandum from the board of education. None of the PPST publications mention predictive-validity studies. Rather, the authors continually stress content validity: Do we want to encourage teacher-trainees who do well on these kinds of questions? Would we want to reject as teachers people who do poorly on these kinds of questions? The PPST authors suggest that the tests are "job relevant" if those in authority judge that 80% or more of the questions are job relevant (ETS, 1987).

Of course, "content validity," like beauty, is in the eyes of the beholder. Others may disagree, but this reviewer sees high content validity in all three tests in the PPST. Consider the following test items cited from the *Bulletin of Information* (ETS, 1986a). The Reading test is replete with questions involving judgments and inferences:

1. The main idea of the passage is that . . .
2. The passage supplies information that would answer which of the following questions?
3. The author of the statement assumes which of the following?

The Mathematics test is heavily conceptual, with a minimum of arithmetic computations:

1. According to the steps in the flowchart above, if the number 5 is the input, the answer printed would be . . .
2. Which of these is NOT a correct way to find 75% of 40?
3. 3.57 is how many times 0.00357?

All of these sample questions canvass adult quantitative thinking. However, there is an ominous note: in that same bulletin, the authors assert that terms such as "integer," "factor," and "prime number" may be used in the test questions (ETS, 1986a, p. 11). If true, such would seem to violate the spirit of a test for teachers. Surely the PPST authors recognize that many excellent teachers would not know the word "integer." Many of those same teachers also would not know the difference between a "prime number" and a hole in the wall, and, moreover, they would not care.

Like the Reading and Mathematics tests, the multiple-choice Writing test is mature and "real life." Clearly, the authors have followed the research of Breland (1977a, 1977b) in his work on the College Board's *Test of Standard Written English*. The Breland-type items on the PPST are of two kinds:

1. He spoke bluntly and angrily to we spectators. No error

    A   B   C   D   E

2. The smallest error has the capability to change the meaning of a whole sentence.

In the first item, the examinee seeks an error in writing and marks choice E on the answer sheet if no error is found. In the second item, the examinee is presented with five choices of phrases to insert above the underline. Examinees are told to "choose the answer that produces the most effective sentence—clear and exact, without awkwardness or ambiguity. Do not make a choice that changes the meaning of the original sentence" (ETS, 1986a, p. 12). For this item, the keyed answer is the simple one: *can change*. These two item types come so close to real-life writing and editing that this reviewer will not quarrel with the test publisher's claim that it is a "writing" test.

The actual writing sample in the second part of the Writing test is derived from an assigned topic. Students are given 30 minutes to address the topic. An example of an assigned topic, taken from the *Bulletin of Information*, follows:

> Imagine that you have been asked, as part of an application for a job, to describe a particular experience—either in or outside a classroom setting—from which you learned something. Write an essay in response to that request. Be sure to include in it what you learned from the experience and whether what you learned changed you or the way you act. (ETS, 1986a, p. 13)

(Is this a well-written assignment? Would Breland endorse its wording? We leave that question to readers of *Test Critiques*.)

Basic statistical data for the PPST are provided in *Test Analysis* (ETS, 1986c)—a detailed analysis of Form 3IPS2 that was administered to user colleges in March 1986. Some of the major findings, shown here in Table 1, can be summarized as follows:

1. Correlations among the three multiple-choice tests are typical for tests of general educational development, averaging around .65 (42% common variance).

2. The multiple-choice tests were perhaps a bit easy for this group, averaging around 68% of the items correct. However, if we believe that the questions are "job relevant," these mean scores are certainly acceptable.

3. The spread-of-scores (SDs) are satisfactory for this group.

4. The reliability estimates for the multiple-choice tests are quite satisfactory; the interrater reliability estimate for the essay portion of the Writing test is disappointingly low.

With regard to the fourth finding, the correction-for-attenuation formula was applied to the data for the two sections of the Writing test:

$$r_{AB} = \frac{r_{ab}}{\sqrt{r_{aa} \cdot r_{bb}}} = \frac{.49}{\sqrt{.83 \cdot .63}} = \frac{.49}{.72} = .68$$

If both sections of the Writing test had been perfectly reliable, the correlation between the two would be estimated to be .68 (46% common variance). But that figure is a function of measurement *theory.* In fact, the actual correlation was .49—a figure that is markedly short of similar correlations reported by Breland (1977a), without corrections for attenuation!

Table 1

Descriptive Data for the Four Scores (N = 6,292)

	Mean	SD	Rel*	Read	Math	Wm-C	Wess
				Intercorrelations			
Reading (Read)	27.8	6.7	.86	...	.64	.72	.43
Mathematics (Math)	27.3	7.3	.88	.64	...	.60	.33
Writing—multiple-choice (Wm-c)	29.6	6.8	.83	.72	.60	...	.49
Writing—essay (Wess)	8.3	1.5	.63	.43	.33	.49	...

* KR-20 for the three multiple-choice tests (ETS, 1986c, p. D); first-rating vs. second-rating for the Writing essay (ETS, 1986c, p. 11).

Two further findings are cited in *Test Analysis:* all three multiple-choice tests are clearly power tests because well over 90% of the examinees completed each test section (ETS, 1986c, p. D). Furthermore, the Standard Error of Measurement of the scale scores is about 2.4 on each scale (ETS, 1986c, p. 5). Given the somewhat similar distributions of scale scores noted previously, this reviewer would suggest that scale score differences of 8-9 points or more could be brought to the student's attention. Thus, to one who scored $R = 173$, $M = 184$, $W = 175$, we would say: "Compared with other students wanting to enter teacher training, you scored about the same on the Reading and Writing tests, but you scored significantly higher on the mathematics test than you did on Reading and Writing." (Total group percentiles for those three scale scores were approximately $R = 27$, $M = 79$, $W = 38$.)

The NIU Studies. In 1985, the College of Education at Northern Illinois University adopted the PPST for experimental use in approving students for teacher-training programs. Cutoffs on the three scores were assigned judgmentally in terms of local concepts of how high a teacher should score in order to become an effective teacher (Livingston & Zieky, 1982). During the first year of PPST testings, approximately two thirds of the students passed all three cutoffs the first time; approximately one third failed to meet one or more of the cutoffs the first time.

After the first year of PPST testings ending in August 1986, data were gathered to compare PPST scale scores with ACT scale scores, sex, age, teaching preference (elementary or secondary), and cumulative grade point average. The data were analyzed using a 13 x 13 correlation matrix,* shown in Table 2. The variables in Table 2 were

1. PPST Scores (4): Reading, Mathematics, Writing, Total.
2. ACT scores (5): English, Mathematics, Social Studies, Science, Total.
3. Major (1): Coded High = 1, Elem = 0 as of December 1986.
4. Sex (1): Coded M = 1, F = 0.
5. Age (1): Age in years as of December 1986.
6. GPA (1): Cumulative grade-point average on a 4-point scale as of December 1986.

Although there were exceptions, the ACT typically was taken early in Grade 12, and the PPST typically was taken around the end of the college sophomore year. Thus, for most students, there was a 2½-year lapse between the ACT and PPST testings.

One other fact to be noted is that the investigators did not expect to find high correlations among the ACT scores, PPST scores, and GPAs because of a restriction of range on each of those three measures: students with low ACT scores or low GPAs would not be in college to take the PPST. That caution aside, Table 2 suggests the following conclusions:

1. PPST Reading correlated most highly ($r = .58$) with the ACT Social Studies score.

2. PPST Mathematics correlated most highly ($r = .44$) with the ACT Mathematics score. However, this was a fairly low correlation that was probably a function of the fact that the ACT Mathematics test is more "academic" and requires more skills in actual arithmetic computations.

3. PPST Writing correlated most highly ($r = .62$) with the ACT English score.

4. The PPST Composite score yielded expected correlations with the ACT Total score ($r = .65$) and with cumulative GPAs ($r = .45$).

The PPST scores were unrelated almost totally to the three variables of teaching level, sex, and age.

There were no great surprises in these studies, except possibly for the relatively low correlation between the two Mathematics scores. Our colleagues generally were pleased to note the near-zero correlations between the PPST scores and the sex and age of the examinees.

Critique

The current (1987-88) fee of the PPST program is $35.00 per examinee. The current prices of the ACT and the College Board SAT programs are well under $15.00

---

*Thanks are extended to Dr. Peter Abrams of Northern Illinois University for his assistance in conducting the 13×13 correlation studies described in this review.

Table 2

13x13 Correlation Matrix including PPST Scores (N = 422)

	MEAN/SD	PPST R	PPST M	PPST W	PPST T	ACT E	ACT M	ACT SOC	ACT SCI	ACT T	MAJ	SEX	AGE	GPA
PPST Reading (PPST R)	176.07/6.57	...	.35	.48	.71	.52	.34	.58	.43	.49	.17	.00	-.02	.36
PPST Mathematics (PPST M)	178.16/9.00	.35	...	.37	.60	.38	.44	.36	.35	.40	.11	.02	.00	.29
PPST Writing (PPST W)	175.84/3.88	.48	.37	...	.72	.62	.34	.40	.43	.44	.03	-.13	-.08	.33
PPST Total (PPST T)	530.07/13.05	.71	.60	.72	...	.68	.63	.58	.61	.65	.15	.01	-.05	.45
ACT English (ACT E)	20.01/4.11	.52	.38	.62	.68	...	.46	.58	.61	.61	.04	-.18	-.14	.32
ACT Mathematics (ACT M)	19.74/6.25	.34	.44	.34	.63	.46	...	.54	.52	.63	.09	.01	-.12	.30
ACT Social Studies (ACT SOC)	18.82/5.90	.58	.36	.40	.58	.54	.36	...	.58	.66	.11	.08	-.06	.30
ACT Science (ACT SCI)	21.80/5.00	.43	.35	.43	.61	.50	.52	.58	...	.67	.19	.11	-.07	.31
ACT Total (ACT T)	80.37/19.28	.49	.40	.44	.65	.61	.63	.66	.67	...	-.11	-.03	-.10	.32
Major—Hi=1, El=0 (MAJ)	0.35/0.48	.17	.11	.03	.15	.04	.09	.11	.19	-.11	...	.46	.24	-.04
Sex—M=1, F=0 (SEX)	0.16/0.37	.00	.02	-.13	.01	-.18	.01	.08	.11	-.03	.46	...	-.21	.07
Age in Years (AGE)	20.56/1.87	-.02	.00	-.08	-.05	-.14	-.12	-.06	-.07	-.10	.24	-.21	...	.03
Cum. Grade Point Average (GPA)	2.72/0.58	.36	.29	.33	.45	.32	.30	.30	.31	.32	-.04	.07	.03	...

per examinee. We need to ponder these pricing facts as we review our data and impressions of the PPST.

First, this reviewer cheerfully endorses the three multiple-choice tests in the PPST as being highly "job relevant." But he also endorses the ACT and the SAT for similar reasons. Would that all of our tests were as real-life as these!

This reviewer is also on record as endorsing the College Board's Test of Standard Written English (TSWE) as an excellent approximation of an individual's actual writing and editing skills (Bauernfeind, 1988). Because the College Board studies were conducted at ETS and because the PPST people have full access to Breland's research and item data, it is not at all clear why the PPST includes an essay exercise when other excellent testing programs, including the ACT and the SAT, do not. This reviewer's arguments against inclusion of the essay exercise are these:

1. Multiple-choice tests in the format of the TSWE come remarkably close to assessing an individual's actual writing and editing skills and do so with much greater reliability than essay tests.

2. Elimination of the essay exercise would save examinees approximately 45 minutes of testing time.

3. Elimination of the essay exercise should markedly reduce the $35.00 fee now charged for the PPST program.

This reviewer urges the PPST people to make available at least one form based on the three multiple-choice tests only. Let each user then decide whether to test students with the multiple-choice tests only or to require an essay as well.

References

Bauernfeind, R. H. (1988). Test of Standard Written English. In D. J. Keyser & R. C. Sweetland (Eds.), *Test critiques* (Vol. VI, pp. 609-614). Kansas City, MO: Test Corporation.

Breland, H. M. (1977a). *A study of college English placement and the Test of Standard Written English.* Princeton, NJ: Educational Testing Service.

Breland, H. M. (1977b). Can multiple-choice tests measure writing skills? *The College Board Review, 103.*

Cooper, C. R. (1977). Holistic evaluation of writing. In C. R. Cooper & L. G. Odell (Eds.), *Evaluating writing: Describing, measuring, judging.* Urbana, IL: National Council of Teachers of English.

Ebel, R. L. (1961). Must all tests be valid? *American Psychologist, 16,* 640-647.

Educational Testing Service. (1985). *Descriptions of the PPST.* Princeton, NJ: Author.

Educational Testing Service. (1986a). *Bulletin of information for the PPST.* Princeton, NJ: Author.

Educational Testing Service. (1986b). *Interpretation guide for the PPST.* Princeton, NJ: Author.

Educational Testing Service. (1986c). *Test analysis for the PPST, Spring 1986 administrations* (Form 3IPS2). Princeton, NJ: Author.

Educational Testing Service. (1987). *Using the PPST as a requirement for program entry: Validation and standard setting at individual institutions.* Princeton, NJ: Author.

Livingston, S. A., & Zieky, M. (1982). *Passing scores.* Princeton, NJ: Educational Testing Service.

Susan Homan, Ph.D.

Associate Professor of Childhood/Language Arts/Reading Education,
University of South Florida, Tampa, Florida.

PREREADING EXPECTANCY SCREENING SCALE

Lawrence C. Hartlage and David G. Lucas. Creve Coeur,
Missouri: Psychologists and Educators, Inc.

Introduction

The Prereading Expectancy Screening Scale (PRESS) is a group diagnostic battery used to predict reading problems in beginning readers as well as to identify children who are ready to begin reading instruction.

The manual does not provide background information about the test's authors, Lawrence C. Hartlage, Ph.D., and David G. Lucas, M.A. They began development of the PRESS in 1973 in response to what they perceived as a need for a quick and comprehensive screening instrument for kindergarten and first-grade children. Because no such instruments were available at the time, they believed that reading problems in children often were not detected until the problems had grown quite serious.

Initially, the PRESS was developed and administered to 44 children in two sections of a team-taught first-grade class. Their mean age was 6.7, and their mean I.Q. was 107. The subjects were administered the auditory sequencing, visual motor skills, auditory spatial skills, visual sequencing, and combined auditory and visual spatial skills sections of the battery. In addition, they were administered the reading section of the Wide Range Achievement Test (WRAT), and two teachers ranked all the children on estimates of reading ability. Teacher ranking and WRAT scores then were correlated to each section of the test. Based on these correlations, the visual motor skills subtest was dropped, and a section on identification of lower- and uppercase letters was added. The norming population for the revised instrument included all children beginning the first grade in Washington Township Schools, Indianapolis, Indiana ($N = 1,384$). These subjects were administered the Metropolitan Readiness Test and a version of the WRAT in addition to the revised PRESS.

The PRESS consists of four tests, each of which are printed on a separate page of the stapled test booklet, a 16-page administration manual, and separate profile sheets for boys and girls. The student test pages are partially typed and partially hand-sketched, giving the test booklet a rather amateur appearance. The test pages are numbered with roman numerals, which are unfamiliar to most first-grade students. The cover sheet provides space for the student's name, birthdate, and sex; date; teacher's name; score; and comments. Students mark their responses directly on the test pages.

Test I, Sequencing, contains 15 items that measure visual sequencing. Students complete a sequence containing a missing element, for example, 1 2 3 4 _____. The second test, Spacial Relations, contains 12 items assessing auditory and visual spatial relationships. The student is directed to look at a stimulus, listen to the teacher read a description of the stimulus, and then circle a correct response. For example, for the first item, the teacher is directed to say, "The square is under the cross. The square is A) under the cross, or B) over the cross. I want you to circle the correct answer." The oral directions are repeated once, and the child responds by circling either "A" or "B" on the student form. In Test III, the 11-item Memory Test, students repeat, after a 1-second time lapse, sequences of numbers and letters spoken by the teacher. The number and letter sequences range in combination from 3 to 6 letters or digits. The fourth test, Letter Identification, contains 14 items requiring students to circle, from a choice of three letters, the letter spoken by the teacher.

Practical Applications/Uses

The PRESS was designed as a diagnostic screening instrument for identifying first-graders with potential reading problems and children ready to begin reading instruction. In light of the current movement to begin reading instruction in Kindergarten, the PRESS also could be used at that level; however, kindergartners were not included in the norming population. Although it was developed for classroom use, the PRESS could be used in a clinical setting as well.

The PRESS is a 25-minute group test administered by a classroom teacher; however, the publishers recommend the presence of another adult to monitor student progress. Administration procedures are clear and relatively simple with the exception of the Spatial Test. While the directions are to the point, the task itself is rather confusing for 6-year-old children. The order in which the four tests are administerd could be altered, but the current arrangement seems the most logical.

The scoring is well explained and relatively simple. After reading the manual, a teacher should be prepared to administer and score the test. Although the manual does not estimate the time required to score the test, the task could probably be accomplished in 5-7 minutes. Interpretation is based on objective scores. The profile sheets allow for the transfer of raw scores to percentiles, permitting ease of interpretation. The PRESS manual explains that students with overall profile scores at the 40th percentile or above should experience little trouble learning to read, whereas students scoring at the 30th percentile or below are in danger of encountering problems with reading instruction.

Technical Aspects

As stated previously, the standardization sample consisted of all the children entering first grade in a township in Indiana (N = 1,384). The subjects were retested 8 months after the initial testing. The total number of subjects available on both testings was N = 1,032, of which 551 were boys and 481 were girls. Split-half reliabilities were computed on all subtests on a randomly selected group of 200 subjects. Reliability coefficients ranged from .80 to .94.

In this reviewer's opinion, the manual's presentation of the standardization information is confusing. Apparently, during the standardization, a research study in which subjects were administered the PRESS and the Metropolitan Readiness

Test also was being conducted. Subjects were then "more or less randomly assigned" (Hartlage & Lucas, 1973, p. 13) to one of three groups, each employing either the phonetic, look-say, or special teaching alphabet (Initial Teaching Alphabet) reading methods. At the end of the school year, subjects were ranked by their teachers for reading skills and individually administered the Wide Range Achievement Test (WRAT).

The manual reports correlations of subtest scores with teaching method as exemplified by WRAT scores. Correlations of subtests with teachers' judgment of reading ability also are reported. This information is not particularly useful and apparently is presented in lieu of validity information. Although the number of subjects in the initial sample was large, the subjects were not necessarily representative of students in other school districts. No information is given concerning the racial or socioeconomic composition of the group. Furthermore, the manual does not explain the decision to include only 200 subjects in the reliability study, and a chart presenting all the reliability coefficients is not included.

Critique

Although the PRESS may have filled a void in the area of reading readiness in 1973, currently many informal assessments for determining prereading status are available. Of these, the PRESS is a rather limited instrument in light of new research on the subject. Because it was published in 1973, the PRESS does not include tests assessing print concepts or oral language development, two variables now recognized as being extremely important to the development of reading skills and, likewise, to the identification of reading problems. In addition, the standardization seems to be inadequate and poorly reported. The greatest advantage of the PRESS is that it is a group instrument, but the time needed for scoring and interpretation appears to counter that advantage.

If the PRESS were revised to include additional subtests in relevant areas, namely, print knowledge and oral language development, and renormed to include kindergartners as well as first-graders, it would have greater utility. Until then, it should be used only as one of several indicators measuring readiness for reading.

References

This list includes text citations and suggested additional reading.

Clay, M. M. (1985). *The early detection of reading difficulties* (3rd ed.). Aukland, New Zealand: Heinemann Publishers.
Downing, J., & Thackeray, D. (1976). *Reading readiness.* London: Hodder and Stoughton.
Hartlage, L. C., & Lucas, D. G. (1973). *Prereading Expectancy Screening Scale.* Jacksonville, IL: Psychologists and Educators, Inc.
Holdway, D. (1979). *The foundations of literacy.* Exeter, NH: Ashton Scholastic.

Eileen Stitt Whitlock Kelble, Ed.D.
Coordinator of Early Childhood Education, College of Education,
University of Tulsa, Tulsa, Oklahoma.

PRESCHOOL AND KINDERGARTEN INTEREST DESCRIPTOR

Sylvia Rimm. Watertown, Wisconsin: Educational Assessment Service, Inc.

Introduction

The Preschool and Kindergarten Interest Descriptor (PRIDE) is a report inventory designed to identify the creative characteristics of young children ages 3-6. A parent, rather than the child, responds to the items on the inventory in order to give a more reliable picture of the child's characteristics. The parent's responses are based on the parent's observations and perceptions of the child's behavior as it occurs in the home environment. Although the PRIDE was designed as a preschool and kindergarten screening instrument for creativity, the manual, the report of the scores, and all materials related to the inventory implicitly state that it should not be used to deny a child entry into a program. The PRIDE was designed to help identify children with creative characteristics who have not been identified already.

Sylvia Rimm, Ph.D., developer and author of the PRIDE, is a psychologist and the director of the Educational Assessment Service in Watertown, Wisconsin and the Family Achievement Clinic in Oconomowoc, Wisconsin. She is the author of several books and articles on gifted and talented education and underachievement. Her work with families and schools in properly placing and handling under-achievers has gained her respect, admiration, and nationwide recognition. She is a leader in the field of gifted and talented education.

The PRIDE, developed in 1983, was preceded by the successful design and validation of the Group Inventory for Finding Creative Talent (GIFT; Rimm, 1976, 1980) for children in Kindergarten through Grade 6; the Group Inventory for Finding Interests I (GIFFI I; Rimm & Davis, 1979) for Grades 6-9; the Group Inventory for Finding Interests II (GIFFI II; Davis & Rimm, 1980) for Grades 9-12; and How Do You Think (HDYT; Davis, 1975) for college students. The PRIDE was normed on a population of 114 children representing urban, suburban, and rural populations. The Hoyt reliability correlation for PRIDE was .92, indicating a high internal consistency.

Test materials, which are simple and straightforward, consist of the inventory to be filled out by the child's parent and the *Manual For Administration* for the teacher or test administrator.

The inventory consists of 50 questions to which the parent responds using a 5-point rating scale ("no" to "definitely"). The questions deal with the child's behavior as observed by the parent's, such as play habits and patterns, relationship with

415

other children, number and kinds of interests, relationship and activity with parent, and likes and dislikes. See Table 1 for sample items.

Table 1

Sample Items for Dimensions of PRIDE

Dimensions	*Item*
Many Interests—High scorers are curious and ask questions. They show high interest in learning, stories, books, and things around them. Low scorers show less curiosity and have fewer interests.	My child seems to like to think about ideas. My child likes to take things apart to see how they work. My child likes to try new things. My child gets very interested in things around her/him.
Independence-Perseverance—High scorers play alone and do things independently. They do not give up easily and persevere even with difficult tasks. Low scorers tend to prefer easier tasks and are more likely to follow the lead of other children.	My child gets interested in things for a long time. My child gets bored easily. My child usually does whatever other children do. My child likes to pick his/her own clothes to wear to school.
Imagination-Playfulness—High scorers enjoy make believe, humor, and playfulness. Low scorers tend to be more serious and realistic.	My child spends much time playing make believe. My child and I make jokes together. My child likes to play out in the rain. My child says things that are funny.
Originality—High scorers tend to have unusual ideas and ask unusual questions. They are inventive in their art and play and tend to think differently than other children. Low scorer's ideas and art work appear to be more typical of children of similar age.	My child asks unusual questions. My child has unusual ideas. My child seems to do things differently than other children. My child often points to unusual things around him/her.

Practical Applications/Uses

The PRIDE is intended to place children "into" programs and not screen them "out of" programs. The target age of the test, 3 to 6 years, reflects its attempt to identify at an early age young children with a high degree of creativity for the purpose of providing them with the kind of intervention that is needed to maintain this creativity. Scores derived from the PRIDE indicate whether a child has characteristics similar to those which are typical of highly creative children. High scores (85-99 percentile) are considered to be good indicators, but low or average scores do not indicate that a child is not creative.

The simplicity of the test lies in the fact that the test administrator needs no previous training other than being able to convey the simple instructions accompanying the test. The person who reports the scores, however, must be totally cognizant of the test's intended purpose, which is clearly stated in the manual and on the computer printout that accompanies the scores.

The teacher or test administrator instructs the parents to print the first and last name of the child; the parents' names; the child's age, grade, and school; and the date on the PRIDE booklet. The administrator then reads the short instructions to the parent. These instructions, which also appear on the front cover of the booklet, simply request the parent to consider the child's interests and activities and the parent's own relationship to the child when responding to the questions. In addition, the parent is asked to compare the child to other children in the same age category. The parent responds by filling in a circle under the desired answer, which is based on the parent's personal observations of the child.

Administrators of the PRIDE are encouraged to respond to any questions parents may have about the meanings of items and terms, pronounce any words that cause parents difficulty, and emphasize that no answers should be left blank. According to the instructions, if parents are indecisive about any item, they should choose an answer that is closest to the way they feel. It is permissible to erase and change any response upon reconsideration. The PRIDE is not a timed test, and completion usually takes between 20-35 minutes.

It is recommended that parents complete the PRIDE at a parent meeting or during a preschool or kindergarten screening for the child. It may not be given in a group setting. The test author does not recommend completing the inventory at home for several reasons. Typically, return rates are low and parents are less likely to attend seriously to the questions in a home environment with its normal and typical distractions.

The PRIDE is computer scored by Educational Assessment Service, Inc. Turnaround time is about 2 weeks; however, for special circumstances, scores can be returned within a week. Scores provided for each student include percentiles and normal curve equivalents (NCEs). NCEs are standard scores with a mean of 50 and a standard deviation of 21.06. The cost for scoring the PRIDE is included in the original cost of the instrument. Special analyses, such as pre- and posttest comparisons, can be requested for small additional costs.

The administration manual emphasizes the importance of using all test scores with caution. More than one identification procedure is recommended for selecting students for creativity programs. Furthermore, it is recognized that although high scores on PRIDE are very useful in the selection process, teacher, parent, and peer nominations also are important and should be considered when reviewing PRIDE scores:

> creativity inventory scores, like achievement test scores or I.Q. scores, should be utilized to screen children "into" a program and not "out" of a program. For example, a student with a high PRIDE score should be included in a program even though the child may not have been selected by a teacher. However, a child who is selected by a teacher as being highly creative should not be eliminated from the program because of a low or average PRIDE score. *Creativity is a*

subtle characteristic which is difficult to identify. PRIDE can help your school in that identification process. (italics added; Rimm, 1983, p. 1)

Technical Aspects

The design of the PRIDE was based on the research of the creative characteristics of preschool and kindergarten children. To establish construct validity, items that describe personality characteristics of creative preschool and kindergarten children as reported in papers by Auerbach (1972); Newland (1976); Johnson (1978); Fuqua, Bartsch, and Phye (1975); and Torrance (1965) were used. The main personality characteristics of these children include many and varied interests, curiosity, independence, perseverance, imagination, playfulness, humor, and originality. Some biographical items also are included.

To establish criterion validity, inventory scores were correlated with outside measures on creativity. Rimm (1983, p. 4) states

> The main validity criterion has been a composite score consisting of teacher ratings of creativeness and experimenter ratings of a picture and a brief dictated story. These criteria . . . were combined and equally weighted before calculating validity correlations with the inventory.

Although teacher ratings have not been considered reliable, teachers are frequently the sole identifiers of creativity in young children. Therefore, the teachers were trained to distinguish between intellectual and creative ability on a scale from 1 to 5, and experimenters rated a picture and a brief, dictated story on a scale from 1 to 5. The composite of these scores was used to obtain the score for the main validity criterion. Criterion-related validity was calculated for three populations. The first population, from St. Aemillian's Preschool and Kindergarten, included 62 children between the ages of 3 and 6. The correlation coefficient between the PRIDE score and the criterion score was .38 ($p < .01$). The second population was based on a Canadian suburban population of 14 children ages 4-5. The correlation coefficient for this population was .50 ($p < .05$). The last population ($n = 18$) was from a suburban community in New York State. The validity correlation for this population was .32.

A factor analysis of young children's creativity characteristics used in the PRIDE resulted in four dimensions: Many Interests, Independence-Perseverance, Imagination-Playfulness, and Originality. These dimensions are described in Table 1. Dimension scores are included in the scoring report for each child.

Critique

Identification of creativity at an early age is important so that the potential for that creativity can be nurtured. Too frequently, the creative young child does not fit into the mold established by our schools today. They are labeled behavior problems and treated as such.

However, the term "creativity" is a loosely defined, broad, multifaceted construct that does not correspond to a precisely defined entity (Anastasi & Schaefer, 1971). Tests of creativity exhibit a low correlation with each other (Sattler, 1982), largely because the items on any inventory used to measure creativity depend on the author's perception of the construct. Also, testing young children is difficult at best, and preschool and kindergarten screening is highly controversial. However, in this reviewer's opinion, the developer of the PRIDE has been cautious and wise. The use of two sets of criteria to determine the criterion validity of the characteristics measured in the test and the subsequent factoring of the characteristics into four dimensions, which in turn are defined extensively, is an advantage. The use of a parent report rather than a report by the child is another positive factor. Most research, except for a current study by Abrahams (1987), indicates that parents were the most reliable indicators of creativity in young children. Unfortunately, because parents today are aware of the phenomena of creativity and because creativity has become a highly desirable trait in our cultural value system, it is possible that the reliability of parents as identifiers of creativity traits in their children is decreasing.

A word of caution in regard to the changing attitude of parents toward their children's abilities is in order here. Young parents increasingly are becoming concerned about obtaining the best of all possible worlds for their children. They are anxious for them to be schooled in the best manner available and for them to operate at peak performance levels which, unfortunately, can lead to overplacement—a concern many child rearing experts and educators share today. This is why the author of the PRIDE strongly states and this reviewer strongly recommends that if the PRIDE is used, it should be used as instructed by the developer, and not as a single measure to identify creative children for the purpose of placing them in a differentiated program. On the other hand, it should never, as stated by Rimm, be used to hold a child back. This reviewer has discussed this test with parents of young children and they feel that it has merit. They do not find the questions too difficult to answer, and they feel that it is worth their effort to complete if it will help their child to be placed correctly in a program that will benefit the child.

The PRIDE reflects its developer's extensive experience with creative children and her expertise in using tests of this nature to assist in the identification of creativity traits and the subsequent placement of the children with these traits into programs suitable for their talents. Rimm's concern for the underachiever is deep and sincere and it has, in this reviewer's opinion, helped her devise an instrument that has considerable merit.

The test's simplicity, its easy administration, its reliability and validity, and, in particular, its use of parent rather than child responses are all positive factors. However, the most exemplary and most refreshing factor is the test author's emphasis that the test should not be used to deny any child admission into a program. In this reviewer's opinion, many tests today are used for that purpose. Rimm's experience with the underachiever and her knowledge that there is a distinct need for children to be placed in programs that meet their individual needs, not the needs of the school system, has helped her devise a test that recognizes its limits. It has merit but should be used in conjunction with other measures such as teacher observation and recommendation when attempting to identify creative potential in young children. It is not a performance measure; it is a measure of potential based on characteristics that have been associated with creativity in children of this age group.

References

This list includes text citations and suggested additional readings.

Abrahams, W. F. (1987, August). *The early years of giftedness (Identification, home and preschool programming, family/preschool partnership)*. Paper presented at the Seventh World Conference on Gifted and Talented Children, Salt Lake City, UT.

Anastasi, A., & Schaefer, C. E. (1971). Note on the concepts of creativity and intelligence. *Journal of Creative Behavior, 5*, 113-116.

Auerbach, A. G. (1972). The bisociative or creative act in the nursery school. *Young Children, 28*(1), pp. 27-31.

Barnes, K. E. (1982). *Pre-school screening: The measurement and prediction of children at risk.* Springfield, IL: Charles C. Thomas.

Davis, G. A. (1975). How do you think? In frumious pursuit of the creative person. *Journal of Creative Behavior, 9*, 74-87.

Davis, G. A., and Rimm, S. (1980). GIFFI II: Group Inventory For Finding Interests. Watertown, WI: Educational Assessment Service.

Fuqua, R. W., Bartsch, T. W., & Phye, G. D. (1975). In investigation of the relationship between cognitive tempo and creativity in preschool age children. *Child Development, 46*, 779-782.

Johnson, J. E. (1978). Mother-child interaction and imaginative behavior of preschool children. *Journal of Psychology, 100*, 123-129.

Newland, R. E. (1976). *The gifted in socioeducational perspective.* Englewood Cliffs, NJ: Prentice-Hall.

Rimm, S. (1976-1980). GIFT: Group Inventory for Finding Creative Talent. Watertown, WI: Educational Assessment Service.

Rimm, S. (1984). The characteristics approach: Identification and beyond. *Gifted Child Quarterly, 28*(4), 181-187.

Rimm, S., and Davis, G. A. (1979). GIFFI I: Group Inventory For Finding Interests. Watertown, WI: Educational Assessment Service.

Sattler, J. M. (1982). *Assessment of children's intelligence and special abilities* (2nd ed.). Boston: Allyn & Bacon.

Tannenbaum, A. J. (1983). *Gifted children: Psychological and educational perspectives.* New York: Macmillan.

Torrance, E. P. (1965). *Gifted children in the classroom.* New York: Macmillan.

John Beattie, Ph.D.
Assistant Professor of Special Education, University of North Carolina at Charlotte, Charlotte, North Carolina.

Kay Haney, Ph.D.
Clinical Assistant Professor of Special Education, University of North Carolina at Charlotte, Charlotte, North Carolina.

Gayle Mayer, Ph.D.
Clinical Assistant Professor of Special Education, University of North Carolina at Charlotte, Charlotte, North Carolina.

PSYCHO-EDUCATIONAL BATTERY

Lillie Pope. North Bergen, New Jersey: Book-Lab.

Introduction

The Psycho-Educational Battery (P.E.B.; Pope, 1976) is a screening instrument designed 1) for early identification of students with learning problems, 2) for individual student assessment, and 3) as an aid in appropriately planning an individual instructional program. The instrument also helps to identify areas in which students need further assessment. By using a structured questionnaire format, the evaluator can identify students' learning styles and interests, as well as their areas of strengths and weaknesses. The P.E.B. may be used with students of all ages, including kindergartners and adults.

The P.E.B. was developed by Dr. Lillie Pope, the director of the Psycho-Educational Center of the Coney Island Hospital in New York. Dr. Pope has worked extensively with students with special needs, especially in the evaluation of these students. In an attempt to make the educational planning process more efficient, she designed the P.E.B. as a criterion-referenced battery that would provide information on individuals' deficits and skills, as well as on individual learning styles. Realizing that individuals are complex organisms influenced by physical, emotional, and social factors as well as by learning experiences, Dr. Pope compiled test items that address student's academic and behavioral performances in light of the interaction of all of these variables. As an individual's performance lies on a continuum, this instrument encourages probing to identify each student's behavior or academic patterns and the appropriate teaching strategies therein.

The test manual provides a description of the P.E.B.'s several components and subsections as well as the philosophical perspective of the test development. The P.E.B. is available in two levels: Level Y (younger level) for children in Grades K-6 and Level O (older level) for examinees ranging from Grade 7 through adulthood. In addition to the manual, the P.E.B. consists of a student recording form (Part A), an evaluator's recording form (Part B), a family, social, and medical history form (Part C), a visual packet (Part D, presented during individual evaluation), and a teacher referral form for each level.

The administration of Level Y (Grades K-6) is initiated by providing each student

421

with a student recording form (Part A). Part A may be administered as a group or individual screening test. Students are asked to draw, copy, color, trace, compute arithmetic problems, and write spelling words. Once the students have completed Part A, the teacher categorizes the results into three piles: the first consists of the test forms of students who had no difficulties and demonstrated competence on all tasks, the third consists of the forms of students who committed obvious errors, and the second is made up of forms of students who cannot be placed in either the first or third pile. Any student previously referred by the teacher but not identified by Part A testing is also placed in the third pile; however, it is noted on the referral form that Part A testing indicated no learning or behavior problems. Every child whose form is placed in the third pile must be further evaluated on an individual basis, using Part B. Students whose forms were placed in the second pile are carefully observed.

Part B consists of 22 specific subtests as well as an overall determination of the student's physical status and appearance. The 22 subtests are 1) Gross Motor Coordination; 2) Fine Motor Coordination; 3) Awareness of Place and Time (interview); 4) Probe of Interests (interview); 5) Knowledge of Left and Right; 6) Tactile and Kinesthetic Perception; 7) Knowledge of Letter Sounds and of Digits; 8) Knowledge of Reversible Words and Tendency to Reverse; 9) Sight Words; 10) Reading Paragraphs: Decoding and Comprehension; 11) Phonics Evaluation (beginning); 12) Spelling; 13) Arithmetic Concepts; 14) Knowledge of Colors and Language of Spatial Relationships; 15) Auditory Word Discrimination; 16) Concept Development, Ability to Classify, and Language Usage; 17) Auditory Memory; 18) Visual Memory; 19) Speech; 20) Phonics Evaluation (continuation); 21) Arithmetic (basic functions); and 22) Behavior and Learning Style. As Part B is administered, it is often necessary to provide additional stimulus materials: paper, pencil, scissors, and a shoestring (Gross and Fine Motor Activities); a sack, a rubber ball, a metal key, sandpaper cubes, and wooden letters (Tactile and Kinesthetic Perception); and crayons (Knowledge of Colors). These materials, which are packaged separately and referred to as Part D, provide stimuli for the following subtests: Knowledge of Letter Sounds and of Digits, Knowledge of Reversible Words and Tendency to Reverse, Sight Words, Reading Paragraphs, Phonics Evaluation, and Visual Memory.

Upon completion of Part B, it is suggested that information concerning the student's health, developmental history, and medical and birth history be collected. This task, referred to as Part C, is accomplished using the Family, Social and Medical History form.

Once the data in Part C have been collected, the evaluator refers to pages 18-20 in the Part B booklet. These pages are summary sheets that list the subtests as well as additional areas of concern. At this time, the evaluator analyzes each student's test results and determines whether the student is competent in each area, needs more practice, or has yet to learn the information.

As noted, the P.E.B. also includes a Teacher Referral form, used by the teacher to identify students whose learning or behavior is of concern. When other students are identified from group testing (Part A) without a completed referral form, the teacher is asked to complete one for each of these students.

Level O (Grades 7-adult) is similar in design to Level Y. The test begins with Part

A, which is administered to either groups or an individual. However, unlike Level Y, students are asked only to draw and copy shapes. After completion of Part A, appropriate students are administered Part B, which consists of 16 specific subtests as well as an overall determination of the student's physical status and appearance. The 16 subtests are 1) Motor Performance (gross motor); 2) Motor Performance (fine motor); 3) Awareness of Place and Time (interview); 4) Probe of Interests (interview); 5) Sight Words; 6) Reading Paragraphs: Decoding and Comprehension; 7) Sight Words (continuation); 8) Inventory of Basic Reading Skill, including Auditory Recognition of Initial Consonants, Auditory Blending, Blend Sounds into Words, Visual Recognition of Consonants, Reading Short and Long Vowel Sounds in Words, Reversal, Auditory Recognition of Blends and Digraphs, Visual Recognition of Blends and Digraphs, and Reading Vowel Combinations Including "r" controlled vowels; 9) Spelling; 10) Concept Development; 11) Auditory Memory (below 4th-grade reading level); 12) Auditory Word Discrimination (4th-grade reading level and below); 13) Orientation in Time (continuation) (those with confused responses to Unit 3); 14) Speech, including Language Development and Articulation; 15) Arithmetic Basic Functions; and 16) Behavior and Learning Style (subtests include items similar to Level Y).

Again, upon completion of Part B, the examiner uses the Family, Social and Medical History form to collect data concerning the student's health, developmental history, and medical and birth history. Next, the evaluator refers to pages 17-20 of the Part B booklet, where the test results are summarized.

Practical Applications/Uses

The P.E.B. is designed for use by teachers, counselors, social workers, principals, or other persons in the school with the designated responsibility of developing educational plans for students. Guidelines for referral to interdisciplinary sources also are included. Directions for administration are clear and direct. Exact procedures for administering the test, descriptions of the materials needed, and explanations and rationales for each task type are well presented throughout each of the test sections. So that definitive decisions regarding future instructional programming can be made, testing suggestions and examples are provided. Extensive directions and suggestions for interpreting the results allow the test administrator to analyze the test results carefully. The order and procedures for completing each form (Teacher Referral form; Family, Social and Medical History form; etc.) are noted and explained carefully. All the materials are easy to read and understand. Administration difficulty is minimal, provided the evaluator is thoroughly familiar with all the materials in the kit. Although not specified, administration time appears to average about 1 hour.

The P.E.B. is designed for use with a wide range (K-adult) of students. No specific categories of students (e.g., learning disabled) are recommended as primary candidates for test administration; consequently, it may be inferred that the test is appropriate for any students identified by teacher referral as performing poorly on the screening portion of the battery or displaying some academic difficulty.

Although no specific recommendations are made regarding specific test settings, the author emphasizes that the student should be as comfortable as possible. Once the student is comfortable, the teacher, counselor, psychologist, or program administrator acting as examiner may administer the test. The only caution provided by the author is that the evaluator be familiar with the test materials.

The P.E.B. is intended to be used as a criterion-referenced assessment measure that compares the student's proficiency against the criteria included in the test. The only areas in which comparisons are made to other students are math, reading, and spelling. In these areas, a crude estimate of grade level is presented in order to prove direction in the selection of appropriately leveled school materials. The author believes that in program planning it is much more important to know a student's precise areas of strength and weakness than how that student compares to another with regard to any particular task. Referrals to other competent professionals is encouraged when necessary.

The author encourages evaluators to probe for specific levels of competence in any assessed skill rather than simply to note the presence or absence of a skill. This assists educators in pinpointing the skill at which instruction should begin, facilitating the development of appropriate and useful education plans. Finally, the author stresses the need for determining how the student learns.

Throughout the test the evaluator is encouraged to administer test sections that are pertinent to the individual student. These sections are selected based on the evaluator's experience with the subject and his or her professional judgment. Performance on each subtest for each level is recorded as *competent, hesitant* (needs more practice), or *inadequate* (has not learned). There is also a comment section provided, which is used for recommending educational strategies and teaching techniques. In summarizing the results of the P.E.B., the evaluator is referred to the summary section at the end of both Level Y and Level O. The summary sections are helpful in synthesizing all test data.

Some concern has been raised by these reviewers regarding the summary section. Throughout the test itself, the items are objectively hand scored. However, a great deal of subjectivity may be required when summarizing results. For more inexperienced educators, decisions regarding competence may be difficult. Although these reviewers recognize that this is an inherent feature of criterion-referenced tests, some caution is warranted. The test author acknowledges this potential problem and strongly encourages inexperienced users to communicate with other professionals.

Technical Aspects

In considering the reliability and validity of the P.E.B., it is critical to review some basic differences between criterion-referenced tests (CRT) and norm-referenced tests (NRT). The criterion-referenced test is not designed to emphasize differences among individuals; instead, it focuses on specific skill development. Consequently, the traditional estimates of reliability frequently used with norm-referenced tests may provide misleading, useless results when used with criterion-referenced tests (Gronlund, 1985). The P.E.B. does not provide any reliability data, which appears to be warranted in this case. Additionally, validity measures often used with norm-referenced tests may not be equally applicable with criterion-referenced tests (Gronlund, 1985). Consequently, the lack of a validity statement in the P.E.B. may be viewed as appropriate. However, some mention of content validity (how adequately the sample of items represents the domain) appears to be warranted. The manual's omission of a statement regarding content validity is of concern with the P.E.B.

Critique

The Psycho-Educational Battery is designed to help improve educational planning by providing data on an individual's deficits, skills, and learning style. The P.E.B. is founded in sound educational philosophy. Unfortunately, it appears to be beyond the scope of any single measure to provide adequately the detailed information necessary to make specific diagnostic suggestions in every academic area. For example, of the 21 arithmetic problems administered in the Arithmetic Basic Functions subtest, only five items assess addition skills. With so few samples of behavior, no accurate diagnosis is possible. It also appears, for the same reason, that specific suggestions made on the basis of the P.E.B. results would be tenuous at best. For example, the arithmetic computation subtest requires the student to answer "4 + 3," "8 + 5," and so on. This, however, provides only one example in each hierarchical level (i.e., sums to 9, sums to 18, etc.). Any instructional decision at this point would require addition, more thorough evaluation.

Also of concern to these reviewers is the test's lack of a content validity statement. This omission raises many questions as to the composition of the test itself. Why did the author select the specific items that appear on the test? Why did she choose the 22 subtests and not others? These questions would be satisfied by an indication of appropriate content validity.

Finally, these reviewers question the overall usefulness of the P.E.B. In analyzing the data provided it appears that additional testing of students likely will be warranted after administration of the P.E.B. Given the length of the test, it is self-defeating to turn around and administer additional instruments. However, to accomplish the P.E.B.'s goal of improving educational planning, such additional testing seems inevitable. The consensus of these reviewers is to use the time required to administer and interpret the P.E.B. for developing a teacher-made device or for administering a more in-depth criterion-referenced test such as the Brigance Diagnostic Inventory of Basic Skills (Brigance, 1977).

References

Brigance, A. H. (1977). *Brigance Diagnostic Inventory of Basic Skills.* North Billerica, MA: Curriculum Associates.

Gronlund, N.E. (1985). *Measurement and evaluation in teaching* (5th ed.). New York: Macmillan.

Pope, L. (1976). *Psycho-Educational Battery.* North Bergen, NJ: Book-Lab.

Brian W. McNeill, Ph.D.
Assistant Professor and Counselor, University Counseling Center and Department of Counseling Psychology, University of Kansas, Lawrence, Kansas.

PSYCHOLOGICAL DISTRESS INVENTORY
Claudia J. Sowa, Patrick J. Lustman, and Dennis J. O'Hara. Charlottesville, Virginia: Claudia J. Sowa, Ph.D.

Introduction

The Psychological Distress Inventory (PDI) is a brief, 50-item instrument designed to assess the constructs of depression, anxiety, somatic discomfort, and stress in college students. The PDI was developed and introduced in 1984 by Claudia J. Sowa, Ph.D. of the University of Virginia, Patrick J. Lustman, Ph.D. of the Washington University School of Medicine, and Dennis J. O'Hara, Ph.D. of St. Joseph's University in Philadelphia (Lustman, Sowa, & O'Hara, 1984). Still in the early stages of its development, the PDI has been used primarily for research. However, Sowa and her colleagues currently are involved in further refining the PDI as a clinical tool and envision its use in clinical settings in the near future.

In developing the PDI, the test authors studied the literature pertaining to stressful life events and addressed what they perceived to be three major deficits in existing instruments designed to measure symptomatology experienced by college students seeking clinical care. First, research in the area of life stress indicates that events perceived as aversive or undesirable are more likely to have a detrimental or negative impact on an individual (Brown & Harris, 1978). Thus, test items addressing events of an aversive nature are more predictive of clinical states like anxiety and depression. Unlike previous instruments that assess both positive and negative events in order to determine an individual's stress level, the PDI excludes items likely to be viewed positively and focuses on items that would be experienced as unpleasant. Second, the literature on life stress also suggests that *ratings* of the undesirability of experienced events are more predictive of psychological distress symptoms than a simple sum of experienced events (Sarason, Johnson, & Siegal, 1978). Therefore, the PDI utilizes an idiographic, event-rating format in which experienced events are rated on a scale from 1 (not aversive at all) to 5 (extremely aversive). The test authors view this format as advantageous when examining and interpreting individual items in individual counseling situations. Third, unlike other psychological distress measures, the PDI includes test items directed specifically towards and experienced frequently by college students.

The development of the PDI utilized both rational/intuitive and empirical methodologies. A total of 136 possible items were derived by asking four psychologists to generate item lists reflecting developmental tasks confronting college students and by using items from previously existing instruments. These items were administered to a sample of college students who were asked to rate them 1) on the 1 to 5

426

scale of aversiveness and 2) for self-reported frequency of occurrence. Correlations then were performed between items and 1) subjects' ratings of their current degree of stress (i.e., high, medium, or low) and 2) self-reported levels of depression, ill health, and anxiety as measured by the Beck Depression Inventory (BDI; Beck et al., 1972), the Cornell Medical Index (CMI; Brodman, Erdman, Lorge, & Wolff, 1949), and the State-Trait Anxiety Inventory (STAI; Spielberger, Gorsuch, and Lushene, 1970), respectively. Three criteria were used to select items: 1) a mean aversiveness rating of greater than 3.2, 2) occurrence reported by 25% of the subjects, and 3) significant correlation with one of the previously mentioned self-report measures. Of the 136 original items, 56 met these criteria.

The scales on the PDI were formed by multiple regressions of four criterion measures on the 56 items; that is, a composite score on the STAI, the BDI, the CMI, and the Negative Change scale of the Life Stress Questionnaire (LSQ; Lustman, Sowa, & Day, 1981). This procedure resulted in the elimination of six items, leaving the PDI with a total of 50 items. Step-down regression analysis on these 50 items then were executed for all four criterion measures, resulting in four scales, each consisting of a series of life events.

The Depression scale consists of 20 items related to reports of self-blame, pessimism, and helplessness (e.g., consistently missing deadlines; rejection by others). The Anxiety scale consists of 19 items related to generalized anxiety, including aspects of both state and trait anxiety (e.g., failing a test; being caught cheating in school). The Somatic Discomfort scale consists of 17 items related to the report of physical symptoms (e.g., unwanted pregnancies; lack of concentration). The Stress scale consists of 20 items related to the amount of negatively perceived experienced life events (e.g., breakup with boyfriend or girlfriend; not graduating on time).

The complete PDI consists of answer forms, a scoring key, and the test manual. Although the two-page questionnaire is self-administered, either individually or to groups, a proctor is needed to oversee administration, and a professional should be available to clarify instructions or respond to questions regarding item content. Because the PDI is designed specifically for use with college students and the difficulty level and content of the items appear to reflect this age level, it is not appropriate for use with other individuals. At this point, the test has not been adapted for use with other groups.

Practical Applications/Uses

Because the PDI is designed to measure the constructs of depression, anxiety, somatic discomfort, and stress in college students, it will be of most use to mental health professionals (i.e., counselors, psychologists, social workers, and psychiatrists) working in university counseling and mental health agencies. The PDI may be used in research investigations of the constructs that it taps in college students (e.g., frequency of depression). Also, as mentioned previously, it has potential for clinical use.

Presently many university counseling and mental health centers work with clients within the context of short-term therapeutic modalities (i.e., 10-15 sessions) in response to both the overwhelming demand for services on campuses and the resulting waiting lists and the recent evidence for the efficacy of short-term therapy models (Gelso & Johnson, 1983). An integral component of many of these short-term models is the initial assessment or intake interview in which the counselor gathers information in relation to a client's areas of concern and needs prior to assignment to a particular mode of counseling (e.g., individual or group) or a particular counselor. Consequently, the issues of accurate problem identification, level of severity, and subsequent case disposition and referral become paramount. Due to both its briefness and the content areas it taps, the PDI appears to have great potential as an information-gathering device within the context of the initial contact with a client. For example, the PDI might be administered easily as part of the routine intake process, subsequently scored by an administrative assistant, and placed in the examinee's records. An intake counselor or administrator responsible for case disposition then would have access to the information tapped by the PDI, which may aid in the decision-making process. Moreover, information gathered via the PDI may be placed in the examinee's file prior to his or her first meeting with an assigned counselor, alerting counselors beforehand to the potential problem areas expressed by clients. On the other hand, administration of the PDI may be left to the discretion of individual counselors, who may use it when concerns surrounding depression, anxiety, somatic complaints, and stress arise in regard to their clients.

The examinee is asked simply to circle the numbers that correspond to events he or she has experienced in the past 12 months. The examinee then is instructed to review the list of circled events and rate how stressful or aversive each event was on a scale of 1 (not at all aversive) to 5 (extremely aversive). Average completion time is about 15 minutes.

Scores on each of the PDI scales are derived simply by summing the ratings for the circled items for the appropriate scale, yielding four subscale scores ranging from 0 to 100, 95, 85, and 100 for the Depression, Anxiety, Somatic Discomfort, and Stress scales, respectively. Although the PDI appears to hold promise for both computer administration and scoring, neither is available at this time.

In regard to interpretation of the PDI, at the simplest level, higher scores on the various scales are indicative of higher levels of depression, anxiety, somatic complaints, and stress. Descriptive statistics for the normative sample of college students, as well as percentile rankings for raw scores on each of the PDI scales, are provided in the manual. For example, if a client's scores on the Depression, Anxiety, Somatic Complaints, and Stress scales are 60, 68, 65, and 77, respectively, the counselor can interpret that the examinee is at the 40th percentile level of depression, the 60th percentile for anxiety, and the 70th percentiles for somatic discomfort and stress. These percentile ranks indicate that more than half of the standardization sample scored lower than the client on the last three scales and may imply that the client is experiencing higher-than-average levels of experienced life events that correlate significantly with symptoms of anxiety, somatic discomfort, and stress. To further specify the nature of a particular client's concerns, individual items may be examined. Although a high level of sophistication and training generally may not be neccessary in order to obtain an estimate of an examinee's levels of reported symptomatology on the PDI scales, examination of the meaning of these scores for a particular individual or follow-up examination of individual items should be performed by a trained mental health professional.

Technical Aspects

Initial reliability studies of the Psychological Distress Inventory indicate adequate levels of test-retest reliability over a 6-week period. Coefficients of .76, .72, .83, and .80 were yielded for the Depression, Anxiety, Somatic Complaints, and Stress scales, respectively. Internal consistency reliability estimates using the Spearman-Brown split-half method yield generally lower estimates of .68, .73, .61, and .64 for the same scales, respectively. According to the authors, these lower estimates "may reflect a variation in the number of experienced events per subject on each half of the test and/or content heterogenity of the test" (Lustman et al., 1984).

In terms of validity, significant correlations have been obtained between the PDI subscales and appropriate corresponding criterion measures. For instance, the correlation between the Depression scale of the PDI and the BDI is .65. There are also some significant, though moderate, correlations between the subscales of the PDI, reflecting item overlap between the scales. This may be expected given the frequent coexistence of symptoms tapped by the PDI (e.g., anxiety and depression).

Consistent, nonsignificant negative correlations between the PDI scales and the short form of the Marlow-Crowne Social Desirability Scale (Strahan & Gerbasi, 1972) suggest that the subscales of the PDI do not reflect subjects' tendencies to respond in a socially desirable manner.

In further examinations of the instrument's validity, the test authors found that students seeking help at a university counseling center scored significantly higher on the Depression and Anxiety scales than those who were not seeking help. Depressed students (as defined by scores above 10 on the BDI) evidenced significantly higher Anxiety, Depression, and Stress scores than nondepressed students. Finally, students reporting higher levels of stress demonstrated significantly higher scores on all PDI scales than those students reporting low stress levels.

Critique

In evaluating the PDI, one must keep in mind that the instrument is still in the early stages of development. To the test authors' credit, they readily acknowledge that more work is needed, especially in relation to the test's potential clinical use. For example, factor analytic studies should be performed to clarify issues regarding the instrument's underlying factor structure and the independence of PDI scales. Lustman et al. (1984) also note that analogue student populations reporting a psychometrically defined depression via BDI scores greater than 10 may *not* exhibit characteristics similar to students meeting more stringent criteria for clinical depression. Additionally, follow-up analysis of PDI data may be necessary in order to establish a person's coping skills or reactions to stressful events before attempting to establish strong etiological connections between scale scores and future illness.

In further refining the PDI, the test authors may want to consider the development of an index designed to evaluate subjects' tendencies to both distort or exaggerate symptomatology (fake bad) and deny problems (fake good). Presumably, once the PDI is developed more fully, a more comprehensive manual will be published utilizing case examples to illustrate the application of the PDI in clinical settings.

In sum, the PDI shows great potential as a result of its brief item format, ability to provide an assessment of multiple problems, and its orientation toward the clinical concerns of college students. Current studies are investigating the use of the PDI in predicting dropout from school, and further collection and refinement of normative data for the instrument is being conducted. With increasing frequency, the PDI appears to be utilized in various investigations of college student stress. Further work by Sowa and her colleagues, as well as others, will determine the future usefulness of the PDI as both a research instrument and a clinical tool.

References

Beck, A.T., Ward, C.H., Mendelson, M., Mock, J.E., & Erbaugh, J.K. (1972). An inventory for measuring depression. *Archives of General Psychiatry, 25,* 57-63.

Brodman, K., Erdman, A.J., Jr., Lorge, I., & Wolff, H.G. (1949). The Cornell Medical Index: An adjunct to medical interview. *Journal of the American Medical Association, 140,* 530-534.

Brown, G.W., & Harris, T. (1978). *Social origins of depression.* New York: Free Press.

Gelso, C.J., & Johnson, D.H. (1983). *Explorations in time-limited counseling and psychotherapy.* New York: Teachers College Press.

Lustman, P.J., Sowa, C.J., & Day, R.C. (1981, August). *Evaluating life changes: Development of the Life Stress Questionnaire.* Paper presented at the meeting of the American Psychological Association, Los Angeles.

Lustman, P.J., Sowa, C.J., & O'Hara, D.J. (1984). Factors influencing college student health: Development of the Psychological Distress Inventory. *Journal of Counseling Psychology, 31,* 28-35.

Sarason, I.G., Johnson, J.H., & Siegal, J.M. (1978). Assessing the impact changes: Development of the life experiences survey. *Journal of Consulting and Clinical Psychology, 46,* 932-946.

Spielberger, D.D., Gorsuch, R.L., & Lushene, R.E. (1970). *Manual for the State-Trait Anxiety Inventory.* Palo Alto, CA: Consulting Psychologists Press.

Strahan, R., & Gerbasi, K.C. (1972). Short homogeneous versions of the Marlowe-Crowne Social Desirability Scale. *Journal of Clinical Psychology, 28,* 191-193.

Phyllis L. Newcomer, Ed.D.

Professor of Education, Beaver College, Glenside, Pennsylvania.

THE PUPIL RATING SCALE REVISED

Helmer R. Myklebust. Orlando, Florida: Grune & Stratton, Inc.

Introduction

The Pupil Rating Scale (Revised): Screening for Learning Disabilities (PRS) is designed to identify children who may have learning disabilities. Built upon what the author describes as a "psycho-neurological-cognitive" frame of reference, it consists of five subscales: Auditory Comprehension and Memory, Spoken Language, Orientation, Motor Coordination, and Personal-Social Behavior. The behavioral items associated with each subscale were developed by the author and his associates from the extensive observation of hundreds of LD children.

Helmer R. Myklebust, the author of the PRS, is a distinguished scholar and respected pioneer in the study of learning disabilities. Among his many publications in the field are a leading text, *Learning Disabilities: Educational Principles and Practices*, written with Doris Johnson and published in 1967, and a report written with Benjamin Boshes entitled *Minimal Brain Damage in Children*, published in 1969. These publications convey Myklebust's emphasis upon neurological dysfunction as the basis of learning disabilities, his concern with interrelated learning modalities termed auditory, visual, kinesthetic, and so forth, and his interest in disorders of inner, expressive, and receptive language.

The PRS was published initially in 1971. The current revision, completed in 1981, reports additional research to establish reliability and validity and presents increased standardization data. Information about the types of behavior included in the test also is provided. The actual items that constitute the PRS have not been altered since the original edition.

The PRS evolved from a pilot study involving 500 children who were not learning efficiently despite good potential and good educational opportunity. The author determined that teacher ratings on an experimental edition of the PRS were as effective in identifying children with learning problems as an elaborate battery of medical, psychological, and educational tests. Thus, the scale was expanded to its current length and used with an original standardization sample of 2,176 (1,138 male and 1,038 female) third- and fourth-grade students from four suburban schools.

The current edition extends the standardization sample to include children from kindergarten through the sixth grade whose ages range from 5 years, 5 months to 14 years. This sample, totaling 1,264 children, included many black and Hispanic subjects as well as many who were socially disadvantaged. The results (i.e., means and standard deviations) obtained with this sample were similar to those obtained with the initial sample. Therefore, the author concludes that the PRS may be used with children in Grades K through 6 and with socially disadvantaged children.

The PRS consists of a package of 50 rating forms and an 84-page manual that contains information about the development of the scale, standardization, reliability, and validity. The manual presents the 24 behavioral characteristics grouped under five major subscale headings (which also are included on the rating forms):

Auditory Comprehension/Memory—comprehending word meanings, following instructions, comprehending class discussions, and retaining information.

Spoken Language—vocabulary, grammar, word recall, story telling/relating experiences, and formulating ideas.

Orientation—judging time, spatial orientation, judging relationships, and knowing directions.

Motor Coordination—general coordination, balance, and manual dexterity.

Personal-Social Behavior—cooperation, attention, organization, new situations, social acceptance, responsibility, completion of assignments, and tactfulness.

Practical Applications/Uses

The Pupil Rating Scale (Revised) has the advantage of being easy to use. It consists of behaviors that are familiar to teachers and important for school success. Therefore, it is logical that teachers can easily recognize those behaviors and rate them accurately. Additionally, a child can be rated in 10 minutes, a commitment of time that is not burdensome to teachers. Scoring also is efficient, consisting simply of the addition of each item rating. Another advantage is that the PRS is not intimidating. Children are assessed in a familiar environment (the classroom), by a familiar person (the teacher) via a procedure that should not alarm them or their parents.

Clearly, it is essential that the PRS be efficient because it is simply a screening instrument. In one sense, its use with older children who have experienced difficulty with academic achievement may be superfluous because their academic failure suggests the possibility of a learning disability and necessitates diagnostic testing. Seemingly, this type of screening instrument is most useful at the preschool, kindergarten, or first-grade level, before the children have experienced academic failure.

The PRS may be administered by teachers, psychologists, counselors, or any person in a position to observe children's behavior in a classroom. The rater simply judges the child's behavior on a scale ranging from a low of 1 to a high of 5. The author states that approximately 10 minutes is required to rate each child but cautions that, to ensure maximum reliability, a rater should have at least 1 month of extensive experience with the children and should not rate more than 30 children per session.

An average rating on each item is 3. Scores of 1 or 2 are below average, and scores of 4 or 5 are above average. A hypothetical average score (3 points per item) per subtest is Auditory Comprehension/Memory—12, Spoken Language—15, Orientation—12, Motor Coordination—9, and Personal-Social Behavior—24. The scale yields a Verbal Score obtained by totalling Auditory Comprehension and Spoken Language—27, a Nonverbal Score from a combination of Orientation, Motor Coordination, and Personal Social Behavior—45, and a Total Score from all subscales—

72. Actual mean scores for the standardization sample were Verbal—28, Nonverbal—49, and Total—77, showing that in fact average children perform close to the hypothetical means.

Cutoff scores to determine a performance indicative of a learning problem have been arbitrarily set at approximately one standard deviation below the mean—verbal scores below 20, nonverbal scores below 40, and total scores below 65. In addition to using the total score, the author recommends the interpretation of verbal, nonverbal, and subscale scores to identify specific deficit areas. For example, a low score on auditory language but not on other scales may be indicative of childhood aphasia or hearing impairment.

Most importantly, the author clearly states that the PRS is a screening device, not a diagnostic instrument. It identifies children at risk of failing in school, not the reasons for failure. Thorough follow-up examinations are necessary to reveal the causal factors. Myklebust, Bannochie, and Killen (1971) report that among 15% of children who failed the PRS, further diagnostic assessment revealed that half of them failed because of learning disabilities and that the remaining half failed because of factors such as mental retardation, visual impairment, and so forth.

Although the author of the PRS is to be lauded for the simplicity and clarity of the actual rating scale, the manual is not particularly well organized and is not always clearly written. For example, critical information regarding the cutoff scores used to denote a potential learning problem is buried well along in the manual (in the middle of a paragraph on page 23).

Although it is of more interest to professionals concerned with test development than to teachers or practitioners, the standardization section of the manual is particularly obscure; the writing is redundant and disorganized. For example, there are references to the original sample without providing the number of subjects, then there is a discussion of 500 subjects in a pilot study who may or may not have been included in the original sample. When the number of subjects in the sample finally is provided, the author gives no data in the text on sex (found in the tables) or, incredibly, on the number of subjects per age or grade. Additionally, although the PRS is represented as being useful for children from a broad demographic distribution, the reader is not told where (i.e., in what section[s] of the country) the test was standardized. Factual omissions of this type are somewhat surprising in the work of an author of this stature.

Technical Aspects

Myklebust reports that the reliability of the Pupil Rating Scale (Revised) was established through a variety of studies. The first type of evidence he presents in the manual is in the findings of the standardization studies where means and standard deviations for Learning Problem and Normal Learner groups were consistent for each group.

Another study of reliability showed the similarity of ratings when 994 of the 1,250 children in the second standardization sample (K through 6) were re-evaluated 1 year later. The author was encouraged by this study because the means by item for the entire group were significantly different when computed by T-tests on only one item, Knowing Directions. On a less encouraging note, dealing with this same study, the author reports the results of analysis of variance that showed greater variability in ratings. However, these results were not adequately explained. Also, subjects are not broken down by age or grade. This omission is unfortunate because there might be greater variation among children at specific ages.

Unfortunately, no test-retest data are provided despite the ease with which such data could be gathered. The author opines that the time interval between ratings would alter the results, but that explanation seems inadequate, because the time between ratings could be kept to a minimum.

Evidence of rater consistency was provided by comparing the ratings of 49 teachers with means obtained from the "initial investigation" (presumably with 2,090 subjects). Because most of the teachers' ratings fell within 1 or 2 points of the original means, the author assumes rater consistency is at a high level. Once again, however, the strategy seems less than adequate. Interrater reliability, whereby various raters rate the same child, was not established. This is an important omission, as the scale's usefulness in identifying at-risk children should not be a reflection of the person using it.

The issue of validity is addressed by comparing the extent to which the PRS identified the same children as deficient as were so identified by a large battery of psychoeducational screening tests. The author reports that "agreement was unusually high, indicating the high validity of the teachers' ratings." Then, in a more complex, and poorly explained study (Myklebust, 1981), 10 subgroups were drawn from children who failed the screening test criteria and a control group that passed. These subjects were given diagnostic tests. Results showed that children designated as learning disabled were rated below true controls. Also, the ratings were accurate enough to differentiate between mild, moderate, and severe learning disabilities.

In an extreme group analysis (Reeves & Perkins, 1976), the PRS accounted for much of the variance in differentiating between LD and normal children, particularly between severe LD children and controls. Regarding specific PRS items, all but Motor Coordination items distinguished between children with moderate learning deficits and controls at high levels of statistical significance. When the groups compared were severely learning disabled and controls, all PRS items differentiated between them.

Concurrent and predictive validity were investigated by Colligan in two separate studies (1977, 1979). In the first, he found that the PRS scale scores and reading readiness scores obtained concurrently on kindergarten children were highly intercorrelated. In a 2-year follow-up study of PRS predictive usefulness, with 26 boys and 29 girls from the original sample, he found significant correlations between the PRS scale scores and criterion measures of reading. Auditory Comprehension/Memory and Spoken Language were most highly related to reading. Also, ratings on the poorest readers were comparable to those obtained on the author's original LD group. Both of these studies are offered as corroborative proof of validity.

A final indication of validity reported in the manual is the author's correlation research. He obtained high positive correlations between measures of reading, spelling, and arithmetic and the PRS scales Auditory Comprehension/Memory and Spoken Language (.42 to .53). Additionally, he found relatively low correlations between the PRS subscales and intelligence measures, demonstrating that, as stated, the PRS is not a measure of learning potential (Myklebust, 1981).

The very critical question of what behavior characteristics are being rated on the PRS is addressed in the manual in a detailed discussion of a variety of factor analyses. A study by the author with the standardization sample of children aged 6 to 14 and one by Bryan and McGrady (1972) with children in the third through sixth grades revealed that the PRS measured only four separate factors rather than five discrete types of behavior. In both studies, the most important factor, accounting for most (72% in the author's study) of the variance, contained the verbal behaviors comprised in the Auditory Comprehension/Memory and Spoken Language scales.

Myklebust chose to interpret these factorial results as supporting the design of the PRS. In fact, however, the consistency of the results suggest that the instrument might be redesigned to reflect four rather than five major areas and scored accordingly. Indeed, the impact of the Auditory Comprehension/Memory and Spoken Language scales is so significant, it might be beneficial to experiment with an abbreviated scale that consists exclusively of those behaviors. It would be a simple matter to determine the extent to which the Verbal Scales identified the same high-risk students as were screened out by the total score.

Critique

In conclusion, three points pertaining to the Pupil Rating Scale (Revised) deserve emphasis. First, when the variety of assessment devices available for use with LD children is considered, the PRS stands out as an inexpensive and useful screening device. Part of its usefulness lies in the fact that it directs teachers' attention to many behaviors that relate to academic achievement. This point is particularly true with regard to the behaviors that comprise the Auditory Comprehension/ Memory and Spoken Language scales. Not only are the behaviors that comprise these two scales highly related to academic achievement, but their significance is understood by teachers. It seems likely that most teachers have a relatively accurate perception of what constitutes the norm for these types of abilities; thus, they can efficiently identify deviance.

In a less positive vein, aspects of the other scales, Orientation, Motor Coordination, and Personal-Social Behavior, although theoretically related to learning disabilities, are less highly correlated with academic achievement and are less valuable predictors of school failure than the verbal skills. Additionally, teachers may have more difficulty recognizing normal and deviant behaviors on these scales. The extent to which teachers are and regard themselves as competent evaluators of these behaviors is an important, viable research question.

The second point pertains to the statistical justification for the PRS. As has been noted, some evidence pertaining to reliability and validity has been reported; however, these data are sparse and limited for an instrument that had been marketed for 10 years (1971 to 1981) before revision. In particular, a greater effort should be made to establish interrater reliability and predictive validity.

The final point pertains to the purpose of the PRS. The consumer is cautioned to remember that the instrument is a screening device, not a diagnostic tool. Essentially, it identifies underachievers. As Wall (1985) pointed out in his review of the PRS, "whether these teacher ratings predict the identification of children as learning disabled needs additional study."

References

Bryan, T., & McGrady, H. (1972). Use of a teacher rating scale. *Journal of Learning Disabilities, 5*, 199-206.

Colligan, R. (1977). Concurrent validity of the Mykelbust Pupil Rating Scale in a kindergarten population. *Journal of Learning Disabilities, 10*, 317-320.

Colligan, R. (1979). Predictive utility of the Myklebust Pupil Rating Scale: A two-year follow-up. *Journal of Learning Disabilities, 12*, 59-62.

Johnson, D., & Myklebust, H. (1967). *Learning disabilities: Educational principles and practices.* New York: Grune & Stratton.

Myklebust, H. (1981). *The pupil rating scale revised.* New York: Grune & Stratton.

Myklebust, H., & Boshes, B. (1969). *Minimal brain damage in children* (Final Rep., Contract No. 108-65-142, Neurological and Sensory Disease Control Program). Washington, DC: U. S. Department of Health, Education & Welfare.

Myklebust, H., Bannochie, M., & Killen, J. (1971). Learning disabilities and cognitive processes. In H. Myklebust (Ed.), *Progress in learning disabilities* (Vol. 2). New York: Grune & Stratton.

Reeves, J., & Perkins, M. (1976). The Pupil Rating Scale: A second look. *Journal of Special Education, 10*, 437-439.

Walls, S. (1985). Review of the Pupil Rating Scale Revised: Screening for Learning Disabilities. In J.V. Mitchell, Jr. (Ed.), *The ninth mental measurements yearbook* (pp. 1250-1251). Lincoln, NE: Buros Institute of Mental Measurements.

R. R. Hutzell, Ph.D.

Clinical Psychologist, VA Medical Center, Knoxville, Iowa.

PURPOSE IN LIFE TEST

James C. Crumbaugh and Leonard T. Maholick. Murfreesboro, Tennessee: Psychometric Affiliates.

Introduction

The Purpose in Life Test (PIL) is used to measure the degree to which an individual experiences life as meaningful, how much an individual feels like "somebody that matters," or how strongly an individual has developed a sense of purposeful direction in life.

James C. Crumbaugh, Ph.D., developed the PIL as part of his work to objectify concepts within the philosophy of Viktor E. Frankl, Ph.D., M.D. Frankl's logotheory postulates that humankind's strongest motive is to find a meaning and purpose in life. This is accomplished, according to logotheory, through the actualization of personally meaningful, self-transcendent values. Thus, Frankl's philosophy is an action- and goal-directed philosophy consonant with Western ideological thought. According to logotheory, existential vacuum ensues when one fails to establish a sense of personal life-meaning. This vacuum, in turn, creates motivation for the individual to strive harder to discover life-meaning or, alternatively, it opens the door to noogenic (existential) neuroses. Crumbaugh originally developed the PIL to be an objective measure of existential vacuum.

Prior to his retirement in 1980, Dr. Crumbaugh served as chairperson of the department of psychology at MacMurray College at Jacksonville, Illinois, then research director at the Bradley Center in Columbus, Georgia, and then staff psychologist at the Veterans Administration Medical Center in Gulfport, Mississippi. He studied logotheory under Dr. Frankl, who was visiting professor at Harvard University's summer school. Frankl invited Crumbaugh to join his seminar and to report on Crumbaugh's research on quantification and validation of logotheory concepts. Crumbaugh since has gained wide respect for his ability to put logotheory's philosophical concepts into practical action.

The development of the PIL is not described in the four-page manual (Crumbaugh & Maholick, 1969). A journal article (Crumbaugh & Maholick, 1964), however, describes the development as follows:

> An attitude scale was specifically designed to evoke responses believed related to the degree to which the individual experienced "purpose in life." The *a priori* basis of the items was a background in the literature of existentialism, particularly in logotherapy, and a "guess" as to what type of material would discriminate patients from non-patients. The structure of all items followed the pattern of a seven-point scale. . . . A pilot study was performed using 25 such items; on the basis of the results half were discarded and new items were substituted. Twenty-two then stood up in item analysis.

Two items were discarded later because they required negative scoring.

The PIL is produced in relatively large print on the front and back of an 11" x17" sheet of paper. Part A comprises three fourths of the space and consists of 20 items structured to conform to the following pattern:

1. I am usually:

1	2	3	4	5	6	7
completely			neutral			exuberant
bored						enthusiastic

Parts B and C take up the remaining space. Part B is a 13-item incomplete sentences test. Part C is an open space with the instruction that the respondent write a paragraph detailing aims, ambitions, goals in life, and progress being made in achieving these. Objective scoring of Parts B and C has not been established, and research with the PIL pertains only to Part A. Many administrators ignore Parts B and C, and at one time a different form of the PIL consisting of only Part A was available, but the current test catalog does not list this form.

Many persons have developed altered versions of the PIL for research purposes. These focus upon Part A only and include one or more of the following: a smaller sample of the 20 items, simplified wording for the two extreme points of each item, 7-point ratings of agreement with only one end point of each item, and dichotomous choice items. A Spanish version of the PIL has been published by Psychometric Affiliates in the past.

The PIL is written for adults. Although the manual states that the instructions are easily understood by most adults and adolescents with a fourth-grade reading level, some of the words may be difficult for younger persons to interpret.

Practical Applications/Uses

One major area of application for the Purpose in Life Test derives from the original intent of the PIL to objectify Frankl's concept of existential vacuum. As noted in the manual, existential vacuum affects more than one half the general population, and noogenic neuroses constitute roughly one fifth of a typical clinical case load. The authors of the PIL suggest that it can be used to detect the presence of existential vacuum so that logotherapy may be instituted where needed. They emphasize that the PIL can be useful in vocational guidance, rehabilitation, and individual counseling of students, neurotic patients, alcoholics, handicapped persons, and retirees. Unfortunately, there is little in the scientific literature except for anecdotal data to support the hypothesis that using the PIL to select patients for logotherapy improves overall outcome. Yet such use of the PIL remains appropriate from a theoretical perspective, and its use in this context is ripe for additional research.

A second practical application of the PIL is as a research tool to measure the degree to which an individual has developed a sense of life-meaning. As one might expect, this application of the PIL has seen more use in professional literature than has the use of the PIL to detect existential vacuum for case selection.

As a research tool, the PIL has been employed in a wide variety of settings, particularly in studies emphasizing existential, humanistic, and theological ideas.

Unfortunately, most research with the PIL has been for master's theses and doctoral dissertations, which often go unpublished and are difficult to locate. However, the current manual includes a two-page bibliography insert that provides some information about the thesis and dissertation research prior to 1980. Journal publications employing the PIL appear in clinical psychology, social psychology, psychological measurement, existentialism, and humanism journals. Publications in theological journals are less prevalent, although much of the unpublished thesis and dissertation research would fit well in such journals.

In the diverse range of settings noted above, the PIL has found application in research with a wide variety of individuals. Samples have included persons with strong religious orientations, retired individuals, neurotics, psychotics, college students, alcohol and drug abusers, criminals, the critically ill, adolescents, delinquents, and others.

Numerous potentially relevant variables may be related to one's sense of life-meaning, and the PIL has been used in many studies to assess such relationships. Among others, the relationships between the PIL and the following variables have been studied: alcohol abuse, anomie, death issues, demographic variables (including socioeconomic status), depression, job satisfaction, mental health adjustment, personality, religious/spiritual issues, retirement, self-actualization, sexual issues, subjective well-being, and time orientation.

Holistic medicine or *wellness* is an area of great potential for new application of the PIL. Most holistic medicine programs include a spiritual component that is considered a vital, but often slighted, aspect of a complete wellness program.

The PIL can be presented in either individual or group settings. The instructions are printed on the answer sheet. The examiner's participation in the testing process, scoring, and interpretation of the scores can be minimal as long as only Part A is employed. There is no time limit, but most individuals can complete Part A in 10 to 15 minutes. Parts B and C require more time to complete, and they are intended for professional interpretation by a clinical psychologist or other person similarly trained in psychological evaluation.

No particular expertise is required of the test administrator when Part A is used. However, the manual notes that the PIL should be employed with caution in any competitive situation where motivation to present a favorable self-image exists; thus it would appear that the test administrator should have some skill at defusing natural tendencies toward competition or social responding.

Scoring of Part A is quite simple and consists of summing the numerical values circled for the 20 items. Scores can range from 20 to 140. Interpretation of the scores is objective. The manual suggests using the mean of "normals" (112) and of patients (92) as cutoff scores; thus, scores above 112 suggest definite feelings of life-meaning, scores below 92 suggest lack of life-meaning, and scores of 92 through 112 are of uncertain definition. For research purposes, raw scores typically are employed for correlational and outcome studies. In both cases, higher raw scores suggest a stronger sense of life-meaning.

Technical Aspects

The four-page manual provides very little information about the technical and statistical aspects of the Purpose in Life Test. The manual provides technical data from only six studies, three of which are unpublished. Although a large number of studies have been conducted with the PIL, many of them are published in rather obscure journals or remain unpublished altogether. The following review of technical material includes data from studies additional to those reviewed in the manual, but given the varied sources of PIL articles, some relevant articles no doubt were missed. These data are statistically significant unless otherwise indicated.

PIL split-half reliability estimates and test-retest reliability estimates appear adequate for a short, paper-and-pencil, self-report scale and suggest that the PIL offers sufficient consistency for its intended use.

Two studies reviewed in the manual, plus several others, present split-half reliability correlations that range from .77 (Spearman-Brown corrected to .87) to .85 (Spearman-Brown corrected to .92). The data were collected largely from students, psychiatric outpatients of mixed diagnoses, hospitalized alcoholics, penitentiary inmates, "high purpose" nonpatients, and active Protestant parishioners (Butler & Carr, 1968; Crumbaugh, 1968; Crumbaugh & Maholick, 1964; Reker, 1977; Reker & Cousins, 1979).

Test-retest reliability is not discussed in the manual, but several studies have been conducted, yielding the following data: a 1-week coefficient of .83 (Meier & Edwards, 1974, $N = 57$ church members); a 6-week coefficient of .79 (Reker & Cousins, 1979, $N = 31$ college students); and a 12-week coefficient of .68 (Reker, 1977, $N = 17$ penitentiary inmates).

Validity assessment of the PIL has been somewhat cumbersome. This may be largely due to having no direct criterion for quantitative experiences of life-meaning against which to validate.

Face validity seems adequate in that the items look like they measure that which is intended. Two of the items (#7, retirement and #15, preparation for death), though face valid for an existential scale, may cause negative reactions by many people.

Regarding convergent validity, there is evidence that the PIL correlates with what Frankl meant by the term "existential vacuum." The sum of 6 quantifiable items from a series of 13 questions (Frankl Questionnaire) used by Frankl to estimate the presence of existential vacuum was found to correlate .68 with the PIL as reported in the manual (from Crumbaugh & Maholick, 1964, $N = 136$ mixed patients and nonpatients) and .56 in a separate study (Meier & Edwards, 1974, $N = 200$ church members). Meier and Edwards (1974) report similar correlations of .68 and .59 from two unpublished dissertations. Although these correlations are statistically significant and in the predicted direction, they are smaller than one might wish if the PIL and the Frankl Questionnaire are expected to measure the same phenomenon.

Regarding concurrent validity, the manual provides additional data (from Crumbaugh, 1968). After two therapy sessions, psychotherapists completed PILs for 50 psychiatric outpatient neurotics with mixed diagnoses. The therapists were instructed to complete the PILs in the way they thought the patient should have filled them out in order to be truthful. The correlation between the therapists' ratings and the patients' actual PIL scores was .38. A similar, earlier study with 39 subjects by Crumbaugh and Maholick (1964) had yielded a correlation of .27, which was not statistically significant. In addition, Crumbaugh constructed a rating scale by which ministers scored 120 parishioners for evidence of life-meaning; a correlation of .47 was found between these ratings and the parishioners' PIL scores. These correlations are in the predicted direction but are disappointingly small and add minor support to the validity of the PIL.

The manual reviews a construct validity study (from Crumbaugh, 1968) in which it was reasoned that if the PIL measures life-meaning, and if certain groups experience greater levels of life-meaning compared to other groups, then those groups expected to have higher levels of life-meaning should receive higher PIL scores. The order of the mean PIL scores of four "normal" populations was predicted correctly: successful business or professional personnel ($M = 118.90$, $SD = 11.31$, $N = 230$); active and leading Protestant parishioners ($M = 114.27$, $SD = 15.28$, $N = 142$); college undergraduates ($M = 108.45$, $SD = 13.98$, $N = 417$); and indigent, nonpsychiatric hospital patients ($M = 106.40$, $SD = 14.49$, $N = 16$). The prediction of the order of means of psychiatric populations was less accurate but did show a predicted drop from neurotics to alcoholics to nonschizophrenic psychotics.

It was further argued in the manual that, within Frankl's logotheory, some but not all psychiatric syndromes evolve from lack of life-meaning. Thus, one would expect some psychiatric patients to have lower than normal feelings of life-meaning. Statistically, this would be reflected in PIL scores of psychiatric patients showing a lower mean and greater variance than that of a normal population. In Crumbaugh's data, both the means and the variances of the patient and the nonpatient populations were different at statistically significant levels (combined "normal" groups: $M = 112.42$, $SD = 14.07$, $N = 805$; combined psychiatric groups: $M = 92.60$, $SD = 21.34$, $N = 346$).

Additional studies replicate the relationship between mean PIL scores and various group memberships. Crumbaugh and Maholick (1964), for example, had earlier shown a significant discrimination between nonpatients ($M = 119$, $N = 105$) and patients ($M = 99$, $N = 120$). Once again there was a progressive decline in mean PIL scores as predicted: "high purpose" nonpatients ($M = 124.78$, $SD = 11.80$, $N = 30$); undergraduate college students ($M = 116.84$, $SD = 14.00$, $N = 75$); psychiatric outpatients, mixed diagnoses ($M = 101.80$, $SD = 22.38$, $N = 49$); patients from a nonprofit psychiatric outpatient clinic, mixed diagnoses ($M = 101.30$, $SD = 18.14$, $N = 50$); and inpatient alcoholics ($M = 89.57$, $SD = 16.60$, $N = 21$). Garfield (1973) also found statistically significant differences in mean PIL scores between several groups: religion graduate students ($M = 119.29$, $SD = 10.01$, $N = 48$); commune inhabitants ($M = 113.43$, $SD = 11.03$, $N = 42$); psychology graduate students ($M = 102.93$, $SD = 17.18$, $N = 50$); professional engineers ($M = 94.26$, $SD = 19.89$, $N = 42$); and ghetto residents ($M = 85.71$, $SD = 24.27$, $N = 40$). Black and Gregson (1973) found statistically significant differences between New Zealand "normals" ($M = 115.07$, $SD = 13.87$, $N = 30$), first-sentence penitentiary inmates ($M = 99.07$, $SD = 18.72$, $N = 30$), and recidivist penitentiary inmates ($M = 86.80$, $SD = 15.35$, $N = 30$).

Correlation of the PIL with existing psychometric instruments can help to clarify the construct of life-meaning as measured by the PIL. Viewing across studies that correlate the PIL with a particular psychometric instrument and across studies that correlate the PIL with separate instruments designed to assess similar personality variables, the PIL has shown a potential, small, positive correlation with the following: absence of depression, extroversion and group achievement, positive attitude toward life at present and in the future, self-acceptance, psychological mindedness, self-control, emotional stability, absence of anxiety, responsibility, and absence of anomie. Studies supporting these relationships are reviewed in the next few paragraphs. (One must be cautioned here that, given the tendency for null results to go unpublished, the relationships described below should receive further evaluation. The relationships appear logical within the framework of logotheory, and they would appear to be fruitful areas for future research.)

The relationship between PIL scores and the MMPI has received attention in four investigations, all reviewed in the manual. All four studies demonstrated a small, negative, but statistically significant relationship with Depression. The only other MMPI scale that showed a statistically significant relationship with the PIL in more than one study was Social Introversion, which showed a negative relationship in two of the studies.

The relationship between the PIL and psychometric instruments other than the MMPI has received less attention. Reker (1977) and Reker and Cousins (1979) report positive relationships between PIL scores and Life Areas Survey scores for attitude toward life at present ($r = .45$, $N = 48$ prison inmates; $r = .65$, $N = 239$ college students) and attitude toward life in the future ($r = .54$; $.41$). Garfield (1973) correlated PIL scores with California Psychological Inventory (CPI) scores for his five groups (ghetto residents, professional engineers, psychology graduate students, commune inhabitants, and religion graduate students), and the manual reports Nyholm (1966) added an additional group (34 nonpatients). Four of these six groups showed a positive, statistically significant correlation between PIL scores and Achievement via Conformance (r ranged from $.49$ to $.71$). Three of the groups showed similar relationships between PIL scores and Self-Acceptance ($.40$ to $.53$) and Psychological-Mindedness ($.47$ to $.59$). Two showed similar relationships with Self-Control ($.46$; $.57$) and with Sense of Well-Being ($.52$; $.67$). Crandall and Rasmussen (1975) and Simmons (1980) studied the relationship between the PIL and terminal values of the Rokeach Value Survey with college students. They replicated a positive relationship between the PIL and Salvation and a negative relationship between the PIL and hedonistic Pleasure.

Viewing accross publications, several personality traits appear to have small but consistent correlations with the PIL. Depression is one such variable. Additional to the MMPI data noted previously, Harlow, Newcomb, and Bentler (1986) showed negative correlations between a variation of the PIL and the Center for Epidemiologic Studies—Depression Scale (CES-D), between the PIL and a 7-item self-derogation scale, and between the PIL and a 5-item suicide ideation scale. The correlations were significant for both men and women and ranged from $-.55$ to $-.79$ ($N = 722$ subjects from a longitudinal study of adolescent growth). In an earlier paper using the same subjects (Newcomb, Harlow, & Bentler, 1985), a correlation of $-.87$ was found between the PIL and experimental scales of life-meaninglessness. Phillips (1980) found negative correlations between the Zung Self-rating Depression Scale (SDS) and 7 of the 20 individual PIL items (r ranged from $-.22$ to $-.29$, $N = 134$ college students).

Again viewing across studies, a relationship appears between the PIL and an emotional stability factor. The manual showed a positive relationship between the PIL and Emotional Stability of the Gordon Personal Profile and Inventory ($r = .43$; $N = 56$ trainee Dominican sisters) as well as a positive relationship with Emotional

Stability on the Sixteen Personality Factor Questionnaire (16PF; $r = .41$). Also on the 16PF, there was a negative relationship between the PIL and Insecurity (-.44) and Neuroticism (-.32). Thauberger, Cleland, and Nicholson (1982) conducted factor anslyses with 16 existential scales ($N = 133$ college students and community residents) and found the PIL to load heavily on a factor labeled Neuroticism. Pearson and Sheffield (1974) report significant correlations for both male and female subjects ($N = 144$ outpatient neurotics with mixed diagnoses) with Neuroticism on the Eysenck Personality Inventory.

The PIL has shown a negative relationship with anxiety. A study reviewed in the manual ($N = 56$ Dominican Sisters in training) found this relationship with the 16PF ($r = -.52$). Yarnell (1971) found it with both "normals" ($r = -.34$, $N = 40$ Air Force men taking university psychology courses) and schizophrenics ($r = -.58$, $N = 40$ male inpatients) through an MMPI Anxiety factor and through a negative relationship between the PIL and the State-Trait Anxiety Inventory (rs ranged from -.37 to -.60).

A sense of self-control also appears to be associated with higher PIL scores. Rotter's Internal-External Locus of Control Scale was correlated negatively (suggesting higher PIL scores are related to "Internalizers") in the studies by Yarnell (1971 $r = -.32$, $N = 40$ normals; $r = -.49$, $N = 40$ schizophrenics) and Reker (1977) ($r = -.71$, $N = 48$ prison inmates). The manual records a positive relationship with self-control as measured by the 16PF ($r = .40$, $N = 56$ Dominican Sisters in training). Simmons (1980) found a relationship with both self-control and responsibility on Rokeach's Value Survey ($N = 99$ college students).

Responsibility appears to be positively correlated with PIL scores. The manual reports a correlation of .39 with responsibility, as measured by the Gordon Personal Profile and Inventory, and a correlation of .37 with conscientiousness, as measured by the 16PF ($N = 56$ Dominican Sisters in training). Reker (1977) reports a correlation of .36 with Plans and Organizes Things on the EPI ($N = 48$ prison inmates).

The manual notes small, negative, but statistically significant relationships between the PIL and the Srole Anomie Scale ($r = -.48$, $N = 94$ male college students; $r = -.32$, $N = 155$ female college students), and similar relationships between the PIL and the Elmore Anomie Scale's General Anomie factor ($r = -.51$, $N = 94$). Garfield (1973) assessed the relationship between the PIL and two anomie scales—Srole's and another by McClosky and Scharr. Garfield reported that correlations were largely nonsignificant within each of his five groups. A subjective view of his means, though, suggests he would have detected a significant negative relationship between the PIL and both of the anomie scales had he correlated his data across all groups.

As noted, these studies of correlations between the PIL and personality variables help to clarify the construct of life-meaning as measured by the PIL. The results do not validate the PIL directly, but most fit predictions generated from the perspective of logotheory and most are what would be expected from a valid measure of the strength of subjective life-meaning.

Another interesting area of study is not reviewed in the manual but supports the validity of the PIL. If the PIL is a valid measure of life-meaning, then PIL scores should increase if an individual participates in an experience that increases feelings of life-meaning. Crumbaugh's (1973) logoanalysis is designed to help individuals increase such feelings. Crumbaugh (1972) administered the PIL to 81 inpatients at the beginning and end of alcoholism treatment. Thirty of the patients were selected to participate in logoanalysis in addition to the regular alcoholism treatment. The 51 remaining patients were not selected and thus served as a comparison group. PIL scores increased more for the alcoholics who participated in logoanalysis than for those who did not. Similarly, Crumbaugh and Carr (1979) found that the PIL scores of subjects who started therapy with existential vacuum increased more after participating in closed-ended logoanalysis sessions than did control patients who did not receive logoanalysis ($N = 25$ alcoholic inpatients per group).

In broad overview, then, there are data that support the validity of the PIL as a measure of the degree to which an individual experiences life as meaningful. Data supporting the validity of the PIL are consonant with predictions made from logotheory, yet many of the relationships are of small magnitude and the support is indirect. At this point, it would appear very useful to conduct a large scale investigation using the multitrait-multimethod matrix, which can provide considerable information to answer questions that remain regarding convergent and discriminant validity of the PIL.

Turning now to psychological norms, the manual does not present data from individuals systematically selected to be representative of any particular population. Rather, data are based on 1,151 cases (from Crumbaugh, 1968) largely influenced by convenience and availability. Approximately 70% of these cases are labeled as "normal," and the remaining 30% are "patients."

The manual states that no consistent relationships have been reported between PIL scores and age, education, intelligence, or gender. There have been several attempts (many of which are not reviewed in the manual) to detect relationships between PIL scores and these variables. Studies of gender effects have produced mixed results. Age, education, and intelligence, on the other hand, often have shown very slight (not statistically significant) but positive relationships with PIL scores. One might suspect that a small, positive relationship between PIL scores and these latter variables would be established if a large sample were obtained to reflect the normal distribution of these variables in the general population.

Relevant to the discussion of norms, Garfield (1973) argues that the PIL is based upon assumptions grounded in white, middle-class, capitalist establishmentarianism that include advocacy of the work ethic, stress upon future orientation, tendancy toward unidimensional conceptualization, advocacy of purposive behavior, and positivity of high levels of stimulation. He additionally argues that the PIL is biased by the following Western philosophical perspectives: acceptance of mind-body dualism, primacy of physical over spiritual existence, and advocacy of process over stasis. Garfield presents some clinical evidence that the PIL items are less appropriate for assessing life-meaning or meaninglessness in populations that deviate from middle-class, Western philosophical thought.

Critique

As a test of the degree to which an individual experiences life as meaningful, Part A of the PIL has much potential for use in noncompetitive situations. It has been used with adults and adolescents in a wide variety of settings. The protocols are hand scored. With little guidance, Part A can be self-administered, self-scored, and self-interpreted easily by most adults within 15 minutes.

The manual needs to be updated and expanded. A considerable service would be done for potential PIL users if the results of all the studies that have employed the PIL were collected and synthesized into a single document. Although the development, technical aspects, and statistical elements of the PIL are not adequately documented, the PIL does remain the most systematically developed and most frequently used test to come out of Frankl's logotheory.

Available reliability scores appear adequate, but reliability estimates in divergent populations need to be established. Criteria regarding the validity of the PIL are hard to identify. The PIL demonstrates face validity in that it "looks like" it measures that which it is purported to measure. Other validity assessments, though indirect, generally do support the validity of the instrument.

The PIL was designed to measure a concept that other psychological instruments do not measure, and thus the PIL would not be expected to show a high correlation with existing psychometric instruments. Although the PIL does show consistent relationships with some other instruments and personality variables, those relationships are expectedly low.

Normative data are notably absent. Although the position that higher PIL scores suggest higher life-meaning might be expected to generalize across many groups, establishment of local norms for cutoff scores rather than those suggested in the manual is highly recommended until more representative data are published.

It must be remembered that the PIL was developed out of Frankl's *logo*philosophy, which suggests that life-meaning is experienced by the actualization of personally meaningful, self-transcendent values. As this philosophy is consonant with Western philosophy and middle-class thought, divergent cultural and subcultural groups may find life-meaning in contexts not addressed by the PIL or may interpret the PIL items differently from the bulk of the samples studied to date. Middle-class America and populations with similar values include a very large number of individuals for whom the PIL is a potentially relevant predictor of the degree to which life-meaning is experienced, but for those many groups that depart from middle-class American values, the generalization of the PIL must be questioned and specific validity studies are warranted.

References

Black, W. A. M., & Gregson, R. A. M. (1973). Time perspective, purpose in life, extraversion and neuroticism in New Zealand prisoners. *British Journal of Social and Clinical Psychology, 12,* 50-60.

Butler, A. C., & Carr, L. (1968). Purpose in life through social action. *Journal of Social Psychology, 74,* 243-250.

Crandall, J. E., & Rasmussen, R. D. (1975). Purpose in life as related to specific values. *Journal of Clinical Psychology, 31,* 483-485.

Crumbaugh, J. C. (1968). Cross-validation of Purpose-in-Life Test based on Frankl's concepts. *Journal of Individual Psychology, 24,* 74-81.

Crumbaugh, J. C. (1972, May). Changes in Frankl's existential vacuum as a measure of therapeutic outcome. *Newsletter for Research in Psychology,* pp. 35-37.

Crumbaugh, J. C. (1973). *Everything to gain: A guide to self-fulfillment through logoanalysis.* Chicago: Nelson-Hall.

Crumbaugh, J. C., & Carr, G. L. (1979). Treatment of alcoholics with logotherapy. *International Journal of the Addictions, 14,* 847-853.

Crumbaugh, J. C., & Maholick, L. T. (1964). An experimental study in existentialism: The psychometric approach to Frankl's concept of noogenic neurosis. *Journal of Clinical Psychology, 20,* 200-207.

Crumbaugh, J. C., & Maholick, L. T. (1969). *Manual of instructions for the Purpose in Life Test.* Murfreesboro, TN: Psychometric Affiliates.

Garfield, C. A. (1973). A psychometric and clinical investigation of Frankl's concept of existential vacuum and of anomia. *Psychiatry, 36,* 396-408.

Harlow, L. L., Newcomb, M. D., & Bentler, P. M. (1986). Depression, self-derogation, substance use, and suicide ideation: Lack of purpose in life as a mediational factor. *Journal of Clinical Psychology, 42,* 5-21.

Meier, A., & Edwards, H. (1974). Purpose-in-Life Test: Age and sex differences. *Journal of Clinical Psychology, 30,* 384-386.

Newcomb, M. D., Harlow, L. L., & Bentler, P. M. (1985, April). *The Purpose-in-Life Test: Exploratory confirmatory, and validational analyses.* Paper presented at the meeting of the Western Psychological Association, San Jose, CA.

Nyholm, S. E. (1966). *A replication of a psychometric approach to existentialism.* Unpublished master's thesis, University of Portland.

Pearson, P. R., & Sheffield, B. F. (1974). Purpose-in-Life and the Eysenck Personality Inventory. *Journal of Clinical Psychology, 30,* 562-564.

Phillips, W. M. (1980). Purpose in life, depression, and locus of control. *Journal of Clinical Psychology, 36,* 661-667.

Reker, G. T. (1977). The Purpose-in-Life Test in an inmate population: An empirical investigation. *Journal of Clinical Psychology, 33,* 688-693.

Reker, G. T., & Cousins, J. B. (1979). Factor structure, construct validity and reliability of the Seeking of Noetic Goals (SONG) and Purpose-in-Life (PIL) tests. *Journal of Clinical Psychology, 35,* 85-91.

Simmons, D. D. (1980). Purpose-in-life and the three aspects of valuing. *Journal of Clinical Psychology, 36,* 921-922.

Thauberger, P. C., Cleland, J. F., & Nicholson, L. (1982). Existential measurement: A factor analytic study of some current psychometric instruments. *Journal of Research in Personality, 16,* 165-178.

Yarnell, T. D. (1971). Purpose-in-Life test: Further correlates. *Journal of Individual Psychology, 27,* 76-79.

Barbara A. Rothlisberg, Ph.D.
*Assistant Professor, Department of Educational Psychology and
Foundations, University of Northern Iowa, Cedar Falls, Iowa.*

Rik Carl D'Amato, Ed.D.
*Assistant Professor and Co-Director, School Psychology Program,
Department of Educational Psychology, Mississippi State University,
Mississippi State, Mississippi.*

RECEPTIVE ONE-WORD PICTURE VOCABULARY TEST

*Morrison F. Gardner. Novato, California: Academic Therapy
Publications.*

Introduction

The Receptive One-Word Picture Vocabulary Test (ROWPVT; Gardner, 1985) is an instrument that attempts to assess one aspect of children's vocabulary development, one-word receptive vocabulary. The measure seeks to provide a baseline of vocabulary development as it relates to pictorial recognition of single words. A stimulus word is provided to the child, and the child must respond by pointing to, verbalizing, or visually fixating on the correct pictorial illustration. The ROWPVT was designed by Morrison F. Gardner, a psychologist at Children's Hospital of San Francisco, as a companion tool to the Expressive One-Word Picture Vocabulary Test (EOWPVT; Gardner, 1979). The two instruments used concurrently could address children's possible preferences for or differences in one-word picture vocabulary skill. Gardner (1985) claimed that the significant difference in the scores of the two instruments when they are administered concurrently suggests "real" vocabulary differences between expressive and receptive abilities (p. 6). Such differences could be used to plan a more comprehensive vocabulary assessment.

The assessment of children's vocabulary has long been an area of interest to educators and psychologists alike. Almost six decades ago, Terman (1918) published research investigating vocabulary as an index of mental ability. Recently, however, the emphasis of research in vocabulary ability has shifted toward an educational or achievement context (D'Amato, Gray, & Dean, 1987). Vocabulary skills appear to be related intimately to reading behaviors as well as to other academic tasks (Sattler, 1982). For example, receptive vocabulary skills have been deemed similar to the verbal comprehension or achievement factor reported on the Wechsler Intelligence Scale for Children-Revised (WISC-R, Wechsler, 1974; D'Amato, Gray & Dean, in press; Hollinger & Sarvis, 1984). In light of the important role it plays in these various disciplines, the interest in the measurement of vocabulary skills should come as no surprise.

The test stimuli appear nonthreatening and consist of 100 spiral-bound test plates, each with four black-and-white, hand-drawn pictures placed horizontally across the page to simplify the visual display. The stimuli, the manual, and 25 individual record forms are packaged in a vinyl-covered folder.

Practical Applications/Uses

The Receptive One-Word Picture Vocabulary Test has been designed to quickly assess receptive one-word pictorial vocabulary and provide data for comparison with its companion tool, the EOWPVT. Because the ROWPVT requires no language expression, it is considered appropriate for children with suspected expressive difficulties (i.e., bilingual, withdrawn, orthopedically impaired, etc.) ranging in age from 2 years to 11 years, 11 months. The manual does not discuss whether a specific level of intellectual functioning is required in order to answer the test items adequately.

A Spanish version is available, but in order to administer it, the examiner must be fluent in Spanish and aware of regional dialectic differences. Unfortunately, the manual provides no further information as to the existence of alternate norms or standardization data on a bilingual/Spanish sample. Clearly, the absence of technical information makes the use of this version as a norm-referenced instrument questionable.

The ROWPVT is administered individually and untimed. However, Gardner indicates that administration time averages 10-15 minutes with scoring accounting for an additional 5 minutes. The manual reports that the examiner should provide the best possible testing environment for the child; therefore, children requiring more time or two sessions to complete the instrument should be accomodated.

The training needed to administer the ROWPVT appears to be rather unrestricted. The only caveat mentioned in the manual is that the examiner should be familiar with the principles of psychoeducational assessment and interpretation. Specialists who should possess such familiarity include speech and language therapists, counselors, and physicians.

Test directions are listed on the child's record form; however, because these instructions are brief, the examiner must be well acquainted with the general instructions in the manual in order to understand the testing procedure fully. The examiner says to the child, "I am going to say a word to you and you are to point to one of the four pictures that is a picture of that word." Three examples are given to assure that the child understands the demands of the task before the actual assessment begins. In addition, the manual indicates that the child should be told that he or she can ask the examiner to repeat words and that as the words become more difficult, the child should be instructed to do his or her best. As is standard with other assessment instruments, the examiner should provide the child with unconditional encouragement during the course of the evaluation.

The examiner should note that the record form of the ROWPVT shows items arranged according to age levels, typically with 10 items per age level. In general, age levels encompass 1 year (e.g., Level 2.0-2.11); however, there are three notable exceptions. The 3-year-old age group is represented by two age levels, Level 3.0-3.5 (9 items) and Level 3.6-3.11 (10 items). Level 9.0-11.11 (21 items) collapses three years into one item grouping. No explanation for this arrangement could be found in the manual.

Testing begins at the child's chronological age level unless the examiner suspects that the child is functioning at a lower cognitive level. The examiner administers a *critical range of items* to the subject from a basal level of eight successive correct items to a ceiling of six incorrect items out of eight. The child is asked to indicate (e.g., through pointing, verbalizing the number, or fixating on the correct picture) which illustration corresponds to the stimulus word provided by the examiner. Should a multiple basal or ceiling occur, the highest basal and/or lowest ceiling is used in the computation of the raw score.

The raw score is computed in the standard fashion by subtracting all errors below the ceiling item, but above the basal, from the number of the last ceiling item. Illustrations demonstrating scoring are provided in the manual and serve as valuable aids to those unfamiliar with such a scoring system. Raw scores then can be converted into four types of derived scores: Language Age scores, Language Standard scores, percentile ranks, and stanine scores. Derived scores have the advantage of removing the test performance from its dependence on the specific test situation and placing it within a more standard context of performance against the standardization group. The quality of the derived scores, then, is dependent upon the composition of the normative sample.

Language Age is similar to the concept of mental age and grade equivalents and, consequently, shares the difficulties of these scoring systems. Language Age scores are based on average raw score performance of given age levels in the norm group. Although the norm group for the ROWPVT spanned the ages of 2 to 11 years, 11 months, the language ages run from 1 year, 7 months to over 15 years, 6 months. Cronbach (1984) condemns such scales as providing misleading information on student performance; that is, the student may be compared to a group to which he may not belong.

Language Standard scores indicate the extent to which a child deviates from the performance of individuals at the same age level. The scale Gardner has chosen for the ROWPVT is commensurate with those of most intelligence measures, with a mean of 100 and a standard deviation of 15. Standard scores are perhaps the most useful of derived scores in that they allow comparisons across tests with comparable means and standard deviations and share equivalent normative populations. The Language Standard scores on the ROWPVT range from 55 to 145.

Percentile ranks and stanines scores are also available to index a child's standing relative to others in his or her age group.

Although Gardner provided some limited instruction on the rationale behind various scoring systems, he did not address the issue of score interpretation in the test manual. Basically, the interpretive information provided referred to how the ROWPVT compares to the EOWPVT. Thus, short of showing where the child falls in performance relative to the standardization sample, no mention was made of the utility of these scores for programming or for additional vocabulary evaluation. At best, the measure was portrayed as an adequate screening device suggesting that the child may or may not display receptive deficits.

Technical Aspects

The technical information provided about a test is paramount in determining the utility and quality of the measure (Anastasi, 1982). Specific information regarding the standardization sample for the Receptive One-Word Picture Vocabulary Test was limited. Gardner reported that standardization was obtained by calibrating the ROWPVT with the EOWPVT in a sample of 1,128 children residing in the San Francisco Bay area. Testing occurred at 14 age levels and was conducted at nine schools and a child development center. Sample sizes ranged from 36 subjects at Level 2.0-2.5 to 132 subjects at Level 10.0-10.11. The manual provides no information as to gender, race, culture, socioeconomic class, or tested intellectual ability. Although the ROWPVT was designed to assess receptive difficulty, the test manual does not appear to mention any attempt by the test author to include children of the various target populations (e.g., withdrawn, bilingual, learning disabled, etc.) in the normative group.

Evidence for the ROWPVT's reliability was limited to a measure of internal consistency at each age level for the standardization sample. Coefficients ranged from .81 (ages 2.6-2.11) to .93 (ages 4.0-4.5) with a median value of .90. This reviewer found somewhat disturbing the reliability statement explaining that "at some age levels the variability of the group was not very large and a number of items had zero variance, either because everyone answered correctly or incorrectly" (Gardner, 1985, p. 25). This comment points to the limited and purportedly homogeneous nature of the standardization sample and suggests poor item discrimination exists for parts of the scale.

Support for the validity of the ROWPVT came from two sources: content and criterion-related validity on a subsample of the standardization group. Content validity was supported by the use of "expert" judges (teachers, speech and language therapists) to review, modify, and evaluate stimuli for the instrument. In addition, it was noted that items were arranged and selected in line with their developmental qualities. "Only those items that met all of the specific criteria were retained for the final form of the ROWPVT" (Gardner, 1985, p. 27).

Criterion-related validity can be provided by establishing support for the test's relationship with nontest behavior or with already well-established instruments. The performance of the ROWPVT in this context was obtained by correlating raw scores on the ROWPVT with raw scores from the Vocabulary subtests of the Wechsler Preschool and Primary Scale of Intelligence (WPPSI) or the Wechsler Intelligence Scale for Children-Revised (WISC-R). It is unclear why Gardner chose to analyze and interpret the relationship between the raw scores of the ROWPVT and those of a Wechsler subtest in preference to standard score data.

Critique

The evaluation of vocabulary abilities in individuals experiencing difficulties with verbal expression is a complex issue and not resolved easily by current assessment practice. Gardner should be commended for attempting to assess components of vocabulary by developing the ROWPVT as a complementary screening instrument to the EOWPVT and, consequently, allowing comparison of expressive and receptive one-word skills. Unfortunately, although the concept behind the ROWPVT is interesting, the resulting measure offers little that enables it to be recommended at this time.

The primary stumbling blocks are most associated with the ROWPVT's technical quality, although they also appear in several other areas. The ROWPVT appears to be a hastily constructed instrument developed to complement the EOWPVT and expand on its useable information. As such, time may not have been available to deal adequately with the complex needs of developing such a scale. For example, although a possible 600 items were reduced to 100 through piloting and consideration of item aspects, the final placement of the 100 items was not cross-validated in a second independent sample. It is apparent from the discussion of the test's reliability that internal coefficients probably were affected adversely by poor item discrimination. Given the test's approach to administration—that of a "critical range" of items—the idea that a number of items displayed no discriminative power is extremely unsettling.

Perhaps the greatest drawback to the measure is the poor standardization sample employed. Tying the norms to those of the EOWPVT did not negate the fact that the San Francisco sample was limited and not representative of the populations for which this test purportedly is designed. Although a Spanish version of the test is available, Spanish children's scores derive little meaning when compared to an English-speaking standardization group. Users of the ROWPVT would need to develop local norms to determine whether children in their population areas actually are having difficulty in the receptive area. In addition, Gardner (1985) suggested that sample age groups were homogeneous in their composition. As a result, administering this test to a child who actually has receptive difficulties may make the child appear to be extremely deviant without just cause.

Moreover, poor standardization also permeated the reports of reliability and validity. The measure of internal consistency, which was the only reliability information provided, was, as expected, quite healthy. Given that the item type is consistent throughout the measure and the concept addressed is limited (one-word receptive vocabulary) the median coefficient of .90 is not that surprising. Data have suggested that homogeneous item types and content promote high internal consistency (Anastasi, 1982; Cronbach, 1984). A more revealing type of reliability would have been test-retest. At this point, it is unknown how stable children's scores would be over time. If the intent of the ROWPVT is to reveal "true" receptive difficulties, the temporal functioning of the test is a critical piece of information that was not addressed.

The validity information provided on the ROWPVT was also unusual. As detailed previously, there appeared to be difficulty with the discrimination of some of the items at certain age levels. Consequently, the issue of content validity as explained in the manual is far from resolved. Indeed, conflicting evidence in the manual revealed that the developmental gradient purported for the items was not as clear-cut as the content validity section suggests.

Criterion-related validity information was also mysteriously lacking. Gardner did not justify adequately his use of raw score correlations of the ROWPVT and the Wechsler Vocabulary subtests. However, the use of WISC-R and WPPSI subtests was highly questionable because the subtests demanded the type of expressive facility that the ROWPVT claims to avoid. To be sure, a better validity test for the ROWPVT would be its comparison with a well-established and similar measure, such as the Peabody Picture Vocabulary Test—Revised (Dunn & Dunn, 1981). In essence, if the ROWPVT cannot prove its cost effectiveness and quality when compared to well-standardized and technically sophisticated instruments of its own genre (e.g., the PPVT-R), it should be avoided in applied settings until appropriate adjustments in its construction can be made.

References

This list includes text citations and suggested additional reading.

Amster, J. B. (1987). Test review: Receptive One-Word Picture Vocabulary Test. *The Reading Teacher, 7*, 452-455.

Anastasi, A. (1982). *Psychological testing* (5th ed.). New York: MacMillan.

Cronbach, L. J. (1984). *Essentials of psychological testing* (4th ed.). New York: Harper & Row.

Cummings, J. A. (1985). Review of Expressive One-Word Picture Vocabulary Test [Lower level]. In J. V. Mitchell, Jr. (Ed.), *The ninth mental measurements yearbook* (pp. 564-566). Lincoln, NE: The Buros Institute of Mental Measurements.

D'Amato, R. C., Gray, J. W., & Dean, R. S. (1987). Concurrent validity of the PPVT-R with the K-ABC for learning problem children. *Psychology in the Schools, 24*, 35-39.

D'Amato, R. C., Gray, J. W., & Dean, R. S. (in press). Construct validity of the PPVT with neuropsychological, intellectual and achievement measures. *Journal of Clinical Psychology.*

Dunn, L. M., & Dunn, L. M. (1981). *Peabody Picture Vocabulary Test-Revised.* Circle Pines, MN: American Guidance.

Gardner, M. (1979). *Expressive One-Word Picture Vocabulary Test: Manual and form.* Novato, CA: Academic Therapy.

Gardner, M. F. (1985). *Manual for the Receptive One-Word Picture Vocabulary Test.* Novato, CA: Academic Therapy.

Goldstein, D. J., Smith, K. B., & Woldrep, E. E. (1986). Factor analytic study of the Kaufman Assessment Battery for Children. *Journal of Clinical Psychology, 42*(6), 890-894.

Hollinger, C. L., & Sarvis, P. A. (1984). Interpretation of the PPVT-R: A pure measure of verbal comprehension? *Psychology in the Schools, 21*, 34-41.

Sattler, J. M. (1982). *Assessment of children's intelligence and special abilities.* Boston: Allyn & Bacon.

Spivak, G. M. (1985). Review of Expressive One-Word Picture Vocabulary Test. In J. V. Mitchell, Jr. (Ed.), *The ninth mental measurements yearbook* (pp. 566-567). Lincoln, NE: The Buros Institute of Mental Measurements.

Terman, L. M. (1918). The vocabulary test as a measure of intelligence. *Journal of Educational Psychology, 9*, 452-466.

Wechsler, D. (1974). *Manual for the Wechsler Intelligence Scale for Children-Revised.* New York: Psychological Corporation.

Wyman E. Fischer, Ph.D.
Professor of Psychology, Ball State University, Muncie, Indiana.

REVISED TOKEN TEST

Malcolm R. McNeil and Thomas E. Prescott. Austin, Texas: PRO-ED.

Introduction

The Revised Token Test (RTT) was developed as "a sensitive and quantifiable test battery for the assessment of auditory processing inefficiencies associated with brain damage, aphasia, and certain language and learning disabilities" (McNeil & Prescott, 1978a, p. xiii). The RTT, which was drafted over a 6-year period, represents an expansion and psychometric refinement of previous versions of the Token Test.

At the time of the RTT's publication in 1978, Dr. Malcolm Ray McNeil, the primary author of the test, was an assistant professor of communication disorders and speech science at the University of Colorado in Boulder. The second author, Dr. Thomas E. Prescott, was the coordinator of the speech pathology department at the Veterans Administration Hospital in Denver, an adjunct professor of speech pathology and audiology at the University of Denver, and assistant professor at the University of Colorado in Boulder (NcNeil & Prescott, 1978b).

The authors undertook the present revision because, in their opinion, previous versions of the Token Test lacked specificity in the standardization of test materials, administration, and scoring, although they were clinically useful and superior to many other aphasia instruments presently on the market. They also believed that the Token Test's applicability for both diagnosis and treatment could be enhanced by increasing its length and introducing a multidimensional scoring system to replace the right/wrong scoring used with previous versions.

The Token Test has had a relatively short but active history. It originally was conceived and published by DeRenzi and Vignolo (1962) and subsequently has undergone at least six modifications in test materials, form and number of items, and suggestions for administration and scoring (McNeil & Prescott, 1978a; Berry, 1973). In its original form, the patient was required to manipulate tokens of different shapes, sizes, and colors according to specific orally administered instructions. The instructions, or command statements, included verbs, nouns, and adjectives progressing in complexity and length throughout the five subtests of the battery. Later versions of the Token Test, developed prior to the RTT, were patterned after the De Renzi and Vignolo test but contained various modifications in 1) size, shape, and color of the tokens, 2) structure of the command statements, and 3) total number of test items. Studies conducted with the original Token Test and the modified or shortened versions that followed indicated that these tests were sensitive to disrupted linguistic processes and were able to differentiate adequately among groups of normals, aphasics, and nonaphasic brain-damaged patients (Lezak, 1983).

As stated earlier, the RTT is an expanded and substantially modified version of the original Token Test by De Renzi and Vignolo. Instead of 62 items distributed among five subtests, the RTT has 100 items arranged into 10 subtests. The first four subtests of the RTT are similar in item structure to the first four subtests of the original Token Test, and RTT subtests V, VI, IX, and X contain command statements that are essentially variations of the complex items contained in Subtest V of the original test (Lezak, 1983). A laterality variant, (i.e., knowledge of left and right), which was not included in the original Token Test, is the main ingredient of the command statements contained in RTT subtests VII and VIII. Items were placed in a specific subtest of the RTT if they were judged to be homogeneous in content and difficulty with other items in the set (McNeil & Prescott, 1978a). Representative command statements (McNeil & Prescott, 1978a, pp. 79-83) from each of the 10 subtests of the RTT follow:

Subtest I (large tokens only): "Touch the green circle."
Subtest II (all tokens): "Touch the little white square."
Subtest III (large tokens only): "Touch the blue square and the black circle."
Subtest IV (all tokens): "Touch the little blue square and the big black square."
Subtest V (large tokens only): "Put the blue circle behind the green square."
Subtest VI (all tokens): "Put the little white square on the big green circle."
Subtest VII (large tokens only): "Put the red circle to the right of the blue circle."
Subtest VIII (all tokens): "Put the little green circle to the left of the big red square."
Subtest IX (large tokens only): "Touch the black square if there is a red circle."
Subtest X (all tokens): "Instead of the big red square touch the big white circle."

The complete RTT kit contains a technical manual, a set of plastic tokens, an administration manual, scoring forms, and a number of sets of response summary forms. The technical manual is comprehensive, containing sections on theory and development, administration and scoring, and interpretation. Validity and reliability data are presented in both textual and graphic detail. Normative and reference group data are included in tabular form for normal controls, left-hemisphere brain-damaged subjects with aphasia, and right-hemisphere brain-damaged subjects without aphasia. The 15-category scoring system, patterned after the Porch Index of Communicative Ability (PICA; Porch, 1967), is described in detail. The set of 20 plastic tokens, used as visual stimuli and requiring some type of manual manipulation for each command statement of all subtests, are of two shapes (circles and squares), two sizes (large and small), and five colors (red, black, white, green, and blue). The administration manual, also duplicated in the technical manual, is brief and contains a diagram for token layout and a complete set of command statements. Minimal instructions specific to each of the 10 subtests also are provided. More comprehensive instructions for administration and scoring are contained in the technical manual. The scoring form for the RTT is quite detailed, providing space for the calculation of mean values for each of the 10 subtests in addition to an overall score. Space also is provided on the scoring form for recording intra-examinee mean values for linguistic elements indicating possible specific difficulty in comprehending command verbs, adjectives relating to size and color, geometric shape nouns, place prepositions, and adverbial clauses. Specific summary forms, included with the kit and reproduced in the technical manual, permit a graphic subtest-by-subtest and total score comparison of an examinee's performance on the RTT in both raw score and percentile score units. Up to three administrations can be recorded on the same form providing both a numerical and visual picture of improvement or deterioration of auditory processing efficiency over time. Other summary forms, also included with the RTT, were designed to present an intra-examinee comparison of proficiency with various linguistic elements.

The RTT is an individually administered test requiring adherence to strictly standardized administration and scoring procedures. All instructions for both examiner and examinee are stated clearly in the technical manual. The RTT has been standardized on groups of normals, right-hemisphere brain-damaged persons, and left-hemisphere brain-damaged aphasics, ranging in age from 20 to 80 years. The authors contend, however, that the test may be appropriate for other age and disability groups as well if the examiner has a clear understanding of the theory underlying the test and its possible limitations (McNeil & Prescott, 1978a). Because the test is intended to detect auditory processing inefficiencies rather than provide a measure of processing facility, normal adult subjects are expected to have minimal dificulty with the RTT. Scores reported on normals during the norming process bear this out.

Practical Applications/Uses

The authors of the Revised Token Test state that the test is "valid for assessing auditory processing in any individual as long as he has adequate peripheral hearing, speaks the language in which the test is administered, and has the concepts in his language repertoire for differentiating 'touch' v. 'pick up' and 'put', colors, shapes, and sizes" (McNeil & Prescott, 1978a, p.31). Research on earlier versions of the Token Test support the notion that the test is capable of identifying aphasic disability as marked by disrupted linguistic processes even when communication ability seems relatively intact (Lezak, 1983; Morley, Lundgren, & Haxby, 1979). The Token Tests appear to have utility in identifying nonsevere aphasic disorders masked by concomitant disabilities resulting from brain damage (Wertz, 1979). Hammeke (1985), however, questions whether this holds true for the RTT as well; he feels that the RTT is a sensitive index to brain injury but doubts whether it is effective in distinguishing between specific linguistic abilities and other cognitive disturbances resulting from brain injury. His concern is based on an analysis of normative data supplied in the RTT manual (McNeil & Prescott, 1978a).

Because most normal individuals who have obtained at least a fourth-grade education can complete the original Token Test with, at most, only a few errors (Lezak, 1983), it is conceivable that the RTT could be used with children as well as adults. Norms for children ages 5-13 reportedly have been collected (Hammeke, 1985), but these data are not included as part of the test materials. At present, the RTT appears most appropriate for adult subjects suspected of having auditory processing inefficiencies associated with brain pathology.

The test authors contend that careful analysis of individual examinee data, generated from the multidimensional scoring system, will provide specific information relating to deficits in 1) auditory attention, 2) auditory short-term memory, 3) linguistic elements, and 4) comprehension of semantic relations. These goals appear achievable despite the fact that the RTT, though clearly able to separate normals from the brain-injured, is unable to separate clearly groups with high incidence of aphasic disturbance (left-hemisphere injury) from those with low incidence (right-hemisphere injury). The test authors appropriately caution that "as long as the tester clearly understands the theory of the test and is conservative in interpretation regarding differentiation of normal from abnormal, the Revised Token Test provides a sensitive index of change and a valuable quantifier of variables important in therapeutic management" (McNeil & Prescott, 1978a, p. 31).

Because of the complex multidimensional scoring system and the examiner sensitivity required to obtain a valid score, only professionals with in-depth training in individual diagnoses and a theoretical background in either psychology or communication disorders should administer and interpret the RTT. A study by Poeck and Pietron (1981) emphasized the importance of examiner sensitivity for tests like the RTT. They found that a slowed presentation or "stretched speech" significantly reduced errors made by aphasic subjects but did not affect scores of patients with right-hemisphere lesions. Although the technical manual provides explicit directions concerning administration and scoring, the complexity of the scoring system, which is intricately intertwined with the administration process, requires that even professionals trained in individual evaluation still will need to be formally trained to use the RTT. Teaching tapes and workshops are available for this purpose. Professionals interested in developing scoring proficiency should contact the authors of the test (McNeil & Prescott, 1978a).

As indicated, the technical manual includes step-by-step administration procedures, including such details as seating arrangement for the examiner and examinee; placement of the test materials (tokens) in relation to the examinee; specific pretest instructions; introductory statements for individual subtests; tempo of administration; and rate, intensity, and prosody of the orally adminstered command statements. Briefly, tokens are placed in a prescribed order in front of the examinee. The examinee is requested to 1) touch one or more tokens, 2) place a token in a specified spatial relation to a second designated token, or 3) demonstrate understanding of a complex adverbial clause by touching the appropriate token. The examiner is permitted to repeat command statements or provide cues when certain specified conditions are met. Such repetitions and cues, in turn, are considered when scoring the items.

Strict adherence to these standardized procedures is necessary to insure both interexaminer and intraexaminer administration reliability. This is a particularly important consideration because repeat administrations may be required to measure any progress that might have resulted from either spontaneous recovery or specifically designed rehabilitation procedures. Research studies dealing with the effectiveness of the RTT as a measure of auditory comprehension or with the effects and nature of receptive aphasia also depend upon rigorous standardization in both administration and scoring. The authors of the RTT were of the opinion that previous versions of the Token Test lacked appropriate standardization, making it difficult to conduct meaningful research. The RTT was intended to correct this deficiency.

The 15-category scoring system of the RTT is described thoroughly in the technical manual. The best possible response to a command statement is credited with a score of 15, and the total lack of response is given 1 point. Carefully qualified gradations, arranged hierarchically according to judgments of pathology gleaned from the literature, are credited points between these numerical extremes. For example, a delay in responding is considered more pathological than vocal-sub-vocal rehearsal but less abnormal than an impulsive response, termed "immediacy," leading to an incorrect solution. Consequently, "delay" would be given a 13; "immediacy" a 12; and "vocal-subvocal rehearsal" a 14. As is evident from this description, the scoring, though difficult, attempts to provide both an objective quantification and a qualitatively meaningful measure. It permits intraexaminee graphing of the 10 subtests at a single administration or a comparison of successive administrations to the examinee during the rehabilitation process.

It is assumed that interpretation of the RTT will be attempted only by professionals with appropriate clinical training and thorough knowledge of the theories underlying auditory processing and brain pathology. Therefore, the interpretation section included within the technical manual provides specific examples rather than an exhaustive treatise relating to communicative disorders. Sample case-response summary profiles are provided with most probable diagnoses and prognoses and suggestions for rehabilitation where appropriate. However, no research data are included in the technical manual to support the validity of the sample profiles as they relate to specific diagnostic/prognostic statements.

Technical Aspects

The authors of the Revised Token Test report evidence of construct, concurrent, and content validity. No studies regarding predictive validity are included in the manual or have, to date, appeared in the professional literature.

The section of the technical manual devoted to construct validity consists primarily of a discussion of auditory information processing deficits, based upon several theoretical formulations that tend to differentiate normals from the brain-injured. The test authors state that the RTT "is designed to isolate and assess the auditory input modality, used along with the gestural output modality and the visual input modality for interaction with the stimuli" (McNeil & Prescott, 1978a, p. 11). They further indicate that the test is capable of detecting the degree of reduction of auditory processing resulting from brain injury and that the scoring system provides a description of the specific type of auditory processing deficit. However, no specific evidence attesting to the validity of these descriptions is provided.

Concurrent validity correlations reported for earlier versions of the Token Test are used as support for the validity of the RTT. The test authors justify this action on the basis that the RTT is similar to the earlier forms. A concurrent validity study (based on a sample of 23 aphasic subjects) specific to the RTT and comparing the 10 subtest scores and overall score with 23 measures of the Porch Index of Communicative Ability (PICA) is reported in the technical manual. Correlations ranging from -.06 to .82 were reported. A correlation of .67 was found between the overall score of the PICA and the overall score of the RTT. Moderate correlations among RTT subtests and PICA measures assessing auditory functions as contrasted with low correlations obtained between RTT subtests and PICA measures of visual functions is viewed by the authors as further evidence of acceptable concurrent validity. To date, no other concurrent validity studies on the RTT have appeared in the literature.

McNeil and Prescott (1978a) present several dimensions of the RTT as evidence of content validity: 1) range of subtest difficulty, 2) number of homogeneous subtest items per subtest, 3) stimulus selection, and 4) task selection. They argue that the 10 subtests, characterized by progressive degrees of complexity, provide a broad enough range of difficulty to differentiate among all levels of dysphasia. They state that the larger number of homogeneous subtest items per subtest in the RTT, as compared with previous versions of the Token Test, protect against random ineffi-ciencies or inconsistencies often displayed by the brain-injured. In other words, random inefficiencies are less likely to be interpreted as auditory comprehension deficits. The authors paid careful attention to the selection of stimulus materials so that the characteristics (shape, color, and size) were counterbalanced carefully within and across subtests in order to insure that the individual subtest scores and the overall score of the RTT were measures of auditory comprehension rather than an indeterminant artifact of one or more stimulus characteristics. Tasks selected for the command statements comprising the 10 subtests were similar to those of earlier versions of the Token Test. Modifications were introduced primarily to increase intrasubtest homogeneity. The test authors further contend that the RTT is a true measure of auditory comprehension that is relatively free from linguistic, educa-tional, and intelligence biases. No specific research evidence, however, is provided to support their opinions. To the contrary, some studies with earlier versions of the Token Test indicate that variables such as educational level and intelligence may be significant correlates (De Renzi & Faglioni, 1978; Silverman, Raskin, Davidson, & Bloom, 1977; Van Dongen & Van Harskamp, 1972).

The authors of the RTT discuss four types of reliability in the technical manual: 1) test-retest, 2) intrascorer, 3) interscorer, and 4) live versus videotaped administra-tions. All of the reliability data reported was based on administrations by three examiners trained by the authors of the RTT over a 3-day period. A test-retest cor-relation of .90, based on five brain-damaged individuals retested within a 2-week period, was obtained. This index represented an average of the 10 individual sub-test correlations calculated via z-transformations of the reliability coefficients.

Intrascorer reliability was determined by having each of the three examiners score three of nine videotaped RTT administrations. The 10 means obtained from these paired protocols were correlated. The correlations obtained were averaged resulting in a *r* of .99.

The three examiners independently scored the same administration of the RTT to obtain a measure of interscorer reliability. The three interscorer reliability coeffi-cients obtained averaged .98.

Reliability coefficients of .95, .96, and .94 were obtained between three admin-istrations scored live and rescored from videotaped replays about 1 week later.

All norm tables included in the technical manual are based on standardization samples of 90 normals, 30 left-hemisphere brain-injured adults judged to be aphasic, and 30 right-hemisphere brain-damaged adults without aphasic symp-toms. Subjects in all three groups were between the ages of 20 and 60 years or older. Specific information relating to each of the sample groups is contained in the tech-nical manual.

A study conducted with the normal sample indicated that age was not a significant determinant of total score. Consequently, the entire sample was considered in determining raw score and percentile norms for RTT subtests and overall score. Considerable information relating to the groups is contained in the technical manual. Eight tables, two figures, and four data listings presented in tabular form provide information such as 1) biographical factors/etiology, site of lesions, and years post-onset of the two pathological sample groups; 2) the relationship of these biographical factors to performance on the RTT; 3) a comparison of the three groups on the basis of overall RTT and individual subtest scores; 4) a comparison of groups relating to performance on the RTT parts of speech, 5) mean time required by each group to complete the total RTT and each individual subtest; 6) the power, expressed in percentages, of the overall RTT score, individual subtests, and parts of speech (i.e., ability to respond to direct and implied commands [verbs], size and color [adjectives], shapes [nouns], placement of tokens [prepositions], etc.), to separate the three norm groups from each other; 7) mean performance of the normal and left-hemisphere brain-damaged group on each subtest and overall score as based on the 15-category scoring system; 8) correlation coefficient matrices for frequency of occurence of scoring categories for the normal and left-hemisphere brain-damaged groups; and 9) percentage of predictability (r^2) scoring-category frequency for the normal and left-hemisphere brain-damaged groups.

Of particular importance is the ability of the RTT to adequately differentiate the three norm groups. Data presented in Table 8 of the technical manual (McNeil & Prescott, 1978a, p. 27) indicate that the overall RTT score correctly separated normals from aphasics 91% of the time, normals from right-hemisphere brain-injured 74% of the time, and right-hemisphere brain-injured from aphasics 55% of the time. This indicates that the RTT is a valid instrument for distinguishing between normal and brain-injured subjects but has questionable validity for separating right-hemisphere brain-injured subjects without aphasia from their left-hemisphere counterparts with aphasia.

Critique

In describing one of the earlier versions of the Token Test on which the RTT is based, Lezak (1983) states, ". . . it is remarkably sensitive to the disrupted linguistic processes that are central to the aphasic disability, even when much of the patient's communication behavior has remained intact. Scores on the Token Test correlate highly with scores on tests of auditory comprehension" (p. 321). Even though the authors of the RTT state that "the RTT is believed to be a valid measure to use with 'language'-impaired populations (e.g., aphasics and learning disabled children)" (McNeil & Prescott, 1978a, p. 11), sufficient research concerning the validity of this claim is lacking. A study conducted by the authors and reported in the technical manual indicates that the RTT accurately differentiates between normals and the brain-injured but not necessarily between brain-injured groups with and without documented aphasia. In reference to this study, Hammeke (1985) states in a review of the RTT ". . . the test is likely sensitive to more generalized functions than auditory comprehension of linguistic informaton (e.g., sustained attention and concentration, temporal sequencing, short-term memory, etc.). This level of sensitivity clouds the identification of specific linguistic deficiencies and reduces the test's ability to classify aphasic disturbances" (p. 1284). Until additional data indicate otherwise, Hammeke's reservation appears justified.

The authors of the RTT are to be commended for their efforts to correct the psychometric weaknesses inherent in earlier versions of the Token Test. They are correct in their assertions that an instrument not standardized properly relative to test items and administration procedures or with unproven reliability may have some clinical utility but cannot serve as an appropriate instrument for research. The technical manual is well written and includes specific information on validity, reliability, standardization of test materials, administration and scoring. It also contains norms, in both raw score and percentile units, for several reference groups. The tables and figures, providing descriptive and statistical data, document the textual material and should prove helpful to both clinicians and researchers. The directions for administration and scoring are generally clear. However, some remaining deficiencies require correcting before the RTT can attain its maximum usefulness. These deficiencies relate specifically to reliability, validity, ease of administration and scoring, and possibly to the prescribed method of token presentation to the examinee.

Although the test authors report high test-retest, intrascorer, and interscorer reliability coefficients, the correlations are based on extremely small samples, making generalizations to other samples or the general population questionable. Additional reliability studies with large sample sizes are needed but have not appeared in the literature yet.

Numerous studies on both children and adult subjects have supported various versions of the Token Test as valid measures of language comprehension (de Simoni & Mucha, 1982; Cole & Fewell, 1983; Whitehouse, 1983; Noll & Randolph, 1978; Coupar, 1976), though others have not found such a clear-cut relationship (Fusilier & Lass, 1984; Lesser, 1976). The authors of the RTT report only one concurrent validity study but argue that the similarity of their test to previous versions of the Token Test is sufficient to insure validity for the RTT also. This may be true, but additional validity studies specific to the RTT are needed nevertheless.

Although previous versions of the Token Test were known for their ease of administration and scoring (Lezak, 1983), the same may not be true for the RTT. Even though the administration manual for the RTT is written clearly, scoring decisions, which must be made at the time of administration, are quite complex and require supervised training sessions to ensure interscorer reliability. The authors of the RTT recommend the use of teaching tapes and attendance at sponsored workshops to develop such proficiency. Because the amount of energy expended in obtaining data must, for practical reasons, always be balanced by the value of the information obtained, it is possible, as Lezak suggests, that not enough additional clinical information may be forthcoming from the RTT to make the extra effort worthwhile. In other words, clinicians may opt for one of the earlier, less time-consuming, versions of the Token Test.

The 15-category scoring system of the RTT, presented by its authors as an ordinal measure and treated statistically (via the calculation of arithmetic means) as an equal-interval measure, is open for criticism. Until both the ordinal nature of the categories (e.g., Does category 13 really represent greater facility with auditory comprehension or less dysphasia than categories with lower numerical values?) and the equal interval nature (e.g., Is the distance, for example, between categories 10 and 12 equal to the distance between categories 4 and 6?) are established clearly, it might be wise to treat the data obtained from the scoring system qualitatively rather than quantitatively. Similar concerns were voiced by previous reviewers of the RTT (Franzen & Golden, 1985; Hammeke, 1985).

The specific token layout for each of the RTT subtests is prescribed clearly in the technical manual. However, it is not stated whether the examiner should use a shield while arranging the tokens in front of the examinee. Because no specific directions are provided, it is assumed that the layout may be unshielded and that the order of layout (right to left or vice versa) is left to the discretion of the examiner. Because Abkarian (1984) found that examinees performed differently under masked (shielded) and unmasked (unshielded) administrations, it might be advisable in future revisions to specify a set procedure for token layout.

In summary, the Token Test, now available in numerous versions, is well established as a test of language comprehension for both children and adults. Its successor, the RTT, is a test with unique features that may provide clinical data and research possibilities beyond that available from the earlier versions. The formal training required to master the cumbersome scoring system of the RTT may limit its use, however.

References

Abkarian, G. G. (1984). Procedural influences on auditory comprehension: A second look at the Revised Token Test. *Journal of Communication Disorders, 17,* 101-108.

Berry, W. R. (1973). *A psychometric reconsideration of the Token Test.* Paper presented at the Third Conference on Clinical Aphasiology, Albuquerque, NM.

Cole, K. N., & Fewell, R. R. (1983). A quick language screening test for young children: The Token Test. *Journal of Psychoeducational Assessment, 1,* 143-153.

Coupar, A. M. (1976). Detection of mild aphasia: A study using the Token Test. *British Journal of Medical Psychology, 49,* 141-144.

De Renzi, E., & Faglioni, P. (1978). Normative data and screening power of a shortened version of the Token Test. *Cortex, 14,* 41-49.

De Renzi, E., & Vignolo, L. A. (1962). The Token Test: A sensitive test to detect receptive disturbances in aphasia. *Brain, 85,* 665-678.

de Simoni, F. G., & Mucha, R. (1982). Use of the Token Test for Children to identify language deficits in preschool age children. *Journal of Auditory Research, 22,* 265-270.

Franzen, M. D., & Golden, C. J. (1985). Review of Revised Token Test. In J. V. Mitchell, Jr. (Ed.), *The ninth mental measurements yearbook* (pp. 1282-1283). Lincoln, NE: Buros Institute of Mental Measurements.

Fusilier, F. M., & Lass, N. J. (1984). A comparative study of children's performance on the Illinois Test of Psycholinguistic Abilities and the Token Test. *Journal of Auditory Research, 24,* 9-16.

Hammeke, T. (1985). Review of Revised Token Test. In J. V. Mitchell, Jr. (Ed.), *The ninth mental measurements yearbook* (pp. 1283-1284). Lincoln, NE: Buros Institute of Mental Measurements.

Lesser, R. (1976). Verbal and non-verbal memory components in the Token Test. *Neuropsychologia, 14,* 79-85.

Lezak, M. D. (1983). *Neuropsychological assessment* (2nd ed.). New York: Oxford University Press.

McNeil, M. R., & Prescott, T. E. (1978a). *Revised Token Test* (manual). Austin, TE: PRO-ED.

McNeil, M. R., & Prescott, T. E. (1978b). *Revised Token Test*. Baltimore, MD: University Park Press.

Morley, G. K., Lundgren, S., & Haxby, J. (1979). Comparison and clinical applicability of auditory comprehension scores on the Behavioral Neurology Deficit Evaluation, Boston Diagnostic Aphasia Examination, Porch Index of Communicative Ability and Token Test. *Journal of Clinical Neuropsychology, 1,* 249-258.

Noll, J. D., & Randolph, S. R. (1978). Auditory semantic, syntactic, and retention errors made by aphasic subjects on the Token Test. *Journal of Communication Disorders, 11,* 543-553.

Poeck, K., & Pietron, H. P. (1981). The influence of stretched speech presentation on Token Test performance of aphasic and right brain damaged patients. *Neuropsychologia, 19,* 133-136.

Porch, B. E. (1967). *Porch Index of Communicative Ability.* Palo Alto, CA: Consulting Psychologists Press.

Silverman, I., Raskin, L. M., Davidson, J. L., & Bloom, A. S. (1977). Relationships among Token Test, age, and WISC scores for children with learning problems. *Journal of Learning Disabilities, 10,* 104-107.

Van Dongen, H. R., & Van Harskamp, F. (1972). The Token Test: A preliminary evaluation of a method to detect aphasia. *Psychiatria, Neurologia, Neurochirurgia, 75,* 129-134.

Wertz, R. T. (1979). The Token Test (TT) (Review). In F. L. Darley (Ed.), *Evaluation of appraisal techniques in speech and language pathology* (pp. 238-242). Reading, MA: Addison-Wesley.

Whitehouse, C. C. (1983). Token Test performances by dyslexic adolescents. *Brain & Language, 18,* 224-235.

Leslie J. Fyans, Jr., Ph.D.
*Research Design and Psychometric Testing, Department of Planning,
Research, and Evaluation, Illinois State Board of Education, Springfield,
Illinois.*

ROGERS CRIMINAL RESPONSIBILITY ASSESSMENT SCALES

*Richard Rogers. Odessa, Florida: Psychological Assessment
Resources.*

Introduction

The Rogers Criminal Responsibility Assessment Scales (R-CRAS) were developed by Richard Rogers to meet the needs of the psychologist involved in a forensic practice within the criminal court system. Prior to the publication of the R-CRAS, perhaps the only empirically based criminal responsibility scales was the Mental State At The Time of Offense Screening Examination (Slobogin, Melton, & Showalter, 1984). As professionals in forensic practice realize, the construct of legal insanity is, at best, poorly linked to the knowledge we have of human behavior. The legal profession has developed at least four different "standards" of legal insanity: the M'Naghten standard (Keilitz & Fulton, 1983), the Irresistible Impulse standard (Goldstein, 1967), the American Law Institute standard (Brooks, 1974), and the Guilty But Mentally Ill standard (Robey, 1978). Each of these standards was designed from a particular historical background, with emphasis on particular components of the law. The R-CRAS was designed by using gradations of severity to standardize the evaluation of clinical information, such that the ambiguity of the decision on criminal responsibility is minimized.

Initially, the R-CRAS focus was primarily for the American Law Institute standard (ALI; the standard used by 24 states), which holds that "A prisoner is not responsible for criminal conduct, if, as a result of mental defect or disease, he lacks substantial capacity either to appreciate the criminality (wrongdoing) of his conduct or to conform his conduct to the requirements of the law" (Weiner, 1980; Rogers, Wasyliw, & Cavanaugh, 1984). However, recent work by Rogers, Seman, and Clark (in press) extends the R-CRAS toward the M'Naghten standard (used in 16 states) and the Guilty But Mentally Ill standard (GBMI; used in 7 states).

The R-CRAS is composed of 30 items (or indicators) designed to measure the psychological and contextual indices surrounding a potential decision regarding insanity. Each of these items is scaled in gradation against the anchor of increasing severity (e.g., 0 for no information; 1 for not present or not applicable; 2 for clinically insignificant; and 3 through 6 for increasing levels of severity). These 30 items are organized into five different subscales:

Patient Reliability, which contains items tapping the reliability and honesty of the client's self-report and involuntary interference with that self-report;

Organicity, which includes indicators of mental retardation, brain damage or disease, intoxification, and the relationship of these to criminality;

Psychopathology, which contains items assessing symptoms of severe affective and cognitive pathologies (e.g., delusions, hallucinations, verbal incoherence, etc.) as well as anxiety and amnesia;

Cognitive Control, presenting items that examine client awareness of the criminality of the act, planning and preparation for the act, as well as any disturbance in thought processes concerning the act; and

Behavioral Control, which has items that assess the magnitude of the client's bizarre behavior, responsible behavior, and reported and estimated self-control.

The R-CRAS also provides methods of aggregating the scores of the items into certain patterns and then linking those patterns to the decision models for each of the three legal standards of ALI, M'Naghten, or GBMI. The manual contains decision flow charts for using the R-CRAS in each of these standards and guiding the interpretation associated with the results.

Practical Applications/Uses

The use of the Rogers Criminal Responsibility Assessment Scales is inherently limited to professionals involved in the interface between law and psychology. The practical use of the R-CRAS will require self-discipline and rigor as well as objectivity with regard to the client, the act, and the legal standards being applied. To use the R-CRAS, an extensive array of information should also be available, such as any previous and relevant psychological/psychiatric reports, police reports, recent mental status examinations, and results from psychological tests such as the MMPI. All of these could be used as resource material for the R-CRAS. Thus, the R-CRAS would not supplant previously accepted documentation in insanity trials, but would practically and systematically integrate it.

Technical Aspects

Recent work by Rogers (1986) discusses validation studies of the Rogers Criminal Responsiblity Assessment Scales conducted across a 2-year period (1981-83) at five different locations with forensic specialization. A cross-validation study based upon 111 client-defendants (Rogers, Seman, & Wasyliw, 1983) estimated the kappa reliabilities for each of the five R-CRAS subscales; the results ranged from .68 and .63 (original sample and cross-validation sample, respectively) to 1.00. Interrater reliability of decisions with regard to insanity were .93 for the original sample and 1.00 for the cross-validation sample. This research also included a discriminant analysis in terms of clients evaluated as sane and insane against the five subscales. Rogers et al. reports a 72% correct classification of sane and 99% classification of insane clients. They report nonsignificant discriminant functions on the subscales by age, sex, type of crime, inpatient vs. outpatient status, and forensic center as evidence of the generalizability of the R-CRAS. Some multivariate statisticians (e.g., M.M. Tatsuoka, personal communication) would question the use of stepwise multivariate functions. This may be due to the potential influence on, for example, Type I and Type II errors.

A 1984 study by Rogers, Wasyliw, and Cavanaugh reports results from two forensic centers on 73 clients. The sample was split into an initial sample of 25, with the remainder being the cross-validation sample. Raw mean differences were found on four of the five R-CRAS subscales. In both samples, the correct classification of sanity versus insanity by the R-CRAS ranged from 89% to 100%.

A review of the R-CRAS by Grisso (1984) indicates a weakness in terms of a lack of factor analysis of its items and subscales. Subsequent research published by Rogers, Seman, and Clark (1985) reports factor analytic results based upon 125 clients. Six factors were identified, accounting for a cumulative 60.6% of the variance: 1) psychotic symptoms/loss of control (29.7%), 2) organic impairment (8.3%), 3) intoxication/amnesia (6.7%), 4) affective response (6.0%), 5) additional organic (5.1%), and 6) patient reliability (4.8%). The first factor, comprising approximately one half of the total variance, dealt with impaired judgment, lack of control, lack of awareness of criminality, and bizarre, psychotic, and disorganized behavior. This research also reported discriminant analyses for levels of correct classifications of sanity and insanity based upon these five subscales, where classification was based on GBMI and M'Naghten legal standards. Other considerations with regard to the development and nature of the R-CRAS are given by Rogers (1986).

Critique

The Rogers Criminal Responsibility Assessment Scales are a major step forward in forensic practice toward systematic and standardized assessment of criminal responsibility. They provide an insightful procedure for examination that is sensitive to many of the potent psychological variables and legal standards that are paramount in these cases. They do so in a thorough, yet lucid, framework, and appear grounded in theory and legal precedent.

However, the ease with which the R-CRAS communicates may exceed its own technical nature and its user group. In terms of its own technical nature, it would be assuring to have factor analytic results based upon larger pools of clients. Six factors appear far too many for a simple structure with 25 items. Unfortunately, none of the factor analytic coefficients are presented; thus, one does not know how "pure" or "strong" these factors are.

If these factors are distinct, one also might question how the discriminant analyses might differ if they were based on factor scores on each of these empirically derived factors rather than the theoretically derived five subscales. Perhaps future factor analyses might lead to reorganizations of the R-CRAS itself. It might be suggested that these studies employ oblique rotations due to the correlations among these variables. (Unfortunately, the correlation matrix among these 25 items also was not presented.)

There also may be too much emphasis placed upon the discriminant analyses. In actual fact, with only two groups (sane vs. insane), this discriminant analysis dissolves into a Hotellings T^2 analysis. One might well expect highly favorable classification rates with only two groups (sane, insane), based on Ns less than 100 and with 25 items. One would need assurance that these relatively high classification rates were not solely based upon capitalization of chance. Furthermore, the strength of the discriminant function is based on the distance between the group means and the magnitude and pattern of the discriminant coefficients. These group means on the discriminant functions and the discriminant coefficients are not presented in any publication mentioned here.

It may be more useful to conduct general linear model ANOVAS on each subscale by groups of sane and insane subjects and incorporate analyses of false positive and false negative classifications based upon considerations of the significance of the results, the power of the tests, and so on. Such works as Cohen (1974) would be guides here. The multiple *t*-test approach should be avoided because the numerous comparisons could lead to chance significance (i.e., Galtons problem). Unfortunately, multiple *t*s are presented in the author's literature. However, the real issue is that with only two categories of classification and many predictors, the likelihood of significant classification is probably increased. Furthermore, generalizability should be analyzed through methods designed for generalizability. This literature has been in existence since Cronbach, Gleser, Nanda, and Rajaratnam (1972) and Fyans (1983). While other techniques such as the Hotellings T^2 as used by Rogers may shed light on generalizability, they are perhaps less definitive.

Finally, there may be a sufficient N to conduct real generalizability analyses (cf., Fyans, 1983; Brennan, 1983). It is a questionable use of a nonsignificant MANOVA to prove up generalizability when the N is so meager. Traditional generalizability analyses would not be as sensitive to these issues and might also allow for several independent variables (sex, location, etc.) and their interactions to be tested in one analysis.

The final flaw of the R-CRAS may be external to it. That is, forensic practice appears to be a quickly growing and interesting domain for many professionals. It may well be that professionals utilizing the R-CRAS should be required to participate in workshops, leading to extensive knowledge of the legal issues/standards and of the relationship of present research and human behavior theory to address those legal issues. Rogers (1986) raises similar concerns. Otherwise, the R-CRAS may be too facilitative, and, employed in unskilled or insensitive hands, could be damaging to the client, the judicial process, and the profession of psychology.

Reviewer's Note: Additional factor analyses of the R-CRAS were conducted by Rogers during August and September of 1987. Rogers (personal communication) suggests that a varimax rotation found five factors that "make sense on a conceptual basis." He titles and defines the five factors in language very similar to his previous work:

Factor I, *Psychopathology—Cognitive Control*, explaining 28.4% of the variance and consisting of items 8, 11, 12, 15, 19, 22, and 24;

Factor II, *Affective/Symptoms—Behavioral Control*, explaining 10.2% of the variance and consisting of items 9, 13, 16, 21, 23, and 24;

Factor III, *Organicity*, explaining 7.9% of the variance and consisting of items 4, 6, and 7;

Factor IV, *Substance Abuse Indices*, explaining 6.8% of the variance and consisting of items 3, 10, 18, and 20; and

Factor V, *Dissimulation*, explaining 5.8% of the variance and consisting of items 1, 23, and negative loading of items 14 and 18.

However, as can be seen by the third factor, less then one third of the variance is accounted for than by the first factor. Perhaps the R-CRAS strongly measures one dimension reliably (i.e., Psychopathology—Cognitive Control), at least from a psychometric view point.

Furthermore, the foregoing dimension analytic evidence is based on a varimax rotation, which presupposes uncorrelated factors and an orthogonal world view (which is not even accepted by Rogers [1987]). Thus, it is an oblique rotation, allowing for the items to correlate empirically to the level that they "may not represent independent factors" and "combine elements of different scales."

An evaluation of the loading from the reference vectors of the oblique rotation indicate a far greater scatter of items to factors. For instance, Factor I now consists of items 13, 19, 22, and 23; Factor II, items 17 and 18; Factor III, items 8, 20, and 24; Factor IV, items 9, 14, and 21; and Factor V, items 2, 6, 12, and 13. Except for Factor III (which might be labeled Behaviorial Control) and Factor IV (which might be labeled an Activation level), this empirical factor structure does not match the ones as published by Rogers (Rogers, Wasyliw, & Cavanaugh, 1984, p. 298).

Rogers (personal communication) argues that the R-CRAS was developed from a "rational theoretic" standpoint to meet the needs of statutory concerns, and thus puts less weight on the empirical results. However, the role of the R-CRAS was to assist in objectifying the decision process and perhaps thus respond to critics such as Ziskin (1981a, 1981b, 1983), who argues against the subjectivity of psychiatric or psychological legal testimony. Thus, one is left in a relative quandary regarding the construct validity of the R-CRAS. It either is meant empirically to objectify a previous subjective decision process, or it is not; it is either composed of independent factors, or it is not. Unfortunately, at this point, the construct validity data, at least from the standpoint of factor analysis, remain unresolved.

References

Brennan, R. (1983). *Element of generalized ability theory.* Iowa City: American College Testing Program.

Brooks, A. D. (1980). *Law, psychiatry, and the mental health system.* Boston: Little, Brown.

Cohen J., & Cohen, P. (1983). *Applied multiple regression/correlation analysis for the behavioral sciences.* Hillsdale, NJ: Lawrence Erlbaum.

Cronbach, L. J., Gleser, G., Nanda, H., & Rajaratnam, N. (1972). *The dependability of behavioral measurements.* New York: John Wiley.

Fyans, L. J., Jr. (1983). *Generalized ability theory.* San Francisco: Jossey-Bass.

Goldstein, A. S. (1967). *The insanity defense.* New Haven: Yale University Press.

Grisso, T. (1984). *Review of Rogers Criminal Responsibility Assessment Scales* (Unpublished draft, Grant MH-37321). Washington DC: National Institute of Mental Health.

Keilitz, I., & Fulton, J. P. (1983). *The insanity defense and its alternatives: A guide to policy makers.* Williamsburg, VA: National Center for State Courts.

Robey, A. (1978). Guilty but mentally ill. *Bulletin of American Academy of Psychiatry, 6,* 374-381.

Rogers, R. (1986). *Conducting Insanity Evaluations.* New York: Van Nostrand.

Rogers, R., Seman, W., & Clark, C. C. (in press). Assessment of criminal responsibility: Initial validation of R-CRAS with the M'Naghten and GBMI standards. *International Journal of Law & Psychiatry.*

Rogers, R., Wasyliw, O. E., & Cavanaugh, J. L. (1984). Evaluating insanity: A study of constructive validity. *Law & Human Relations, 8,* 293-303.

Slobogin, C., Melton, G. B., & Showalter, C. R. (1984). The feasibility of a brief evaluation of mental state at the time of the offense. *Law & Human Behavior, 8,* 305-320.

Spitzez, R. G., & Endicott, J. (1978). *Schedule of Affective Disorders and Schizophrenia.* New York: Biometrics Research.

Tatsuoka, M. M. (1971). *Multivariate analysis.* New York: John Wiley.

Weiner, B. A. (1980). Not guilty by reason of insanity: A sane approach. *Chicago Kent Law Review, 56,* 1057-1085.

Ziskin, J. (1981a). *Coping with psychiatric and psychological testimony* (Vol. I). Beverly Hills: Law & Psychology Press.

Ziskin, J. (1981b). *Coping with psychiatric and psychological testimony* (Vol. II). Beverly Hills: Law & Psychology Press.

Ziskin, J. (1983). *Coping with psychiatric and psychological testimony* (3rd ed.). Beverly Hills: Law & Psychology Press.

David P. Lindeman, Ph.D.

Director, Parsons Regional Early Intervention Program, and Research Associate, University of Kansas, Bureau of Child Research at Parsons Research Center, Parsons, Kansas.

SEED DEVELOPMENTAL PROFILES

Karin Schuelke, Beverly Harris, Marcia Blum, Pam Schenfeld, Bambi Dixey, Kathy Molinski, Barbara Sanborn, Genny Rosenburg, Judy Nelson, Francie Sinton, and Darcy Bauer. Denver, Colorado: Sewall Rehabilitation Center.

Introduction

The Sewall Early Education and Development Program (SEED) Developmental Profiles, published by the Sewall Rehabilitation Center, are an informal, criterion-referenced assessment tool (K. Schuelke, personal communication, May 21, 1987) designed to assess the skills of children between the ages of 4 weeks and 6 years of age. These assessment profiles were developed specifically to provide a functional appraisal of a handicapped or "at risk" child's social and emotional, gross-motor, fine-motor, adaptive and reasoning, speech and language, and self-help skills. Following the administration of this assessment, the results can be used as a standard from which to assess a child's development, as the basis for the development of an individual education plan (IEP), or as a criterion-referenced measure to assess a child's eduational process.

The present version (Schuelke et al., 1985) of the SEED Developmental Profiles is a revised form of the original assessment developed in 1976 by Joan Herst, M.S.; Shelia Wolfe, O.T.R.; Gloria Jorgenson, R.P.T.; and Sandra Pallen, M.A., C.C.C.-S.P. Both versions of the assessment were developed with input from a team of specialists in the areas of speech and language, occupational therapy, physical therapy, child development, and special education. The 1976 form was developed for use as a screening instrument (Matson, 1985) under a contract between the Sewall Rehabilitation Center in Denver, Colorado, and the U.S. Office of Education's Bureau of Education for the Handicapped (Algozzine, 1985).

The profiles originally were developed in 1976 in response to the authors' review of existing evaluation tools, which indicated a lack of accurate and detailed assessment instruments for informal functional assessment of young children. Furthermore, the authors felt that due to their structured nature, standardized tests "do not fairly assess the handicapped child's ability" and that they often penalize a child by setting specific time limits or requiring particular types of responses. The authors' disillusionment with existing assessment instruments resulted in the development of the SEED Developmental Profiles. A number of items selected from a variety of standardized tests were consolidated in form, format, and content to provide a means for procuring a detailed, realistic, and complete picture of the

469

child's abilities. The 1985 version of the profiles 1) reflects revisions made in the major sources from which the SEED is a consolidation, 2) suggests more extensive uses for the instrument, and 3) expands the upper end of the profiles to represent skills in the developmental age range of 48 to 72 months.

The testing materials, which are purchased directly from the Sewall Rehabilitation Center, can be acquired in two formats. Both formats are contained in spiral-bound booklets that are easy to manipulate during test administration. The first format is a complete testing booklet that includes an introduction to the assessment and background information; lists of materials needed during administration; the actual profiles or testing items; a composite graph and individual subtest graphs to plot initial testing results, as well as developmental progress over time; and the source list of tests from which the SEED was developed. The second format simply provides the profiles (test items) and graphs that constitute the necessary materials for test administration and a record form for an individual child.

In its present form, the SEED Developmental Profiles contain 860 items that are divided into six subtests: 1) Social/Emotional (133 items), 2) Gross Motor (155 items), 3) Fine Motor (101 items), 4) Adaptive/Reasoning (204 items), 5) Speech/Language (154 items), and 6) Self-Help (113 items). The six subtests, which are divided into skill areas, provide the examiner with information regarding the child's developmental level in each particular area or domain. Table 1 presents the subtests and skill areas with specific information concerning the number and grouping of items and the age ranges for each area. Test items within each skill area are grouped according to developmental age levels. Items below the 1-year age level are grouped in 4-week intervals. Items between the 1- and 2-year level are represented in 3-month intervals, and items between the 2- and 6-year level are grouped in 6-month intervals. There is some variation as to the exact age levels at which items for the various skill areas appear and, consequently, in the number of items for each skill area. Nevertheless, when the results of the subtests are combined, the examiner is presented with a profile of the child's overall development.

Each subtest contains an introduction that outlines the basic rationale for the subtest, factors for consideration during the subtest's administration and scoring, and a list of any additional materials needed to complete the subtest. The first subtest is the Social and Emotional Developmental Profile. The authors point out that due to the nature of social/emotional development, careful observation of the child is required. In addition, these skills often can not be assessed accurately during the initial evaluation. However, specific suggestions are made in reference to behaviors on which to focus during the evaluation. Among these are 1) manner of separation from parents, 2) reaction to strange people or environments, 3) degree of independence, 4) interest in the surroundings and play materials, and 5) general affect, such as demonstrated degree of trust, level of activity or restlessness, and length of attention span.

The second profile, Gross Motor, assesses the development of large-muscle movement and responses to changes in body position. The rationale guiding this subtest maintains that general development progresses from gross, uncoordinated reflexive patterns to more refined and controlled movement. Additionally, it is suggested that this assessment not take the place of an in-depth evaluation in areas such as reflex development, muscle tone, ataxia, or athetosis. The critical factor to

Table 1

Relation of Subtests, Items, and Age of the SEED Developmental Profiles

Subtests and Skill Areas	No. of Age Levels Assessed	No. of Items	Age Range
Social/Emotional			
Socialization	25	54	4 weeks—72 months
Differentiation of Self & Others	15	24	8 weeks—66 months
Play	24	55	12 weeks—72 months
Gross Motor			
Body Responses	11	24	4 weeks—10 months
Head Control	6	16	4 weeks—5.5 months
Sits	9	13	4 weeks—10 months
Creeps	7	9	4 weeks—8 months
Stands	19	27	4 weeks—72 months
Walks	14	26	40 weeks—72 months
Developmental Skills	12	40	52 weeks—72 months
Fine Motor			
Grasp—Prehension Patterns	24	68	4 weeks—72 months
Release Patterns	13	33	28 weeks—60 months
Adaptive/Reasoning			
Visual Regard & Pursuit	16	31	4 weeks—30 months
Activity with Rattle/Bell	9	21	4 weeks—9.2 months
Activity with Blocks/Cup	23	48	12 weeks—72 months
Developmental Concepts	19	44	32 weeks—72 months
Activity with Crayon	13	32	48 weeks—72 months
Activity with Formboard/ Puzzle	13	28	44 weeks—72 months
Speech/Language			
Auditory Awareness	4	8	4 weeks—3.7 months
Gestures	9	11	4 weeks—15 months
Receptive Language	19	43	28 weeks—72 months
Expressive Language	25	67	4 weeks—72 months
Auditory Response & Memory	11	25	18 weeks—72 months
Self-Help			
Feeding Behaviors & Skills	21	55	4 weeks—72 months
Toileting & Hygiene	12	21	4 weeks—60 months
Dressing Skills	15	37	4 weeks—72 months

be considered during administration of this subtest is that a child's performance is not simply a matter of pass/fail. Quality of performance (i.e., the amount of labor the child exerts in completing the task or the appearance of an excess amount of effort) is important.

The third subtest, Fine Motor, focuses on the development of the muscles in the arms and hands. The authors stress that the importance of fine-motor skill "cannot be understated" (Schuelke et al., 1985, p.23). This profile is designed to delineate the developmental progress of highly complex skills. Specifically, the areas assessed are reaching, grasping, prehension, and manipulation of objects. As with the gross-motor subtest, the examiner is cautioned in regard to the performance of the child. Again, quality, rather than performance or nonperformance, is at issue. Other cautions, such as influences of visual acuity, visual-perceptual ability, and observation of tremors or uncoordinated movements are mentioned.

Adaptive/Reasoning, the fourth subtest of the assessment, attempts to tap the child's ability to interact with his or her environment. Again, items are categorized along a developmental continuum and examine the adaptation and organizational skills of the child. Many of the items appear to correlate with items in the fine-motor subtest but are differentiated based on the examiner's focus on the child's ability to solve the problem, rather than on quality of movement. As with the previous profiles, the examiner is cautioned that many other factors may affect the child's performance. For example, learning style, affective development, response to sensory input, and the ability to follow verbal directions may be influencing factors. Furthermore, the examiner is cautioned to assess carefully his or her attitude during the administration of this profile. The author suggests that few tasks have a right or wrong result and that, if given the opportunity, a child may solve or adapt to a problem in a unique fashion. Therefore, a child should be provided ample opportunity to make adjustments to the problem or situation.

The fifth profile, Speech/Language, measures speech and language development. The manual points out that language skills are not simply a matter of speech production, but also involve many other skills, such as receiving input and integrating that information to produce an output, which may take the form of an action, gesture, or more formal system such as speech. This profile is divided into the areas of 1) auditory awareness, 2) gestures, 3) receptive language, 4) expressive language, and 5) auditory response/memory. The auditory awareness section is designed to isolate skills ranging from reactions to sound to comprehension of language. The skills clustered in the gesture category assess nonverbal speech, as well as social interaction with others in the environment. Receptive language is defined as the child's ability to understand and respond to input from the environment; expressive language is output in the form of sounds, gestures, and real words.

The final subtest is Self-Help. Self-help skills reflect the child's ability to engage in self-maintenance activities. The manual suggests that this profile may be affected greatly by the motor skills of the child and the expectations of the family; that is, by the level at which the parents expect the child to assist or demonstrate the skills. Futhermore, the manual cautions that the social and cultural background of the family can influence the results. The profile begins with feeding behaviors and proceeds to hygiene, toileting, and dressing behaviors. Because some items on this subtest may be difficult to observe in a testing situation unless there is ample opportunity to observe the child over a period of time, and only then if significant rapport has been established, many of the items may need to be scored by parental report. As with the other subtests, quality of performance is highlighted.

Practical Applications/Uses

The SEED was designed in such a fashion that with proper review, knowledge of the profiles, and practice it can be administered by any professional or paraprofessional who has experience working with young, handicapped children. Test items are completed by direct observation of the child or, in some cases, by the reports of parents or significant others. The manual suggests strongly that appropriate specialists (i.e., physical therapists, speech clinicians, occupational therapists, etc.) be consulted during the analysis of the results and during the development of the individualized educational plan.

Although the subtests in the profiles may be administered independently of each other, the manual advises that the entire set of subtests be administered in order to achieve a complete picture of the child's strengths and weaknesses. Because all areas of the assessment are interrelated—some of the items are tied closely to one another and items across subtests or areas may be assessed concurrently—the subtests are not administered in any specific sequence. For example, a ball can be used to test a number of skills, among them to 1) elicit the name of the item, 2) measure social interaction, and 3) check understanding of directions. This interrelatedness again highlights the importance of the examiner's familiarity with all of the profiles in order to insure a composite of the child's development and ease of test administration.

Guidelines for administration of the assessment, room arrangements, method of assessment, charting of student performance, and reporting of the results are discussed briefly in the manual. Administration of the assessment should be conducted in a room in which the child is comfortable. It should have adequate space for performing the gross-motor tasks and a table for the fine-motor and adaptive tasks. Scoring of test items and graphing of student performance is done by hand directly in the testing booklet. Items, which may be scored as they are presented, simply are marked with a " + " when passed and a "-" when failed. For many items, criteria for the performance to be observed are often unclear and left to the subjective judgment of the examiner. Also, as points of reference, the examiner is prompted to write comments in the booklet as to the quality of the child's performance.

Unlike many skills assessments, no information is provided regarding an appropriate place at which to begin or end the administration of test items. Generally, an appropriate starting point would be associated with the child's chronological age or estimated developmental age based upon his or her primary handicap, with the test terminated after the child fails a number of items. Although basal and ceiling levels are described in general terms, they do not appear to enter into any decision as to the beginning or ending point for item administration. The manual simply suggests that "all items in a specific profile should be administered" (Schuelke et al., 1985, p.2).

The manual does present guidelines for charting the performance of the child and suggestions for presenting the results of the assessment in narrative as well as graphic form. The authors recommend that, in the narrative, terms such as *gaps* and *splinter skills* be used to comment on skills as they may skew the basal and ceiling levels across and within the subtests. A *gap* represents a skill that is missed in a category, although the child completes higher-level items within the same category; *splinter skills* are those wherein several items in a row are missed with the exception of one at a relatively advanced developmental level. Other than this guidance, little information is provided in terms of SEED interpretation.

This assessment is administered individually. Because there is an extremely large number of items to be administered and because the child's age or handicap could make him or her difficult to evaluate, testing may be labor intensive. There are no guidelines or estimates provided regarding administration time. Because many factors might affect the length of each subtest's administration, the overall administration time may be quite variable. Among these factors would be 1) the examiner's familiarity with the test items, 2) his or her ability to establish rapport with the child, 3) the child's age and type of handicap, 4) the child's attention span, and 5) the span or number of items assessed in each skill area.

Technical Aspects

There are many criteria against which the utility and technical soundness of an assessment instrument can be judged. Among these are its purpose, design, development procedures, and function in the assessment process (i.e., a screening instrument or a diagnostic instrument). The criteria on which tests commonly are judged are the rigors they undergo during development, as well as on information presented in the manual or elsewhere supporting the development process. Information of this type generally adds credibility to statements regarding interpretation of results and intended uses of the instrument.

The SEED Profiles are intended to be a functional assessment instrument; that is, one in which a child's performance is examined in reference to the skills he or she demonstrates in relation to common developmental milestones. Although the authors intended to create an informal measurement tool, they still are required to present a number of psychometric elements that are vital to the assurance of the test's utility. No criteria for inclusion of items, reliability measures, validity of the items, norms, or standardization procedures are reported in the manual. Furthermore, no information as to any attempt to validate the age levels at which items are included is given. It only can be assumed that the age levels for the items included in the SEED Profiles reflect the age levels of the test(s) from which the items were selected. As such, the test does not meet sufficient criteria for the development of an assessment instrument because it relies solely on the expertise of professionals to determine the inclusion of test items. In addition, omission of these elements leads one to question critically the procedures, standards, and practices during the development of the instrument.

Critique

The SEED Developmental Profiles were designed to provide a comprehensive assessment of the strengths and weaknesses of young handicapped children in various skill domains in order to develop individualized educational programs. Considerable time and effort would be required to become totally familiar with and proficient at administration of the profiles. Nevertheless, if one made the investment to prepare adequately, a working knowledge of developmental markers or skills that should be demonstrated by a child could be acquired. Furthermore, if a user followed the guidelines of administering all the items in a specific profile, a substantial amount of time would be spent with a child and knowledge in reference to a child's adaptive behavior should be gained. Such knowledge could provide a point of reference for beginning program development or for decisions regarding other types of assessment that might be required.

Although the profiles are comprehensive in their approach to the assessment of a child's development, a number of problems, or areas of concern, remain. The first is the sparse nature in which information is provided to the prospective test administrator. This is of special concern because the test is designed to be administered by paraprofessionals. Although persons at this level may have extensive knowledge and experience in working with handicapped children, their expertise at test administration may be limited. It is quite troublesome that the information and guidance provided by the manual often leaves the user to rely upon prior experience or knowledge of the testing process. Clearer descriptions and examples containing more details of general administration procedures and use of materials are needed.

Second, detailed information is needed to provide stringent criteria for the administration and scoring of test items. Many items need clearly stated methods of presentation and criteria for performance to avoid misuse or misinterpretation of results. For example, it is questionable that intra- or interrater reliabilty would be consistent for items such as "understands taking turns," "aware of strange situations," "regards blocks and cup," or "drops toy on purpose." Unobservable concepts such as understanding, awareness, regard, and purposefulness are at best vague and pose extremely difficult decisions in determining a child's performance. Clearly stated criteria for performace, such as "drops toy when requested" or "orients head and visually gazes at block," would be of greater benefit to the user.

A final concern involves the nature of the test. Although the test is designed to be an informal evaluation, the user should be assured that the instrument has the basic psychometric properties of reliability and validity. Because there is no evidence of the availability of these technical data, the usefulness and generality of the assessment results could be questioned.

References

Algozzine, B. (1985). Review of SEED Developmental profiles. In J.V. Mitchell, Jr. (Ed.), *The ninth mental measurements yearbook* (pp. 1337-1338). Lincoln, NE: Buros Institute of Mental Measurements.

Matson, J. L. (1985). Review of SEED Developmental Profiles. In J.V. Mitchell, Jr. (Ed.), *The ninth mental measurements yearbook* (p. 1338). Lincoln, NE: Buros Institute of Mental Measurements.

Schuelke, K., Harris, B., Blum, M., Schenfeld, P., Dixey, B., Molinski, K., Sanborn, B., Rosenburg, G., Nelson, J., Sinton, F., & Bauer, D. (1985). *S.E.E.D. Developmental Profiles.* Denver: Sewall Rehabilitation Center.

Ronald C. Pearlman, Ph. D., CCC-A/Sp
Graduate Associate Professor of Audiology, School of Communication, Howard University, Washington, D.C.

A SCREENING TEST FOR AUDITORY PROCESSING DISORDERS

Robert W. Keith. San Antonio, Texas: The Psychological Corporation.

Introduction

A Screening Test for Auditory Processing Disorders (SCAN) is used to detect dysfunction of the central nervous system in processing auditory information. Children with auditory maturation delays and difficulties in perceiving speech should perform poorly on the SCAN. Even children with normal intelligence and hearing sensitivity may have difficulty in learning due to a central auditory disorder. The SCAN is intended for children between the ages of 3 and 11 years.

Robert W. Keith, Ph.D., the author of the SCAN, is the Director of the Division of Audiology and Speech Pathology in the Department of Otolaryngology at the University of Cincinnati Medical Center. He is the editor of two books and the author of several papers on central auditory disorders. As Dr. Keith states in the examiner's manual, he developed the SCAN because ". . . a review of the auditory test materials available at the time of this writing finds that too many are poorly recorded and standardized. Many of them are not easily adapted for use in the child's school situation, and so testing is not generally available" (Keith, 1986, p.iii). Standardization of the SCAN began in 1985, and the test was published in 1986.

SCAN materials include a rather extensive examiner's manual providing information on development, standardization, administration, scoring, interpretation, and remedial implications of the SCAN. One audiocassette with the same stimuli on both sides is provided so that the tape can be replayed by turning the cassette over without having to rewind the tape. Several record forms, which include space for audiometric and tympanometric data, and the child's name, school, age, grade, sex, and so forth are provided also. There is also a section that questions whether English is the child's primary language. A stereo cassette player that is able to provide a signal to two "quality" headphones is required but not included with the SCAN. The second headphone allows the examiner to monitor the auditory stimuli while the child, wearing the first headphone, is responding orally to the stimuli. The original equipment used to standardize the test included a Realistic® Stereo-Mate SCP-16 cassette player and two Realistic® Nova-34 Mini Stereo Headphones.

Three subtests comprise the current version of the SCAN: 1) Filtered Words, 2) Auditory Figure Ground, and 3) Competing Words. A fourth subtest, Auditory Fusion, was included in the original test battery but was eliminated because it ". . . was consistently easy for students and . . . it did not contribute to the test"

(Keith, 1986, p. 42). The vocabulary for all three tests was selected from the Spache readability word lists and the Kindergarten Phonetically Balanced word lists. Words familiar to children in Grades 1-3 were sought.

In the Filtered Words subtest, monosyllabic words are low-pass filtered at 1000 Hz with a filter roll-off of 32 dB/octave. In this type of test, the high frequency sounds are eliminated. If the child does poorly on this subtest, he or she would be expected to have difficulty in environments where the auditory signal is distorted. Examples of distorted auditory environments include listening to poor quality audio tapes, trying to understand teachers who speak toward the chalkboard (not the class) when writing on the board, or trying to understand a teacher who has an accent or speaks rapidly (Keith, 1986, p. 32).

The Auditory Figure Ground subtest uses monosyllabic words with a multi-talker-babble background. Words are 8dB (s/n +8) above the babble. Peaks of the babble noise are compressed so that they do not vary more than 2 dB. Children who have difficulty with this subtest will do poorly when listening in a noisy environment, due to an attention-deficit disorder or an auditory processing disorder or both (Keith, 1986, p. 32).

In the Competing Words subtest, two lists of 25 monosyllabic word pairs are presented to the examinee's right and left ears at the same time. Onset times between words are matched to within 5 milliseconds. The child repeats both words, repeating the word heard in the right ear first and the left ear second for the first half of the test. For the last half of the test, the process is reversed. The child first repeats the word heard in the left ear and then the one heard in the right ear. If the child does not know the concept of right and left, he or she should be directed to hold his or her hand over the ear receiving the word that is to be repeated first. According to Keith, the Competing Words subtest indicates maturation or developmental level (Keith, 1986, p. 33).

The examiner's manual recommends that the SCAN be administered by speech/language pathologists, audiologists, and others trained to identify and design special education programs for children and learning problems.

Practical Applications/ Uses

The SCAN is meant for the child who seems to have a learning disability that has not been delineated. The child tries to do the classwork but becomes frustrated with his or her inability to keep pace with classmates. Frustration with inadequate school achievement may lead to behavior and discipline problems. One possible cause of the child's difficulty to learn may be poor interpretation of the auditory stimuli he or she constantly receives. The child may seem to have a hearing difficulty and may not follow directions. The child's auditory reception is inconsistent and, at times, he or she may seem to be in his or her own little world. When given a pure tone or simple auditory discrimination task, the child's test results are normal. He or she may have had a history of ear infections. Spelling and reading skills are usually below grade level. Often the child will need to have auditory information repeated and will frequently say "what?"

Children with multiple articulation errors may not be candidates for the SCAN. Furthermore, if a child does not have normal hearing, the SCAN should not be administered. SCAN results are invalid if the child has even a temporary hearing loss, as in the case of a middle-ear infection. The examiner's manual suggests screening the child's hearing with a pure tone audiometer at 500, 1000, 2000, and 4000 Hz prior to administering the SCAN. The manual does not state the cutoff threshold in dB needed to obtain a valid score on the SCAN. Tympanometry is recommended in addition to the pure tone test. The first page of the scoring sheet has a space for hearing and tympanometric results.

Test administration requires the examiner to be familiar with articulation substitution, omission, distortion, and development. Examples of the test scoring show the use of phonemic transcription to indicate the errors made by the child. Because a word is scored either correct or incorrect, it is the examiner's responsibility to differentiate whether a word is acknowledged correctly but repeated incorrectly by the child due to an articulation error or whether the word is totally incorrect. In a very real sense, then, proper scoring of the test relies on the auditory discrimination ability of the examiner. Unless the examiner has special training, a speech/language pathologist or audiologist are perhaps the only professionals, by training, equipped to administer the SCAN.

The test site should be a quiet room, free of distractions or disruptions. As stated earlier, a stereo cassette player and two headphones are required for the test. The volume control of the tape recorder should be set at the child's most comfortable listening level. A 1000 Hz tone of 10 seconds duration is recorded on the tape in order to set the intensity level of the test. If a VU meter is available, the tone should be set to "0" on the VU meter for the right and left channels of the tape recorder. When there is no VU meter on the tape recorder, the tone is increased by both channels until the tone sounds as if it is balanced in the center of the head for a normal hearing individual. Directions on taking the test, as well as practice items for each one of the subtests, are recorded on the tape. Children are encouraged to guess if they are not sure of an answer.

All three subtests can be administered in about 20 minutes. Children respond to the test stimuli by repeating the stimulus words. Keith believes that "an imitative auditory processing task is a pure test of auditory processing abilities because it de-emphasizes the cognitive and memory aspects of audition" (Keith, 1986, p. 2). Keith states that an oral response also avoids the cross-modality and cognitive aspects of a picture-pointing response to a word. Children are encouraged to guess if they are not sure of an answer. In the Filtered Words and Auditory Figure Ground subtests, 25 stimuli are presented to the right ear followed by 24 stimuli to the left ear. In the Competing Words subtest both of the child's ears are stimulated simultaneously with different words. The child is to repeat the word presented to the right ear first for 25 stimuli, and then the word presented to the left ear first for 25 stimuli.

If the child repeats a word correctly, the examiner circles the "+" next to the correct response. If the response is wrong, the "-" is circled. There is no partial credit given when the child says part of a word, but if a word is said in the reverse order on the Competing Words subtest, credit still is given for each word. A line is drawn through omitted words. When a substitution is made, the substituted word is written (in phonetics if necessary) next to the correct answer.

A raw score is obtained by summing the correct responses for each ear when necessary, and then adding the right and left ears for a total subtest score. All three subtests then are totaled for a composite raw score of the entire SCAN test. Total SCAN scores are expressed as normalized scores with a mean of 100 and a standard deviation of 15. Percentile ranks and age equivalents are provided for raw score conversion. The greater the deviation from 100, the more abnormal the score. Degrees of normalcy, rather than a "cutoff" score, are used to determine a subject's auditory processing problems. A score of 85 would be -1 standard deviation from the mean and requires analysis of the subtest scores. A score of 55 would be -3 standard deviations from the mean, yielding a 0.1 percentile rank, and very abnormal. Subtest scores are normalized standard scores with a mean of 10 and a standard deviation of 3.

During the test, the examiner notes the child's reaction to the instructions, attention span, conduct, speed of response, and so forth. General observations of the child's behavior are noted on the last page of the scoring sheet. A right versus left "ear advantage" is computed by computing right ear minus left ear. When the difference is a positive number, there is a right ear advantage; if negative, a left ear advantage is indicated. According to Keith, abnormal ear advantage results from poor maturation of the auditory system, reverse cerebral dominance for language, or a neurologic disorder.

Technical Aspects

Standardization of the SCAN took place in 1985. The sample consisted of 1,034 normally achieving children between the ages of 3 and 11 from across the United States. Keith warns that the small sample size of the 3rd- and 4th-grade-level children makes interpretation of the norms for those grades imprecise. Variables considered when selecting subjects included geographic region, age, sex, race, ethnic group, socioeconomic status, and community size. Keith compares various variables of the SCAN with the 1980 United States census. He believes the sample used on the SCAN is similar to the census for the total number of males and females and the number of individuals of Spanish origin. The number of blacks was about 6% greater in the SCAN sample than in the census population. About 65% of the the SCAN's sample was from communities larger than 25,000 people and of different socioeconomic status levels.

A study of the internal consistency of the SCAN was conducted looking at age groups and using Cronbach's alpha. About 41% of the subtests by age groups produced a consistency coefficient above .80. The remaining 59% of the coefficients were below .80 with most of the poorer coefficients occurring in the Filtered Words subtest, followed by the Auditory Figure Ground subtest.

Test-retest reliability studies were conducted using a subsample of the total standardization sample after a 6-month period. The sample consisted of 68 students from the first and third grades. Only test-retest reliability coefficients in the moderate range were found.

Specificity of the SCAN subtests were found to be "ample" using Kaufman's classification. Only the Auditory Figure Ground test, when used with 5-year-old subjects, showed less than "ample" specificity.

Sensitivity studies of the SCAN will require a great deal more research. Dr. Keith states that the SCAN "is too new to have complete validation studies, even in our own facility" (personal communication, June 8, 1987). He did receive information from audiologist Jane A. Rudy at the Children's Medical Center-Dayton, Ohio. Of the children Ms. Rudy tested with the SCAN, 84% of the SCAN scores agreed with a failing score on one or more sections of a total test battery consisting of the Staggered Spondaic Word Test (SSW), the Willeford Central Auditory Test Battery, and the Goldman-Fristoe-Woodcock Auditory Skills Battery.

Critique

Because the SCAN is so new (published in 1986), there have been few studies involving the test. Although the test tries to eliminate cross-modality responses, the child with poor articulation or the shy child will be difficult to analyze. If the test is to be used by professionals without a background in speech-language-pathology or audiology, an optional pointing procedure would be very helpful. As a screening procedure, the SCAN is relatively fast (compared to such tests as the SSW), easy to score, and easy to administer. The need for documentation indicating the child has normal hearing at the time of the test cannot be overemphasized.

References

Goldman, R., Fristoe, M., & Woodcock, R. (1974). *Goldman-Fristoe-Woodcock Auditory Skills Battery*. Circle Pines, MN: American Guidance Service.

Katz, J. (1968). *Staggered Spondaic Word Test*. Vancouver, WA: Precision Acoustics.

Keith, R. W. (1986). *SCAN A Screening Test for Auditory Processing Disorders*. San Antonio, TX: The Psychological Corporation.

Willeford, J. (1968). *Willeford Central Auditory Test Battery*. Ft. Collins, CO: Author.

Roger D. Carlson, Ph.D.
Associate Professor of Education, Willamette University, Salem, Oregon.

SCREENING TEST OF ACADEMIC READINESS
A. Edward Ahr. Skokie, Illinois: Priority Innovations, Inc.

Introduction

The Screening Test of Academic Readiness (STAR) is a group-administered, paper-and-pencil test used for highlighting strengths and weaknesses related to school readiness. The STAR is designed to provide a "broad preventive approach in screening preschool and kindergarten children for possible problems at an early age" (Ahr, 1966, p. 40). The test is intended for use with children ranging from 4.0-6.5 years.

The STAR was developed by A. Edward Ahr, Ed.D., who received his doctorate at Loyola University in Chicago in 1966. Dr. Ahr has served as an elementary school teacher and college professor, as well as a school psychologist. Currently, Dr. Ahr is a supervising school psychologist for Niles Township Department of Special Education in Illinois. Dr. Ahr, who paid special attention to adapting the STAR for use with educationally disadvantaged preschoolers, has developed special administration and scoring instructions for that population.

The test form is a 50-page top-bound booklet. Only one item is presented per page. Each page is a different color so that, in a group administration, the examiner can direct all children to turn to a common page.

The STAR is composed of 50 items in eight subtests: 1) Picture Vocabulary, 2) Letters, 3) Picture Completion, 4) Copying, 5) Picture Description, 6) Human Figure Drawing, 7) Relationships, and 8) Numbers. In the Picture Vocabulary subtest, the examiner reads the names of objects, and the child matches the names to pictures of the objects. The Letters subtest asks the child to identify letters by their names. The child identifies missing parts of drawn objects in the Picture Completion subtest. The Copying subtest requires the subject to copy a circle, triangle, and square, and in the Picture Description subtest, the child identifies items possessing various qualities (such as "is made of glass," "has feathers"). In the Human Figure Drawing subtest, the child draws a man. The Relationships subtest requires the child to identify relationships among and between objects. The Numbers subtest involves the identification, matching, sequencing, and counting of numbers.

Children mark the answer booklet as oral instructions are given. The test is not graded in difficulty; items of various degrees of difficulty are interspersed. A one-page record form is used to summarize scoring.

Practical Applications/Uses

The STAR can be useful to both educators and school psychologists in the initial preschool screening of children. The test is intended to serve as a screening device that will lead to further diagnosis of children for initial remediation or treatment in particular areas prior to school entry. It also can be used to identify students who are particularly strong in various areas and who are ready for early school entry. The STAR can be used for program evaluation as well. In addition, the test can help to establish selection and placement criteria within a given school district. Further, the STAR, used as a research instrument, can determine the characteristic patterns of response relating to various demographic characteristics.

By adapting the administration and scoring procedures, it is possible to use the STAR with educationally disadvantaged children. When used with these children, three scores can be derived: 1) current academic readiness level, 2) potential academic ability, and 3) experiential deficit index. The test has been administered to groups ranging in size from 18 to 72 children. When it is administered to "educationally disadvantaged" preschoolers, Ahr recommendes that groups contain no more than 15, and preferably only 10, students.

No special level of training is specified for administering the STAR, scoring the items, or interpreting the results. The examiner should be familiar with the test items before attempting to administer them, however. In order to establish rapport with examinees and make them feel secure, it is recommended that the STAR be given following an activity period. An activity period also should separate the first from the second half of the test. The manual suggests several activities for these intervals.

The examiner gives instructions orally, and children are required to either 1) mark one of four or more choices or 2) print, copy, or draw a response to the item. Although liberal time limits are given for each item, it is estimated that administration of the entire test will take no longer than 1 hour.

Hand-scoring procedures are presented clearly in the manual, in which correct answers are given along with the point values to be awarded with each. The greatest difficulties in scoring involve judgments necessary for the assignment of points to drawing items. The manual does present guidelines for scoring these subjective items, however. Bonus points are given for correctly completing designated series of items. The test author states that hand-scoring of the STAR takes no longer than 6 minutes. Machine- and computer-scoring are not available.

Scores are interpreted using a table, in the manual, that associates raw scores with chronological age (from 4.0-6.5 years). Deviation IQ equivalents that are based on a norming study of the test are given. Average raw score ranges for each subtest as associated with various chronological ages are given.

Technical Aspects

The STAR was standardized on 1,500 middle- to upper-middle class, suburban, white, preschool and kindergarten children between the ages of 4.0 and 6.5 years. According to the test author, the sample is "presently being expanded to include 'educationally disadvantaged' children" (Ahr, 1966, p. 40). The manual reports that reliability for various groups ranges from .87 to .93. In a study of 14 preschoolers using a 4-week test-retest interval, reliability was estimated at .87. Reliability was reported as .91 for a group of preschoolers ($n = 20$) using an 8-week test-retest interval. In another study, reliability for 90 preschoolers was reported as .88 using Kuder-Richardson formula 21 and as .91 using Hoyt's ANOV. A study involving

Head Start children (*n* = 137) showed reliability as .90 using Kuder-Richardson formula 21.

The author reports concurrent validity of .72 with the Stanford-Binet, Form L-M (*n* = 90 preschoolers). Predictive validity is reported at .67 with the Stanford-Binet, Form L-M (*n* = 50 preschoolers) and at .76 with the Metropolitan Readiness Tests (*n* = 391 kindergartners).

Several independent studies have been conducted assessing various characteristics of the STAR. Hutton (1970) found that STAR scores correlated significantly with 1) end of first grade reading performance, 2) reading grade point average, 3) arithmetic grade point average, and 4) passing or failing first grade. The individually administered Slosson Intelligence Test and the Sprigle School Readiness Test (Sprigle, 1965) did not correlate any higher with four academic performance criterion variables or with teacher ratings than the group-administered STAR, thus making it the most economical and efficient of the three school readiness screening tests to administer.

Telegdy (1974) performed a factor analysis of the STAR. Three factors were identified: (I) Visual-Perceptual-Motor Function, (II) Language Comprehension, and (III) Abstraction of Essential Characteristics (attention to details). The following year, Telegdy (1975) found that the subtest of the STAR that was most predictive of Wide Range Achievement Test scores and Gray Oral Reading Test scores was Letters. In most respects, Telegdy found that the STAR's total scores also were adequate predictors of the criterion measures. Although he later found a modest correlation (*r* = .45) with the Peabody Picture Vocabulary Test (PPVT), Telegdy (1976) found a 21-point discrepancy between the mean deviation IQs obtained on the STAR and those obtained from the PPVT (the STAR yielded the highest). However, Telegdy cautioned users to be wary of the high deviation IQs yielded by the STAR because the PPVT correlated well with the WISC and Stanford-Binet.

In comparing results of a prekindergarten-administrated STAR with end-of-kindergarten Metropolitan Readiness Test scores and teachers' academic rankings, Tsushima, Onorato, Okumura, and Sue (1983) found only one very low correlation (*r* = .24) between the STAR Numbers subtest and the Metropolitan Readiness Test performance and a low but significant correlation between the STAR and teacher rankings (*r* = .23).

Hollinger and Kosek (1985) found prekindergarten-STAR scores to correlate .29, .38, and .40 with the WISC-R Verbal, Performance, and Full Scale IQ scores, respectively, obtained in subsequent years. STAR IQ scores ranged from 96 to 153, and Full Scale WISC-R scores ranged from 89 to 144. In terms of using the STAR for selection into a gifted program, Hollinger and Kosek report that when a criterion score of 116 is employed, a 72% accuracy rate is obtained.

Critique

It is important and instructive to note that the results of the STAR are not intended to be definitive. The author states that an important use of test results is "to help locate children with learning problems or social and emotional difficulties early so referral and remediation or treatment can be instituted prior to school entry (Ahr, 1966, p. 31). School personnel may need a quick, group, screening instrument such as the STAR that requires little in terms of response competence, but it is important that the results not be considered conclusive.

From the point of view of public school personnel, it seems evident that an instrument like the STAR, which attempts to reference the competencies it measures to the demands made in schools, is needed. The difficulty that remains unaddressed is that the STAR, as well as other such school readiness tests (e.g., Gesell Preschool Test [Haines, Aimes, & Gillespie, 1980], Gesell School Readiness Test [Ilg, Aimes, Haines, & Gillespie, 1978]), was constructed in a nexus of assumptions and is imbedded with a multitude of value judgments about development and the "goodness of fit" between child and school. (See Carlson [1985a] and Carlson [1985b] for a discussion of those assumptions.)

The most significant deficit of the STAR in this regard concerns Ahr's assumptions about scaling. Although certainly not unique to the STAR, the assignment of numbers of points to items is arbitrary and camouflages quality of performance encapsulated within a binary unit. In addition, the number of items that sample a particular competency seem limited and culture-bound. Nowhere in the manual is there an explanation of how the items were chosen and the rationale for their selection. This is particularly significant because, in most cases, the number of items for each subtest is what defines the number of points possible, and, thus, governs the weight assigned to the subtest as it contributes to the overall score. In general, there seems to be no theoretical structure that has guided item development or combinations of items.

Again, without apparent theoretical or quantitative justification, bonus points are awarded for getting a majority of certain series of items correct. The author is in effect weighting perfect or near-perfect performance on a given series of items more heavily than performance of a lesser sort. Such a practice represents very arbitrary and superficial psychometric decision-making. Nowhere in the manual are the assumptions or rationale justifying such a practice stated. The assignment of bonus points, in effect, creates two different weights for a series of items: 1) a "base" weight and 2) a heavier weight determined by bonus points when the respondent answers most or all the items correctly. Weighting must always be justified; if it is not, psychometric justification remains mysterious and the meaning of numbers becomes lost. Scores, too, become numbers void of meaning. The total score on the STAR, then, represents the cumulation of varying performances on tasks of varying sorts, the weights of which vary depending upon the quality of performance on given series of items. Quite different mixes of items, as well as bonus weightings, can contribute, then, to two identical scores or deviation IQs.

The STAR yields deviation IQ equivalents based on norming studies of the test, as well as average raw score ranges associated with various chronological ages for each subtest. Singularly, however, a deviation IQ serves no real purpose because it does not refer to local norms or local school success. Without local norms to provide information about what skills are needed to perform well in school, deviation IQs seem to serve no purpose; they leave the examiner no standards upon which to interpret performance. In contrast to total scores and the derived deviation IQ, individual subtest norms would be of most help for identifying examinee's strengths and weaknesses and for remediation. Although Ahr has provided raw score ranges for the subtests, he has fallen short by not specifying how the ranges were derived or what they represent, making them of limited usefulness. The most powerful support that the derived deviation IQ has is that the author reports that it correlates with the Stanford-Binet, Form L-M. However, the accuracy of its IQs are questionable given the range of results obtained by other independent investigators (Telegdy, 1976; Hutton, 1970; Tsushima et al., 1983; Hollinger & Kosek, 1985).

On the surface, the quality of the drawings upon which responses to test items depend is poor, and some items are ambiguous as to what they represent or what is expected of the child.

Although the STAR appears to have face validity in that it contains items that *look like* items contained on other intelligence tests, the weak psychometric assumptions upon which the test is based make the mechanics suspect; therefore, the test becomes little more than a rough portrayal of performance. On the positive side, however, the STAR attempts to go beyond the Gesell School Readiness Test and the Gesell Preschool Test by providing a normative outcome. It is unfortunate that the author has failed to attach meaning to test performance through careful selection of items and graphic portrayal, attention to psychometrics and scaling, and aiding the user in interpretation of outcomes.

References

Ahr, A. E. (1966). *Screening Test of Academic Readiness (STAR): Examiner's manual.* Skokie, IL: Priority Innovations.

Carlson, R. D. (1985a). Gesell Preschool Test. In D. J. Keyser & R. C. Sweetland (Ed.), *Test critiques (Vol. II,* pp. 310-313). Kansas City, MO: Test Corporation.

Carlson, R. D. (1985b). Gesell School Readiness Test. In D. J. Keyser & R. C. Sweetland (Ed.), *Test critiques (Vol. II, pp. 314-318).* Kansas City, MO: Test Corporation.

Haines, J., Ames, L. B., & Gillespie, C. (1980). *The Gesell Preschool Test manual.* Lumberville, PA: Modern Learning Press.

Hollinger, C. L., & Kosek, S. (1985). Early identification of the gifted and talented. *Gifted Child Quarterly, 29,* 168-171.

Hutton, J. B. (1970). Relationships between pre-school screening test data and first grade academic performance for head start children. *Dissertation Abstracts International, 31,* 395B. (Order No. 70-9268)

Ilg, F. L., Ames, L. B., Haines, J., & Gillespie, C. (1978). *Gesell School Readiness Test.* New York: Programs for Education.

Sprigle, H. A. (1965). *Sprigle School Readiness Screening Test.* Jacksonville, FL: Psychological Clinic and Research Center.

Telegdy, G. A. (1974). A factor analysis of four school readiness tests. *Psychology in the Schools, 9,* 127-133.

Telegdy, G. A. (1975). The effectiveness of four readiness tests as predictors of first grade achievement. *Psychology in the Schools, 12,* 4-11.

Telegdy, G. A. (1976). The validity of IQ scores derived from readiness screening tests. *Psychology in the Schools, 13,* 394-396.

Tsushima, W. T., Onorato, V. A., Okumura, F. T., & Sue, D. (1983). The predictive validity of the STAR: A need for local validation. *Educational and Psychology Measurement, 43,* 663-665.

R. R. Hutzell, Ph.D.
Clinical Psychologist, VA Medical Center, Knoxville, Iowa.

SEEKING OF NOETIC GOALS TEST
James C. Crumbaugh. Murfreesboro, Tennessee: Psychometric Affiliates.

Introduction

The Seeking of Noetic Goals Test (SONG) was created to measure the strength of an individual's motivation to discover a sense of life-meaning. It is one of two tests developed by James C. Crumbaugh, Ph.D., from the orientation of Logotheory. The other test, the Purpose in Life Test (PIL; Crumbaugh & Maholick, 1969) was developed earlier to measure the degree to which an individual already has established a sense of life-meaning.

Prior to his retirement in 1980, Dr. Crumbaugh served as chairperson of the department of psychology at MacMurray College in Jacksonville, Illinois, research director at the Bradley Center in Columbus, Georgia, and staff psychologist at the Veterans Administration Medical Center in Gulfport, Mississippi. He studied Logotheory under its founder, Viktor E. Frankl, M.D., Ph.D., while Frankl was visiting professor at Harvard University's summer school in 1961 under the invitation of Dr. Gordon W. Allport. Frankl, in turn, invited Crumbaugh to join his seminar and to report on Crumbaugh's research on quantification and validation of Logotheory concepts. Crumbaugh has gone on to become widely respected for his ability to put Logotheory's philosophical concepts into practical action.

Frankl's theory postulates that humankind's strongest motive is to find a meaning and purpose in life. When one fails to establish a sense of personal meaning in life, a vacuum (existential vacuum) ensues. This vacuum can create motivation for the individual to strive harder to discover life-meaning or can open the door to neuroses.

It is out of this philosophy that Crumbaugh developed the PIL and his own therapeutic procedure, logoanalysis (cf. Crumbaugh, 1973). A logical next step for him was to develop a scale that would predict an individual's motivation to discover that which logoanalysis was designed to encourage.

The resulting SONG, intended for adult populations, is quite simple and straightforward. It consists of 20 statements (e.g., "I think about the ultimate meaning of life") that are responded to along a 7-point Likert-type frequency scale ranging from "never" (1) to "constantly" (7). The order of the response choices is reversed for nine items to break up response sets. The directions, printed on the SONG, are to "circle the number which most nearly represents your true feeling." The SONG is produced in relatively large print on one side of an 11" x 17" sheet of paper.

486

Practical Applications/Uses

Finding a viable meaning in life appears to be increasingly difficult in today's world, with more persons suffering from feelings of emptiness and loss of direction (Wolman, 1975); thus, it would appear that the number of candidates for meaning-oriented interventions would be on the increase. The Seeking of Noetic Goals Test was designed to help select candidates likely to benefit from meaning-oriented therapies, particularly Logotherapy. The essential component of Logotherapy is the surfacing of the individual's personal value hierarchy so that the hierarchy can have a direction-giving effect. It would follow, then, that the SONG could have potential use in conjunction with other values-oriented interventions such as values clarification.

The validity of the SONG to select candidates for intervention, as will be shown later, does not lend substantial support to its use as a selection device on a case-by-case basis. As a result, the primary use of the SONG at this time should be research. To date, limited research has been conducted on the psychometric qualities of the instrument itself. Some additional research has employed the SONG in studies of the relationships between motivation to discover life-meaning and other psychological states.

Much of the original work with the SONG was done with college student and alcoholic inpatient populations. Additionally, the SONG has been used with health care professionals and with members of various religious communities. If further normative data become available, then the SONG would seem useful for a wide variety of adult populations. It should be particularly useful to researchers in the areas of religion and mental health. The SONG would appear to have potentially substantial use as well with adolescents who are making important decisions about what to do with their lives.

Although no special administration settings are required for the SONG, the manual appropriately suggests that the instrument not be employed in competitive situations. Most of the items are quite transparent, and any intention to dissimulate could be accomplished easily. On the other hand, the simplicity and face validity of the items might increase the accuracy of the responses when self-honesty (rather than competition) is the respondent's goal.

The examiner's participation in the testing process typically is minimal. The SONG can be administered either individually or in groups. Most adults understand what is required without any instructions beyond the directions printed on the SONG. Frequently, less than 10 minutes is required for administration, although there is no time limit.

Adding the 20 numbers that the examinee has circled produces the SONG's score. Interpretation of the score is equally simple. Those with higher scores are purported to have higher motivation to discover a personal life-meaning.

Technical Aspects

A detailed description of the development of the Seeking of Noetic Goals Test has never been published. The manual directs the test user to the basic reference for the SONG, which was published in the *Journal of Clinical Psychology* (Crumbaugh, 1977).

In the development of the SONG an initial item pool was cast into a Likert-type format and reduced to a total of 20 items, selected to correlate only moderately with each other but correlating substantially with the sum of all the items. The size and source of the initial item pool and the specific details of the analyses were not reported.

Normative data, in the usual sense, are not established for the SONG. Rather, a score that is halfway between the mean of a group consisting largely of college students and a group consisting largely of alcoholic inpatients is provided as a "normative cutting score." Crumbaugh argues that the group consisting largely of college students would have less motivation to discover life-meaning (because they already have more life-meaning) than would the group consisting largely of alcoholic inpatients.

The generalizability of the normative cutting score is indeterminable. Crumbaugh (1977) provides mean SONG scores for several groups of college students and several groups of alcoholic inpatients, and these scores are surprisingly similar within the two respective categories. Reker and Cousins (1979) collected data from another group of college students and presented a mean (and standard deviation) almost identical to Crumbaugh's. But Hutzell, Laughlin, and Booth (1987) studied a group of alcoholic inpatients similar to Crumbaugh's group and found a mean score significantly lower than Crumbaugh's patients' mean score and almost identical to Crumbaugh's normative cutting score.

The issues of psychometric reliability and validity receive cursory attention in the manual. In addition to Crumbaugh's (1977) basic reference for the SONG, technical information can be gained from a report by Yarnell (1972) and the article by Reker and Cousins (1979).

Crumbaugh (1977) assessed reliability by correlating the odd and even items of the SONG in a combined sample of 158 persons that included methadone patients, seminary students, general college students, and alcoholic inpatients. The coefficient obtained was .71, Spearman-Brown corrected to .83. Reker and Cousins (1979), with 248 college students in an introductory psychology course, similarly found a split-half coefficient of .76, Spearman-Brown corrected to .87. Reker and Cousins assessed test-retest reliability over a 6-week period for 31 of their subjects and obtained a coefficient of .78. Each of these reliability coefficients appears adequate.

Yarnell (1972) added to the construct validity of the SONG by administering it to 40 "normal" males and 40 schizophrenic males along with the Shipley Institute of Living Scale (Vocabulary), Rotter's Internal-External Locus of Control Scale, the PIL, the MMPI, the State-Trait Anxiety Inventory, and the Kuder Preference Record. The data indicate that the SONG is positively related to anxiety and depression and negatively correlated with ego strength and purpose in life.

Similar to Yarnell, Crumbaugh (1977) and Reker and Cousins (1979) found a moderate but statistically significant negative relationship between the SONG and the PIL. Crumbaugh (1977) notes that such a relationship is to be expected, but the actual correlation coefficients should be small if the SONG and PIL assess different domains. The correlations were small, ranging from -.27 to -.52.

Further evidence that the SONG is not simply a negatively scored alternate form of the PIL comes from Reker and Cousins's (1979) factor analysis of all of the PIL and

SONG items together for their sample of 248 students. They extracted 10 factors, of which 4 came largely from the SONG and 5 came largely from the PIL with a relatively small degree of overlap. (The 10th factor consisted of two SONG items and two PIL items). Reker and Cousins proceeded to factor analyze the SONG separately and again detected four primary factors, which they labeled Existential Vacuum, Goal Seeking, Search for Adventure, and Futuristic Aspirations.

Further evidence of construct validity is Crumbaugh's (1977) verification of his prediction that patient populations would score higher than "normals." He argued that emotional illness would tend to destroy life-meaning and thus increase the need to find it.

Reker and Cousins (1979) extended the relationship between the SONG and sense of life-meaning to include the degree to which life itself is viewed as positive or negative. They administered the Life Areas Survey, which is a 14-item, 5-point semantic differential scale, to their sample of 248 students. A moderate, negative, but statistically significant correlation (-.30) was detected between the SONG and satisfaction with present life.

An important clinical use of the SONG would be its use in conjunction with the PIL to select clients for meaning-oriented therapy such as logotherapy. Unfortunately, the data assessing the validity of the SONG for such purposes have been disappointing. Crumbaugh (1977) suggested that alcoholic inpatients with high SONG and low PIL scores would improve more during the course of logotherapy than would patients with low SONG and high PIL scores. In a study of 43 available patients, ratings of the degree of recovery at the end of logotherapy did not show a statistically significant relationship with the two scoring combinations. Hutzell et al. (1987) studied the relationship between SONG scores and the decision to participate in logotherapy. They evaluated 53 inpatient alcoholics who had completed the SONG, had validly completed the MMPI, were assigned to Logotherapy, and either came to an introductory Logotherapy session plus went on to complete the course of therapy or came to the same introductory session but declined to participate in the therapy. Forty-one of the subjects completed therapy, and 12 declined to participate. No significant relationship was found between the SONG score and participation status ($r = .01$).

Critique

The Seeking of Noetic Goals Test holds a unique position among assessment instruments in that it is the only widely available psychological instrument designed to measure the strength of an individual's motivation to discover a sense of life-meaning. Its administration is simple and straightforward and does not require a professionally trained test administrator. It is equally simple to score, and interpretation of the resulting scores is very direct.

The two-page manual of instructions provides minimal information. The manual lacks the detailed reliability, validity, and normative data that one would expect from the manual of a psychometric instrument. Even when the rather limited existing literature on the SONG is taken into account, additional validation work and normative information is necessary to determine if the instrument can be of widespread, general use. At this point, one should be particularly cautious when interpreting results from adolescents or children.

The current data do not support the use of the SONG for selecting individual clients for meaning-oriented groups. With additional validation, though, the SONG may prove useful in larger scientific studies of the relationships between motivation to discover a sense of life-meaning and other potentially relevant variables.

Although the SONG does not appear to have lived up to the author's original expectations, it remains a valuable contribution to the scientific literature, particularly the Logo- and Meaning-Oriented literature. It is a valuable first step toward objective measurement of the extremely elusive variable described as motivation to discover meaning and purpose in life.

References

Crumbaugh, J. C. (1973). *Everything to gain: A guide to self-fulfillment through logoanalysis.* Chicago: Nelson-Hall.

Crumbaugh, J. C. (1977). The Seeking of Noetic Goals Test (SONG): A complementary scale to the Purpose in Life Test (PIL). *Journal of Clinical Psychology, 33,* 900-907.

Crumbaugh, J. C., & Maholick, L. T. (1969). *Manual of instructions for the Purpose in Life Test.* Munster, IN: Psychometric Affiliates.

Hutzell, R. R., Laughlin, P. R., & Booth, B. M. (1987). *Alcohol treatment recidivism: Effects of logotherapy.* (Final report of Veterans Administration Health Services Research & Development Field Program Project No. RRH-001). Iowa City, VA: Health Services Research & Development Field Program.

Reker, G. T., & Cousins, J. B. (1979). Factor structure, construct validity and reliability of the Seeking of Noetic Goals (SONG) and Purpose in Life (PIL) tests. *Journal of Clinical Psychology, 35,* 86-91.

Wolman, B. B. (1975). Principles of interactional psychotherapy. *Psychotherapy: Theory, Research and Practice, 12,* 149-159.

Yarnell, T. (1972). Validation of the Seeking of Noetic Goals Test with schizophrenic and normal Ss. *Psychological Reports, 30,* 79-82.

Mark A. Mishken, Ph.D.
Adjunct Assistant Professor, Graduate School of Business, Department of Management, Pace University, New York, New York.

SELF-DESCRIPTION INVENTORY
Edwin Ghiselli.

Introduction

The Self-Description Inventory (SDI) is used for the measurement of 13 traits that comprise, according to Edwin Ghiselli, the test author, elements of managerial talent. The book in which the Self-Description Inventory (SDI) and the description of early studies that were conducted using it originally were published in *Explorations in Managerial Talent* (Ghiselli, 1971), now out-of-print. Prior to its publication in this book, the SDI had been utilized in numerous research studies, and was obtained from Dr. Ghiselli by colleagues who were free to make copies and administer the instrument for research purposes. *Explorations in Managerial Talent* was written as both a research treatise on what makes a successful manager and a technical manual of the SDI; therefore, this review will assess the information the book provides about the test. In general, this review of the Self-Description Inventory was written and the instrument's items and scoring keys are published here (see Figures 1 and 2) in order to provide the instrument and evidence for its validity to those conducting research.

Edwin E. Ghiselli received his Ph.D. from the University of California at Berkeley in 1936 and was a professor of psychology there from 1939-73. He authored numerous books and articles in the field of industrial-organizational psychology and was past president of Division 14 of the American Psychological Association. He was a fellow of APA and received an A.B.P.P. Diploma from Division 14. Division 14 gives an annual award for research design in his honor. Dr. Ghiselli died in 1980.

Ghiselli thought that by holding organizational variables constant (e.g., levels of management, organizational structure, technological factors, etc.), then traits and abilities could be considered prime determinants of success. Or, in Ghiselli's own words, managerial talent is "the human quality which is important to success in the managerial occupation" (Ghiselli, 1971, p. 17). Ghiselli's approach to managerial talent would seem to contrast with the more "modern" concept of contingency approaches (e.g., Fiedler, 1967; Vroom & Yetton, 1973), in which environmental and situational variables determine how managers should behave to be successful. It is thought by some (Gannon, 1979; Bass, 1981), however, that Ghiselli's approach may complement, rather than contradict, these approaches because he proposes the existence of enduring traits inherent in the person. Hence, this in combination with

This reviewer would like to thank Platies Young-Brown for assistance in putting this manuscript together.

Figure 1. The Self-Description Inventory

The purpose of this inventory is to obtain a picture of the traits you believe you possess, and to see how you describe yourself. There are no right or wrong answers, so try to describe yourself as accurately and honestly as you can.

In each of the pairs of words below, check the one you think most describes you.

1. ____ capable
   ____ discreet

2. ____ understanding
   ____ thorough

3. ____ cooperative
   ____ inventive

4. ____ friendly
   ____ cheerful

5. ____ energetic
   ____ ambitious

6. ____ persevering
   ____ independent

7. ____ loyal
   ____ dependable

8. ____ determined
   ____ courageous

9. ____ industrious
   ____ practical

10. ____ planful
    ____ resourceful

11. ____ unaffected
    ____ alert

12. ____ sharp-witted
    ____ deliberate

13. ____ kind
    ____ jolly

14. ____ efficient
    ____ clear-thinking

15. ____ realistic
    ____ tactful

16. ____ enterprising
    ____ intelligent

17. ____ affectionate
    ____ frank

18. ____ progressive
    ____ thrifty

19. ____ sincere
    ____ calm

20. ____ thoughtful
    ____ fair-minded

21. ____ poised
    ____ ingenious

22. ____ sociable
    ____ steady

23. ____ appreciative
    ____ good-natured

24. ____ pleasant
    ____ modest

25. ____ responsible
    ____ reliable

26. ____ dignified
    ____ civilized

27. ____ imaginative
    ____ self-controlled

28. ____ conscientious
    ____ quick

29. ____ logical
    ____ adaptable

30. ____ sympathetic
    ____ patient

31. ____ stable
    ____ foresighted

32. ____ honest
    ____ generous

(cont.)

Figure 1. The Self-Description Inventory *(cont.)*

In each of the pairs of words below, check the one you think least describes you.

33. ____ shy
    ____ lazy

34. ____ unambitious
    ____ reckless

35. ____ noisy
    ____ arrogant

36. ____ emotional
    ____ headstrong

37. ____ immature
    ____ quarrelsome

38. ____ unfriendly
    ____ self-seeking

39. ____ affected
    ____ moody

40. ____ stubborn
    ____ cold

41. ____ conceited
    ____ infantile

42. ____ shallow
    ____ stingy

43. ____ unstable
    ____ frivolous

44. ____ defensive
    ____ touchy

45. ____ tense
    ____ irritable

46. ____ dreamy
    ____ dependent

47. ____ changeable
    ____ prudish

48. ____ nervous
    ____ intolerant

49. ____ careless
    ____ foolish

50. ____ apathetic
    ____ egotistical

51. ____ despondent
    ____ evasive

52. ____ distractible
    ____ complaining

53. ____ weak
    ____ selfish

54. ____ rude
    ____ self-centered

55. ____ rattle-brained
    ____ disorderly

56. ____ fussy
    ____ submissive

57. ____ opinionated
    ____ pessimistic

58. ____ shiftless
    ____ bitter

59. ____ hard-hearted
    ____ self-pitying

60. ____ cynical
    ____ aggressive

61. ____ dissatisfied
    ____ outspoken

62. ____ undependable
    ____ resentful

63. ____ sly
    ____ excitable

64. ____ irresponsible
    ____ impatient

Figure 2. Scoring Keys for the Self-Description Inventory

The following lists give the correct responses for each of the various scales. The first number is the item number and the last number is the weight or score of the item. T means the top adjective of the pair is the correct response, and B means the bottom adjective is correct.

Supervisory Ability	Intelligence	Initiative	Self-Assurance	Decisiveness
4 B2	3 B4	3 B3	2 B2	1 T2
5 T2	4 B2	9 T2	7 B1	8 T1
14 B3	8 T2	11 B3	11 B1	9 T2
15 B3	9 B1	12 B2	12 T2	10 B1
21 T2	10 B2	17 B3	13 T1	12 T2
23 T3	12 T2	19 B2	16 B2	16 T2
25 T3	13 T2	21 B3	18 T2	19 T2
27 T3	16 B4	25 T5	20 T1	22 T2
30 T2	19 B2	32 T2	22 B1	24 T2
31 B3	22 B1	33 B3	24 T2	26 T2
33 B1	24 T1	35 B3	25 T2	30 T1
34 T2	25 T3	47 B3	26 T1	34 T3
35 T4	27 T1	53 T3	27 B1	38 T1
36 B1	34 B1	57 B2	30 B1	42 B1
41 T3	35 B1	59 B3	31 B2	45 B1
42 T2	37 B2	60 T5	33 B2	50 T2
44 B1	39 T2	61 T4	37 T1	53 T1
49 B2	40 B2		38 B1	57 T1
50 T2	41 B4		41 B2	60 T2
51 T2	42 T2		42 B1	61 T2
54 T1	43 T1		43 T2	63 T1
56 B3	45 T1		46 T1	
60 T2	46 B3		50 T2	
61 T2	47 B1		51 T2	
	48 B2		53 T2	
	50 T3		56 B1	
	52 B1		57 T1	
	53 T2		58 T1	
	54 T3		59 B2	
	55 T4		60 T2	
	58 T2		62 T1	
	59 T1			
	60 B1			
	61 B1			
	62 T1			
	64 T2			

(cont.)

Figure 2. Scoring Keys for the Self-Description Inventory *(cont.)*

Masculinity/ Femininity	Maturity	Working Class Affinity	Achievement Motivation	Need for Self- Actualization
5 B1	1 B1	2 B2	1 B1	3 B2
6 T2	2 B3	4 B1	2 B3	8 T2
11 B1	6 T4	9 T2	3 B3	11 B1
12 B1	8 B1	12 T1	6 T4	12 T2
18 T1	10 B2	13 B2	7 T2	14 B2
23 B2	12 B2	21 T1	20 B4	21 B2
24 B1	13 T1	25 B2	25 T3	26 T1
29 T1	15 B1	31 T2	26 T3	33 B2
30 B1	16 T2	34 B2	27 T3	36 T1
32 T1	18 T1	42 B1	32 B3	49 B1
33 B1	20 B4	43 T1	41 B5	56 B1
34 B1	21 T1	44 B2	47 B2	60 T1
36 B1	22 B4	45 B2	49 B4	
38 T1	28 T1	52 T2	50 T3	
39 B1	33 B1	54 T2	53 T6	
40 T1	34 B1	60 T2	55 T6	
46 B1	35 B3	63 B2	59 B4	
48 T1	37 B1		61 T2	
52 B1	38 B3		63 T3	
55 T1	40 T3		64 T2	
59 B1	43 T1			
60 T1	46 T2			
64 T1	48 B3			
	59 B1			
	60 T4			
	61 T2			
	63 T2			

Need for Power	Need for High Financial Reward	Need for Security
7 T1	6 B1	3 T2
12 B1	13 T1	7 B1
18 T1	16 B1	8 T1
20 B1	22 T1	11 T2
24 T1	29 T1	12 B1
30 B1	57 T1	14 T1
33 B1	59 T1	18 B1
34 T1	60 B3	20 T1
35 B2		21 T1
37 T1		27 B3
42 T1		31 T2
48 T1		36 B1
51 T1		37 B2
58 T1		45 T1
59 T2		49 T1
63 B2		53 B1
64 T1		57 T1

how managers should act in certain situations and/or in different environments may provide a more total picture of managerial success and behavior.

An important segment of *Explorations in Managerial Talent* deals with the importance of each trait measured by the SDI and its contribution to managerial talent. Ghiselli set three criteria for inclusion of a trait: 1) it should differentiate managers from line supervisors and workers, 2) it should differentiate more successful managers from less successful managers, and 3) the relationship between the trait and job success should be higher for managers than it is for line supervisors and workers. Ghiselli administered the SDI to 306 middle managers, 111 supervisors, and 238 line workers. He hypothesized that the more important the trait or ability, the more it would differentiate between the three groups. Generally, the results depict the means for each trait being in the expected direction, with managers scoring highest (or lowest, e.g., need for higher financial reward), supervisors second, and workers third, except in those traits that are not considered important (see Figure 3). The correlations with job performance were also highest for managers, next highest for supervisors, and lowest for workers. It must be stated that no significance levels were reported, with some differences appearing negligible. The results are depicted in various tables and charts showing means, standard deviations, and correlations between the trait and ability scores and measures of job performance. (Unfortunately, some of the data are hard to decipher, due partly to the mislabeling of Figures 9 and 11 in the book.) Based on this research, Ghiselli rated the importance of 13 traits. The traits, reported in order of importance, are shown in Figure 3. How these findings relate to the specific percentages listed in Figure 3 is not readily apparent.

According to Ghiselli, the 13 traits or "elements" of talent fall into three main categories: 1) abilities (supervisory ability, intelligence, and initiative), 2) personality (self-assurance, decisiveness, masculinity-femininity, maturity, and working-class affinity), and 3) motivation (achievement motivation, need for self-actualization, need for power, need for high financial reward, and need for security). It is upon these traits or elements of talent that Ghiselli based the 13 subscales of the SDI.

The SDI, then, was constructed using a purely empirical method, as opposed to the so called rational approach (Travers, 1951). That is, during test construction, items were included without any theoretical underpinnings concerning how they would differentiate individuals (e.g., individuals with high self-assurance vs. those with low self-assurance); rather, they were included according to how each criterion group differed in its responses to the items.

The methodology of test construction generally consisted of administering pairs of adjectives to the criterion groups that had been rated to possess a certain level of a trait or ability. The paired adjectives for which there were differential responses among the levels of each trait or ability were included in each of the 13 scales with "appropriate weights" attached. A total of 64 pairs of adjectives were included in the inventory.

The criterion groups for each of the 13 scales did not consist of the same subjects for each of the scales. Some groups consisted of undergraduates, some of employed people, and others of a combination of employed and college students. The critera against which there were differential responses to the pool of paired adjectives

Figure 3. The Relative Importance of the 13 Traits to Managerial Talent

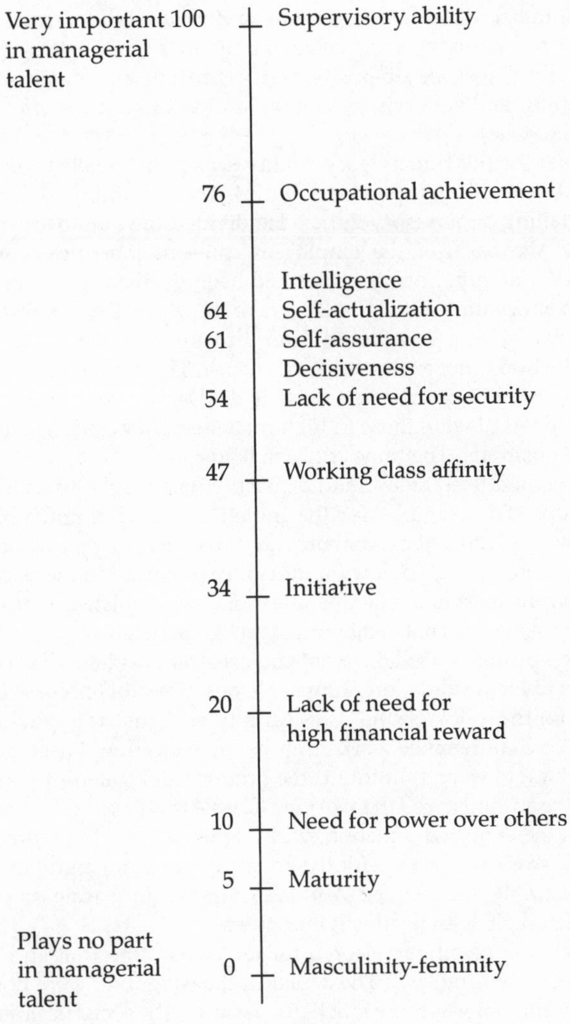

Very important 100 ⊥ Supervisory ability
in managerial
talent

76 ⊥ Occupational achievement

Intelligence
64 Self-actualization
61 Self-assurance
Decisiveness
54 Lack of need for security

47 ⊥ Working class affinity

34 ⊥ Initiative

20 ⊥ Lack of need for
high financial reward

10 ⊥ Need for power over others

5 ⊥ Maturity

Plays no part
in managerial   0 ⊥ Masculinity-feminity
talent

consisted of median splits in a motivation scale, (need for self-actualization, initiative, need for power, need for high financial reward and need for security) high and low scorers in an intelligence test (intelligence, males and females (masculinity-femininity), different age groups and relative differential responses for each age group (maturity), sociometric ratings in a work setting (working class affinity), different levels or job positions in organizations, (decisiveness, achievement motivation, and supervisory ability), and self-ratings of effective behaviors (self-assurance scale).

For example, for the Supervisory Ability scale, 210 subjects were used, 105 of whom had jobs involving supervisory duties. The remaining 105 subjects did not have jobs entailing supervisory duties. He divided the employees into these two categories on the basis of the employees' present job titles, contending that employees without supervisory duties had been deemed as lacking supervisory abilities by their organizations. Couldn't it be, however, that a substantial amount of nonsupervisory workers would later be promoted to supervisory work? Ghiselli's deduction is not explained in any depth. The same method was employed for the construction of other scales, such as the Decisiveness scale. Here, the two criterion groups used were those in high management positions and those in low management positions. The same criticism holds.

A second method used answers to a questionnaire as the basis for forming the criterion groups. For example, for the Initiative scale 324 undergraduates (note population) were given a questionnaire that consisted of a pairing of needs. Those who scored extremely high or low on initiative were used in the selection of items for the scale. Information about this questionnaire, which was used to develop more than one scale, was not dealt with in any great detail.

Ghiselli then properly used different subjects from his item development or criterion groups to form validation groups. He describes this process for each scale. For instance, for the trait Working Class Affinity, sociometric popularity votes were obtained for 185 maintenance workers at the same location. Based on these scores (votes), the workers were split into three groups (high, medium, and low popularity) and were administered the Working Class Affinity scale. The results showed differences in the expected direction, with higher sociometric scores corresponding to higher Working Class Affinity scores (though no significance level was given). Interestingly, the average Working Class Affinity scale score for 24 work groups correlated .48 with productivity ratings.

The SDI items generally are scored for more than one trait and are weighted differentially (see scoring key). The details of the weighting were not discussed at length by Ghiselli, nor was there much discussion of the social desirability of items. He passed over that rather lightly, saying that both adjectives in each pair were of equal value. The multiple scoring of items also may be evident in some of the scale intercorrelations (e.g., Achievement Motivation and Intelligence, $r = .63$); however, with ranges of .00 to .63, they are not as high as one would expect using this type of test construction.

Practical Applications/Uses

The SDI is intended as a research tool for the measurement of traits associated with managerial performance. It can be used in innumerable educational, organizational, and work settings. In addition, the instrument has and can be utilized in masters theses and doctoral dissertations as well as other research studies directed by professionals trained in psychometrics, industrial-organizational psychology, management, and other related areas.

The SDI is geared to subjects ages 18 years and older who are participants in specific research efforts in organizational and educational environments.

The SDI can be administered in a group setting or individually. The examinee simply checks one adjective in each of the pairs that the subject thinks most describes him- or herself (first 32 pairs) and the one from each pair that least describes him- or herself (second 32 pairs). The instrument takes approximately 15 minutes to administer, although there is no time requirement. The instructions are contained on the inventory. The test can be administered by persons without formal training in psychometrics or psychology.

The scoring for each scale consists of adding up the points associated with the adjective checked. For each scale, the pair number is given, with either *T* standing for the top adjective or *B* standing for the bottom adjective, with the amount of points attached to the adjective. It may be wise to code each response on computer sheets and then score the instrument by computer when there are a substantial number of inventories. Although hand scoring is not complex, it is time consuming.

The scores are arrived at objectively by adding the points or weights attached to the adjectives making up each scale. The results should be utilized by grouped data. This instrument is not meant to be utilized as an individual assessment tool. Because the instrument probably should be renormed, individual score interpretation is not recommended. The person who interprets the results should have advanced training in industrial organizational psychology, psychometrics or related areas.

Technical Aspects

No information is given regarding the reliability of the Self-Description Inventory. This is due, perhaps, to Ghiselli's prejudice for the heterogeneous makeup of traits: "The history of psychology shows that as we understand human behavior more and more, we move away from static to dynamic theoretical models" (cited in Mishken, 1973, p. 19). Still, one may wonder why a simple test-retest reliability coefficient was not reported.

The SDI has been used in numerous research studies (Korman, 1967a, 1967b, 1967c, 1967d, 1968, 1969; Mishken, 1973) with statistically significant results. Generally, these studies utilized self-esteem (as measured by the SDI) as a moderator variable in different situations including vocational choice, job satisfaction, and the prediction of job performance. More recently, SDI results have been correlated with assessment center ratings and job performance indices of store managers and district managers in a large retail food organization, showing more than moderate results (see Tables 1 and 2; Lawrence Fogli, personal communication, October 11, 1985). For the assessment center data, the Supervisory Ability, Self-Assurance, Working Class Affinity, and Need for Self-Actualization scales produced the most significant correlates. For job performance criteria, Supervisory Ability, Achieve-

Table 1

Job Performance Assessment and Self-Description Inventory Correlations

	Supervisory Ability	Intelligence	Initiative	Self-Assurance	Decisiveness	Masculine/Feminine	Maturity	Working Class Affinity	Achievement Motivation	Need for Self-Actualization	Need for Power	Need for High Financial Reward	Need For Security
1. Salary (n=121)	.01	-.01	.14	-.03	.11	.01	.05	.12	.07	.15**	.20**	-.06	-.04
2. Bonus (n=116)	.01	-.09	-.05	.02	.14	-.08	.06	.09	.06	-.03	-.09	-.06	-.01
3. Discretionary Bonus (n=35)	.24	.12	-.17	-.41*	.05	.15	-.21	.00	-.14	-.17	.16	.39*	-.19
4. Total Salary (n=121)	.06	-.01	.10	-.04	.13	.01	.03	.13	.08	.17**	.10	-.08	-.11
5. Ranking into Thirds (1=better) (n=114)	.05	.07	.04	.01	.16**	-.20**	-.01	-.02	.05	.09	-.18**	-.17**	-.22*
SUPERVISORY RATINGS: (n=128)													
6. Oral Communication and Presentation	.25*	.20*	.10	.15**	.13	-.05	.00	.00	.22*	.12	.01	.10	-.21*
7. Written Communication	.34*	.35*	.21*	.23*	.10	-.01	.05	.08	.28*	.23*	-.02	.05	-.25*
8. Planning and Organizing	.15**	.03	.05	-.02	.08	-.12	.05	.13	.12	.10	-.06	-.07	-.18**
9. Delegation and Control	.08	.05	.11	.07	.10	-.10	.11	.14	.14	.07	-.08	-.04	-.19*
10. Staffing, Developing and Coaching People	.19**	.11	.12	.11	.13	-.06	.02	.08	.21*	.10	-.05	-.05	-.13

(cont.)

Table 1 (Cont.)

Job Performance Assessment and Self-Description Inventory Correlations

	Supervisory Ability	Intelligence	Initiative	Self-Assurance	Decisiveness	Masculine/Feminine	Maturity	Working Class Affinity	Achievement Motivation	Need for Self-Actualization	Need for Power	Need for High Financial Reward	Need For Security
11. Internal Contacts	.24*	.09	.13	.06	.16**	-.01	.02	.16**	.19**	.16**	-.08	-.11	-.14
12. External Contacts	.22*	.03	.11	.05	.07	-.02	.13	.16**	.17**	.09	.03	.02	-.20*
13. Leadership of Others	.13	-.02	.04	-.05	.09	-.05	.04	.09	.10	.01	-.03	.00	-.17**
14. Problem Analysis and Decision Making	.16**	.03	.09	.04	.10	.00	.04	.03	.08	.17**	-.04	.01	-.18**
15. Innovation and Resourcefulness	.23*	.05	.05	.04	.15**	-.10	-.01	.09	.10	.18**	.05	.04	-.29*
16. Monitoring Business Indicators	.22*	.12	.15*	.10	.12	-.05	.06	.10	.20*	.18**	-.06	-.04	-.27*
17. Knowledge of Business and Operations	.15**	.04	.08	.01	.13	.06	.08	.12	.17**	.21*	-.00	-.07	-.28**
18. Overall Performance Effectiveness	.20*	.06	.06	.05	.10	-.08	.00	.09	.15**	.11	-.08	-.06	-.22*

*p < .01
**p < .05
Source: Lawrence Fogli (personal communication, October 1985), CORE Corporation, 367 Civic Drive, Suite 12, Pleasant Hill, CA 94523

Table 2

Assessment Center Data and Self Description Inventory Correlations

	Supervisory Ability	Intelligence	Initiative	Self-Assurance	Decisiveness	Masculine/Feminine	Maturity	Working Class Affinity	Achievement Motivation	Need for Self-Actualization	Need for Power	Need for High Financial Reward	Need For Security
1. Assessment Center Rating (1=better)	-.22	-.23	-.26	-.24	-.02	-.05	-.14	.25	-.41*	-.39*	-.16	.14	.23
2. Assessment Center Ranking (1=better)	-.36**	-.21	-.25	-.30**	-.06	-.11	-.05	.23	-.36**	-.26	-.12	.09	.18
ASSESSMENT CENTER DIMENSIONS:													
3. Problem Analysis	.12	.14	.13	.18	.11	-.08	.08	-.29**	.19	.35**	.08	-.05	-.13
4. Decisiveness	.21	-.03	-.02	.06	.00	.00	.08	-.40*	.10	.27	-.11	-.19	.04
5. Problem Solving	.13	.25	.16	.25	.00	.11	-.01	-.37*	.15	.44*	.03	-.09	-.12
6. Planning	.35**	.14	.24	.27	.06	-.03	.03	-.43*	.33**	.37*	-.04	-.06	-.14
7. Delegation	.27	.23	.21	.36*	-.04	.15	.04	-.41*	.23	.17	.18	.01	.03
8. Managerial Control	.38**	.10	.18	.32**	-.02	.14	.12	-.54.*	.21	.36**	.10	-.25	-.27*
9. Oral Communications	.26	.29**	.21	*.57*	.32**	-.19	.22	-.21	.49*	.07	-.04	-.44*	-.19
10. Written Communications	.22	.19	.10	.17	-.09	.10	-.06	-.20	.12	.16	.15	.04	.00
11. Human Relations Skills	.32**	.08	.25	.15	.21	.04	.14	-.30**	.07	.23	.02	-.11	-.02

(cont.)

Table 2 *(Cont.)*

Assessment Center Data and Self Description Inventory Correlations

	Supervisory Ability	Intelligence	Initiative	Self-Assurance	Decisiveness	Masculine/Feminine	Maturity	Working Class Affinity	Achievement Motivation	Need for Self-Actualization	Need for Power	Need for High Financial Reward	Need For Security
12. Leadership	.10	.10	.04	.26	.13	.07	.16	-.41*	.22	.21	.11	-.24	.00
13. Self-Confidence	.22	.17	.02	.38*	.22	-.12	.10	-.30**	.26	.08	.07	-.35**	-.20
14. Stress Tolerance	.23	.22	-.09	.32**	.24	-.03	.05	-.27**	.24	.18	-.13	-.10	.02

*p < .01
**p < .05
Source: Lawrence Fogli (personal communication, October 1985), CORE Corporation, 367 Civic Drive, Suite 12, Pleasant Hill, CA 94523

ment Motivation, Need for Self-Actualization, and Need for Security scales produced the highest number of significant correlates.

One should note that although women were included in the norms presented in the book and were used in the development of some of the scales and validation studies, this was not done consistently. It seems to this reviewer that more research using the SDI should be aimed at studying gender similarities and differences as well as relevant norms.

Critique

In conclusion, although the psychometrics (i.e., reliability, weighting of items, etc.) are not perfect, the SDI still shows as much potential as it did when Ghiselli first began using it in the late 1950s. Utilization of the SDI would be enhanced by 1) the integration of current contingency and individually oriented trait approaches to leadership and 2) research into which managerial traits work best in various types of organizations and organizational settings.

References

Bass, B. M. (1981). *Stogdill's handbook of leadership.* New York: Free Press.

Fiedler, F. E. (1967). *A theory of leadership effectiveness.* New York: McGraw-Hill.

Gannon, M. J. (1979). *Organizational behavior.* Boston: Little Brown.

Ghiselli, E. E. (1971). *Explorations in managerial talent.* Pacific Palisades, CA: Goodyear.

Ghiselli, E. E. (1972). Comment on the use of moderator variables. *Journal of Applied Psychology, 56*(3), 270.

Korman, A. (1967c). Self-esteem as a moderator of the relationship between self-perceived abilities and vocational choice. *Journal of Applied Psychology, 51,* 65-67.

Korman, A. (1967a, August). *Ethical judgments, self-perceptions and vocational choice.* Paper presented at the 75th Annual Convention of the American Psychological Association, Washington, DC.

Korman, A. (1967d, April). *Some correlates of satisfaction as moderated by self-esteem.* Paper presented at the annual meeting of the Eastern Psychological Association, Boston.

Korman, A. (1967b). Relevance of personal need satisfaction for overall satisfaction as a function of self-esteem. *Journal of Applied Psychology, 51,* 533-538.

Korman, A. (1968). Self-esteem as a moderator in vocational choice: Republicans and extensions. *Journal of Applied Psychology, 52,* 484-490.

Mishken, M. A. (1973). *Self-esteem as a moderator of the relationship between job ability and job performance.* Unpublished doctoral dissertation, University of Tennessee, Knoxville.

Travers, R. M. W. (1951). Rational hypotheses in the construction of tests. *Educational and Psychological Measurement, 11,* 128-137.

Vroom, V. H., & Yetton, P. W. (1973). *Leadership and decision-making.* Pittsburgh, PA: University of Pittsburgh Press.

Robert L. Heilbronner, Ph.D.

Clinical Psychologist, Section of Neuropsychology, Department of Neurosurgery, HCA-Presbyterian Hospital, Oklahoma City, Oklahoma.

George K. Henry, Ph.D.

Postdoctoral Fellow in Clinical Neuropsychology, Department of Psychiatry and Behavioral Sciences, University of Oklahoma Health Sciences Center, Oklahoma City, Oklahoma.

THE SINGLE AND DOUBLE SIMULTANEOUS STIMULATION TEST

Carmen C. Centofanti and Aaron Smith. Los Angeles, California: Western Psychological Services.

Introduction

The Single and Double Simultaneous Stimulation Test (SDSST) is designed to examine specific somatosensory functions and to determine the accuracy with which subjects can identify single and double simultaneous tactile stimulation applied to the cheek and/or hand. It was developed by Carmen C. Centofanti and Aaron Smith at the University of Michigan and has been employed in diagnostic studies of children and adults with suspected disease or injury to the central nervous system and for the assessment of patients with documented lesions.

The SDSST was developed by its authors in 1975. It was part of the first author's doctoral dissertation at the University of Michigan under the direction of the second author, a well-known and widely published neuropsychologist.

Normative standards for the SDSST are based on the performance of 431 caucasian adults within the age range of 18-75 years (Centofanti, 1975). All subjects were volunteers and lived in one of two industrial communities, Madison Heights, Michigan, or Florence Township, New Jersey. No subject in this group had a history of vascular or infectious disease, trauma, or other central nervous system pathology. There were 223 females and 208 males, approximately equally represented in the various age subgroups. The manual contains descriptive data for the sample in subgroups, including hand dominance and percent of subjects with abnormal scores. In the standardization group, only 11 subjects (9 females, 2 males) made three or more errors; all of them were right-handed and 55 years of age or older. Subjects in the 65-74 age group ($N=9$) made significantly more errors than those in any other age group. All six age groups had at least 19% of their subjects making at least one error; four of the six age groups had subjects with two errors. Only the 55 and over age groups had subjects with three or more errors on the first trial.

Practical Applications/Uses

Various single and double stimulation tests of somatosensory functions have been used in neurological and neuropsychological examinations and research in

505

the past. They have been shown to have diagnostic value in identifying patients with slowly evolving neoplasms and other types of lesions. Several studies have demonstrated their prognostic value with hemispherectomy patients (Smith & Sugar, 1975), aphasia patients (Smith, 1972), and patients recovering from vascular, traumatic, and other resolving lesions (Smith, 1971, 1972, 1974, 1975). The most popular sensory stimulation test used in neuropsychology is the sensory-perceptual examination commonly used in combination with the Halstead-Reitan Neuropsychological Battery. This examination includes single and double simultaneous stimulation in the visual, auditory, and tactile modalities.

The SDSST is administered with the subject seated either in a chair or in a bed. When seated in a chair, the person's hands are placed palms down and over the knees, not touching each other. When sitting up in bed, the hands are placed palms down over each thigh. The examiner is seated facing the subject. To begin testing, the examiner says, "Please close your eyes and put your hands on your knees (thighs). I am going to touch you and you show me where you were touched." Tactile stimuli are then applied by one or two simultaneous light and brisk strokes with the index fingers on the cheek and/or back of the hand. Both sides of the body are tested. The subject indicates where he or she was touched by pointing to the area rather than by a verbal response. If the patient is unable to do this, a verbal response may be used. For patients who are aphasic, instructions should emphasize pointing or some other type of manual gesture. If the subject fails to report two stimulations on either of the first two stimuli presentations, the examiner asks, "Were you touched anywhere else?" and notes accordingly. No more questions are asked after the first two stimuli presentations.

The front of the SDSST test sheet contains spaces for demographic information (name, age, sex, date of birth, and years of education) and the back side is used for writing down the subject's responses and documenting the types of error categories and patterns. Stimuli are presented in blocks of 20 in the order provided on the SDSST record sheet. If there are no errors during the presentation of the first 20 stimuli, the testing is terminated. If one or more errors are made on the first trial, the second trial is administered in order to document fluctuations in performance. The record form also provides the opportunity to record the results of a third and fourth trial. An abnormal performance on the SDSST is three or more errors on the first trial.

There are three types of errors an individual can make: extinction, displacement, and adjunction. Extinction errors are scored when the patient perceives only one of two bilateral simultaneously applied stimuli or fails to respond to single unilateral stimulation. A displacement error occurs when the subject indicates an area stimulated that is different from the cheek and/or hands in single or double simultaneous stimulation. Adjunction errors are scored when an additional stimulation is reported (i.e., two in a single or three in a double simultaneous stimulation). If the errors are restricted to only the right or left hand and/or cheek, the results are regarded as evidence of a lateralized sensory deficit. If they occur on both sides, the results are considered evidence of a bilateral sensory deficit or involvement of the haptic mechanisms in the brain or spinal cord, which mediate lateralized somatosensory functions.

Technical Aspects

Reliability generally refers to the sensitivity or accuracy and repeatability of measurement. In the SDSST manual, no information is provided on the instrument's reliability. With respect to validity, a test is considered valid if it measures what it purports to measure. Validity is not an all-or-none phenomenon, but rather a matter of degree. Psychological tests must ultimately be concerned with three types of validity: 1) predictive or criterion-related, 2) content, and 3) construct.

In the case of the Single and Double Simultaneous Stimulation Test, no information is available on predictive validity; that is, who does and does not have brain damage at the time the test is administered. Although predictive validity is a major issue in applied psychology and education, content and construct validity are of major importance in neuropsychology.

Content validity of a test instrument is often a product of item selection and test construction. Although personal values determine the relative emphasis on different types of content, it is apparent that the SDSST focuses solely on the tactile sensory modality. However, sensory suppressions may also occur via auditory and visual modalities. Thus, by eliminating auditory and visual domains as sources of potential inattention or suppression, the authors have restricted the range of sample items. This could render the SDSST susceptible to a preponderance of false negatives (i.e., erroneously indicating no brain damage). In fact, in one validity study cited in the manual, 44% of 172 patients with confirmed cerebral dysfunction did not meet the SDSST criterion of three or more errors necessary for correct classification as brain damaged.

Several other critical points about this study are worth mentioning. Although 172 of the subjects had confirmed brain damage, the authors did not identify the method of confirmation or provide information on neurological diagnosis. The authors do report that 41% of the 172 subjects were aphasic. This may have diminished the subjects' ability to comprehend fully or understand the nature of the test because of the nature of their language deficit and may have inflated the number of false positives. Although 101 of the 172 were nonaphasics, no breakdown is provided as to localization of lesion. Patients with right parietal lesions may have selective visuospatial deficits (e.g., unilateral neglect) that may interfere with their ability to accurately point to body parts touched by the examiner, especially on the left side of the body. This could exist independently of any specific sensory deficit yet appear as a suppression or extinction error, resulting in a false positive. Also, stimulation of nonhomologous body parts (face and contralateral hand), as recommended in the procedure outlined for the SDSST, does not address the phenomenon of rostral dominance, when the brain-damaged subject is more likely to report facial contact than hand stimulation; similar results are often found in animal studies of recovery of sensory orientation following brain damage (Teuber, 1978).

To the extent that the SDSST alone measures the construct, brain damage, or correlates with one or more other observable variables or tests shown to be sensitive to brain damage, construct validity is provided or assumed. This is also referred to as concurrent validity. The extent to which the SDSST discriminates between different groups is termed discriminant validity. Both discriminant and construct validity of the SDSST remain equivocal. For example, in a separate validity study of 110 hospitalized psychiatric patients (cf. Centofanti & Smith, 1979), the SDSST classified 56% as brain damaged on the basis of three or more errors. This hit rate was identical to that reported in the authors' validity study of 172 confirmed brain-damaged patients. Thus, the SDSST does not appear to discriminate very well between known brain-damaged subjects and subjects with a psychiatric illness. In a third and final validity study of the SDSST with institutionalized retardates (cf. Centofanti & Smith, 1979), SDSST errors directly correlated with lower IQs. Thus, SDSST errors appear to be influenced by both intelligence and psychiatric illness or emotional factors. This confounds interpretation and renders the SDSST questionable as a valid instrument for screening for organicity. Even though some evidence for concurrent validity exists, as the authors state, the SDSST is "especially useful when combined with the Symbol Digit Modalities Test" (p. 3), they provide no data for each of the tests' individual contributions to diagnostic accuracy.

Critique

Although the Single and Double Simultaneous Stimulation Test purports to be a sensitive and valid test of brain damage that is easy to administer and score, it appears to be influenced by a number of variables that render suspect its utility as a screening device for organicity. Variables that may artificially inflate the number of SDSST errors include psychiatric illness and emotional factors, intelligence, age, language ability, and possibly motivation. Thus, the SDSST should only be interpreted in the context of a more comprehensive test battery given by a professional trained in the administration and interpretation of neuropsychological tests.

References

Centofanti, C. (1975). Selected somatosensory and cognitive test performances as a function of age and education in normal and neurologically abnormal adults (Doctoral dissertation, University of Michigan, Ann Arbor). *Dissertation Abstracts International, 36,* 3027B.

Centofanti, C., & Smith, A. (1979). *The Single and Double Simultaneous (Face-Hand) Stimulation Test (SDSST).* Los Angeles: Western Psychological Services.

Smith, A. (1971). Objective indices of severity of chronic aphasia in stroke patients. *Journal of Speech and Hearing Disorders, 36,* 167-207.

Smith, A. (1972). Dominant and nondominant hemispherectomy. In W. L. Smith (Ed.), *Drugs, development, and cerebral function.* Springfield, IL: C.C. Thomas.

Smith, A. (1973). *Symbol Digit Modalities Test.* Los Angeles: Western Psychological Services.

Smith, A. (1974). *Related findings in neuropsychological studies of patients with hemispherectomy and of stroke patients with chronic aphasia.* Paper presented at Second Annual Meeting of the International Neuropsychological Society, Boston.

Smith, A. (1975). Neuropsychological testing in neurological disorders. In W. J. Friedlander (Ed.), *Advances in neurology, Vol. 7,* (pp. 49-110). New York: Raven Press.

Smith, A., & Sugar, 0. (1975). Development of above normal language and intelligence 21 years after left hemispherectomy. *Neurology, 25,* 813-818.

Teuber, H. (1978). The brain and human behavior. In R. Held, H. W. Leibowitz, and H. L. Teuber (Eds.), *Handbook of Sensory Physiology: Vol. 7. Perception.* Berlin: Springer-Verlag.

Edwin E. Wagner, Ph.D.
Professor of Psychology, The University of Akron, Akron, Ohio.

SLOSSON INTELLIGENCE TEST

R. L. Slosson. East Aurora, New York: Slosson Educational Publications, Inc.

Introduction

The Slosson Intelligence Test (SIT) is an individually administered test of mental ability. It is designed to be quickly given, scored, and interpreted by practically any "responsible person" concerned with assessing intelligence. The SIT is an example, par excellence, of the congruent approach to test construction inasmuch as its validity is deliberately and exclusively linked to its high correlation with the Stanford-Binet Intelligence Scale: Form L-M.

The SIT consists of the basic instructional manual, a technical booklet, expanded 1985 norm tables, and tablets of scoring blanks. The Slosson Oral Reading Test is included in the kit as a recommended complement to the SIT. A sample cassette is available, which exemplifies proper oral administration procedures, and an SIT item analysis kit (manual and score sheets) can also be obtained.

The test is made up of 194 observations, tasks, and oral questions arranged in ascending order of difficulty and encompassing an age range of ½ month to 27 years. There are a few performance questions, but, for the most part, the SIT is highly verbal. The questions are untimed, but administration is relatively brief, varying from 10 to 30 minutes.

Responses are recorded on a scoring sheet as they occur, thus saving time. A basal level of 10 consecutive passes is established, and testing is concluded after 10 successive failures. Instructions are clear-cut, and scoring is objective.

The SIT was first published by Slosson in 1963. A new version, touted as a "revision," was developed by Armstrong and Jensen (hereafter referred to as the "authors") and published in 1981. It is quite similar to the original SIT but reports deviation rather than ratio IQs to bring the scores more closely in line with the renormed 1972 Stanford-Binet.

Practical Applications/Uses

As advocated by the authors, the SIT is eminently practical and can be used in almost any assessment situation where a valid IQ is required. Purportedly, it can be properly administered and interpreted by teachers, principals, guidance counselors, special education and learning disability teachers, psychologists, psychometricians, and social workers (Slosson, 1983). The SIT is recommended for use in estimating the IQs of infants (down to age 2), children, public school students, college students, mental patients, retardates, the blind, individuals with minimal brain dysfunctions, and subjects with learning disabilities.

In general, the SIT is represented as a viable, across-the-board alternative to the Stanford-Binet. However, as concluded in a critical review by Stewart and Jones (1976), the SIT is best used for screening purposes and "interpretation of an individual's SIT IQ should not go beyond relatively broad categorizations . . ." (p. 377).

Technical Aspects

The authors of the Slosson Intelligence Test do not deem it necessary to define the nature of intelligence in a theoretical sense. Because the SIT was deliberately constructed to correlate highly with the SB, it is assumed, pragmatically, that whatever the criterion test measures, the SIT does, too. In fact, the SIT is lauded by the authors, inaccurately, as a parallel form of the SB.

The reliability of the SIT has been consistently reported in the .90s. The validity is also in the .90s, and this is quite impressive, provided one is willing to accept the basic "piggyback" rationale that a high correlation with the SB is, per se, proof of the test's usefulness. Also, various studies can be cited that establish the SIT's concurrent validity in terms of substantial correlations with the WAIS, WISC, and other tests.

During the original standardization of the SIT, over 4,000 subjects spread over various geographic areas were tested. Subsequent renorming, however, was based on only 1,109 cases confined to the New England states. The new norms make use of the equipercentile procedure whereby the scores of the SIT are equated to those of the SB at each age level. Consequently, an IQ derived from the SIT can be expected to approximate that which would have been obtained from the SB. Jensen and Armstrong's research indicates that 75% of the time the SIT IQ should not deviate from the SB IQ by more than $\pm$ 4.86 points. However, more empirical research is needed in order to establish whether these parameters will hold up for specific groups at different age levels.

Little confidence can be placed in the newer item analysis data. As noted by Bohning (1980), remediation based on differential scores across eight categories is appealing at face value but cannot be taken seriously until relevant technical information is provided regarding the development, reliability, and uniqueness of these item clusters.

Critique

The Slosson Intelligence Test is a versatile instrument, but one of its major advantages, ease of administration, is also a potential source of misuse. Almost anyone involved in evaluation can give the test, and the manual and tape are presented in a disarmingly breezy, upbeat manner that accentuates the positive features of the SIT. There is a danger that unsophisticated, untrained, and unwary test users could draw unwarranted conclusions due to lack of professional training.

The SIT is probably best regarded as a convenient screening test with the caveat that "tests developed for screening should be used only for identifying test takers who may need further evaluation" (American Educational Research Association, American Psychological Association, & National Council on Measurement in Education, 1985, p. 43).

References

American Educational Research Association, American Psychological Association, & National Council on Measurement in Education. (1985). *Standards for educational and psychological testing*. Washington, DC: American Psychological Association.

Bohning, G. (1980). Item analysis of Slosson Intelligence Test: A review. *Psychology in the Schools, 17,* 339.

Slosson, R. L. (1983). *Slosson Intelligence Test (SIT) and Oral Reading Test (SORT)* (2nd ed.). East Aurora, NY: Slosson Educational Publications, Inc.

Stewart, K. D., & Jones, E. C. (1976). Validity of the Slosson Intelligence Test: A ten-year review. *Psychology in the Schools, 13,* 372-380.

Tim Roberts, Ed.D.
Associate Professor of Special Education, East Texas State University, Commerce, Texas.

SOCIAL INTELLIGENCE TEST
F. A. Moss, T. Hunt, K. T. Omwake, and L. G. Woodward. Washington, D. C.: Center for Psychological Service.

Introduction

The Social Intelligence Test is a now-dated attempt to assess several components of social interactions. In three editions, this timed paper-and-pencil instrument purportedly measures the judgment, information, and memory related to an individual's ability to respond to social situations (Moss, Hunt, Omwake, & Woodward, 1955). Approximately 80 reference citations focus on either the construct of social intelligence or the Social Intelligence Test; the majority of these citations occur prior to 1960.

The Social Intelligence Test was developed in an effort to address E. L. Thorndike's suggestion, in the 1920s, of the need to create a distinction between abstract, mechanical, and social intelligence (Taylor, 1949; Thorndike, 1940). The test was first published in 1927 under the name George Washington University Social Intelligence Test; a second form was published in 1929 (Moss et al., 1955). A revised form, designated "first edition," was published in 1937 and consisted of the "better" items from the two earlier versions (Moss et al., 1955). This first edition and the current second edition (which dates from 1949) are titled George Washington University Series: Social Intelligence Test.

In its present form, the Social Intelligence Test consists of a universal manual (1955), scoring keys, and three test editions: Revised Form, Second Edition (Moss et al., 1949); Short Edition (Moss, Hunt, & Omwake, 1947); and SP (Special) Edition (Moss, Hunt, & Omwake, 1978). The Short and SP Editions consist of subtests selected from the Revised Form, Second Edition. Each edition of the test has similar directions for group administration.

Limited information is presented describing the normative sample. All subjects came from the mid-Atlantic states, although no indication is given regarding gender, ethnic/racial, or SES distributions. However, an atypical distribution seems to be present, due in part to the large number of college students reported on in the 1955 test manual.

Each test edition (i.e., Revised Form, Second Edition; Short Edition; and SP Edition) has a different normative sample and grade range. Apparently, no control for age differences was addressed in the development of these test editions. A second normative sample focuses on the category of employed adults, whose occupations are classified as professional, managerial, and skilled (Moss et al., 1955). However, no distribution for these classifications is given.

The Revised Form, Second Edition provides normative comparisons for ninth- and tenth-graders ($N = 718$), high school seniors ($N = 984$), college freshmen

(N = 2,270), college graduate students (N = 236), and employed adults (N = 1,275). Normative data for the Short Edition report high school seniors (N = 340), college freshmen (N = 863), and college upperclassmen (N = 450). The SP Edition reports high school seniors (N = 812), college freshmen and sophomores (N = 1,200), and employed adults (N = 2,830).

Scoring procedures are similar for each test edition. Raw scores for each subtest are totaled and converted to percentile ratings. A summary table for subtest scores is provided on the Revised Form and Short Editions, but not on the SP Edition. Each summary table has three columns: total possible, score, and rating. Neither the manual nor the conversion table explains a procedure for converting subtest raw scores to percentile ratings, and some confusion may exist over the concept of percentile ratings. The standard scores provided, percentile ratings, are percentile points, which are based on the percent of the sample members that have scores less than a specified value (Ferguson, 1981); therefore, percentile ratings range from 0-100. This use of percentile points also allows for quartile and decile score interpretation.

Revised Form, Second Edition. The Revised Form, Second Edition has five subtests. The testing session begins with a single-sheet stimulus consisting of 12 names and faces. Subjects are directed to remember these names and faces for future test reference. After 4 minutes, the names-and-faces sheet is collected and the test booklets distributed. Subjects are asked to provide identification data (e.g., name, age, sex, occupation, etc.) on the front of the test booklet and then to follow the instructions for each subtest, start a new subtest upon the completion of the previous one, and work as quickly as possible. Forty-five minutes are allowed to complete the exam.

Judgment in Social Situations (30 items) asks the subject to respond to a statement by selecting one of four possible alternatives. Items can be grouped into three broad categories of business relations, interpersonal relations, and social concerns (Hunt, 1928). Many of the test stimuli are laced with masculine pronouns. Each correct response is awarded 1 point, with a maximum score of 30 possible.

The second subtest, Recognition of the Mental State of the Reader, has 18 items. Subjects read each item, selecting the one of four alternatives that reflects the emotional/mental state depicted by the passage. The maximum score for this subtest is 36.

Memory for Names and Faces (12 items), the third subtest, utilizes the single-sheet stimulus of 12 names and faces that is presented at the beginning of the testing session. These 12 names and faces are mixed in among an additional 13 names and faces. From this newly created matrix of 25 pictures, subjects are required to associate a name with one of four alternative pictures. All of the pictures are vintage 1940s and consist of young Anglo males. Each correct multiple-choice association is awarded 2 points, for a maximum of 24 points.

Observation of Human Behavior, subtest four, consists of 50 true/false items that could be described as axioms, myths, and vagaries. The scoring procedure is right minus wrong, and no penalty is given for omitted items or items marked both T and F. The maximum score for this subtest is 50.

The last subtest of the Revised Form, Second Edition is entitled Sense of Humor (20 items). Directions instruct subjects to select one of four suggested statements that will make the "best joke." Each correct answer is worth 2 points. This subtest is by today's standards prejudicial and chauvinistic: approximately one half of the stimuli consist of racial- and sexist-directed stereotypic statements.

Short Edition. Four subtests constitute the Short Edition, modified with regard to subtest format, test stimuli, directions, or the number of items presented. The manual provides no information relative to the effects of these modifications on reliability or validity.

Directions to the examinees are similar to those in the Revised Form; however, the amount of time allowed to complete the Short Edition is reduced to 40 minutes. The Memory for Names and Faces subtest is eliminated along with the single-sheet stimulus. Scoring procedures are similar to those of the Revised Form, and a percentile rating conversion table is presented in the manual.

Judgment in Social Situations, the first subtest of the Short Edition, is modified with regard to directions and the order of stimuli presented; in addition, new items replace two original stimuli from the Revised Form. The replaced items involve 1) a shift from a religious situation to a personal ethic viewpoint and 2) the substitution of an interpersonal/management judgment for a political/personal item. Another modification includes a change in directions, which results in response changes. The Revised Form asks subjects to indicate a letter designation in a four-choice alternative format; on the Short Edition, examinees are asked to check (ν) the "most nearly correct" alternative. The final modification of this subtest involves an alteration in the stimulus sequence (e.g., item #12 in the Revised Form becomes item #10 in the Short Edition). This altered sequence occurs six times in a set of 30 stimuli.

Subtest two, Recognition of the Mental State of the Speaker, consists of 18 items worth 2 points each. The primary modifications involve format, directions, and response rather than test items. Format changes account for the majority of the directional and response alterations. Instead of the Revised Form's four-choice format, the Short Edition employs a "matching" format. Eighteen emotional/mental state descriptors are alphabetized and numbered, divided into two columns of 8 and 10 respectively. Examinees are instructed to write the number of the emotional/mental state descriptor in the parentheses corresponding with each stimulus. The directions also advise the subject that some emotional/mental state descriptors may not be represented among the stimuli, while others may be represented more than once.

The third subtest in the Short Edition is Observation of Human Behavior. This subtest is reduced from 50 to 40 items, but an explanation is not given for item retention/deletion. The format changes to two columns of 20 items, each preceded by a T or F. This in turn results in modifications of the directions, which in this case instruct the subject to circle rather than write T or F.

The last subtest of the Short Edition is the 15-item Sense of Humor subtest, in which items, format, and directions are modified. The racially biased items and one other are eliminated (but again, no criteria are given for item deletion). Two columns of 7 and 8 stimuli are used. The subtest retains its stem and four-choice format; however, each choice in the shortened version is identified by a number (i.e., 1, 2, 3, 4) rather than a letter (e.g., a, b, c, d) designation. A score of up to 30 points is possible.

SP Edition. The current SP Edition (copyright 1978) is a revision of the original (dating from about 1955). General directions are similar to those for the Revised Form and Short Editions. The total testing time with this edition is 30 minutes.

Several alterations were made in the Judgment in Social Situations subtest for the SP Edition; however, no information is provided regarding the impact these alterations have on the constructs of the test. The most significant alteration was the elimination of sexist language in 13 items. Directions for this subtest are identical to those found in the Short Edition. Each correct item is credited with 1 point, with a maximum of 30 points possible.

Observation of Human Behavior consists of 50 true/false items. The format and directions in this subtest are consistent with those in the Short Edition, but the presentation reverts to the original items found in the Revised Form (1949). In addition, one question has been altered to reflect a specific point of reference. As with the first subtest in this edition, no information is given regarding the impact of these changes on the internal consistency of the subtest.

Practical Applications/Uses

The stated purpose of the Social Intelligence Test is to measure certain factors related to social interactions through subtests designed to examine judgments, information, and memory (Moss et al., 1955). Historically this test has been used as a prediction variable for business testing programs (Bruce, 1953), and the authors suggest it as a means for selecting and placing employees in jobs that involve interpersonal relationships (Moss et al., 1955). Further, Moss et al. indicate that the test may be helpful in identifying deficiencies in students presenting adjustment problems.

The Social Intelligence Test can be administered by a variety of qualified personnel (e.g., vocational counselors, guidance counselors, personnel managers, teachers, etc.). Time duration is the primary responsibility for the proctor. The test is easily administered, to groups or individuals, and could be considered self-administering. The manual clearly indicates the proctor's role and the directions to be given.

Scoring the test can be completed in approximately 30 minutes, depending on the edition used. Raw scores for each subtest are determined and entered onto a summary table. Three subtests (Recognition of the Mental State of the Speaker, Memory for Names and Faces, Sense of Humor) utilize a weighted score, obtained through total correct multiplied by 2. No justification or support is given for this procedure.

To assist in test scoring, each edition has answer keys that correspond to the test's format. These keys are relatively easy to follow and use, with the exception of the one used for the Short Edition. This key does not identify any subtest stimuli by number. The scorer must either number each subtest answer or consistently monitor the key's and subtest's configuration; therefore, scoring the Short Edition would seem to promote a higher error frequency.

The total raw score for each edition is converted to a percentile rating. These ratings are based on intervals of five to allow for decile and quartile interpretations. These ratings may facilitate some interpretation error, though this may benefit the user, as percentile ratings explain a percentage instead of a ranking. The use of these ratings would allow a variety of qualified personnel to interpret the test scores.

Technical Aspects

Serious technical problems surround the Revised Form of the Social Intelligence Test and its two companion tests. The most obvious problem relates to reliability. The authors report a total of seven reliability studies, with one each conducted on the current Revised Form, Second Edition (1949), the Short Edition (1947), and the SP Edition (1955). No current reliability information is reported for the revised form, SP Edition (1978).

Only one test-retest reliability study has been conducted (circa 1927), using Form 1 on 100 college sophomores ($r = .83$). Another study (circa 1930) found a moderate correlation ($r = .83$) between Forms 1 and 2. The remaining reliability data consist of content samplings utilizing split-half studies. Of these split-half studies reported in the 1955 test manual, only three focus on the current versions of the test, excluding the 1978 SP Edition. Split-half reliabilities for the Revised Form, Second Edition; the Short Edition; and 1955 SP Edition were $r = .85$, $r = .81$, and $r = .72$, respectively. These reliability studies were conducted on 100 college freshmen, 150 high school seniors, and 120 adult high school graduates. The reliability data provided do not conform to the ranges of the normative groups represented on the raw score conversion tables. Additionally, given established standards (Guilford & Fruchter, 1978; Nunnally, 1978; Kelley, 1927), it would appear the three editions of the Social Intelligence Test have moderately acceptable reliability on the samples reported, though these data may not generalize to the population at large.

Validity studies focusing on the Social Intelligence Test seem to be of an equivocal nature. Moss et al. (1955) reported that "validity studies have demonstrated conflicting and ambiguous results" (p. 2). No indication of content validity procedures and/or item development and analysis were provided by the authors. The validity studies reported in the manual generally reflect diagnostic validity, rather than factor identification and construct validity.

In the original validation study, Hunt (1928) found a high correlation ($r = .61$) between test scores and ability rating of employees. Differences were also reported between the Social Intelligence Test scores of college students involved in extracurricular activities and those not involved. However, the significance of this validation study seems to be a description of the groups, rather than the location of significant differences between or within these groups.

Flemming and Flemming (1946) also reported a significant criterion validity index betwen judgments of sale performance made by company executives and test findings. Kaess and Witryol (1955), examining the Memory for Names and Faces subtest, found that though it possibly measured a useful skill, it lacked sufficient generality for widespread practical application.

Questions regarding validity of the Social Intelligence Test have been raised in several studies (Bruce, 1953; Taylor, 1949; Thorndike, 1940; Thorndike, 1936; Thorndike & Stein, 1937). Moss et al. (1955) reported correlation coefficients between the Social Intelligence Test and various tests of abstract reasoning (i.e., verbal intelligence) tests. The mean correlation coefficient between these measures was .55. It would appear that a significant amount of common factor space is shared among verbal and social intelligence tests, a position supported by Thorndike and Thorndike and Stein. Specifically, Thorndike and Stein (1937) reported factors consisting of verbal ability and speed. These researchers identified some evidence for a social intelligence factor, though this factor was not independent of the primary verbal ability factor (Taylor, 1949; Thorndike, 1940; Thorndike & Stein, 1937). Bruce (1953), predicting the effectiveness of factory foremen, found the SP Edition highly correlated with the Otis Self-Administering Test of Mental Ability and tests of reading, spelling, computation, and so on. Therefore, the Social Intelligence Test contributed minimally as a criterion variable.

Critique

Clearly, the Social Intelligence Test and the measured construct of social intelligence are equivocal. While technical aspects of the test (i.e., reliability, validity, normative data, etc.) fail to demonstrate consistent support for any of the three editions, a far more serious breach surrounds the 1949 Revised Form's Sense of Humor subtest. As noted previously, this subtest violates current social consciousness and respectability in its depiction of blacks and women. While the racially oriented "jokes" are eliminated from the Short Edition, a more important concern is the presence of racial and/or sexist humor on the companion versions. These shortcomings would appear to preclude using this test and its two companions. Consumers may consider utilizing the 1978 SP Edition, although manual and norms are over 30 years old now. Further, additional information has not been provided indicating what effects, if any, its non-sexist language had on the constructs of the test.

In summary, the validity of the Social Intelligence Test and the construct of social intelligence appear highly suspect. Thorndike's (1940) initial review of the test has been supported by Bruce (1953) and Taylor (1949). Additionally, Keating (1978) questions the validity of the social intelligence construct and concludes that the putative social intelligence domain is empirically incoherent. The assessment of social intelligence, as measured by existing tests, would appear to overlap with the abilities and skills measurements of other tests of abstract (verbal) intelligence and academic abilities. It would further appear that any attempt to measure social competence must be performed through carefully controlled observations.

References

Bruce, M. (1953). The prediction of effectiveness as a factory foreman. *Psychological Monographs: General and Applied, 67*, 1-17.

Ferguson, G. (1981). *Statistical analysis in psychology and education*. New York: McGraw-Hill.

Flemming, E., & Flemming, C. (1946). A qualitative approach to the problem of improving selection of salesmen by psychological tests. *Journal of Psychology, 21*, 127-150.

Guilford, J., & Fruchter, B. (1978). *Fundamental statistics in psychology and education*. New York: McGraw-Hill.

Hunt, T. (1928). The measurement of social intelligence. *Journal of Applied Psychology, 12*, 317-334.

Kaess, W., & Witryol, S. (1955). Memory for names and faces: A characteristic of social intelligence? *Journal of Applied Psychology, 39*, 457-462.

Keating, D. (1978). A search for social intelligence. *Journal of Educational Psychology, 70*, 218-223.

Kelley, T. (1927). *Interpretation of educational measurement.* Yonkers-on-Hudson, NY: World Press.

Moss, F., Hunt, T., & Omwake, K. (1947). *George Washington University Series: Social Intelligence Test (Revised Form, Short Edition).* Washington, DC: Center for Psychological Service.

Moss, F., Hunt, T., & Omwake, K. (1978). *George Washington University Series: Social Intelligence Test (Revised Form, SP Edition).* Washington, DC: Center for Psychological Service.

Moss, F., Hunt, T., Omwake, K., & Woodward, L. (1949). *George Washington University Series: Social Intelligence Test (Revised Form, Second Edition).* Washington, DC: Center for Psychological Service.

Moss, F., Hunt, T., Omwake, K., & Woodward, L. (1955). *George Washington University series: Social Intelligence Test manual.* Washington, DC: Center for Psychological Service.

Nunnally, J. (1978). *Psychometric theory* (4th ed.). New York: McGraw-Hill.

Taylor, H. (1949). Social Intelligence Test: George Washington University Series. In O. K. Buros (Ed.), *The third mental measurements yearbook* (p. 197). Highland Park, NJ: The Gryphon Press.

Thorndike, R. (1936). Factor analysis of social and abstract intelligence. *Journal of Educational Psychology, 27*, 231-233.

Thorndike, R. (1940). Social Intelligence Test. In O. K. Buros (Ed.), *The 1940 mental measurements yearbook* (pp. 1253). Highland Park, NJ: The Gryphon Press.

Thorndike, R., & Stein, S. (1937). An evaluation of the attempts to measure social intelligence. *The Psychological Bulletin, 34*, 275-285.

Patricia L. Hollingsworth, Ed.D.

Director, University School for Gifted Children, The University of Tulsa, Tulsa, Oklahoma.

SOI-LEARNING ABILITIES TEST: SCREENING FORM FOR GIFTED

Mary Meeker and Robert Meeker. Los Angeles, California: Western Psychological Services.

Introduction

The SOI-Learning Abilities Test: Screening Form for Gifted (Form G) is a seven-page, multiple-item, paper-and-pencil test consisting of 12 of the 26 subtests of the SOI-LA. The subtests deal with a combination of three types of intellectual functioning, first identified by J.P. Guilford: content abilities, operation abilities, and product abilities.

The various Structure of Intellect Learning Abilities Tests (SOI-LA), including Form G and forms for children ages 3 to 7, were developed by Mary Meeker, Ed.D., and Robert Meeker, Ed.D. The basic SOI-LA Forms A and B and the Career Form consist of 26 factorial abilities. From these, a select group of 12 subtests comprise the gifted screening tests. The subtests were first identified as predictors of giftedness by established procedures using the SOI Analyses (Meeker, 1969) of the Stanford-Binet Intelligence Scale: Form L-M (Terman, Merrill, & Thorndike, 1973). Secondly, tests were developed from 1962 through 1973 to validate the factors. The SOI-Learning Abilities Test: Screening Form for Gifted (Form G) includes two subtests to assess figural and semantic divergent or creative production. Thus, Form G is intended to assess a variety of cognitive abilities, including two creative abilities.

All SOI-LA tests are based on the multifactorial model of intelligence developed over a period of 20 years by J. P. Guilford (1959, 1967, 1977). Each of the 26 subtests is identified by a three-letter code (trigram) and a name and is designed to measure a factor of intelligence in the Guilford cube. By combining the 26 subtests, the 14 Structure of Intellect general intellectual abilities can be obtained.

Guilford's model, known as the SI model, is the product of extensive factor analytic research (1966, 1977). One dimension of the cube represents the mental *operations* or processess (capital letters indicate SOI abbreviations): Cognition, Memory, Evaluation, coNvergent production, and Divergent production. Another dimension represents the types of intellectual *content* operated on: Figural, Symbolic, and seMantic. The third dimension represents *products* (levels of informational complexity): Units, Classes, Relations, Systems, Transformations, and Implications. These are the 14 SOI general intellectual abilities.

The 12 SOI learning abilities assessed in Form G are 1) Comprehension of Figural Units (CFU), which measures visual closure with the recommended time limit of 3 minutes for gifted; 2) Comprehension of SeMantic Units (CMU), which measures vocabulary understanding with the time limit of 5 minutes; 3) Comprehension of SeMantic Relations (CMR), which measures verbal relations or analogies, 5 minutes; 4) Comprehension of SeMantic Systems (CMS), which measures comprehension of complicated verbal information, 5 minutes; 5) Memory for Symbolic Visual Units (MSUv), which measures memory for visual symbols, 5 minutes; 6) Memory for Symbolic Visual Systems (MSSv), which measures memory for visual sequence, 5 minutes; 7) Memory for Symbolic Auditory Units (MSUa), which measures memory for auditory symbols, 5 minutes; 8) Memory for Symbolic Auditory Systems (MSSa), which measures memory for auditory sequence, 5 minutes; 9) Problem Solving with Symbolic Transformations (NST), which measures speed of word recognition, 10 minutes; 10) CoNvergent Production with Symbolic Implications (NSI), which measures ability to utilize logic and reasoning, 5 minutes; 11) Divergent Production with Figural Units (DFU), fluency and originality of ideas, 5 minutes; 12) Divergent Production with SeMantic Units (DMU), creativity with words and ideas, 5 minutes.

Dr. Mary Meeker, a clinical psychologist, began in 1962 to apply Guilford's theory of intelligence to school settings. Meeker, who had studied with Guilford, became disillusioned with conventional IQ scores because they gave so little information for teachers to use. What followed was 13 years of research and development prior to the 1975 publication of the SOI-LA. In that same year, Dr. Meeker was selected by the U. S. Office of Education as one of the five most outstanding social psychologists in the United States for her research on the specific kinds of intellectual abilities needed for successful school learning. She is the author of numerous books and articles and has lectured widely in the United States and abroad.

Her husband, Dr. Robert Meeker, began his professional career as a human factors scientist at System Development Corporation and later worked extensively with computer-based survey research programs at the University of California-Los Angeles. His principal work in education has been with the development of computer programs and training materials for the SOI Institute.

The authors continue to develop and refine the SOI products. New York City is using the new SOI Screening Kit, which has norms on 22,000 kindergarten through third-grade students who speak over 60 different languages. The SOI materials have been used extensively in the United States and in many foreign countries.

The SOI-LA tests, based on those used by Guilford and his colleagues, were scaled to be age appropriate for students in Grades 2 through 12, as well as for adults. Of the 120 abilities in the Guilford SI model, 26 were selected by Dr. Meeker by means of a series of research studies investigating how each related to school achievement.

Because Form G is derived directly from the SOI-LA, the majority of normative studies concentrated on the SOI-LA Forms A and B. Thus, norms for Form G are the same as for Forms A and B. To insure against cultural bias, a pilot norming was conducted based on a sample of whites, blacks, and Hispanics from inner-city and suburban elementary schools. These data were used as preliminary norms for the tests published in 1975. In 1980, the SOI-LA Forms A and B were renormed. The test was standardized on a sample of 2,014 elementary students and stratified by grade levels, geographical locations, ethnic backgrounds, and city and rural residencies.

In addition, approximately 2,000 students, Grades 7 to 12, and approximately 500 adults contributed to the norming data. Since that time, many school districts have established their own averages on their gifted populations.

The Primary Forms, Reasoning Readiness and the Process and Diagnostic Forms, are designed to be developmentally appropriate for use with students in Kindergarten through second grade and are thus in a larger visual format than the other forms. The Process and Diagnostic Form contains 11 subtests from the basic forms of the SOI-LA. These forms can be used for identification of younger gifted children. Both forms can be administered to a group or to an individual. The SOI-LA: Screening Form for Atypical Gifted is comprised of subtests from Form A on which culturally and linguistically diverse populations have made scores in the gifted range. It is designed to identify students who are usually penalized and excluded from gifted programs because of their weaknesses in the semantic abilities. The Screening Form for Atypical Gifted offers a full range of nonsemantic (figural and symbolic) abilities for testing. Students selected in this manner will need a gifted program that is designed specifically to meet their strengths, rather than the traditional, semantically oriented program (Meeker & Meeker, 1986). SOI training materials are devised to be used to develop individual and group programs. The SOI-LA is also available in French, Spanish, and German.

The test profile is printed on the front of the test booklet using standard deviations. There are columns of scores at these points for each subtest, each grade (2 through 6), the intermediate level, and the adult level. There are rows of percentiles and evaluation levels. In addition, stanines are provided. The profile is easily graphed and used. Within the correct age group band, the examiner circles the nearest number that corresponds to the subtest raw score. The circles are then connected, and the profile can be read using the evaluation level on the far left. The evaluation levels are based on the following percentile equivalents: 94%, gifted; 84%, superior; 66%, high average; 50%, average; 34%, low average; 16%, limiting; 6%, disabling.

Practical Applications/Uses

From a practical point of view, the SOI-Learning Abilities Test: Screening Form for Gifted can be useful to any professional who wants more information and direction for curriculum instruction than is provided by unitary IQ scores or grade-level achievement tests. The SOI assessment materials are unique in that Form G gives an extensive profile of abilities. These abilities are directly keyed to SOI instructional materials and strategies for each specific ability tested.

Form G can be used for the selection and placement of students in gifted programs. The test yields a useful individual profile of 12 intellectual abilities showing specific strengths and weaknesses that can be shared with parents, teachers, and the student. The manual (Meeker, Meeker, & Roid, 1985) provides a table of the number of subtest scores in the gifted and superior range that are suggested as criteria for identification of students for gifted programs. For example, if a school district takes the top 3% of students, a fifth-grader needs to have scored in the gifted range on 7 of the 12 subtests.

The next logical step after using Form G in selection for a gifted program would be to use the profile to plan individual educational programs. Training materials come in two forms: 1) sourcebooks of lessons for group instruction and 2) self-help modules for individual instruction. Each self-help module, designed to train one SOI ability, comes in an attractive workbook form that most students enjoy using. None of the training materials is similar in stimuli to any test items. The SOI training materials can be adapted to meet specific program goals. For example, different programs could be devised depending on whether the program goal was enhancing academic achievement, developing creative abilities, or maintaining cognitive strengths.

In addition to being used for assessment and training, Form G can be used for accountability and evaluation. After the initial assessment and prescribed training, reassessment on the SOI-LA will judge whether the program goals were met. The creativity subtests allow predictive correlations with the SOI Rating Scale.

Form G can be used as a means of communicating the individual goals and progress of the program to students, teachers, and parents (Meeker & Meeker, 1986). There are report forms to translate students' profiles into readable and understandable language for parents and administrators—all children have intelligence; the SOI communicates "what kind."

The usefulness of Form G in the school setting can be equaled in other settings also. The instrument's power to assess and prescribe for specific intellectual abilities makes it useful in a clinical setting to discover a variety of learning-related problems.

Form G is a group test that also can be administered individually. It was designed as a cost-effective way to identify giftedness. The test can be used as a screening device for entrance into gifted programs, as a general cognitive assessment for the gifted, as a guide for program and curriculum planning, and as a gifted program evaluation instrument. The manual specifically states that all of the SOI-LA tests are assumed to be part of a larger collection of data from a variety of sources. In other words, Form G is not intended to be the sole instrument to be used in making an identification or placement decision. It is recommended that other tests, grades, products (e.g., books, inventions, stories, plays, etc.), motivation, and talent also be used for selection consideration. If children have been prescreened by other selection devices, the four Memory tests can be omitted. A unique aspect of all the SOI tests is that they not only identify intellectual strengths and weaknesses, but they also are keyed to prescribed curriculum training materials and specific training methods, which can be used for either remediation or enrichment.

Form G can be used in a school, clinic, or home setting by school administrators, teachers, school psychologists, psychometricians, clinical psychologists, and counselors. The test is appropriate for assessing gifted abilities of both children and adults. By the use of written instructions, as on the Career Form, Form G can be used for hearing-impaired or deaf persons. Form G is not appropriate for "low-socioeconomic, non-linguistic-proficient students" (Meeker, Meeker, & Roid, 1985, p.67). Such students or adults should be tested on Forms A and B or the Atypical Gifted Screening Form.

Form G can be administered by trained teachers, other professionals, or paraprofessionals. A classroom is an appropriate setting for the test because it generally contains the requisite chairs and tables, at a comfortable writing height, and access to seeing and hearing the examiner. A test booklet and soft-leaded pencils need to be provided for each person taking the test.

The examiner needs to read all directions thoroughly and practice giving the test prior to administering it. The examiner determines the number of testing sessions and the sequence, although following the sequence in the test booklet is strongly suggested. In order to administer the test, the examiner needs the manual, the stimulus cards for the Memory subtests, a test booklet, and a stopwatch or wrist-watch with a second hand. Before beginning the administration, the examiner explains the nature and purpose of the test. Next, the examiner reads the instructions from the manual, which is clear and easy to understand. (Although there is a detailed table of contents, it would be helpful to have an index.) Approximately 1 hour, 15 minutes is required to administer Form G. The subtests' time limits are indicated in the Introduction section of this review.

The examiner leads students through the test by reading instructions, providing visuals when required, and timing each subtest. The test can be given to students in Grades 2 through 12 (and to adults). Most of the subtests are articulated, beginning with easy items and then moving to the more difficult. Examinees write directly in the test booklet. Scantron sheets are not used because of immature vision functions in children through age 12. The test booklet contains a variety of figural, symbolic, and semantic material. To maintain interest, some answers are to be circled, some are to be written, some are drawn, and some have small boxes to be filled in below the answers. Because this is a test of a variety of intelligences, there is variety in the test items and in the modes of responses.

An hour of reading and practice with the manual will prepare a person to determine both subtest scores and general ability scores. The instructions for scoring are clearly presented and, for most of the subtests, are easy to follow. Only the two subtests to assess Divergent Production (DFU and DMU) are complex. The other 10 subtests can be scored easily and quickly. The scoring criteria for each of these subtests is included on the transparent scoring keys. Once having read the instructions, the scoring of these 10 subtests takes only about 15 or 20 minutes. After obtaining subtest scores, general ability scores can be computed with a calculator using the formula provided in the manual.

The Divergent Production subtests have explicit, though fairly complex, directions. Four factors—fluency, set change, transformation, and originality—are considered in scoring Divergent Production of Figural Units, a test that requires subjects to draw. Divergent Production of Semantic Units, which requires subjects to write a story, is scored on fluency and originality. Points are given for each of the factors assessed and are totaled for a raw score for both of the subtests. General abilities scores for these subtests are computed in the same manner as for the other subtests.

Although the Divergent Production subtests do require judgment in scoring, interrater reliability scores have been high, ranging from .91 to .95 (Thompson & Andersson, 1983). Most classroom teachers would be able to competently score both DFU and DMU, but it would be time-consuming if there were many students taking the test—both subtests can take 30 minutes to score. Although all SOI-LA tests can be hand scored, computer analysis is available from SOI Systems, Vida, Oregon, and is advisable to anyone who feels insecure about scoring the test. However, it can be beneficial to a classroom or resource room teacher to score the subtests by hand because of the information and insights into individual intellectual functioning gained by doing so. This can help the teacher move from assessment into prescriptive training materials by becoming more aware of individuals' needs.

Technical Aspects

Reliability, the degree to which test scores are dependable, consistent, and repeatable, is reported from a variety of sources. Divergent Production subtests require a degree of scoring judgment, thus necessitating estimates of interrater reliability. Interrater reliability correlations for three raters of 205 sixth-grade students ranged from .91 to .95 (Thompson & Andersson, 1983). Test-retest reliability coefficients were calculated during the normative study for Forms A and B and used the subjects previously described. Initially, half of the subjects were given Form A ($N = 514$) and half Form B ($N = 507$). Within 2 to 4 weeks subjects were retested. Half of those initially given Form A were retested on Form B and the other half on Form A. Those who were initially given Form B were retested using the same design conversely. Test-retest correlations range from a low of .46 for Memory of Symbolic Systems-Auditory and Divergent Figural Units to a high of .88 for Convergent Production of Symbolic Transformations. Alternate forms correlations range from a low of .43 for Memory of Symbolic Systems-Auditory to a high of .84 for Convergent Production of Symbolic Transformations.

Although some of the individual subtests' correlations are lower than one might wish (Coffman, 1985)—notably memory and divergent production—subtests by themselves are not intended to be interpreted in isolation (Meeker & Meeker, 1980). Pupil-related educational decisions are to be made within the context of other information that the test provides, such as indications of the 14 general intellectual abilities. Test-retest correlations for the general abilities range from a low of .65 for Divergent Production to highs of .90 for Cognition and .91 for Symbolic Abilities. Alternate form correlations for the general abilities range from a low of .63 for Divergent Production to highs of .87 for Cognition and .89 for Symbolic Abilities. For both test-retest and alternate form correlations, a large majority of the correlations are in the .70s and .80s.

Decision-consistency indexes assess the degree to which a test provides stable information in situations in which dichotomous decisions are made. This would be particularly important reliability information to one using Form G to identify students for a gifted program. Using the original normative sample (Meeker, Meeker, & Roid, 1985) students identified as gifted on the initial test were compared to students who were likewise identified on the second administration of the test. The decision-consistency reliability ranged from .83 to .85 in the normative samples for Grades 2 to 6.

In deciding whether a test has a sufficiently high reliability coefficient, the most important consideration is the gravity of the decision to be made on the basis of the test score (Wesman, 1976). Because the manual clearly states that 1) the SOI-LA forms are not to be used in isolation of other SOI information, 2) the test is to be used in conjunction with other assessment instruments, and 3) the assessment instrument is to be used with the SOI training materials, in this reviewer's opinion there is little risk in the decisions made using the SOI-LA forms. Considering the evidence and the intent and purpose, there should not be any serious concerns regarding the reliability of Form G.

The manual provides evidence concerning content validity, criterion validity, (both concurrent and predictive), and construct validity. Additionally, decades of research by Guilford (1967) provide evidence of the representativeness of the cognitive and learning abilities in the SOI model. Many studies by Meeker (1963, 1965, 1966, 1969) have established the link between Guilford's SI factors and school learning.

Three types of evidence of criterion validity are presented in the manual: 1) diagnostic, 2) concurrent, and 3) predictive utility with school achievement and teacher ratings. Several studies reported that students selected for gifted programs on measures other than the SOI-LA also scored in the gifted range on the SOI-LA tests. Pearce's (1983) study ($n = 59$) of high-achieving students found a significant correlation between the WISC-R Full Scale IQ and the SOI-LA CMS ($r = .48$, $p < .01$) and NST ($r = .31$, $p < .05$) subtests. Gore (1980) reported significant differences ($p = .006$ and $p = .001$) when comparing junior high identified-gifted students with regular classroom students on two SOI-LA subtests (DFU and DMU). Thompson, Alston, Cunningham, and Wakefield (1978) compared reading and math achievement of 145 fourth- and fifth-graders on the Iowa Tests of Basic Skills with hypothesized correlates of reading and math on the SOI-LA, the reading correlation was .59 ($F = 7.9$, $p < .05$), and the math correlation was .83 ($F = 16.26$, $p < .05$). Johnson (1979) found 22 significant ($p < .05$) correlations in a comparison of the SOI-LA to the Peabody Individual Achievement Test. In Stenson's 1982 study of 339 elementary students the correlation between Form G and the WISC-R showed a multiple correlation of .337 ($p < .05$).

A number of studies are presented that provide confirmatory factor analysis of the SOI-LA subtests. Maxwell (1984) reported all SOI-LA memory subtests to have significant factor loadings (lambda coefficients) on their respective factors. Roid (1984) found all 9 figural subtests to load on the figural factor, 10 of 13 symbolic subtests were confirmed, and 3 of 4 semantic subtests were confirmed. In addition, extensive documentation of construct validity evidence is provided for the Guilford model.

The new SOI-LA test manual (1985) has solved several of the problems detailed by Coffman (1985) and Leton (1985). The percentiles in the manual now correspond with those on the test profile. The manual provides the standard error of measurement for 40 subtest and general abilities. For 35 out of 40 of these the standard error of measurement is smaller than the width of the profile category.

In terms of use validity (Leton, 1985), the new manual provides studies showing the effects of SOI training on specific abilities. Robert Meeker (1979) conducted a study ($N = 157$) concerning the development of divergent production with gifted students in Grades 4 through 6. Students were pretested and posttested on SOI-LA divergent production subtests, with the experimental group receiving SOI training in divergent production while the control group received instruction in convergent production. Statistically significant differences ranging from $p < .05$ to $p < .01$ were reported in five of the eight categories of divergent production.

In another study (Hengen, Keith, & Bessai, 1982), children in Grades 4 through 6 were pretested and posttested on the SOI-LA and the Wide Range Achievement Test (WRAT). The experimental group ($n = 47$) received SOI instruction but the control group ($n = 49$) did not. The results showed significant improvement (p values not reported in the manual) by the experimental group in 11 of the 24 SOI-LA subtests, 9 out of 14 of the general abilities, and in all three subject areas (reading, arithmetic, and spelling) on the WRAT.

Buisman's (1981) study, involving students from Grades 1 through 8 ($N = 40$), was to determine the effect of SOI training on learning abilities and achievement. Students were pretested and posttested on the SOI-LA and given the California Achievement Test (CAT) at the end of the school year. Students given SOI training showed statistically significant gains in 18 of the 26 learning abilities. A correlation of .72 was reported between the CAT and the SOI-LA.

The SOI-LA Test Manual (1985) makes it clear that all of the SOI-LA tests are to be used to give information and training to persons concerning their learning abilities. Form G is not intended to be used as the sole instrument for identification or labeling but rather for focused feedback and educational training.

Every effort seems to have been made by the authors of the manual (Meeker, Meeker, & Roid, 1985) to provide complete and thorough evidence concerning validity and reliability in accordance with the Standards for Educational and Psychological Testing (American Educational Research Association, American Psychological Association, National Council on Measurement in Education, 1985). Although there was concern in the past (Coffman, 1985), the present evidence for validity and reliability seem quite adequate for the purpose and intent of the instrument.

Critique

The SOI-Learning Abilites Test: Screening Form for Gifted is an important and useful tool, particularly for schools wanting to identify gifted students and develop a program for them. With Form G, schools can assess strengths and weaknesses of intellectual functioning and develop a program suited to individual needs. This unique aspect of the SOI-LA materials has been noted by J. P. Guilford (1977). In his popular classic *Way Beyond the IQ*, Guilford writes:

> Mary N. Meeker . . . has probably done most to apply SI concepts to educational problems. . . . She has not only developed new SI tests for children but also workbooks of exercises for different SI abilities. . . . She has founded the SOI Institute for fostering the use of SI in education. (p. 184)

Form G has the advantages of 1) a readable manual, 2) test administration that requires a minimum of special equipment and preparation, 3) a relatively short time period for administration, 4) inexpensive test and training materials, 5) interesting test and training booklets, 6) adequate evidence of reliability and validity, 7) moderately easy scoring, 8) two ways to profile score, and 9) training materials that can remediate weaknesses and/or enhance strengths.

The Divergent Production subtests require some judgment in scoring. However, if this proves to be too difficult, or too time-consuming, the test can be sent to SOI Systems for scoring. As with any instrument, Form G needs to be used as instructed, with the intent and purpose of the authors in mind (i.e., it should be used in conjunction with other instruments in identifying gifted students, and the subtest scores are not to be used in isolation of other test information). If Form G is used as it was intended, it is a unique, invaluable educational instrument.

References

This list includes text citations and suggested additional readings.

American Educational Research Association, American Psychological Association, & National Council on Measurement in Education. (1985). *Standards for educational and psychological testing*. Washington, DC: American Psychological Association.

Buisman, J. (1981). *SOI training related to achievement*. Unpublished manuscript, Sherwood School District, Sherwood, OR. (Available from SOI Systems, Box D, Vida, OR 97488)

Coffman, W. E. (1985). Structure of Intellect Learning Abilities Test. In J. V. Mitchell, Jr. (Ed.), *The ninth mental measurements yearbook* (pp. 1486-1488). Lincoln, NE: Buros Institute of Mental Measurements.

Gore, D. (1980). *SOI training in memory and creativity*. Unpublished manuscript, Greater Albany School District, Albany, OR. (Available from SOI Systems, Box D, Vida, OR 97488)

Guilford, J. P. (1959). Three faces of intellect. *American Psychologist, 14,* 469-479.

Guilford, J. P. (1966). Intelligence: 1965 model. *American Psychologist, 21,* 20-26.

Guilford, J. P. (1967). *The nature of human intelligence*. New York: McGraw-Hill.

Guilford, J. P. (1977). *Way beyond the IQ*. Buffalo, NY: Creative Education Foundation.

Hengen, T., Keith, C., & Bessai, F. (1982). *Identification and enhancement of giftedness in Canadian Indians*. Unpublished manuscript, The Regina Catholic Schools, Regina, Saskatchewan. (Available from SOI Systems, Box D, Vida, OR 97488)

Johnson, B. (1979). *The relationships between the Structure of Intellect Learning Abilities Test (SOI-LA) test scores and the Peabody Individual Achievement Test (PIAT) scores of educationally handicapped junior high students*. Unpublished doctoral dissertation, University of Denver.

Leton, D. A. (1985). Structure of Intellect Learning Abilities Test. In J. V. Mitchell, Jr. (Ed.), *The ninth mental measurements yearbook* (pp. 1488-1489). Lincoln, NE: Buros Institute of Mental Measurements.

Meeker, M. (1963, August). *The NSWP behavior samplings in the Binet*. Paper presented at the annual meeting of the American Psychological Association, Philadelphia.

Meeker, M. (1965). A procedure for relating Stanford-Binet behavior samplings to Guilford's structure of intellect. *Journal of School Psychology, 3*(3), 26-36.

Meeker, M. (1966). *Immediate memory and its correlates with school achievement*. Unpublished doctoral dissertation, University of Southern California, Los Angeles.

Meeker, M. (1969). *The structure of intellect: Its interpretation and uses*. Columbus, OH: Charles E. Merrill.

Meeker, M., & Meeker, R. (1980). *SOI inservice training manual*. El Segundo, CA: SOI Institute.

Meeker, M., & Meeker, R. (1986). The SOI system for gifted education. In J. S. Renzulli (Ed.), *Systems and models for developing programs for the gifted and talented* (pp. 194-215). Mansfield Center, CT: Creative Learning Press.

Meeker, M., Meeker, R., & Roid, G. H. (1985). *Structure of Intellect Learning Abilities Test manual*. Los Angeles, CA: Western Psychological Services.

Meeker, R. (1979). Can creativity be developed in gifted? *Roeper Review, 1,* 17-18.

Pearce, N. (1983). A comparison of the WISC-R, Raven's Standard Progressive Matrices, and Meeker's SOI Screening Form for Gifted. *Gifted Child Quarterly, 27,* 13-19.

Roid, G. H. (1984). Construct validity of the figural, symbolic, and semantic dimensions of the Structure-of-Intellect Learning Abilities Test. *Educational and Psychological Measurement, 44,* 697-702.

Stenson, C. M. (1982). Note on concurrent validity of Structure of Intellect Gifted Screener with Wechsler Intelligence Scale for Children-Revised. *Psychological Reports, 50,* 552.

Terman, L. M., Merrill, M. A., & thorndike, R. L. (1973). *Stanford-Binet Intelligence Scale: Manual for the third revision.* Chicago: Riverside Publishing.

Thompson, B., Alston, H. L., Cunningham, C. H., & Wakefield, J. A., Jr. (1978). The relationship of a measure of structure of intellect abilities and academic achievement. *Educational and Psychological Measurement, 38,* 1207-1210.

Thompson, B., & Andersson, B. V. (1983). Construct validity of the divergent production subtests from the Structure of Intellect Learning Abilities Test. *Educational and Psychological Measurement, 43,* 651-655.

Wesman, A. G. (1976). Reliability and confidence. In W. A. Mehrens (Ed.), *Readings in measurement and evaluation* (pp. 35-44). New York: Holt, Rinehart & Winston.

William T. Martin, Ph.D.
Director of Psychological Services, Winfield State Hospital and Training Center, Winfield, Kansas.

STAFF BURNOUT SCALE FOR HEALTH PROFESSIONALS

John W. Jones, Park Ridge, Illinois: London House Press.

Introduction

The Staff Burnout Scale for Health Professionals (SBS-HP; Jones, 1980b, 1980c) was designed to measure the degree of burnout among health professionals, primarily those in the medical professions. The SBS-HP was based on definitions of burnout of Maslach (1976), Maslach and Pines (1977), and Pines and Maslach (1978). The SBS-HP is an objective, paper-and-pencil instrument that employs a rating scale method that examinees use to rate their perceived degree of agreement with the instrument items. A single Burnout score and a Lie score, which the examiner uses to ascertain the degree to which the examinee may be attempting to "fake good," are obtained. Jones (1980c, p. 2) described the SBS-HP as an instrument designed to assess adverse cognitive, emotional, behavioral, and physiological reactions that comprise the 'burnout syndrome' defined by Maslach and Pines." Jones (1980b) noted that in a series of validity studies that he reviewed in 1981, the SBS-HP scores reliably correlated with items such as turnover, absenteeism, tardiness, physical illness, job errors and patient neglect, job dissatisfaction, substance abuse, and stress as related to work shift.

The SBS-HP is published by London House, Inc., of Park Ridge, Illinois. The author of the test, John W. Jones, Ph.D., C.A.C., was an instructor with the Psychology Department, DePaul University, Chicago, Illinois, when he developed the test. Dr. Jones was associated with London House Management Consultants of Park Ridge, Illinois, through his research with the test. Jones has presented two papers addressing the topic of staff burnout. The first, presented at the 52nd annual meeting of the Midwestern Psychological Association in St. Louis, Missouri, was titled *The Staff Burnout Scale: A Validity Study* (Jones, 1980c). Later, at the first annual Chicago Conference on Behavior Analysis, Jones (1980a) presented a paper titled *Environmental and Cognitive Correlates of Staff Burnout: Implications for Rational Restructuring*. Specific early strategies on the development of the SBS-HP were not available to this reviewer.

Although the results of a wide-scale normative study do not appear to be available at this time, normative development of the SBS-HP appears to be continuing. A bibliographical listing of 18 normative studies using the SBS-HP was supplied to this reviewer. Seven of these studies were conducted by the test author, eight were either master's theses or doctoral dissertations, and three were miscellaneous studies. Several other studies have been published by London House. Jones (1980b) suggests that SBS-HP users develop local norms for the test.

The SBS-HP is a 30-item instrument printed on an 11"x17" sheet folded to make an 8½"x11" booklet. The examinee uses a 6-point Likert-type scale (agree very much, agree pretty much, agree a little, disagree a little, disagree pretty much, and disagree very much) to rate each of the 30 items. The examinee writes his or her name, social security number, and occupational position on a separate page of the booklet.

A four-page preliminary test manual contains sections on purposes of the test, administration instructions, validity, preliminary norms, a scoring key for the 20 burnout items, and a key for the 10 lie items. This reviewer found no information indicating the availability of a scoring template or other scoring device.

The manual does not provide age norms or suggest examinee groups. However, based upon the stated purpose of the SBS-HP and the nature of several validity and reliability studies examined by this reviewer, the instrument appears to be suited for employed health-care professionals. Given the qualifications required of such personnel, this reviewer assumes at least a high school education is necessary to meet the reading level difficulty of the instrument. The SBS-HP is essentially a 20-item, one-factor scale (based upon several subfactors that purportedly contribute to the majority of the variance of the total scale). A separate set of statements randomly placed in the instrument includes 10 lie items.

Practical Applications/Uses

The SBS-HP appears to be another instrument for the measurement of the burnout construct. For example, London House publishes the Staff Burnout Scale for Police and Security Personnel (SBS-PS) and the Job Burnout Scale. In a technical report, Jones (1982b) noted that empirical research on the phenomenon of burnout was in its infancy. Jones (1980b) also noted that such areas as physical and emotional exhaustion, which included subdimensions of a poor attitude about one's job, lowered self-concept, and loss of concern for clients, were some of the areas examined with regard to burnout. Negativity toward one's job included such attitudes as discouragement, pessimism, irritation with patients, and projecting one's problems onto the "system," among others. As a burnout scale, then, the SBS-HP purports to measure in an empirical manner the aforementioned constructs of an individual's job perceptions. The SBS-HP also was noted by the test publisher, London House, to have been used to assess counterproductivity, dishonesty and theft, the taking of unauthorized work breaks, job turnover, absenteeism, tardiness, and other variables (Jones, 1980c; Jones, 1981a, 1981b; Jones, 1982a, 1982b).

The SBS-HP might be used potentially by employee assistance counselors in an organization, in organizational research, for screening for stress levels during training of medical-care personnel, and in outpatient or inpatient psychiatric settings. It does not appear to be suited for use in secondary education. At this point in time, the major utility of the SBS-HP would appear to be in further research of the various constructs that purportedly comprise the syndrome of burnout.

As cited in several of the reprints supplied by the test publisher, characteristic users of the SBS-HP would include nurses, mental health technicians or aides, alcoholism and drug abuse counselors, hospital emergency room personnel, geriatric counselors and workers, oncology personnel, and others employed in related high-stress occupations. Essentially, the instrument seems most useful in either the health professional training environment or in the applied work setting of these individuals. It may not be appropriate for use with individuals who have significant insights into the construct of burnout, as test items appear to be transparent.

The instrument may be administered individually or in a supervised group setting. It also could be used as a "blind survey" technique of an organization's employees. As such, employees would take the survey in an anonymous manner, thereby promoting more valid responses. Based upon the American Psychological Association's ranking of psychological tests, the SBS-HP would appear to be categorized as a Level A test, indicating that trained personnel, such as secretaries, supervisors, psychological or management internists, company trainers, or similar persons could administer the SBS-HP effectively. Because the SBS-HP is a self-administering instrument, requiring about 5-15 minutes to complete, the examiner functions as a psychometric monitor in the testing process. The instrument is a self-contained booklet; there are no separate answer forms. Instructions (two brief sentences) are printed at the top of the answer portion of the booklet. A profile sheet was not available to this reviewer.

The SBS-HP is hand-scored. Without templates or other scoring devices (such as carbon sets), the scoring is cumbersome, but relatively foolproof. The six responses are assigned the following numerical values: 1) agree very much, 7; 2) agree pretty much, 6; 3) agree a little, 5; 4) agree a little, 4; 5) disagree pretty much, 3; 6) disagree very much, 2. The examiner assigns the numerical values of the 30 test items according to the responses marked by the examinee. The scores of the 20 items that are included on the burnout scale are summed, yielding the Burnout score. Burnout scores can range from 20 (no burnout) to 140 (severe burnout). Scoring of the Lie scale is a little more complicated. For the 10 items designated as the Lie scale, each of five designated items that had responses with a raw score of "1" are given a Transformed score of "1" each. Any of the other five items are given a Transformed score of "1" if the raw score was "7." The Transformed scores from the Lie scale are summed. Total scores can range from between 0 to 10. High scores on the Lie scale indicate an attempt by the examinee to "fake good."

Interpretation of an examinee's scores is rather straightforward. Objective scoring of the answers provides a total raw score ranging from 20 to 140 on the Burnout scale. Objective scoring of the Lie scale can produce raw scores ranging from 0 to 10. Interpretation of the relation of these scores to a norm is based upon comparing means and standard deviations to other studies. The manual noted that the mean Lie scale score was about 2.74 (SD = 2.12) with means ranging from about 51.1 (SD = 16.8) to about 62.6 (SD = 20.2) for the Burnout scale score. The number of study participants ranged from 34 to 185 among six studies (Jones, 1980b). In another summary study, Jones (1980c) indicated that SBS-HP mean Burnout scores for three groups (Ns ranging from 8 to 23; total N = 49 for the three groups), a mean SBS-HP raw score was 59.0 (SD = 28.3). In a study of 31 nurses at an emergency room trauma center in one hospital, Jones (1981a) noted that the mean Burnout scale score was about 62.6 (SD = 20.2; range about 27 to 101).

Technical Aspects

The SBS-HP appears to have been developed upon theoretical constructs supported by the accumulation of evidence in the literature regarding burnout. However, based upon the findings of studies supplied this reviewer by the test publisher (Jones, 1980a, 1980b, 1980c; Jones, 1981a, 1981b; Jones 1982a, 1982b) there did not appear to be adequate evidence of either convergent or discriminant validity. All of the studies had a small number of subjects, and concurrent or predictive evidence did not appear to be adequately presented.

Jones (1980c) studied 49 hospital health professionals in Chicago using the SBS-HP and a variety of criterion variables. Patient to staff ratios were correlated at .42, $(P \leq .05)$. The results of other comparisons were as follows: 1) SBS-HP versus absenteeism ($r = 52; p < .01$), 2) SBS-HP versus tardiness ($r = .37; p < .05$), 3) SBS-HP versus desires to look for a new job ($r = .49; p < .01$), 4) SBS-HP versus disciplinary action received ($r = .40; p < .01$), 5) SBS-HP versus alcohol use ($r = .57; p < .01$), and 6) SBS-HP versus errors made by nurses ($r = .42; p < .10$). In another comparison of Albert Ellis' 11 irrational beliefs, the SBS-HP was compared with the irrational beliefs of 38 nurses in an Alabama hospital. Of the 11 Pearson product-moment correlations ranging from $-.08$ to $+ .50$, only 6 were significant at the $p < .05$ level or above.

In another study by Jones (1981a), which examined responses among 31 nurses in a hospital trauma unit, the SBS-HP was compared with another London House scale, the Personnel Selection Inventory (London House Management Consultants, 1980). Results indicated that dishonesty scores correlated with nurses taking excessive work breaks at .33 ($p \leq .05$). Comparisons between burnout and excessive work breaks was .47 ($p = .005$). Finally, dishonesty scores correlated with burnout scores revealed a correlation of .24 ($p \geq .05$).

In a comparison of the SBS-HP, the Job Descriptive Index (JDI; Smith, Kendall, & Hulin, 1969), and London House's Personnel Security Inventory (London House Management Consultants, 1977) among anonymous responses of 33 staff nurses at an emergency trauma room in a Chicago hospital, Jones (1981a) found that no relationship existed between theft admissions and job satisfaction. However, the SBS-HP was correlated at .45 ($p \leq .01$) to the admission of overall theft. Interestingly, it was found that about 85% of the nurses had admitted to theft, with about 74% of those nurses admitting to theft of drugs.

Among an anonymous study of 89 employees from two Chicago area hospitals, Jones (1982a) compared responses on the London House Employee Attitude Inventory (London House Management Consultants, 1981). Subscales included one on job burnout, theft knowledge/suspicion, theft attitudes, drug abuse, and dissatisfaction. Theft attitudes were correlated about .56 ($p \leq .01$) with theft admissions. In addition, a correlation of about .21 ($p \leq .05$) was obtained between theft admissions and burnout. Theft attitudes were correlated about .33 ($p \leq .01$) with burnout scores.

There was no evidence in any of the reviewed articles relating to the SBS-HP that the scale had been compared to The Maslach Burnout Inventory (Maslach & Jackson, 1981). Although burnout has been related to stress in the literature, there was no evidence that the SBS-HP had been compared with any instrument measuring the stress construct. None of the concurrent validity correlations for any of the comparisons were very high, and p-values were similarly marginally significant.

Jones (1980a, 1980b) reported a Spearman-Brown split-half reliability coefficient of .93 for the SBS-HP. Because validity is a function of reliability and vice versa, the validity estimate based upon a reliability coefficient of .93 would be about .96; however, this estimate may be somewhat optimistic due to the nature of the studies reported. Item versus total score on the SBS-HP was reported to be at or equal to a p-level of .001 with mean item-with-total correlations being about .71, ranging from .59 to .82. There was no mention of test-retest reliability studies to examine the effects of temporal factors upon an examinee's responses.

Critique

This reviewer was disappointed by the lack of well-controlled studies in both the initial development and subsequent validation of the SBS-HP, as reported in the preliminary test manual and other studies. Additionally, there was a significant lack of any evidence of experimental control, such as random selection of subjects or the use of control groups. There was a strong suggestion that "available subject pools" were used to conduct the studies reported.

Although Jones' and London House's work with the burnout construct appears to have been based upon valid constructs of the various phenomena encompassing stress and burnout, these concepts did not appear to have been developed empirically with the SBS-HP. The majority of the printed evidence supplied by the publisher on this instrument apparently was generated "in-house." The other studies were based on graduate student theses and dissertations.

Although the SBS-HP is a test by all usual definitions, the preliminary test manual was significantly lacking in information that complied with APA standards for the development of psychological tests. Test-retest reliability studies could not be found among the materials reviewed. To use the Spearman-Brown formula for determining split-half reliability is to assume that the Likert-type items on the scale do, in fact, contribute to *summated ratings.* Summated ratings are the summing of ordinal scale rating values to arrive at a single score for a subscale or entire test. There is an assumption that the items contained in the construct are homogeneous. However, Ghiselli, Campbell, and Zedeck (1981) noted that the Likert-type rating method falls under the "Law of Comparative Judgment," which requires the examinee to compare one possible response with another.

Although Jones (1980b) mentioned that four factors on the SBS-HP had been found to have eigenvalues equal to or greater than 1, only one of the factors, Job Dissatisfaction, which was measured by seven test items, contributed to any significant portion of the total test variance (about 75%). It was noted that these items seemed to be related to health service provider's negative beliefs and/or feelings about the workplace (i.e., unadaptive behavior). Three other factors contributed only about 6.3%, 9.7%, and 8.7% to the total variance of the burnout dimension. Generally, factor loadings of less than .40 are not significant enough to be labeled major factors. Although an imbedded index of "fake good" or "fake bad" is important for a personality instrument, about one third of the SBS-HP's items ($n = 10$) form the Lie scale, an amount that appears to be excessive. On the other hand, if the remaining 20 burnout items had highly significant factor loadings, then the number of Lie scale items would be no problem.

Considering the very small number of subjects used in the SBS-HP studies, and their heterogeneity of study settings, the issue of validity generalization is essentially moot due to an apparent lack of research controls and the small number of subjects used in the studies. As such, attempts to examine these studies through the multitrait-multimethod process probably would not be productive. If the studies had been controlled better and had used larger numbers of subjects, then meta-analysis methods (e.g., Glass, McGaw, & Smith, 1981) would be appropriate to determine the research's general significance. On the other hand, there does seem to be somewhat of a scoring trend among the various groups used for normative purposes: most of the scores appeared to have a mean of about 50 to 70 for the various groups. However, the standard deviations of the mean scores of the sample groups tended to be rather large, which suggested significant dispersion of raw scores over the score range. There would be a lot of false positives and false negatives with such scores.

In the event that the raw scores in all of the studies reported were combined into a heterogeneous group of "health professionals" encompassing a variety of occupational situations, then these combined data might be used to develop an experimental edition of a profile sheet that examiners could use to better relate an individual's scores to local scores. Again, without data based upon random sampling, it would be difficult to generalize across health settings.

The SBS-HP seems to be a good "common-sense" instrument. It has inherent appeal, and, if used as a management consulting tool for "ice-breaking" activities, it would seem to have utility. However, due to the SBS-HP's seeming lack of predictive validity, this reviewer would be distressed to see the instrument used as an indicator of potential or present burnout, or as a predictor of absenteeism, drug use, theft, or other organizational problems. Discriminant and convergent validity studies, as well as some well-designed reliability studies, need to be conducted. Based upon the materials studied by this reviewer, the SBS-HP should be used only as a research instrument at this time. To make concurrent or predictive decisions based upon SBS-HP scores would be premature at best.

References

Ghiselli, E. E., Campbell, J. P., & Zedeck, S. (1981). *Measurement theory for the behavioral sciences.* San Francisco: Freeman.

Glass, G. V., McGaw, B., & Smith, M. L. (1981). *Meta-analysis in social research.* Beverly Hills, CA: Sage.

Jones, J. W. (1980a, September). *Environmental and cognitive correlates of staff burnout: Implications for rational restructuring.* Paper presented at the First Annual Chicago Conference on Behavior Analysis, Chicago.

Jones, J. W. (1980b). *Preliminary test manual: The Staff Burnout Scale for Health Professionals (SBS-HP).* Park Ridge, IL: London House.

Jones, J. W. (1980c, May). *The Staff Burnout Scale: A validity study.* Paper presented at the 52nd Annual Meeting of the Midwestern Psychological Association, St. Louis.

Jones, J. W. (1981a). Attitudinal correlates of employee theft of drugs and hospital supplies among nursing personnel. *Nursing Research, 30,* 349-351.

Jones, J. W. (1981b). Dishonesty, burnout, and unauthorized work break extensions. *Personality and Social Psychology Bulletin, 7,* 406-409.

Jones, J. W. (1982a). *Dishonesty, staff burnout, and employee theft* (Tech. Rep. No. E2). Park Ridge, IL: London House.

Jones, J. W. (1982b). *Measuring staff burnout* (Tech. Rep. No. E3). Park Ridge, IL: London House.

London House Management Consultants. (1977). *The Personnel Security Inventory.* Park Ridge, IL: London House.

London House Management Consultants. (1980). *The Personnel Selection Inventory.* Park Ridge, IL: London House.

London House Press. (1981). *The Employee Attitude Inventory.* Park Ridge, IL: Author.

Maslach, C. (1976). Burned out. *Human Behavior, 5,* 16-22.

Maslach, C., & Jackson, S. (1981). *The Maslach Burnout Inventory.* Palo Alto, CA: Consulting Psychologists Press.

Maslach, C., & Pines, A. (1977). The burnout syndrome in the day care setting. *Child Care Quarterly, 6,* 100-113.

Pines, A., & Maslach, C. (1978). Characteristics of staff burnout in mental health settings. *Hospital and Community Psychiatry, 29*(4), 233-237.

Smith, P. C., Kendall, L. M., & Hulin, C. L. (1969). *The measurement of satisfaction in work and retirement: A strategy for the study of attitudes.* Chicago: Rand McNally.

J. Stanley Ahmann, Ph.D.

Distinguished Professor of Education, Professor of Psychology, and Professor of Professional Studies, Iowa State University, Ames, Iowa.

STANFORD ACHIEVEMENT TEST: 7th EDITION

Eric F. Gardner, Herbert C. Rudman, Bjorn Karlsen, and Jack C. Merwin. San Antonio, Texas: The Psychological Corporation.

Introduction

For over 60 years, the successive editions of the Stanford Achievement Test (SAT) have been widely used as norm-referenced, objectively scored measures of elementary and junior high school student achievement, especially in the basic skills. The seventh edition, published in 1982, is the most recent major effort to improve this test battery and to strengthen its role as the centerpiece of the Stanford Achievement Test Series. This series of three integrated test batteries is designed to measure student achievement in basic learning areas from kindergarten through the 12th grade, beginning with the Stanford Early School Achievement Test: 2nd Edition for kindergarten and first grade and ending with the Stanford Test of Academic Skills: Second Edition for secondary school students and those entering community colleges. The SAT is administered to students in the middle group, specifically Grades 1.5 to 9.9. All levels and all forms of these three test batteries have been normed and scaled together, thereby facilitating longitudinal evaluation of growth in achievement and out-of-level testing.

All test items included in the seventh edition of the SAT are new. The large national sample used in the item tryout program (described later in this review) was drawn with considerable care. In order to improve the likelihood that the sample was representative of the national school population, variations in socioeconomic level, school system enrollment, and geographic region were considered. Approximately 50 school systems were included in the tryout program and, as a group, they were very similar to the total U.S. school system in terms of school enrollment and regional representation.

Standardization of the seventh edition of the SAT took place during the 1981-82 school year. About 250,000 students from 300 school districts participated in the fall standardization, with about 200,000 students tested as a part of the spring standardization. Again, the samples were selected through the use of a stratified random sampling technique and demographically strongly resembled the national pattern.

These research efforts established the degree of equivalence between two alternate and equivalent forms of the SAT, forms E and F, at each of the six levels included in the battery and permitted the development of extensive tables of norms, including: 1) scaled scores, 2) individual student percentile ranks and stanines, 3) grade equivalent scores, 4) normal curve equivalent scores, and

536

5) small and large group percentile ranks and stanines. Norm tables were developed for identifying the relative performance of individual students for each of the six levels for a national sample including both public and nonpublic schools and for a nonpublic school sample including only church-affiliated and private schools. Furthermore, tables of group norms were computed, but for a national sample only. All in all, the standardization of the SAT was extensive, systematic, and technically sound.

The Stanford Achievement Test Series contains a total of 10 levels, each overlapping somewhat the grades of adjacent levels:

Stanford Early School Achievement Test (SESAT)

      SESAT 1: K.0 - K.9
      SESAT 2: K.5 - 1.9

Stanford Achievement Test (SAT)

      Primary 1: 1.5 - 2.9
      Primary 2: 2.5 - 3.9
      Primary 3: 3.5 - 4.9
  Intermediate 1: 4.5 - 5.9
  Intermediate 2: 5.5 - 7.9
     Advanced: 7.0 - 9.9

Stanford Test of Academic Skills (TASK)

      TASK 1: 8.0 - 12.9
      TASK 2: 9.0 - 13.0

The position of the SAT in the Stanford Achievement Test Series is crucial; it is the linchpin of the program. That it be a high quality instrument is essential.

The fundamental focus of the Stanford Achievement Test: 7th Edition is on the basic skill areas: reading, listening, language, and mathematics. For the two lowest levels (Primary 1 and 2), the reading subtests are *Word Study Skills, Word Reading,* and *Reading Comprehension; Word Reading* is excluded at all other levels. At the advanced level, only reading comprehension is tested.

The listening subtests are *Vocabulary and Listening Comprehension* at all six levels. Administration of these subtests requires that the teacher read a short story or paragraph (usually not longer than five or six sentences) to the students, following which several (usually not more than three or four) multiple-choice questions about the passage are read aloud, with a short pause after each to allow the students to respond. Some questions only require examinees to recall what was heard; others require them to draw inferences on the basis of what was heard.

For the two lowest levels of the SAT, the only language subtest is a spelling test. A subtest devoted to this area is offered at the other four levels as well, along with a language subtest. An optional writing test is also available for these four levels, in which writing samples are obtained from the students and evaluated holistically by classroom teachers. The total writing assessment is designed to assess four different kinds of informative writing: describing, narrating, explaining, and reasoning. For each type of writing, a separate writing sample is obtained that addresses a particular topic announced by the teacher. The categories used for evaluating each sample are general merit, quantity and quality of ideas, effectiveness of organization, and wording. The evaluation of writing mechanics is optional.

The mathematics subtests of the SAT emphasize three skill areas at all six levels: concepts of number, mathematical computation, and mathematical applications. It should be noted that, for those who do not wish to administer the entire test battery, separate test booklets are available for mathematics, as well as reading, listening comprehension, and the writing assessment.

In addition to the basic skills subtests, the SAT offers a subtest measuring achievement in environment for the two lowest levels and subtests in science and social science for the remaining four levels. The environment subtests measure social aspects (e.g., understanding community, family, transportation, communications) and natural aspects (e.g., understanding plants and animals) of the environment. The science subtests include questions about both physical and biological science, whereas the social science subtests sample student knowledge of geography, history, sociology, political science, and economics.

Student responses to selected items from the language, mathematics, science, and social science subtests are combined to compute a "using information" score. This score is based on test items that require examinees to demonstrate their reference skills (e.g., using a dictionary and interpreting tables, graphs, charts, and various visual materials such as maps and signs in science and social science). A multiple-choice item format is used throughout the SAT, with the exception of the writing assessment program, of course.

Practical Applications/Uses

The instructions for administering and scoring the SAT are very complete, which is typical for standardized achievement tests of this type. Following a general orientation to the test battery, the test administrator is carefully instructed with respect to both the preliminary planning required and the actual steps of the test administration. No details are overlooked.

To help improve the quality of the measurements made, attention is given to possible student fatigue during testing. To this end, administration of each subtest in a separate sitting is advised, although giving two subtests in one sitting (with a rest period of 10 minutes in between) is possible. This means that there could be as many as 11 sittings, with the possibility that these are preceded by a practice test several days before the first subtest is administered.

The students fill out the required personal information on the separate answer sheet under the guidance of the test administrator. Both MRC and NCS machine-scorable answer folders are available, as well as a hand-scorable answer folder.

The test administrator (usually a classroom teacher) does not need to have special training, but must be thoroughly familiar with the procedures contained in the *Directions for Administering*, which are provided for each level of the test battery. Indeed, it would be beneficial for the examiner to take the tests him- or herself in order to become fully acquainted with them.

In view of the many subtests included in the complete battery of the SAT, it is not surprising that testing a fourth-grade class, for instance, requires 4 hours, 55 minutes. The basic battery, which excludes the science and social science subtests, requires 4 hours, 5 minutes for administration. If the optional writing assessment along with the Otis-Lennon School Ability Test—a scholastic aptitude test recommended for use with the SAT—are added, the test time is no doubt long enough to require a dozen or more short sittings in order to avoid excessive student fatigue.

This investment of time yields a great deal of information. For example, testing students in the fourth grade using the Primary 3 complete battery will produce 11 subtest achievement scores plus the "using information" score, along with a total score for each of the four basic skill areas (i.e., a separate score for reading, listening, language, and mathematics). Moreover, there is a composite score for the entire basic battery and another for the complete battery.

These results can be reported as scaled scores, national and local percentile ranks, national and local stanines, grade equivalents, and normal curve equivalents. Fall, midyear, and spring national norms are available for each grade level. Additionally, within almost every subtest area there are several content clusters, which are groups of test items measuring a specific type of achievement. Scored separately, these clusters can be helpful when attempting to identify a student's specific strengths and weaknesses in a particular content area. For example, the following shows the Primary 3 content clusters for each subtest except the vocabulary subtest, for which none is listed.

Word Study Skills: structural analysis; phonetic analysis—consonants; phonetic analysis—vowels.

Reading Comprehension: textual reading; functional reading; recreational reading; literal comprehension; inferential comprehension.

Listening Comprehension: retention; organization.

Spelling: sight words; phonetic principles; structural principles.

Language: conventions; language sensitivity; reference skills.

Concepts of Number: whole numbers and place value; fractions; operations and properties.

Mathematics Computation: additions with whole numbers; subtraction with whole numbers; multiplication with whole numbers; division with whole numbers.

Mathematics Applications: problem solving; geometry/measurement; graphs and charts.

Social Science: geography; history and anthropology; sociology; political science; economics; inquiry skills.

Science: physical science; biological science; inquiry skills.

To analyze a student's level of success for a content cluster, the score of that student on the designated test items is converted to a "performance category" (i.e., it is classified as below average, average, or above average as compared to the scores of a national sample of students in the same grade). The same type of analysis can be prepared for groups of students such as a self-contained class.

To examine student test performance even more closely, the response of each student to each test item within a content cluster can be made available. A response code designates the student's answer as correct or, if incorrect, identifies the wrong answer selected. This item-by-item information provides a detailed picture of a student's pattern of success and failure. The same type of report can be provided for an entire class, school, or district. The percent of the students in a specified group answering a question correctly can be compared to the percent (*p*-value) of the national sample that did so. Finally, a summary of the number of students selecting each option of a multiple-choice question can also be prepared (at extra cost).

Interpretation of these data is facilitated by a booklet containing an index of instructional objectives that is published for each level of the SAT. The booklet contains a description of the behavior measured by each test item, along with the percent of the national sample that answered the question correctly at fall and spring test administrations. These descriptions of behavior can then be compared to local instructional objectives to estimate the degree to which the test item content is relevant and subsequently to estimate the degree of importance to be attached to the ability of a student and/or a class to answer the test item successfully. Hence, objective-referenced interpretations are possible.

It is anticipated that this type of information, voluminous as it is, would be useful as one component in a multifaceted effort to improve student achievement and even to understand better the relative effectiveness of the teaching strategies used. These are extremely ambitious goals, and if they are to be attained, a very serious effort would have to be made over a substantial period of time with a wide variety of information sources available.

Technical Aspects

Attempting to build a reasonably short, objectively scored, nationally useful battery of achievement tests that measures student proficiency in the major skill areas across virtually nine grade levels is indeed a formidable task. In the first place, a high degree of congruence must exist between the content and behaviors embedded in the test and what is taught in schools nationally. In other words, the test battery is not sufficiently valid if its content relevance is modest.

In the case of the Stanford Achievement Test: 7th Edition, an analysis of curriculum materials and research literature was made, and five or six of the most widely used textbook series in each subject area were studied—a common approach for tests of this type. Then test specifications were developed that in effect outlined each subtest. After review and revision of the specifications, appropriate test items based on them were constructed, field-tested, reviewed and edited, and, when found to be promising, included in the extensive tryout program to determine appropriateness of format, level of difficulty, discriminating power, possible ethnic and sex bias, and the like.

The degree of reliability of the subtests of the SAT is represented by three different approaches: 1) the internal consistency method, using the Kuder-Richardson formulae 20 and 21, 2) the alternate forms method, in which data from Form E and Form F are correlated, and 3) the consistency-over-time method, in which student performance in the fall is correlated with performance in the spring. The first two are probably the most useful in a reliability review.

As expected, the reliability coefficients tend to be somewhat higher for older students (Grades 8 and 9) than younger students (Grades 1 and 2) and for scores based on many test items (total scores for subtests) than for scores based on fewer test items (content cluster scores). In any event, reliability coefficients were respectable for the subtests, often clustering around .90, whereas those for content cluster scores were lower and varied widely. Some content cluster scores that were based on only three test items were as low as .11 and .19 for students in the third grade. Obviously, data from such small groups of items should not be used in a definitive manner. On the other hand, the reliability coefficients of content cluster scores based on 20 or more test items reached .80 and beyond in some instances.

The intercorrelations among the various subtests of the SAT and the correlations between these subtests and the Otis-Lennon School Ability Test (OLSAT) present a picture that is difficult to interpret. First of all, the intercorrelations between various pairs of subtests tend to be large. In the case of the Primary 3 level (Form E) administered to students in the fourth grade, the lowest intercorrelation coefficient among the subtests was .45 (Spelling and Listening Comprehension) and the highest was .86 (Science and Social Science). About half (28 out of 55) fell within the .60-.69 interval. At the same time, the correlation coefficients between the various subtests and the OLSAT ranged between .63 and .76. These findings raise an interesting question about the degree of construct validity of the subtests of the SAT and, for that matter, the OLSAT. Moreover, can the OLSAT be used jointly with the SAT to estimate over-, normal, and under-achievement with a decent degree of accuracy? These questions may not be answerable, but it is clear that adding the OLSAT to a testing program including the SAT will not increase appreciably the fund of information available.

Critique

The Stanford Achievement Test: 7th Edition is augmented by an unusually large and attractive set of supporting materials. For example, the directions for administering the SAT are well crafted, the indices of instructional objectives should be most helpful to those who wish to study the degree of content relevance of the test battery, and the occasional papers entitled *The Stanford Bulletin from the Test People* are well written, pertinent statements—each of a few pages—focusing on topics of interest to SAT users. Also, the eight *Stanford Special Reports* provide valuable data of a more technical nature that enlarge upon the information included in the *Technical Data Report*.

Not to be overlooked are the special materials developed to help parents, teachers, and administrators understand better the nature of the SAT and some practical uses of the data it yields. There are two much-needed publications for parents, *Previews for Parents* and *Understanding Test Results: A Guide for Parents*. For teachers, the *Handbook of Instructional Strategies* and *Guides to Classroom Planning* are available, both of which contain suggestions as to how data from the SAT, along with a wide variety of additional information, might be used to improve instruction and to strengthen the decision-making process needed for effective classroom planning.

The *Handbook of Instructional Strategies* contains separate sections for each of the subtests and describes instructional strategies in each area that are teacher-directed and use materials constructed by the teacher or students. In contrast, the *Guides to Classroom Planning* addresses the possible uses of data from the SAT along with information from other less formal measures (e.g., teacher-made tests and anecdotal records) when the classroom teacher must make decisions about such activities as pacing instruction, instructional grouping, diagnosis of learning levels, retention, and promotion. Illustrative data are presented.

The *Guide for Organizational Planning* is designed to assist school administrators in the process of making better management decisions with regard to such tasks as evaluating the effectiveness of educational programs, planning a district-wide testing program, and interpreting student achievement levels to the general public.

It must be understood that there is considerable risk associated with the dissemination of materials such as these, especially those designed for teachers and school administrators. The suggestions and guidelines for making decisions about an instructional program might be adopted uncritically, and various important local factors would be ignored. Of course, there are no all-purpose recipes for solving instructional problems that can be designed on the basis of the findings generated by an achievement test battery—even one with strong technical qualities. Yet these test findings need to be considered more seriously than is often the case when teachers and administrators make decisions about instructional problems. Therefore, it is reasonable that general suggestions such as those included in these publications be promulgated, even in the absence of compelling empirical data regarding associated validity considerations.

Underusing achievement test data may be as troubling as overusing it. Common sense laced with caution may make it possible for data from a high quality test battery like the SAT to be added to information from many nontest sources, thereby producing a successful effort to solve some of our instructional problems by triangulation.

On balance, the Stanford Achievement Test: 7th Edition is a worthy successor to all preceding editions of this well-regarded achievement test battery. It has been developed with care by a group of distinguished authors. Nevertheless, as in the past, it faces strong competition from other achievement test batteries that also are technically sound. Undoubtedly it will meet this challenge; as a part of the Stanford Achievement Test Series it is in a strong position.

References

This list includes text citations and suggested additional reading.

Airasian P. W. (1985). [Review of the Stanford Achievement Test, Forms E and F]. *Journal of Educational Measurement, 22,* 163-167.

Antonak, R. F., King, S., & Lowy, J.J. (1982). Otis-Lennon Mental Ability Test, Stanford Achievement Test, and three demographic variables as predictors of achievement in grades 2 and 4. *Journal of Educational Research, 75,* 366-373.

Bower, R. (1983). Matching Stanford Achievement Tests to local objectives. *Spectrum, 1,* 40-43.

Caskey, W. E., Jr., & Larson, G. L. (1983). Relationship between selected kindergarten predictors and first- and fourth-grade achievement test scores. *Perceptual and Motor Skills, 56,* 815-822.

Crowell, D. C., Hu-pei Au, K., & Blake, K. M. (1983). Comprehension questions: Differences among standardized tests. *Journal of Reading, 26,* 314-319.

Gardner, E. F., Callis, R., Merwin, J. C., & Rudman, H. C. (1982). *Stanford Test of Academic Skills* (2nd ed.). San Antonio, TX: The Psychological Corporation.

Gardner, E. F., Rudman, H. C., Karlsen, B., & Merwin, J.C. (1982). *Stanford Achievement Test* (7th ed.). San Antonio, TX: The Psychological Corporation.

Linn, R. L. (1986). Educational testing and assessment: Research needs and policy issues. *American Psychologist, 41,* 1153-1160.

Madden, R., Gardner, E. F., & Collins, C. S. (1982). *Stanford Early School Achievement Test* (2nd ed.). San Antonio, TX: The Psychological Corporation.

Otis, A. S., & Lennon, R. T. (1979). *Otis-Lennon School Ability Test.* San Antonio, TX: The Psychological Corporation.

Powers, S., Thompson, D., Azevedo, B., & Schaad, O. (1983). The predictive validity of the Stanford Mathematics Test across race and sex. *Educational and Psychological Measurement, 43,* 645-649.

Suddick, D. E., & Bowen, C. L. (1981). Longitudinal study of mathematics scores of the Stanford Achievement Test. *Psychological Reports, 48,* 284-286.

Suddick, D. E., & Bowen, C. L. (1982). Longitudinal study of reading scores of the Stanford Achievement Test. *Perceptual and Motor Skills, 54,* 369-370.

Watkins, E. O., & Wiebe, M. J. (1984). Factorial validity of the Stanford Achievement Test for first-grade children. *Educational and Psychological Measurement, 44,* 951-954.

Jean Spruill, Ph.D.
Associate Professor and Director, Psychological Clinic, The University of Alabama, Tuscaloosa, Alabama.

STANFORD-BINET INTELLIGENCE SCALE, FOURTH EDITION

Robert L. Thorndike, Elizabeth P. Hagen, and Jerome M. Sattler. Chicago, Illinois: The Riverside Publishing Company.

Introduction

The Stanford-Binet Intelligence Scale: Fourth Edition (Stanford-Binet/Fourth Edition) is an individually administered intelligence test based on a three-level hierarchical model of cognitive abilities. Taking into consideration the ways that clinicians and educators have used previous editions of the Stanford-Binet, the developers of the hierarchical model were influenced by current theories and research in cognitive psychology. Figure 1 illustrates the model and the tests that measure each of its factors. At the top of the model is *g*, a *general reasoning* factor used by individuals to solve problems they have not been taught to solve. The next level divides *g* into three broad factors: Crystallized Abilities, Fluid-Analytic Abilities, and Short-Term Memory. The Crystallized Abilities factor represents the cognitive factors necessary to acquire and use information to deal with verbal and quantitative concepts in order to solve problems. Highly influenced by education, crystallized abilities also represent more general verbal and quantitative problem-solving skills acquired through a variety of learning experiences, both formal (school) and informal. At the third level of the model, Crystallized Abilities are further divided into verbal and quantitative reasoning. The fluid-analytic abilities factor represents the cognitive skills necessary to solve new problems involving nonverbal or figural stimuli. General life experiences are considered more important than formal education in the development of these abilities. Fluid-analytic abilities are measured by abstract/visual reasoning.

The third factor at the second level of the cognitive abilities model, Short-Term Memory, is included because of its positive relationship to more complex tasks of cognitive performance. This is a measure of the individual's ability to retain information until it can be stored in long-term memory and to hold information drawn from long-term memory so that the individual may use it for solving problems. Short-Term memory is not further differentiated at the third level of the model.

The term *intelligence* has been replaced by *cognitive development*. The terms *intelligence, IQ* and *mental age* are not used in reference to the Fourth Edition anywhere in the administration or technical manual. Instead of IQ, the term *Standard Age Score* (SAS) is used. The Fourth Edition is intended to provide a clinically useful profile of an individual's cognitive abilities in addition to the overall level of cognitive development. Another goal was to de-emphasize verbal skills from previous editions. To accomplish these goals, a wide variety of item types (tests) were used

544

Figure 1

Model of the theoretical structure of the Stanford-Binet Intelligence Scale, Fourth Edition

$$g$$
General Reasoning

CRYSTALLIZED ABILITIES		FLUID-ANALYTIC ABILITIES	SHORT-TERM MEMORY
Verbal Reasoning	*Quantitative Reasoning*	*Abstract/Visual Reasoning*	____
Vocabulary	Quantitative	Pattern Analysis	Bead Memory
Comprehension	Number Series	Copying	Memory for Sentences
Absurdities	Equation Building	Matrices	Memory for Digits
Verbal Relations		Paper Folding & Cutting	Memory for Objects

to assess each area of cognitive abilities in the theoretical model. Although the Fourth Edition is a revision of earlier editions of the test, "it is a thoroughly modern test in terms of content, scales, and testing procedures" (Thorndike, Hagen, & Sattler, 1986b, p.3).

Extensive reviews of the Binet scales can be found in the various editions of the *Mental Measurements Yearbook* (e.g., Buros, 1972), and only a brief sketch of their voluminous history is given here.

The history of modern intelligence testing began with the publication of the first Binet test in 1905 in Paris, France. Alfred Binet, collaborating with Theophilus Simon, devised 30 objective tests with the goal of differentiating mentally retarded from normal children. Binet revised his test twice before his death in 1911. Refinements of the test included grouping the tests into age levels, introducing the concept of mental age, eliminating tests that were heavily dependent upon reading and writing, and including adults in the standardization sample.

The Binet-Simon scales received wide acceptance in the United States. In 1916, Lewis Terman of Stanford University published the American version of the test (Terman, 1916). Major changes included the use of the intelligence quotient (IQ), adopted from Wilhelm Stern's work, and perhaps the first attempt at a truly representative sample for standardization. The 1916 edition was called the Stanford Revision and Extension of the Binet-Simon Intelligence Scale. Although inadequate by modern standards, the 1916 version was standardized on 1,000 children and 400 adults, representing a very large sample for that time.

The 1937 revision by Terman and his colleague at Stanford, Maude Merrill, resulted in two forms of the test, Form L and Form M, and improved standardization. There were still problems with the test, namely, in the scoring and difficulty level of items, lack of adequate discrimination at the upper end of the intelligence distribution, a heavy emphasis on verbal and rote memory items, and, although improved over prior versions, still inadequate standardization.

A third revision of the Stanford-Binet was published in 1960. Although Terman's name appears on the revision, it was carried out primarily by Maude Merrill, because Dr. Terman died in 1956. The best subtests from the L and M Forms of the 1937 scale were incorporated into a single scale; some tests were relocated, dropped, or rescored; items were updated; a new group of children was used to check changes in test difficulty; scoring principles were clarified; and age 18 rather than 16 was used as a ceiling level. Also, the Deviation IQ was substituted for the ratio IQ (Terman & Merrill, 1960). Although the revision was an improvement over earlier versions, many of the criticisms of the 1937 edition remained true of the 1960 revision.

In 1971, Robert Thorndike undertook a revision of the norms of the 1960 Form L-M. Unlike the earlier versions, which sampled only white, English-speaking individuals, the 1972 norms included black, Mexican-American, and Puerto Rican-American English-speaking subjects (Terman & Merrill, 1973). Insufficient information is provided in the publication of the 1972 norms to adequately assess the representativeness of the normative sample.

The 1972 revision was only an updating of the test norms; prior to this Fourth Edition, the test itself had not been revised since 1960. The present revision of the test was undertaken because of the vast social and cultural changes in the United States that have taken place since 1960. "These changes, together with new research in cognitive psychology," led Thorndike, Hagen, and Sattler (1986c, p. 8) to revise Form L-M. These authors of the Fourth Edition are all well-known psychologists with a great deal of expertise in the areas of intelligence and test development.

Although the Fourth Edition is markedly different from Form L-M, it has retained some of its predecessor's features. Cognitive abilities related to progress in school continue to be a major factor. The Fourth Edition covers the same age range, requires the examiner to establish a basal and ceiling for each subject, and includes a few of the items from Form L-M. However, these features seem to be the only similarities between the two tests. In keeping with the development of other individual tests of intelligence (e.g., the Wechsler scales and the Kaufman Assessment Battery for Children), the Fourth Edition has a point scale format rather than an age scale format and is devised so that an individual's pattern of abilities, as well as overall cognitive development, can be assessed. A wide variety of item types (tests) are used to measure each area of cognitive ability.

To be included in the Fourth Edition, each item type had to be 1) a measure of verbal, abstract/visual, or quantitative reasoning, or short-term memory, 2) reliably scored, 3) relatively free of ethnic or gender bias, and 4) able to function over a relatively wide age range. Many new item types were constructed, and in 1979 the authors began preliminary tryouts of the items. After the tryouts, the item types were reduced to 23 for possible inclusion in the final test. Items were arranged according to difficulty level within each type, and field-test booklets were prepared. Because 23 tests were too many to give to any one person, three forms were prepared. Each form (A, B, and C) had at least two tests that were designed to measure each of the four cognitive areas being tested.

Initial testing was done on a small group of subjects during 1981-82. Based on the results of the field testing, additional revisions were made, and a second field test was carried out on a larger sample during late 1983 and early 1984. After the second field trial, the additional revisions in administration and scoring took place, and the final form of the test was developed. At each stage of development, examiners were asked to critique the items, and standard item analyses and Rasch analyses were run on each item. A panel of judges made decisions about any potential ethnic and gender biases of items. In the final test, 15 item types were retained. The field tryouts and development of the test appear to have been done very well.

Standardization was carried out from January through July, 1985. The standardization sample of 5,013 people was stratified according to six variables: geographic region, community size, race/ethnic group, gender, parental occupation, and parental education. School districts that met the stratification variables were selected throughout the country. Children in representative classes were sent home with parental permission slips for testing. Each examiner was responsible for locating children (from those whose parents gave permission) who met certain specifications of the sampling design. The technical manual (Thorndike et al., 1986c) describes how examiners were selected and trained and gives examples of how specific subjects were selected for testing. Under-school-aged subjects were children enrolled in some type of preschool or day care or were siblings of school-aged children tested. Adult subjects aged 18 to 23 were siblings of school-aged children. The standardization sample originally included a group of subjects 24 to 32 years of age but did not include them in the normative data because they were not a representative sample.

The standardization sample was roughly representative of the 1980 U.S. census population for all variables except parental occupation and parental education. The sample was overrepresented for the college-graduate-or-beyond education group and for the managerial/professional occupation group. This was taken into consideration and the sampling bias presumably reduced by weighting the data in the development of the scale scores. The weighting of scores assumed the differences among the subjects were only quantitative, not qualitative, and this is a questionable assumption. The weighting procedure was described only in general terms. The authors state: "Each child from an advantaged background was counted as only a fraction of a case (as little as 0.28), while each child from a less advantaged background was counted as more than one case" (Thorndike et al., 1986c, p. 25). What is meant by "advantaged" and "disadvantaged" backgrounds is not explained, nor is the weighting procedure further described.

The Fourth Edition comes in a sturdy carrying case that contains the test materials, the administration and scoring manual, and the technical manual. The record book must be ordered separately. Most of the test items are contained in four item books arranged in an easel-kit format that allows the examiner to see the test directions and scoring key at the same time the child sees the items. In addition to the item books, there are four boxes containing test materials. Two boxes contain blocks; in one are the nine green blocks used in the copying test and in the other are black and white cubes with varying designs to be used in the Pattern Analysis test. The third box contains beads used in the Bead Memory test, and the last box contains dice used in the Quantitative test. Additional materials are a modern picture of a child and a new version of the Form Board.

For each of the 15 tests of the Fourth Edition, the items are arranged in levels of increasing difficulty designated by the letters A through Y. Each level has two items of approximately equal difficulty. For each test, the examiner must establish a basal and a ceiling age, and the administration manual gives detailed instructions on doing so (Thorndike, Hagen, & Sattler, 1986b, p. 12).

Testing with the Fourth Edition uses a multistage format. The Vocabulary test, administered first, is a routing test to determine the entry level for all other tests. Entry level for the Vocabulary test is determined by the examinee's chronological age. Using the results of the Vocabulary test, the examiner determines the entry level for the remaining tests by using the "Entry-Level Chart" on the back cover of the record booklet. The rows of the chart are the chronological ages, and the columns are the highest pair of vocabulary items administered. The intersection of the appropriate row and column determines the entry level for the remaining tests.

Test 1: Vocabulary. The Vocabulary test contains 46 items; the first 14 are pictures of common objects, and the remaining 32 are words the examinee is asked to define. This test is administered to all examinees.

Test 2: Bead Memory. The examinee is shown a pattern of beads for 5 seconds and is then asked to reproduce the pattern from memory by using beads of three colors and four shapes. Bead Memory is administered for all entry levels; the sample items and tasks required differ depending upon entry level. For example, entry levels A-G (items 1-10) merely require the examinee to point to the beads they have been shown. Entry levels H-Q (items 11-26) require the examinee to copy a pattern of beads.

Test 3: Quantitative. The Quantitative test requires examinees to place counting blocks (dice) in the counting-blocks tray to match, count, add, subtract, or form logical series of numbers for items 1-12. For items 13-40, the examinee is asked to solve quantitative problems presented either visually or orally. The Quantitative test is administered at all entry levels, and examinees may use paper and pencil in answering. There are no sample items.

Test 4: Memory for Sentences. In the Memory for Sentences test, the examinee repeats sentences after the examiner. Sample items are given, and the test is administered to all entry levels. The examiner is instructed to read at a steady pace and to drop his or her voice at the end of each item.

Test 5: Pattern Analysis. This is the only test that is timed; it is administered to all entry levels and has sample items except for Levels A-C. The first six items use the Form Board; items 7-42 use the black and white cubes. For some items, the examiner uses the cubes to construct a model that the examinee then copies. For other items, the examinee is shown a picture that he or she is expected to copy.

Test 6: Comprehension. The first six items require the examinee to point to various body parts on a card showing a picture of a child. Items 7-42 require a verbal response to questions about common events (e.g., "Where do people buy books?"). Comprehension is administered at all entry levels; there are no sample items.

Test 7: Absurdities. The Absurdities test is administered only for examinees with entry levels A-L. The examinee is asked to tell what is wrong with absurd pictures; there are examples for entry levels A-B.

Test 8: Memory for Digits. The examinee is read a series of digits and asked to repeat them, first forwards, then backwards. Digits are read at the rate of one per second, and the examiner's voice is dropped at the end of each item. This test is administered only to examinees with entry levels I or above.

Test 9: Copying. This test is given only for entry levels A-J; no sample items are administered. For items 1-12, the examinee is asked to duplicate the examiner's design made from blocks. Items 13-28 require the examinee to draw the designs shown. Erasing is allowed.

Test 10: Memory for Objects. This test is administered to examinees with an entry level of I or above. Examinees are shown pictures of one or more objects and then asked to identify the objects in the correct order of their appearance. Both order and choice of object must be correct. Objects are shown at the rate of one per second, and sample items are given.

Test 11: Matrices. In this test, the examinee is asked to fill in the missing object in a series, first using a multiple choice format (items 1-22) and then using a written format (items 23-26). The examinee must figure out the rule or pattern to determine the correct response.

Test 12: Number Series. The examinee is shown a series of numbers that are arranged according to a certain rule. The examinee is asked to supply the next two numbers according to the rule. This test is given only to examinees with entry levels of I or above. The examinee may use paper and pencil, and both numbers must be correct to receive credit. Although the test is not timed, a limit of 2 minutes per item is suggested.

Test 13: Paper Folding and Cutting. This test is administered only to examinees with entry levels M or above. For the sample items, the examiner folds and cuts pieces of paper, and the examinee, using a multiple-choice format, picks the picture that demonstrates how the paper would look if unfolded. For the test items, the examinee looks at a series of pictures that illustrate a piece of paper being folded and cut, then picks the picture that shows how it would look when unfolded.

Test 14: Verbal Relations. This test is administered only to examinees with entry levels M or above. The examiner names four things, and the examinee tells how the first three things are alike and how they are different from the fourth.

Test 15: Equation Building. In this test, given to entry levels M or above, the examinee is given several numbers and operational symbols (+, -, =, etc.) to use in building a meaningful mathematical relationship. This test is not timed, but a limit of 2 minutes per item is suggested. The examinee may use paper and pencil in figuring out the answers.

No more than 13 of the above tests are given to any one examinee. The complete battery ranges from 8 to 13 tests, depending upon the age of the examinee and the entry level established by the Vocabulary routing test. There are several abbreviated versions recommended by the authors, and these are detailed in the administration manual (Thordike et al., 1986b, pp 34-36).

The record booklet is 40 pages long and at first glance appears quite intimidating. The first page contains information about the examinee and his or her scores on the test. The administration manual gives clear guidelines for completing the front page and obtaining the Standard Age Scores. Five Standard Age Scores are obtained, one for each of the four areas measured (Verbal Reasoning, Abstract/Visual Reasoning, Quantitative Reasoning, and Short-Term Memory) and a Test Composite. The Composite SAS is a deviation score; it appears to be simply another name for the Deviation IQ score in the previous editions of the Stanford-Binet.

The second page contains a profile analysis, which is used to determine the strengths and weaknesses of the examinee's cognitive abilities. The last page in the booklet is the Entry-Level Chart. Pages 3-39 are for recording verbatim the examinee's answers in the spaces provided for each test. Each test has the entry level items marked, and the sample items, if any, that accompany each entry level are designated. The raw score for a test is the item number of the highest item administered minus the number of items that were failed.

Practical Applications/Uses

The Stanford-Binet Intelligence Scale, Fourth Edition was constructed as a measure of the cognitive abilities of individuals from 2 to 23 years of age. According to the authors, additional purposes of the test are 1) to differentiate between mentally retarded and learning disabled students, 2) to aid in understanding why a particular student is having difficulty in learning in school, and 3) to identify gifted students.

Although the Fourth Edition covers the age range from 2 to adult, the use of the term *adult* is misleading. Normative data are available only for individuals 18 to 23 years of age, limiting the usefulness of the Fourth Edition for adults. It is most appropriate for the purposes given previously, which relate primarily to school-aged children.

At the present time it is not clear if the Fourth Edition will be useful in institutions for the mentally retarded. The lowest SAS that can be obtained for any area or composite is 36, making the test inappropriate for individuals classified in the severe or profound range of retardation.

The Fourth Edition is designed to be individually administered by an experienced examiner in a setting that is free from distractions. This test may be used by several professional groups, among them school psychologists, educational specialists, and clinical psychologists. To administer and interpret the test, graduate-level training in test administration in general and in the Fourth Edition in particular is necessary. The administration manual gives many helpful suggestions, such as how to 1) tailor tests to specific individuals, 2) develop and maintain rapport, and 3) test preschool children.

One of the major criticisms of Form L-M was the difficulty in administration. Unfortunately, difficulty in administration also is likely to be a criticism of the Fourth Edition. Much practice with the Fourth Edition is needed to achieve a smooth transition from test to test.

Instructions for administration appear in the item books and, in general, are clear and easy to follow. Finding the correct starting place is not always easy; the tabs in the item books are not labeled. The administration guide recommends administering the tests in a certain order, changing the order only if circumstances dictate. Although the order of administration of the tests may be changed if necessary, the order of items within each test should never be changed.

Whereas most administration instructions are clear and easy to follow, there is a problem in the administration of the Bead Memory test. In Bead Memory, items 1-10 involve one type of response and items 11-42 another. Individuals starting with item 11 (or above) are given sample items. The problem arises when an examinee starts the test below item 11 and continues beyond it. The directions do not specify the administration of a sample item or an explanation that the task has now changed. In fact, the directions explicitly state that sample item 1 is given only to those examinees starting the test at entry levels H-K (Item Book 2, p. 15). Without an explanation of the task and a sample item, examinees starting below entry level H will have difficulty understanding the changed nature of the task. Apparently, at workshops on administering the Fourth Edition, examiners are being told to give Sample 1 to all examinees who are administered item 11, regardless of their entry Level. Surely, later printings of Item Book 2 (Bead Memory) will correct this problem.

The administration of the Fourth Edition is lengthy. To keep the testing time to 60 to 90 minutes, the authors recommend that the examiner use one of the batteries of tests suggested for different age groups; however, the most reliable and valid measure of intellectual functioning is obtained with the complete battery. Thus, for some individuals, testing may have to be split over two sessions. Other reasons for not administering an abbreviated battery are given below.

Instructions for scoring the responses are contained in the item books, generally on the same page as the instructions for administering the items. Only 5 of the 15 tests require using an expanded scoring guide from the administration manual. Few problems with scoring items are noted; one exception is the scoring for the Pattern Analysis test. In this test, the examiner shows the subject a picture of a design to copy. The examiner also sees the design that the subject is supposed to make. However, for items numbered 26, 28, 29, 30, Sample 5, 31, 32, 33, 34, 35, 36, 37, and 42, the pattern seen by the examiner as the "supposedly" correct answer is wrong—typically, the patterns are rotated 180x either vertically or horizontally. Presumably, this error will be corrected in future printings of Item Book 1. The record booklet also shows the pattern for each item, and these illustrations are accurate, so the examiner can use the record book to score the item.

Potential problems also exist in scoring the Equation Building test. Although the scoring key in the item book lists many correct solutions, there are other correct solutions not listed. An examiner who forgets that there are other correct answers, or whose mathematical skills are limited, may incorrectly score some answers.

Using the tables in the administration manual, raw scores are transferred to the front page and converted to Standard Age Scores. Although scoring of individual items is easy, obtaining test raw scores, test SASs, Area SASs, and the Composite SAS is more complex; therefore, to ensure accuracy, scoring should be gone over at least twice.

All test raw scores are converted to SASs with a mean of 50 and a standard deviation of 8. The Area and Composite SASs have a mean of 100 and a standard deviation of 16. The standard deviation was kept at 16 to make comparisons with Form L-M scores easier and to maintain historical continuity with Form L-M. The technical manual also has a table converting the scores to a distribution with a mean of 100 and a standard deviation of 15 in order to facilitate comparisons with scores from tests with a mean of 100 and a standard deviation of 15.

For older and/or very bright examinees, some tests have a ceiling that is much too low. For example, a perfect score on Copying could result in an SAS of 55 for a 12- to 13-year-old child. An opposite problem occurs with the mentally retarded subjects. In testing subjects ranging in ability from mild to moderate retardation, this reviewer has had difficulty in establishing a basal level for several tests. In addition, when starting the Bead Memory test at the level dictated by the Vocabulary routing test, this reviewer has had to go backwards an average of 5 levels to establish a basal.

Interpretation is based on objective scoring and clinical judgment. At the present time, the information provided for the interpretation of the Fourth Edition is inadequate. The *Expanded Guide for Interpreting and Reporting Fourth Edition Results* (Thorndike, Hagen, & Sattlerr, in press) has not been published, although it has been advertised since the publication of the test. There is very little information, other than technical data, in either the administration or technical manual to assist in interpreting the results. The technical manual does provide descriptive classifications for the Composite SAS. The terminology is similar to that associated with the IQ scores of Form L-M, except that the terms *Mentally Retarded* and *Slow Learner* have replaced *Mentally Defective* and *Borderline Defective*, respectively. In order to conform more closely with current classification practices, the score ranges for the categories were changed. In fact, when converted using a scale with a standard deviation of 15 (Table D.1 in the technical manual), the score ranges are the same as those used in the Wechsler scales and the K-ABC.

It has become increasingly common to report confidence intervals around standard (IQ) scores. Confidence intervals are not presented in the technical manual; however, if one knows how, they can be calculated easily from the standard errors of measurement. In addition to using confidence intervals around SAS scores, interpretation of test performance is aided by determining the examinee's strengths and weaknesses. Table F.2 in the technical manual (Thorndike et al., 1986c, p. 134) presents the differences between SASs for the areas and composite required for statistical significance at the 15% and 5% levels of confidence. However, these values were computed without taking into consideration the fact that multiple comparisons are being made. Although the problems involved in making multiple pairwise tests of significance have been recognized for years, it is only recently that test developers and clinicians have considered the problems caused by multiple comparisons when doing a profile analysis of test scores.

Sattler (1982) and Silverstein (1982), in their calculations of the differences required for significance among Wechsler subtests, used the Bonferroni *t* inequality to control for the increased error rate caused by multiple comparisons on the same data. Similarly, Naglieri (1982) corrected the differences required for significance for the McCarthy Scales using the Bonferroni *t*. Kaufman and Kaufman (1983), in their development of the K-ABC, made the appropriate corrections for their comparisons among test scores. The failure of the developers of the Fourth Edition to apply currently accepted statistical procedures in their analysis of differences represents a major error.

Technical Aspects

Reliability of the scores varies as a function of the different tests and the different age groups. For the most part, the reliability coefficients reported are based on Kuder-Richardson Formula 20 (KR-20) and are well within the typical range for individual tests of intelligence. However, as pointed out in the technical manual, to use the KR 20 formula it is necessary to assume that all items below the basal level were passed and all items above the ceiling level were failed. Because the assumption of failure above the ceiling level is not likely to be strictly met, the resulting correlation coefficients may be somewhat inflated and therefore should be considered as upper bound estimates of reliability.

As would be expected, the most reliable score, at all ages, is the Composite SAS; reliability coefficients ranged from .95 to .99. Reliability coefficients for the area scores depended upon the number of tests making up the score and ranged from .80 to .97. Not surprisingly, the more tests given in a particular area, the more reliable the Area SAS. The individual test reliabilities, with the exception of Memory for Objects, are in the .80s and .90s. Memory for Objects is also a short test (14 items), which contributes to its lower reliability. Reliability coefficients tend to be higher for older age groups.

Test-retest reliability was computed for two samples, ages 5 and 8. For most of the tests, the coefficients reported, although lower than the internal consistency measures based on KR-20, are in the acceptable range. Exceptions are the very low test-retest correlations of .28 for the Quantitative test, .46 for the Copying test, and .51 for the Quantitative Reasoning area. The authors speculate that a restricted range of scores in their small sample of 8-year-olds ($n=57$) resulted in the low retest correlations. However, the authors did not correct for restricted range, nor did they explain why the range of scores may have been restricted. It would have been helpful to see some test-retest data for adolescents and adults.

With the exceptions mentioned, the reliabilities reported in the technical manual (Thorndike et al., 1986c, pp. 38-49) compare favorably with those reported for the WISC-R, WAIS-R, K-ABC, and other individually administered tests.

Standard errors of measurement are reported for the various age groups for each test, each area score, and the composite score. Standard errors appear to be of reasonable magnitude given the reliability of the tests and considering that they are based on scales with a standard deviation of 8.

To understand what the Fourth Edition measures, the authors focused their research on three areas: 1) the internal structure of the test (i.e., factor analysis and intercorrelations among tests); 2) correlations between the Fourth Edition and other measures of intellectual ability; and 3) the performance of individuals identified by other measures as being high or low in intellectual ability.

A variant of confirmatory factor analysis was used to investigate the internal factor structure of the Fourth Edition. The results of the factor analysis were reported "to provide positive support for the rationale underlying the battery" (Thorndike et al., 1986c, p. 55). Unfortunately, the authors do not explain how this "variant of confirmatory factor analysis" was carried out, making it difficult to evaluate the accuracy of their statement. Except for the loadings on the *g*, or general factor, the factor loadings reported in the technical manual are smaller than desirable. Clearly, each test's highest loading is on the g, or general ability factor. Not surprisingly, Vocabulary had the highest loading on g (.76), whereas Memory for Objects had the lowest (.51). Some tests (e.g., Memory for Sentences) show loadings on factors other than those proposed by the model, and other tests fail to load on the group factors dictated by the model (e.g., Matrices and Copying). Test specificity for most tests is high, indicating the tests are strong measures in their own right.

When additional factor analyses were computed on various age groups, the model did not fare well. According to the theoretical model, four factors in addition to the g factor should occur. Separate analyses were conducted for three age groups. For the age group 2 through 6, only two factors in addition to g emerged: a Verbal factor and an Abstract/Visual factor. The Short-Term Memory and Quantitative factor did not appear. Because only one quantitative test is given at these ages, a common Quantitative factor was not expected. There are two memory tests given at ages 2 through 6; Bead Memory had a high loading on the Abstract/Visual factor, and Memory for Sentences had a high loading on the Verbal factor. The Quantitative test also loaded on the Abstract/Visual factor; in fact, its factor loading of .25 was higher than the factor loadings of two of the tests (Pattern Analysis and Copying) that supposedly are measuring abstract/visual reasoning.

Factor analysis for the age group 7 through 11 resulted in the identification of three factors: Verbal, Memory, and Abstract/Visual. A Quantitative factor did not occur, even though two quantitative tests were given in this age range. Only three of the four memory tests loaded on the Memory factor. Bead Memory had a factor loading of .05 on Memory, and Memory for Sentences had a loading of .19 on the Verbal factor.

The four factors predicted by the theoretical model did emerge for "age group 12 through 18-23." However, the Quantitative factor was not very strong; factor loadings were .10 for the Quantitative test, .20 for Number Series, and .40 for Equation Building. Matrices also had a weak loading on the Abstract/Visual factor, perhaps because it had such a high loading on g.

Some support for the theory underlying the construction of the Fourth Edition is found in a study by Keith and his colleagues at the Lindquist Center at the University of Iowa (Keith, Cool-Hauser, Novak, White, & Pottebaum, 1987). Keith et al. used a LISREL VI computer program to perform a first order, confirmatory factor analysis (CFA) using the entire standardization sample and the same three age groups used by the authors of the Fourth Edition in their factor-analytic studies. The CFA approach is somewhat different from programs typically used and is thought to provide a stronger test of the underlying structure of a test than does exploratory factor analysis. Keith et al. conducted a systematic series of analyses, checking the fit of various models to the proposed structure of the Fourth Edition. Based on their analyses, Keith et al. found some support for the four factors underlying the structure of the Fourth Edition. However, neither theirs nor Thorndike et al.'s factor-analytic procedure found a Memory factor for the ages 2 through 6 group.

The conclusion by Thorndike et al. (1986c) that the factor structure of the Fourth Edition both conformed to the theoretical framework used to construct the test and provided good support for the theoretical rationale underlying the test is questionable. Clearly, the results support the existence of a strong *g* component; the results are less supportive of the second level of the model in general, especially for the younger age groups.

The technical manual reports a number of studies conducted to obtain data on the validity of the Fourth Edition. Most of the subjects selected for the studies were not part of the standardization sample. Two kinds of samples were used for these studies: 1) groups of examinees designated by their schools or institutions as members of an exceptional group (gifted, mentally retarded, or learning disabled), and 2) nonexceptional samples consisting of groups of subjects that were in a regular educational setting.

When comparing the Fourth Edition Composite SAS with the Full Scale IQ scores of the Wechsler tests (WISC-R, WAIS-R, and WPPSI) and the Mental Processing Composite score of the K-ABC, the correlations ranged from .81 to .91, and the differences between the mean SASs and Full Scale IQ or Mental Processing Composite scores ranged from a low of .4 (K-ABC) to a high of 5.0 (WPPSI). These results were for the nonexceptional samples and clearly indicate that, for these groups, the Fourth Edition Composite SAS is equivalent to other test scores.

The authors had predicted certain patterns of correlations between the Fourth Edition SASs and the various standard scores of the other tests. The results were mixed. Overall, the studies were well done and the results generally supportive of this aspect of the validity of the Fourth Edition.

Using exceptional samples, that is, samples identified by their school system or institution as "gifted, learning disabled, or mentally retarded," the Fourth Edition was compared to the Form L-M, WISC-R, WAIS-R, and K-ABC. Two samples of gifted students were studied, one using the WISC-R and one using Form L-M. Using gifted students restricted the range of scores on the two tests, and hence reduced the possible correlation between the two. An additional factor that might reduce the correlation is the different emphasis on verbal skills for the two tests, particularly as gifted individuals are generally considered to be higher in verbal skills than in other areas. Nevertheless, the correlation of .27 between the Composite SAS of the Fourth Edition and the IQ score of Form L-M is quite low, particularly considering that the sample size was 82. The mean Composite SAS was 121.8 (SD = 9.0); the mean Form L-M IQ was 135.3 (SD = 9.7). The difference in overall scores is probably a function of the differing verbal emphasis between the two tests.

When comparing the Fourth Edition with the WISC-R in a sample of 19 gifted students, the mean Composite SAS was 116.3 (SD = 16.4), and the Mean Full Scale IQ was 117.7 (SD = 12.1). The correlation between the Composite SAS and the Full Scale IQ score was .69. It is interesting that the mean WISC-R IQ score for this "gifted" sample was 117.7, only slightly more than one standard deviation above the population mean IQ—hardly what is typically thought of as gifted.

Using learning disabled subjects to compare the Fourth Edition with the WISC-R, Form L-M, and the K-ABC, the authors found that the Fourth Edition Composite SAS scores are comparable to scores obtained on the other tests. The mean differences between the Fourth Edition SAS and the total standard scores for the other tests ranged from 1.7 to 3 points; the correlations between the Composite SAS and the total standard scores for the other tests ranged from .66 to .87. Specific predictions concerning patterns of relationships between the various test scores were only partially supported; however, the overall results indicate that the Fourth Edition is comparable to other tests for this group of subjects.

Studies using subjects identified as mentally retarded showed similar results to the studies using learning disabled subjects. In general, when testing groups of retarded subjects, the Fourth Edition Composite SAS is very similar to the Form L-M IQ score (50.9 vs. 49.5 respectively) and the WISC-R Full Scale IQ score (66.2 vs. 67.0 respectively). When compared to the WAIS-R Full Scale IQ Score of 73.1, the mean Composite SAS was 63.8, which is almost 10 points lower. However, the difference between the WAIS-R and Fourth Edition scores is not surprising; the WAIS-R overestimates the IQ scores of the retarded (Spruill & Beck, in press), and the Fourth Edition has a slightly lower floor than the WAIS-R.

At present, there is little published information about the Fourth Edition in the professional literature, primarily because of the short time since its publication; however, there are many studies being done. Carvajal and Weyand (1986) compared the General Purpose Abbreviated Battery of the Fourth Edition with the WISC-R and found a correlation of .783 between the Composite SAS and the Full Scale IQ score. It is likely that the correlation would be even higher had they given the complete Fourth Edition. Carvajal and his colleagues (Carvajal, McVey, Sellers, Weyand, & McKnab, 1986) also compared the General Purpose Abbreviated Battery of the Fourth Edition to the Peabody Picture Vocabulary Test-Revised (PPVT-R) and found a correlation of .601; this is similar to correlations found between the PPVT and other individual measures of intelligence.

Several studies comparing the Fourth Edition and various other measures of intellectual ability were reported at the 1987 meeting of the National Association of School Psychologists. Livesay and Mealor (1987) presented the results of their study comparing the scores obtained on the Fourth Edition with Form L-M using a sample of subjects referred for evaluation for the gifted programs in their schools. Their results were similar to those found by the authors when studying gifted children; the IQ scores on the Form L-M were significantly higher than the Composite SAS (130 vs. 122), and the correlations between the two tests were all equal to or higher than those reported for similar studies in the technical manual. One reason for Livesay and Mealor's higher correlations could have been a broader range of scores in their sample. Because the subjects were referred for a determination of eligibility for a gifted program, they probably had a broader range of scores than would be found in a group of subjects identified as gifted.

Krohn and Lamp (1987), using a sample of low income preschool children, compared the K-ABC, Form L-M, and the Fourth Edition. There were no significant differences among the means of composite scores for the total group on all three measures. In general, the correlations found by Krohn and Lamp were slightly lower than those reported in the technical manual, probably because of the low socioeconomic status of the subjects.

Sims (1987), in her investigation of the relation between the Fourth Edition and the WISC-R, concluded that the Fourth Edition was a strong measure of general reasoning ability, providing a Composite SAS very similar to the WISC-R Full Scale IQ. The correlations between the two tests, except for Composite SAS and Full

Scale IQ, were smaller than those reported in the technical manual for similar studies.

In conclusion, there is a strong theory underlying the development of the Fourth Edition. The extent to which the test matches the theory is yet to be determined. The evidence presented thus far is inconclusive. In particular, the factor-analytic results are not adequately explained, and the factor structure of the theoretical model is not supported for all age groups. Correlational data between the Fourth Edition Composite SAS and the other major individual tests of cognitive ability is acceptable; however, specific predicted patterns of results between the various scores of the Fourth Edition and other tests have not always been borne out. Clearly, validity of the Fourth Edition as a measure of *g*, or a general reasoning, cognitive ability factor, has been adequately demonstrated. At the present time, the validity of the other levels of the theoretical model is undetermined. Much more needs to be done in this area to ascertain what other factors are measured by the test.

Critique

During the last two decades, the Stanford-Binet Intelligence Scale, Form L-M has declined in popularity (Lubin, Larsen, & Matarazzo, 1984), and the reviews in the various editions of the *Mental Measurements Yearbook* (e.g., Buros, 1972) have become increasingly negative. The development of the Fourth Edition was clearly aimed at regaining the Stanford-Binet's prominence in the area of assessing cognitive abilities. The Fourth Edition is, in most respects, a completely new version of a very old test. Unfortunately, the test appears to have had significant problems from the start.

The publication of the Fourth Edition seemed somewhat rushed. In retrospect, it would have been better to have delayed the publication date. The test was published without accompanying technical data to allow the user to judge the appropriateness and technical adequacy of the instrument. This is a violation of Standard 5.1 in the *Standards for Educational and Psychological Testing* (American Educational Research Association, American Psychological Association, & National Council on Measurement in Education, 1985). Even more serious than the failure to provide technical information is the fact that the first printing of the *Guide for Administering and Scoring the Fourth Edition: Stanford-Binet Intelligence Scale* (Thorndike et al., 1986b) contained errors in the norms tables. The errors were corrected, and a new administration manual was sent to all known purchasers of the test. Hopefully, no examiners are still using the original administration manual. The *Expanded Guide for Interpreting and Reporting Fourth Edition Results* is not yet available. The test has been out for well over 1 year, and materials crucial to its use still are not available.

However, the problems with the Fourth Edition started earlier than the development of the manuals. One consistent criticism of the Binet test and its subsequent revisions has been inadequate standardization. This continues to be a problem. In the Fourth Edition, the standardization sample contained a larger percentage of high-socioeconomic-status subjects than in the population at large. The adequacy of the weighting procedure used to correct for sample bias in the data is not clear. The Fourth Edition also does not include adults over age 23 in its normative data.

There are problems associated with the administration of the Bead Memory test and in the scoring of the Pattern Analysis test. Furthermore, the statistical data presented in the technical manual do not fully support the factor structure of the model underlying the construction of the test.

The National Association of School Psychologists did not endorse the use of the Fourth Edition when it was first published because of the lack of accompanying technical information. As of the time of this writing (June, 1987), some states have yet to approve the test for use in making decisions about placement in educational programs.

In spite of the preceding negative comments, the Fourth Edition is an exciting addition to the array of tests used to measure intellectual abilities. It has a strong theory based on recent research in cognitive psychology and provides a broader coverage of the cognitive skills of examinees than previous editions. The administration is more flexible than is true of many tests, and children, especially younger ones, find the items challenging and fun.

Although less emphasis on verbal skills is present in the Fourth Edition, the Stanford-Binet continues to be a very good assessment of cognitive skills related to academic progress. In spite of numerous problems, the Fourth Edition will be around for many years—after all, it is still the Binet.

References

This list includes text citations and suggested additional reading.

American Educational Research Association, American Psychological Association, & National Council on Measurement in Education. (1985). *Standards for educational and psychological testing.* Washington, DC: American Psychological Association.

Buros, O. K. (Ed.). (1972). *The seventh mental measurements yearbook.* Highland Park, NJ: The Gryphon Press.

Carvajal, H., & Weyand, K. (1986). Relationships between scores on Stanford-Binet IV and Wechsler Intelligence Scale for Children-Revised. *Psychological Reports, 59,* 963-966.

Carvajal, H., McVey, S., Sellers, T., Weyand, K., & McKnab, P. (1986). Relationships between scores on the General Purpose Abbreviated Battery of Stanford-Binet IV, Peabody Picture Vocabulary Test-Revised, Columbia Mental Maturity Scale, and Goodenough-Harris Drawing Test. *The Psychological Record, 1,* 127-130.

Holden, R. H. (1984). Stanford-Binet Intelligence Scale: Form L-M. In D.J. Keyser & R.C. Sweetland (Eds.), *Test Critiques: Volume I* (pp. 603-607). Kansas City, MO: Test Corporation of America.

Kaufman, A. S., & Kaufman, N. L. (1983). *Kaufman Assessment Battery for Children.* Circle Pines, MN: American Guidance Service.

Keith, T. Z., Cool-Hauser, V. A., Novak, C. G., White, L. J., & Pottebaum, S. M. (1987). *Confirmatory factor analysis of the Stanford-Binet Fourth Edition: Testing the theory—test match.* Unpublished manuscript.

Kennedy, W. A. (1973). *Intelligence and economics: A confounded relationship.* Morristown, NJ: General Learning Press.

Krohn, E. J. & Lamp, R. E. (1987, March). *Validity of K-ABC and Binet-Fourth Edition for low income preschool children.* Paper presented at the annual convention of the National Association of School Psychologists, New Orleans.

Livesay, K., & Mealor, D. J. (1987, March). *A comparison of the Stanford-Binet: Fourth Edition and Form L-M for gifted referrals.* Paper presented at the annual convention of the National

Association of School Psychologists, New Orleans.

Lubin, B., Larsen, M., & Matarazzo, J. (1984). Patterns of psychological test usage in the United States: 1935-1982. *American Psychologist, 39*, 451-454.

Naglieri, J. A. (1982). Interpreting the profile of McCarthy Scale indexes: A revision. *Psychology in the Schools, 19*, 49-51.

Sattler, J. M. (1982). *Assessment of children's intelligence and special abilities* (2nd Ed.). Boston: Allyn & Bacon.

Silverstein, A. B. (1982). Pattern analysis as simultaneous statistical inference. *Journal of Consulting and Clinical Psychology, 50*, 234-240.

Sims, L. (1987, March). *Concurrent validity of the new Stanford-Binet and WISC-R.* Paper presented at the annual convention of the National Association of School Psychologists, New Orleans.

Spruill, J., & Beck, B. (in press). Comparison of the WAIS and WAIS-R: Different results for different IQ groups. *Professional Psychology: Research and Practice.*

Terman, L. M. (1916). *The measurement of intelligence.* Boston: Houghton Mifflin Company.

Terman, L. M., & Merrill, M. A. (1937). *Measuring intelligence.* Boston: Houghton Mifflin Company.

Terman, L. M., & Merrill, M. A. (1960). *Stanford-Binet Intelligence Scale.* Boston: Houghton Mifflin Company.

Terman, L. M., & Merrill, M. A. (1973). *Stanford-Binet Intelligence Scale: 1972 norms edition.* Boston: Houghton Mifflin Company.

Thorndike, R. L., Hagen, E., & Sattler, J. (1986a). *Stanford Binet Intelligence Scale: Fourth Edition.* Chicago: The Riverside Publishing Company.

Thorndike, R. L., Hagen, E., & Sattler, J. (1986b). *Guide for administering and scoring the fourth edition: Stanford Binet Intelligence Scale.* Chicago: The Riverside Publishing Company.

Thorndike, R. L., Hagen, E., & Sattler, J. (1986c). *Technical manual: Stanford-Binet Intelligence Scale, Fourth Edition.* Chicago: The Riverside Publishing Company.

Thorndike, R. L., Hagen, E., & Sattler, J. (in press). *Expanded guide for interpreting and reporting fourth edition results.* Chicago: The Riverside Publishing Company.

Glenn E. Snelbecker, Ph.D.
Professor, Psychological Studies, Temple University, Philadelphia, Pennsylvania.

Michael J. Roszkowski, Ph.D.
Research Psychologist, Research and Evaluation Department, The American College, Bryn Mawr, Pennsylvania.

STRESS EVALUATION INVENTORY

Raymond W. Kulhavy and Samuel E. Krug. Champaign, Illinois: Institute for Personality and Ability Testing, Inc.

Introduction

The Stress Evaluation Inventory (SEI) is a brief self-rating instrument used to measure stress resulting from one's career, family, and personal-social circumstances. The SEI was *not* designed to be administered alone, and is not available for purchase by itself; rather, together with the Sixteen Personality Factor Questionnaire (16PF), it forms the basis for a stress assessment and stress management package called the Individualized Stress Management Program (ISMP). The authors and publisher indicate that they are offering a stress management system that uses the information from these two instruments to design personalized prescription plans for stress reduction. Given this "captive role" for the SEI, this review will examine the ISMP as well as the SEI as an integral instrument.

The authors of the SEI and the related stress management program are Drs. Raymond W. Kulhavy and Samuel E. Krug. Dr. Kulhavy received his Ph.D. in educational psychology from the University of Illinois in 1971. He has served as a professor at Arizona State University since 1971 and as director of the Laboratory for the Study of Intellectual Processes (also at Arizona State University) since 1975. His areas of interest include the study of thinking processes and personality. Dr. Krug received his Ph.D. in personality psychology from the University of Illinois in 1971. He worked with the Institute for Personality and Ability Testing (IPAT) as director of the Test Services Division and is currently president of MetriTech, Inc., a company he founded in 1982.

The SEI form contains 60 items, but only 30 items are actually used for scoring purposes at present. These 30 items now comprising the SEI are first-person statements presenting situations or symptoms considered to be indicative of stress (e.g., "My current job makes great demands in terms of my performance"). The statements, distributed equally into three scales—Career Stress, Family Stress, and Personal-Social Stress—permit one of the following responses: "a. frequently applies to me," "b. sometimes applies to me," or "c. does not apply to me."

These 30 items were derived from an initial pool of 60 statements developed from a search of the literature and the suggestions of 12 clinicians regarding potentially

560

stressful experiences. A two-step process was used to select these 30 items. First, an item analysis was conducted on the 60-item pool to identify items that loaded on all three of the scales in order to eliminate statements that were not specific to one given source of stress (i.e., career, family, or personal-social). Second, the remaining items were used to assess 20 persons undergoing a stress reduction program. The clinicians conducting the program were asked to rate the participants on the relative importance of career, family, and personal-social circumstances as the cause of the individual's stress. The 10 items on each scale that had the highest correlation with the clinicians' ratings were retained.

As stated previously, the present SEI forms contain 60 (rather than only 30) items. These additional 30 items represent an ongoing attempt by IPAT and the authors to improve the SEI. Apparently, these items were developed after the first version of the SEI was first distributed commercially, and they were obtained separately from the 30 items that were originally reviewed and rejected. Following procedures used in selecting the initial SEI items, these new items were developed in an attempt to increase the range of situations tapped, to try to get better items, and to improve instrument reliability. Although the additional 30 items were not being fully utilized at the time this review was prepared, the IPAT procedures apparently make reference to these items when selecting exercise chapters (described in more detail later in this review) for a client's personalized stress management program.

The SEI authors and IPAT staff note that, in contrast with the "general causative factors" that are the focus in other stress analysis and stress management programs, they have opted to rely more on an "individual differences" model in formulating their approach

> one that recognizes the interaction between personality structure and stress responses. . . . As far as we know, the program described here is the only stress management approach which blends together test information with a systematically designed training model.
> The research effort that ultimately led to the development of the Individualized Stress Management program began by analyzing the relationship between personality programming and distress. We then developed a series of instructional modules designed to help reduce stress. In doing so we attempted to match the content of these modules to our best understanding of the relationship between stress sources and individual reactions. When the modules were completed, we field tested the entire program and obtained participant evaluations of its effectiveness. (Kulhavy & Dee-Burnett, 1984, p. 2)

As noted previously, the ISMP is based on the 16PF and the SEI. The 16PF is one of the most widely used comprehensive personality tests, especially with "normal" populations. Because the 16PF has been reviewed elsewhere in this series (Wholeben, 1985), only a few comments are needed here. Designed to assess normal adult personality attributes, the 16PF measures levels of assertiveness, emotional maturity, shrewdness, tension, self-sufficiency, and 11 other attributes designated as "primary" personality traits.

Kulhavy and Krug indicate that they reviewed the research literature and conducted research to identify personality predictors of good versus bad stress management. *Tech Report #1, Revised* (Kulhavy & Dee-Burnett, 1984), dated November 29, 1984, describes a study in which the 16PF and the SEI were administered to 270 people (about half males and half females) from a cross-section of occupations. Attempts were then made to use the 16PF results to predict the SEI scales. On average, a linear combination of 16PF factors could explain 30% of the reliable variance in the SEI. The 16PF Factor L (Suspiciousness) and Factor O (Insecurity) could predict stress in the career, family, and personal-social areas. Three other personality factors, however, were related to only one specific source of stress. That is, Factor Q_4 (Tension) predicted stress arising from family relationships, whereas Factor I (Sensitivity) predicted stress due to personal-social circumstances, and Factor G (Persistence) predicted stress originating in one's career. These relationships were replicated with another sample of 756 subjects.

These results, along with other research findings linking the 16PF to stress handling capabilities, served as the basis for the development of individualized stress reduction exercises. Each of the three sources of stress has four exercises. For career-related stress, the exercises are entitled a) Planning to Make Your Time Work for You, b) Learning to Compete for Important Things, c) Enjoying the Challenge of Work, and d) Smoothing Out Your Working Day. The exercises for stress stemming from one's family situation are a) Reviewing and Adjusting Important Life Areas, b) Thinking Straight About Long Term Plans, c) Seeing Yourself in a Better Light, and d) Easing Close Relationships. Stress due to personal-social circumstances has exercises named a) Getting Through to Other People, b) Standing Up for Your Personal Rights, c) Dealing with Little Things That Bother You, and d) Getting Your Thoughts Together. The exercises were developed by psychologists familiar with the content of the 16PF. The interventions were field tested on 206 adults; feedback from this sample indicated that they viewed the interventions positively.

The exercises consist of instructions and activities and contain what are called Personal Behavior Goals. Based on elevations occurring in the person's 16PF scores, there are two types of goals: those associated with *undesirable* 16PF factors, and those associated with *desirable* 16PF characteristics. If the respondent's personality is characterized by traits that are likely to cause stress, these flaws are identified and suggestions are offered on how to overcome such behavioral tendencies. For example, if the individual scores above the 85th percentile on the Q_1 (Radicalism) factor of the 16PF, the following behavioral goal is contained in the intervention unit called Planning to Make Your Time Work for You: "You have an inquiring, analytic mind and you tend, quite frequently, to be critical of traditional solutions. As a consequence, you may spend too much time looking for novel approaches to problems. However, you can reduce many time pressures by recognizing that not every problem requires a novel solution." The second type of Personal Behavior Goal, associated with personality traits that are conducive to stress management, is intended to "reinforce awareness" of that trait. Here is one example: "The fact that you usually approach situations calmly is an important asset in competitive activities" (Tech Report #, p. 1). Four Personal Behavior Goals have been developed for each intervention unit (exercise).

The results of the analysis of the 16PF, SEI, and the intervention procedures are presented in a 10-chapter booklet. Chapters 1 to 3 and chapter 10 are the same in all the booklets, irrespective of the particular pattern of scores on the 16PF and the SEI. These chapters provide an introduction to the ISMP (chapter 1), a definition of stress and a discussion of its effects (chapter 2), and a discussion of general stress reduction techniques (chapter 3). Chapter 10 is entitled "Where Do You Go From Here" and discusses what needs to be done after the booklet is read.

Chapters 4 through 9, on the other hand, are individualized to conform to the test results. Chapter 4 presents feedback on one's performance on the SEI, identifying the relative contribution of career, family, and personal-social areas to one's overall distress level. Each of the chapters numbered 5 through 9 consists of selected intervention plans, which are reflected in the chapter titles.

The sequence of these intervention modules (i.e., chapters) is determined by the relative severity of the person's stress on the Career, Family, and Personal-Social scales of the SEI. Three exercises (intervention modules), presented specifically as chapters 5 through 7, are concerned with the individual's greatest source of stress. For example, if the Career area is indicated as the highest source of stress for someone, that person's chapters 5 through 7 would focus on reducing stress arising from Career concerns. Similarly, someone else would have chapters 5 through 7 deal with Personal-Social topics if the Personal-Social score was higher than Career or Family on the SEI scales. Then, chapter 8 consists of guidelines for dealing with stress arising from the second highest SEI source of distress. Chapter 9 describes one intervention plan addressing the third highest SEI source of distress.

Chapters 5 through 9 contain a similar format, though addressing different areas. For each of these chapters, the first page provides an overview and time estimate for the client to read and to follow instructions provided in that chapter. The second page contains at least one Personal Behavior Goal that is relevant for that chapter, along with suggestions about the stress management implications of the goal. Also, the client is encouraged either to add additional goals or to substitute another related goal on which to focus in completing exercises provided in the chapter. The remaining pages contain descriptions of topics relevant to the particular area of stress management, a series of exercises to be completed, and other instructions for using the information provided in the chapter.

Practical Applications/Uses

IPAT and the authors of the Stress Evaluation Inventory have tried to maintain professional control over use of SEI and ISMP, especially by attempting to market their system only to qualified professionals. Page 16 of the 1987 IPAT catalog of psychological assessment instruments, computer interpretive services, and books contains these statements: "Specialized training to present IPAT's Individualized Stress Management Program is necessary. Attendance at one of the Stress Management Training Seminars is the preferred method of qualification." Psychologists and other professionals with counseling skills who complete their 1-day training program are designated as "IPAT Stress Management Consultants." The 1-day training program focuses on 1) the features of the ISMP; 2) the development, presentation, and administration of the materials; 3) the benefits a participant in the program can experience; and 4) marketing and delivery techniques. Participants receive a training manual for the Stress Program Coordinator, several technical reports (five at the time of this writing), audiotapes of sessions, training materials for clients, and their own Individualized Stress Management Program based on their SEI and 16PF results.

The ISMP apparently has been offered in business, education, health care, government, and private practice settings. Information to develop a client's Individualized Stress Management Program is obtained by IPAT Stress Management Consultants, who can administer the SEI and the 16PF to clients individually or in groups. The protocols are submitted to IPAT for scoring. IPAT provides test scores and interpretations, an *Individualized Stress Management* booklet, and a professional summary to the consultant for additional discussions with clients.

The suggested method for conducting sessions with clients is as follows. Session #1: approximately 2 hours—overview of stress and administration of the two diagnostic instruments (SEI and 16PF). Session #2: (conducted after the test materials are received from IPAT) approximately 2 hours—explanation of the workbook and discussion of the assigned stress management activities. Session #3: (after the client has read the booklet materials and completed the training activities) approximately 2 hours—review of results from the workbook activities, exploration of the client's experiences and insights, and consideration of further stress management plans.

Technical Aspects

Internal consistency reliability for the SEI is reported (Tech Report #5, p. 1), based on the responses of 827 persons drawn from enlisted military personnel, blue collar workers in the construction industry, and white collar workers. The reported alphas are .59 for the Career Stress Scale, .72 for the Family Stress Scale, and .76 for the Personal-Social Stress Scale. The total score reliability was .83.

The authors factor analyzed the responses from the same sample of 827 persons to demonstrate the SEI's construct validity. Three-, five- and eight-factor solutions were studied. The five-factor solution was considered to be the most appropriate. The first three of these five factors were unidimensional and homogeneous and were interpreted as reflective of the rationally derived scales (Career, Family, Personal-Social stress). However, three of the career items also defined a factor that seems to suggest Job Dissatisfaction Distress. The fifth factor consists of several items from the Career and Family Scales and is believed to measure stress due to a sense of obligation.

Critique

The question of what constitutes stress and how best to measure it has been debated at length. Thus, in order to evaluate the Stress Evaluation Inventory properly as a stress measure, it is really necessary to consider the instrument within the context of existing approaches to the measurement of stress.

It is noteworthy that the SEI, along with virtually all available stress measurement instruments, focuses on the *negative* aspects of stress and thus emphasizes "managing stress" by reducing the adverse impact of unpleasant experiences. Selye (1956, 1974; see also Snelbecker, 1983), who is credited with inventing the term "stress" in psychology and medicine, took the position that ongoing life experiences involve stress as a matter of course. Some forms of stress are *negative*, representing "threats" to the individual, whereas other forms of stress are *positive*, representing "challenges" to the individual. Selye proposed that we use the term *distress* to refer to negative instances of stress, and *eustress* for positive instances of stress. His publications provide a basis for addressing eustress as well as distress, but many subsequent authors have focused mainly on distress.

Instruments and programs that are designed to focus exclusively or mainly on distress typically enable clients to deal with threats and various sources of distress. With an exclusive or primary emphasis on distress, stress management programs provide little or no means either for identifying eustressful instances or for managing challenging experiences constructively. Because the SEI contains very little consideration for eustress and the management of challenges, this review will follow the convention of focusing mainly on distress, including the convention of treating *distress* and *stress* virtually as synonyms.

One can differentiate between physiological and *psychological* measures of stress. Obviously, the SEI falls into the psychological category of instruments. The most popular means of psychologically gauging the level of stress is through self-reports about one's exposure to stressful circumstances. The SEI and the ISMP involve self-reported measures and interventions for managing distress.

Self-report stress instruments typically have been concerned with one of the following questions: 1) whether the individual has recently experienced "major life events" or 2) the extent to which the person is subjected to "daily hassles" (Kanner, Coyne, Schaefer, & Lazarus, 1981; Lazarus, DeLongis, Folkman, & Gruen, 1985). Major life events are acute in nature and include such things as the death of a loved one, a divorce, loss of a job, moving to a new house, a promotion, and so on. Both unpleasant and pleasant events are considered. This approach to measuring the amount of stress in one's life is typified in the classic Social Readjustment Scale developed by Holmes and Rahe (1967). In contrast, "daily hassles" (sometimes also known as "microstressors" or "chronic role strains") are relatively less serious but are generally experienced on a day-to-day basis. Generally, hassles constitute a dissatisfaction with various aspects of one's life (e.g., not having one's accomplishments recognized by one's boss). The Hassles Scales (Kanner et al., 1981) is an example of the latter type of stress measure. At this time, it is not really clear whether major life events or daily hassles induce the more detrimental form of stress.

The content of the items in the SEI primarily reflects "daily hassles," although at least three of the items are of the major life events type (i.e, job change, new family responsibility, a change of residence). Many of the "daily hassles" items in the SEI can also be viewed as a list of symptoms of what Selye termed "strain" (see Freedman, 1985). As such, the SEI seems to be a hybrid measure of stress rather than one that can be neatly compartmentalized into the "life events" or "daily hassles" categories. It is probably prudent to measure both types of stressful situations for a comprehensive picture of a person's stress, and the authors of the SEI are wise to include both in their instrument.

On the other hand, the item response format for some of the "major life event" items could be improved somewhat. Presently, the scoring for these items is the same as for the "daily hassles" items, namely "frequently applies to me," "sometimes applies to me," or "does not apply to me." These answers are not good options for answering items 19, 20, and 26 on the SEI. For these three items, a two-option response format seems more appropriate (i.e., "applies to me" or "does not apply to me"). The same two-option item response format could also be used with item 29.

A number of other items on the SEI could perhaps be answered more precisely with a different set of options than those currently available or, alternatively, by editing the content of the items themselves. Because the answer options are in terms of event frequency, such words as "constantly" (item 2) and "often" (item 32) should be eliminated to minimize confusion for respondents. If the frequency-type modifiers are retained in the stem, the response options should be changed to the two-item response format suggested previously.

Another unresolved issue in the measurement of stress is the degree to which a person's cognitive appraisal of the situation needs to be considered in judging its stressfulness (Cohen, 1986; Dohrenwend & Shrout, 1986; Green, 1986; Lazarus et al., 1985; Lazarus & Folkman, 1986; Sowa, Lustman, & Day, 1986). One school of thought maintains that only the occurrence or nonoccurrence of the event needs to be determined. If the event occurred, then it is judged to be stressful to the same degree for all persons. In contrast, according to a second school of thought, one not only needs to check whether the event occurred but also how stressful that event was for the respondent. In other words, it is not the situation itself that is stressful, but, rather, it is the interpretation of that situation that determines its stress level. For instance, the death of an uncle may be much more stressful if the person was well known by the respondent than if the two people had never met.

The issue of how much "subjective weight" needs to be assigned to a particular situation continues to be an important issue in the construction of both "major life events" and "daily hassles" scales. Some suggest, for example, that the personal interpretation of a situation should receive greater consideration in the measurement of "daily hassle" type events than in the measurement of "major life events"; the reasoning is that daily situations are more ambiguous. However, not all researchers agree on this. Examples of both measurement approaches can be found in the literature. Among "major life event" scales, the previously mentioned Holmes and Rahe scale (1967) does not consider the subjective appraisal of the situation, whereas The Life Experiences Survey (Sarason, Johnson, & Siegel, 1978) does take this into account. Among "daily hassle" type scales, the Hassles Scale (Kanner et al., 1981) does not make provision for assessing whether an event is perceived to be stressful independently of its occurrence, but the Perceived Stress Scale (Cohen, Kamarck, & Mermelstein, 1983) does make such provision.

The SEI only measures whether a particular situation applies to a respondent's life circumstances. The model underlying the SEI thus seems to assume that all the situations presented are in fact a source of distress to the person and that each situation is equally stressful to everyone. In this respect, its creation appears to have been guided by the first school of thought.

In the opinion of these reviewers, stress is not purely a characteristic of the environment or of the individual, but an interaction of the two. Therefore, preferable response formats would be those that allow one to obtain information indicating whether 1) the situation described in the item applies to the respondent and, 2) if it applies, how much distress it is causing. We believe that this approach to the measurement of stress is more precise and that including this additional feature would make the SEI a more precise instrument. It is possible that, to some individuals, some SEI items may be "applicable" (in the sense that the hassle or major life event has occurred), but the client may not really experience distress from this event. For instance, it is conceivable that for some persons having a job that "makes great demands" on them is not a source of distress. In fact, it could even be a source of eustress, in that the person sees such "demands" as challenges rather than threats.

The reported internal consistency reliability of the SEI subscales is disappointing, especially for the Career Stress subscale. Only the SEI total score meets Nunnally's (1978) recommended reliability criterion of .70 for use with groups, and none of the SEI's scales meet his minimal criterion of .90 for practical applications with individuals. In fairness, however, Nunnally's standards are quite difficult to meet when one is dealing with personality-type scales and could constitute a target, but probably unattainable, ideal for instruments like the SEI. Reliability problems are not uncommon with personality-type measures in general, including measures of stress. There is work being done by IPAT and the authors on 30 experimental items (noted at the beginning of this review). Theoretically, if the SEI is made longer by adding at least some of these items, the internal consistency reliability of the subscales should improve.

Perhaps a test-retest reliability index would provide more encouraging results. There is the opinion that evaluating a scale's reliability by its homogeneity is a false and outmoded idea (Cattell, 1986). Internal consistency reliability is viewed by some as particularly noncritical for "life event" scales. As Hurst, Jenkins, and Rose (1978) note, there is no reason to assume that the death of a friend should be correlated with job change. Hurst et al. believe that the comprehensiveness and relevancy of life event inventories are more important considerations in judging the value of a scale.

Further work also needs to be conducted to demonstrate the SEI's validity. The technical report discussing the SEI's validity only deals with the scale's factor structure (a form of construct validity). Some evidence for the SEI's criterion-related validity may be found in the technical report describing the SEI's relationship to the 16PF, but more evidence is needed. It would be extremely instructive to learn about this instrument's relationship with other measures of stress as well as about the SEI's ability to predict some of the negative physical outcomes associated with stress (e.g., hypertension, heart attacks, strokes, etc.)

The literature on stress contains studies linking specific personality traits to stress. Some of these studies have used the 16PF as the personality instrument but have measured stress with instruments other than the SEI. The predictive validity of Factors L (Suspiciousness), O (Insecurity), and Q_4 (Tension) have been demonstrated in a study by Duckitt and Broll (1982). Factor-analyzing the 16PF into six second-order factors and relating these second-order factors to the Langer Index (a 22-item measure of psychological strain), these authors found that two second-order 16PF factors were related to stress, namely Extroversion and Anxiety. The second-order Anxiety factor was defined by factors L, O, Q_4, and C. These results provide additional support for the associations identified above between the SEI and factors L, O, and Q_4. However, it should be noted that Duckitt and Broll attributed the 16PF Anxiety factor/Langer Index correlation to a conceptual overlap between the items on the two scales. A similar conceptual overlap may contribute to the 16PF/SEI correlations.

The preceding portion of this section was concerned with the SEI as an instrument. For a full evaluation of the SEI, it is also necessary to consider the stress management system that it supports. It may be best to start by comparing the ISMP with other stress management programs currently available. Freedman (1985) reports that most stress management programs simply inform the participants about the debilitating consequences of stress, working under the implicit assumption that such knowledge by itself will somehow produce beneficial transformations in the person's behavior. According to Freedman, few stress management programs teach the person how to acquire the coping skills needed to reduce stress effectively. In this respect, the ISMP appears to surpass most other programs.

Kirkpatrick (1975) proposes that stress management programs can be evaluated on four levels: 1) the emotional reactions of the participants to the program immediately after its completion; 2) the level of mastery of the material presented; 3) the degree to which the participants apply this knowledge; and 4) if applied, what effects the program has on the physiological symptoms of stress. Freedman (1985) contends that there are very few comprehensive evaluations of stress management programs that examine Kirkpatrick's four levels. Most evaluations of stress management programs typically focus on data regarding the participants' immediate reactions to the program, which Freedman believes informs us primarily about the entertainment value of the program. Tech Report #2, dated March 28, 1983, notes positive feedback from clients who used early versions of the ISMP. Krug (personal communication, July 30, 1986) indicated that IPAT and the SEI authors have been collecting details about client reactions to the SEI and the ISMP and that a report of findings may be forthcoming in a technical report. Hopefully, further work will continue and will go beyond the first level identified by Kirkpatrick (1975).

References

Cattell, R. B. (1986). The 16PF personality structure and Dr. Eysenck. *Journal of Social Behavior and Personality, 1,* 153-160.

Cohen, S. (1986). Contrasting the Hassles Scale and the Perceived Stress Scale: Who's really measuring appraised stress? *American Psychologist, 41,* 716-718.

Cohen, S., Kamarck, T., & Mermelstein, R. (1983). A global measure of perceived stress. *Journal of Health and Social Behavior, 24,* 385-396.

Dohrenwend, B. P., & Shrout, P. E. (1986). "Hassles" in the conceptualization and measurement of life stress variables. *American Psychologist, 40,* 780-785.

Duckitt, J., & Broll, T. (1982). Personality factors as moderators of the psychological impact of life stress. *South African Journal of Psychology, 12,* 76-80.

Freedman, A. M. (1985). Stress-management training. In W. R. Tracey (Ed.), *Human resources management and development handbook* (pp. 1221-1240). New York: AMACOM.

Green, B. L. (1986). On the confounding of "hassles" stress and outcome. *American Psychologist, 41,* 714-715.

Holmes, T. H., & Rahe, R. H. (1967). The Social Readjustment Rating Scale. *Journal of Psychosomatic Research, 11,* 213-218.

Hurst, M. W., Jenkins, C. D., & Rose, R. M. (1978). The assessment of life change stress: A comparative and methodological inquiry. *Psychosomatic Medicine, 40,* 126-141.

Kanner, A. D., Coyne, J. C., Schaefer, C., & Lazarus, R. S. (1981). Comparison of two modes of stress measurement: Daily hassles and uplifts versus major life events. *Journal of Behavioral Medicine, 4*, 1-39.

Kirkpatrick, D. (1975). *Evaluating training programs.* Madison, WI: American Society of Training and Development.

Lazarus, R. S., DeLongis, A., Folkman, S., & Gruen, R. (1985). Stress and adaptational outcomes: The problem of confounded measures. *American Psychologist, 40,* 770-779.

Lazarus, R. S., & Folkman, S. (1986). Reply to Deutsch and Green. *American Psychologist, 41,* 715-716.

Nunnally, J. C. (1978). Psychometric theory (2nd ed.). New York: McGraw-Hill.

Sarason, I. G., Johnson, J. H., & Siegel, J. M. (1978). Assessing the impact of life changes: Development of the Life Experiences Survey. *Journal of Consulting and Clinical Psychology, 46*, 932-936.

Selye, H. (1956). *The stress of life.* New York: McGraw-Hill.

Selye, H. (1974). *Stress without distress.* Philadelphia: Lippincott Company.

Snelbecker, G. E. (1983). Managerial stress and strategic planning for company stress management programs. In W. Schiemann (Ed.), *Managing human resources/1983 and beyond* (pp. 264-285). Princeton, NJ: Opinion Research Corporation.

Sowa, C. J., Lustman, P. J., & Day, R. C. (1986). Evaluating stressful life events. *Educational and Psychological Measurement, 46*, 353-358.

Wholeben, B. E. (1985). Sixteen Personality Factor Questionnaire. In D. J. Keyser & R. C. Sweetland (Eds.), *Test critiques* (Vol. IV, pp. 595-605). Kansas City, MO: Test Corporation of America.

Technical Reports for the SEI and the Individualized Stress Management Program:

Kulhavy, R., & Dee-Burnett, R. (1984). *Tech Report #1, Revised (November 29, 1984): The 16PF as a guide to predicting and managing stress.* Champaign, IL: Institute for Personality and Ability Testing.

Tech Report #2 (March 28, 1983): Field test summary. Champaign, IL: Institute for Personality and Ability Testing.

Tech Report #3, Revised (November 29, 1984): A check on level of individualization in the selection and sequencing of book chapters. Champaign, IL: Institute for Personality and Ability Testing.

Tech Report #4, Revised (November 29, 1984): Structure and design of the Personal Behavior Goals module. Champaign, IL: Institute for Personality and Ability Testing.

Tech Report #5 (May 25, 1983): Construction and validation of the Stress Evaluation Inventory (The SEI). Champaign, IL: Institute for Personality and Ability Testing.

Jerry B. Hutton, Ph.D.

Professor of Special Education, East Texas State University, Garland, Texas.

THE STRESS RESPONSE SCALE

Louis A. Chandler. Pittsburgh, Pennsylvania: Louis A. Chandler, Ph.D.

Introduction

Psychological stress is experienced when an individual perceives an event or situation as threatening and/or when needs are unmet. The Stress Response Scale (SRS) was developed to assess the impact of stress on the behavioral adjustment of children. By rating the child on 40 items, parents and teachers may provide information that assists the diagnostician, counselor, or psychologist in measuring the magnitude of maladjustment as well as describing the typical behavioral pattern for the child in response to stress. The assessment of the child's behavioral adjustment is actually the fourth step in the Stress Assessment System (Chandler, 1985a), following the identification of stressors; the exploration of the child's perception of those stressors; and the assessment of the impact of stress on the child's health, school, and social functioning. The fourth step, assessing the child's behavioral reaction to stress, involves the administration of the SRS. Scores on the SRS are expressed as a total T-score, total percentile, and T-scores for each of the five subscales: Acting Out, Passive-Aggressive, Overactive, Repressed, and Dependent.

Louis A. Chandler, Ph.D., author of the SRS, received his training at the New School for Social Research of Duquesne University and at the University of Pittsburgh. He is currently Clinical Associate Professor in the Department of Educational Psychology at the University of Pittsburgh and is director of the Psychoeducational Clinic. His previous experience includes work as a staff psychologist with the Exceptional Children's Program of the Allegheny Intermediate Unit in Pittsburgh. He is the author of two books on stress, *Children Under Stress* and *Assessing Stress in Children*, and has written several articles. Work on the SRS dates back to at least 1979 when an unpublished manuscript (Chandler, 1979) contained the report of a factor analytic study of the structure of behavioral disorders, replicated later by Piso (1981). The most recent SRS manual is a 1986 revision, titled *The Stress Response Scale for Children* (Chandler, 1986). The latest research on the SRS that was available at the time this review was being conducted was dated 1986 (Chandler & Shermis, 1986). The SRS protocol and response charts are dated 1982.

The SRS consists of a manual, a rating form, and separate profile charts for boys and girls. The rating form is a single sheet of paper with both sides used to print the 40 items. The items consist of one- to four-word descriptors such as "Worries" and "Poor attitude toward school." Teachers or parents rate each item according to the estimated frequency of occurrence by marking one of six categories of response ranging from "never" to "always." The profile charts have one side for graphing the child's profile, and the other side may be used for scoring.

The SRS was normed on 947 children (459 boys and 488 girls), 5 through 14 years of age. The children attended schools in six school districts in western Pennsylvania. Students assigned to special education classes were excluded. Tables for converting raw scores to T-scores and percentiles are given separately for boys and girls at three age levels: 5 to 6 years, 7 to 12 years, and 13 to 14 years.

Raw scores are summed to represent a total score and subscores for each of the five subscales. The values awarded for each item may range from 0 (never) to 5 (always). Eight of the items are scored in reverse order, and these items are designated with an "R" on the scoring sheet. A few of the items are included in more than one of the subscales. For example, item 23 ("Cares about school work") is reversed in scoring and occurs in both the Acting Out and Passive-Aggressive subscales. The raw scores may be converted to T-scores and plotted on a profile chart to represent behavior patterns graphically. The potential for clerical errors in transferring scores from the protocol to the scoring sheet is lessened by using the Apple II series computer program in scoring. The computer program also provides a printout of a descriptive report of the SRS results, including the profile type and the incidence of the profile type in the normative sample and in a clinical sample.

Practical Applications/Uses

According to Chandler (1985a), the Stress Response Scale consists of descriptors that were taken from psychological reports of children referred to the Psychoeducational Clinic at the University of Pittsburgh. The descriptors were thought to represent four primary types of behavioral patterns exhibited by mildly or moderately emotionally disturbed children, and factor analytic studies confirmed the primary types, except one type (Impulsive), which was subdivided by factor analysis into two (Acting Out and Overactive). The resulting five types may have diagnostic utility in a school or clinical setting. Further, the SRS may assist counselors or other school personnel in elementary and junior high schools in screening students for possible referral or intervention. Because the number of items to be rated is less than some behavior rating scales, teachers may be more accommodating to requests to provide information regarding their observations of students. Parents may also rate the child using the SRS. This may facilitate the comparison of perceptions of the child's behavior at home with behavior at school. On a very general level, this assists in comparing assessment and intervention needs in two major settings, at home and at school. The private practitioner may wish to consider the SRS as a part of the routine battery of instruments used to assess children referred due to behavior problems.

Administration and scoring are clear and simple. The 40 descriptors are brief, concise, and uncomplicated. The dimensions (or subscales) represent symptom clusters that are fairly easily understood due to their common names: Acting Out, Passive-Aggressive, Overactive, Repressed, and Dependent. The author notes that the SRS was developed to be used with children who have possible emotional adjustment problems rather than seriously emotionally disturbed (psychotic), mentally retarded, or learning disabled children. Children with mild or moderate emotional disturbance probably compose the bulk of the caseload of private practitioners and psychoeducational clinics. Chandler (1985a) reports that over half of a sample of 100 children referred to the Psychoeducational Clinic at the University of Pittsburgh were classified as expressing emotional adjustment reactions.

The author indicates that it takes about 10 minutes to rate a child on the SRS. This amount of time seems realistic. Scoring without the computer program takes another 5 to 10 minutes. Thus, within 15 to 20 minutes a profile may be developed on a student. This feature of the SRS not only contributes to its value as an initial assessment tool but may also make it practical to use in assessing a child's response to counseling or special intervention. Chandler (1985b) has also provided guidelines for intervention. Of course, as suggested by Chandler (1986), the SRS should be used in combination with other assessment data in making decisions about children. Although not stated in the manual, assessment personnel should probably be the ones to make the greatest use of the SRS.

Technical Aspects

Chandler (1986) reports information regarding the validity of the Stress Response Scale under the headings of construct validity, content validity, factorial validity, discriminant validity, and criterion-related validity. The construct validity of the SRS is based on the selection of the conceptual model, which contains four hypothesized response patterns to stress. Drawing from the taxonomy of coping responses proposed by Moos and Billings (1982) and taking the position that emotional adjustment problems are maladaptive responses to stress (Klerman, 1979), Chandler supports his decision to view the child's maladaptive behavior as an ineffectual effort to cope with stress. The four ineffectual response patterns are expressed in the conceptual model as repressed, dependent, passive-aggressive, and impulsive. These types are derived from behavior that is extreme on two dimensions of personality: passive versus active and introversion versus extraversion (Chandler, 1983). Chandler (1986) notes the similarities between the four response patterns and selected coping responses from Moos and Billings (1982): the dependent response is similar to resigned acceptance, the impulsive response is like emotional discharge, and so on. Also, Chandler, Shermis, and Marsh (1985) reported similarities between four of the SRS categories and four selected diagnoses from the *Diagnostic and Statistical Manual of Mental Disorders* (DSM-III; American Psychiatric Association, 1980): Impulsive (Acting Out) and Conduct Disorder, Aggressive Type; Passive-Aggressive and Oppositional Disorder; Impulsive (Overactive) and Attention Deficit Disorder; and Repressed and Anxiety Disorders. Considering that construct validity refers to the extent to which test performance can be interpreted in relation to theoretical constructs, the SRS appears fairly sound.

The items for the SRS were developed from expected characteristics derived from the four response patterns, characteristics appearing in the psychological reports of clinic-referred children and in the literature on childhood behavior disorders. The descriptors were organized under their respective response patterns (repressed, dependent, passive-aggressive, and impulsive). The first factor analysis (Chandler, 1979) of SRS scores was performed using the scores of 120 clinic-referred children. This analysis found a five-factor solution accounting for 62% of the variance. The factors were very similar to the predetermined assignment of items to categories of response patterns. Items in the category labeled Impulsive formed two factors, which were called Acting Out and Overactive. Piso (1981) used a larger group of children ($N = 376$) classified as nonreferred and found a similar factor structure. Another factor analytic study ($N = 167$) reported a five-factor solution accounting for 64% of the variance (Chandler & Lundahl, 1983). The confirmatory factor analyses support the construct validity of the SRS. Further, the content validity of the SRS appears adequate, with the item selection following sound procedures.

Studies related to criterion-related validity are reported in the manual. The SRS differentiates between clinical samples of children and nonreferred samples (Chandler, 1979, 1983; Kemmerer, 1984; Krotec, 1982; Piso, 1981). Further, Chandler and Shermis (1986) analyzed the SRS ratings of 857 randomly selected, nonreferred children and 84 clinic-referred children in order to identify and compare profile types. The two profiles identified most frequently among the clinic-referred group were labeled Acting Out and Repressed. Mixtures of low-level responses were found to be fairly evenly distributed within the nonreferred sample. Chandler and Shermis (1985) submitted the ratings of the children in the normative sample to a cluster analysis in order to obtain data on typical behavior patterns. They found an overactive rating to be the typical area elevated among boys and girls across the three age groupings. Another study (Chandler, Shermis, & Marsh, 1985) compared SRS subscale scores with the diagnoses assigned to a sample of clinic-referred children using the DSM-III. The SRS ratings were shown to reduce the error in predicting diagnostic group membership. The ratings of children by teachers predicted the diagnostic groups better than those of parents. Thus, the evidence for the criterion-related validity of the SRS is good.

Reliability involves the determination of the internal consistency as well as the stability of the instrument. The SRS has good internal consistency with a reported coefficient alpha of .94 (Chandler, 1986). Two studies show good test-retest stability of the SRS using a 1-month interval. In the first study (Chandler, 1986), 25 students in regular classrooms were rated by teachers, and the two ratings yielded a correlation coefficient of .86. In the second study (Mramor, 1986), teachers rated 68 elementary students, and the correlation coefficient for the total score was .87. Coefficients for the subscales were: Acting Out, $r = .83$; Passive-Aggressive, $r = .83$; Overactive, $r = .72$; Repressed, $r = .80$; and Dependent, $r = .73$. The stability of ratings made by different raters within or between settings is not reported.

Critique

The Stress Response Scale has some very apparent strengths that should be noted. It was developed as the fourth step in the Stress Assessment System, as mentioned earlier, and is the outgrowth of the author's continuing interest in stress as a major mediating variable in the maladaptive behavior of children. The SRS may be used by parents and teachers to rate the behavior of children, and the ratings may assist practitioners who make judgments concerning the need for further assessment and intervention at school and/or at home. Its limited number of items (40), ease of scoring, available computer scoring program, and easily understood constructs all contribute to its appeal as a practical tool. Several studies support the validity and reliability of the SRS.

A major limitation of the SRS is the lack of norms established on a national sample. Presently, SRS users are restricted to comparing obtained ratings with the normative sample of children in western Pennsylvania. Also, information regarding the ethnicity of children in the normative sample is not given in the manual. Some critics might react to the response scaling of the SRS. The SRS calls for a six-step response to each item, with the judgment ranging from "never" to "always." Edelbrock (1983) notes that untrained raters such as teachers and parents may have difficulty making the fine distinctions involved in response formats with more than three choices. However, the six-step response scaling of the SRS is only a minor criticism in light of the good reliability and validity of the instrument.

One of the advantages of a behavior rating scale that has only a few items is that it facilitates the assessment of behavior across settings. Multiple ratings across school settings require the cooperation of more than one teacher, and brief scales meet less resistance. The SRS may help the diagnostician, counselor, or practitioner view the child's behavior patterns in a variety of settings. However, in order to be of benefit in assessing behavior across settings, data need to be furnished regarding the reliability of the SRS when rated by two raters within the same setting as well as different raters in different settings. More information regarding differences between teacher-rated and parent-rated behavior would be helpful. It may be that ratings by parents require separate norms for interpretation.

References

American Psychiatric Association. (1980). *Diagnostic and statistical manual of mental disorders* (3rd ed.). Washington, DC: Author.

Chandler, L. A. (1979). *A classification scheme for behavior disorders.* Unpublished manuscript, University of Pittsburgh.

Chandler, L. A. (1983). The Stress Response Scale: An instrument for use in assessing emotional adjustment reactions. *School Psychology Review, 12,* 260-265.

Chandler, L. A. (1985a). *Assessing stress in children.* New York: Praeger.

Chandler, L. A. (1985b). *Children under stress: Understanding emotional adjustment reactions* (2nd ed.). Springfield, IL: Charles C. Thomas.

Chandler, L. A. (1986). *The Stress Response Scale for Children.* Unpublished manuscript, University of Pittsburgh. (Available from Louis A. Chandler, The Psychoeducational Clinic, University of Pittsburgh, 5D Forbes Quadrangle, Pittsburgh, PA 15260)

Chandler, L. A., & Lundahl, W. T. (1983). Empirical classification of emotional adjustment reactions. *American Journal of Orthopsychiatry, 53,* 460-467.

Chandler, L. A., & Shermis, M. D. (1985). Assessing behavioral responses to stress. *Educational and Psychological Measurement, 45,* 825-844.

Chandler, L. A., & Shermis, M. D. (1986). Behavioral responses to stress: Profile patterns of children. *Journal of Clinical Child Psychology, 15,* 317-322.

Chandler, L. A., Shermis, M. D., & Marsh, J. (1985). The use of the Stress Response Scale in diagnostic assessment. *Journal of Psychoeducational Assessment, 3,* 16-29.

Edelbrock, C. (1983). Problems and issues in using rating scales to assess child personality and psychopathology. *School Psychology Review, 12,* 293-299.

Kemmerer, A. (1984). *A comparison of the Stress Assessment System data of clinic-referred and nonreferred children.* Unpublished doctoral dissertation, University of Pittsburgh.

Klerman, O. L. (1979). Stress, adaption, and affective disorders. In J. E. Barrett (Ed.), *Stress and mental disorder* (pp. 151-160). New York: Raven Press.

Krotec, S. C. (1982). *A comparison of behavior, personality, and academic variables of learning disabled, emotionally disturbed, and normal adolescents.* Unpublished doctoral dissertation, University of Pittsburgh.

Moos, R. H., & Billings, A. G. (1982). Conceptualizing and measuring coping responses and processes. In L. Goldberger & S. Brenitz (Eds.), *Handbook of stress: Theoretical and clinical aspects* (pp. 212-230). New York: The Free Press.

Mramor, N. (1986). *The measurement and instruction of stress management techniques with elementary-school students.* Unpublished doctoral dissertation, Saybrook Institute, San Francisco.

Piso, C. N. (1981). *A revision, factor analysis, and concurrent validity study of a children's behavior scale.* Unpublished doctoral dissertation, University of Pittsburgh.

R. A. Bornstein, Ph.D.
Neuropsychology Laboratory, Department of Psychiatry, The Ohio State University, Columbus, Ohio.

L. J. Suga, Ph.D.
Neuropsychology Laboratory, Department of Psychiatry, The Ohio State University, Columbus, Ohio.

SYMBOL DIGIT MODALITIES TEST

Aaron Smith. Los Angeles, California: Western Psychological Services.

Introduction

The Symbol Digit Modalities Test (SDMT) is a symbol manipulation task that involves the association of meaningless geometric designs with written and/or oral number responses. The SDMT is similar to the Digit Symbol subtest of the Wechsler intelligence scales with some modification of the nature of the required response. The test requires the examinee to learn associations between nonsense designs and numbers and to provide those numbers in response to the particular design stimuli. The SDMT is timed (90 seconds) and requires the subject to substitute numbers (written or oral) in response to randomly presented designs. Similar to the Digit Symbol subtest, successful performance of the SDMT would appear to require visuoperceptual and graphic (writing) skills as well as visuomotor coordination, learning abilities, motor persistence, sustained attention, and response speed. The SDMT is purported by its author to be useful as a screening device for cerebral dysfunction in both adults and children. Whatever utility the SDMT may have in this context is very likely related to the multifaceted demands of the test.

The SDMT was developed by Aaron Smith, Ph.D., professor of neuropsychology and education at the University of Michigan. Dr. Smith is one of the early contemporary investigators in the field of clinical neuropsychology. The development of the SDMT appears to have evolved from his early experiences and work, as well as from the observation that symbol substitution tests were generally (but nonspecifically) sensitive to the effects of cerebral dysfunction. An underlying assumption was that this sensitivity arose from test demands related to the mechanisms and capacities of both cerebral hemispheres, as well as the integration of these activities.

Normative and experimental work with the SDMT began in 1965. The test was initially published by Western Psychological Services in 1973, and the manual was revised in 1982. The test materials have not been modified. Normative samples include a large number of children from Nebraska ($N = 3,680$), adults from New Jersey and Michigan ($N = 420$), and a second Michigan sample ($N = 887$). The age range for these samples is 8 to 75 years. The initial development of the test included both oral and written administration procedures, but alternate forms of the test are otherwise not available. Because the test involves manipulation of geometric figures and numbers, there are few language specific demands. Therefore, taking the test does not require proficiency with English, apart from the level required to understand test directions. Normative studies indicated differences in children's performance related to sex, and the manual provides separate norms for the appropriate ages. Separate norms are also provided for the two administration procedures (oral and written). Normative studies found no differences in performance between men and women, so tables are provided based on age group (18-24, 25-34, 35-44, 45-54, 55-64, and 65 and older) as well as educational level (12 years or less, 13 years or more). In the standardization tables, classification rates associated with various cutoff scores are presented.

The answer sheet pictures eight rows of double boxes divided horizontally at the midline with 15 boxes per row. In the top of each box is a geometric design; nine different designs are arrayed randomly. The boxes below the midline are empty. At the top of the page is a key in which each of the designs in the upper box is assigned a single digit number (1-9) in the lower box.

Practical Applications/Uses

The Symbol Digit Modalities Test was designed and developed as a dichotomous screening measure for the identification of cerebral dysfunction in children and adults. Many of the validity studies suggest that this goal is accomplished in that patients with cerebral dysfunction in fact do perform poorly on the SDMT. This appears to be true for patients with a variety of neurological disorders and, as acknowledged by the test authors, likely relates to the multiple demands of the task. However, as the test authors state, this same sensitivity means that test performance may also be depressed by other factors, and, therefore, "SDMT scores *alone* should not be considered diagnostically definitive" (Smith, 1982, p. 4). In the context of diagnostic studies of suspected cerebral dysfunction, the manual points out the value of the test "in conjunction with standardized intelligence and other carefully selected tests" (Smith, 1982, p. 2); the most widespread use of the SDMT has been in the context of a comprehensive neuropsychological test battery (The Michigan Neuropsychological Test Battery; Smith, 1975). The SDMT is also purported to be of value in the prediction or early identification of reading disabilities in children.

For the written version of the test, the examiner draws the subject's attention to the key at the top of the page and points out the unique association between numbers and geometric designs. Attention is then drawn to the eight rows in which the lower boxes are empty. The subject is then instructed to match each figure with the corresponding number, and then write that number in the appropriate box. Ten practice trials are given. Each subject is then encouraged to work as quickly as possible and is allowed 90 seconds to complete as many items as possible. If performance on this written task falls below 1 standard deviation (based on age, sex, or educational means) the administration of the oral version is recommended.

The instructions for the oral version are substantially condensed. The subject is instructed to repeat the task, but to vocalize the correct numbers associated with the designs (rather than writing them). No practice trials are given, and the examiner records the subject's responses. The test stimuli are identical. The written version is suggested to be given first, although no specific recommendation is made regarding retest interval. There are no additional subtest scores. The manual contains a relatively coarse method for classifying performance in relation to normative means and standard deviations and also provides some interpretive guidelines for evaluation of children and adult test scores.

Administration is easily accomplished within 5 minutes by a neuropsychologist or trained psychometrist, and scoring can be quickly and easily completed in a few minutes. Hand scoring is the only method available; this involves counting the number of substitutions attempted and the number of correct items. The scoring criteria and interpretation of scores based on proposed cutoffs are presented in a straightforward manner. Specific interpretations of patterns of performance are provided only for children. Scores approximately 1.5 standard deviations below the appropriate age or education group mean are reported as "indicative of possible cerebral dysfunction" in adults (Smith, 1982, p. 6). For adults, the manual also provides some gradient for scoring in relation to deviation from the expected group mean. There is some inconsistency between interpretive comments for children versus adults. For example, scores approximately 1 standard deviation below expected means are described as a low score for adults but as a markedly abnormal score for children. Again the manual is careful to emphasize that "subnormal SDMT scores alone cannot be used as evidence of brain damage" and that such scores indicate the need for further examination (Smith, 1982, p. 4).

The SDMT is used most appropriately when individually administered in conjunction with other tests. The manual also suggests the potential utility of the SDMT as a group-administered screening test with children. Although the test clearly has merit in the context of a comprehensive neuropsychological examination, its lack of specificity raises some concern as to its applicability as an accurate screening device for learning difficulties. Public Law 94-142 requires test specificity in any instrument used for educational planning; this would appear to mitigate against the utility of the test for widespread screening purposes.

Technical Aspects

The normative data base of the Symbol Digit Modalities Test includes a large number of children and adults. In adults, the manual reports that test-retest reliability studies have suggested that both written and oral versions of the test have adequate reliability with a mean 1-month interval (.80 and .76, respectively). In adult aphasic patients (treated and untreated), there was little change over a nearly 2-year interval (Smith, Champoux, Leri, London, & Muraski, 1972). The manual does not include sufficient information to judge the merits of this study; several important points were not mentioned, such as duration of symptoms and duration and nature of treatment. If the SDMT was insensitive to the effects of a therapy that was effective for treatment of aphasia, this could imply an important weakness of the test. The correlation between the written and oral forms of the SDMT (see Smith, 1982, p. 10) appears adequate in adults (.82 and .78 in two studies).

A significant omission in the manual is the lack of adequate descriptive information about the nature of the adult normals used for the normative and reliability studies. In one of the reliability studies, the mean age was 34.8, with a mean education of 16.2 years. In the larger of the normative/reliability samples ($N = 887$), the mean age was 33.6 with a mean educational level of 15.2 years. The number of subjects in the age/education subgroups in the manual's Table 6 and the basis for subdivisions are not provided. The meaning of the statistical differences on age and education is somewhat obscure in view of the arbitrary basis of these divisions. There is no evidence presented that Grade 12 is the appropriate point for subgrouping, or that only two education groups are needed. Test-retest reliability correlations are not provided for children.

The reliability studies are methodologically flawed by the failure to use a counterbalanced design. The manual suggests that the written form was given first in all studies to approximate the normal clinical use of the SDMT. However, if the equivalence of the test versions is to be established, such counterbalanced studies would seem to be critical. Furthermore, the reliability studies with children contain additional weaknesses that could compromise the value of this test with children. The issue of practice effects in children is discussed and addressed (in part) by examining the relationship or duration of test-retest interval with the difference in scores obtained from the oral and written versions. The manual reports that the mean oral scores were consistently higher, regardless of duration of interval. However, the duration of this interval is not specified and could have a bearing on the interpretation of clinically derived scores.

It is possible, for example, that these scores were all within a single test session lasting perhaps a few hours, thus providing a very narrow range. In clinical practice following a group-administered screening, it could be several days before an initially poor score on the written version might be followed up with the administration of the oral version. This difference in intertest interval could well lead to comparatively lower scores on the oral version and could be interpreted to represent a significant problem. The manual provides no guidelines as to recommended interval, and no specific guidelines are given for group administrations where the above possibility is more likely to occur.

There is an additional problem with the normative data for children. The data in the manual's Table 1 indicate that the mean scores for the oral version are based on administration when this is the only version given, not as a retest as recommended by the manual. Thus, in accordance with the stated administration procedures, the norms for the oral version would not appear to be directly applicable.

A considerable number of studies have documented the sensitivity of the SDMT to cerebral dysfunction arising from a variety of neurological disorders including commissurotomy (Campbell, Borgen, & Smith, 1981), cerebrovascular disease (Burkland & Smith, 1967), brain trauma (Smith et al., 1972), Huntington's disease (Whelan, Berker, Campbell, & Smith, 1980), Cushing's disease (Whelan, Schteingart, Startkman, & Smith, 1980), chronic brain lesions (Centofanti, 1975; Smith & Centofanti, 1975), and dementia (Pfeffer et al., 1981). Additional studies (Hartlage, Mains, Kovach, & Stovall, 1981) have examined the effects of anticonvulsant medications on SDMT performance. Studies attempting to discriminate between patients with psychiatric disorders versus "organics" have produced mixed results. Initial, unpublished studies by Watson (1970) found schizophrenic patients to perform worse than patients with brain dysfunction, but later studies (Watson, Gasser, Schaefer, Buranen, & Wold, 1981) from the same group of investigators found the opposite result. In the examination of depressed patients, this same group of investigators (Watson, Davis, & Gasser, 1978; Watson, et al., 1981) has shown 1) that the SDMT is an important variable in discriminating between those patients and others with cerebral dysfunction and 2) that high classification rates can be obtained. The manual describes these studies, but does not contain adequate information to assess the validity of the findings. Furthermore, the SDMT scores used to arrive at such classifications are not presented, so that it is impossible to compare those experimental results with cutoff scores proposed in the SDMT manual.

Furthermore, the studies do not appear to use the same criteria for determining abnormality. For example, in the manual's discussion of the study of patients with chronic lesions (p. 14), it is suggested that 1 to 1.5 standard deviations from the mean be considered suggestive of cerebral dysfunction, whereas the discussion of patients with Cushing's disease (p. 15) reports the percentage of patients with scores 1 standard deviation below. Because the manual (p. 6) suggests a criterion of 1.5, it is not clear what the latter study is demonstrating. In the study of anticonvulsant medications, no information regarding mean performance is presented.

In contrast to the numerous studies documenting the sensitivity of the SDMT in adults, the test is poorly documented in regard to its purported use as a predictor of reading difficulties. One study (Hutton, 1973) reports that SDMT and reading proficiency scores were correlated. This, of course, has little to do with prediction, and no studies are cited that actually demonstrate that SDMT scores are at all predictive. With several groups of brain-damaged or retarded children, Rathbun and Smith (1982) found scores lower than age-equated norms. Although this may demonstrate the sensitivity of the test to impaired brain function in children, it does not support the role of the SDMT in prediction or early identification of reading problems.

Critique

The Symbol Digit Modalities Test is intended to be used as a screening measure for the identification of cerebral dysfunction and the prediction or early identification of reading problems in children. Although it is clear that the test is sensitive to cerebral dysfunction, other factors can influence performance. The manual is quite correct in cautioning that scores from the SDMT (regardless of severity) cannot be used *in isolation* to make any definitive statement regarding cerebral dysfunction. The SDMT appears to have some value as a nonspecific screening measure, which due to its complexity is sensitive to a variety of factors. However, its greatest value would appear to be in the context of a complete psychological and neuropsychological examination interpreted by a competent neuropsychologist. In fact, the author is to be commended for his cautious approach in emphasizing the need for additional testing. In regard to the SDMT's ability to predict reading problems, there is simply no evidence that this is the case. Lack of clarity in the manual and the failure to relate empirical research findings with proposed clinical use of the test are additional obstacles.

The SDMT is a complex task that appears to require a multitude of basic and higher level psychological abilities. It appears to be a nonspecifically sensitive test that, when used in conjunction with a comprehensive neuropsychological test battery, would be of value in the psychometric examination of the effects of cerebral dysfunction. The SDMT would not appear to be of value in the prediction of reading problems at its present stage of documentation. In view of the considerable degree of similarity between the SDMT and the Digit Symbol subtest of the Wechsler intelligence scales, it is not clear that any incremental validity ensues from the use of this test when an age-appropriate Wechsler intelligence scale is also administered.

References

Burkland, C. W., & Smith, A. (1967). Unpublished manuscript.

Campbell, A. L., Jr., Bogen, J. E., & Smith, A. (1981). Disorganization and reorganization of cognitive and sensorimotor functions in cerebral commissurotomy. *Brain, 104*, 493-511.

Centofanti, C. C. (1975). *Selected somatosensory and cognitive test performances as a function of age and education in normal and neurologically abnormal adults.* Unpublished doctoral dissertation, University of Michigan, Ann Arbor.

Hartlage, L. C., Mains, M. R., Kovach, R., & Stovall, K. (1981). *Anticonvulsant medication as a determinant of neuropsychological test profiles.* Paper presented at the meeting of the International Neuropsychological Society, Atlanta.

Hutton, S. G. (1973). *Reading difficulty: Early and economic identification of children at risk.* Unpublished doctoral dissertation, University of Capetown, South Africa.

Pfeffer, R. I., Kuroskai, T. T., Harraha, C. H., Jr., Chance, J. M., Bates, B., Detels, R., Files, S., & Butzke, C. (1981). A survey diagnostic tool for senile dementia. *American Journal of Epidemiology, 114*, 515-527.

Rathbun, J., & Smith, A. (1982). Comment on the validity of Boyd's validation study of the Hooper Visual Organization Test. *Journal of Consulting and Clinical Psychology, 50*, 281-283.

Smith, A. (1975). Neuropsychological testing in neurological disorders. In W. J. Friedlander (Ed.), *Advances in Neurology* (Vol. 7). New York: Raven Press.

Smith, A. (1982). *Symbol digit modalities test.* Los Angeles: Western Psychological Services.

Smith, A., & Centofanti, C. C. (1975). Two economic tests of cognitive and somatosensory functions for the detection of cerebral dysfunction. *Neurosciences Abstracts, 1*, 510.

Smith, A., Champoux, R., Leri, J., London, R., & Muraski, A. (1972). *Diagnosis, intelligences and rehabilitation in chronic aphasics. Final report* (Social and Rehabilitation Service Grant No. 14-P-55198/5-01). Ann Arbor: University of Michigan.

Watson, C. G., Davis, W. E., & Gasser, B. (1978). The separation of organics from depressive with ability- and personality-based tests. *Journal of Clinical Psychology, 34*, 393-397.

Watson, C. G., Gasser, B., Schaefer, A., Buranen, C., & Wold, J. (1981). Separation of brain-damaged from psychiatric patients with ability and personality measures. *Journal of Clinical Psychology, 37*, 347-353.

Whelan, T., Berker, E., Campbell, A., & Smith, A. (1980, August). *Comparative vulnerability of neuropsychological test functions in Huntington's disease (HD).* Paper presented at the meeting of the American Psychological Association, Montreal.

Whelan, T. B., Schteingart, D. E., Starkman, M. N., & Smith, A. (1980). Neuro-psychological deficits in Cushing's syndrome. *Journal of Nervous and Mental Disease, 168*, 753-757.

Francis X. Archambault, Ph.D.

Professor and Head, Department of Educational Psychology, University of Connecticut, Storrs, Connecticut.

A TEACHER ATTITUDE INVENTORY

Joanne Rand Whitmore. East Aurora, New York: D.O.K. Publishers, Inc.

Introduction

The Teacher Attitude Inventory (TAI) was developed to provide "a reliable indicator of [elementary school] teacher attitudes regarding philosophical issues and contrasting educational practices" (Whitmore, 1985, p.4). Responses to the TAI are suggested to be indicative of an underlying philosophical orientation toward either a "traditional" or an "experimental" teaching style, which influences teachers' decision-making and teaching practices.

The inventory was developed at the Stanford Center for Research and Development in Teaching as part of an inservice project designed to encourage elementary school teachers to individualize instruction and to experiment with alternative approaches to teaching. Based on her review of the literature, Whitmore (1985) concluded that existing inventories were not specific enough to assess the extent to which these project goals were being met. Moreover, she felt that existing instruments were also too lengthy to allow for efficient administration and to encourage teacher cooperation. Thus, 40 items would be developed for immediate application in the low SES black schools cooperating in the project and for later application in any district or community. With this goal in mind, "statements believed to reflect basic attitudes that would influence either classroom decisions and practices or interest in opportunities for professional growth and development were compiled" (Whitmore, 1985, p. 8). How they were compiled and how they became the TAI is not evident from the test manual.

The TAI contains 24 Likert-type items equally distributed across four subscales: *Controlling* versus Releasing, *Rigidity* versus Flexibility, *Individualism* versus Group-orientation, and interest versus disinterest in *Professionalism*. In the manual, Whitmore suggests that

> Higher scores across subscales would report preference for more pupil-centered, individualized, flexible, and innovative teaching behavior, and probably greater interest in opportunities for professional growth and contribution. Lower scores would suggest a tendency to prefer teaching behavior that is more teacher oriented (controlling), large group-oriented, and bound to traditional methods and content, and to favor a strong administration and minimal teacher participation in professional activities. (Whitmore, 1985, p.9)

The reviewer wishes to express his deep appreciation to Ms. Lesley Welsh, a doctoral student at the University of Connecticut, for her valuable assistance in drafting and editing this review.

The total score is intended to distinguish between teachers who employ traditional or experimental styles.

Practical Applications/Uses

According to Whitmore, the TAI can provide elementary school administrators with useful information for evaluating the extent to which the philosophical positions of an individual teacher or group of teachers conform to desired philosophical or pedagogical attitudes. It also may be used to cluster teachers with either compatible or contrasting educational preferences, to determine the extent to which the self-reported educational perspectives of the teaching faculty are suitable to the adoption of proposed innovations, to determine if faculty development is needed, and to determine the extent to which teacher attitudes or opinions have changed over time. In addition, teachers may also use the TAI to examine their own beliefs and opinions concerning their teaching styles.

As with other self-report inventories, a climate of trust and openness must be established in order to elicit accurate responses to the TAI. In the manual for the inventory, Whitmore offers some good advice for establishing such trust. She also provides clear directions for test administration and scoring. In actuality, however, scoring may be somewhat confusing because 13 of the 24 items must be reverse-scored. Machine scoring is not available.

Interpretation of TAI scores also is questionable. The test author notes that extreme caution must be exercised by interpreters of the results and that a lack of field testing and normative data further limit score interpretation. Nonetheless, as stated previously, the test manual suggests that teachers who achieve high scores (probably of above 100) are inclined to be flexible and pupil directed. On the other hand, low scorers (those with scores below about 85) are thought to be least willing to experiment and tend to be more traditional in their teaching styles. Mid-range scores (86-99) are even more difficult to interpret. Whitmore suggests that in cases in which interpretation is difficult, one might analyze the subscores; however, she also informs the reader that there is "no strong evidence indicating that it would be especially beneficial to do so" (Whitmore, 1974, p.11).

Technical Aspects

The test manual claims that the TAI "is moderately reliable and the total score provides the most reliable information" (Whitmore, 1985, p.9). Although the manual provides little data to support this claim, a technical report written by the author states that measures of internal consistency for the subscales are quite low, ranging from -.08 to .70 across samples. Alphas for the total scale, on the other hand, were between .76 to .84.

Factor analyses failed to support the four hypothesized subscales:

> The experimenter could not make sufficient sense of the four resulting factors to advocate the use of different scales with confidence. . . . Furthermore, the fact that the four factors account for only 35 percent of the total variance raises questions about possible measurement error. Certainly more comparative data are needed for further analyses before a decision about scales can be made. (Whitmore, 1974, pp. 25-26)

Although the TAI was published 11 years after Whitmore recognized the need for further validation studies, there is no evidence to suggest that further scale validation has taken place. Criterion-related validity was assessed by correlating TAI scores with performance ratings obtained by naive observers. According to the author, "data was available for only 22 teachers and the results showed a large amount of variance within the two groups and a lack of consistent differences between groups" (Whitmore, 1985, p.10).

The author also employed informal judgments to support the instrument's validity: "The self-reports from individuals were apparently accurate, since they were consistent with the experimenter's informal observations" (Whitmore, 1974, p.29). In addition, multiple *t*-tests were used to test the significance of the informally observed differences between groups and yielded significant differences for total score and several of the subscales. However, in view of the difficulties associated with multiple *t*-tests (significance due to chance), this support is suspect also.

Critique

The Teacher Attitude Inventory has a number of serious shortcomings, including a) problems with the items themselves, b) the assignment of items to subscales, c) the scoring of the instrument, and d) the reliability and validity of both the subscales and the total score.

The major problem with the items is that the two statements comprising each stem are assumed to be dichotomous; that is, agreement with one presupposes disagreement with the other. The author states, however, that "error may have occurred in the original pairing of statements assumed to be dichotomous, or in the assignment of items to subscales" (Whitmore, 1974, p. 14). Our examination of the statements leads us to concur with this warning. For example, the statement "Most children in the grade that I teach are capable of increasing their responsibility for self-evaluation and self-discipline . . ." was assumed to be antithetical to "Pupils cannot be expected to assume responsibility for self-discipline and evaluation before secondary level; until then the teacher must assume most responsibility for discipline and evaluation." Similar problems exist with other pairs of statements.

Whitmore concluded that she could not make sufficient sense of the factor structure to advocate the use of subscores. Nonetheless, the TAI scoring sheet clearly displays the four factors and provides directions for assigning items to them. As there is no evidence to suggest that such scoring is warranted, test users should not be led to believe that this method of scoring is acceptable. Although the author focuses primarily on the total test score when discussing score interpretation, she provides no evidence for the score cutoffs suggested for differentiating among teaching styles. Furthermore, her suggestion that "one might look at the subscales for more specific information" (Whitmore, 1985, p. 14) about total scores is misleading and unsubstantiated.

There is little evidence to suggest that the TAI truly measures teaching style. The only validity data provided in the manual were based on correlations of TAI scores with observations of the performance of 22 teachers, and these data are not convincing. Clearly, more concurrent and predictive validating data from a larger sample are required. Reliability data, including stability estimates, also should be obtained from a large sample of subjects. The test author appears to have recognized the need for such data as early as 1974: "At this stage, there is not enough evidence of the reliability and validity of the instrument for its use to be extended beyond that of qualified researchers desiring to obtain more data" (Whitmore, 1974, p. 42). It is indeed unfortunate that such data had not been acquired prior to test publication in 1985. Because they were not, this reviewer recommends that the test be used only in a research context, and even then only with extreme caution.

References

Whitmore, J. R. (1974). *A Teacher Attitude Inventory: Identifying teacher positions in relation to educational issues and decisions* (Research and Development Memorandum No. 118). Stanford, CA: Stanford University, Stanford Center for Research and Development in Teaching.

Whitmore, J. R. (1985). *A Teacher Attitude Inventory (TAI): Identifying teacher positions in relation to educational issues and decisions* (manual). East Aurora, NY: United Educational Services, Inc.

John F. Schmitt, Ph. D.
Associate Professor of Communicative Disorders, The University of Alabama, Tuscaloosa, Alabama.

TEST FOR AUDITORY COMPREHENSION OF LANGUAGE-REVISED

Elizabeth Carrow-Woolfolk. Allen, Texas: DLM Teaching Resources.

Introduction

The Test for Auditory Comprehension of Language—1985 Revised Edition (TACL-R) is the revision of a test first published in 1971 and updated in 1973. The TACL-R is designed to test the understanding of the literal meaning of selected features of spoken English syntax, morphology, and semantics in children from 3 to 10 years of age. Guidelines also are given for using the test with adults. The TACL-R is not designed to test comprehension of the pragmatic functions of language or the rules of language as a tool for social interaction and discourse.

Elizabeth Carrow-Woolfolk, the author of the TACL-R, is also the author of the Carrow Auditory-Visual Abilities Test (1981) and the Carrow Elicited Language Inventory (1974). Additionally, she has co-authored the book *An Integrative Approach to Language Disorders in Children* (Carrow-Woolfolk & Lynch, 1982) and has written numerous articles on language disorders. She is a speech-language pathologist and Fellow of the American Speech-Language-Hearing Association, was professor and head of the program in speech pathology and audiology at the University of Texas, and has served as editor of the *Journal of Speech and Hearing Disorders*.

The 1973 edition of the TACL was the standard in assessment of auditory comprehension of language by speech-language pathologists. A 1979 survey of 84 university programs in speech-language pathology by Muma, Webb, and Muma showed that the TACL was the most frequently used test of both receptive and expressive language abilities. Numerous reviewers (e.g., Darley & Spriestersbach, 1978; Miller, 1979; Plante, 1984; Wiig & Semel, 1984) concluded that the TACL was strong in its conceptualization, design, and utility, but that it also had limitations. Among the 21 limitations of the TACL enumerated by both reviewers and the TACL-R manual are the effect of guessing on test performance; the need for simplification of the design of the test form; the need for simpler administration procedures, such as not having to administer the entire test to each child; the need for more complex constructions among the test items; the need to extend the age range of the test; and the need for significantly more than 200 subjects in the standardization sample. Furthermore, the 1973 edition of the TACL had guidelines but no norms for the assessment of Spanish-speaking children. These guidelines are absent from the revised edition of the test, except for an explanation that testers should not expect meaningful differences between scores of the norm group and those of minority subjects on the TACL-R.

Carrow-Woolfolk incorporated all but one of the 21 changes suggested by reviewers and summarized in the TACL-R manual. In chapter 6 of the manual there is a rationale for not changing the test from three pictures per test plate to four, which would have reduced the guessing rate from .33 to .25 for each stimulus. The author states that having three pictures arranged horizontally on each page elicits more reliable performance than having four pictures in a grid format because research has shown that subjects prefer the bottom two pictures in such a grid.

The TACL-R consists of a 185-page examiner's manual, a test booklet with three black-and-white drawings per plate, and a four-page individual record form. The manual is well written but would be improved by the addition of a subject index. The test norms are in 6-month intervals from 3 to 10 years. The test has 120 items divided into three subsections: word classes and relations (nouns, verbs, adjectives, and adverbs); grammatical morphemes (noun-verb agreement, verb number and tense, and noun number and case); and elaborated sentences (embedded sentences, and partially and completely conjoined sentences). Within each section, the items are of increasing difficulty. The TACL-R is an easy test to administer and one that most children enjoy taking. Out of three pictures on each plate, the correct one shows the meaning of either a word, morpheme, or syntactic structure spoken by the examiner. The two foils on each plate either both contrast (semantically or syntactically) with the correct response or there is one decoy and one contrast. The child simply points to the picture corresponding to the stimulus. The page facing each test plate has the test stimulus written for the examiner so that all examiners administer identical stimuli.

The individual record form is constructed carefully. On the first page, the examiner records demographic information, computes the child's chronological age in years and months, and records a total score and scores for the three subsections. Pages 2 and 3 fold open to show three preliminary item examples, basal and ceiling rules, lines for scoring each test item, and boxes for adding the total correct in each section. On page 4 the examiner graphically displays three profiles: standard score by age-level norms, standard score by grade-level norms, and age-equivalent score.

Practical Applications/Uses

The Test for Auditory Comprehension of Language—1985 Revised Edition is valuable in identifying children with problems in understanding various morphologic, syntactic, and semantic rules of the English language system. Furthermore, the test goes beyond identification by allowing the examiner to 1) understand the nature of the language comprehension deficit and 2) generate information that is useful in planning for language intervention. Once therapy is initiated, the TACL-R is an effective tool for measuring the child's progress and the efficacy of the intervention program. As with the previous version of the test, the TACL-R should prove to be the test of choice for many researchers who need valid and reliable information on the language comprehension abilities of their subjects.

There are multiple settings in which to use the TACL-R. It is effective in assessing language comprehension in venues such as public schools, hospitals, rehabilitation centers, and private practice. Additionally, the test enjoys widespread use by both clinicians and researchers in college and university speech and hearing centers. In their initial and periodic testing, nationally certified and/or state-licensed speech-language pathologists are the professionals who most often use the TACL-R. Educational diagnosticians, psychologists, and other qualified personnel may also use the test, according to the author.

The norms for the TACL-R are a considerable improvement over the TACL norms. The norm group had 1,003 children, with between 97 and 104 in each subgroup. This essentially meets the criterion of 100 per significant cell (Winer, 1971). Norms for the test group are given for percentile ranks, four normalized standard scores (z-scores, T-scores, deviation quotients, and normal curve equivalents), and age equivalent scores. There are norms for the general population, for grade levels K-4, as well as nonnormalized standard scores with which to verify progress when performance on the test is very good or very poor.

The TACL-R is appropriate for all children who pass one of the preliminary sample items. It is inappropriate for children who fail all three samples or for those whose lack of attention or other inappropriate behaviors indicate that continued testing would invalidate the findings. In the test manual, there are descriptions of the test's use with normal and aphasic adults, hearing-impaired children, and mentally retarded children. Data in the manual show that the test retains its split-half and test-retest reliabilities quite well with these groups.

Proper administration of the TACL-R first requires careful reading of the examiner's manual. There also is an excellent 42-minute instructional videotape with an introduction by the author and a demonstration of administration and scoring. The publisher provides the videotape and invites testers to copy it so that all who administer the test can view the tape. Prospective test users should practice a minimum of three test administrations to ensure valid and reliable administration, scoring, and interpretation. Practice should take an average of 4 to 6 hours. The examiner also must obtain information from an audiologist regarding auditory acuity before administering the test. The test is designed to be given individually in a well-lighted room that is free of distracting background noise. The examiner and subject sit at a table across from each other. The test book is on the table with pictures facing the child and instructions facing the examiner. The examiner places the individual record form in a position for easy scoring but out of the child's direct vision. The scoring details in the manual are comprehensive and clearly written. Testing takes between 10 and 20 minutes.

Before a test session, the examiner completes the demographic information on the individual record form and computes the child's chronological age in years and months. The examiner first administers the sample items, giving the verbal stimulus and having the subject point to the picture that represents it. The examiner discontinues testing if a child fails all three practice items. Children under five years of age may receive a repetition of the stimulus if there is no response within 10 seconds. If there is hesitation, the examiner should provide encouragement for the subject to respond. On the individual record form, a slash is placed through the number of the item to which the child pointed, or through "NR" if no response is given.

Each 40-item test section has stimuli ordered serially for difficulty. Testing within a section begins at the level corresponding to the child's chronological age and continues uninterrupted from the point at which the child establishes a basal of four consecutive correct responses; if one of the first four items is failed, testing reverts to the next lower age level. The child receives one credit for all items below the basal and none for items above the ceiling. Failure of three consecutive items constitutes a ceiling. When a ceiling is established, testing proceeds to the next section.

Computer scoring is available from the publisher. The computer program is called *COMPUSCORE* and is configured for the Apple IIe, II+, IIc, and the IBM PC. By entering raw scores, one may obtain printouts of all of the scores and profiles for the TACL-R. There also is a data base management system for saving data, reconfiguring it by classifications (such as type of disorder), and tallying descriptive statistics for comparisons, the establishment of local norms, and research purposes.

The number correct, including a credit of one point for all items below the basal, is the raw score for each section. The total raw score is the sum of the raw scores for Sections I, II, and III. This number and all subsequently determined scores are entered on page 1 of the individual record form. The manual has two tables of norms, one for age level and one for grade level, which the examiner uses to interpret raw scores. The tables have percentile ranks, standard scores (z, T, deviation quotients, and normal curve equivalents) by age and grade level, standard errors of measurement, and age equivalent scores. After placing the percentile rank for each section and for the total score on page 1 of the record form, the examiner uses guidelines in the manual to decide which standard score to use. The standard errors of measurement (SEM) for each standard score in the age level norms are in Table 4 of the manual. They tell the examiner the range of scores in which the true score of the subject likely falls. The examiner computes the upper and lower limits of the confidence interval by adding and subtracting the SEM from the standard score. The confidence interval then is placed on the record, followed by age equivalent scores for the entire test and for each subtest. The examiner enters an index number on the record form when using nonnormalized scores for a subject whose raw score is below the 1st or above the 99th percentile.

The manual and videotape clearly demonstrate the guidelines for interpreting TACL-R results. Clinical judgment does play an important part in interpreting what the test findings mean with respect to the subject's strengths and weaknesses in the receptive language abilities tapped by the TACL-R. For example, although an examiner determines age equivalent scores, the author cautions that such scores often are misleading and should not be used to label language disorders. She also warns against interpreting age scores as meaning that a child is performing like a child of another age level. Age equivalent scores reveal performance but not the reasons responsible for the performance. They are part of the TACL-R because governmental regulations and third-party reimbursement agencies often require them, not because it is always appropriate to use them. For a subject whose background is similar to that of the subjects in the norm group, Carrow-Woolfolk says to use a cutoff score 1.5 standard deviations below the mean for the subjects' age level. The cutoff can be higher or lower, however, for persons of low socioeconomic status and/or from other groups whose performance is expected to be better or poorer than average. These guidelines may seem somewhat vague and uncomfortable to the inexperienced test user, but those who are experienced will appreciate the care with which test interpretation guidelines are presented in order to avoid the kinds of misuses of data that so often are associated with tests of language development.

Among the appendices in the manual are a progress summary form to record the data from multiple administrations of the test to one subject over a period of time; data for nonclinical adult samples; procedures for determining the statistical significance of position responding; and alternative stimulus items to use for informal assessment, but not with the TACL-R norms. The appendices are important additions to the test manual.

Technical Aspects

In developing the Test for Auditory Comprehension of Language—1985 Revised Edition, the author employed a stratified random sampling technique to select the sample of 1,003 subjects (501 males and 502 females) from 20 states representing four geographical regions of the United States. To stratify for geographical area, 1981 U.S. Census Bureau data were used; this procedure resulted in a sample that was approximately 75% urban and 25% rural. Community size, gender, age, family occupation, and ethnic origin were the other factors used to stratify subjects at each age level. The detailed text and tables show that, for each factor, there were no fewer than 97 subjects at each of the 10 age levels.

In chapter 9 of the manual, there is a discussion of the reliability of the TACL-R. Carrow-Woolfolk computed the standard error of measurement (SEM) to determine how much a subject's test score is expected to vary because of factors inherent in the test itself. The range of SEMs for the 488 subjects in grade levels K-4 was 1.29 to 3.32; the range was 1.25 to 3.73 for the 1,003 subjects in $\frac{1}{2}$-year age levels between 3 and 10 years. The individual SEM values are very small compared with mean scores, thus attesting to the stability and precision of the TACL-R scores for the norm group sample.

Split-half (odd-even) reliability was computed using the Spearman-Brown formula. The reliabilities for sections I to III and the total score for the 488 children in Grades K-4 were .91, .91, .93, and .96, respectively. The reliabilities for the 10 age groups ranged from a low of .88 to a high of .97, with an average of .95. The lowest correlations were for the upper four age levels. This is not surprising because children in those groups generally had high scores, and one or two errors by a child substantially decreased the split-half correlation coefficient for the group. The TACL-R clearly has acceptable internal consistency as measured by split-half reliability.

Test-retest reliability was assessed on 129 subjects within 4 weeks of the first administration. The respective reliabilities for sections I-III and total scores were .90, .91, .89, and .95. Again, the data were quite stable and reflective of the test's internal consistency.

Neither inter- nor intra-examiner reliability is addressed in the manual. According to McCauley and Swisher (1984), this is a violation of the test-selection criterion that test authors should demonstrate empirically the reliability of examiners in giving the test. However, the test administration and scoring procedures are specific, simple, and independent of subjective judgment; thus, it is assumed that an examiner who studies the manual and practices the test at least three times prior to administering it will be reliable as compared to himself and other examiners. A discussion of this in the manual would seem appropriate, especially because the 35 examiners who tested the standardization sample represented a number of different areas of specialization (early childhood education, psychology, nursery school and day-care staff), as well as a variety of settings (schools, private practice, and advanced training programs).

Lieberman and Michael (1986) state that the content validity of a test is the result of "careful content sampling from a relevant content domain, not the evaluation and interpretation of scores" (p. 72). Furthermore, Messick (1980) presents two requisites for content validity: content relevance and content coverage. The former refers to the clarity of definition of the test domain; the latter refers to how well the sample of test items represents the domain. The TACL-R appears to be sensitive to these content validity criteria. The items in each of the three sections represent relevant content domains of the lexicon, morphology, and syntax of the English language, and the degree of emphasis in each subsection generally reflects the frequency with which those items occur in the language. The manual reports the results of numerous studies and reviews of the 1973 TACL in which analyses of individual items, the sequence in which test items were presented, and the placement of items into their respective categories showed the test to have generally good content validity. Carrow-Woolfolk states that because every valid suggestion regarding the overall test construction and the validity of individual items on the TACL was incorporated in the updated test, the TACL-R should be considered to have improved content validity. This conclusion appears to be tenable.

To demonstrate the construct validity of the TACL-R, Carrow-Woolfolk describes how each section of the test reflects Brown's (1973) model of semantic roles and modulations, sentence modalities, and embedded and coordinated sentences. The model for the test does not represent all aspects of language content, form, and use (Bloom & Lahey, 1978), but it was not designed to do so. For example, the TACL-R does not elicit information on the child's understanding of the many functions and social-interactional uses of language. Tables in the manual show that 1) there are statistically significant correlations ($p < .001$) between overall test performance and performance on sections I-III for the norm group; 2) for the various age groups, there is consistently better performance on easier items than on more difficult items; 3) performance on more difficult items improves with age; and 4) understanding of the most difficult items is evident in the highest age levels. Carrow-Woolfolk further states that the TACL-R has construct validity because it was given to 234 communicatively impaired subjects, after which t ratios were computed between the scores of these subjects and those in the norm group. Persons with disorders primarily of language comprehension or language expression scored significantly poorer than the norm group, and those with articulation (speech-sound) disorders did not differ significantly from the norm group. As noted previously, Lieberman and Michael (1986) would not agree that these findings prove the construct validity of the TACL-R.

To assess concurrent criterion-related validity, correlations were computed between the TACL and TACL-R scores of 140 children. This procedure is appropriate because the TACL is the criterion most similar to the TACL-R, but more importantly, because of the demonstrated validity of the TACL. The correlations ranged from .71 to .86, and all were significant beyond an alpha level of .001. The highest correlation (.86) was between the total scores of the two tests. Correlations also were significant ($p < .01$) between the TACL-R and several other receptive language measures, namely, the Sequenced Inventory of Communication Development (Hedrick, Prather, & Tobin, 1975), the Peabody Picture Vocabulary Test (Dunn, 1965), and the Auditory Association, Auditory Reception, and Grammatic Closure subtests of the Illinois Test of Psycholinguistic Abilities (Kirk, McCarthy, & Kirk, 1968).

Predictive criterion-related validity data are not yet available. At the time of this writing, the TACL-R had been published for less than 2 years. In the manual, Carrow-Woolfolk indicates that the determination of predictive validity is an ongoing process and that additional studies are under way.

Critique

The Test for Auditory Comprehension of Language—1985 Revised Edition can be considered both a valid and reliable test for determining a subject's knowledge of word classes and relations, grammatical morphemes, and elaborated sentences. The test is easy to administer, score, and interpret, in large measure because of the detailed and clearly written manual and the excellent instructional videotape. In revising the TACL, the author was sensitive to the many suggestions for improvement offered in a number of published reviews; she acted upon virtually every suggestion. The norms for the TACL-R represent a significant improvement over those of the TACL, and the administration and scoring are easier because of the new rules for computation of a basal and ceiling for each subject. These rules shorten the test administration time considerably in a majority of situations. The descriptive information and tables on selection of standard scores are most helpful in reducing the possible intimidation of some potential test users, such as students in training, who might want to use the test despite their lack of extensive knowledge of psychometric characteristics. The improvements of the TACL-R over previous versions of the test make it a valuable tool for clinical and research purposes.

An area of concern with the TACL-R is the lack of detailed information on item selection. The author states that she received comments regarding problems with about 25 items on the TACL, but that the comments were too numerous to mention in the manual. At least a summary of the comments and how they were addressed in the revised version of the test would have been helpful for prospective test users in determining item adequacy. A second area of concern is the lack of information on the 35 examiners in 26 sites who standardized the test. We are not told the selection criteria for the administrators, such as national certification and/or state licensure in their areas of specialization. This is particularly important because the group was eclectic, including not only speech-language pathologists, psychologists, early childhood specialists, and "assessment specialists," but also nursery school and day-care center staff. At the very least, the qualifications of the nursery school and day-care center personnel to assess language development of children are open to question.

Users of the TACL-R should note the author's clear statements about what this test does and does not measure. It is not a test of all of the components of the English language system, or of all of the important areas of language comprehension;

unfortunately, some users report the results of the test as if it were. Of course, misuse of the test is not the fault of the author. Furthermore, when the test is given and interpreted according to Carrow-Woolfolk's well-defined guidelines, it is the best commercially produced measure of selected areas of language comprehension available today.

References

Bloom, L., & Lahey, M. (1978). *Language development and language disorders*. New York: John Wiley & Sons.

Brown, R. (1973). *A first language: The early stages*. Cambridge, MA: Harvard University Press.

Carrow, E. (1973). *Test for Auditory Comprehension of Language*. Allen, TX: DLM Teaching Resources.

Carrow, E. (1974). *Carrow Elicited Language Inventory*. Allen, TX: DLM Teaching Resources.

Carrow, E. (1981). *Carrow Auditory-Visual Abilities Test*. Allen, TX: DLM Teaching Resources.

Carrow-Woolfolk, E., & Lynch, J. (1982). *An integrative approach to language disorders in children*. New York: Grune & Stratton.

Darley, F. L., & Spriestersbach, D. C. (1978). *Diagnostic methods in speech pathology* (2nd ed.). New York: Harper & Row.

Dunn, L. M. (1965). *Peabody Picture Vocabulary Test*. Minneapolis: American Guidance Service.

Dunn, L. M., & Dunn, L. M. (1981). *Peabody Picture Vocabulary Test-Revised*. Circle Pines, MN: American Guidance Service.

Hedrick, D. L., Prather, E. M., & Tobin, A. R. (1975). *Sequenced Inventory of Communication Development*. Seattle: University of Washington Press.

Hedrick, D. L., Prather, E. M., & Tobin, A. R. (1984). *Sequenced Inventory of Communication Development-Revised*. Seattle: University of Washington Press.

Kirk, S. A., McCarthy, J. J., & Kirk, W. D. (1968). *Illinois Test of Psycholinguistic Abilities: Revised Edition*. Urbana, IL: University of Illinois Press.

Lieberman, R. J., & Michael, A. (1986). Content relevance and content coverage in tests of grammatical ability. *Journal of Speech and Hearing Disorders, 51*, 71-81.

McCauley, R. J., & Swisher, L. (1984). Psychometric review of language and articulation tests for preschool children. *Journal of Speech and Hearing Disorders, 49*, 34-42.

Messick, S. (1980). Test validity and ethics of assessment. *American Psychologist, 35*, 1012-1027.

Miller, J. F. (1979). [Review of the Test for Auditory Comprehension of Language]. In F. L. Darley (Ed.), *Evaluation of appraisal techniques in speech and language pathology.* (pp. 71-74). Reading, MA: Addison-Wesley.

Muma, J. R., Webb, P. H., & Muma, D. B. (1979). Language training in speech-language pathology and audiology: A survey. *ASHA, 21*, 467-473.

Plante, E. M. (1984). Nonlinguistic influences on the Test for Auditory Comprehension of Language. *NSSLHA Journal, 12*, 3-11.

Wiig, E. H., & Semel, E. (1984). *Language assessment and intervention for the learning disabled* (2nd ed.). Columbus, OH: Charles E. Merrill.

Winer, B. J. (1971). *Statistical principles in experimental design.* (2nd ed.). New York: McGraw-Hill.

E. Wayne Holden, Ph.D.
Assistant Professor of Psychology, Auburn University, Auburn, Alabama.

J. Bart Hodgens, Ph.D.
Assistant Professor, Division of Adolescent Medicine, Department of Pediatrics, University of Alabama at Birmingham, Birmingham, Alabama.

TEST OF DIABETES KNOWLEDGE

Suzanne Bennett Johnson. Gainesville, Florida: Suzanne Bennett Johnson, Ph.D.

Introduction

The Test of Diabetes Knowledge: General Information and Problem Solving (TDK; Johnson, 1979; Johnson, Lewis, & Alexander, 1981; Pollak & Johnson, 1979) is a series of assessment instruments designed to evaluate knowledge about and behavioral skills for adhering to specific components of outpatient medical regimens used to treat insulin-dependent diabetes mellitus. The importance of accurate knowledge and well-developed behavioral skills in adherence to medical prescriptions for treating diabetes and the absence of well-standardized, reliable, and valid measures were the basis for the construction of the tests. The tests assess general knowledge about insulin-dependent diabetes, regimen-specific problem solving, insulin injection skills, serum glucose monitoring skills, and urine testing skills. This battery of tests provides a comprehensive set of cognitive-behavioral measures for evaluating skills that are purportedly related to regimen adherence in insulin-dependent diabetic patients.

The Test of Diabetes Knowledge was developed by Suzanne Bennett Johnson, Ph.D., and her colleagues in a research program that has extended for nearly a decade at the University of Florida Health Sciences Center. The measures were devised specifically for use in research on diabetic children and adolescents and their families. The measures have been employed in a number of investigations conducted over the last several years and have been administered to children, adolescents, and their parents in research and clinical settings (Johnson, 1984). Dr. Johnson has published extensively in the area of chronic childhood illness while developing and directing the psychological component of a specialized inpatient treatment program for diabetic children and adolescents at the University of Florida Health Sciences Center.

The general information and problem-solving components of the test initially were developed in the late 1970s based upon the earlier work of Etzwiler and associates (Etzwiler & Robb, 1972; Etzwiler & Sines, 1962). This early research underscored the importance of knowledge about illness factors and treatment in the effective self-management of juvenile diabetes. Initially, Johnson and her colleagues (1982) developed a pool of self-report multiple-choice items that represented the broad spectrum of problems confronting diabetic youth. From this large pool of potential items, 39 general information and 36 problem-solving items were selected based upon response agreement between physicians and nurses actively working in the area of juvenile diabetes. The content of the items covered issues in the areas of diet, insulin, urine testing, insulin reaction, illness, exercise, and emotions. Observational procedures for evaluating urine testing skills and insulin injection skills also were constructed (Pollak & Johnson, 1979). These observational measures were developed by dividing each regimen adherence skill into component parts and devising pass-fail scoring criteria for each component.

The self-report items (general information and problem-solving) and observational measures were administered to a sample of 151 diabetic children and adolescents to develop initial normative data and to evaluate the psychometric characteristics of the tests (Johnson et al., 1982). In addition, 179 parents of the children participating in the study completed the general knowledge and problem-solving measure.

The results of this study revealed differences in performance as a function of demographic variables. Both parents and children obtained more correct responses on the general knowledge than on the problem-solving items, and girls were more accurate overall than boys. Age effects also were found on the self-report measure with children in the 6- to 8-year-old age range scoring more poorly than older children. Mothers were more knowledgeable than either fathers or children. The duration of the child's diabetes did not appear to be related to performance on the self-report measure. Variability was displayed across the content areas measured by the test although items in the areas of urine testing, illness, and illness related emotions were the most difficult for the parents and children to complete successfully.

Results obtained with the urine testing and insulin injection tests largely paralleled the results from the self-report measure. Males were less accurate than females on both tests and made more errors directly leading to an incorrect urine testing reading or incorrect insulin administration. Age effects also were obtained. Younger children displayed greater inaccuracy on both tests and committed more errors on the insulin injection test that potentially could lead to an inaccurate injection. Again, the duration of the disease was not related significantly to performance on either behavioral skills test.

The third component of the behavioral skills tests—The Chemstrip Skills Demonstration Test—was published at a later date (Johnson, Lewis, & Alexander, 1981). Normative data on this test and information regarding its psychometric characteristics were collected in an unpublished master's thesis (Brody, 1985). The general trends displayed by the other skills tests with respect to sex and age differences were found with this test as well. Females exhibited greater accuracy than males and older children, and adolescents (ages 10-15) were more accurate than younger children (ages 6-9).

Further research with independent populations of diabetic children and adolescents (Gilbert et al., 1983) has confirmed the presence of the sex and age effects obtained in the initial normative investigation. Johnson (1984) has interpreted these results as indicating the importance of cognitive developmental level in successful diabetes self-care. The observational tests have not been revised since their initial introduction into the literature. A revision of the general information and problem-solving tests has recently become available through the test author.

The general knowledge and problem-solving components of the TDK are assessed by a single 75-item self-report measure (Johnson, 1979). Written instructions at the beginning of the test indicate that the instrument is designed to measure knowledge about diabetes, as well as specific responses to the disease. Furthermore, respondents are encouraged to read each item carefully before responding and to restrict their answers to one alternative per question. The author provides a scoring key with the test materials.

The three skills testing portions—urine testing (34 items), insulin injection (24 items), serum glucose monitoring (15 items)—each require a set of specific regimen-related materials. These materials are placed in random order before the child. Once rapport is established and it is ascertained that the child has either completed these procedures by him- or herself or has observed others completing the procedures, the individual tests are administered. For each test, the respondent is instructed to demonstrate the appropriate skill as well as he or she can. The examiner observes the respondent's performance and scores the responses as either pass or fail using a multiple-item behavior rating scale. As noted previously, the rating scales break each regimen-adherence skill down into specific component steps, and the manual for each test provides detailed scoring criteria for each item within each of the three skills tests.

Practical Applications/Uses

The three broad aspects of diabetes care assessed by the Test of Diabetes Knowledge—general information, problem solving, and skill—cover a wide range of content areas. The general information and problem-solving components of the test include items dealing with diet, insulin, urine testing, insulin reactions, illness, exercise, anxiety/excitement, and miscellaneous issues. The behavioral skills components assess the ability to perform serum glucose monitoring, urine glucose monitoring, and preparation for self-injection.

The TDK was designed primarily as a research tool with the goals of yielding standardized, reliable information that could be validated against relevant criteria and that assessed knowledge from more than one area or domain. The test has been used to describe the nature and development of diabetes knowledge among child and adolescent populations, parental knowledge, and the relationship of parent-patient knowledge (Johnson, 1986; Johnson et al., 1982). Another important use of the TDK has been to assess the impact of education and self-management programs (Brody, 1985; Harkavy et al., 1983). The test is designed for use by health-care professionals with knowledge of the relevant aspects of diabetes care. Because of its research focus, the TDK has been used primarily in settings with significant numbers of children and adolescents with diabetes, for example, in summer camps and large outpatient clinics. However, the test appears to be useful for any health-care professional involved in the management of insulin dependent diabetes and interested in obtaining standardized and reliable information about self-care.

The Test of Diabetes Knowledge has been administered to patients ranging in age from 6-18 years. However, the reading level of the general information and problem-solving components may be too high for patients ages 6-9. The test's authors report the reading level to be 16-18 years (Grades 10-11) for the general information test and 10-12 years (Grades 6-7) for the problem-solving test. When diabetes-related technical terms are removed from the test items, the reading level drops to 11-12 years (Grade 7) for the general information test and 8-9 years (Grade 4) for the problem-solving test. The author recommends reading the test to younger children or older children with suspected reading problems (S.B. Johnson, personal communication, August 19, 1987).

The use of the Test of Diabetes Knowledge with adult patients diagnosed with insulin-dependent diabetes has not been explored. Examination of scores obtained by parents of child and adolescent patients does not indicate ceiling effects on any components of the test, suggesting that the test may be useful for diabetic adults. An exception is the problem-solving component, which contains items pertaining to child and adolescent activities. It should be emphasized, however, that the TDK is *not* designed for patients with non-insulin-dependent diabetes mellitus, which is the most common adult form of the disease.

The general information and problem-solving questionnaires can be administered within a group format and do not require examiner expertise. They are scored easily for the percentage of correct items by dividing the number of items correctly completed by the total number of items. The behavioral skills tests require one-on-one administration and an examiner familiar with conducting and reading the specific tests. Explicit definitions are provided for scoring each behavioral item as pass-fail. Useful tips also are provided to facilitate observation. The test's authors have classified items from each of the skills tests to identify those items directly resulting in an inaccurate serum/urine glucose reading or failed insulin injection. Percentage correct scores are computed for each of the skills tests by dividing the number of items correctly completed by the total number of items. The total time required for administration of the TDK is less than 1 hour. The general information and problem-solving questionnaires require approximately 30 minutes and the skills tests 5-10 minutes each.

Interpretation of the test is based on comparing the derived objective scores to established normative data. Descriptive indices are provided for each component of the test on a total standardization sample and in each of four age groups: 6-9-year-olds, 10-12-year-olds, 13-15-year-olds, and 16 years and older. Descriptive indices include means, standard deviations, and a range that is plus or minus one standard deviation from the mean.

Technical Aspects

The reliabilities of the assessment instruments included in the Test of Diabetes Knowledge have been evaluated in several investigations (Brody, 1985; Gilbert et al., 1982; Harkavy et al., 1983; Johnson et al., 1982). Emphasis has been placed on examining the internal consistency of the self-report measures and the interrater agreement of the behavioral skills tests. Attempts also have been made to evaluate the relationships between the measures included in the test battery. Test-retest reliability has not been examined specifically for any of the five measures.

The internal consistencies for the general information and problem-solving tests have been reported to be quite high, ranging from .84 to .90 when calculated using the Spearman-Brown prophecy formula (Harkavy et al., 1983; Johnson et al., 1982). The correlations between these two self-report measures have been somewhat lower. Johnson and associates (1982) reported that the measures correlated .62 for parents and .72 for children. The magnitude of the relationship between children's scores on the general information and problem-solving components were at a similar level (.65) in another investigation (Harkavy et al., 1983). These results indicate that a substantial amount of shared variance exists but also that unique variance is being tapped by the two self-report components of the test. Summary scores that were computed for eight conceptually derived content areas, irrespective of their membership within the general information or problem-solving components, were intercorrelated in the initial published investigation (Johnson et al., 1982). The average correlation obtained from this analysis for children was .57 and for parents was .38. The conceptually derived content areas assessed by the test are at least somewhat interdependent but do not overlap to the extent that excessive item redundancy exists.

Interrater agreements for the behavioral skills tests also have been in the high range. To evaluate interrater agreement, two observers independently rated the same child in approximately 25% of the sample investigated in the original study published on the TDK (Johnson et al., 1982). Interrater agreement was 88% for the urine testing test and 93% for the insulin injection test. A similar interrater agreement rate (93%) for the insulin injection test was obtained by Gilbert and colleagues (1982) although somewhat higher interrater agreement indices for both tests (urine testing=96%, insulin injection testing=97%) were reported by Harkavy and associates (1983). Furthermore, Brody (1985) reported a 95% interrater agreement rate for the serum glucose monitoring test. The urine testing and insulin injection tests are also relatively independent of each other and the self-report measures. The correlation between the insulin injection test and urine testing test was reported to be within the low range and the correlations between the behavioral skills tests and self-report measures also were reported to be within the low range (Harkavy et al., 1983; Johnson et al., 1982).

In comparison to reliability, the validity of the TDK is not as well documented. Harkavy and associates (1983) stated that the content validity of the general information and problem-solving components were supported by the selection procedure used to generate the items. That is, agreement between two physicians and a nurse on the correct response alternative for each item provided support for the the content valididty of the self-report portions of the test. Although this is a conceptually accurate conclusion, the methodology used to select items can be questioned with respect to its utility in validly evaluating item content. Significant and substantial positive correlations (>.90) were reported between performance on the serum glucose monitoring test and daily readings of serum glucose levels obtained via a glucometer at a diabetes summer camp (Brody, 1985). No other concurrent or predictive validity data are available for the set of measures.

The measures included in the Test of Diabetes Knowledge, however, have been reported to be sensitive to interventions designed to improve children's and adolescents' skills for coping with diabetes. In a systematic analysis of a therapeutic diabetes summer camp, general information, problem solving, urine testing, and insulin injection skills improved significantly as a result of daily education and practice (Harkavy et al., 1983). As noted previously, these changes were dependent on the sex and ages of the children assessed. Older children improved substantially more than younger children across all areas as a function of the camp experience. Moreover, females displayed greater improvement than males. Similarly, insulin injection performance improved substantially as a function of videotaped instruction (Gilbert et al., 1982), and serum glucose monitoring improved significantly as a function of daily diabetes education and practice conducted at a diabetes summer camp (Brody, 1985) in other investigations. These changes followed the same pattern with respect to age and sex already noted.

Age differentiation provides support for the construct validity of the measures included in the TDK. The absence of control groups in the pre- and post-evaluations of therapeutic and educational interventions, however, places substantial limits on the applicability of the changes displayed as indices of construct validity.

Critique

The authors of the Test of Diabetes Knowledge should be commended for taking on the task of developing a set of standardized and reliable disease-specific measures that evaluate both behavioral skills and multiple domains of disease knowledge. Measures such as these are clearly important to the further refinement of research strategies and effective clinical interventions in the rapidly developing area of pediatric psychology. Disease-specific rather than global psychological constructs appear to be important to the further understanding of long-term therapeutic adherence in chronic childhood illness.

Continued development of these measures by the test authors and the collection of data on their utility outside of the setting in which they were developed seems to be an appropriate and warranted direction for the TDK. In its current form, the test package may need revision as it reflects its research origins and may not be perceived as user friendly by professionals employing the measures in clinical settings. Moreover, an alternate form of the general information and problem-solving questionnaire with diabetes-related technical terms removed is not included in the test package. The most critical research questions to address in the future are 1) further information on criterion-related validity, 2) examination of test-retest reliabilities, and 3) evaluation of the role of reading level, especially with respect to reported age differences on the general information and problem-solving components.

Although the set of measures clearly has utility for researchers and potentially for clinicians, a number of issues should be kept in mind for those opting to use the TDK at this point in time. Rapid developments in the technology for treating insulin-dependent diabetes mellitus have resulted in changes in the self-care regimen. Electronic readings of serum glucose obtained via glucometer are currently available, and the cost of such equipment for use on a daily basis is becoming increasingly affordable. In addition, the utility of urine testing on a regular basis has been challenged in some circles. With technological advances likely to continue, the measures as they currently stand may need periodic revision to keep pace with technology-based improvements in self-care.

References

Brody, H.J. (1985). *An evaluation of blood-glucose monitoring skills among children with diabetes.* Unpublished master's thesis. University of Florida, Gainesville, FL.

Etzwiler, D.D., & Robb, J.R. (1972). Evaluation of programmed education among juvenile diabetics and their families. *Diabetes, 21,* 967-971.

Etzwiler, D.D., & Sines, L.K. (1962). Juvenile diabetes and its management: Family, social and academic implications. *Journal of the American Medical Association, 181,* 94-98.

Gilbert, B.O., Johnson, S.B., Spillar, R., McCallum, M., Silverstein, J.H., & Rosenbloom, A. (1982). The effects of a peer-modeling film on children learning to self-inject insulin. *Behavior Therapy, 13,* 186-193.

Harkavy, J., Johnson, S.B., Silverstein, J., Spillar, R., McCallum, M., & Rosenbloom, A. (1983). Who learns what at diabetes summer camp. *Journal of Pediatric Psychology, 8,* 143-153.

Johnson, S.B. (1979). *Manual for the general information and problem solving components of the Test of Diabetes Knowledge.* Unpublished manuscript, University of Florida, North Florida Regional Diabetes Program, Gainesville.

Johnson, S.B. (1984). Knowledge, attitudes, and behavior: Correlates of health in childhood diabetes. *Clinical Psychology Review, 4,* 503-524.

Johnson, S.B. (1986, March). *Knowledge, attitudes, and behavior in childhood diabetes: Developmental considerations.* Paper presented at the meeting of the Society of Behavioral Medicine, San Francisco, CA.

Johnson, S.B., Lewis, C., & Alexander, B.B. (1981). *Administration and scoring manual for Chemstrip Skills Demonstration Test.* Unpublished manuscript, University of Florida, North Florida Regional Diabetes Program, Gainesville.

Johnson, S.B., Pollak, T., Silverstein, J.H., Rosenbloom, A.L., Spillar, R., McCallum, M., & Harkavy, J. (1982). Cognitive and behavioral knowledge about insulin-dependent diabetes among children and patients. *Pediatrics, 69,* 708-713.

Pollak, R.T., & Johnson, S.B. (1979). *Administration and scoring manual for the Test of Diabetes Knowledge skills demonstration.* Unpublished manuscript, University of Florida, North Florida Regional Diabetes Program, Gainesville.

Ruth G. Thomas, Ph.D.
Associate Professor and Head, Division of Home Economics Education, Department of Vocational and Technical Education, University of Minnesota, St. Paul, Minnesota.

TEST OF PRACTICAL JUDGMENT, FORM 62

Alfred J. Cardall. Montreal, Quebec: Institute of Psychological Research, Inc.

Introduction

The purpose of the Test of Practical Judgment—Form 62 (TPJ) is to measure problem-solving ability in everyday life. The test manual indicates that the TPJ measures 1) the ability to draw logical conclusions in business and social situations, 2) mental balance between applied intelligence and potential, and 3) preferential mind set toward things or people in problem solving. The manual indicates that the TPJ will also provide information upon which to base inferences about relative degree of drive or compulsion.

The TPJ provides two subscale scores, one measuring factual judgment and the other measuring social judgment or empathy. Judgment is defined as the act of selecting the best alternative from several possible courses of action at a particular time. The TPJ is intended to measure the ability to exercise judgment in general life situations, as opposed to judgments involving underlying technical skills in a specialized field.

Factual judgment is defined as the ability to select the best of several alternatives in solving a factual problem and involves 1) the use of applied intelligence in understanding a problem in order to determine possible action alternatives, 2) the ability to project and compare the possible consequences of each alternative, and 3) the selection of the best alternative over the other possibilities. Skill (or know-how) is differentiated from factual judgment, the former being identified as doing a thing in a uniform way across situations and the latter involving the use of a skill in different ways depending on the situation. The complexity of the judgment situation is perceived to increase as the number and variety of alternatives increases, as the projection of consequences becomes more complex, and as the time required for proof is more remote.

Social judgment is defined as the ability to select the best of several possible alternatives in complex social situations by employing empathy, or the imaginative projection of one's consciousness into the personality of another in order to understand him or her better. Social skill is identified as ranging from social conformity to manipulative or motivational social skills, whereas social judgment is required in complex situations where subtle nuances are involved.

The TPJ was developed by Alfred J. Cardall, Ed.D. The current form of the test was produced in 1961, and represents a refinement of earlier versions developed in 1941 and 1948. Construction of the original (1941) version was based on "an extensive study of specific personality requirements as they were revealed in a broad range of occupational descriptions" (Cardall, 1961, p. 1). This study is not referenced in the manual. The factors that were determined to occur most frequently in the occupational descriptions were practical judgment, business judgment, and common sense. These were believed to be synonymous factors, and the term chosen to represent the single factor was *practical judgment*.

Consultation with vocational psychologists and personnel representatives was done to determine what practical judgment is and how it might be measured. The manual reports that there was a consensus that practical judgment involves the recognition of a problem in terms of possible alternatives and action and the ability to select the best of these alternatives. A multiple-choice format was identified as an appropriate way to test the ability to exercise judgment in choosing among a set of possible action alternatives.

The 1941 version of the test included 48 items that resulted in a single score. A short form of the test was developed in 1948, which provided two factors or clusters of items and two scores, one factual and one social. Although experimental use revealed that the two subscores had relatively low intercorrelation and suggested that the construct measured by the test was multidimensional, the number of items in each subscale was too small to yield acceptable reliabilities. Consequently, when the test was revised in 1961, new items were added. This revision produced Form 61 of the test and brought reliability coefficients for the two subscales to a higher level than that of the original version total score. The scoring method was also changed to a simpler procedure, employing a simple unit count of correct responses rather than the more complex weighted scoring procedure used in the original version. Another modification of the scoring procedure entailed scoring all four possible alternative responses rather than just the best three, resulting in the current test and procedure.

Information provided in the manual concerning the norm populations is not detailed. A norms table is provided with the indication that norms are based on a general business population. There is no information regarding what business sectors are represented in the norm group or on the experience level, sex, age, educational level, or positions of those who comprised the norm population or sample.

The TPJ is composed of 60 multiple-choice items. There are four response alternatives for each item. The first 8 items require selection of the best response from among the four alternatives. The remaining 52 items require selection of the best response followed by ranking of second and third best responses from among the remaining three response choices. Items are of two types. One type is the presentation of a situation with action alternatives as responses. The second item type is a stem that requires a response that provides reasons for the existence of a condition, state, action, or purpose. Responses are written on the test form itself. A handscoring key and a four-page manual accompany the test.

The examiner's role in the testing process is distribution of the test and making sure that the examinees understand the instructions and record their starting and stopping times on the test forms. Although not explicitly recommended, the examiner might go through the two sample items in order to introduce the test.

The TPJ is intended for mature adults of all ages who are employed or seeking employment in business and industry settings, entering college, or undergoing clinical evaluation.

Test difficulty is not reported. The TPJ contains two subscales, factual judgment and social judgment or empathy. Thirty-four items comprise the factual judgment subscale and the remaining 26 items comprise the social judgment or empathy subscale. In addition to the two subscale scores, a total score is also obtainable but is not used in the test score profile.

The manual suggests that stanines are the most useful raw score equivalents, and it includes stanine equivalents in a table of score norms for generating stanine score profiles of test results for individuals. A stanine is a type of standard score based on a normal curve. The norms table includes stanines for score ranges for the two subscale scores, in addition to the total score and the score ranges on the Cardall-Miles Mental Alertness Test (Cardall, 1961). The Cardall-Miles Mental Alertness Test is no longer available. The manual recommends that by using an intelligence test in conjunction with the TPJ, an individual's stanine for learning rate may be combined in a profile with the person's stanine for factual judgment and social or empathic judgment to yield a three-factor profile of mental structure: learning rate, factual judgment, and empathic judgment.

Practical Applications/Uses

The Test of Practical Judgment—Form 62 is designed to measure an individual's problem-solving ability in everyday life. This capacity, as differentiated from general intelligence, is described in the manual as an important factor in success, achievement, or status. The construct measured by the TPJ is purported to be particularly relevant in situations that require problem solving and getting along with people. The manual indicates that the TPJ differentiates an individual's orientation between things or people and can detect potential imbalance between common sense and a theoretical learning rate. The manual also states that when this test is used with an intelligence test, the resulting profile reflects mental balance and indicates the relative drive and social maturity of an employee.

The TPJ is intended for use with adults in business and industry, young adults entering college, and for the clinical evaluation of adults. In business and industry, the TPJ has been used to select and promote executives, supervisors, and sales personnel and for the general screening, selection, and placement of employees who will be involved in work requiring thinking, planning, or getting along with people. The TPJ is also recommended as a means of appraising present employee assignments by matching the employee's test profile with their job demands. Examples report the use of such profiles for the selection of management personnel in various companies representing a range of industries. The manual further indicates that profile cutoff points for screening job applicants for positions can be determined from written job descriptions but that doing so may require professional guidance. The manual suggests that cutoff points should be set for each individual situation and work context.

Use of the TPJ is reported for guidance and admissions situations as a measure of maturity and readiness for college. With respect to clinical uses, the manual indicates that, when used with an intelligence test, the TPJ aids in identifying the nature and extent of psychoses, specifically, mental and emotional imbalance.

Professions that are reported to use the test include personnel officers, guidance and college admissions officers, counselors, psychologists, and psychiatrists. A potential new application for the test may be research, particularly in light of the growing body of knowledge in cognitive science that indicates that expertise entails more than factual knowledge and general intelligence. It is increasingly apparent that the ability to use knowledge and employ one's intelligence in making judgments regarding a particular problem or situation in a particular context is a key element in expertise. It appears that this capacity for judgment, as differentiated from general intelligence, was recognized long ago by Cardall in developing this test.

The types of subjects for which this test is intended are normal and psychologically abnormal adults. The TPJ is not appropriate for children, who would lack the language development and experience required by this test, or for adolescents, who would lack understanding of the employment context underlying many of the items. The TPJ would be appropriate for adults of average reading ability and above.

The TPJ might be adapted for situations other than work and employment situations. Many of the items are factual questions requiring judgment about reasons for certain business practices. Others are questions concerning interpersonal actions in a supervisory or co-worker situation. The same types of judgment could be tapped by similar types of questions that incorporate nonbusiness contexts such as family situations, political situations, friendship and peer group situations, and so on.

The TPJ may be administered in group or individual settings. The examiner does not need special qualifications beyond being able to clearly communicate instructions for taking the test. However, special qualifications are needed for interpreting test results, which should be done by someone well versed in psychometrics and in psychological assessment. The TPJ is not difficult to administer. The most challenging task in administering the test is likely to be making sure that examinees understand that there are two different types of items requiring different kinds of responses. Instructions for the items in which examinees are to select and rank three of the four alternative responses to each question should be gone over especially carefully in introducing the test. There is no time limit for taking the test. Most people are reported to take about 30 minutes.

A hand-scoring key is provided with the TPJ. Scoring necessitates being alert to the correct order of each response and to counting the fourth alternative as a correct response if it has been left blank and if the three other alternatives have clearly been designated as first, second, and third choices. A test user could learn to score this test in a short time by actually scoring several sample tests. It should not take more than about 5 minutes to score a test, once proficiency is attained. The most difficult scoring tasks are watching for three designated choices in order to count the fourth alternative correct and being sure that the correct ranking has been indicated for each response alternative.

Scoring involves obtaining a simple count of the number correct. Two test keys are used, one for the factual judgment items and one for the social judgment items. The procedure involves laying the appropriate key alongside the test items and noting correct and incorrect responses. This procedure is followed by counting the number correct for each item on the two subscales and entering the raw subscale score in the appropriate box at the upper right hand corner of the test. The stanine raw score equivalent is then determined from the norms table provided in the test manual and entered in the score box behind each subscale raw score.

Interpretation is based on objective scores using profile cutoff points that have been established for personnel selection or promotion decisions. Although general guidelines for cutoff points are provided in the manual for supervisory and secondary level management positions, it is recommended that cutoff points be set for each situation and employment context. The establishment of cutoff points from job descriptions is discussed, but not in enough detail for a person without experience or professional training in the procedure to do it.

As stated previously, analysis may be performed on the relationships between intelligence test scores, factual judgment, and social or empathic judgment. The balance between these three scores is examined to make interpretations about orientations toward people or things and mental and emotional balance. For example, the manual states that factual judgment scores that are considerably lower than intelligence scores indicate a tendency to rationalize problems rather than study them. The reverse pattern indicates compulsiveness in problem solving and reliance on authorities rather than one's own thought processes. Profiles with high learning rate and factual judgment scores tend to show low empathy scores. Empathy scores that are higher than factual judgment scores indicate reliance on political skills over factual exploration: extremely skewed profiles in this direction indicate a tendency to distort problems into one's preferential mind-set.

Although the cutoff score interpretation approach requires experience and/or professional training, once these are acquired, interpretation of test results is straightforward and could be done by personnel without special background in psychology or psychometrics. Interpretation of the TPJ based upon score interrelationships, however, requires professional skill and understanding of psychology and psychometrics and would also seem to require considerable experience in using the test.

Technical Aspects

Although validity and reliability studies are mentioned in the manual in part of a general discussion of these psychometric test properties, no specific studies are referenced, nor is a general overview provided of pertinent findings. However, consistency reliability information is provided. The cutoff coefficient used for item retention on the subscales was .20. Split-half reliability coefficients for each subscale are .92 for Factual Judgment and .88 for Empathic Judgment. The total score split-half coefficient is .87. These reliability coefficients represent an average of split-half coefficients obtained from random samples of 300 cases drawn from various undescribed populations.

A moderate intercorrelation (.39) is reported between the Factual Judgment and the Empathic Judgment subscale scores, based on a sample of 300 undescribed cases. Despite this moderate correlation between the subscale scores, the manual recommends the use of the subscale scores rather than the total score, based on a statistical test for independence of the two subscales.

Moderate correlation coefficients are reported between the Cardall-Miles Mental

Alertness Test and the Factual Judgment subscale (.55) and the Empathic Judgment subscale (.43), based on what is described as a general population sampling of 300 cases.

Studies bearing on predictive validity are mentioned in relation to the total score on the original version of the test. The criterion used in one of these studies was position level within the management structure of the company. The manual provides little discussion of the specific criteria that the test is intended to predict. A passage reporting minimal profile cutoff points that have been established by specific companies for various types of positions refers to a "success profile" criterion. Subjects in one of these studies included management personnel in a food subsidiary of an airline. Other studies involved a large finance company and banks with extended branch systems. No descriptive information about experience level, sex, or age of participants in these studies is provided.

Critique

The importance of being able to measure the capacity of practical judgment, as distinguished from general intelligence, academic ability, and so on, is receiving considerable attention in the cognitive science literature (e.g., Sternberg & Wagner, 1986). The capacity of practical judgment, as defined by Cardall, would seem to be of increasing importance in a work world and a society that are moving in the direction of increasing complexity, where the ability to make fine but critical distinctions is significant. Further, it is well established that interpersonal problems are more often the cause of work-related problems than are problems with technical job skills, which reinforces the importance of the Emphatic Judgment scale in the Test of Practical Judgment—Form 62. At the same time, it should be observed that the TPJ needs considerable development with respect to norm groups, validity studies, and interpretation guidelines, not to mention the need for a general revision of both the test and the manual to remove sexist language (e.g., "personnel men"; man-development"; a man's general success"; [Cardall, 1961, pp. 1, 3] "why is a man superior to a productive machine?" [Cardall, 1980, p. 1]) and sexist role images ("Why are food products advertised in women's magazines?—To aid women in planning meals" [Cardall, 1980, p. 1]). The test has apparently been reprinted several times (1976, 1977, 1980) without giving attention to such revision.

Information about norm groups provided in the manual is incomplete. At a most basic level, the size of the norm group upon which the table of norms is based is not reported. More specific descriptions of norm groups concerning age, sex, educational and experience levels, and types of industry is needed for the user to determine comparability of his or her own group with those on which this test was normed. It would be most desirable to have norms reported separately by gender and for different industries and types of positions. From a psychometric standpoint, more information about sampling procedures employed in assembling norm groups would be important for assessing the degree to which it was a systematic process likely to produce unbiased results as opposed to a set of convenience samples likely to contain a number of unknown biases. If norm groups for this test have been largely male, there is considerable question about the appropriateness and soundness of applying them in interpreting the test scores of females.

The manual should also provide instructions for the two types of responses test items require. The first eight items have instructions for selecting one alternative response. Items 9 through 60 are preceded by another set of directions for choosing and ranking three of the four alternative responses. No instructions are provided as to whether the test administrator should give both sets of instructions at the outset or ask everyone to stop after completing item 8 and wait for further instructions. Sample items of each type are provided on the test to help assure that the examinee will understand the instructions and follow them accurately. The speed with which the test is completed is mentioned as indicating speed of judgment, and information concerning this variable is obtained on the test, but no guidelines for interpreting this information are provided.

The use of an intelligence test in conjunction with the TPJ is given major attention in the manual but insufficient information is provided to guide the test user in deciding which intelligence test to use. Norms for specific intelligence tests, besides the Cardall-Miles Mental Alertness Test, are not provided. This is a particularly serious limitation, given the unavailability of the Cardall-Miles Mental Alertness Test. The manual should be updated and revised to include norms for existing intelligence tests. Theoretical norms are given with the suggestion that they can be used for other intelligence tests. Such general norms are much less desirable than specific norms for specific intelligence tests.

More explicit guidance in establishing cutoff profiles based on job descriptions is needed. Currently, there is not enough information provided concerning how this can be done, and yet the manual implies that, with experience, it can be learned. Guidelines for the interpretation of test results suggest that the test user could possibly interpret test results for an individual's score profile interrelationships. However, not enough information is provided to allow test users who are not professional psychologists and who have not had extensive experience with this test to do this. More explicit information about qualifications needed for interpretation of the test results is needed.

The construct this test is intended to measure, practical judgment, is discussed relatively extensively in the manual. The discussion relates practical judgment capacity to success, achievement, and status in management positions in business settings. Yet, construct validity is not discussed, nor is evidence of it presented. Predictive validity is discussed, and some evidence is presented in its support, but the criterion variable and its measures are unspecified for the TPJ. Consequently, exactly what it is that cutoff profiles will predict if used in personnel selection is unclear. More study and evidence of construct and predictive validity are needed in order to evaluate this test as a tool for promotion, selection, and job reassignment decisions. No validity data are presented for use of the test as a measure of maturity and readiness for college or for detecting psychoses, two of the purported uses of the test.

In addition to internal consistency reliability information, evidence regarding stability reliability is equally relevant and should be provided. Reporting of the standard error of measurement would also be helpful in making judgments about test reliability.

In summary, the construct that the TPJ claims to measure has much promise in assessing and understanding human performance. Given the information available about the test at the present time, however, it is difficult to ascertain whether the TPJ does measure practical judgment as it is defined in the test manual. It is also unclear 1) what specific performance characteristics relevant to a work context could be expected from individuals scoring at various levels on this test and 2) what the test will predict regarding college performance or attrition or if it will detect psychoses in clinical situations. It would seem that the TPJ could measure a promising and significant construct not easily captured in a paper-and-pencil test and that it is worthy of the suggested revisions to remove sexist language and the necessary research required to validate its construct validity and predictive power. Until such validation studies are done, users are likely to experience difficulty in interpreting its results. Those who want to consider using the test should definitely plan to do norming and validation studies for their own situation before relying on the TPJ as a categorizing or predictive tool.

References

Cardall, A. (1961). *Manual: Test of Practical Judgment, Form 62*. Yardley, PA: Cardall Associates.

Cardall, A. (1980). *Test of Practical Judgment, Form 62*. Montreal: Institute of Psychological Research, Inc.

Sternberg, R., & Wagner, R. (Eds.). (1986). *Practical intelligence: Nature and origins of competence in the everyday world*. Cambridge, England: Cambridge University Press.

Robert H. Bauernfeind, Ph.D.

Professor of Education, Northern Illinois University, DeKalb, Illinois.

TEST OF STANDARD WRITTEN ENGLISH

The College Board. New York, New York: The College Board Publications.

Introduction

The Test of Standard Written English (TSWE) is an objective (multiple-choice) test of editing skills designed for use with high school seniors and college freshmen. Students' raw scores are converted to scale scores ranging from a low of 20 to a high of 60+.

Initial forms of the TSWE were developed at Educational Testing Service in the early 1970s. The newer forms, carrying 1980s copyrights, consist of 50 five-choice items. These newer forms are distributed by the College Board in two ways. First, TSWE can be given along with the College Board SATs through the national Admissions Testing Program of the College Board. Second, TSWE can be purchased by high schools and colleges for local institutional use.

The test materials are visual and oriented to appropriate use of the English language. Thus, we should expect no adaptations for individuals who are visually impaired, nor translations into other languages.

The TSWE test booklet consists of 12 pages—a cover page, one page of directions, a page showing a sample answer sheet, six pages of test questions, and three pages that are blank. The sample test sent for review is printed black-on-white—not especially attractive, but certainly readable.

The TSWE consists of two item types. The first, called "Usage," has the student attempt to identify an error in a short sentence. Example:

Prefabricated housing <u>is</u> economical <u>because</u>
                        A   B
<u>they reduce</u> labor costs <u>considerably.</u> <u>No error</u>
     C*   D   E

This kind of skill is tested in items 1-25 and 41-50.

The second item type, called "Sentence Correction," asks the student to select which of five parallel phrases best completes a short sentence. Example:

<u>Eddie was as angry as Linda was</u> when he discovered that thieves had stripped her car.

  *(A) Eddie was as angry as Linda was
  (B) Eddie had anger like Linda's
  (C) Eddie's anger was like Linda was
  (D) Eddie's anger was as great as Linda
  (E) Eddie had an anger as great as Linda

This kind of skill is tested in items 26-40.

TSWE will be quite easy for students who are good writers, but difficult for those who have not tuned in to the "logical flow" of good English and/or who are poor readers.

The examiners' responsibilities simply involve making certain that each student understands the test problems and the use of the separate answer sheet. The examiners also are responsible for calling time after exactly 30 minutes.

Practical Applications/Uses

The authors intend that TSWE will be used with high school seniors and with college freshmen. Noting that many colleges provide two semesters of English composition courses in the freshman year, Breland (1977a) suggested that TSWE scale scores be used to divide incoming freshmen into four groups:

Category 1. Students scoring 60+ might be exempted from freshman composition, provided they also pass an actual essay-writing test provided by the local college's English faculty. (Breland's studies found that only 4% of college-bound high school seniors scored 60+ on the TSWE.)

Category 2. Students scoring 45-59 might take only the second of the two writing courses. (Breland's studies found that 46% of high school seniors scored in the 45-59 range.)

Category 3. Students scoring 35-44 would be required to take both writing courses. (Breland's studies found that 27% of high school seniors scored in the 35-44 range.)

Category 4. In Breland's (1977a) opinion, students scoring below 35 would need remediation *before* enrolling in a college writing course. (Breland's studies found that 23% of high school seniors scored in the 20-34 range.)

Individual colleges will, of course, wish to make categories based on their own experience, but Breland's suggestions give each a model from which to make local adaptations.

TSWE is a group test, very easily administered. In a single English class it could be administered just by the teacher. In an all-school testing program, there should probably be one proctor for every 30 students. The 30-minute time limit means that TSWE can be easily administered in the usual 40-50 minute high school or college class period.

The College Board provides a complete scoring service. In addition, the authors provide an answer key (one correct answer per item) that is readily adaptable for hand-scoring or machine-scoring. One disturbing aspect of the TSWE program is that the authors use the antiquated correction-for-guessing formula in arriving at each student's final raw score. While TSWE was developed and normed at Educational Testing Service, so also were the Pre-Professional Skills Tests:

> Your score on any of the PPST is based on the number of questions you answer correctly. . . . On these tests, there is no subtraction for incorrect answers. Therefore, if you are uncertain about the answer to any of the multiple-choice questions . . . you should guess at the answer rather than not respond at all. You risk nothing by guessing; but if you do not respond, you lose the chance to raise your score by making a lucky guess. (Educational Testing Service, 1985, p. 8)

And the Law School Admission Test:

> Answer every question—guess if necessary. When we score your answer sheet, we will give you credit for every correct answer but deduct nothing for wrong ones. (Educational Testing Service, 1974, p. 12)

The right hand knoweth not. . . .

The correction-for-guessing formula imposed on TSWE scores is disturbing for at least two reasons. First, it is unnecessary and probably counterproductive (Bauernfeind, 1978). Secondly, it requires *three* additional steps in the scoring process, making it extremely error-prone in high schools and colleges that do their own scoring. Schools using the College Board's scoring services will have to live with the correction-for-guessing procedures. But for schools doing their own scoring, this reviewer strongly urges use of the kinds of directions cited above, to be followed by local norms and local cutoffs on the items-right raw scores.

TSWE is a straightforward test of ability to recognize correct standard English. It is further intended to distinguish among students who need help in this skill. It is *not* intended to distinguish among students whose command of standard English is "better than average" (College Board, 1984b).

Technical Aspects

The TSWE package sent to test reviewers contains a strange assortment of materials. We are given (in my opinion) an excellent test of editing skills and a summary of Breland's (1977a) interesting article. The Breland article makes repeated references to 50% common variance with ratings of actual essays, thus suggesting correlations around .70 between TSWE scores and essay ratings. Such correlations, not corrected for attenuation, would be remarkably high. But, we are given no scoring key, no information about the 20-60+ scale, no information about the percentile rank norming group, no information about interform equating studies, and not a single report of a reliability coefficient.

The review package *does* tell us the following:

1. The percentile rank scores "are based on the scores of candidates who took the TSWE in 1977-78" (College Board, n.d.). The word "candidates" suggests that the percentile rank scores are based on scores of college-bound high school seniors.

2. TSWE is fairly easy. The norms suggest a median number-correct raw score of about 33 for Form WWE6, and about 34 for Forms WWE1 and WWE2.

3. The wide spread of raw scores indicates that these scores are quite reliable. This reviewer estimates the KR-21 reliability for the percentile-rank norms group to be .88. However, there is a skew to the left in the raw score distribution. This means that the percentile-rank scores are more reliable with low-scoring students, less reliable with high-scoring students.

Reviewers who write for a conventional technical report are sent only a copy of the original project report by Breland (1977b). Although the Breland report is *not* a conventional technical report, it does show some interesting correlation data involving TSWE scores:

	r	N
SAT Verbal	.73	7013
SAT Mathematical	.53	7013
Essay, early in Freshman year	.63	770
Essay, later in Freshman year	.58	766
Essay, total of three	.72	261
TSWE Posttest (Fall)	.83	789
TSWE Posttest (Spring)	.84	493

Source: Breland (1977b, pp. 38, 47)

In "Essay, total of three," the TSWE scores were correlated with the sums of ratings of three student essays gathered at different times during the college freshman year. Correlations with TSWE posttests constitute reliability studies of a sort, but there were intervening treatments between the TSWE testings, making it very difficult to interpret these data.

The Breland (1977b) study was focused on *uses* of the TSWE, not on the test itself. Nowhere in the Breland report does one find an explanation of the 20-60+ scale, a description of the percentile rank norming group, or a single report of an internal consistency reliability coefficient.

One final effort by this reviewer to obtain such basic technical information brought forth a 225-page technical report that had been published in 1984 (College Board, 1984a). This totally unpublicized technical handbook is one of the best I have ever seen. Among its revelations:

1. The TSWE scale *is* equated to the SAT 200-800 scale, except that TSWE deletes the final zero on the SAT scale. Thus, TSWE 40 is equivalent to SAT 400, TSWE 50 is equivalent to SAT 500, and so on (p. 20).
2. The TSWE is fairly easy for good editors. Rather than following a bell-shaped curve, raw scores are somewhat bunched at the top, skewed at the bottom. For this reason it was not possible to set the scores to the bell-shaped 20-80 scale; rather, the bunched high scores are simply given a scale value of 60+ (p. 20).
3. Each new form is anchored to previous forms using conventional linear methods. SAT Verbal scores are also referenced in scaling new forms of the TSWE (p. 20).
4. TSWE correlations with SAT Verbal and SAT Mathematical have been running slightly higher than those reported in the Breland study. Across eight studies, TSWE scores showed a median correlation of .78 with SAT Verbal, .64 with SAT Mathematical (p. 81).
5. In studies in 25 colleges, TSWE scores showed a mean correlation of .37 with grades in English courses (pp. 166-167). Note, however, that "English courses" often include skills other than writing/editing.
6. TSWE is essentially a power test. Typically around 85% of students complete the entire test (pp. 80-81).
7. TSWE reliabilities are quite satisfactory for the groups tested. Across eight studies the median KR-20 coefficient was .88; across six studies the median test-retest coefficient was .82 (p. 82).

An incidental note: While 500 was at one time the mean scale score for 12th--graders taking the SAT, the mean had fallen to V = 427, M = 468 during the 1977-78 school year (Educational Testing Service, 1978). The mean of those two means is 447.5, which accounts for the fact that the median 20-80 scale score on TSWE is about 44, not 50 (College Board, 1984b).

Critique

This reviewer would offer three suggestions to the authors of the TSWE. First, redesign the review package to make clear the fact that the original Breland study and a full technical handbook are in print and available. Secondly, develop number-correct norms for the benefit of high schools and colleges doing their own scoring. In addition, update the percentile rank norms. While the 1977-78 norms are probably very close to current norms (because the SAT means are very similar), it would look better to students and test critics if the norms appeared to be current.

However, those considerations are peripheral. The "heart" of any test embodies the test items themselves, and the heart of TSWE is in good shape. Although each prospective test user will have to judge the "content validity" of the TSWE, this reviewer considers it very high. The items reflect real-life communication problems: the authors do *not* test for *laid/lain, can/may, better/best,* or other "errors" that in fact do *not* interfere with clear communications. While the task is clearly one of "editing," the test items come so close to real life that I will not quarrel with those who call this a test of "writing skills."

The test is relatively easy. It focuses on low-scoring students and the kinds of errors they make. The raw scores of good editors are bunched at the top of the test; these high-scoring students will be told that they are good with standard English and, as suggested by Breland, they might be excused from one or more college English composition courses.

I gave this 50-item test to five people who work as "copy editors" for newspapers, magazines, test publishers, or book publishers. The raw scores were 49, 49, 49, 50, and 50. Four of the five volunteered that this is "a very good test." The usefulness of the TSWE could be greatly enhanced if the authors would generate raw score distributions for successful copy editors, successful secretaries, and successful writers (James Reston and Mike Royko types). These latter kinds of data could be extremely helpful to high school and college students who want something more than a norm-referenced score within a local norms group. In short, TSWE may have a much greater variety of uses than its authors have proclaimed.

References

Bauernfeind, R. H. (1978). *Building a school testing program* (2nd ed.). Bensenville, IL: Scholastic Testing Service.

Breland, H. M. (1977a). *Can multiple-choice tests measure writing skills?* (The College Board Review No. 103). New York: College Board.

Breland, H. M. (1977b). *A study of college English placement and the Test of Standard Written English*. Princeton, NJ: Educational Testing Service.

Cohn, S. J. (1985). Review of the College Board Scholastic Aptitude Test and Test of Standard Written English. In J. V. Mitchell (Ed.), *The ninth mental measurements yearbook* (pp. 360-362). Lincoln, NE: The Buros Institute of Mental Measurements.

College Board. (n.d.). *Guidelines for TSWE administrations.* New York: Author.

College Board. (1980). *Test of Standard Written English—Form M-WWE6.* New York: Author.

College Board. (1984a). *The College Board technical handbook for the Scholastic Aptitude Test and Achievement Tests.* New York: Author.

College Board. (1984b). *Information for candidates about the Test of Standard Written English.* New York: Author.

Cronbach, L. J. (1985). Review of the College Board Scholastic Aptitude Test and Test of Standard Written English. In J. V. Mitchell (Ed.), *The ninth mental measurements yearbook* (pp. 362-364). Lincoln, NE: The Buros Institute of Mental Measurements.

Educational Testing Service. (1974). *Law school admission bulletin 1974-1975.* Princeton, NJ: Author.

Educational Testing Service. (1978). *1977-1978 update for Table 2 in* On Further Examination. Princeton, NJ: Author.

Educational Testing Service. (1985). *Bulletin of information for the Pre-Professional Skills Tests of Reading, Writing, and Mathematics.* Princeton, NJ: Author.

Scarvia B. Anderson, Ph.D.
Adjunct Professor of Psychology, Georgia Institute of Technology, Atlanta, Georgia.

TESTS OF BASIC EXPERIENCES, SECOND EDITION

Margaret H. Moss. Monterey, California: CTB/McGraw-Hill.

Introduction

The Tests of Basic Experiences, Second Edition (TOBE 2) were designed to assess the skills and general knowledge of children in preschool, kindergarten, and first grade. There are two overlapping levels of the TOBE 2, K and L, and four tests at each level: Mathematics, Language, Science, and Social Studies. According to the test publisher, CTB/McGraw-Hill, the TOBE 2 test results can help educators evaluate program effectiveness and student progress and determine instructional needs and methods. CTB/McGraw-Hill effectively dismisses the TOBE 2 as a "readiness" test, as well as the ambiguities that label introduces. It "is neither a reading readiness test nor a test of readiness to learn. Instead, it measures factors that contribute to readiness" (CTB/McGraw-Hill, 1978, p. 1).

The TOBE 2 is a revision of an earlier series of tests by the same name developed by Margaret H. Moss and CTB/McGraw-Hill in the late 1960s. The TOBE has its origins in two instruments developed in the mid-1960s: Items of Space end Location, which Moss used in research for her doctoral dissertation at George Washington University, and Test of Basic Information, used in Cooperative Research Project No. 619, U. S. Office of Education. Like the TOBE 2, the first edition of the TOBE was "designed to measure the differences in children's awareness of the world around them" with the assumption that "children's experiences and learning opportunities vary considerably" (CTB/McGraw-Hill, 1978, p. 1).

The purposes of the revision were to 1) update the test content, 2) develop winter and spring norms (the first edition had only fall norms), 3) increase the difficulty of Level L (students were "topping out" on the old Level L), 4) improve the format and response method, 5) add a practice test (the first edition included only a few demonstration items to help children learn the mechanics of test taking), 6) eliminate the General Concepts test, and 7) reduce slightly the number of items in each of the remaining tests.

The development of the TOBE 2 involved three major steps: 1) item development, 2) pretesting, and 3) item selection. Item development began with the definition of the content of the four subject areas: mathematics, language, science, and social studies. The item pool included items from the first edition, as well as new items. Special attention was given to the quality of the illustrations. Items also were screened for their "translatability" into Spanish.

The reviewer wishes to acknowledge the helpful assistance of Dr. John S. Helmick.

Pretesting involved about twice as many items as needed for the final forms of the tests. Items were assembled into eight "books" and "administered to a national sample of approximately 3,200 students in kindergarten and first-grade classes during the second week of February in 1977. Students in kindergarten were administered Level K, while students in first grade were administered Level L. . . . To minimize the effects of practice on test results, the order of administering the books for a given subject area was alternated" (CTB/McGraw-Hill, 1979, p. 3). According to CTB/McGraw-Hill, consideration was given to the ethnic, cultural, and sex content of the items and to dialects in the Spanish-language version.

Final selection of items was based on statistical and content criteria, including 1) an "acceptable" point biserial correlation with total score, 2) a positive "select index," indicating that an item should make a positive contribution to the reliability of the test, 3) satisfactory distribution of choices among the distracters and difficulties among the items, 4) good coverage of the skills areas, and 5) a further check for possible bias.

According to the test publisher, "the population to which the TOBE 2 norms are intended to apply includes all public school districts with total enrollments of eleven or more students, and which are listed in the 1970 Census School District Data Tapes and all Catholic schools listed in The Official Catholic Directory 1975" (CTB/McGraw-Hill, 1979, p. 16). For the public school sample, the national population of public school districts (the first-stage sampling units) was stratified into 56 cells on the basis of geography, size of enrollment in Grades 1-8, and a demographic index (socioeconomic data from the 1970 census). A 57th cell consisted of districts in the five largest cities. Schools (the second-stage sampling units) were selected proportionately from these cells. All students in Kindergarten and Grade 1 in the participating schools were tested. The Catholic school sample was stratified by region (Mideast plus Great Lakes, and the rest of the U.S.) and by enrollment (two size categories). The first-stage sampling unit was the diocese or archdiocese.

There are also references to a Head Start sample, but the test publisher does not provide information about how it was selected. Moreover, no information about the number of public or Catholic schools that were selected but were unwilling or unable to participate is provided. A School Characteristics Questionnaire was sent to the 31 Head Start centers and 125 elementary schools that did participate in the study, presumably to check how well the distribution of their characteristics conformed to the distribution of characteristics in the original population. Such a comparison is not included in the technical report, however. The report does summarize the characteristics of the 27 Head Start Centers and 112 elementary schools returning the questionnaire, along with each institution's geographic region, age groups served (for Head Start centers), and community types (for elementary schools). For example, of the 112 elementary schools that returned the questionnaire, 29 were classified as rural farm, 18 as rural nonfarm, 29 as suburban, 21 as urban, and 5 as inner city. The community type for 10 schools was listed as unknown.

There is a marked difference between the characteristics of the Head Start centers and the elementary schools. For example, the percentages of children in the centers and the schools who were black, Hispanic, and "other" were, respectively, about 40, 8, and 52 for the centers, and about 12, 6, and 82 for the schools. The percentage of Head Start families receiving AFDC was 53; the corresponding percentage for the schools was 17. These discrepancies discourage comparisons of the performance of the prekindergarten (all Head Start), kindergarten, and first-grade groups.

The norming studies involving the aforementioned samples were conducted in October 1977 and April 1978. Level K was administered to four groups of children: 1) 465 prekindergartners (all Head Start) (spring), 2) 4,925 kindergartners (fall), 3) 2,563 kindergartners (spring), and 4) 3,336 first graders (fall). Level L was administered to 1) 2,553 kindergartners (spring), 2) 3,463 first graders (fall), and 3) 7,046 first graders (spring). According to the technical report, 550 prekindergartners enrolled in Head Start programs participated in the spring norming. The discrepancy between 550 and 465 is not explained. Were some children's test results discarded? And, if so, what were the bases for eliminating them?

According to officials at CTB/McGraw-Hill, a new childhood assessment system serving the same educational levels as the TOBE 2 is being readied for publication in 1989-90. It is not anticipated, however, that the TOBE 2 will be withdrawn from the market at that time.

Each test at each level takes about 40 minutes to administer. In addition a 20-minute practice test is available. Hand-scorable test booklets are available for all the tests and machine-scorable booklets may be used with the Mathematics and Language subtests. The tests can be administered by regular classroom teachers or aides. Directions for administering the tests in Spanish and English are provided.

The four subtests at both Level K and Level L are designed to measure the following areas:

1. Mathematics: order of numbers; counting; geometry; time and money; weight, volume, and linear measurement; properties and operations; fractions.

2. Language: visual discrimination, alphabet knowledge, initial sounds, final sounds, rhyming, space and location, verb tense, sentence sense, context, comprehension, reading terminology.

3. Science: electricity and magnetism; force, motion, and mechanics; light, optics, and sound; chemistry; earth science and astronomy; animal reproduction and development; animal behavior and characteristics; plant life.

4. Social Studies: geography and travel, environment and use of natural resources, human relations and behavior, occupations and the world of work, sociocultural geography, history and chronology, health and safety, money and consumer behavior.

This is a rather awesome list of topics when one considers that the test takers are from 4 to 7 years old and that each test is only 26 items long! However, the items do seem to bear on the intended subjects, albeit at an appropriately low level. For example, a question assessing knowledge of electricity and magnetism on the Science subtest asks children which of four pictured devices must have electricity to work; an animal reproduction and behavior question on the same subtest asks children to mark the one that changes into a butterfly; and a history and chronology question on the Social Studies subtest asks children to identify the pictured event that happened a long time ago.

As the test publisher indicates, the Mathematics and Language tests include skills that are not learned readily without some kind of systematic instruction. On the other hand, the Science and Social Studies tests deal with topics that children might observe for themselves, be told about, or hear about from stories. Obviously, children's exposure to these topics will vary according to their environments, including preschool experiences and family support.

Practical Applications/Uses

Although the publisher recommends that the TOBE 2 be used for both program and individual evaluation, the former suggests itself more strongly. Tests in the TOBE 2 battery would seem to be good candidates for program evaluation in situations in which group results are of interest and the test(s) selected is appropriate for the educational program under examination. The TOBE 2 might be used to 1) monitor any changes in the skills or knowledge of entering pupils over time, 2) provide a basis for historical comparisons (e.g., of the performance of pupils in a new program with the performance of pupils in prior years), and 3) provide same-year comparisons of the performance of two or more groups of pupils (perhaps exposed to different educational programs).

However, users must remember that the TOBE instruments are survey tests. Because there are only 26 items in each test and only a *very few* items touch on each of the skills or topics a test covers, it would be difficult to draw educationally meaningful conclusions about the "academic" strengths and weaknesses of individual children, conclusions that would help guide instructional placement or prescription. If a child misses the *one* item on the Level L Language that requires recognition of a rhyme, what are the instructional implications? If the child also misses the *one* item on the same test requiring recognition of the sound of the last letter in a word, can one conclude that the child needs special drills on "final sounds"?

The Language test *might* be used as a sort of "structured interview" with an individual child in order to gain insights about how he or she approaches language tasks. In order to gain knowledge about their "bilingualism," children from Hispanic homes might be given both the English version and the Spanish version of the test. However, when the tests are used in these ways, interpretation necessarily would be judgmental or clinical. The publisher's norms would not be very helpful!

The TOBE 2 should present no significant administration problems for classroom teachers or for trained aides or volunteers. The examiner's manuals (one for each level) are detailed with respect to the examiner's responsibilities, use of proctors, breaks, arrangement of the testing room, what children should be told about the testing, purpose of the Practice Test, amount of flexibility allowed in the oral directions, when a test should be considered nonvalid, and similar topics. The directions for administering each test are similarly detailed and easy to follow. Directions for administering the questions in Spanish are presented in a separate publication and not included in the regular examiner's manuals.

The Practice Test takes about 20 minutes to administer, and each of the regular subtests takes about 40 minutes. The publisher recommends that only one subtest be given per day or, if children attend classes for a full day, perhaps one subtest could be administered in the morning and one in the afternoon.

Each item for every test at both levels consists of four pictures. Children respond to the questions, which are read aloud by the examiner, by blackening a small circle under the picture. Two kinds of test booklets are available: hand-scorable (available for all four tests at both levels) and machine-scorable (available for only the Mathematics and Language tests at both levels). At the lower level (K), the hand-scorable test booklet has one item per page, and the machine-scorable version has two. At the upper level (L), the test booklets are identical with two items per page. With rare, and probably insignificant, exceptions, the test items show an orderly progression in difficulty as judged by percent correct across grade levels; that is, on Level K, first graders tested in the fall performed better than kindergartners tested in the spring, who in turn performed better than kindergartners administered the test in the fall or prekindergartners who underwent testing in the spring. A similar pattern characterized the item difficulties for Level L.

Raw scores can be converted into percentile ranks, normal curve equivalents (NCEs), stanines, and scale scores. Separate norms are available for each of the tests, for the Mathematics-Language combination, and for the total battery as follows: 1) Level K—prekindergarten (spring), kindergarten (fall, winter, and spring), and first grade (fall); Level L—kindergarten (spring) and first grade (fall, winter, and spring). The publisher states that Language/Mathematics score reports, including NCEs, are especially useful for Title 1 reporting.

The basic reporting form is the Class Evaluation Record, which allows space for indicating whether each student's answer to each question on a given test is right or wrong, raw scores, percentile ranks, stanines, and NCEs. For recording and reporting individual pupil scores, teachers can use the removable cover of the test booklet or order individual evaluation records. When the Mathematics and Language tests are machine-scored by the publisher, the Class Record Sheets group and summarize item responses ("rights") by skill area and also include school and district mean scores for answer forms processed together. Optional services available from the publisher include frequency distributions of scores by school and district.

For both levels and all administrations, the Mathematics test was the most difficult (in terms of number of items answered correctly). For example, for the prekindergarten sample, the average raw score was 12.34, compared with 15.14 for Social Studies. On Level L, first graders tested in the spring had average raw scores of 20.42 on Mathematics and 22.29 on Language. (All tests at both levels contain 26 items each.) The Mathematics test also showed the greatest variations in scores. For example, the standard deviation for Mathematics was 6.3 for prekindergartners, compared to 4.8 for Social Studies. On Level L, the standard deviation for first-graders tested in the spring was 5.2 for Mathematics and 3.8 for Social Studies. There were smaller differences among the difficulty levels of the other tests. The Science test tended to be somewhat easier for the younger children and the Language test (Level L) for the older children.

Those who would place a great deal of stock in individual scores on these tests would do well to examine the differences in interpretations that can attend differences of even 2 raw score points. For example, suppose a first-grader takes the Level L Language test in December and answers 19 items correctly. According to the publisher's tables, a raw score of 19 places the child at the 25th percentile (edging the bottom quarter of the norms group) and corresponds to a stanine of 4 ("slightly below average"). Now suppose that holiday thoughts had not crowded the child's head for a few moments and the child answered two more items correctly. A raw score of 21 corresponds to the 43rd percentile and pushes the child into stanine 5, which is generally interpreted as "average."

To be fair to the author and publisher of the TOBE 2, few if any standardized group tests for such young children provide more than limited help in accurately identifying childrens' levels of knowledge and skills or diagnosing educational needs. The kinds of testing that would be necessary for such purposes are not practical for such young children in a group setting.

Technical Aspects

The intercorrelations among scores on the four tests are high for all eight testing instances reported. No correlation is below .40; only two are below .50; some correlations are in the .70s. In general, Mathematics scores correlate highest with Language scores, Language scores highest with Social Studies or Science scores, and Social Studies and Science scores with each other. There is obviously a large general verbal ability component in all the tests. The uniformly high relationship between Social Studies and Science scores reflects the great similarity in the content of the tests. In a review of the earlier TOBE, Thurber suggested "that studies be carried out to ascertain the factor structure" (1973, p. 281). No information on formal structural analyses of the TOBE 2 is presented in the technical data book.

The publisher suggests a Mathematics-Language combination score (and offers a scoring service for that pair of tests), as well as a Total Battery score based on all four tests. The lowest correlation between the Mathematics-Language score and the Total Battery score is .92; the other nine are all .94 or .95!

The publisher's norms were developed from raw score distributions for each level, grade, and time of testing. Distributions from each district were weighted to ensure proportional representation for each substratum of the design and combined to form the national distributions. (Presumably Catholic dioceses were treated as districts, although the manual does not make this clear.

A polynomial of best fit was calculated for each distribution, and raw score to percentile rank and Normal Curve Equivalent (NCE) conversions were obtained. The process of determining the final "best fit" also made use of scales derived from Thurstone's absolute scaling procedure, although according to the publisher, their lack of linearity led to the decision not to publish the resulting scale scores (CTB/McGraw-Hill, 1979, p. 24). Raw to scale scores conversions are presented in Tables 46 and 55 in the *Norms and Technical Data Book*.

The October and April distributions produced empirical norms for fall and spring; straight-line interpolation was used to determine January norms. Levels K and L were equated *separately* for Kindergarten in April and for Grade 1 in October.

A fall administration to 2,500 kindergarten children comparing the hand-scorable and machine-scorable editions of Level K showed no educationally significant differences.

Kuder-Richardson Formula 20 (K-R 20) was used to estimate the internal consistency (reliability) of the tests. K-R 20s for the separate tests were in the high .70s or low .80s. The combination scores, of course, yielded higher K-R 20 coefficients: .88-.90 for the Mathematics-Language combination at Level K, .87-.88 for the same combination at Level L; .92-.94 for the Total Battery score at Level K, .93 at Level L.

Standard errors of measurement (SEM) for the individual tests hovered around 2.0, with the Level L Mathematics test having the highest SEM.

Critique

These are nice little tests for survey and program evaluation purposes. As a reviewer of the earlier TOBE said, they "should be of particular use in isolating *general types* of difficulties in *classes* at the start of a school year, especially in classes for disadvantaged children or those of subnormal capacity. However, the use of TOBE as a quasi-diagnostic battery for individual children is questionable . . ." (Proger, 1972, p. 184).

The items are relatively unambiguous and, especially in the Mathematics and Language tests, seem to address topics considered important in most school curricula for the age levels for which TOBE 2 is intended. The usefulness of the Science and Social Studies tests (and the separation of the items into *two* tests) is less obvious, but perhaps the tests have relevance for some early learning programs. Most of the pictures are readily interpretable (not an easy feat), with the exception of coins (which are always difficult to represent pictorially) and some of the natural phenomena (e.g., the moon).

TOBE 2 does not seem to suffer from the unfortunate sex stereotyping that characterized the first TOBE (Diamond, 1978), and population subgroups probably will not be offended by the content of the tests. The tests are easy to administer and score and do not require measurement professionals for the interpretation of the scores.

If school systems want to compare children's scores with the scores of children from other schools around the country (not necessary for many evaluation purposes), the TOBE 2 norms present some problems. They are old. The mixture of actual and interpolated distributions leaves something to be desired. The Head Start and school samples are not comparable. Although there are directions for administering the tests in Spanish, there are no specific guidelines for determining which children should be tested in Spanish and apparently no materials available for interpreting the scores from these administrations. The description of the equating of the levels should make users cautious about attempting to compare scale scores on Level K with scale scores on Level L (as in a pretest-posttest situation). In other words, the tests are better than the data about them.

References

CTB/McGraw-Hill. (1978). *Level L examiner's manual; Level K examiner's manual*. Monterey, CA: Author.

CTB/McGraw-Hill. (1979). *Norms and technical data book*. Monterey, CA: Author.

Diamond, E. E. (1978). Tests of Basic Experience. In O. K. Buros (Ed.), *The eighth mental measurements yearbook* (pp. 119-120). Highland Park, NJ: The Gryphon Press.

Proger, B. B. (1972). Tests of Basic Experiences: Review. *Journal of Special Education, 6*, 179-184.

Thurber, S. (1973). Tests of Basic Experiences: Review. *Journal of School Psychology, 11*, 280-281.

John J. Venn, Ph.D.

Associate Professor of Special Education, University of North Florida, Jacksonville, Florida.

VULPÉ ASSESSMENT BATTERY

Shirley German Vulpé. Downsview, Ontario: National Institute on Mental Retardation.

Introduction

The Vulpé Assessment Battery is a comprehensive developmental assessment instrument for handicapped children from birth through six years of age. It is criterion-referenced and administered individually. The Vulpé provides a detailed measure of performance in all the major developmental learning areas. Measures of performance in several areas typically not found on developmental scales also are included: organizational skills, the environment, basic senses and functions, developmental reflexes, muscle strength, motor planning and balance. Because there is a large number of items in each learning area, the Vulpé is effective with all types of handicapped children, including children with multiple and profound handicaps. The assessment information derived from the Vulpé is valuable for measuring a child's present levels of performance and progress over time. Individual program plans also can be developed using all or parts of the Vulpé. Educators, therapists, psychologists, and other professionals who serve handicapped children have found the Vulpé to be a practical tool for gathering assessment information that can be translated into meaningful programming.

The original version of the Vulpé was published in 1969 by the National Institute on Mental Retardation in Canada. The manual was titled *Home Care and Management of the Mentally Retarded Child* and the test was called the Assessment Battery. The initial purpose of the manual was to supplement a series of training seminars on care and management of mentally retarded children sponsored by the Canadian Association for the Mentally Retarded. Shirley German Vulpé wrote the original manual and assessment battery based on her extensive practical experience with children as an occupational therapist and her knowledge of the research on child development. The revised edition is over twice as long as the 1969 edition.

The second edition was developed in response to consistent demand over several years. It was published in 1977 and reprinted with minor revisions in 1978, 1979, and 1982. The current edition includes new sections on cognitive processing, reflexes, and environmental assessment. Other revisions entail extending the age range down to birth, expanding the language section, and providing more detail for specific skills at each age level. Reference citations supporting statements about child development and references for each item on the scale have been furnished as part of the new edition. The revisions were necessary due to the explosion of information on developmental assessment, changes in the assessment process, and the need for new ways to assess severely and profoundly handicapped children.

The complete Vulpé Assessment Battery can be purchased in one 397-page paperback volume, the *Vulpé Assessment Battery: Developmental Assessment, Performance Analysis, and Individualized Programming for the Atypical Child*. The book includes 1) an introduction to and thorough review of developmental assessment and programming, 2) the assessment battery, and 3) a developmental reflex test. Scoring pads for scoring individual children, the developmental reflex test, sections of the assessment, and spirit masters of specific pages are also available separately. A list of the materials and equipment needed to administer the battery is included in the Appendix. No equipment is provided with the test itself. Most of the needed materials and equipment are available in a typical classroom for young children.

The extent of the examiner's direct participation in the assessment process varies according to the needs of the child and the purpose of the evaluation. Although direct administration of items by the examiner is encouraged, it is not required. The purpose of the evaluation is to obtain information about the child, an objective that can be met by gathering some information indirectly from reports from the child's mother, teacher, or other primary caregiver as long as the source is reliable and is noted by the examiner.

The battery is composed of eight subtests and five supplemental tests. The eight subtests address 1) basic senses and functions, including vision, hearing, touch, proprioception/kinesthesia and other central nervous system operations; 2) gross-motor behaviors; 3) fine-motor behaviors; 4) receptive and expressive language behaviors; 5) cognitive processes and specific concepts; 6) the organization of behavior, including attention and goal orientation, behavior control, problem-solving and learning patterns, and dependence/independence; 7) activities of daily living; and 8) the environment, including the setting and the caregiver. The five supplemental tests include 1) an analysis of posture and mobility; 2) a developmental reflex test written by Janet Wilson; 3) functional tests of muscle strength; 4) functional tests of motor planning; and 5) functional tests of balance. The items on the tests are arranged chronologically according to the earliest age at which they appeared on reviewed assessment scales.

The answer form, the Performance Analysis/Developmental Assessment, is long because each item is named, described in detail, scored, and interpreted on the answer form itself. The test, including the supplements, contains over 1,300 items. However, only those items that fall within the child's current age level of functioning are administered during an evaluation; therefore, each child is tested using only a small portion of the 1,300 items on the test.

Two types of profiles are available: a graph of the child's functioning and a program profile. After all the items have been recorded, the graph is made by joining all of the scale score dots together to show strengths and weaknesses across and within developmental areas. The program profile, a form that is completed after the evaluation, outlines individual intervention goals, materials needed, the child's responses to the activity, and comments.

Practical Applications/Uses

The Vulpé Assessment Battery provides a system for gathering the information required to develop an in-depth, individualized learning program for a child. The "whole" child is evaluated in relation to environment and central nervous system function. The battery is designed to examine more than categorical disabilities or weaknesses in isolated skills.

The Vulpé is not a traditional test but a diagnostic tool that provides information for developing intervention goals and activities. In addition to its use as a developmental scale, it is helpful as a programming guide, program evaluation tool, and training manual. In addition, the battery has been an effective assessment and curriculum guide for members of interdisciplinary teams using various parts of the battery. The team members combine their results and use them to establish priorities for intervention.

The Vulpé is used extensively in schools, home-training programs, diagnostic clinics, therapy programs, and institutions. It also is used widely in professional training programs due to the detailed information and references on developmental theory and practices that are included in the manual. Professionals using the test include diagnosticians, special educators, home intervention specialists, occupational therapists, physical therapists, language therapists, and psychologists.

The subjects for whom this test is appropriate include handicapped infants and children, especially those with severe and/or multiple handicaps. Although the test originally was designed for mentally retarded children, it has widespread applicability to other handicapped children, including those who are physically or perceptually impaired. Items that measure verbal language or visual skills must be modified before the test can be used with the deaf or the blind. The flexibility of the test makes it easy to modify items, and items that cannot be modified can be eliminated. Because there are so many items, elimination of a few does not alter the accuracy of the assessment.

The most suitable setting for administering the test is a quiet room with a table and a chair of appropriate size for the child and a play area in which the child can perform the activities required for the gross-motor items. If a very young child is being tested, a mat for him or her to lie on should be available. Materials should be accessible but not in the way. Vital assessment information also can be gathered in natural, familiar settings in the child's own environment, such as a classroom, bedroom, living room, and so on. If the evaluation is conducted in the home, the setting must be prepared in advance. The parents should be instructed carefully and materials must be brought into the home. Unlike the procedures used with some standardized, norm-referenced tests, the participation of the primary caregivers (i.e., parents, teacher, etc.) is encouraged as an important part of the diagnostic process.

Examiners are not limited to one professional group such as psychologists. However, evaluators using the test should be familiar with developmental theory and practices and experienced in administering developmental tests. In addition, examiners should be thoroughly familiar with the Vulpé before adminstration and should practice administering and scoring the items before evaluating a child. The test is not appropriate for administration by paraprofessionals such as teacher aides, secretaries, or proctors.

Certain parts of the battery may be omitted depending on the professional background of the examiner. For example, the developmental reflex subtest should be administered only by a professional who has training and experience in reflex testing and related areas. Likewise, the Vulpé should not be given to a blind child, for example, by an evaluator who has no training or experience with blind children.

The steps involved in administration are 1) prepare the testing environment and equipment, 2) establish rapport with the child, 3) begin the battery at the child's approximate developmental age level in each learning area and administer the items by observation or report, 4) score each item and make comments about the child's performance and modifications when appropriate, 5) establish a basal level by administering only a few items at each age level as long as the child passes all the items, and 6) when the child misses an item, administer every item in that area until the child no longer successfully completes any items.

This procedure can be altered as necessary to meet the needs of the individual child. Adaptability is an intrinsic attribute of the evaluation process. The complete Vulpé does not need to be administered. For example, a language therapist may administer only the receptive and expressive language and the oral reflex sections. A behavioral specialist may use only the behavior organization section. A program consultant may use the assessment of the environment section, which evaluates the setting, the caregiving personnel, teaching and behavior management techniques, and related information. On the other hand, the complete test is utilized if the child requires a comprehensive diagnostic evaluation and a detailed prescriptive program.

Although the large number of items on the test is an advantage in terms of gathering as much information as possible about a child, it makes the test more difficult to administer compared to most developmental scales which have fewer items. One of the problems encountered by evaluators who are inexperienced in developmental assessment is deciding when to modify the administration procedures. Because the adminstration process is flexible, there are no hard and fast rules for deciding which parts of the test to administer or which items to score by report rather than through direct observation. These decisions usually are not difficult for the experienced evaluator.

The time required to administer the Vulpé is variable. It takes about an hour to administer the test to a young child. Administration time can be much shorter if only part of the test is given. On the other hand, the time can be much longer if all of the battery is administered and most of the items are scored by direct observation.

The method of scoring the Vulpé is very different from that used for most other tests. The goal in presenting the task in each item is to find out what skills the child has at whatever level they may be present. If the child does not perform an activity as described, the evaluator changes the method of task presentation by modifying the materials, level of assistance, language cues, or setting to see if the child can perform the task under specific circumstances. A structured process for changing the way each item is presented is provided with the Vulpé Performance Analysis System (VPAS), a 7-step scoring procedure. The scoring procedure uses the following seven criteria to rate the child's performance on each item presented. Criteria 3-7 indicate the child can perform the task at some level of competence.

1. *No.* The child does not perform the activity and either shows no interest in the task or is not attentive at all to the task.
2. *Attention.* The child shows either intermittent attention or sustained interest in the task but is still unable to perform it.

3. *Physical Assistance.* The task is performed when it is modified so that the child is given either physical support or other physical modifications.
4. *Social/Emotional Assistance.* The task is performed if changes are made in the manner of relating to the child (e.g., increasing reassurance) or a change is made in the personnel administering the task (e.g., the parent).
5. *Verbal Assistance.* The child can perform the task if given increased verbal cues or other verbal assistance.
6. *Independent in Familiar Situations.* The child consistently performs the task in familiar settings with familiar people and material.
7. *Transfer.* The child can perform the task consistently in more than one environment and using unfamiliar media or materials.

This scoring procedure is not complex in that it follows the developmental approach normally used in teaching new skills to children. A helpful description of how to use the 7-step scoring system with different types of items is provided. However, the use of a system with seven criteria for scoring each item is difficult to apply to certain types of behavior. Some language behaviors, for example, are either performed by the child or not. Thus a multistep scoring procedure does not apply easily to these items. Likewise, certain social interaction behaviors do not lend themselves to multiple-scoring criteria because a child either exhibits these behaviors or does not exhibit them. In such cases, the item can be scored using criteria 6 or criteria 2, criteria that correspond to a pass and a fail, respectively, on traditional pass-fail scoring systems.

Despite the nontraditional procedures used to score the battery, instructions are presented clearly. It takes approximately 1 hour to learn how to score the test, and once mastered takes less than 20 minutes for actual scoring.

Interpretation is based upon objective scores for some items and internal clinical judgment for others. Objective scoring is used with items such as those involving number concepts, for example, counting objects. Clinical judgment is used with items such as the primitive reflexes. The level of difficulty in using clinical judgment to interpret the results from items such as the primitive reflexes is high. Overall, advanced training and experience is required to interpret the results of some parts of the assessment.

Detailed information explaining how to interpret the Vulpé, including a well-written chapter on designing individual program plans and several complete samples of individual programs, is provided. The manual also describes how to transfer the assessment results into an individual program.

Technical Aspects

The Vulpé Assessment Battery is a criterion-referenced assessment instrument. The developmental age scores derived from the test are used to measure the progress of individual children by comparing their performance over time but are not used to compare children to each other. An individual child's subtest scores are compared to each other in order to yield a profile of relative strengths, weaknesses, gaps in development, and splinter skills across the learning areas.

No research studies on the validity of the Vulpé were reported in the manual, but two reports of pilot studies on reliability were given. One of the reliability studies was described very briefly. This study used the Vulpé to examine two methods of instructing parents in home-based physical therapy activities for children with developmental disabilities. The other study used the Vulpé Performance Analysis System (VPAS) to score the performance of handicapped children on various developmental tasks. Five raters proficient in the use of the VPAS simultaneously scored the performance of five handicapped children on various items from the battery. One rater interacted with the child while the others observed through one-way glass. The results indicated that there were no significant differences in the scoring of the raters across children and tasks. The small sample size, however, limits the extent to which these results can be generalized to all evaluators and to other children.

The structure of the test is similar to other published developmental scales. The test items were derived from a review of 188 references. These references included norm-referenced developmental tests such as the Bayley Scales of Infant Development (Bayley, 1969) and criterion-referenced developmental tests such as Assessment in Infancy: Ordinal Scales of Infant Psychological Development (Uzgiris & Hunt, 1972). A bibliography of 116 references on developmental reflexes is included in the appendix of the manual. The references, which are keyed to individual test items, provide evidence indicating that the Vulpé has excellent content validity.

Critique

The Vulpé Assessment Battery fulfills the purpose for which it was designed. It is a comprehensive assessment and programming guide for young children with special learning needs. The content includes the traditional developmental learning areas and subtests that measure neurological development; therefore, it is also an excellent tool for professionals who use the neurodevelopmental approach. In addition, the Vulpé is valuable in programs using an interdisciplinary team approach.

The Vulpé is not as useful when test scores that provide general information about levels of functioning are needed. The length of the battery and the complex 7-step measuring system make it difficult to obtain global scores. Likewise, the Vulpé is not at all suited for use as a screening test that takes a brief time to administer and score.

The Vulpé contains aspects that make it particularly effective for pre-service and in-service training. It includes a sequential approach to teaching about the developmental assessment process. Specific information is provided about how to analyze the interaction between skill areas and how to determine the way in which a child is processing information. Methods of assessing perceptual and neurological functioning also are emphasized.

Overall, the Vulpé is an excellent assessment instrument that contains many unique features. Best of all, evaluations using the Vulpé provide diagnostic information about the activities a child can perform and a child's emerging skills rather than simply testing to determine a child's developmental delays.

References

Bayley, N. (1969). *The Bayley Scales of Infant Development.* New York: The Psychological Corporation.

Uzgiris, I. C., & Hunt, J. (1972). *Assessment in Infancy: Ordinal Scales of Psychological Development.* Champaign, IL: Illinois Psychological Development Laboratory, University of Illinois.

Vulpé, S. G. (1982). *Vulpé Assessment Battery.* Toronto: National Institute on Mental Retardation.

Jan Faust, Ph.D.

Staff Psychologist, Children's Hospital at Stanford University, Palo Alto, California.

C. Eugene Walker, Ph.D.

Professor of Pediatric Psychology, Department of Psychiatry and the Behavioral Sciences, University of Oklahoma Health Sciences Center, Oklahoma City, Oklahoma.

WAHLER PHYSICAL SYMPTOMS INVENTORY

H. J. Wahler. Los Angeles, California: Western Psychological Services.

Introduction

The Wahler Physical Symptoms Inventory (WPSI) is an instrument designed to measure the degree of physical or somatic complaints endorsed by an individual. Wahler (1968) designed the inventory to specifically include those complaints considered to be exclusively somatic in composition, eliminating items of a psychological nature. H. J. Wahler developed the questionnaire in the late 1960s in an attempt to aid professionals in their delineation or differential diagnosis of physical and/or psychological problems. The original report of Wahler's work was published in 1968 as the Physical Symptoms Inventory (PSI). This publication represented a culmination of several years of data collection while the author was Director of Outpatient Psychological Services for University Hospitals at Ohio State University.

The WPSI is composed of 42 items that were written following careful analysis of the medical symptoms sections (A-K) of the Cornell Medical Index (CMI; Brodman, Erdmann, Lorge, & Wolff, 1949) and the Minnesota Multiphasic Personality Inventory Hypochondriasis Scale (MMPI/Hs; Dahlstrom, Welsh, & Dahlstrom, 1975; Hathaway & McKinley, 1940). In order to generate items for inclusion in the WPSI, 138 CMI items and all of the MMPI Hypochondriasis (Hs) scale items were first categorized by two judges with respect to the physical system/sensations to which these items belonged. Next, a phrase was written that denoted the general character of the physical complaint in each category. Following this, two psychiatrists, four psychologists, and two social workers judged whether the items tended to be common complaints of individuals and whether or not they were diagnostically important. Those physical complaints denoted by at least 75% of the raters as rare in occurrence and/or as clinically insignificant were eliminated. Finally, the judges "listed additional somatic complaints they considered diagnostically important for psychiatric patients." The criterion of 50% agreement among raters was utilized in determining whether these complaints were considered further. As a result of the preceding item selection procedures, a total of 42 "general" physical complaints were obtained.

It should be noted that the CMI was originally designed to provide cursory information for the physician concerning the overall medical problem, including physical as well as psychological components. Wahler (1968), however, stated that he developed his physical symptom checklist to measure the degree or intensity of somatic complaining as well as the specificity or diffuseness of the complaints. Hence, he selected items that pertained only to distress in somatic systems as opposed to abnormalities in psychological systems, and he constructed the instrument in such a manner that the patient or subject is able to endorse the frequency of each symptom's occurrence, thus obtaining individual differences with respect to what Wahler believes is the "intensity" of the symptoms reported.

In order to determine whether a response bias could occur as a result of the sequencing of the physical complaints, items were ordered in three different ways. In addition, three different forms of scaling were utilized to detect possible differences if the scales were formatted by frequency of symptoms, severity of symptoms, or duration of symptoms. The items were presented in six different versions to male and female college students and veterans applying for psychotherapy at a Veterans Administration (VA) hospital. All three rating formats (frequency, intensity, and duration) were correlated at .85 or above and no significant differences were detected among the mean scores and variances obtained from the three different-ordered presentations of the items. Because the correlations were very high, Wahler arbitrarily selected one of the item orders and used the frequency scaling format for the final version because he felt that subjects would find it easier to respond to this format. Wahler decided to use "anchoring phrases" to determine frequency of the scale value beginning with 0 (almost never), 1 (about once a year), 2 (about once a month), 3 (about once a week), 4 (about twice a week), and 5 (nearly every day).

Thus, the WPSI test form includes a list of the 42 symptoms in columns arranged so that for each symptom, the subject can circle one of the numbers mentioned above (0-5) to indicate the frequency of occurrence for himself or herself. Further, there is a space at the bottom of the WPSI form for the patient to indicate a physical problem not included on the WPSI. Because this is a self-report inventory, the examiner's participation in the testing process is minimal. The examinee is instructed to read the directions and complete the inventory. Wahler (1973) has provided a simple phrase for examiners to repeat should the examinee need encouragement or reassurance. The phrase is as follows: "You are being asked to complete a questionnaire to show how often you are bothered by some common physical troubles" (Wahler, 1973). The WPSI was designed for those individuals with a sixth- to eighth-grade reading ability, and Wahler contends that people with Intelligence Quotients as low as 80 may be able to complete the inventory without assistance.

Practical Applications/Uses

The intent of the WPSI is to provide the examiner with information about the frequency of occurrence of a patient's or subject's somatic complaining behavior. Psychological sequelae were purposefully excluded in order to obtain a purer measure of somatic symptomatology. However, Wahler (1973) argues for the WPSI's usefulness when it is used in conjunction with the Wahler Self-Description Inventory (WSDI), an inventory designed to measure the degree to which respondents differentially emphasize favorable or unfavorable characteristics of their self-evaluations.

Although the WPSI has been primarily promoted as a screening instrument, researchers have incorporated the WPSI as an outcome measure in scientific endeavors. While some researchers have used the WPSI in drug studies (cf. Bianchi, Fennessy, Phillips, & Everitt, 1974; Bianchi & Phillips, 1972; Norman, Burrows, Bianchi, Maguire, & Wurm, 1980), others have relied upon the WPSI in psychological studies (cf. Aubuchon & Calhoun, 1985; Hampton & Tarnasky, 1974). Further, Weissmann (1984) cited the WPSI as an instrument potentially useful for assessment in legal cases in which emotional or physical pain and suffering are arguable issues.

The WPSI was designed to be administered in mental health and medical settings. Physicians and "mental health professionals and workers in the different areas of guidance and counseling" have been identified as appropriate WPSI administrators (Wahler, 1973). As mentioned previously, it appears that researchers have broadened this range of administration by incorporating the WPSI in scientific research. Presumably subjects who have at least a sixth-grade reading ability and with IQs of at least 80 are able to successfully complete the WPSI. However, because normative data have not been obtained on individuals younger than 16 years of age, results of questionnaires completed by teenagers and young adolescents may be used as an aid to interviewing but not specific interpretation. Further, the examiner may wish to read the WPSI to those individuals unable to see or read but who can grasp the weighted scale concepts. Unfortunately this raises the issue of examinee response bias. In fact, Wahler (1973) has stated that the WPSI is subject to response bias when it is sent home with the young examinee (subject to coaching by parents). However, with non-court-referred adults, no differences were found between means and variances of scores and item preferences of those inventories completed at home or in a clinic/office over a 5-year period (Wahler, 1973).

Wahler (1973) offers an alternative use of the WPSI involving retrospective completion of the inventory based upon the best and worst periods of a patient's life. The examinee is requested to think about a period in his or her life during which he or she has either felt the most "secure and free of distress" or the most "insecure and distressed." The patient completes the WPSI based on this best and/or worst period.

Scoring of the WPSI is relatively easy. The scorer is to eliminate those items that were omitted or rated twice and to designate a questionnaire as invalid if more than eight items are omitted or rated twice. If the completed questionnaire is deemed valid, the examiner subtracts the number of omits and double ratings from the total number of items; next, all of the remaining ratings are summed and then divided by the number of items on which they are based (total number minus omitted and double-marked items). This final number constitutes the WPSI score. In order to make these scores more meaningful, raw scores can be converted into deciles, and a decile table is provided in the WPSI manual. The deciles and their corresponding raw scores are derived from normative data based on the psychiatric patient's responses to the WPSI. Women tended to score higher and have a wider range than men; consequently, the normative decile will often be different for different-sex responders who have the same raw scores. There is also a table in the manual that reveals distributions of the WPSI scores (in intervals) for junior and senior high school students, although standard deciles are not provided.

Technical Aspects

Wahler Physical Symptoms Inventory reliability information was obtained from several different sources. First, internal consistency was derived from two student populations, a group of male patients with physical disabilities and a group of people applying for Social Security compensation. Except for the physically disabled male sample, each group was subdivided into female and male groups. The Kuder-Richardson Formula 20 was utilized to determine the internal stability of the WPSI based on men's and women's performances. Using the different subject samples previously noted, the WPSI internal consistency proved to be excellent, ranging from .85 to .94.

Wahler also measured the reliability of the WPSI based on test-retest administrations. The rationale for such an administration is unclear, given the purpose for which the WPSI was constructed. Because the instrument was constructed to measure the presence of somatic complaints at the time of testing, it is conceivable and highly likely that complaints would change over time, given the remission or acquisition of physical problems. It is not surprising, then, that reliability coefficients varied considerably with the passage of time (ranging from .45 to .94). Furthermore, it is difficult to compare test-retest reliability differences across groups because each group had a different time interval between testing. In addition, Wahler combined both males and females in his computation of test-retest reliability scores for one of his student populations and for the psychiatric population, but segregated by gender the other student population.

With respect to the validation of the WPSI, Wahler utilized an unusual approach to concurrent validation of the instrument. He based such validation on scores obtained from groups who would be "expected" to report and emphasize somatic symptomatology. Not surprisingly, those individuals with physical disabilities, psychiatric problems, and those who had applied for Social Security disability demonstrated greater complaining behavior when compared to the student samples. Although the use of this contrasting-groups design is appropriate, it is questionable how well it was conducted with respect to the WPSI because a true medical (e.g., ill) population was not utilized. Concurrent validation of the WPSI would have been better evaluated utilizing a physically ill sample as well as a more psychometrically sound instrument or objective measure (e.g., medical records) than by the inferential conclusions based on the characteristics of the samples Wahler utilized.

Wahler did obtain correlations between the WPSI and all subscales of the MMPI, most of which were significant at the < .05 level. However, it is not surprising that the strongest relationship was between the MMPI Hs and Hy subscales and the WPSI, given the fact that the WPSI is comprised partly of items derived from study of the content of the MMPI Hypochondriasis subscale. Once again, the WPSI would have been better validated had different instruments or objective measures been utilized.

Finally, Wahler reports in his WPSI manual that no relationship between age and

WPSI scores was discerned. He did, however, find significant differences with respect to an inverse relationship between WAIS Verbal IQs and WPSI scores. He explained this relationship as suggesting that people with poor verbal skills, and hence lower educational level, tend to endorse physical complaints to a greater degree. This may very well be the case; however, because the examiner has not been given any indication of numerical values for actual IQ levels, it is difficult to make conclusions about the inflated WPSI scores based on lowered intelligence.

Finally, Wahler obtained significant correlations between his Self-Description Inventory, a measure designed to determine self-concept in individuals, and the WPSI. The rationale for such correlations is unclear; as a result, it was also unclear whether Wahler was attempting to obtain convergent or divergent validation data.

Critique

Some of the positive attributes of the WPSI can be discerned by its test construction. These include the actual design of the inventory, which includes not only the number and type of physical problems the patient believes he or she experiences but also the frequency of problem occurrence. Furthermore, the utilization of anchoring phrases enables the patient to relay unambiguously perceived somatic difficulties. Additionally, item selection was obtained in a fairly comprehensive and conservative fashion, thereby reducing the inclusion of irrelevant items. This is further reflected by the high internal consistency exhibited by the instrument. Additionally, Wahler addressed and attempted to eliminate the problem of response bias based on presentation of the order of symptoms, although no rationale was given for the selection of the particular order of symptoms finally used.

The WPSI is very easy to administer and may be completed by a wide range of individuals with differing intelligence levels and educational backgrounds. Moreover, the instrument may be utilized in a number of settings for a number of purposes. The WPSI also is easy to score, and some comparative norms are provided.

Problems with the WPSI include those issues inherent in a self-report instrument, such as response biases (e.g., social desirability). Also, although Wahler refers to the WPSI as a measure of intensity of somatic complaining behavior, the scale is formatted only to account for the frequency of physical symptoms. Furthermore, normative data are scant with respect to age groups below 16 years.

A major weakness of the WPSI is the lack of comparison groups for psychologically normal adults (other than college students) as well as for individuals who have documented physical illnesses. The two physical illness groups used were a rehabilitation group and a Social Security disability claim group. Similarly, the predictive power of the WPSI is unclear, especially given the lack of data exploring this facet of the test. Consequently, it is unreasonable to suggest, as Wahler (1973) has, that the WPSI can identify conversion disorders and other somatic-psychological problems. This is particularly questionable given that the WPSI was designed to identify somatic difficulties and not to predict or delineate psychological ones. Wahler's (1973) allusion to the inherent predictive power of the WPSI is particularly confusing because he stresses throughout the manual the need to account for "adjunctive information" when making specific diagnoses, but the type of data needed and methods by which to collect this adjunctive information were neither clarified nor explained.

Updated normative data are needed because the last collection of normative data occurred approximately 20 years ago. In light of Wahler's endorsement of the utilization of the WPSI as a retrospective physical state questionnaire (i.e., completion of the WPSI based on a previous "best" and "worst" period in the patient's life), reliability data need to be gathered on the durability of the WPSI when making retrospective comparisons. In addition, more refined, updated, and comprehensive validity studies are required. The validity and purpose of utilizing the WPSI with Wahler's other scale, the WSDI, as opposed to other instruments and measures, are unclear. Although it would appear the WPSI has the potential to be a valid scale, due to the lack of evidence and data one cannot assume that the WPSI is, in fact, valid. Future studies need to address these issues.

The WPSI appears to be an internally consistent self-report instrument. It has the merit of being an excellent quick-screening device, but it must not be used to the exclusion of other data the practitioner or scientist has available, nor should it be weighted too heavily when considering all other patient data.

References

Aubuchon, P. G., & Calhoun, K. S. (1985). Menstrual cycle symptomatology: The role of social expectancy and experimental demand characteristics. *Psychosomatic Medicine,* 47(1), 35-45.

Bianchi, G. N., Fennessy, M. R., Phillips, J., & Everitt, B. S. (1974). Plasma level of diazepam as a therapeutic predictor in anxiety states. *Psychopharmacologia, 35,* 113-122.

Bianchi, G. N., & Phillips, J. (1972). A comparative trial of doxepin and diazepam in anxiety states. *Psychopharmacologia, 33,* 86-95.

Brodman, K., Erdmann, A., Jr., Lorge, I., & Wolff, H. G. (1949). The Cornell Medical Index, an adjunct to medical interview. *Journal of the American Medical Association, 140,* 530-534.

Dahlstrom, W. G., Welsh, G. S., & Dahlstrom, L. E. (1975). *An MMPI handbook: Vol. II. Research applications* (rev. ed.). Minneapolis: University of Minnesota Press.

Hampton, P. T., & Tarnasky, W. G. (1974). Hysterectomy and tubal ligation: A comparison of the psychological aftermath. *American Journal of Obstetrics and Gynecology, 119*(7), 949-952.

Hathaway, S. R., & McKinley, J. C. (1940). A multiphasic personality schedule (Minnesota): Construction of the schedule. *Journal of Psychology, 10,* 249-254.

Norman, T. R., Burrows, G. D., Bianchi, G. N., Maguire, K. P., & Wurm, J. (1980). Doxepin plasma levels and anxiolytic response. *International Pharmacopsychiatry, 15,* 247-252.

Wahler, H. J. (1973). *Wahler Physical Symptoms Inventory manual.* Los Angeles: Western Psychological Services.

Wahler, H. J. (1968). The Physical Symptoms Inventory: Measuring levels of somatic complaining behavior. *Journal of Clinical Psychology, 24,* 207-211.

Weissman, H. N. (1984). Psychological assessment and psycho-legal formulations in psychiatric traumatology. *Psychiatric Annals, 14*(7), 517-529.

John Beattie, Ph.D.

Assistant Professor, Department of Curriculum and Instruction, The University of North Carolina at Charlotte, Charlotte, North Carolina.

WELLER-STRAWSER SCALES OF ADAPTIVE BEHAVIOR: FOR THE LEARNING DISABLED

Carol Weller and Sherri Strawser. Novato, California: Academic Therapy Publications.

Introduction

The Weller-Strawser Scales of Adaptive Behavior: For the Learning Disabled (WSSAB) are a pair of separate scales used to assess the adaptive behavior of two age groups of learning disabled students in four areas: 1) social coping, 2) relationships, 3) pragmatic language, and 4) production. The Social Coping subtest assesses the manner in which a student deals with environmental situations. The Relationships subtest is concerned with how the student relates to other individuals. The Pragmatic Language subtest does not evaluate language itself but instead evaluates how language is used in social or communicative situations. Finally, the Production subtest, rather than concentrating specifically on *what* is produced, assesses *how* the student produces information. The scales yield student profiles of either mild to moderate or moderate to severe adaptive behavior problems in each of these four areas.

The WSSAB are designed for use specifically with LD students, and the authors bring excellent credentials and experience to the task of developing a test evaluating the adaptive behavior needs of this population. Carol Weller is currently an associate professor in the Department of Special Education at the University of Utah. She has recently served as president of the Division for Learning Disabilities within the Council of Exceptional Children. She has gained vast experience with LD students as a classroom teacher and has also served as director of the Northeast Indiana Instructional Resource Center for Handicapped Children and Youth.

Sherri Strawser is currently a school psychologist in the Jordan County (Utah) School District. She has served as an adjunct instructor at several universities as well as a materials consultant for the Northeast Indiana Instructional Resource Center for Handicapped Children and Youth. She too has extensive experience with LD students and has held the position of president of the Indiana Division for Children with Learning Disabilities.

Within the WSSAB the four areas of social coping, relationships, pragmatic language, and production are evaluated for 6- to 12-year-olds with the Elementary Scale and 13- to 18-year-olds with the Secondary Scale. By identifying how an LD student functions in a certain environment, the Social Coping subtest should help the teacher in preparing a more suitable environmental structure for that specific student. On the Elementary Scale, the Social Coping subtest uses eight items to assess the student's use of time and space, the effect of change on the student's behavior, how he or she handles worry or depression, and how the student deals with criticism and problem-solving situations. The Secondary Scale uses nine items to evaluate the student's understanding of cause/effect relationships, how he or she follows directions, the ability to orient oneself in space, the acceptance of change and disorganization, and the student's habit of dress.

The manner in which the LD student interacts with others is evaluated in the Relationships subtest. The examiner is provided with an understanding of the problems of social interaction experienced by the LD student as well as a plan of instruction relative to this social interaction. The Elementary Scale has eight items that evaluate a student's relationships with peers and authority figures, the manner in which a student seeks entertainment, the student's expectations of the behavior of others, and the implications of possessiveness in a relationship. The Secondary Scale considers the types of relationships that the student has formed, but is also concerned with the expectation and acceptance of the behaviors of the other members of those relationships. This scale uses several items to measure the use of profanity, the degree of involvement in social activities, and the social awareness of the LD student.

The Pragmatic Language subtest evaluates the problems that a learning disabled student might have in using language without consideration for the specific language problems. In other words, the subtest measures how well the student uses language in a social setting. The Elementary Scale has nine items evaluating the use of verbal language, the use of nonverbal language, the need for verbalization, the use of gestures and humor, and the ability to use language to either cover up or make obvious the learning deficits that may exist. The Secondary Scale has nine items that assess the degree to which a student understands voice inflection and jokes, the impact of gestures, nonverbal communication or body language as related to the meaning of language, the importance of others' feelings, and the use of slang.

The final subtest, Production, evaluates *how* the student produces, rather than *what* he or she produces. In so doing, the Elementary Scale has 10 items that measure how tasks are performed, how past learning is transferred to new situations, how the student uses work time, and how he or she responds to various activities. The Secondary Scale uses 10 items to evaluate how group size affects learning, how a student shifts tasks, the organization of work space, the need for stimulation to continue working, and the desire to participate in tasks and task strategies.

Practical Applications/Uses

The Weller-Strawser Scales of Adaptive Behavior: For the Learning Disabled were developed to consider two distinct populations of LD students. The Elementary Scale is for use with students aged 6 through 12 who have been previously diagnosed as learning disabled, and the Secondary Scale is designed for similarly diagnosed students aged 13 to 18. In an attempt to assess the adaptive behavior needs of these populations adequately, the authors identify seven purposes for the WSSAB: 1) to discriminate between mild to moderate and moderate to severe adaptive behavior problems of LD students; 2) to measure the degree to which students who were previously diagnosed as learning disabled function in the adaptive areas of social coping, relationships, pragmatic language, and production; 3) to provide a profile of adaptive behavior skills in an attempt to focus on specific adaptive areas; 4) to provide information to a wide variety of personnel, including the classroom teacher, the educational diagnostician who may interact with the student, and the student him- or herself; 5) to identify adaptive behaviors most commonly found among LD students; 6) to enhance the placement options of LD students; and 7) to identify specific behaviors of LD students that can be modified through specialized programming and environmental modifications.

The WSSAB are completed after observing the student in the classroom setting. The authors emphasize that as many observations as necessary should be conducted to provide a complete and thorough knowledge of the student's adaptive behaviors. Because the WSSAB have been standardized on learning disabled students, the authors stress that the test is most appropriately used only with them.

The administration of the WSSAB is quite simple. After observing the student in the classroom setting for a sufficient number of times, the examiner completes the rating form. The authors state that the examiner may contact the student's parents to clarify any questions. The procedures for using a forced-choice format are clearly described in the manual. The user must select one of two descriptions that best describes the adaptive behavior of the student. Several examples of completed forms are provided. The length of time needed to complete the administration will vary depending on the number of observations required. In the experience of this reviewer, the amount of time needed to complete the rating form itself usually ranges from 10 to 20 minutes.

The WSSAB manual provides step-by-step instructions for the scoring of the WSSAB. The scores are ranked numerically, and a profile score is determined. Initially, students are determined to have mild to moderate or moderate to severe problems in adaptive behavior. Upon completion of this broad ranking, the examiner is referred to a section in the manual that provides further interpretation of the scores.

The authors provide 16 potential profiles that may result from the administration of the WSSAB. These profiles are interpreted with specific attention given to both the Elementary and Secondary levels. After the results of the WSSAB have been obtained, the manual suggests programming recommendations for each of the 16 profiles (e.g., class placement, areas of instruction, and suggestions for seating). Once this has been completed, the examiner is referred to another section of the manual where several remedial activities are provided for each of the profiles. The authors offer specific instructional ideas that likely will be of great value to the practitioner.

Technical Aspects

In developing a pool of items for the Weller-Strawser Scales of Adaptive Behavior: For the Learning Disabled, the authors interviewed special education teachers; however, the number of teachers is not specified in the manual. Upon completing the interviews and establishing an item pool, the authors then determined that the items generally fell into one of the four categories of Social Coping, Relationships,

Pragmatic Language, or Production; items were then assigned to one of the categories. Experts in the field of learning disabilities then read each of the selected items to determine its appropriateness.

Teachers of LD students were then asked to complete checklists on as many students as they could. This resulted in ratings for 236 students between the ages of 6 and 18. Of these 236 students, 154 were of elementary age (i.e., 6 through 12) and 82 were of secondary age (i.e., 13 to 18). Items were then rated according to their measure of severity of adaptive behavior deficits, resulting in 35 items each for the Elementary and the Secondary Scales.

Internal consistency and interrater estimates of reliability are provided for the WSSAB from the ratings of the same 236 students. The internal consistency was determined by the split-half method, and reliability checks were made for both the Elementary Scale and the Secondary Scale. Scores ranged from .94 to .96 on the Elementary Scale, and from .95 to .96 on the Secondary Scale. The total reliability rating was .99 for both the Elementary and Secondary Scales. Interrater reliability estimates, indicating agreement in the rankings of two examiners, were based on the ratings of 63 elementary students and 19 secondary students. Correlations of .88 for the Elementary Scale and .89 for the Secondary Scale were determined.

Several measures of validity were utilized. Content validity was assured by selecting test items representing a wide range of adaptive behaviors of LD students. Correlations among subtests were used as the initial measure of construct validity. Correlations of each subtest score to the total score ranged from .87 to .97; for subtest scores and total scores to severity ratings, the range was .93 to .99. Diagnostic validity, established by demonstrating that the scales differentiate groups of students with specific characteristics, was determined to be appropriate using Z-scores. All of these analyses were completed on the 236 students involved in the development of the scales.

Critique

The Weller-Strawser Scales of Adaptive Behavior: For the Learning Disabled are designed to assist teachers in identifying adaptive behavior skill deficits in students diagnosed as learning disabled. Additionally, the WSSAB provide guidelines and instructional activities directed at the adaptive behavior weaknesses identified. In so doing, this instrument is a valuable tool for special educators. Learning disabled students often enter the out-of-school environment without the appropriate social and social language skills necessary for adequate functioning. This often results in failure to establish or maintain friendships, or failure to secure or keep a job. It is clear that the information derived from the WSSAB would be extremely useful for the classroom teacher, particularly because the information is provided only after the examiner is familiar with the examinee.

The WSSAB are easy to administer and can be completed in a short period of time. The manual is basically very clear and helpful. The major concern involves the small number of subjects used to develop the scales. Two hundred thirty-six subjects over such a broad age range is a very small sample. However, the instrument was not developed as a placement device. The authors repeatedly emphasize that the WSSAB are to be used only after students have been diagnosed as learning disabled and that the scales are useful only for this specific population.

This reviewer has used the WSSAB and found them to be helpful in pinpointing specific adaptive behavior skill deficits. It should be emphasized, however, that the scales are not designed to stand alone. Their use, along with observational, norm-referenced, and informal assessment data, will provide a strong basis upon which to develop an individual program for learning disabled students.

References

Weller, C., & Strawser, S. (1981). *Weller-Strawser Scales of Adaptive Behavior for the Learning Disabled*. Novato, CA: Academic Therapy Publications.

Lee N. June, Ph.D.
Professor and Director, Counseling Center, Michigan State University, East Lansing, Michigan.

WORK ATTITUDES QUESTIONNAIRE

Maxene S. Doty and Nancy E. Betz. Columbus, Ohio: Marathon Consulting and Press.

Introduction

The purposes of the Work Attitudes Questionnaire (WAQ) are to assess whether an individual's degree of commitment to work is high or low and to distinguish between various types of committed workers (addicted, dedicated, confused, or uninvolved).

The WAQ was developed by Maxene S. Doty and Nancy E. Betz. It is the product of Doty's master's thesis and doctoral dissertation. These studies focused on levels of commitment to work and their implications. Dr. Doty is currently a staff psychologist in the counseling center at the University of Maine, Orno. Dr. Betz, a faculty member in the department of psychology at Ohio State University, has written fairly extensively in the area of vocational psychology. She assisted in the WAQ's overall construction and chaired Dr. Doty's dissertation committee.

The WAQ was constructed utilizing the rational-empirical approach to instrument development. That is, items that seemed logically or intuitively related to the constructs of worker commitment and psychological health were originally assembled, reduced, and further refined after review by judges and eventually administered to a normative group of 93 male managers. As a result of item-scale correlational analyses, the final version of the WAQ contains 45 items as compared to the initial pool of 80 items. As of this review's publication, no revisions in the instrument have been made and no other forms have been developed.

The WAQ is a fairly simple and straightforward instrument to complete. Its 45 items are contained in a four-page booklet. The instrument consists of two subscales—Work Commitment (23 items) and Psychological Health (22 items). The respondents are instructed to indicate their answers by circling in the booklet the degree (strongly disagree, disagree, uncertain, agree, and strongly agree) to which they believe each of the statements describes their behavior and feelings. No answer or profile sheets accompany the instrument.

No specific age group is mentioned in the manual, but based on the items and the constructs the questionnaire purports to measure, it seems primarily appropriate for persons 18 years and older who are in the work force. Though no reading level is reported, the test appears to be written at an eighth-grade level.

Practical Applications/Uses

The WAQ is offered only as a research instrument. As such, it measures two levels of work commitment and distinguishes between various types of commitment to work. Because it lacks the proper psychometric studies, it is *not* presented

640

as an instrument for general clinical use, thereby limiting its overall application. However, given the types of areas it seeks to assess it theoretically has wide *potential* applications. If certain validity and reliability issues were addressed more adequately, the WAQ could be applicable for clinical assessment in the areas of worker adjustment and occupational satisfaction and dissatisfaction. The attempt to measure concepts such as workaholics and Type A behavior is a very attractive feature of the instrument. Hence, a wide range of mental health and career development specialists could find it attractive.

The instrument is potentially appropriate for experienced workers. It also could be used to assess the work attitudes of persons about to enter the world of work. Such use could allow counselors to assess and discuss with potential workers their attitudes and feelings toward work as well as areas that might develop into problems based on their attitudes and feelings about work. Though not mentioned in the manual, the instrument could be administered with minor adaptations, to blind persons and other handicapped individuals, given that no tasks of manipulation or manual dexterity are required.

The WAQ may be administered individually or in group settings. Given the simplicity, brevity, and conciseness of the instructions, it may be administered by almost anyone who is familiar with the general principles of test administration. No time limit or range is mentioned in the manual; however, considering the length of the instrument and the type of instructions, examinees should be able to complete it within 15-20 minutes.

The instructions for scoring are fairly straightforward. However, six items are reverse-scored, detracting from the ease of scoring. It would have simplified matters if those items originally had been worded differently or, at minimum, if those items had been asterisked in the appendices of the manual as a reminder of the reverse-scoring procedure. Approximately 10-15 minutes is required to derive the three possible scores: Work Commitment, Psychological Health, and Total. The total score is derived by adding the weights (values) of each of the 45 items. Adding the weights of the 23 items comprising the Work Commitment subscale results in a Work Commitment score. Likewise, the Psychological Health score is derived by adding the 22 items of which the Psychological Health subscale is comprised. No item is scored on both subscales. Neither machine nor computer scoring is available through the publisher. However, given the small number of items in the instrument and the relative ease of scoring, such methods could be adopted easily by the user.

Interpretation of the scale scores is based on objective data (norms). However, given the paucity of normative data, the interpretation of high and low scores for each of the subscales basically is left to clinical judgment. The overall level of difficulty in interpreting the results is not high, assuming a moderate amount of training in vocational psychology and familiarity with concepts such as Type A behavior and workaholics. Hence, a person with a master's degree in counseling or psychology should be able to interpret the instrument adequately and properly.

Technical Aspects

Only internal consistency reliability is reported in the manual. The internal consistency reliability coefficients for the Total Score, Commitment, and Health scales are .90, .80, and .85, respectively. These are within the respectable range. No test-retest reliability data are reported. Evidence for two types of validity is presented, concurrent and construct. Concurrent validity data are based on statistically significant positive correlations between the WAQ scales, the Career Salience scale (Greenhaus, 1971), and subjects' reports of the number of hours worked per week. The evidence for construct validity is based on examining the relationship between low Commitment scores and high Health scores on the WAQ. The relationships were observed to be in the hypothesized direction; that is, none of the managers studied obtained low Commitment scores in combination with high (defined as unhealthy) Health scores. Normative data are reported for 93 male managers from a pharmaceutical firm. Cutoff scores for judging levels of commitment to work and psychological health are suggested based on this sample.

The data on reliability, validity, and norms leave much to be desired. Although technical information that is presented is supportive of the instrument, the data are too meager.

Critique

The WAQ, while conceptually attractive, fails from a psychometric standpoint. It seems to have been published prematurely, without adequate preliminary supporting data. M. S. Doty (personal communication, June 30, 1987) indicated that no revisions or additional work on the instrument currently are planned. Because the instrument has a 1981 publication date and only one additional validation study has been conducted by the authors (Doty, 1984), and no other studies were found by this reviewer through a literature search, the instrument's viability is questionable. In her review of the instrument, Tenopyr (1985) suggested "that those who plan to use the inventory for further research do so with extreme caution."

In sum, the instrument contains four major flaws. First, in general, reliability and validity data supporting the instrument are sparse. Second, norms are reported in the manual for males only. In this reviewer's opinion, this is inexcusable given the variety of issues that have been raised in the literature over the last 2 decades concerning diverse populations and the specific issues often surrounding them. Third, the norms apply only to managers. Because managers are only a small domain of workers, the possibility of comparisons is limited severely. Fourth, all validation studies reported in the manual appear to have been conducted with a single sample of 93 male managers. Even in the early stages of an instrument's development, more extensive validation is necessary before placing an instrument on the market. Though female norms are reported in the dissertation (Doty, 1984), such data are unpublished and not easily accessible to possible users of the instrument; they should have been included in the manual.

In the publication *Standards for Educational and Psychological Testing* (American Educational Research Association, American Psychological Association, & National Council on Measurement in Education, 1985), it is stated that "validity is the most important consideration in test evaluation" (p. 9). On this criterion the instrument is extremely deficient. Additional validation studies clearly are needed. At minimum, the following four issues should be addressed. First, a more diverse norming sample that includes at least females but also racially diverse groups should be developed. Second, reversing the labeling of the Psychological Health subscale should be considered. At present, high scores on the scale are to be interpreted as indicating unhealthy attitudes, which is conceptually confusing. Third, a profile sheet must be developed. This would facilitate the interpretive process. Finally, the reverse-scoring procedure for the items scored in that manner should be eliminated. At the very least, the procedure should be referred to in the appendices as a reminder to those scoring the instrument.

On the positive side, the WAQ conceptually is a contribution to the literature and field of psychology. Nevertheless, its flaws must be corrected in order for it to reach its potential and to meet the minimum standards for test construction. Unless additional validation work is done on the instrument soon and its manual updated, the WAQ should be removed from the market even as a research instrument.

References

American Educational Research Association, American Psychological Association, & National Council on Measurement in Education. (1985). *Standards for educational and psychological testing.* (1985). Washington, DC: American Psychological Association.

Doty, M.S. (1984). High levels of commitment to work and dimensions of achievement motivation among women and men in management. *Dissertation Abstracts International, 45,* 1945b.

Doty, M.S., & Betz, N.E. (1981). *Manual for the Work Attitudes Questionnaire.* Columbus, OH: Marathon Consulting and Press.

Greenhaus, J. H. (1971). An investigation of the role of career salience in vocational behavior. *Journal of Vocational Behavior, 1,* 209-216.

Tenopyr, M.L. (1985). Review of Work Attitudes Questionnaire. In J. V. Mitchell, Jr. (Ed.), *The ninth mental measurements yearbook.* Lincoln, NE: The Buros Institute of Mental Measurements.

Don C. Locke, Ed.D.

Professor and Head, Department of Counselor Education, North Carolina State University at Raleigh, Raleigh, North Carolina.

THE WORLD OF WORK INVENTORY

Robert E. Ripley, original version; Karen S. Hudson and John W. Hudson, revisions. Scottsdale, Arizona: World of Work, Inc.

Introduction

The World of Work Inventory (WOWI) was developed as a multiple-purpose instrument to be used in career decision-making, career exploration, job placement, and training-program selection. The author began work on the inventory in 1961 when he recognized the need for an instrument that would be relevant to the actual jobs at which most people would be working. The focus of the inventory was then, as it is now, on self-exploration and job relevancy. The inventory was designed to provide a single instrument that would 1) focus on the more important phases of career choice and identity, 2) provide a method for systematic vocational counseling, and 3) provide more effective personal educational planning and improved job satisfaction.

Robert E. Ripley, author of the inventory, holds a Ph.D. from the University of Minnesota. Prior to joining the counselor education faculty at Arizona State University, he was a consultant to the U.S. Department of Labor, where he saw the need for such an assessment instrument and began development of the WOWI. The WOWI rights were obtained by John W. Hudson and Karen S. Hudson in 1975. John Hudson (Ph.D., Ohio State University) is Professor of Sociology at Arizona State University.

The WOWI is one of the most carefully and elaborately constructed instruments published. When it was first published in 1973, 12 years of observation and research had been devoted to its development. The methods employed in its construction were coordinated with the latest theories in career development, which focused on individual differences, life stages, career patterns, occupational ability patterns, and job satisfaction (Super, 1953).

The 516-item inventory is divided into four parts. Part A, Identifying Information, includes spaces to record a subject's name, date of birth, gender, identification number, choice of two occupational areas, and choice of two most-liked school subjects.

Part B, Career Interest Activities, contains 238 items in a forced-choice format requiring subjects to respond whether they like, dislike, or are neutral to each of the job-related activities presented. Data from this section yield scores in 17 homogeneous categories. In addition, the 17 Occupational Exploration Worksheets may be used with the profile results. These worksheets expand the 17 basic areas into 117 career families. Through exploration of the 17 career families, subjects can investigate all 20,000 occupations in the WOWI represented in the *Dictionary of Occupa-*

tional Titles (DOT; U.S. Department of Labor, 1977). The items used in the scale are based on job analysis and activities that persons actually perform on the job, not on individual personality traits.

Part C, Vocational Training Potentials, includes 98 items, placed in the inventory according to degree of difficulty, which measure six aptitude-achievement areas. These six scales are 1) Verbal (ability to read and comprehend words), 2) Numerical (ability to perform basic mathematical functions), 3) Abstractions (potential to solve problems without the assistance of words or numbers), 4) Spatial-Form (potential for visualizing and thinking in three dimensions), 5) Clerical (potential for organizing information, including the ability to perform sequential reasoning) and 6) Mechanical-Electrical (potential for constructing, operating, and repairing machinery and understanding physical forces). Four of the six scales attempt to measure factors similar to those found in Volume II of the DOT, including Verbal, Numerical, Spatial, and Form Perception.

In Part D, Job Satisfaction Indicators, 180 items measure 12 job temperament areas: 1) versatile, 2) adaptable to repetitive work, 3) adaptable to performing under specific instruction, 4) dominant, 5) gregarious, 6) isolative, 7) influencing, 8) self-controlled, 9) valuative, 10) objective, 11) subjective, and 12) rigorous.

In addition, these 12 areas, like the scales in Part B, attempt to measure factors similar to those listed in Volume II of the DOT. The interpretation manual provides information for each job temperament area for the specific vocations within a career family in which a person may be most effective and have the greatest job satisfaction.

Practical Applications/Uses

The WOWI was designed to provide an assessment of an individual's aptitude/achievement in several career-related areas (i.e., numerical, abstraction, mechanical-electrical); temperament factors related to job satisfaction; and interest in specific career families. The instrument is written at an eighth-grade reading level, with a modified version available for the deaf and the mentally handicapped. In addition, a computer version (IBM-PC), as well as a version in Spanish, are also available.

All versions may be used in career education/development classes or research, vocational guidance in high school or junior college, and employment or training placement. Counselors and employment agency personnel should find the WOWI useful in work with clients in their settings.

The WOWI is a four-page, paper-and-pencil instrument in a reuseable, machine-scorable format. The instrument cannot be hand scored. In addition to the NCS/Interpretive Scoring System report, a profile may be generated from the WOWI floppy disk, which may be used on any IBM or compatible system.

Items in parts B, C, and D of both the inventory booklet and the answer sheet are arranged in clusters of five to reduce errors in recording answers. The use of large type and blank space makes the inventory booklet easy to read. The inventory is designed to be self-administered and may be completed in one, two, or three sittings. The average time for completion of the entire inventory is reported as 120 minutes. Materials included with the WOWI inventory are the interpretation manual, the mini interpretation manual, and 17 Occupational Exploration Worksheets.

Test scores are reported on the inventory profile according to norms for age and educational level. The inventory profile may be used with the 17 Occupational Exploration Worksheets. A narrative interpretative summary that explains the various scales and scores and offers suggestions on appropriate jobs and occupations is available. An instructional cassette tape, designed to guide the counselor through an interpretation of the WOWI, is also available. When 100 or more inventories are submitted for scoring, a statistical summary with group norms may be purchased separately.

Technical Aspects

The WOWI has undergone several systematic steps to determine validity of the individual items, scales, and total profile analysis. The items within each scale were developed from the job analysis and job descriptions presented in the DOT. Personal job analyses and job descriptions were collected and each item was written around job activities and tasks in order to maintain job relevancy. The scales were reviewed by four or five judges in each of the 17 career families after the item clusters were developed and grouped. The judges had to agree unanimously on the use of an item before its final placement in a particular scale. After revisions, a final form was developed for the instrument. A stratified sample by age, gender, educational level, minority-group membership, and occupational groupings was used to determine the inter-item, intrascale correlations and the interscale correlations.

Of the 7,280 inter-item, intrascale correlations computed for the 17 Career Interest Activities scales, the 12 Job Satisfaction Indicator scales, and the 6 Vocational Training Potential scales, 91% of the correlations were statistically significant, leading to the conclusion that each item within a given scale was converging on the same construct and that each scale was internally consistent with respect to the construct it was measuring.

Interscale correlations of the 35 different scales also are reported (Hudson & Hudson, 1986). The Pearson product-moment coefficients suggested high correlations between various related Career Interest Activities areas, lower correlations between Career Interest Activities and Job Satisfaction Indicator areas, and even lower correlations between Career Interest Activities areas and Vocational Training Potentials. The coefficients suggested that the three areas are measuring different factors.

To determine the overall reliability of each scale, the sample was composed of approximately 51% males and 49% females ranging in age from 13-65 and whose formal education levels ranged from Grades 7-17. Afro-Americans, Chicanos, native Americans, Asian-Americans, and Caucasians were represented in the sample. Major subpopulations included university, junior and technical college students, vocational counseling and vocational rehabilitation clients, public and private high school students, and male prison inmates. All geographical areas of the United States were represented.

Coefficient alphas were computed to determine the internal consistency of the various scales. A reliability coefficient was computed for each scale; no summing over was computed because each scale is homogeneous. The coefficient alpha method provided the mean reliability of all possible split-half reliabilities ($r = .80$). The range of reliability coefficients was .94 to .83 on Career Interests; .83 to .72 on Job Satisfaction Indicators; and .65 to .24 on Vocational Training Potentials. Some dates indicating that profiles are fairly stable ($rs > .72$) over a 6-week period are presented (Neidert, 1986).

It seems unsound that more test-retest data have not been collected on an instrument that has been on the market nearly 15 years. One would hope that the publishers would systematically collect data on the long-term stability of the WOWI scores and profiles for without demonstrated stability use of the inventory is unwarranted.

Critique

Ripley, Hudson, and Hudson have undertaken an ambitious project in developing the inventory. Although the instrument seems to have been constructed according to acceptable standards, subsequent studies of validity and reliability are obviously lacking. After 15 years, only one study reporting reliability is available. The validity studies seem to have been based on convenience samples. Given some of the low intercorrelations between the dimensions, valid interpretations may be based more on faith than on demonstrated validity. The WOWI has some potential uses, but users must interpret results with these cautions in mind.

References

This list includes text citations and suggested additional reading.

Hudson, J.W., & Hudson, K.S. (1986). *Inter-scale correlations of career interest activities, job satisfaction indicators, and vocational training potentials of a stratified sample of women and men.* (Available from World of Work, Inc., 2923 North 67th Place, Scottsdale, AZ 85251)

Locke, D.C. (1977). Test review: World of Work Inventory. *Measurement and Evaluation in Guidance, 10,* 62-64.

Neidert, G.P.M. (1986). *Summary of test-retest reliability results for the World of Work Inventory.* (Available from World of Work, Inc., 2923 North 67th Place, Scottsdale, AZ 85251)

Neidert, G.P.M., Hudson, K.S., Ripley, R.E., & Ripley, M.J. (1984). *World of Work Inventory interpretation manual and guide to career families.* Scottsdale, AZ: World of Work, Inc.

Ripley, R.E. (1975). *World of Work administration manual.* Tempe, AZ: World of Work, Inc.

Super, D. (1953). A theory of career development. *American Psychologist, 8,* 185-190.

U.S. Department of Labor, Employment and Training Administration. (1977). *Dictionary of occupational titles.* Washington, DC: U.S. Government Printing Office.

INDEX OF TEST TITLES

INDEX OF TEST PUBLISHERS

Academic Therapy Publications, 20 Commercial Boulevard, Novato, California 94947; (415) 883-3314—[II:621; IV:172, 213, 278, 627; V:237, 244, 396; VI:447, 635]

American College Testing Program, (The), 2201 North Dodge Street, P.O. Box 168, Iowa City, Iowa 52243; (319)337-1000—[I:11]

American Foundation for the Blind, 15 West 16th, New York, New York 10011; (212)620-2000—[IV:390]

American Guidance Service, Publisher's Building, Circle Pines, Minnesota 55014; (800) 328-2560, in Minnesota (612)786-4343—[I:322, 393, 712, 715; III:99, 304, 480, 488; IV:327, 368, 704; V:172; VI:153]

American Orthopsychiatric Association, Inc., (The), 1775 Broadway, New York, New York 10019; (212)586-5690—[I:90]

American Psychiatric Association, *American Journal of Psychiatry,* 1400 K Street, N.W., Washington, D.C. 20005; (202)682-6000—[III:439]

American Psychological Association, *Journal of Consulting and Clinical Psychology,* 1200 Seventeenth Street, N.W., Washington, D.C. 20036; (202)955-7600—[V:198, 412]

American Society of Clinical Hypnosis, *The American Journal of Clinical Hypnosis,* 2250 E. Devon, Ste. 336, Des Plaines, Illinois 60018; (312)297-3317—[V:447]

American Testronics, P.O. Box 2270, Iowa City, Iowa 52244; (319)351-9086—[III:164]

Ann Arbor Publishers, Inc., P.O. Box 7249, Naples, Florida 33941;(813)775-3528—[VI:26]

ASIEP Education Company, 3216 N.E. 27th, Portland, Oregon 97212; (503) 281-4115—[I:75; II:441; V:86]

Associated Services for the Blind (ASB), 919 Walnut Street, Philadelphia, Pennsylvania 19107; (215)627-0600—[II:12]

Aurora Publishing Company, 1709 Bragaw Street, Ste. B, Anchorage, Alaska 99504; (907)279-5251—[V:444]

Australian Council for Educational Research Limited, (The), Radford House, Frederick Street, Hawthorn, Victoria 3122, Australia; (03) 819 1400—[IV:560]

The Barber Center Press, 136 East Avenue, Erie, Pennsylvania 16507; no business phone— [VI:127]

Behar, Lenore, 1821 Woodburn Road, Durham, North Carolina 27705; (919)733-4660— [V:341]

Behavior Science Systems, Inc., Box 1108, Minneapolis, Minnesota 55440; no business phone—[II:472; V:252, 348]

Bloom, Philip, 140 Cadman Plaza West, Apt. 3F, Brooklyn, New York 11201; no business phone—[VI:48]

Book-Lab, 500 74th Street, North Bergen, New Jersey 07047; (201)861-6763 or (201)868-1305—[V:209; VI:421]

Brink, T.L., 1044 Sylvan, San Carlos, California 94070; (415)593-7323—[V:168]

Bruce, (Martin M.), Ph.D., Publishers, 50 Larchwood Road, Larchmont, New York 10538; (914)834-1555—[I:70; IV:496; V:529]

Cacioppo, (John T.), Department of Psychology, University of Iowa, Iowa City, Iowa 52242; no business phone—[III:466]

Callier Center for Communication Disorders, The University of Texas at Austin, 1966 Inwood Road, Dallas, Texas 75235; (214)783-3000—[IV:119]

Center for Child Development and Education, College of Education, University of Arkansas at Little Rock, 33rd and University, Little Rock, Arkansas 72204; (501)569-3422—[II:337]

Center for Cognitive Therapy, 133 South 36th Street, Room 602, Philadelphia, Pennsylvania 19104; (215)898-4100—[II:83]

Center for Epidemiologic Studies, Department of Health and Human Services, 5600 Fishers Lane, Rockville, Maryland 20857; (301)443-4513—[II:144]

Center for Psychological Service, 1511 K Street N.W., Suite 430, Washington, D.C. 20005; (202)347-4069—[VI:512]

Chandler, Louis A., Ph.D., 5D Forbes Quadrangle, Pittsburgh, Pennsylvania 15260; (412)624-1244—[VI:570]

Chapman, Brook & Kent, 1215 De La Vina, Suite F, Santa Barbara, California 93101; (805) 962-0055—[IV:183]

Childcraft Education Corporation, 20 Kilmer Road, Edison, New Jersey 08818; (800)631-5652—[IV:220]

Clinical Psychology Publishing Company, Inc., 4 Conant Square, Brandon, Vermont 05733; (802)247-6871—[III:461]

Clinical Psychometric Research, 1228 Wine Spring Lane, Towson, Maryland 21204; (301) 321-6165—[II:32; III:583]

The College Board Publications, 45 Columbus Avenue, New York, New York 10023; (212)713-8000—[VI:120, 609]

College Hill Press, Inc., 4284 41 St., San Diego, California 92105; (619)563-8899—[III:293]

Communication Research Associates, Inc., P.O. Box 11012, Salt Lake City, Utah 84147; (801)292-3880—[I:707; III:669]

Communication Skill Builders, Inc., 3130 N. Dodge Blvd., P.O. Box 42050, Tucson, Arizona 85733; (602)323-7500—[II:191, 562; V:118]

Consulting Psychologists Press, Inc., 577 College Avenue, P.O. Box 60070, Palo Alto, California 94306; (415)857-1444—[I:34, 41, 146, 226, 259, 284, 380, 482, 623, 626, 663, 673; II:23, 56, 113, 263, 293, 509, 594, 697, 729; III:35, 51, 125, 133, 349, 392, 419; IV:42, 58, 132, 162, 570; V:141, 189, 226, 303, 556; VI:29, 87, 97]

C.P.S., Inc., Box 83, Larchmont, New York 10538; no business phone—[I:185; III:604]

Creative Learning Press, Inc., P.O. Box 320, Mansfield Center, Connecticut 06250; (203) 423-8120—[II:402]

Croft, Inc., Suite 200, 7215 York Road, Baltimore, Maryland 21212; (800)638-5082, in Maryland (301)254-5082—[III:198]

CTB/McGraw-Hill, Publishers Test Service, Del Monte Research Park, 2500 Garden Road, Monterey, California 93940; (800)538-9547, in California (800)682-9222, or (408)649-8400 —[I:3, 164, 578; II:517, 584, 780; III:186; IV:79, 238; V:406, 494; VI:149, 615]

Curriculum Associates, Inc., 5 Esquire Road, North Billerica, Massachusetts 01862-2589; (800)225-0248, in Massachusetts (617)667-8000—[III:79]

Dean, Raymond S., Ph.D., Ball State University, TC 521, Muncie, Indiana 47306; (317)285-8500—[VI:297]

Delis, (Dean), Ph.D., 3753 Canyon Way, Martinez, California 94553—[I:158]

Devereux Foundation Press, (The), 19 South Waterloo Road, Box 400, Devon, Pennsylvania 19333; (215)964-3000—[II:231; III:221; V:104]

Diagnostic Specialists, Inc., 1170 North 660 West, Orem, Utah 84057; (801)224-8492—[II:95]

DLM Teaching Resources, P.O. Box 4000, One DLM Park, Allen, Texas 75002; (800)527-4747, in Texas (800)442-4711—[II:72; III:68, 521, 551, 726; IV:376, 493, 683;

V:310; VI:80, 586]

DMI Associates, 3604 Lansdowne, Cincinnati, Ohio 45236—[VI:115]

D.O.K. Publishers, Inc., 71 Radcliffe Road, Buffalo, New York 14214; (716) 837-3391—[II:211; VI:303, 582]

Economy Company, (The), P.O. Box 25308, 1901 North Walnut Street, Oklahoma City, Oklahoma 73125; (405)528-8444—[IV:458]

Educational Activities, Inc., P.O. Box 392, Freeport, New York 11520; (800)645-3739, in Alaska, Hawaii, and New York (516)223-4666—[V:290; VI: 249]

Educational and Industrial Testing Service (EdITS), P.O. Box 7234, San Diego, California 92107; (619)222-1666—[I:279, 522, 555; II:3, 104, 258; III:3, 215; IV:199, 387, 449; V:76]

Educational Assessment Service, Inc., Route One, Box 139-A, Watertown, Wisconsin 53094; (414)261-1118—[II:332, VI:415]

Educational Development Corporation, P.O. Box 45663, Tulsa, Oklahoma 74145; (800) 331-4418, in Oklahoma (800)722-9113—[III:367;VI:244]

Educational Testing Service (ETS), Rosedale Road, Princeton, New Jersey 08541; (609) 921-9000—[III:655; VI:404]

Educators Publishing Service, Inc., 75 Moulton Street, Cambridge, Massachusetts 02238-9101; (800)225-5750, in Massachusetts (800)792-5166—[IV:195, 611; VI:188, 392]

Elbern Publications, P.O. Box 09497, Columbus, Ohio 43209; (614)235-2643—[II:627]

El Paso Rehabilitation Center, 2630 Richmond, El Paso, Texas 79930; (915)566-2956—[III:171, 628]

Elsevier Science Publishing Company, Inc., 52 Vanderbilt Avenue, New York, New York 10017; (212)867-9040—[III:358]

Essay Press, P.O. Box 2323, La Jolla, California 92307;(619)565-6603—[II:646; IV:553]

Evaluation Research Associates, P.O. Box 6503, Teall Station, Syracuse, New York 13217; (315)422-0064—[II:551; III:158]

Family Stress, Coping and Health Project, School of Family Resources and Consumer Sciences, University of Wisconsin, 1300 Linden Drive, Madison, Wisconsin 53706; (608)262-5712—[VI:10, 16]

Foreworks, Box 9747, North Hollywood, California 91609; (213)982-0467—[III:647]

Foundation for Knowledge in Development, (The), KID Technology, 11715 East 51st Avenue, Denver, Colorado 80239; (303)373-1916—[I:443]

G.I.A. Publications, 7404 South Mason Avenue, Chicago, Illinois 60638; (312)496-3800—[V:216, 351]

Grune & Stratton, Inc., Orlando, Florida, 32887-0018; (800)321-5068, (305)345-4500—[I:189; II:819; III:447, 526; IV:523; V:537; VI:52, 431]

Guidance Centre, Faculty of Education, University of Toronto, 252 Bloor Street West, Toronto, Ontario, Canada M5S 2Y3; (416)978-3206/3210—[III:271]

Halgren Tests, 873 Persimmon Avenue, Sunnyvale, California 94807; (408)738-1342—[I:549]

Harding Tests, Box 5271, Rockhampton Mail Centre, Q. 4701, Australia; no business phone—[IV:334]

Harvard University Press, 79 Garden Street, Cambridge, Massachusetts 02138; (617)495-2600 —[II:799]

Hilson Research Inc., 82-28 Abingdon Road, P.O. Box 239, Kew Gardens, New York 11415; (718)805-0063—[VI:265]

Hiskey, (Marshall S.), 5640 Baldwin, Lincoln, Nebraska 68507; (402)466-6145—[III:331]

Hodder & Stoughton Educational, A Division of Hodder & Stoughton Ltd., P.O. Box 702,

Mill Road, Dunton Green, Sevenoaks, Kent TN13 2YD, England; (0732)50111—[IV:256]

Hodges, Kay, Ph.D., Duke University Medical Center, P.O. Box 2906, Durham, North Carolina 27710; (919)684-3044—[VI:91]

Humanics Limited, 1182 W. Peachtree Street NE, Suite 201, Atlanta, Georgia 30309; (602) 323-7500—[II:161, 426]

Humanics Media, 5457 Pine Cone Road, La Crescenta, California 91214; (818)957-0983—[V:522, 524; VI:76]

Industrial Psychology Incorporated (IPI), 515 Madison Avenue, New York, New York 10022; (212)355-5330—[II:363]

Institute for Personality and Ability Testing, Inc. (IPAT), P.O. Box 188, 1602 Coronado Drive, Champaign, Illinois 61820; (217)352-4739—[I:195, 202, 214, 233, 377; II:357; III:139, 246, 251, 319, 567; IV:595; V:283; VI:21, 359, 560]

Institute for Psycho-Imagination Therapy, c/o Joseph Shorr, Ph.D., 111 North La Cienega Boulevard #108, Beverly Hills, California 90211; (213)652-2922—[I:593]

Institute for Psychosomatic & Psychiatric Research & Training/Daniel Offer, Michael Reese Hospital and Medical Center, Lake Shore Drive at 31st Street, Chicago, Illinois 60616; (312)791-3826—[V:297; VI:387]

Institute of Psychological Research, Inc., 34, Fleury Street West, Montreal, Quebec, Canada H3L 1S9; (514)382-3000—[II:530; VI:601]

Instructional Materials & Equipment Distributors (IMED), 1520 Cotner Avenue, Los Angeles, California 90025; (213)879-0377—[V:109]

International Universities Press, Inc., 315 Fifth Avenue, New York, New York 10016; (212)684-7900—[III:736]

Jamestown Publishers, P.O. Box 6743, 544 Douglass Avenue, Providence, Rhode Island 02940; (800)USA-READ or (401)351-1915—[V:212]

Jastak Associates, Inc., 1526 Gilpin, Wilmington, Delaware 19806; (302)652-4990—[I:758, 762; IV:673; VI:135]

Johnson, Suzanne Bennett, Ph.D., Shands Teaching Hospital, Psychiatry Service, Childrens's Mental Health Unit, Box J-234, J.H.M.H.C., University of Florida, Gainesville, Florida 32610—[VI:594]

Jossey-Bass, Inc., Publishers, 433 California Street, San Francisco, California 94104; (415) 433-1740—[III:395]

Kent Developmental Metrics, 126 W. College Avenue, P.O. Box 3178, Kent, Ohio 44240-3178; (216)678-3589—[III:380]

Kovacs, Maria, Ph.D., University of Pittsburgh, School of Medicine, Western Psychiatric Institute and Clinic, 3811 O'Hara Street, Pittsburgh, Pennsylvania 15213-2593; no business phone—[V:65]

Krieger, (Robert E.), Publishing Company, Inc., P.O. Box 9542, Melbourne, Florida 32901; (305)724-9542—[III:30]

Ladoca Publishing Foundation, Laradon Hall Training and Residential Center, East 51st Avenue & Lincoln Street, Denver, Colorado 80216; (303)296-2400—[I:239]

Lafayette Instrument Company, Inc., P.O. Box 5729, Lafayette, Indiana 47903; (317)423-1505—[V:534]

Lake, (David S.), Publishers, 19 Davis Drive, Belmont, California 94002; (415)592-7810—[II:241]

Learning House, distributed exclusively by Guidance Centre, Faculty of Education, University of Toronto, 10 Alcorn Avenue, Ontario, Canada M4V 2Z8—[VI:66, 70, 73]

Lewis, (H.K.), & Co. Ltd., 136 Gower Street, London, England WC1E 6BS; (01)387-4282—[I:47, 206, 595; IV:408]

P.O. Box 26240, San Diego, California 92121; (619)578-5900—[I:559; II:765; VI:109]

University of Illinois Press, 54 E. Gregory Drive, Box 5081, Station A, Champaign, Illinois 61820; institutions (800)233-4175, individuals (800)638-3030, or (217)333-0950—[I:354; II:543; V:32]

University of Minnesota Press, 2037 University Avenue S.E., Minneapolis, Minnesota 55414; (612)373-3266. Tests are distributed by NCS Professional Assessment Services, P.O. Box 1416 Minneapolis, Minnesota 55440; (800)328-6759, in Minnesota (612)933-2800—[I:466]

University of Vermont, College of Medicine, Department of Psychiatry, Section of Child, Adolescent, and Family Psychiatry, 1 South Prospect Street, Burlington, Vermont 05401; (802)656-4563—[I:168]

University of Washington Press, P.O. Box 85569, 4045 Brooklyn Avenue N.E., Seattle, Washington 98105; (206)543-4050, business department (206) 543-8870—[II:661, 714]

Valett, (Robert E.), Department of Advanced Studies, California State University at Fresno, Fresno, California 93740; no business phone—[II:68]

Variety Pre-Schooler's Workshop, 47 Humphrey Drive, Syosset, New York 11791; (516) 921-7171—[III:261]

Vocational Psychology Research, University of Minnesota, Elliott Hall, 75 East River Road, Minneapolis, Minnesota 55455; (612)376-7377—[II:481; IV:434; V:255; VI:350]

Walker Educational Book Corporation, 720 Fifth Avenue, New York, New York 10019; (212) 265-3632—[II:689]

Western Psychological Services, A Division of Manson Western Corporation, 12031 Wilshire Boulevard, Los Angeles, California 90025; (213)478-2061—[I:315, 338, 511, 543, 663; II:108, 430, 570, 607, 723, 826; III:145, 255, 282, 340, 402, 415, 615, 714, 717; IV:15, 33, 39, 259, 274, 300, 351, 382, 440, 501, 565, 606, 649; V:9, 73, 83, 378, 382, 425, 458, 549; VI:60, 260, 505, 519, 576, 629]

Wilmington Press, The, 13315 Wilmington Drive, Dallas, Texas 75234; (214)620-8531—[VI:383]

Wonderlic, (E.F.), & Associates, Inc., P.O. Box 7, Northfield, Illinois 60093; (312)446-8900—[I:769]

World of Work, Inc., 2923 North 67th Place, Scottsdale, Arizona 85251; (602)946-1884—[VI:644]

Wyeth Laboratories, P.O. Box 8616, Philadelphia, Pennsylvania 19101; (215)688-4400—[V:499]

Zung, (William W.K.), M.D., Veterans Administration Medical Center, 508 Fulton Street, Durham, North Carolina 27705; (919)286-0411—[III:595]

INDEX OF TEST AUTHORS/REVIEWERS

669

SUBJECT INDEX

Marriage and Family: Family

Marriage and Family: Premarital and Marital Relations

Neuropsychology and Related

Personality: Adolescent and Adult

Personality: Child

Personality: Multi-level

Research

EDUCATION

Academic Subjects: Business Education

Academic Subjects: English and Related: Preschool, Elementary, and Junior High School

Academic Subjects: English and Related—Multi-level

Academic Subjects: Fine Arts

Academic Subjects: Foreign Language & English as a Second Language

Academic Subjects: Mathematics—Basic Math Skills

Academic Achievement and Aptitude

Reading: High School and Above

Reading: Multi-level

Sensorimotor Skills

Special Education: Gifted

Special Education: Learning Disabled

Teacher Evaluation

BUSINESS AND INDUSTRY

Aptitude and Skills Screening

Clerical

Computer

Intelligence and Related

Interests

ABOUT THE EDITORS

Daniel J. Keyser, Ph.D. Since completing postgraduate work at the University of Kansas in 1974, Dr. Keyser has worked in drug and alcohol rehabilitation and psychiatric settings. In addition, he has taught undergraduate psychology at Rockhurst College for 15 years. Dr. Keyser specializes in behavioral medicine—biofeedback, pain control, stress management, terminal care support, habit management, and wellness maintenance—and maintains a private clinical practice in the Kansas City area. Dr. Keyser co-edited *Tests: First Edition, Tests: Supplement*, and *Tests: Second Edition* and has made significant contributions to computerized psychological testing.

Richard C. Sweetland, Ph.D. After completing his doctorate at Utah State University in 1968, Dr. Sweetland completed postdoctoral training in psychoanalytically oriented clinical psychology at the Topeka State Hospital in conjunction with the training program of the Menninger Foundation. Following appointments in child psychology at the University of Kansas Medical Center and in neuropsychology at the Kansas City Veterans Administration Hospital, he entered the practice of psychotherapy in Kansas City. In addition to his clinical work in neuropsychology and psychoanalytic psychotherapy, Dr. Sweetland has been involved extensively in the development of computerized psychological testing. Dr. Sweetland co-edited *Tests: First Edition, Tests: Supplement*, and *Tests: Second Edition*.